T0389184

Handbook of Flexible and Stretchable Electronics

Handbook of Flexible and Stretchable Electronics

Edited by

Muhammad Mustafa Hussain

King Abdullah University of Science and Technology, Saudi Arabia

and

University of California Berkeley, USA

Nazek El-Atab

King Abdullah University of Science and Technology

Saudi Arabia

CRC Press

Taylor & Francis Group

Boca Raton London New York

CRC Press is an imprint of the

Taylor & Francis Group, an **informa** business

Cover photo credit to Anastasia Khrenova.

CRC Press
Taylor & Francis Group
6000 Broken Sound Parkway NW, Suite 300
Boca Raton, FL 33487-2742

Printed on acid-free paper

International Standard Book Number-13: 978-1-138-08158-1 (Hardback)

Visit the Taylor & Francis Web site at
http://www.taylorandfrancis.com

and the CRC Press Web site at
http://www.crcpress.com

Contents

SECTION III Systems and Applications

Preface

Electronics technology has enabled today's digital world where connectivity is increasingly considered as one of the basic needs for humanity. Moore's law has driven this amazing growth of electronics technology in the areas of computation, communication, and infotainment. Most of the crystalline materials which are dominantly used as active functional materials that show superior electronic, radio frequency (RF), and optoelectronic performance, manufacturability, reliability in state-of-the-art electronics are still physically rigid and brittle. Therefore, in the last 3 decades, there was an extensive exploration in the areas of physically compliant (flexible and stretchable) materials for their potential use in a new class of emerging electronics: flexible and stretchable electronics. One of the major lab innovations to have commercial success from such an effort is organic light emitting diode (OLED)-based displays. As the focus has initially been on display technology, in the last 2 decades, significant efforts are underway to develop flexible and stretchable electronics for applications focusing on healthcare, sensing, photovoltaic, and battery technology. Many conceived and some demonstrated applications are entirely the resultant of such effort and advances in the area of flexible and stretchable electronics. While there is no doubt that the dedicated focus has been on naturally flexible and stretchable materials-based electronics and low-cost garage fabrication inspired low-temperature, low-cost process technologies, there is an increased interest in the physically compliant version of the traditional crystalline materials-based electronics. Thousands of research papers, hundreds of patents or patent applications, and tens of new starts-ups are natural outcomes of such concerted and sustained effort. As we see more and more interest and practical applications in this fascinating area, it is necessary to have a comprehensive collection of variety of related as well as discrete topics/areas of flexible and stretchable electronics. And in that regard, the motivation and the objective of this handbook are simple—to offer a comprehensive account of flexible and stretchable electronics.

This could not have been realized without the support of the publishing house CRC Press by Taylors & Francis Group. Undoubtedly, they have global leadership in compiling handbooks on important and timely topics. I would also like to thank the contributing authors for their dedicated efforts to enrich this collection. They have made this handbook unique. I have had the pleasure of working with many brilliant students, post docs, and colleagues from whom I have personally learnt a lot about this field. That knowledge and experience have helped me to edit this handbook. And finally, I would like to thank Dr. Nazek El-Atab for helping me stitch together the last portion of this handbook.

I am sincerely hopeful that this comprehensive handbook will be a great resource of knowledge and inspiration for countless electronics enthusiasts. More importantly, this area of flexible and stretchable electronics often draws the interest from a multidisciplinary community. And therefore, it will definitely be a treasure trove of knowledge for many others irrespective of their disciplinary and foundational backgrounds and preparations, respectively.

Nothing is perfect and to disclaim due to the veracity and the volume of the area of display technology, we intentionally have omitted it from this handbook. My personal favorite computational logic devices are missing more than their formation, the materials, and processes that are critical for them, and they have been captured very well here.

This handbook is dedicated to my family: my parents, my wife, and my children who will one day contribute to and will be benefited by this exciting area of flexible and stretchable electronics.

Editors

Prof. Muhammad Mustafa Hussain received his BSc in Electrical and Electronic Engineering from Bangladesh University of Engineering and Technology, Dhaka, Bangladesh, in 2000. Then he completed an MSc degree in Electrical Engineering from the University of South California, Los Angeles, CA, in 2002. Next he earned another MSc and PhD in Electrical and Computer Engineering from the University of Texas at Austin, in 2004 and 2005, respectively. Currently, he is a professor of Electrical Engineering at King Abdullah University of Science and Technology (KAUST), Thuwal, Saudi Arabia. He was recently a visiting professor in EECS of University of California, Berkeley. Before joining KAUST, Prof. Hussain was a program manager of the Emerging Technology Program in SEMATECH, Inc. Austin, Texas. A regular panelist of US NSF grants reviewing committees, he is the fellow of American Physical Society (APS), Institute of Physics, UK, and Institute of Nanotechnology, UK, IEEE Electron Devices Society Distinguished Lecturer, editor-in-chief of *Applied Nanoscience* (Springer-Nature), editor of *IEEE Transactions on Electron Devices*, and an IEEE Senior Member. He has served as first or corresponding author in 90% of his 2300+ research papers (including 27 invited reviews, 32 cover articles, and 150 journal papers). He has more than 60 issued and pending US patents. His students are serving as faculty and researchers in MIT, Stanford, UC Berkeley, Caltech, Harvard, UCLA, Yale, Purdue, TSMC, Intel Corporation, KACST, KFUPM, KAU, and DOW Chemicals. Scientific American has listed his research as one of the Top 10 World Changing Ideas of 2014. *Applied Physics Letters* selected his paper as one of the Top Feature Articles of 2015. He and his students have received 39 research awards including IEEE Region 5 Outstanding Individual Achievement Award 2016, World Technology Award Finalist in Health and Medicine 2016, TEDx 2017, Outstanding Young Texas Exes Award 2015 (UT Austin Alumni Award), US National Academies' Arab-American Frontiers of Sensors 2015, 2016, DOW Chemical Sustainability Challenge Award 2012, etc. His research has been highlighted extensively in international media like in *Washington Post*, *Wall Street Journal* (*WSJ*), *IEEE Spectrum*, etc. His research interest is to expand the horizon of CMOS electronics and technology for futuristic applications.

Dr. Nazek El-Atab received her BSc degree in Computer and Communications Engineering from the Rafik Hariri University, Lebanon, in 2012, her MSc degree in Microsystems Engineering from the Masdar Institute of Science and Technology, Abu Dhabi, UAE, in 2014, and her PhD degree in Interdisciplinary Engineering from the Masdar Institute in 2017. Currently, she is a post-doctoral research fellow with Professor Muhammad Mustafa Hussain at the MMH labs at KAUST. Her current research focuses on the design and fabrication of futuristic electronics.

She has received several awards for her research, including the 2015 For Women in Science Middle East Fellowship by L'Oreal-UNESCO, Best Paper Award in the Micro/Nano-systems section at the UAEGSRC 2016 conference, the 2016 IEEE Nanotechnology Student Travel Award in Japan, the 2017 International Rising Talents Award by L'Oreal-UNESCO, the 2018 "Rafik Hariri University" Alumni Award, and was portrayed in the 2019 "Remarkable Women in Technology" by UNESCO. She has published over 30 papers in international peer-reviewed scientific journals and conference proceedings and

2 book chapters, and has 4 pending US patents. She has served as a Jury member for the 2017 For Women in Science Middle East Fellowship by L'Oreal-UNESCO and a member of the Panel of Experts for the 2019 For Women in Science Awards by L'Oreal-UNESCO. Her research has been highlighted extensively in international media including The National, America's Navy, Office of Naval Research, Emirates News Agency, Skynews Arabia TV, Khaleej Times, Monte Carlo Paris, the Health Medicine Network, etc.

Contributors

Md. Kawsar Alam
Department of Electrical and Electronic Engineering
Bangladesh University of Engineering and Technology
Bangladesh, India

Joseph B. Andrews
Department of Electrical and Computer Engineering
Duke University
Durham, North Carolina

Nouha Alcheikh
King Abdullah University of Science and Technology
Thuwal, Saudi Arabia

Sushmee Badhulika
Department of Electrical Engineering
Indian Institute of Technology Hyderabad
Hyderabad, India

Saidur R. Bakaul
Materials Science Division
Argonne National Laboratory
Argonne, Illinois

Pavel Bolshakov
University of Texas at Dallas
Richardson, Texas

Jorge A. Cardenas
Department of Electrical and Computer Engineering
Duke University
Durham, North Carolina

Richard A. Chapman
University of Texas at Dallas
Richardson, Texas

Huanyu Cheng
Department of Engineering Science and Mechanics, and Materials Research Institute
The Pennsylvania State University
University Park, Pennsylvania

Canan Dagdeviren
MIT Media Lab
Massachusetts Institute of Technology
Cambridge, Massachusetts

Tae Hoon Eom
Department of Chemical and Biochemical Engineering
Dongguk University-Seoul
Seoul, South Korea

Aaron D. Franklin
Department of Electrical and Computer Engineering
and
Department of Chemistry
Duke University
Durham, North Carolina

Jeong In Han
Department of Chemical and Biochemical Engineering
Dongguk University-Seoul
Seoul, South Korea

Suk-Wong Hwang
KU-KIST Graduate School of Converging Science and Technology
Korea University
Seoul, Korea

Aftab M. Hussain
Center for VLSI and Embedded Systems Technologies (CVEST)
International Institution of Information Technology (IIIT)
Hyderbad, India

Mahnaz Islam
Department of Electrical and Electronic Engineering
Bangladesh University of Engineering and Technology
Bangladesh, India

Sherjeel M. Khan
King Abdullah University of Science and Technology
Thuwal, Saudi Arabia

Chang-Hyun Kim
Department of Electronic Engineering
Gachon University
Seongnam, Republic of Korea

Ramendra Kishor Pal
Department of Mechanical and Aerospace Engineering
Rutgers University
Piscataway, New Jersey

Gwan-Jin Ko
KU-KIST Graduate School of Converging Science and Technology
Korea University
Seoul, Korea

Arwa Kutbee
King Abdulaziz University
Jeddah, Saudi Arabia

Inkyu Lee
Department of Electrical and Computer Engineering
University of Wisconsin-Madison
Madison, Wisconsin

Joong Hoon Lee
KU-KIST Graduate School of Converging Science and Technology
Korea University
Seoul, Korea

Juhwan Lee
Department of Electrical and Computer Engineering
University of Wisconsin-Madison
Madison, Wisconsin

Tongfen Liang
Department of Mechanical and Aerospace Engineering
Rutgers University
Piscataway, New Jersey

Zhenqiang Ma
Department of Electrical and Computer Engineering
University of Wisconsin-Madison
Madison, Wisconsin

Aaron D. Mazzeo
Department of Mechanical and Aerospace Engineering
Rutgers University
Piscataway, New Jersey

Joanna M. Nassar
Hopkins Marine Station
Stanford University,
Pacific Grove, California

Jhonathan Prieto Rojas
Electrical Engineering Department
King Fahad University of Petroleum and Minerals
Dhahran, Saudi Arabia

Nadeem Qaiser
King Abdullah University of Science and Technology
Thuwal, Saudi Arabia

Manuel Quevedo-Lopez
University of Texas at Dallas
Richardson, Texas

Rodolfo A. Rodriguez-Davila
University of Texas at Dallas
Richardson, Texas

Anna Root
Department of Mechanical and Aerospace Engineering
Rutgers University
Piscataway, New Jersey

Parikshit Sahatiya
Department of Electrical Engineering
Indian Institute of Technology Hyderabad
Hyderabad, India

Galo Torres Sevilla
Empa Swiss Federal Laboratories for Materials Science and Technology
Duebendorf, Switzerland

Rinky Sha
Department of Electrical Engineering
Indian Institute of Technology Hyderabad
Hyderabad, India

Sohail Faizan Shaikh
King Abdullah University of Science and Technology
Thuwal, Saudi Arabia

Atif Shamim
IMPACT Lab, Computer, Electrical and Mathematical Sciences and Engineering (CEMSE) Division
King Abdullah University of Science and Technology (KAUST)
Thuwal, Kingdom of Saudi Arabia

Devendra Singh
National University of Singapore
Singapore

Mohammad Vaseem
IMPACT Lab, Computer, Electrical and Mathematical Sciences and Engineering (CEMSE) Division
King Abdullah University of Science and Technology (KAUST)
Thuwal, Kingdom of Saudi Arabia

Irmandy Wicaksono
Massachusetts Institute of Technology
Cambridge, Massachusetts

Chadwin D. Young
University of Texas at Dallas
Richardson, Texas

Xiyue Zou
Department of Mechanical and Aerospace Engineering
Rutgers University
Piscataway, New Jersey

I

Fundamentals

1

Plastic Electronics

Chang-Hyun Kim

1.1 Introduction

A general way that the term 'organic' is perceived by the public is that it describes something that is close to the nature and living matters. Even without scientific knowledges, this kind of impression seems enough for us to imagine one important property of organic materials; the mechanical flexibility. Organic semiconductors are indeed soft and flexible materials, much like biological systems around us such as tissues, organs, skins, and many other parts of our body. There is a fundamental reason that organics are ideally positioned for flexible or even stretchable electronics (which will be explained in this chapter) [1,2], and this characteristic is a basis for creating a wide range of unconventional device concepts that provide a physical form factor that is not easily achievable by traditional semiconductors [3–5].

From a scientific viewpoint, organic semiconductors are carbon-based molecules with varying sizes that have a special chemistry that makes them a tunable electric conductor (thus satisfying the classical notion of semiconductors). This chemical common ground will be also dealt with hereinafter. Suffice it to say that such a somewhat 'loose' definition of a materials class translates into a practically unlimited number of possible molecular structures. As the field steadily advances, the material toolbox has exploded, thanks to the consistent synthetic efforts and new applications that validate and guide these efforts [6–14]. Now it seems safe to believe that the once critical stability and lifetime issues have been largely overcome, and the materials developments are now more devoted to obtaining target-specific properties, such as color purity of light-emitting materials, the structural order that facilitates carrier movement, and sensitivity to the environment for harvesting wasted energies. In brief, the material's exceptional tunability is another key merit that adds to the flexibility, and it is paving the way to a new generation of multi-functional platforms.

Organic or plastic electronics is a field where optically or electrically active molecules or polymers replace classical semiconductors like silicon to build technological applications. Many will agree that the organic light-emitting diodes (OLEDs) represent a first success story of organic semiconductors, the products based on which have been commercialized in the late 2000s [15]. Partly owing to this momentum, the last decade has witnessed impressive boosts in the research and developments of organic-based technologies. Materials, processing, devices, systems, and theories are all under increasing global investigation, and flexible displays, radio-frequency identification (RFID) tags, energy-, or healthcare-related products are likely to become the next driving force to convince the market. In this context, summarizing the vast field of organic electronics into a single text is a difficult task. I will nonetheless take this opportunity to set an easily understandable rationale behind this fast-developing, multi-faceted area, particularly reflecting its relevance to flexible applications. Above all, I will elaborate the defining features of organic semiconductors that distinguish themselves from other semiconducting materials. To this end, I will not attempt to be exhaustive, and selected examples will be used to highlight these key features. This approach will allow one to quickly digest the past and present of plastic electronics. Ideally, one will be able to eventually find answers to questions like 'why are organic semiconductors promising for flexible devices?' and 'how will the field change over the coming years?'

1.2 Brief History

The first notation of organic semiconductors dates back to the 1940s (Figure 1.1) [16]. This and related early observations were mostly focused on the photoconductivity in a certain type of molecules, which have not been previously considered as an electrical conductor. In 1963, Pope, Kallmann, and Magnante's experiments on the electroluminescence (EL, generation of light by an electric field) in

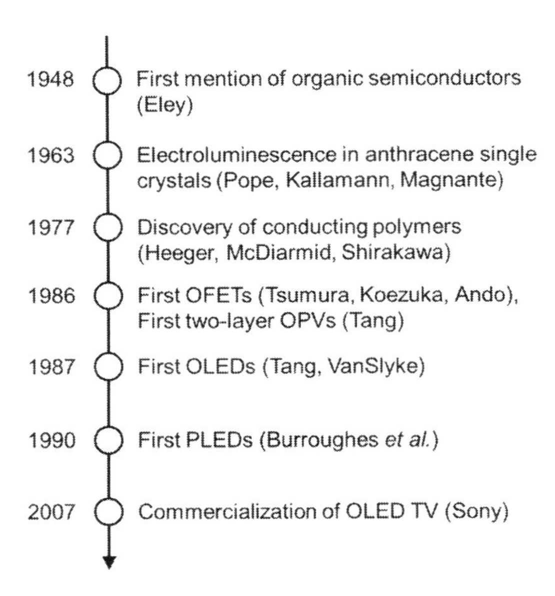

FIGURE 1.1 Historical developments of organic electronics.

anthracene single crystals were reported [17]. As it was the first observation of dc-voltage-induced emission from an organic material, it can be considered as the first step toward the OLED technology. While it was an important historical achievement, the required high voltage (up to 400 V) and sophisticated growth of bulk crystals were among major limitations. It is the seminal work by Shirakawa et al. reported in 1977 that is widely believed as the 'opening' of the field of organic electronics [18]. Here, the authors demonstrated the vapor-phase doping of polyacetylene by several halogen atoms. The iodine doping resulted in the increase in conductivity by a factor of 10^7 upon increasing the vapor pressure, reaching the maximum of 38 S/cm at room temperature. Among a multitude of key implications of this experiment, we note that the systematic doping, which is crucial for any semiconductor, was clearly shown to be possible, further justifying the appellation 'organic semiconductor.' Other now well-known notations 'conducting polymer' and 'synthetic metal,' also originated from it, as the material could enter into the metallic regime of conductivity. The Nobel Prize in Chemistry in 2000 was awarded to the main authors of this paper, Heeger, MacDiarmid, and Shirakawa, for their discovery of conducting polymers. In the 1980s, improved understanding and controllability of materials led to the demonstrations of thin-film devices with huge application potentials. The year 1986 saw the first macromolecular polythiophene field-effect transistors [19] and the efficient two-layer solar cells (using copper phthalocyannine and perylene tetracarboxylic derivative) [20], reported by Tsumura et al. and Tang, respectively. In 1987, Tang and VanSlyke reported a green OLED using a 60-nm 8-hydroxyquinoline aluminum (Alq_3) film as an EL medium [21]. Since then, these three devices, organic field-effect transistors (OFETs), organic photovoltaics (OPVs), and OLEDs, have been under intensive investigation, and they constitute the major building blocks of the current organic technology (see Section 1.5 for more details). The work by Burroughes et al. published in 1990 showed that solution-phase deposition of poly(p-phenylene vinylene) (PPV) precursors can make efficient OLEDs [22]. This paper is credited for the first polymer OLEDs (PLEDs) and provided an implication for large-area displays fabricated by low-cost printing. As already mentioned, the OLEDs are the first successful consumer products, now available as an ultra-thin, light-weight, bright, and energy-efficient display in mobile phones and TVs. For OLEDs, clear market need to replace bulky liquid-crystal displays in emerging applications ideally met the timely development of the technology. In view of recent research outcomes, we may soon be able to see more organic-based products, either in conventional or new applications.

1.3 Semiconducting Properties

1.3.1 Concept of π-Conjugation: A Molecule Carrying Electricity

Now we try to understand how an organic molecule can be a semiconductor. A good starting point is the basic concept in organic chemistry, π-conjugation [23]. Often illustrated with the example of benzene, when a carbon atom has three neighboring atoms to make covalent bonds with, one *s* and two *p* orbitals of this carbon are hybridized into three sp^2 orbitals. Three of the four valence electrons are shared through these orbitals forming strong σ bonds. The remaining fourth electron resides in an unhybridized *p* orbital and makes a weak π bond with the adjacent carbon's fourth electron. As the geometrical overlap between the *p* orbitals is small, the shared π electrons are not tightly bound to the parent atoms, but are delocalized throughout the entire molecule. It justifies the free choice between single and double bonds in writing the molecular structure, and even a ring often replaces double bonds to clarify the equivalence of all carbon-carbon bonds. This 'delocalization' of electrons already gives us a sense of electric conductor. If we inject an electron into this kind of molecule, it may freely move around all the atoms in this molecule.

A more direct analogy to conventional semiconductors is given by Figure 1.2a, which also introduces a few useful terminologies. When we pick up two same atoms and let them approach each other, the Pauli exclusion principle forces two equivalent atomic orbitals to split into one bonding and one antibonding molecular orbital that belongs to the pair. Because the σ bonds are strong, the energetic

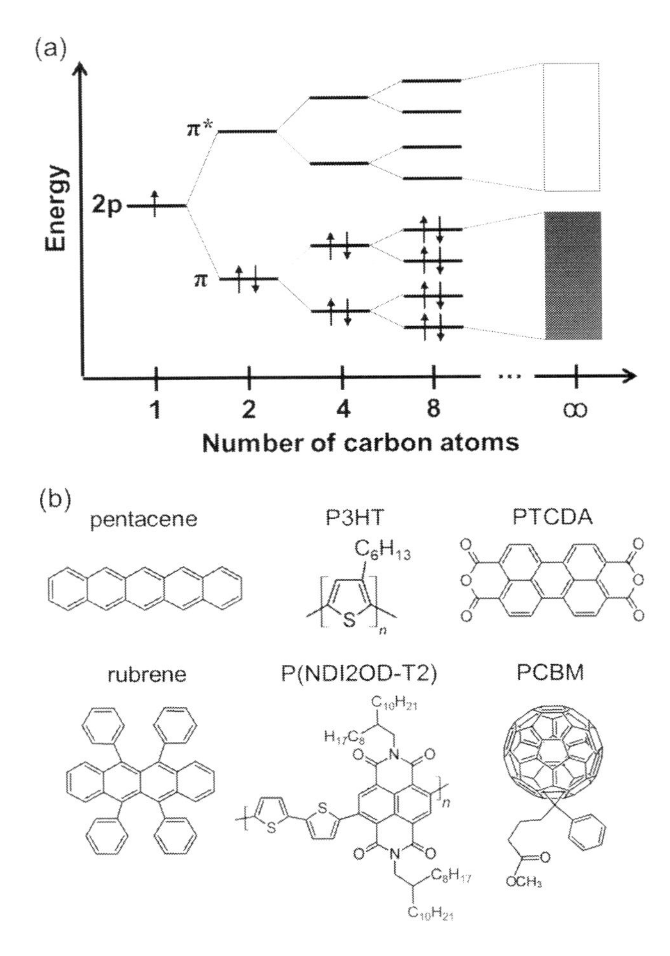

FIGURE 1.2 (a) Illustration of the energy levels and formation of energy bands in conjugated carbon-based molecules. Here, π and π^* denote the bonding and antibonding molecular orbital, respectively. (b) Chemical structure of well-known organic semiconductors.

distance between their bonding and antibonding orbitals is large. This makes the π bonds-originated molecular orbitals the frontier orbitals, the highest occupied molecular orbital (HOMO) and the lowest unoccupied molecular orbital (LUMO). As illustrated in Figure 1.2a, when we increase the number of carbons, the energy splitting at some point gives rise to the formation of continuous HOMO and LUMO bands. It resembles the formation of conduction and balance band in inorganic semiconductors.

Figure 1.2b shows the chemical structure of a small number of organic semiconductors. Some of them are small molecules (or oligomers), others are polymers, and they all have different motifs and/or functional groups. As mentioned in Section 1.1, now there are a huge number of materials available, and the detailed synthetic principles will not be described here. To know more about these rules, readers may refer to the reviews on that topic [7,8,11–14]. Another article contains a short explanation useful for understanding the notation of n- and p-type semiconductors [24]. Here, for the initially stated purpose, it should be sufficient to note that all organic semiconductors (including those in Figure 1.2b) feature substantial alternating single-double bonds in a carbon-based framework, which is the signature of π-conjugation. In brief, not all organics can be a semiconductor, but only those with extended conjugation meet the basic requirement, because only these molecules can carry or transport an electron (or hole) through the connected LUMO (or HOMO).

1.3.2 Hierarchical Understanding: From Molecule to Device

We have just rationalized the possible electron movement or delocalization in certain types of organics. However, this concept is only strictly valid at a single molecule level. A practical device such as OFET or OPV contains a molecular 'thin film' having a thickness of the order of 10–100 nm, and its width and length are generally larger and can vary widely depending on applications (from submicron to cm or above). We should therefore bridge the molecular understanding to macroscopic features, because such a film gathers a huge number of molecules. Then, interfaces with other layers (e.g., electrodes, dielectric) have to be taken into account to fully conceptualize the device characteristics.

Figure 1.3a illustrates the crystal structure of dinaphtho[2,3-b:2′,3′-f]thieno[3,2-b]thiophene (DNTT), a high-performance p-type semiconductor [25]. Because molecules form such a crystal via weak van der Waals force, the electronic coupling between molecules is small. Therefore, the formation of HOMO and LUMO bands throughout the entire solid is unlikely; charge carriers may be delocalized within a single molecule, but are localized at this molecule when seen macroscopically. Together with the strong tendency to polarize electron clouds around a charged molecule

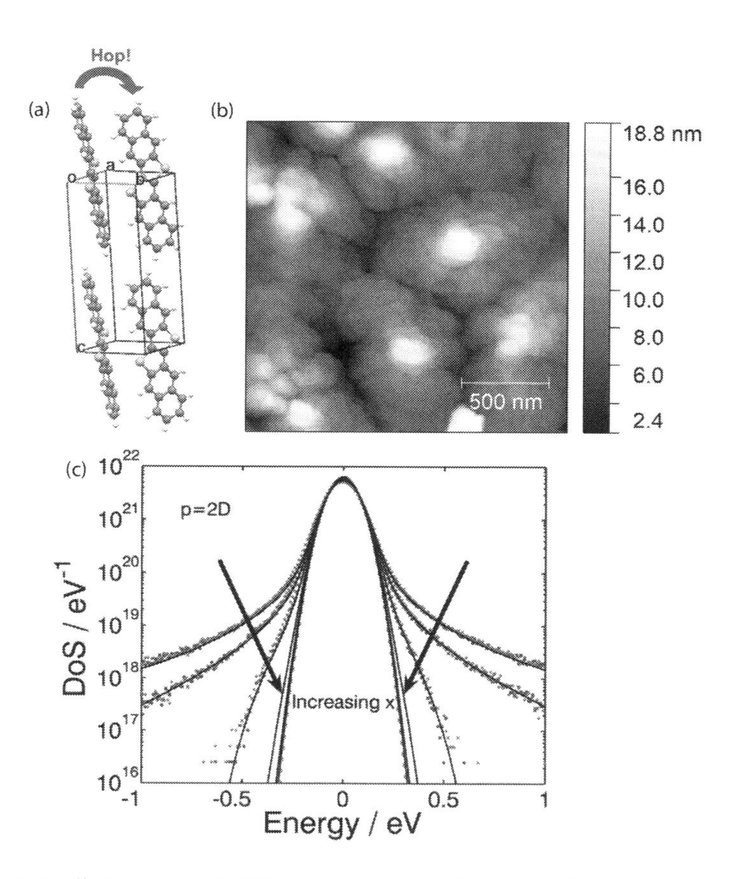

FIGURE 1.3 (a) Unit-cell structure of DNTT crystal, where the intermolecular hopping process is graphically illustrated. (b) Atomic-force microscope image of a polycrystalline DNTT thin film (~20 nm). (c) DOS in an organic semiconductor with increasing distance from the insulator surface (x = 0–0.5 nm). The dots indicate numerical simulation results and the solid lines are from the proposed analytical model. The dashed line shows the intrinsic (unbroadened) Gaussian DOS with a characteristic width of 60 meV. (Reprinted with permission from Richards, T. et al., *J. Chem. Phys.*, 128, 234905, 2008. Copyright 2008 by the American Institute of Physics.)

and self-trap this charge [6], it frustrates long-range transport. Instead, a carrier should move from a molecule to another via hopping, a thermally activated tunneling process. Such intermolecular transport is less efficient than intramolecular transport, and therefore often becomes the limiting factor of (macroscopically averaged) carrier mobility. At a higher level, the atomic-force microscope (AFM) image of a DNTT film in Figure 1.3b provides an insight. For polycrystalline materials, hopping carriers should cross abrupt grain boundaries from time to time when traveling a substantial distance. These boundaries are detrimental to macroscopic transport as they function as a high energy barrier. To sum up, the chemical structure of a molecule is not solely responsible for electrical properties; the knowledges about the film structure (e.g., intermolecular distance, grain shapes and sizes) are also critical.

The fact that a device contains several layers makes the interface effects important. The electrode interface will be discussed in the next section, and here we focus on the insulator interface. When we deposit an organic film, molecules arriving on a surface have a certain degree of freedom before being crystallized or solidified [26]. Because of the weak molecule-molecule force, the substrate-molecule interaction may easily become dominant. Therefore, the final film structure can be strongly dependent on the chemical groups, surface energy, and roughness of the substrate [27–29]. Another important aspect regarding the interface is its electrostatic influence. The density of states (DOS) in an organic material is approximated by a Gaussian distribution that reflects intrinsic disorder, and an energetically narrow DOS facilitates the hopping conduction [30]. Richards et al. showed that, due to the static dipolar disorder in a dielectric, the DOS of the semiconductor in contact with that dielectric can be locally modified [31]. Their representative modeling data are given in Figure 1.3c. Here, the dielectric material has a fixed dipole moment (p) of two Debye (D), and x denotes the distance from the insulator surface to a region in the semiconductor. When x is close to zero, we clearly observe a DOS that is broadened due to the dielectric's polarity. At larger x, this effect vanishes and the DOS recovers its intrinsic Gaussian shape. It was therefore shown that a low dielectric constant insulator is preferred not to stretch the DOS at the critical current-carrying interface, and thus not to degrade the carrier mobility of a given semiconductor.

1.3.3 Junction Behavior

Another characteristic of an organic semiconductor is that it is often used without (intentional) doping [32]. A key concept in solid-state device physics is the formation of a depletion region (or space-charge region) at a metal-semiconductor junction. Here, we recall that the depletion width is inversely proportional to the square root of doping density. Because of the absence of doping, the depletion width at a metal-organic junction can be very large. We may choose a metal whose work function is close to the HOMO or LUMO of the organic material in question to avoid the formation of a Schottky barrier. However, organics feature interface defects and Fermi-level pinning. Therefore, making an ohmic contact is not straightforward [33,34]. An overall consequence of minimal doping and a non-vanishing barrier is that a thin organic layer can be easily fully depleted (i.e., the depletion width exceeds the film thickness), and the efficiency of carrier injection from electrodes becomes critical.

For diode systems, such as single-layer rectifying diodes, OLEDs, and OPVs, the metal-insulator-metal model (MIM) is applicable [35]. As in Figure 1.4a, the organic semiconductor is represented with straight energy profiles. It is reasonable to regard the semiconductor as a dielectric-like medium which provides the conduction paths (HOMO and LUMO) that can be utilized by the carriers injected from external sources (e.g., electrodes or light). For OFETs, the contact resistance (R_c) can be comparable to or even dominant over the channel resistance (R_{ch}). Figure 1.4b shows the hole distribution inside the semiconductor of a p-type OFET obtained by finite-element drift-diffusion simulation. It is similar to our previous data [36], but here we added trap sites. The hole density at the channel is high because of the capacitive charge accumulation in the on state. However, the zones near the source and drain

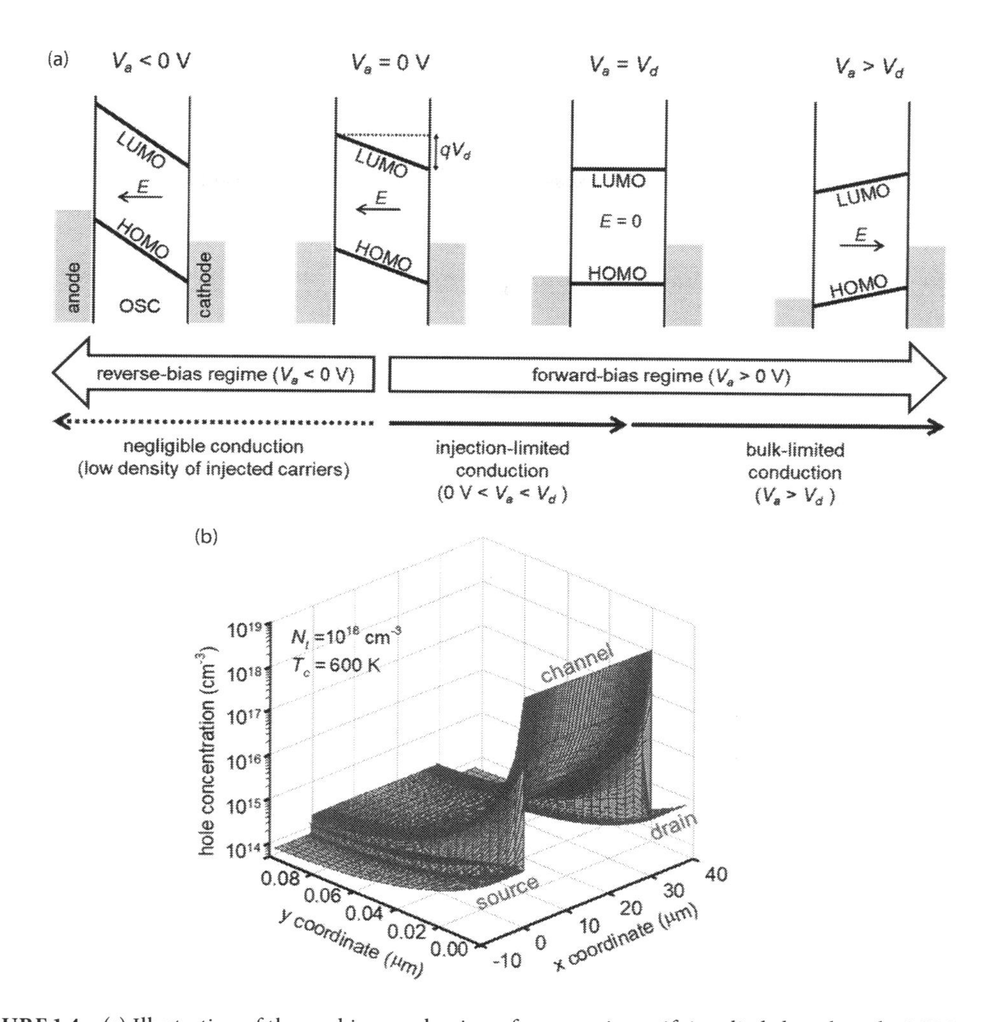

FIGURE 1.4 (a) Illustration of the working mechanism of an organic rectifying diode based on the MIM model (OSC: organic semiconductor, *E*: electrical field, V_d: built-in potential, V_a: applied voltage at the anode). (b) Hole distribution in a switched- on p-type OFET calculated by 2D finite-element simulation (N_t: total trap density, T_c: trap distribution temperature).

contacts suffer from much fewer carriers available; these zones become the bottleneck of the current and the origin of R_c [37]. The effects of R_c on OFETs will be further discussed in Section 1.6.3.

1.4 Routes to Flexible Systems

1.4.1 First Consideration: Materials' Softness

A recent review by Root et al. comprehensively analyzed mechanical properties of organic semiconductors [38]. As illustrated in that article, there are several multi-scale mechanisms that contribute to the dissipation and release of mechanical stress, arising from the structural complexity of an organic or polymeric film. While the chain entanglements, packing motifs, and macroscopic morphologies should all play a significant role at different levels, the weak van der Waals force between molecules seems to be the most common and fundamental reason that accounts for the intrinsic softness and excellent flexibility of organic electronic materials.

The inorganic semiconductor solids constructed entirely by covalent bonds (e.g., Si) or ionic bonds (e.g., ZnO) would have limited deformability because of the strength and stiffness of these chemical bonds. In contrast, the van der Waals bonds holding molecules together can be viewed as highly stretchable springs, which allow materials to deform substantially. A figure-of-merit describing a material's mechanical elasticity is Young's modulus, which is defined as the ratio of tensile strength to tensile strain. It is practically measured as the slope in the stress-strain curve before the yield point. It describes how well a material can deform in the direction of applied stress before being irreversibly damaged. Silicon has a Young's modulus of 130 GPa, while many organic semiconductors have a value of the order of 1 GPa [38,39]. It seems that many materials can be made quite flexible by reducing some of their physical dimensions. For instance, an ultra-thin silicon membrane can make highly flexible devices [40]. Nonetheless, the fundamental modulus indicates that soft organic materials have a clearly competitive edge, which makes them well positioned for flexible applications with less tight dimensional constraints.

1.4.2 Low-Temperature Processability

Having accepted the materials' own mechanical softness as a fundamental limit, we now again need to think hierarchically about the multi-layer structure of a functional device. Actually, we now take the substrate into our consideration, a material that is inactive in most cases, but serves as the physical support to mount a device on. For a final device to be truly flexible, not just the semiconductor needs to be flexible, but ideally all the constituent materials have to as well. Especially, the chosen active material's processing temperature is a critical factor that may screen the use of certain substrates.

The weakly-bound nature of organic molecules makes it possible to process them with a remarkably low energy budget. Organic semiconductors can be thermally evaporated at a moderate temperature or can be dissolved in a liquid solvent to be entirely solution-processed. Some as-deposited films may require a mild thermal treatment (e.g., for solvent evaporation), yet the maximum processing temperature is around 200°C or lower, in general cases. This level of temperature is relevant to the use of nontraditional substrates, such as plastic, textile, or paper, which are not only flexible, but also provide an intriguing playground for highly creative platforms.

A hint of organic-based future electronics may be found in Figure 1.5. Shown in Figure 1.5a is the fully additive-printed display backplane, developed by Arias et al. [41,42]. Pixel switches in information displays are one of the important applications of OFETs. Such an additive approach can make the fabrication simpler and potentially faster than traditional microfabrication that involves repeated cycles of growth and patterning, not to mention its alignment with a major future trend of customized manufacturing. In ideal cases, printing can be done in ambient air, thus eliminating the needs for ultraclean environments and vacuum-based tools. This aspect, together with the possible use of high-throughput techniques (e.g., roll-to-roll printing) is a key driver for low-cost applications [43]. Figure 1.5b shows the organic circuits built on an ordinary banknote, reported by Zschieschang et al. [44]. This demonstration combined several chemically and thermally harmless deposition methods to preserve the intactness of this somewhat delicate substrate. In addition to their possible immediate use as anticounterfeit technologies, such devices can be placed in the broader context of paper electronics that forms a main branch in biodegradable and environmentally friendly electronics [45,46]. Kaltenbrunner et al. developed the extremely thin OPVs shown in Figure 1.5c [47]. These devices were fabricated on a 1.4-μm-thick plastic, and this wrapping-around-a-hair demonstration emphasizes their exceptional flexibility and bending stability. These OPVs could also accommodate a tensile strain of over 300%. The authors put forward the specific weight (in the unit of W/g) as the key figure-of-merit of their solar cells. Their OPVs showed an unprecedented value of 10 W/g, outperforming all other established technologies. To generalize, the performance of organic devices may not compare particularly favorably to that of inorganic devices, but this article clearly shows that there can be a unique opportunity for organic electronics. Returning to the article, the uniqueness in this case was the generated power per unit weight (while the power-conversion efficiency, PCE, itself was not dramatic), which may appeal to systems such as unmanned aerial vehicles.

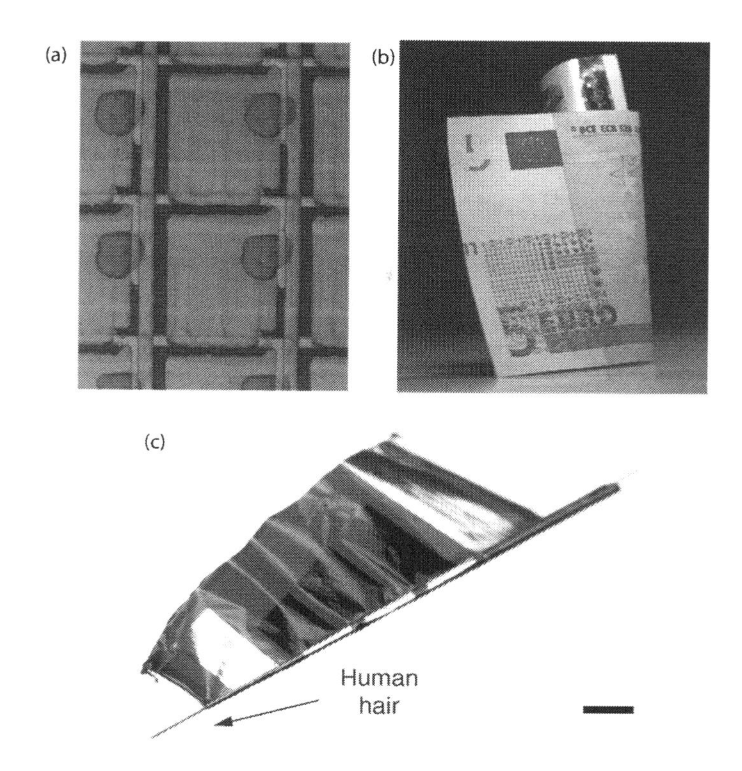

FIGURE 1.5 (a) All-additive-printed OFETs that serve as a pixel driver in an electrophoretic display. (From Street, R.A.: Thin-Film Transistors. *Advanced Materials*. 2009. 21. 2007–2022. Copyright Wiley-VCH Verlag GmbH & Co. KGaA. Reproduced with permission.) (b) A banknote with arrays of OFETs and logic inverters. (From Zschieschang, U. et al.: Organic Electronics on Banknotes. *Advanced Materials*. 2011. 23. 654–658. Copyright Wiley-VCH Verlag GmbH & Co. KGaA. Reproduced with permission.) (c) Extreme bending flexibility demonstrated by wrapping a solar cell around a 35-μm-radius human hair (scale bar: 2 mm). (Reprinted with permission from Macmillan Publishers Ltd. *Nat. Commun.*, Kaltenbrunner, M. et al., 2012, copyright 2012.)

1.5 Devices

It is expected that other chapters of this book will deal with many of the state-of-the-art devices based on organic semiconductors. In this section, basic structures and working mechanisms of three representative devices, OFETs, OLEDs, and OPVs will be only briefly introduced.

A general structure of an OFET is drawn in Figure 1.6. An OFET is structurally similar to other thin-film transistors (TFTs), and so is also called as an organic TFT or OTFT. The term OFET can, however, cover both thin-film channel devices and single-crystal devices, the latter being widely employed to explore fundamental charge-transport mechanisms [48]. When considering a p-type channel, an applied negative gate voltage (V_G) larger than the threshold voltage (V_T) leads to the capacitive accumulation of holes at the semiconductor-dielectric interface, thus creating a conducting channel. Then, a negative drain voltage (V_D) drifts these holes from the source and to the drain, generating a drain current (I_D). Therefore, this device is behaviorally similar to a metal-oxide-semiconductor field-effect transistor (MOSFET) in that the gate controls the source-drain conductivity through capacitive coupling. A major difference is that an OFET works in accumulation mode, while a MOSFET operates in inversion mode. It implies the difficulty in physically defining V_T [49]. OFETs are used as an electrical switch and amplifier, finding their applications in displays, logic gates, analog circuits, and active-matrix imagers [50]. Furthermore, their chemical sensitivity toward certain molecules and the possibility of surface

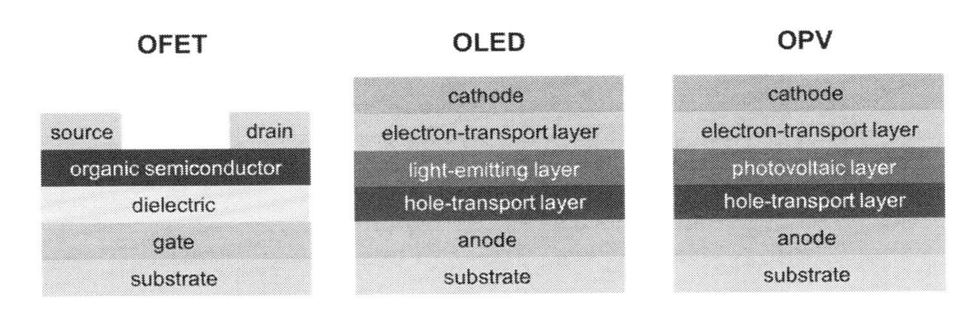

FIGURE 1.6 Structure of widely investigated organic electronic devices.

functionalization are combined to make OFETs a good sensing platform [51,52]. Also, the OFETs are widely used as a testbed for estimating semiconducting properties (e.g., charge-carrier mobility) of a newly synthesized molecule.

An OLED is generally built as a multi-layer stack as shown in Figure 1.6. This device converts an electrical current to light. When we apply a positive voltage to the anode (with the cathode grounded), holes are injected from the anode and electrons are injected from the cathode into the organic layer(s). There are dedicated hole and electron transport layers that facilitate the balanced injection of these two types of charges into the emission layer. In this central layer, a pair of electron and hole meets to form a radiative exciton, the recombination of which creates a photon. This photon eventually passes through a transparent electrode (e.g., indium-tin oxide, ITO) for us to see the light. As the energy of emitted photon is associated with the HOMO-LUMO gap of the emitter, synthetic approaches can be applied to fine-tune the colors or to make devices emit non-visible wavelengths. As outlined in Section 1.1, OLEDs have been already commercialized, and the market is under continuous growth. In addition to the traditional display applications, OLEDs can be used as an excitation source in biosensors and sheet-type lighting devices [53,54], which are some of their possible future orientations. For simplicity, Figure 1.6 only includes a single transport and emissive layer, but devices may have more than one layer for each function to exhibit optimum performances.

An OPV cell has a structure quite similar to that of an OLED, as shown in Figure 1.6, and its operating mechanism is the reverse of that of an OLED; this device converts light to electricity. While a planar-junction OPV (similar to a *p-n* junction) can be realized, the so-called bulk-heterojunction architecture is particularly well suited for organic semiconductors [55]. In this structure, electron-donating and -accepting organic materials are deposited from a mixed solution or by co-evaporation, forming a tightly inter-digitated single active layer. The nanoscale phase separation in a bulk-heterojunction film is a strategy for overcoming a short exciton diffusion length in organic materials [56]. A light absorption in the active layer gives rise to the formation of a bound electron-hole pair (exciton), and the intermixed structure allows most of these excitons to diffuse safely to a donor-acceptor interface and subsequently get dissociated. This exciton dissociation creates a free hole and a free electron, which reach their respective electrode through a transport layer. OPVs, thanks to the flexible form factors, transparency, and lightness, are promising technologies for wearable electronics and building-integrated energy sources [57].

1.6 Topics and Trends in Research

Plastic electronics has experienced particularly rapid growth over the past few years. This section will briefly outline some of the remarkable progresses that have a broad impact on the field. Some of them addressed the methodological issues in traditional analysis, and others concern the possibilities for new material combination and drastically new device concepts. Overall, they strengthen our knowledges about technologically relevant materials and devices for us to build an applicable design principle and employ a viable engineering approach. Therefore, this section may show a blueprint for the future of organic electronics technology.

1.6.1 Re-defining the Structure-Property Relationship

As explained in Section 1.3.2, the microstructure of a film strongly affects the semiconducting property, and how we manipulate the molecular arrangements can influence the device performance dramatically even in the case of using the same semiconductor [58]. In this context, the structure-performance relationship has been one of the biggest questions of the field, and still forms an important research motivation.

One of the traditional beliefs was that the crystalline morphology is preferred, because we may expect better intermolecular electronic coupling in such a situation. Venkateshvaran et al., however, rationalized how a nearly amorphous polymer film can show a practically disorder-free transport, thus proposing a new molecular design principle [59]. Based on the observation of a surprisingly high field-effect mobility in certain donor-acceptor copolymers, these authors compared OFETs made of several crystalline and non-crystalline polymers. Among them, indacenodithiophene-co-benzothiadiazole (IDTBT) showed evidences for the extremely low energetic disorder despite lacking long-range crystallinity. For instance, the power-law dependence of I_D on the gate field was estimated in the saturation regime, using the proportionality between I_D and $(V_G - V_T)^\gamma$. The parameter γ is therefore 2 in the ideal case where there is no gate enhancement of mobility associated with the trap filling [32]. Figure 1.7a shows

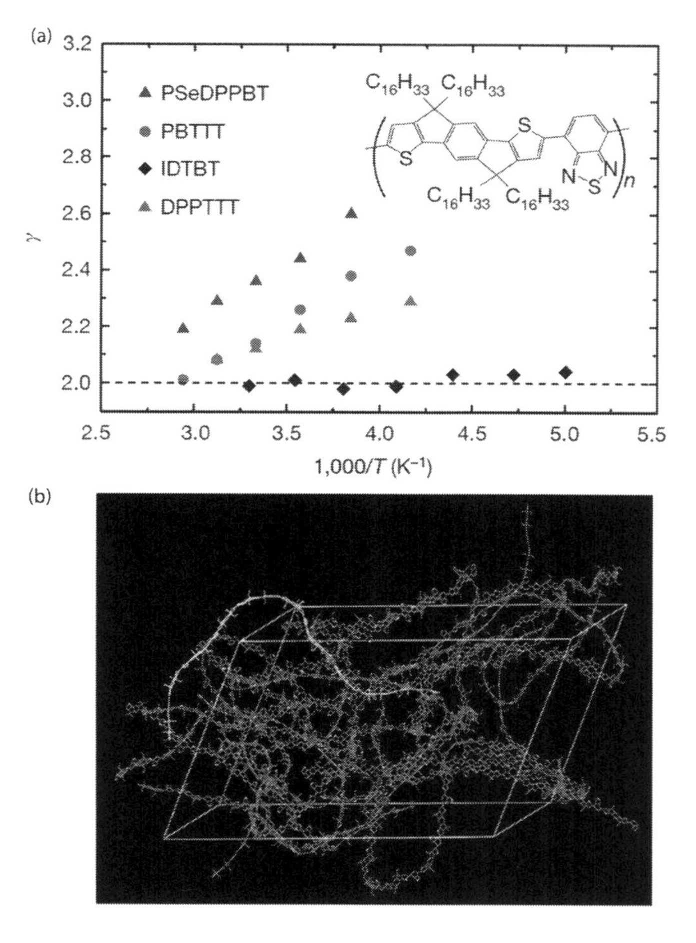

FIGURE 1.7 (a) The disorder-related exponent γ plotted as a function of $1000/T$ for the four different semiconducting polymers tested in an OFET structure. (b) Simulation of the backbone conformation of IDTBT in the amorphous phase. A single chain from the simulated unit cell has been highlighted in bright yellow. (Reprinted with permission from Macmillan Publishers Ltd. *Nature*, Venkateshvaran, D. et al., 2014, Copyright 2014.)

that IDTBT approaches this ideality while other polymers had larger γ values and temperature (T) dependence. Along with additional evidences of low disorder from the field-effect-modulated Seebeck coefficient and Urbach energy measurements, molecular dynamics simulations were conducted to identify the origin. It was found that IDTBT features a highly planar torsion-free backbone conformation which is resilient to side-chain disorder. As shown in Figure 1.7b, even in a fully amorphous phase, the material accommodates disorder through backbone bending, while keeping its planarity. This was in sharp contrast to other simulated polymers, which showed significant broadening of the DOS when becoming amorphous.

Our group and other groups reported on related evidences that the crystallinity itself may have less importance in certain cases than the interconnectivity of domains and the efficiency of intrachain transport [60,61]. This implies a new paradigm of material development, where we may choose between the materials that are non-sensitive to the microstructure and those with structural tunability.

1.6.2 Novel Heterostructures and Devices

The processing versatility and soft contact of organic materials facilitate hybridized use of other materials in developing new functionalities. Mixed inorganic and organic transistors and solar cells represent some of the broadly employed approaches [62–64]. More recently, heterostructures between organic semiconductors and two-dimensional (2D) materials have received growing attention.

We proposed for the first time the vertical graphene-C_{60} heterojunction in a low-voltage high-performance switching device [65]. This architecture combined the electric-field-tunable work function of graphene and the highly injection-limited transport of an organic material. As shown in Figure 1.8a, a graphene/C_{60}/Al diode was mounted on an ITO/Al_2O_3 gate/dielectric substrate, initially forming a large electron-injection barrier from graphene to C_{60}. We can make this graphene less blocking and more injecting by applying a positive V_G to indium-tin oxide, and it allowed us to demonstrate a wholly vertical transistor that works with a driving voltage as low as 200 mV. An inverter circuit connecting C_{60} and pentacene heterojunctions was also operated as shown in Figure 1.8b. Thanks to the soft van der Waals contact and the resulting clean interface, we were able to extract the change of electron barrier from graphene to C_{60} by using the Arrhenius plot in Figure 1.8c, which approached a high value of 300 meV. Later, related device structures have been widely reported, and it was also shown that the additional doping or surface treatment can lead to the delicate tuning of the graphene-organic junction devices [66–69]. Our group also recently proposed a simple device model based on an equivalent circuit method [70]. These devices can be positioned within the broader context of mixed dimensional van der Waals systems [71,72]. As both organic semiconductors and 2D materials show further advances, we will be able to develop more hybrid device concepts that synergistically utilize the key strengths of these materials.

1.6.3 Contact Issues and Parameter Reliability

As organic electronics enters into technological maturity, there is an increasingly intensive debate on the research practices and performance measurement procedures [73]. For OPVs, the discussion has been focused on the possible overestimation of PCE originating from the fabrication and measurement steps that do not perfectly follow the standard [74]. For OFETs, the field-effect mobility is a relevant parameter to extract. While several alarming comments have arisen [75,76], it seems still difficult to standardize the mobility extraction flow partly due to the absence of a standard model that applies to a wide variety of devices [32].

Recently, Bittle et al. reported on a detailed analysis of a mobility overestimation issue in OFETs [77]. They fabricated rubrene single-crystal OFETs which are relatively free from the complicating disorder inside the semiconductor. By doing so, they were able to focus on the contact property

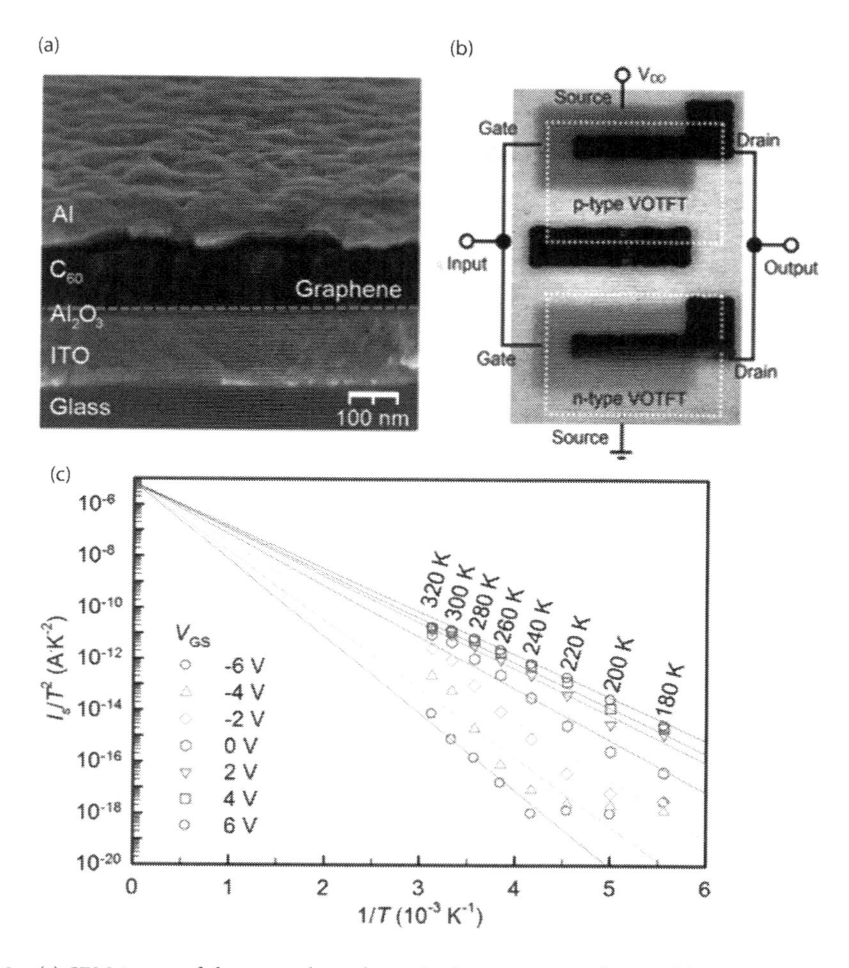

FIGURE 1.8 (a) SEM image of the vertical graphene-C_{60} heterojunction device. (b) Optical micrograph of the inverter circuit made of p- and n-type vertical devices. (c) Arrhenius plot of the temperature dependence of the saturation current, used for the extraction of the injection barrier height. (Reprinted with permission from Hlaing, H. et al., *Nano. Lett.*, 15, 69–74, 2015. Copyright 2014 American Chemical Society.)

that eventually leads to the parameter overestimation. As shown in Figure 1.9a, the fabricated devices showed a particular behavior; there were two distinct slopes in the square-root plot of the saturation regime transfer characteristic. While this behavior has been already widely observed, these authors clarified the ambiguity in selecting only one slope to provide the sole representation of the device. For this purpose, they re-analyzed a number of previously reported data, and plotted a device's peak mobility versus aggregate mobility, extracted from the low and high V_G region, respectively. As shown in Figure 1.9b, the ratio between these two was sometimes quite significant, exceeding a factor of 5. This further raised the question of which value to take if we are supposed to choose only one. These authors utilized the impedance spectroscopy to decouple R_C and R_{ch}. As shown in Figure 1.9c, R_{ch}, which represents the intrinsic semiconducting property, clearly dominates the measured current at high V_G. It means that the aggregate mobility better represents the true mobility in this particular device. In the lower V_G region, R_C was initially high, but rapidly decreased as V_G increased. This fast reduction in R_c explains the increased slope in the transfer curve and the overestimation of the mobility obtained from this region.

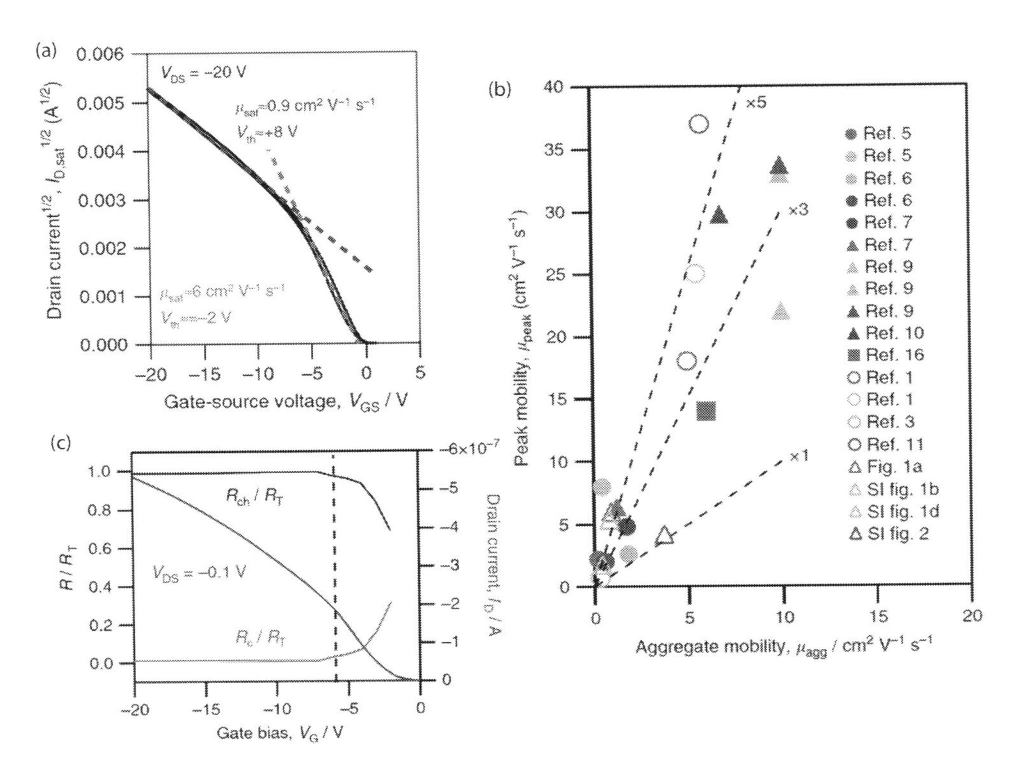

FIGURE 1.9 (a) Square-root transfer curve of a rubrene single-crystal OFET in the saturation regime, featuring an ambiguity in extracting the field-effect mobility. Parameters separately extracted from the two fitting lines are given. (b) Plot of the peak mobility versus aggregate mobility extracted from the low and high V_G regions of various reported OFETs, respectively. (c) The change in relative magnitude of channel and contact resistances as a function of V_G in the linear regime of the rubrene OFET. (Reprinted with permission from Macmillan Publishers Ltd. *Nat. Commun.*, Bittle, E.G. et al., 2016, Copyright 2016.)

It was a clearly meaningful message to the community. The performance improvement has been a major motivation, especially at the early stage of development. This and related revelations show that now we need to care more about the robustness of the methods that we use to estimate this performance, to precisely understand the current status of the field and to orient our research efforts toward right directions.

1.6.4 Mixed Conductivity: Ions in Action

While the established devices (e.g., OFETs, OLEDs, and OPVs) rely on the manipulation of electronic charges (i.e., electrons and holes), certain organic materials have been found to be able to conduct both electronic and ionic charges [78]. This so-called mixed conductivity has been one of the most intensive topics in recent years, as this property is expected to maximize the uniqueness of organic materials in opening up many new application areas.

Inal et al. proposed the benchmarking of organic mixed conductors, comprehensively reflecting the understanding of these materials accumulated over the past few years [79]. The widely used structure of an organic electrochemical transistor (OECT) is shown in Figure 1.10a. This device resembles an OFET, but the major differences are that a gate dielectric is replaced by an electrolyte and that mixed conductivity is required for a channel in an OECT. As shown in the illustrations in Figure 1.10b, ionic transport within the electrolyte is driven by an applied V_G, which eventually forces a certain type of ions to penetrate into the channel and chemically dope it. While an OFET operates via capacitive coupling between

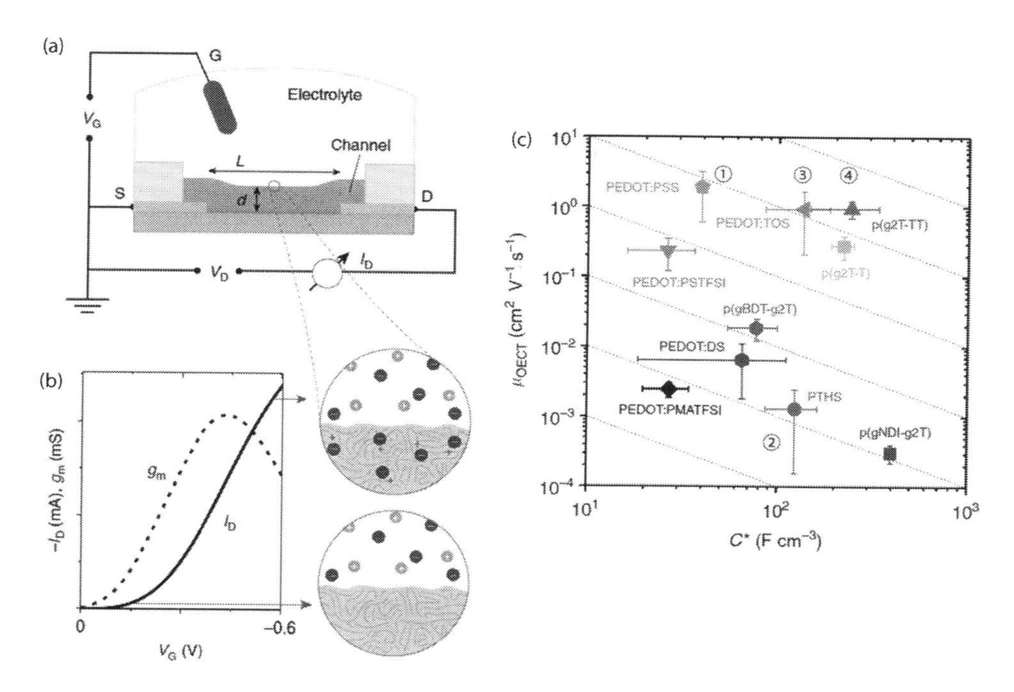

FIGURE 1.10 (a). General structure of an OECT with physical dimensions noted. *L*: channel length, *d*: channel thickness. (b) Representative transfer and g_m plot for a p-type accumulation mode OECT ($V_D < 0$ V). The illustrations show the penetration of anions into the channel at a high negative V_G that stabilizes hole in the semiconductor. (c) The OECT mobility versus volumetric capacitance plot of the ten reported materials. Circled numbers are for the purpose of discussion in the original paper. (Reprinted with permission from Macmillan Publishers Ltd. *Nat. Commun.*, Inal, S., 2017, Copyright 2017.)

gate and channel, I_D of an OECT is modulated electrochemically, thus a very low-voltage operation is possible (less than 1 V), as indicated in the transfer curve in Figure 1.10b. Also, combining this narrow voltage window and a high on-state current of a conducting polymer, a high transconductance ($g_m = dI_D/dV_G$) is obtainable. The OECT mechanism was not well documented until recently, and the concept of volumetric capacitance (C^*) was then introduced to clarify this. The capacitance term in an equation for a field-effect device is a unit-area capacitance of the dielectric, because of the formation of a 2D charge sheet at dielectric interface. In contrast, in an OECT, there is no electrostatic coupling, but the whole volume of the film reacts to V_G electrochemically. It allows us to regard the channel itself as a capacitor, whose capacitance scales with the physical volume of the film. In addition, C^* allows for the extraction of a (electronic) mobility from an OECT's transfer characteristic. These authors focused on this aspect of OECTs. They suggested the product of mobility (μ_{OECT}) and C^* as a new figure of merit relevant to organic mixed conductors, as this parameter addresses both how large the ion-induced charging capability is and how well the material conduct electronic charges. Also, from a mathematical point of view, both of these two factors are linear scaling factors of I_D, so their contributions are equivalent in the static operation [80]. Figure 1.10c shows the plot of μ_{OECT} and C^* for OECT materials. It is clear from this type of plot which material will result in a larger I_D and g_m given the equal effective biasing conditions.

Poly(3,4-ethylenedioxythiophene):polystyrene sulfonate (PEDOT:PSS) has long been the material of choice for OECTs. The improved molecular understanding is translating into the design principles for new materials [81–83], and such benchmarking seems to be useful to further this understanding. The availability of improved electronic-ionic active materials will also significantly contribute to the realization of organic bioelectronics [84,85].

1.7 Summary and Outlook

This chapter has briefly reviewed the vibrant field of plastic electronics. The worldwide research ignited by the somewhat unexpected electronic property of carbon-based materials has led to the tremendous scientific discoveries over the last 3 decades. If the established OLED display technology has proved the technological viability of these materials, the quest for the next big thing seems relentless. I have identified three important aspects that the field will most likely address in the near future. Firstly, creation of niche markets will be necessary. The large-area low-cost processability of materials is particularly attractive for disposable and customizable applications. Electronic food packages, point-of-care sensing kits, wearable identification tags, and card-integrated displays might be some of these applications where organic materials are not required to provide the best possible long-term performance, but to carry out simple tasks with clear cost advantages. Secondly, the ambient electronics will be a key motive. People will continue to enjoy a convenient life surrounded by electronics, but may want to feel less of its presence. The materials' lightness, mechanical softness, and optical transparency through hybrid design underlines a competitive edge. Skin-patch-type health monitors, building-integrated photovoltaics, window information displays, and indoor energy harvesters are possible examples. Finally, fundamental physics and chemistry of materials will remain a valuable theme. As many works cited in this chapter illustrated, the technological development has been underpinned by the revelation of new properties or the renewed interpretation of known mechanisms. Despite inherent complexity and variety, the continuing efforts to establish governing principles will eventually give a clearer view of how a device works, and they will broadly back both academic and commercial activities. Some of the basic understanding that this chapter has tried to deliver might serve as a useful reference for such activities, helping especially those who are new to the field.

Acknowledgments

This work was supported by the National Research Foundation of Korea (NRF) grant funded by the Korean government (MSIT) (NRF-2019R1C1C1003356).

References

1. T. Sekitani and T. Someya. 2010. Stretchable, large-area organic electronics. *Adv. Mater.* 22:2228–2246.
2. A. L. Briseno, R. J. Tseng, M.-M. Ling, et al. 2006. High-performance organic single-crystal transistors on flexible substrates. *Adv. Mater.* 18:2320–2324.
3. H. Klauk (ed.). 2006. *Organic Electronics: Materials, Manufacturing and Applications.* Weinheim, Germany: Wiley-VCH.
4. H. Klauk (ed.). 2012. *Organic Electronics II: More Materials and Applications.* Weinheim, Germany: Wiley-VCH.
5. J. A. Rogers, T. Someya, and Y. Huang. 2010. Materials and mechanics for stretchable electronics. *Science* 327:1603–1607.
6. G. Horowitz. 2004. Organic thin film transistors: From theory to real devices. *J. Mater. Res.* 19:1946–1962.
7. C. R. Newman, C. D. Frisbie, D. A. da Silva Filho, et al. 2004. Introduction to organic thin film transistors and design of n-channel organic semiconductors. *Chem. Mater.* 16:4436–4451.
8. A. Facchetti. 2007. Semiconductors for organic transistors. *Mater. Today.* 10:28–37.
9. G. Dennler, M. C. Scharber, and C. J. Brabec. 2009. Polymer-fullerene bulk-heterojunction solar cells. *Adv. Mater.* 21:1323–1338.
10. A. C. Arias, J. D. MacKenzie, I. McCulloch, et al. 2010. Materials and applications for large area electronics: Solution-based approaches. *Chem. Rev.* 110:3–24.

11. J. E. Anthony, A. Facchetti, M. Heeney, et al. 2010. N-type organic semiconductors in organic electronics. *Adv. Mater.* 22:3876–3892.
12. D. Gendron and M. Leclerc. 2011. New conjugated polymers for plastic solar cells. *Energy Environ. Sci.* 4:1225–1237.
13. K. Takimiya, S. Shinamura, I. Osaka, et al. 2011. Thienoacene-based organic semiconductors. *Adv. Mater.* 23:4347–4370.
14. C. B. Nielsen, S. Holliday, H.-Y. Chen, et al. 2015. Non-fullerene electron acceptors for use in organic solar cells. *Acc. Chem. Res.* 48:2803–2812.
15. B. Geffroy, P. Le Roy, and C. Prat. 2006. Organic light-emitting diode (OLED) technology: Materials, devices and display technologies. *Polym. Int.* 55:572–582.
16. D. D. Eley. 1948. Phthalocyanines as semiconductors. *Nature* 162:819.
17. M. Pope, H. P. Kallamann, and P. Magnante. 1963. Electroluminescence in organic crystals. *J. Chem. Phys.* 38: 2042–2043.
18. H. Shirakawa, E. J. Louis, A. G. MacDiarmid, et al. 1977. Synthesis of electrically conducting organic polymers: Halogen derivatives of polyacetylene, (CH)x. *J. Chem. Soc. Chem. Commun.* 16: 578–580.
19. A. Tsumura, H. Koezuka, and T. Ando. 1986. Macromolecular electronic device: Field-effect transistor with a polythiophene thin film. *Appl. Phys. Lett.* 49:1210–1212.
20. C. W. Tang. 1986. Two-layer organic photovoltaic cell. *Appl. Phys. Lett.* 48:183–185.
21. C. W. Tang and S. A. VanSlyke. 1987. Organic electroluminescent diodes. *Appl. Phys.Lett.* 51:913–915.
22. J. H. Burroughes, D. D. C. Bradley, A. R. Brown, et al. 1990. Light-emitting diodes based on conjugated polymers. *Nature* 347:539–541.
23. W. Brown and T. Poon. 2004. *Introduction to Organic Chemistry.* 3rd ed. Hoboken, NJ: Wiley.
24. C.-H. Kim and I. Kymissis. 2017. Graphene–organic hybrid electronics. *J. Mater. Chem. C* 5:4598–4613.
25. U. Zschieschang, F. Ante, D. Kälblein, et al. 2011. Dinaphtho[2,3-b:2′,3′-f]thieno[3,2-b]thiophene (DNTT) thin-film transistors with improved performance and stability. *Org. Electron.* 12:1370–1375.
26. R. Ruiz, D. Choudhary, B. Nickel, et al. 2004. Pentacene thin film growth. *Chem. Mater.* 16:4497–4508.
27. S. Steudel, S. De Vusser, S. De Jonge, et al. 2004. Influence of the dielectric roughness on the performance of pentacene transistors. *Appl. Phys. Lett.* 85:4400–4402.
28. S. Y. Yang, K. Shin, and C. E. Park. 2005. The effect of gate-dielectric surface energy on pentacene morphology and organic field-effect transistor characteristics. *Adv. Funct. Mater.* 15:1806–1814.
29. S. A. DiBenedetto, A. Facchetti, M. A. Ratner, et al. 2009. Molecular self-assembled monolayers and multilayers for organic and unconventional inorganic thin-film transistor applications. *Adv. Mater.* 21:1407–1433.
30. S. Jung, C.-H. Kim, Y. Bonnassieux, et al. 2015. Injection barrier at metal/organic semiconductor junctions with a Gaussian density-of-states. *J. Phys. D.-Appl. Phys.* 48:395103.
31. T. Richards, M. Bird, and H. Sirringhaus. 2008. A quantitative analytical model for static dipolar disorder broadening of the density of states at organic heterointerfaces. *J. Chem. Phys.* 128:234905.
32. C.-H. Kim, Y. Bonnassieux, and G. Horowitz. 2014. Compact dc modeling of organic field-effect transistors: Review and perspectives. *IEEE Trans. Electron Devices* 61:278–287.
33. N. Koch. 2007. Organic electronic devices and their functional interfaces. *ChemPhysChem* 8:1438–1455.
34. C. Van Dyck, V. Geskin, and J. Cornil. 2014. Fermi level pinning and orbital polarization effects in molecular junctions: The role of metal induced gap states. *Adv. Funct. Mater.* 24: 6154–6165.
35. C. H. Kim, O. Yaghmazadeh, D. Tondelier, et al. 2011. Capacitive behavior of pentacene-based diodes: Quasistatic dielectric constant and dielectric strength. *J. Appl. Phys.* 109: 083710.

36. C. H. Kim, Y. Bonnassieux, and G. Horowitz. 2011. Fundamental benefits of the staggered geometry for organic field-effect transistors. *IEEE Electron Device Lett.* 32:1302–1304.
37. C. H. Kim, Y. Bonnassieux, and G. Horowitz. 2013. Charge distribution and contact resistance model for coplanar organic field-effect transistors. *IEEE Trans. Electron Devices* 60:280–287.
38. S. E. Root, S. Savagatrup, A. D. Printz, et al. 2017. Mechanical properties of organic semiconductors for stretchable, highly flexible, and mechanically robust electronics. *Chem. Rev.* 117:6467–6499.
39. T. Someya, Z. Bao, and G. G. Malliaras. 2016. The rise of plastic bioelectronics. *Nature* 540:379–385.
40. J. A. Rogers, M. G. Lagally, and R. G. Nuzzo. 2011. Synthesis, assembly and applications of semiconductor nanomembranes. *Nature* 477:45–53.
41. R. A. Street. 2009. Thin-film transistors. *Adv. Mater.* 21:2007–2022.
42. A. C. Arias, J. Daniel, B. Krusor, et al. 2007. All-additive ink-jet-printed display backplanes: Materials development and integration. *J. Soc. Inf. Disp.* 15:485–490.
43. B. Kang, W. H. Lee, and K. Cho. 2013. Recent advances in organic transistor printing processes. *ACS Appl. Mater. Interfaces* 5:2302–2315.
44. U. Zschieschang, T. Yamamoto, K. Takimiya, et al. 2011. Organic electronics on banknotes. *Adv. Mater.* 23:654–658.
45. D. Tobjörk, and R. Österbacka. 2011. Paper electronics. *Adv. Mater.* 23:1935–1961.
46. M. Irimia-Vladu. 2014. 'Green' electronics: Biodegradable and biocompatible materials and devices for sustainable future. *Chem. Soc. Rev.* 43:588–610.
47. M. Kaltenbrunner, M. S. White, E. D. Głowacki, et al. 2012. Ultrathin and lightweight organic solar cells with high flexibility. *Nat. Commun.* 3:770.
48. V. Podzorov. 2013. Organic single crystals: Addressing the fundamentals of organic electronics. *MRS Bull.* 38:15–24.
49. S. Jung, C.-H. Kim, Y. Bonnassieux, et al. 2015. Fundamental insights into the threshold characteristics of organic field-effect transistors. *J. Phys. D-Appl. Phys.* 48: 035106.
50. G. Gelinck, P. Heremans, K. Nomoto, et al. 2010. Organic transistors in optical displays and microelectronic applications. *Adv. Mater.* 22:3778–3795.
51. C. Zhang, P. Chen, and W. Hu. 2015. Organic field-effect transistor-based gas sensors. *Chem. Soc. Rev.* 44:2087–2017.
52. L. Kergoat, B. Piro, M. Berggren, et al. 2012. Advances in organic transistor-based biosensors: From organic electrochemical transistors to electrolyte-gated organic field-effect transistors. *Anal. Bioanal. Chem.* 402:1813–1826.
53. Y. Cai, R. Shinar, Z. Zhou, et al. 2008. Multianalyte sensor array based on an organic light emitting diode platform. *Sens. Actuator B-Chem.* 134:727–735.
54. H. Sasabe and J. Kido. 2013. Development of high performance OLEDs for general lighting. *J. Mater. Chem. C* 1:1699–1707.
55. B. Kippelen and J.-L. Bredas. 2009. Organic photovoltaics. *Energy Environ. Sci.* 2:251–261.
56. C.-H. Kim, J. Choi, Y. Bonnassieux, et al. 2016. Simplified numerical simulation of organic photovoltaic devices. *J. Compt. Electron.* 15:1095–1102.
57. K. Forberich, F. Guo, C. Bronnbauer, et al. 2015. Efficiency limits and color of semitransparent organic solar cells for application in building-integrated photovoltaics. *Energy Technol.* 3:1051–1058.
58. C. H. Kim, H. Hlaing, F. Carta, et al. 2013. Templating and charge injection from copper electrodes into solution-processed organic field-effect transistors. *ACS Appl. Mater. Interfaces* 5:3716–3721.
59. D. Venkateshvaran, M. Nikolka, A. Sadhanala, et al. 2014. Approaching disorder-free transport in high-mobility conjugated polymers. *Nature* 515:384–388.
60. R. Noriega, J. Rivnay, K. Vandewal, et al. 2013. A general relationship between disorder, aggregation and charge transport in conjugated polymers. *Nat. Mater.* 12:1038–1044.
61. K. Yu, B. Park, G. Kim, et al. 2016. Optically transparent semiconducting polymer nanonetwork for flexible and transparent electronics. *Proc. Natl. Acad. Sci. U. S. A.* 113:14261–14266.

62. K. D. G. I. Jayawardena, L. J. Rozanski, C. A. Mills, et al. 2013. 'Inorganics-in-organics': Recent developments and outlook for 4G polymer solar cells. *Nanoscale* 5:8411–8427.
63. S. Sung, S. Park, W.-J. Lee, et al. 2015. Low-voltage flexible organic electronics based on high-performance sol–gel titanium dioxide dielectric. *ACS Appl. Mater. Interfaces* 7:7456–7461.
64. V. Pecunia, K. Banger, and H. Sirringhaus. 2015. High-performance solution-processed amorphous-oxide-semiconductor TFTs with organic polymeric gate dielectrics. *Adv. Electron. Mater.* 1:1400024.
65. H. Hlaing, C.-H. Kim, F. Carta, et al. 2015. Low-voltage organic electronics based on a gate-tunable injection barrier in vertical graphene-organic semiconductor heterostructures. *Nano Lett.* 15:69–74.
66. Y. Liu, H. Zhou, N. O. Weiss, et al. 2015. High-performance organic vertical thin film transistor using graphene as a tunable contact. *ACS Nano* 9:11102–11108.
67. G. Oh, J.-S. Kim, J. H. Jeon, et al. 2015. Graphene/pentacene barristor with ion-gel gate dielectric: Flexible ambipolar transistor with high mobility and on/off ratio. *ACS Nano* 9:7515–7522.
68. J. S. Kim, B. J. Kim, Y. J. Choi, et al. 2016. An organic vertical field-effect transistor with underside-doped graphene electrodes. *Adv. Mater.* 28:4803–4810.
69. S. Parui, M. Ribeiro, A. Atxabal, et al. 2017. Graphene as an electrode for solution- processed electron-transporting organic transistors. *Nanoscale* 9:10178–10185.
70. C.-H. Kim, H. Hlaing, and I. Kymissis. 2016. A macroscopic model for vertical graphene-organic semiconductor heterojunction field-effect transistors. *Org. Electron.* 36:45–49.
71. D. Jariwala, T. J. Marks, and M. C. Hersam. 2016. Mixed-dimensional van der Waals heterostructures. *Nat. Mater.* 16:170–181.
72. Y. Lui, N. O. Weiss, X. Duan, et al. 2016. Van der Waals heterostructures and devices. *Nat. Rev. Mater.* 1:16042.
73. K. D. G. I. Jayawardena, L. J. Rozanski, C. A. Mills, et al. 2015. The true status of solar cell technology. *Nat. Photon.* 9:207–208.
74. E. Zimmermann, P. Ehrenreich, T. Pfadler, et al. 2014. Erroneous efficiency reports harm organic solar cell research, *Nat. Photon.* 8:669–672.
75. H. Sirringhaus. 2014. 25th anniversary article: Organic field-effect transistors: The path beyond amorphous silicon. *Adv. Mater.* 26:1319–1335.
76. D. Choi, P.-H. Chu, M. McBride, et al. 2015. Best practices for reporting organic field effect transistor device performance. *Chem. Mater.* 27:4167–4168.
77. E. G. Bittle, J. I. Basham, T. N. Jackson, et al. 2016. Mobility overestimation due to gated contacts in organic field-effect transistors. *Nat. Commun.* 7:10908.
78. D. C. Martin and G. G. Malliaras. 2016. Interfacing electronic and ionic charge transport in bioelectronics. *ChemElectroChem* 3:686–688.
79. S. Inal, G. G. Malliaras, and J. Rivnay. 2017. Benchmarking organic mixed conductors for transistors. *Nat. Commun.* 8:1767.
80. J. Rivnay, P. Leleux, M. Ferro, et al. 2015. High-performance transistors for bioelectronics through tuning of channel thickness. *Sci. Adv.* 1:e1400251.
81. C. B. Nielsen, A. Giovannitti, D.-T. Sbircea, et al. 2016. Molecular design of semiconducting polymers for high-performance organic electrochemical transistors. *J. Am. Chem. Soc.* 138:10252–10259.
82. A. Giovannitti, C. B. Nielsen, D.-T. Sbircea, et al. 2016. N-type organic electrochemical transistors with stability in water. *Nat. Commun.* 7:13066.
83. C. M. Pacheco-Moreno, M. Schreck, A. D. Scaccabarozzi, et al. 2017. The importance of materials design to make ions flow: Toward novel materials platforms for bioelectronics applications. *Adv. Mater.* 29:1604446.
84. J. Rivnay, R. M. Owens, and G. G. Malliaras. 2013. The rise of organic bioelectronics. *Chem. Mater.* 26:679–685.
85. C. Liao, M. Zhang, M. Y. Yao, et al. 2015. Flexible organic electronics in biology: Materials and devices. *Adv. Mater.* 27:7493–7527.

2

Materials—Flexible 1D Electronics

Aftab M. Hussain

2.1 Introduction

Material science has made great progress in the last century. We have understood the nature of the most fundamental building blocks of matter and have modeled their properties with quantum theory. Included in this revolution is the realization that not all materials necessarily occupy a 3D lattice. There are certain materials that form stable 1D and 2D structures. The 1D structures consist of long chains of molecules that are only a few atomic distances in diameter. A major turning point in the study of 1D materials was the discovery of carbon nanotubes in 1991. Over the past 2 decades, many such materials have been discovered or synthesized that have diameters less than 100 nm and lengths in the order of centimeters. Further, various studies have been focused on the electronic, material, thermal, and optical properties of such materials.

Truly 1D materials are those that have a single atomic chain several atoms in length. However, such materials are rarely stable and are mostly analyzed theoretically for understanding the properties of 1D materials. In general, most materials of interest have a diameter of several atomic distances, however, because their lengths are so large compared to their diameter, they are classified as *quasi* 1D materials. These terms are mostly used interchangeably, and in this chapter, we will use the term '1D materials' to refer to both 1D and quasi 1D materials. These materials are also commonly referred to as 'nanowires,' 'nanotubes,' 'nanorods,' etc. depending on their structure.

The ratio of length to diameter, called the aspect ratio of a 1D material, varies from 100 to 10^7. The high aspect ratio and the peculiar chemical bonding required to keep it stable gives some unique properties to the 1D materials. Firstly, because of extremely low thickness in two dimensions, the flexural rigidity of these materials is very low in these planes (Rogers et al. 2011). Thus, they are easily flexed in these planes. Secondly, several of these materials provide unique electronic and material characteristics. Thus, for a given combination of electrical and mechanical properties, there is generally a selection of 1D materials to choose from. Thirdly, their large surface to volume ratio enables high sensitivity to surface interactions, thus, enabling their application as sensor systems. Lastly, these materials are stable over a large range of temperatures. Hence, with these characteristics, these materials can be ideal for flexible electronics applications.

In the next section, we will discuss several well studied 1D materials. These materials have been proven to be useful for many flexible electronics applications. These applications will be discussed in the subsequent section.

2.2 Various 1D Materials

2.2.1 Carbon Nanotubes

Carbon nanotubes (CNTs) are the most studied 1D material system. Carbon nanotubes are made of sp^2 hybridized carbon atoms arranged in a hexagonal lattice structure. They are generally described as graphene sheets rolled into cylinders (Figure 2.1a). CNTs can be single-walled (SWNTs), with only one atomic layer of carbon atoms forming the cylinder, double-walled (DWNTs), with two such cylinders, or multi-walled (MWNTs), wherein multiple cylinders of atomic carbon are nestled within each other (Figure 2.1b). Much work has been done in the past 2 decades in advancing the synthesis and applicability of CNTs and some CNT applications are finally showing true promise. In this section, we will look at the synthesis of various kinds of CNTs, and how their structure and properties are unique providing multiple avenues of potential applications, which will be discussed in detail in the next section.

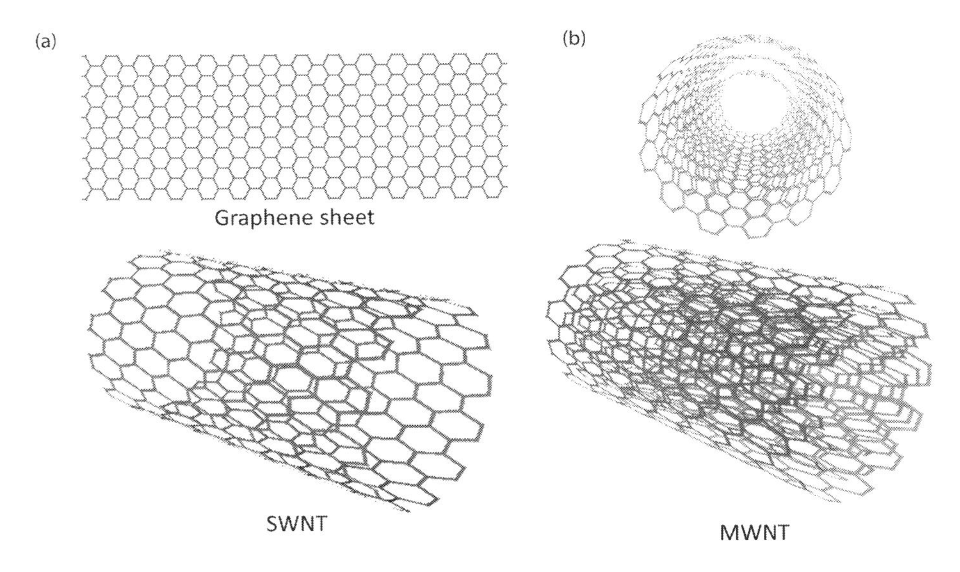

FIGURE 2.1 (a) Single graphene sheet with planar hexagonal lattice structure rolled into a SWNT. (b) Multiwall NTs (MWNTs) consist of many coaxial SWNTs.

2.2.1.1 Synthesis

Numerous laboratory scale processes have been reported to synthesize both SWCNTs and MWCNTs. A suitable process for synthesis is chosen based on the particular application in question. In particular, the size, diameter, number of walls, and orientation considerations dictate the preferred method of synthesis. The most famous synthesis process is the arc discharge process which was used by Iijima in his landmark paper in 1991 (Iijima 1991). In this process, an electric arc is formed between two carbon (graphite) electrodes at a fixed distance from them both. The arc is generally formed at low pressure (in the presence of inert gases) with a current around 50–100 A and voltage at 30–40 V. This creates a high temperature and vaporizes the carbon atoms at the tip of the electrodes. These vapors then condense as CNTs and other carbon-based materials such as fullerenes. A simple arc discharge synthesis system is shown in Figure 2.2a. Because the electrodes are consumed during the process of CNT formation, the electrodes have to be steadily shifted to maintain a constant distance, and have to be eventually replaced. The arc discharge process has been extensively used and studied to obtain both SWNTs, DWNTs and MWNTs (Shi et al. 2000, Sugai et al. 2003, Michael 2007). We suggest the review by Arora et al. for detailed taxonomy and review of arc discharge synthesis of CNTs (Arora and Sharma 2014). One of the major challenges with this process is the accumulation of other carbon-based materials along with the CNTs, thus entailing a subsequent, often cumbersome, purification process. Thus, several studies have been dedicated to obtain a solution to this problem using either in-situ modifications to the process (Roslan et al. 2016, Yatom et al. 2017) or post synthesis treatments (Hokkanen et al. 2016). Another key synthesis process is the laser ablation which involves the use of lasers to create high temperature to vaporize carbon targets and obtain CNTs on condensation. The laser ablation process is similar to the arc discharge method in that it also creates many other carbonaceous by-products apart from CNTs. Studies have shown that including metal nanoparticles in the graphite target can give rise to the growth of SWNTs. The quality and quantity of the nanotubes produced with this method can be controlled to some extent using the laser power, duration, and wavelength (Chrzanowska et al. 2015). Das et al. provide a detailed review of these two processes and discuss the question of optimizing these for industry applications (Das et al. 2016).

Although arc discharge is one of the earliest methods used for CNT synthesis, it is not suitable for large scale synthesis mainly due to other carbon-based materials synthesized in the process and due to low throughput of the process. For industrial scale manufacturing, chemical vapor deposition (CVD) is the most commonly used process. In this process, a hydrocarbon gas is flown over a transition metal catalyst in a furnace at a high temperature. Typically, a quartz tube furnace is used, and the hydrocarbon gas is diluted with argon or hydrogen. The temperature ranges from 500°C–1200°C, and

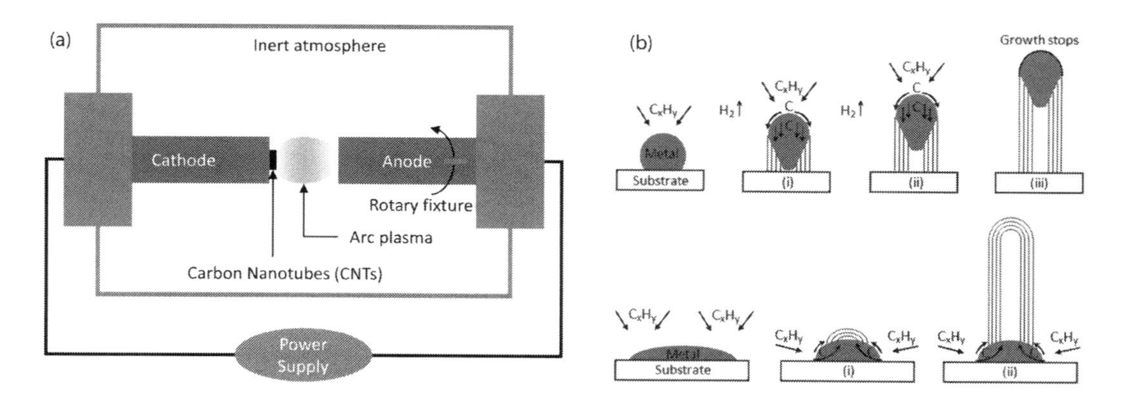

FIGURE 2.2 (a) Schematic of an arc discharge furnace. (b) Mechanisms for CVD growth of CNTs. Top: tip-growth process, bottom: base-growth process. (From Kumar, M., Carbon nanotube synthesis and growth mechanism, in *Carbon Nanotubes—Synthesis, Characterization, Applications*, ed. Yellampalli, S., Ch. 8, InTech, Rijeka, Croatia, 2011.)

the pressure is kept at a few Torr range. During the process, the hydrocarbon breaks into hydrogen and carbon at high temperatures at the metal catalyst surface, and the carbon atoms are dissolved into the catalyst. Once the solubility limit for carbon in the metal is reached, the carbon atoms are precipitated as crystalline nanotubes. This general process is the basic CVD process for CNT synthesis and has been thoroughly studied with many variations. A schematic illustration of the basic process is shown in Figure 2.2b. The most commonly reported hydrocarbon gas used in the synthesis is methane (Kong et al. 1998, Colomer et al. 2000, Su et al. 2000, Harutyunyan et al. 2002), however, studies have also reported synthesis using ethane (Louis et al. 2005), ethylene (Lyu et al. 2004), acetylene (Couteau et al. 2003), and benzene (Tian et al. 2004). Other gases reported in literature for CNT growth include carbon monoxide (Zheng et al. 2002) and ethyl alcohol (Maruyama et al. 2002). On the other hand, the most commonly used catalysts are iron, nickel, and cobalt. These metals have a high solubility of carbon and a high melting point which facilitates the large range of temperatures to be used for the synthesis. Many other transition metals have been reported in literature, including Pt, Au, Pd, Ag, Sn, Mo, Al, Mg, Cu, and Cr. Generally, metal nanoparticles are used as a catalyst which increases the surface area for reaction and offers control of CNT diameter using nanoparticle size, as typically one nanoparticle supports the growth of one CNT. Although this poses a challenge of fabricating or depositing metal nanoparticles on a given substrate for CVD growth of CNTs, this control is a distinguishing factor for this process. Similarly, the CVD process affords much more control in terms of temperature, pressure and gas flow variation, carbon precursor variation, metal catalyst variation, and so on. Indeed, CVD also produces far less carbonaceous by-products compared to other synthesis processes and thus requires less purification steps. For a comprehensive review of catalytic CVD for CNT synthesis, we recommend a recent review by Shah and Tali (2016). Yan et al. also provide a detailed review of all the three synthesis processes discussed (Yan et al. 2015).

2.2.1.2 Structure

CNTs consist of sp^2 hybridized carbon atoms σ-bonded together in a planar hexagonal structure as shown in Figure 2.1a. This planar structure is commonly observed in graphite, wherein these planar graphene sheets are held together by weak van der Waals forces. In this structure, each carbon atom is bonded to three other carbon atoms with a single covalent bond with bond angles of 120° and bond lengths of 0.142 nm each. The remaining π electrons form orbitals out-of-plane of the bond plane. Because sp^2 hybridized orbitals are ideally planar, the curved structure of a CNT causes σ–π rehybridization which results in slightly out-of-plane σ-bonds and more delocalized π electrons outside the tube. In general, this nanotube structure is extremely stable and is seen to have very few defects, however, in some cases, pentagons, heptagons, and vacancies are seen to be incorporated in the hexagonal structure (Suenaga et al. 2007). These localized 'defects' in the nanotube structure also locally change π electron density and change the electronic properties of the nanotubes (Charlier et al. 1996, Charlier 2002).

A single-wall carbon nanotube can be pictured as a cylinder formed by rolling a planar graphite sheet. However, for a given sheet of hexagonally tessellated plane, there can be many ways of forming the cylinder based on the vector used for 'folding' the plane. This vector is represented on the graphite sheet plane using indices n and m for the two graphite lattice vectors $\boldsymbol{a}_1$ and $\boldsymbol{a}_2$ as shown in Figure 2.3.

Thus, vector $\boldsymbol{C}$ is defined as,

$$\boldsymbol{C} = n\boldsymbol{a}_1 + m\boldsymbol{a}_2. \tag{2.1}$$

The length of vector $\boldsymbol{C}$ forms the circumference of the resulting nanotube. Thus, the diameter of the nanotube is given by:

$$d = \frac{C}{\pi} = \left(\frac{a}{\pi}\right)\sqrt{n^2 + nm + m^2}, \tag{2.2}$$

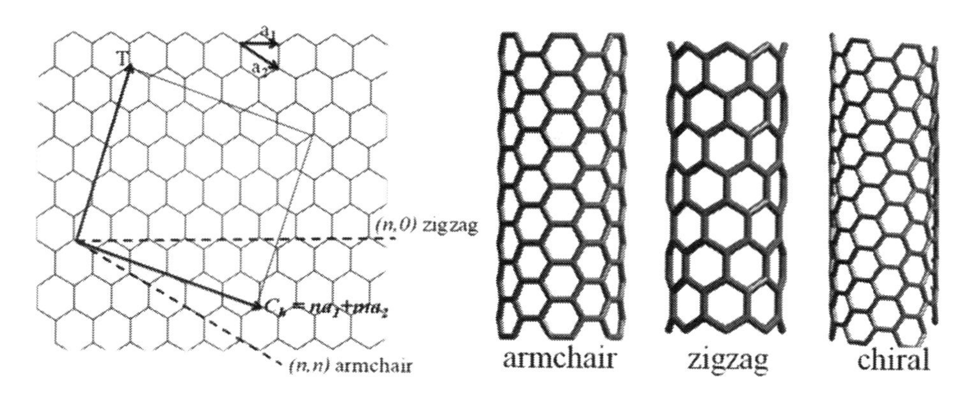

FIGURE 2.3 Formation of carbon nanotubes by rolling a graphene sheet. Depending on the indices n and m, different nanotube structures are obtained. (From Choudhary, V. and Gupta, A., Polymer/carbon nanotube nanocomposites, in *Carbon Nanotubes—Polymer Nanocomposites*, ed. Yellampalli, S., Ch. 4, InTech, Rijeka, Croatia, 2011.)

where a is the magnitude of vectors $\boldsymbol{a}_1$ and $\boldsymbol{a}_2$. The value of the lattice parameter a is 0.246 nm, which can be obtained from the C–C bond length and has been confirmed by many experiments and theoretical calculations. The angle of vector $\boldsymbol{C}$ with respect to vector $\boldsymbol{a}_1$ is called the chiral angle of the nanotube. In terms of n and m, the chiral angle is given by:

$$\theta = \tan^{-1}\left[\frac{\sqrt{3}m}{m+2n}\right]. \tag{2.3}$$

In the special case of $m=0$, $\theta=0°$. Thus, vector $\boldsymbol{C}$ is in the direction of vector $\boldsymbol{a}_1$, and the nanotube obtained is called a zigzag nanotube. In the case of $m=n$, $\theta=30°$, the nanotube obtained is called an armchair nanotube. Any angle apart from these angles produces a 'chiral nanotube.'

2.2.1.3 Properties

The peculiar structure of CNTs is responsible for their unique electronic, mechanical, and thermal and optical properties. The electronic properties of CNTs are the most studied and reported in literature. Because the nanotubes are modeled as a rolled graphite sheet, many of the properties are closely related to graphite sheet properties. However, the rolling of sheets causes a periodic boundary condition to be imposed on the wave vector, quantizing it along the circumference (or the direction of vector $\boldsymbol{C}$). This causes certain nanotubes to be completely metallic, while a significant band gap opens up in other nanotubes, which is tunable based on their chirality index (n, m). In general, two-thirds of SWNTs are semiconducting, while one-third are metallic. This ratio, predicted by theoretical analysis, has been confirmed by experimental observations (Cambré et al. 2010). The band gap of the semiconducting nanotubes is around 0.4–0.7 eV (Wilder et al. 1998). Also, even in case of metallic zigzag SWNTs, the curvature of nanotubes and the resulting σ–π rehybridization opens up a small band gap of around 0.04–0.08 eV (Matsuda et al. 2010). These results have been calculated and reported for SWNTs, hence, when considering the properties of MWNTs, the coupling between the individual coaxial nanotubes has to be taken into account.

A key characteristic of nanotubes is the presence of delocalized π electrons, which provides excellent electrical conductivity to the nanotubes. Early studies have shown the resistivity of metallic CNTs to be 90 μΩ-cm (Kane et al. 1998). Because of the presence of a relatively defect-free lattice, the π electrons can achieve near ballistic conduction under an applied electric field. Even in the case of semiconducting nanotubes, most MWNTs tend to be semimetallic due to the inter-tube coupling. Hence, many current and potential applications are related to exploiting this high conductivity in CNT forests. However, these characteristics pertain to single SWNTs or MWNTs, and the electrical properties of a bulk sample or a large thin film of nanotubes depend greatly on the fabrication process, sample quality, and alignment. Along with high electrical conductance, the presence of π electrons also accords phenomenal thermal conductivity to CNTs. Reported values range widely, again pertaining to the quality of the sample, however, single CNTs have been reported to have theoretical thermal conductivity of more than 6000 W/mK (Berber et al. 2000).

The other most studied and reported aspect of CNTs is their mechanical strength. The sp^2 carbon-carbon σ-bond is stronger than the sp^3 bond in a diamond, thus making CNT tougher than diamond in the tube-axis. Elastic modulus of more than 1 TPa has been reported for CNTs along the tube axis (Treacy et al. 1996, Krishnan et al. 1998). The modulus values do not depend on chirality, however, they are dependent on the diameter of the nanotubes. Also, MWNTs tend to have larger modulus values because of the presence of van der Waals interactions between the tubes adding to the moduli of the individual SWNTs. Lastly, it is worth mentioning the thermal and chemical stability of CNTs. Again, because of the strong C–C bonding, CNTs tend to be relatively inert. CNTs have been shown to be biocompatible for the short-term applications (Silvano et al. 2006), while long-term biocompatibility is being investigated (Smart et al. 2006). This is an important property because many of the potential applications being considered for CNTs involve wearable or implantable devices.

2.2.2 Metal Nanowires

Nanowire structures of metals have gained attention in the past 2 decades mainly for flexible and possible transparent electrode applications. In particular, silver nanowires (AgNWs) have been well studied and reported for many reasons. In bulk form, silver is the most conductive metal known to man, followed by copper and gold (Matula 1979). Further, silver nanowires are relatively easily synthesized and are more stable than nanowires of other metals. Lastly, silver is considerably more affordable than platinum and gold, thus opening up opportunities for large scale applications. The most promising process for synthesizing AgNWs on a large scale is the polyol process. The general process consists of precipitating metal nanocrystals from metal salts in a reducing bath along with a capping agent to restrict the growth of metal nanostructures in a single direction to form nanowires. In the case of silver, the reducing bath is that of a polyol (mostly ethylene glycol) and polyvinyl pyrrolidone is used as a capping agent. Silver nitrate ($AgNO_3$) can be used as a metal precursor (Sun et al. 2002, Whitcomb et al. 2016). This process is of particular importance because it can be scaled up merely by scaling the reducing bath, all the chemicals involved in the process are relatively low-cost, and the process can be done at a relatively low temperature of around 150°C. The process affords some control over the shape and size of the nanowires by changing the relative concentration of the chemicals and temperature, or by addition of $PtCl_2$ or NaCl as nucleating seeds (Lin et al. 2014, 2015). Several other processes have been reported for synthesis of metal nanowires, prominent among them are the hard-templating process, UV irradiation process, hydrothermal process, and electrothermal process. We recommend the review by Daniel et al. for details on silver nanowire synthesis (Daniel et al. 2013). Other than silver nanowires, copper, and gold nanowires have shown promise. However, copper nanowires tend to form native oxide in air and can lose conductivity at contact points which necessitates processes to protect them (Chen et al. 2014), and gold nanowires are expensive to make. Hence, silver nanowires seem to be the most promising metal nanowire candidate for large scale applications. In general, metal nanowires have the same electrical properties as the bulk metals unless the nanowire diameters are small enough for quantum confinement effects to be observed. These nanowires are generally used as a network to form conducting thin films (Figure 2.4).

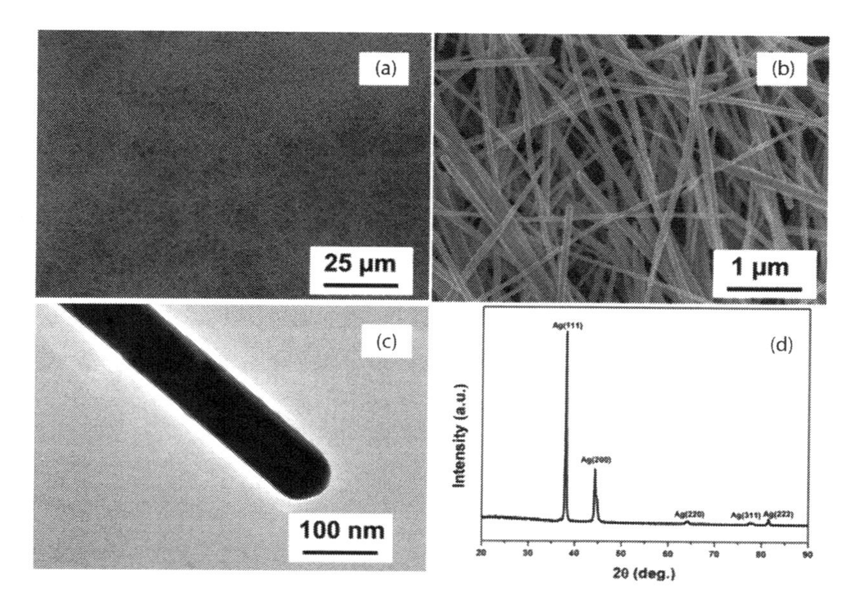

FIGURE 2.4 Silver nanowires (AgNWs): (a, b) Scanning electron micrographs (SEMs), (c) Transmission electron micrograph (TEM), and (d) X-ray diffraction pattern. (Reprinted from *J. Mater. Sci. Technol.*, 31, Huang, Q. et al., Separation of silver nanocrystals for surface-enhanced Raman scattering using density gradient centrifugation, 834–839, Copyright 2015, with permission from Elsevier.)

2.2.3 Semiconducting Nanowires and Nanotubes

Semiconducting 1D materials have the potential to be used for fabrication of ultra-small electronic devices such as field effect transistors (FETs). This potential has sparked a great interest in the synthesis, properties, and applications of semiconducting 1D materials, particularly, silicon nanowires (SiNWs). In general, CVD is most commonly used to synthesize semiconducting nanostructures. A precursor vapor such as silane (SiH_4) can be broken down at high temperatures (>800°C) at particular nucleating sites to form 1D nanostructures. However, because of the requirement of very high temperatures for CVD, modifications such as the vapor-liquid-solid (VLS) method have been introduced to obtain condensation at much lower temperatures. In general, the VLS process involves the use of a liquid metal alloy droplet to absorb atoms from a precursor vapor, and the atoms are eventually precipitated in the form of a nanowire once supersaturation is reached. In particular, Au-Si eutectic liquid droplets are commonly used to synthesize SiNWs. These droplets are formed in-situ by annealing substrates deposited with a thin film of Au in the presence of silicon precursor vapor. The thickness of the gold thin film determines the size of the droplets formed, and hence the diameter of the nanowires synthesized. Because Au-Si metal alloy has a melting point of 363°C at eutectic point (at around 19% silicon) (Anantatmula et al. 1975), the synthesis process can be done at a much lower temperature compared to a simple CVD process. A similar process called the vapor-solid-solid (VSS) method is said to be employed when the reaction temperature is lower than the eutectic melting point of the alloy, and the alloy exists as solid nanoparticles. These processes are also used to obtain III-V semiconductor nanowires, however, in this case, the precursor gas is obtained directly by evaporating a solid source. Referred to as the solid source CVD (SSCVD) process, this method reduces the cost of procuring, transporting, and storing highly toxic and volatile precursor gases (Fang et al. 2014).

A major challenge in the growth of semiconducting nanowires is the control of the crystal plane orientation of the resulting nanowire. This is of particular importance because most of the electrical, chemical, and physical properties of the nanowire depend on it. In case of VLS and VSS methods, the preferential plane of crystallization of semiconducting nanowires depends on the droplet size of metal-semiconductor alloy and the temperature of growth. Apart from the crystalline orientation, it is important to control the position of the nanostructures on the substrate to be able to use them in electronic applications. The selective area epitaxy (SEA) method is one of the ways of obtaining semiconducting nanostructures with a controlled orientation and position on the substrate. Epitaxial deposition is the process of depositing crystalline thin films on a crystalline substrate of same or similar lattice using precursor gases. In case of SEA, a thin film mask layer is used to protect areas where growth is undesirable. The thin film can be patterned and etched using advanced lithography methods to obtain nanoscale trenches. Epitaxy using such a masked substrate provides nanostructures with well-defined crystal orientation, diameter, and pitch. Such a process has been demonstrated to obtain controlled nanostructures of both silicon (Fahad et al. 2014) and III-V semiconductors (Pratyush Das et al. 2013, Kruse et al. 2016, Berg et al. 2017).

2.2.4 Other 1D Materials

Apart from the materials already discussed, many more have been shown to exist in stable 1D and quasi 1D structures. These include metal oxide nanowires of copper (Filipič and Cvelbar 2012), zinc (Greene et al. 2006), cobalt (Rakhi et al. 2012), tungsten (Gu et al. 2002), indium (Wan et al. 2006), and tin (Dai et al. 2002). Both VLS and VSS processes discussed above have been successfully employed to fabricate these metal oxide nanowires, apart from methods such as solution processing, thermal oxidation, and thermal evaporation (Shen et al. 2009). Metal oxide nanowires can be insulating, semiconducting, or conducting, depending on the metal in question. Another compound studied extensively in 1D nanostructure form is boron nitride (BN). Both nanowires and nanotubes have been reported for boron nitride, but boron nitride nanotubes (BNNTs) are the most commonly studied and reported. Again, CVD is a key method used for synthesis of BNNTs, although other methods such as arch discharge and laser ablation have been reported (Ahmad et al. 2015). BNNTs are cylindrical structures of hexagonal boron nitride (h-BN) with diameters ranging from 1 to 100 nm. They are semiconductors with a large bandgap and have intrinsic magnetization dependent on their chirality index.

2.3 Flexible Electronics Based on 1D Materials

2.3.1 Circuits

The most common application of 1D materials such as CNTs, SiNWs, BNNTs, and even metal oxide nanowires has been as FETs. This application gained significance after the physical limits of scaling started to threaten the continuation of Moore's law in silicon electronics. Field effect transistors based on 1D nanostructures were considered the electronic devices of the future. However, challenges relating to alignment, large-scale integration, and contact formation did not allow this scenario to materialize. Even so, 1D nanostructure devices still promise to be of interest, particularly, after the rise of wearable electronics for Internet of things (IoT) applications. With better control on synthesis, purification, alignment, and integration, larger flexible electronic circuits are being reported based on 1D materials with continually increasing performance.

One of the landmark studies in the application of CNTs as transistors was published by Shulaker et al. (2013). Although the CNT 'computer' reported in this work was not flexible, this was the first time a large number of CNT FETs (178 transistors) had been interconnected to form a working circuit. Chen et al. reported a 4-bit adder using 140 top gate CNT FETs in 2016 (Chen et al. 2016a). Although these are still far from a billion-plus transistors routinely fabricated using silicon complementary metal-oxide semiconductor (CMOS) technology, these efforts signify the first of many such works with increasing CNT transistor density and reliability. In 2016, Zhang et al. reported flexible circuits using top gate CNT FETs with an impressive I_{on}/I_{off} ratio of 10^5 and mobility of 15 cm^2/V-s (Zhang et al. 2016d). Similar success has been obtained with SiNWs as well. Yun et al. reported the fabrication and characterization

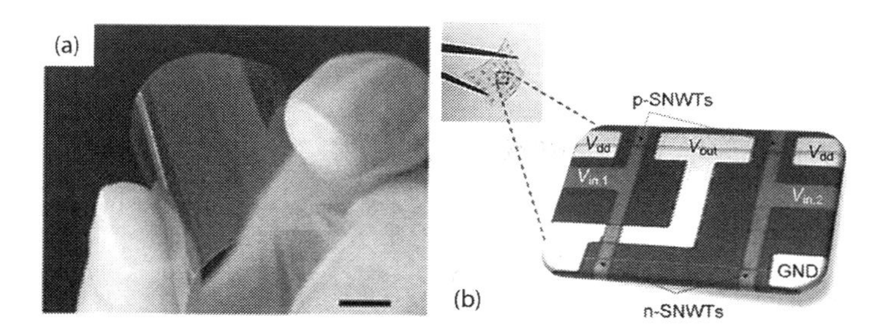

FIGURE 2.5 (a) Photograph of an all-carbon device fabricated on a flexible PEN substrate (scale bar, 10 mm). (Reprinted with permission from Macmillan Publishers Ltd. *Nat. Commun.*, Sun, D.M. et al., 2013, Copyright 2013.) (b) Flexible SiNW FET-based NAND circuit. (With kind permission from Springer Science+Business Media: *Nano Res.*, Nanowatt power operation of silicon nanowire NAND logic gates on bendable substrates, 9, 2016, 3656–3662, Yun, J. et al.)

of SiNW-based flexible NAND logic gates with supply voltage as low as 0.8 V (Yun et al. 2016). Navaraj et al. demonstrated a flexible array of tactile sensors based on silicon nanowires with high-κ Al_2O_3 gate dielectric and Pt/Ti electrodes (Taube Navaraj et al. 2017). Recently, indium oxide-based transistors have been reported to have I_{on}/I_{off} ratio of 10^6 and mobility of 218.3 cm^2/V-s (Liu et al. 2017a) (Figure 2.5).

2.3.2 Electrode Applications

CNTs and metal nanowires have both been shown to have very high conductivity and flexibility. These properties make them ideal candidates for formation of flexible conductive thin films. However, to obtain two-dimensional thin films from these 1D and quasi 1D structures, a network or mesh of many nanostructures is required. Ideally, a network of nanowires/nanotubes would contain them in regular mesh so that their properties can be repeatably obtained. However, this is extremely difficult to achieve, and most of the work in literature has focused on a 2D mesh or 3D composite of randomly oriented NW/NTs with varying thicknesses, diameters. and lengths. This makes their properties hard to analyze.

One of the commonly applied mathematical frameworks to study these random electrode structures is the percolation theory (Hu et al. 2004, Pfeifer et al. 2010). The theory models the behavior of randomly connected clusters in a graph. Given a particular random distribution of CNTs or AgNWs on a substrate or in a polymer composite, the probability of obtaining conductive pathways from point A to point B can be analyzed using the percolation theory. It is obvious that a large number of possible conductive pathways between two points will decrease the apparent electrical resistance of the thin film. Additionally, according to percolation theory, there is a minimum density of 1D nanostructures required to 'just' obtain conductivity, i.e., to obtain at least one conductive pathway between any two points on the thin film. This limit is called the percolation limit. The conductivity of the thin film increases drastically as the NT/NW density is increased beyond the percolation limit. However, the increase in conductivity quickly saturates and a subsequent additional increase in nanostructure density does not yield the same conductivity increase. The same theory holds true for 3D nanostructure polymer composites as well (Bauhofer and Kovacs 2009) (Figure 2.6).

While flexibility is a key enabler for many applications involving 1D nanostructure thin films, their unique advantage lies in their stretchability. They are ideal for applications requiring thin films that maintain their electrical properties even under large strains. Dielectric elastomer actuators (DEAs) are one such example. Broadly, DEAs are capacitors made of complaint dielectrics (typically polymers such as silicones or acrylics), that produce lateral motion when an electric field is applied (O'halloran et al. 2008). However, to be able to continually apply the electric field, the electrode material also needs to undergo strain without cracking and losing resistivity. Thin films based on percolation networks of 1D nanostructures have been extensively used to fabricate DEAs (Curdin et al. 2016, Duduta et al. 2016,

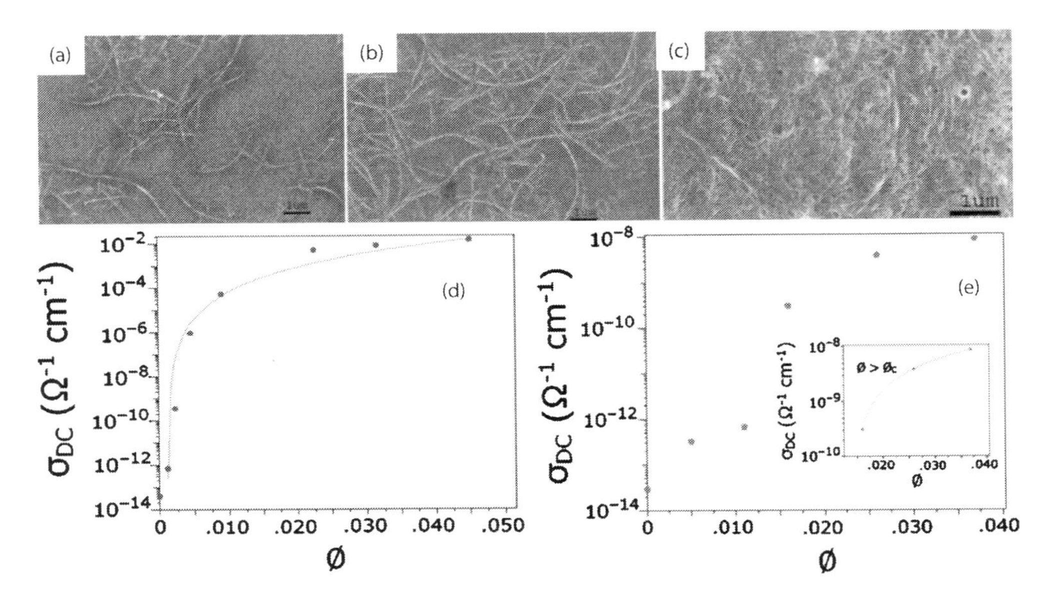

FIGURE 2.6 (a–c) SEM images of networks with increasing concentration of CNTs. (Reprinted with permission from Hu, L. et al., *Nano Lett.*, 4, 2513–2517, 2004. Copyright 2014 American Chemical Society.) (d) and (e) The variation in the dc electrical conductivity (σ_{dc}) with volume fraction (ϕ) for (d) SWNTs and (e) MWNTs dispersed in a polymer matrix. (Reprinted with the permission from Pfeifer, S. et al., *J. Appl. Phys.*, 108, 024305, 2010. Copyright 2010 by the American Institute of Physics.)

Park et al. 2017). Other applications of stretchable electrodes using 1D nanostructures reported in literature include field effect transistors (Chortos et al. 2016, Shlafman et al. 2016), supercapacitors (Cherusseri et al. 2016, Lv et al. 2016, Yu et al. 2016), biomedical applications (Hosseini et al. 2016, Kim et al. 2016), and sensors (Choi et al. 2017). These applications are based on the fact that individual 1D nanostructures merely slip with respect to each other when the thin film or composite is strained, thus, affording continuous conductivity while straining. Further, when such a thin film is strained, the individual nanostructures are not strained themselves, thus, the modulus of the thin film is relatively low and is generally close to the polymer matrix encompassing them. Lastly, it should be pointed out that several studies have reported creating thin films by combining CNTs and metal nanowires to obtain hybrid electrodes with improved performance metrics (Jun et al. 2017, Lee et al. 2017b).

Apart from flexibility and stretchability, a conductive network of a 1D nanostructure has another interesting and desirable property: transparency. It is evident from the optical images of thin films of 1D nanostructures that a certain percentage of substrate is not covered by the nanostructures. These gaps are sufficient to allow a percentage of light to be transmitted through the thin film. Traditionally, indium tin oxide (ITO) thin films have been used to obtain transparent and conductive thin films. However, with an increase in indium prices and concerns about its abundance, there has been a great push towards replacing them with thin films of 1D materials (Chen et al. 2016b). A myriad studies have reported various percentages of transmittance and haze for a given concentration of CNTs or metal nanowires (Hwang et al. 2016) (Figure 2.7).

2.3.3 Optoelectronic Applications

Semiconducting thin films of SW-CNTs can have a large range of band gaps depending on their chirality index. These band gaps are close to the photon energy of solar irradiation, thus enabling use of semiconducting CNT thin films in photovoltaic (PV) devices. When photons of a sufficient energy are absorbed

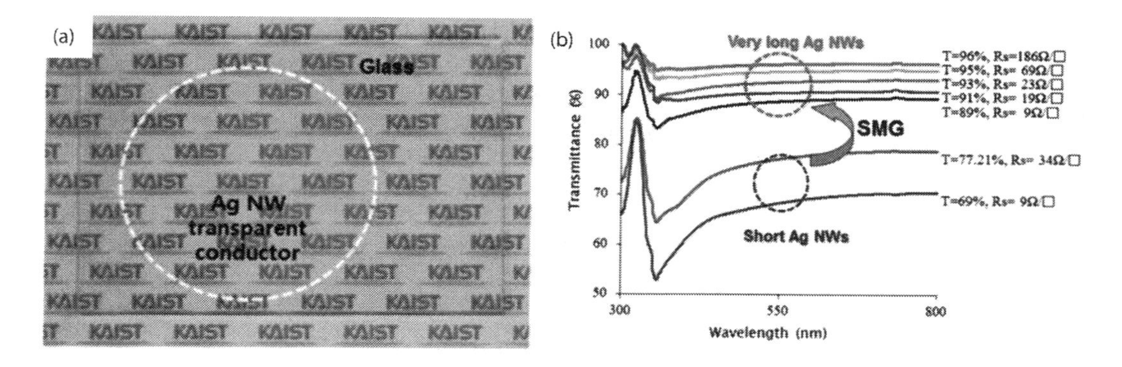

FIGURE 2.7 (a) Optical image of a transparent and conductive AgNW thin film on glass. (b) Transmittance of AgNW thin films versus wavelength for short AgNWs and long AgNWs synthesized using successive multistep growth (SMG) process. (Lee, J. et al., *Nanoscale*, 4, 6408–6414, 2012. Reproduced by permission of The Royal Society of Chemistry.)

by a semiconductor, an exciton (electron-hole pair) is formed, which is separated into the two carriers by the built-in potential, converting light intensity to electric current. Based on this phenomenon, SWNTs can be used to make ideal diodes with high power conversion efficiency (Lee 2005). However, single CNT-based PV devices are difficult to fabricate on a large scale and offer very low current to be of practical use. Thus, efforts are made to fabricate PV structures using CNT thin films. The main challenge here is to separate semiconducting CNTs from metallic ones, which cannot be used for PV applications. Even after this is achieved, semiconducting CNT thin films are generally used in a heterojunction configuration with another material (Wang et al. 2017a, Wei et al. 2007) or as a polymer composite (Hosseini and Kouklin 2016). However, some effort has been made to use CNTs both as a photoactive material and an electrode to transport the carriers, thus making all carbon PV cells (Klinger et al. 2012, Liu et al. 2017c). Liu et al. presented a comprehensive review on the state-of-the-art of CNT-based PV (Liu et al. 2016).

Similar to CNTs, other semiconducting 1D nanostructures such as silicon nanowires (SiNWs) and III-V nanowires have been extensively studied for PV applications (Priolo et al. 2014, Zhang et al. 2016a). These 1D structures offer some distinct advantages compared to CNT thin films, such as, they can be grown scalably to a high degree of purity and crystallinity using CVD, they can be oriented according to need, they can be doped in-situ, and they can be integrated with existing CMOS processes. However, this process makes these materials expensive compared to CNT thin films. The semiconductor nanowires are generally grown in the form of arrays to form large area solar cells capable of producing currents usable in practical applications. In 2013, Wallentin et al. reported solar cells based on indium phosphide (InP) nanowire arrays, grown using epitaxial process with in-situ doping, that achieve efficiencies up to 13.8% (Wallentin et al. 2013). Recently, a GaAs nanowire array-based solar cell was reported by Åberg et al. to have efficiency of 15.3% (Åberg et al. 2016).

2.3.4 Battery Applications

Lithium ion batteries (LIBs) have revolutionized the portable electronics industry with their high energy density, slow charge loss, and long life. These batteries generally consist of a lithium metal oxide cathode, a carbonaceous anode (graphite), and a lithium salt electrolyte. During charging, the lithium atoms in the cathode ionize and diffuse to the anode to be absorbed there in a process called intercalation. Generally, graphite is used to intercalate lithium ions owing to its chemical stability, high conductivity, and theoretical intercalation or gravimetric capacity of 372 $mAhg^{-1}$. However, lithium absorption causes graphite to expand, reducing the practical intercalation capacity. Thus, CNTs are being considered as anode materials for Li-ion batteries because they provide all the advantages of graphite along with potentially large intercalation capacity. Many studies have reported the use of CNTs as anode materials for flexible li-ion batteries (Zhang et al. 2017). The legacy approaches for incorporating CNTs in flexible lithium ion batteries have been adequately summarized in the review by Sehrawat et al. (2016). Recent attempts to integrate CNT-based anodes into flexible LIBs have revolved around methods such as incorporating dopants (Pan et al. 2016, Yehezkel et al. 2017), constructing innovative nanostructures with CNTs (Ahmad et al. 2016, Chiwon et al. 2016, Yehezkel et al. 2016), or creating composites (Yao et al. 2016, Yin et al. 2018).

In general, silicon has the highest gravimetric capacity of 4200 $mAhg^{-1}$ among anode materials other than lithium metal (Zhang et al. 2016b). This makes silicon a promising anode candidate, however, silicon expands substantially due to the absorption of such a large amount of lithium, thus causing mechanical failure. This concern can be mitigated to some extent by structuring silicon in creative ways, as a result, SiNWs have been studied thoroughly as LIB anode material. Chan et al., in 2007, demonstrated a LIB with SiNWs as anode, showing that nanowire architecture can successfully circumvent the problem of silicon expansion (Chan et al. 2007). The stable intercalation capacity was demonstrated to be 3500 $mAhg^{-1}$ for 0.2°C and 2100 $mAhg^{-1}$ for 1°C, which is very high compared to carbon-based anodes. Many recent studies have improved on the capacity or stability even further using novel methods. Wang et al. have demonstrated a capacity of 3500 $mAhg^{-1}$ for 500 cycles using carbon-coated SiNW on carbon fabric as anode (Wang et al. 2017b). Salvatierra et al. reported a flexible battery based completely on nanostructure electrodes—silicon nanowire and graphene nanoribbon anode and lithium cobalt oxide nanowire cathode (Salvatierra et al. 2016). Germanium nanowires have also been considered as an anode material for LIBs. In 2008, Chan et al. reported GeNW-based anode with a capacity close to 1000 $mAhg^{-1}$ for 20 cycles (Chan et al. 2008). Recently, Chang et al. reported a composite anode with Ge/Cu nanowire mesh electrode, demonstrating a capacity of 1153 $mAhg^{-1}$ at a discharge rate of 0.1°C (Chang et al. 2017), while Gao et al. reported a composite anode comprising of GeNWs and graphene and carbon nanotubes (GCNTs) with a capacity of 1315 $mAhg^{-1}$ for 200 cycles (Gao et al. 2017).

2.3.5 Sensor Applications

FETs based on 1D nanostructures are generally reported for single devices to focus on the properties of the device and material itself. Generally, a single 1D nanostructure is placed on an insulating substrate with metal contacts patterned using advanced lithography. An electric field is applied using either a back gate or a patterned top gate to modulate the resistance of the nanostructure. Although these simple devices may not be useful for complex circuit fabrication, the single NW/NT FETs are generally excellent sensor devices. Most commonly, the NW/NTs are functionalized with a specific biological or chemical receptor molecule. These receptors are designed to bond with a specific chemical, biological marker, protein, or virus. Once the marker is attached to the receptor, the electrical potential of the structure is slightly changed, which can be detected from the change in electrical properties of the FET. 1D nanostructures are ideal candidates to fabricate FET-based sensors because they are both nanostructured, which provides for high sensitivity, and are long along a certain axis, which provides for ease of fabrication. Because of the high sensitivity of these sensors, extremely sensitive devices, which can respond to only a few molecules of a particular chemical in a solution, are possible. Further, because of their small size, they can be integrated into a small chip, with different functionalizations, so as to detect a host of chemicals simultaneously (Figure 2.8).

SiNWs, CNTs, and metal oxide nanowires are common 1D nanomaterials used for FET sensors (Zhang et al. 2016c). Shehada et al. demonstrated the use of SiNW FETs to sense voltaic organic compounds (VOCs) linked with various diseases (Shehada et al. 2016). As many as six different biological molecules were coated on these SiNWs to optimize their interaction with particular VOCs. The small size of SiNW FETs can be used for chemical analysis of micro-droplets, as reported by Schütt et al. (2016). A large area device was fabricated using arrays of up to 10^3 aligned nanowires per FET. Use of a large number of parallel NWs increases the output current, thus increasing signal fidelity.

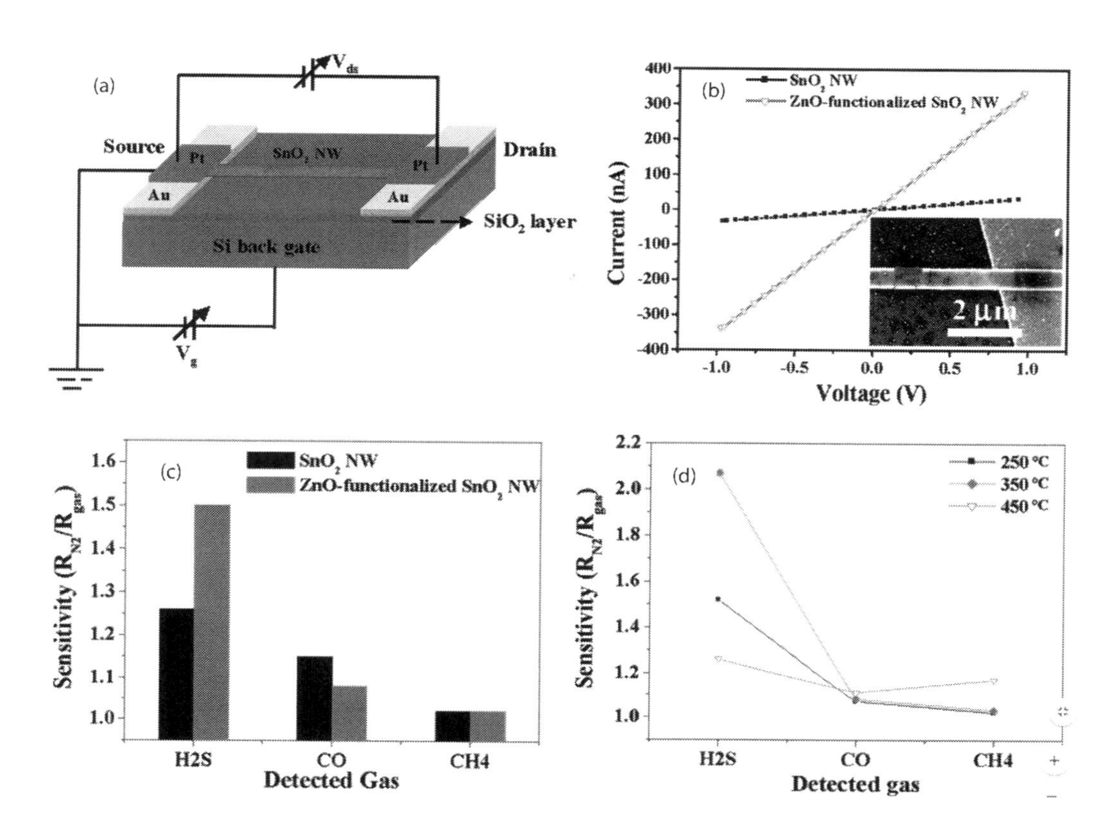

FIGURE 2.8 (a) Schematic illustration of a single SnO_2 NW-based FET. (b) I-V curves of the single SnO_2 NW-based device before and after ZnO deposition. The inset is the corresponding SEM image of the single SnO_2 NW device after ZnO deposition. The deposition thickness of ZnO is about 10 nm. (c) Comparison of gas-sensing sensitivity of the pure and ZnO-functionalized SnO_2 NW-based sensor to three detected gases. (d) Sensitivity to three detected gases of the ZnO-functionalized SnO_2 NW-based sensor at different operation temperatures. The concentration of detected gases is 500 ppm, the operation temperature is 250°C, and the fixed bias is 1 V. (Reprinted with permission from Kuang, Q. et al., *J. Phys. Chem. C*, 112, 11539–11544, 2008. Copyright 2008 American Chemical Society.)

Carbon nanotubes have been reported as sensors for many chemical and physical parameters (Meyyappan 2016). CNT-based FETs for detection of harmful organophosphorus pesticides (OPs) have been reported by Bhatt et al. (2017). Electrolyte-gated CNT FETs have also been reported for direct detection of enzymes using surface functionalization (Melzer et al. 2016). Seichepine et al. reported a pH sensing array of 1024 CNT FETs fabricated on a fully processed complementary metal-oxide semiconductor (CMOS) microsystem (Seichepine et al. 2017). An important application of CNT network-based thin films as sensors is their use for detecting strain (Lee et al. 2017a). Because the application of strain creates larger distances between adjacent nanotubes, thus requiring electrons to traverse new pathways, the resistivity of the thin film is dependent on strain. This effect can be utilized to fabricate extremely sensitive strain sensors with a very large dynamic range (Ryu et al. 2015). Strain sensors based on a similar concept have also been reported using AgNWs (Kim et al. 2015).

Sensors based on metal oxide nanowires have recently gained momentum due to improvements in fabrication and integration strategies (Comini 2016). Hsu et al. reported a ZnO nanowire based relative humidity (RH) sensor with resistance modulation with RH (Hsu et al. 2017). Using indium zinc oxide (IZO) nanowires, Liu et al. reported a pH sensor on a flexible polyethylene terephthalate (PET) substrate with sensitivity of 57.8 mV/pH which is close to the Nernst limit (Liu et al. 2017b). Li et al. fabricated zinc tin oxide (ZTO) nanowire-based light sensors that showed light-to-dark current ratio up to 6.8×10^4, along with a fast response, and excellent stability (Li et al. 2017). ZTO nanowires were decorated with ZnO quantum dots to enhance the photocurrent and sensitivity. Meng et al. reported a single SnO_2 nanowire-based device for electrical sensing of very small concentrations of NO_2 (100 ppb). Pulsed self-Joule-heating of the suspended device enabled pico-joule operation of the sensor (Meng et al. 2016). Several recent review papers have reported summarized in detail the use of nanostructure devices as sensors (Trung and Lee 2016, Han et al. 2017, Yang et al. 2017).

2.4 Challenges

Many innovative techniques have been developed to fabricate 1D nanostructures, leading to the interesting applications discussed in the previous section. While these applications motivate further efforts in nanowire and nanotube research, there remain some challenges that are limiting the potential for their application in consumer products. One of the key challenges with nanostructure-based devices is their cost of production. In some cases, such as indium zinc oxide, the high cost originates from the materials themselves, however, in the case of commonly available elements like carbon and silicon, the cost is associated with the fabrication process. Some form of CVD is commonly used for the large scale growth of nanowires. This process involves the use of large furnaces capable of going to very high temperatures, sophisticated vacuum systems and gas delivery systems. In most cases, because the precursor gases are either toxic or combustible or both, the facility growing nanowires or nanotubes has to be thoroughly vetted for personnel safety. All these overheads drive up the cost per kilogram of nanostructured material produced. Further, the throughput of nanomaterial growth is generally low. Although the volumetric requirement of nanomaterials is small due to their small footprint, a throughput compatible with mass consumer consumption remains elusive. This again drives up the costs due to lack of scalability.

Even with recent and hopefully future advances in the synthesis process driving down costs, there remains the challenge of integrating these nanowires and nanotubes into usable devices, circuits, or systems. One of the key concerns with nanostructure growth is the repeatability of the properties of the synthesized material. With every batch of carbon nanotubes having different diameter and length, hence different electrical characteristics, it is very difficult to fabricate reliable large scale devices or systems, even if these differences are very small. The success of the silicon CMOS and the consumer electronics industry in general is due to the reliable fabrication processes used therein. Processes of similar reliability are needed in order to achieve the true potential of devices and systems based on 1D nanostructure materials.

Indeed, even with the problem of cost, throughout and reliability solved, there remains the issue of obtaining usable bulk materials, composites, or thin films from 1D nanostructures that retain their incredible properties. For example, it is well known that CNTs have near ballistic electronic transport and the strongest bonding known to man. However, this does not translate into CNT fabrics that have the same properties. Indeed, thin films and composites based on CNTs are highly conductive as seen in the previous section, however, these conductivities are severely limited due to the contact resistance between the adjacent nanotubes in the percolation network. Similarly, the tensile strength and Young's modulus shown by CNT-based composites is severely limited due to the composite matrix. While CNTs and other nanostructure materials enhance certain properties of their host, such as conductivity or strength, these benefits are only incremental and in most cases cannot be justified due to high costs of these materials and the presence of other alternatives.

2.5 Summary

In summary, 1D nanostructures demonstrate a wide array of unique properties that can be used in specific applications in consumer electronics, material science, and biological applications. In the past 2 decades, there has been a tremendous increase in the knowledge associated with synthesis, properties, integration, and applications of these materials. The progression in research interest has gone from studying properties of a few synthetic or naturally occurring materials, to large scale synthesis, integration, and application. As this trend continues, it can be expected that 1D nanomaterial-based devices and materials will be in the market shortly. Even with all the challenges discussed in the previous section, the road ahead for 1D nanostructure materials looks promising. The potential of myriad applications already demonstrated in the laboratory setting has fueled research into solving the challenges associated with these materials. For example, already, the CNT market is expected to be a few billion USD by 2020 and is projected to continue to growth at a healthy rate. As market size increases, production will be scaled up resulting in lower costs. Thus, new applications of these materials will become viable due to the lower cost of production, thus increasing demand and market size. This virtuous circle of demand and supply is already underway for CNTs and is expected to start for other 1D nanomaterials. Thus, a future with devices, sensors, circuits, and systems enhanced with the extraordinary capabilities of 1D nanomaterials is just beyond the horizon.

References

Åberg, I., Vescovi, G., Asoli, D., Naseem, U., Gilboy, J. P., Sundvall, C., Dahlgren, A., Svensson, K. E., Anttu, N. & Björk, M. T. 2016. A GaAs nanowire array solar cell with 15.3% efficiency at 1 sun. *IEEE Journal of Photovoltaics* 6: 185–190.

Ahmad, P., Khandaker, M. U., Khan, Z. R. & Amin, Y. M. 2015. Synthesis of boron nitride nanotubes via chemical vapour deposition: A comprehensive review. *RSC Advances* 5: 35116–35137.

Ahmad, S., Copic, D., George, C. & De Volder, M. 2016. Hierarchical assemblies of carbon nanotubes for ultraflexible Li-Ion batteries. *Advanced Materials* 28: 6705–6710.

Anantatmula, R. P., Johnson, A. A., Gupta, S. P. & J. Horylev, R. 1975. The gold-silicon phase diagram. *Journal of Electronic Materials* 4: 445–463.

Arora, N. & Sharma, N. N. 2014. Arc discharge synthesis of carbon nanotubes: Comprehensive review. *Diamond and Related Materials* 50: 135–150.

Bauhofer, W. & Kovacs, J. Z. 2009. A review and analysis of electrical percolation in carbon nanotube polymer composites. *Composites Science and Technology* 69: 1486–1498.

Berber, S., Kwon, Y.-K. & Tománek, D. 2000. Unusually high thermal conductivity of carbon nanotubes. *Physical Review Letters* 84: 4613–4616.

Berg, A., Caroff, P., Shahid, N., Lockrey, M. N., Yuan, X., Borgström, M. T., Tan, H. H. & Jagadish, C. 2017. Growth and optical properties of $In_xGa_{1-x}P$ nanowires synthesized by selective-area epitaxy. *Nano Research* 10: 672–682.

Bhatt, V., Joshi, S., Becherer, M. & Lugli, P. 2017. Flexible, low-cost sensor based on electrolyte gated carbon nanotube field effect transistor for organo-phosphate detection. *Sensors* 17: 1147.

Cambré, S., Wenseleers, W., Goovaerts, E. & Resasco, D. E. 2010. Determination of the metallic/semiconducting ratio in bulk single-wall carbon nanotube samples by cobalt porphyrin probe electron paramagnetic resonance spectroscopy. *ACS Nano* 4: 6717–6724.

Chan, C. K., Peng, H., Liu, G., Mcilwrath, K., Zhang, X. F., Huggins, R. A. & Cui, Y. 2007. High-performance lithium battery anodes using silicon nanowires. *Nature Nanotechnology* 3: 31.

Chan, C. K., Zhang, X. F. & Cui, Y. 2008. High capacity Li ion battery anodes using Ge nanowires. *Nano Letters* 8: 307–309.

Chang, W.-C., Kao, T.-L., Lin, Y. & Tuan, H.-Y. 2017. A flexible all inorganic nanowire bilayer mesh as a high-performance lithium-ion battery anode. *Journal of Materials Chemistry A* 5: 22662.

Charlier, J.-C. 2002. Defects in carbon nanotubes. *Accounts of Chemical Research* 35: 1063–1069.

Charlier, J. C., Ebbesen, T. W. & Lambin, P. 1996. Structural and electronic properties of pentagon-heptagon pair defects in carbon nanotubes. *Physical Review B* 53: 11108–11113.

Chen, B., Zhang, P., Ding, L., Han, J., Qiu, S., Li, Q., Zhang, Z. & Peng, L.-M. 2016a. Highly uniform carbon nanotube field-effect transistors and medium scale integrated circuits. *Nano Letters* 16: 5120–5128.

Chen, D., Liang, J. & Pei, Q. 2016b. Flexible and stretchable electrodes for next generation polymer electronics: A review. *Science China Chemistry* 59: 659–671.

Chen, Z., Ye, S., Stewart, I. E. & Wiley, B. J. 2014. Copper nanowire networks with transparent oxide shells that prevent oxidation without reducing transmittance. *ACS Nano* 8: 9673–9679.

Cherusseri, J., Sharma, R. & Kar, K. K. 2016. Helically coiled carbon nanotube electrodes for flexible supercapacitors. *Carbon* 105: 113–125.

Chiwon, K., Eunho, C., Rangasamy, B. & Wonbong, C. 2016. Three-dimensional free-standing carbon nanotubes for a flexible lithium-ion battery anode. *Nanotechnology* 27: 105402.

Choi, T. Y., Hwang, B.-U., Kim, B.-Y., Trung, T. Q., Nam, Y. H., Kim, D.-N., Eom, K. & Lee, N.-E. 2017. Stretchable, transparent, and stretch-unresponsive capacitive touch sensor array with selectively patterned silver nanowires/reduced graphene oxide electrodes. *ACS Applied Materials & Interfaces* 9: 18022–18030.

Chortos, A., Koleilat, G. I., Pfattner, R., Kong, D., Lin, P., Nur, R. et al. 2016. Mechanically durable and highly stretchable transistors employing carbon nanotube semiconductor and electrodes. *Advanced Materials* 28: 4441–4448.

Choudhary, V. & Gupta, A. 2011. Polymer/carbon nanotube nanocomposites. In *Carbon Nanotubes—Polymer Nanocomposites*, ed. Yellampalli, S., Ch. 4. Rijeka, Croatia: InTech.

Chrzanowska, J., Hoffman, J., Małolepszy, A., Mazurkiewicz, M., Kowalewski, T. A., Szymanski, Z. & Stobinski, L. 2015. Synthesis of carbon nanotubes by the laser ablation method: Effect of laser wavelength. *Physica Status Solidi (b)* 252: 1860–1867.

Colomer, J. F., Stephan, C., Lefrant, S., Van Tendeloo, G., Willems, I., Kónya, Z., Fonseca, A., Laurent, C. & Nagy, J. B. 2000. Large-scale synthesis of single-wall carbon nanotubes by catalytic chemical vapor deposition (CCVD) method. *Chemical Physics Letters* 317: 83–89.

Comini, E. 2016. Metal oxide nanowire chemical sensors: innovation and quality of life. *Materials Today* 19: 559–567.

Couteau, E., Hernadi, K., Seo, J. W., Thiên-Nga, L., Mikó, C., Gaál, R. & Forró, L. 2003. CVD synthesis of high-purity multiwalled carbon nanotubes using $CaCO_3$ catalyst support for large-scale production. *Chemical Physics Letters* 378: 9–17.

Curdin, B., Samuele, G., Hatem, A. & Gabor, K. 2016. Inkjet printed multiwall carbon nanotube electrodes for dielectric elastomer actuators. *Smart Materials and Structures* 25: 055009.

Dai, Z. R., Gole, J. L., Stout, J. D. & Wang, Z. L. 2002. Tin oxide nanowires, nanoribbons, and nanotubes. *The Journal of Physical Chemistry B* 106: 1274–1279.

Daniel, L., Gaël, G., Céline, M., Caroline, C., Daniel, B. & Jean-Pierre, S. 2013. Flexible transparent conductive materials based on silver nanowire networks: A review. *Nanotechnology* 24: 452001.

Das, R., Shahnavaz, Z., Ali, M. E., Islam, M. M. & Abd Hamid, S. B. 2016. Can we optimize arc discharge and laser ablation for well-controlled carbon nanotube synthesis? *Nanoscale Research Letters* 11: 510.

Duduta, M., Wood, R. J. & Clarke, D. R. 2016. Multilayer dielectric elastomers for fast, programmable actuation without prestretch. *Advanced Materials* 28: 8058–8063.

Fahad, H. M., Hussain, A. M., Torres, G. a. S., Banerjee, S. K. & Hussain, M. M. 2014. Group IV nanotube transistors for next generation ubiquitous computing. *SPIE Defense + Security*. SPIE, 7.

Fang, M., Han, N., Wang, F., Yang, Z.-X., Yip, S., Dong, G., Hou, J. J., Chueh, Y. & Ho, J. C. 2014. III-V nanowires: Synthesis, property manipulations, and device applications. *Journal of Nanomaterials* 2014: 14.

Filipič, G. & Cvelbar, U. 2012. Copper oxide nanowires: A review of growth. *Nanotechnology* 23: 194001.

Gao, C., Kim, N. D., Villegas Salvatierra, R., Lee, S.-K., Li, L., Li, Y. et al. 2017. Germanium on seamless graphene carbon nanotube hybrids for lithium ion anodes. *Carbon* 123: 433–439.

Greene, L. E., Yuhas, B. D., Law, M., Zitoun, D. & Yang, P. 2006. Solution-grown zinc oxide nanowires. *Inorganic Chemistry* 45: 7535–7543.

Gu, G., Zheng, B., Han, W. Q., Roth, S. & Liu, J. 2002. Tungsten oxide nanowires on tungsten substrates. *Nano Letters* 2: 849–851.

Han, S.-T., Peng, H., Sun, Q., Venkatesh, S., Chung, K.-S., Lau, S. C., Zhou, Y. & Roy, V. a. L. 2017. An overview of the development of flexible sensors. *Advanced Materials* 29: 1700375.

Harutyunyan, A. R., Pradhan, B. K., Kim, U. J., Chen, G. & Eklund, P. C. 2002. CVD synthesis of single wall carbon nanotubes under 'Soft' conditions. *Nano Letters* 2: 525–530.

Hokkanen, M. J., Lautala, S., Shao, D., Turpeinen, T., Koivistoinen, J. & Ahlskog, M. 2016. On-chip purification via liquid immersion of arc-discharge synthesized multiwalled carbon nanotubes. *Applied Physics A* 122: 634.

Hosseini, T. & Kouklin, N. 2016. Carbon nanotube–polymer composites: Device properties and photovoltaic applications. In *Carbon Nanotubes-Current Progress of Their Polymer Composites*, ed. Berber, M. R. & Hafez, I. H. Rijeka, Croatia: InTech.

Hosseini, V., Gantenbein, S., Avalos Vizcarra, I., Schoen, I. & Vogel, V. 2016. Stretchable silver nanowire microelectrodes for combined mechanical and electrical stimulation of cells. *Advanced Healthcare Materials* 5: 2045–2054.

Hsu, C.-L., Su, I. L. & Hsueh, T.-J. 2017. Tunable Schottky contact humidity sensor based on S-doped ZnO nanowires on flexible PET substrate with piezotronic effect. *Journal of Alloys and Compounds* 705: 722–733.

Hu, L., Hecht, D. S. & Grüner, G. 2004. Percolation in transparent and conducting carbon nanotube networks. *Nano Letters* 4: 2513–2517.

Huang, Q., Shen, W., Tan, R., Xu, W. & Song, W. 2015. Separation of silver nanocrystals for surface-enhanced Raman scattering using density gradient centrifugation. *Journal of Materials Science & Technology* 31: 834–839.

Hwang, C., An, J., Choi, B. D., Kim, K., Jung, S.-W., Baeg, K.-J., Kim, M.-G., Ok, K. M. & Hong, J. 2016. Controlled aqueous synthesis of ultra-long copper nanowires for stretchable transparent conducting electrode. *Journal of Materials Chemistry C* 4: 1441–1447.

Iijima, S. 1991. Helical microtubules of graphitic carbon. *Nature* 354: 56–58.

Jun, K.-W., Kim, J.-N., Jung, J.-Y. & Oh, I.-K. 2017. Wrinkled graphene–AgNWs hybrid electrodes for smart window. *Micromachines* 8: 43.

Kane, C. L., Mele, E. J., Lee, R. S., Fischer, J. E., Petit, P., Dai, H. et al. 1998. Temperature-dependent resistivity of single-wall carbon nanotubes. *EPL (Europhysics Letters)* 41: 683.

Kim, K. K., Hong, S., Cho, H. M., Lee, J., Suh, Y. D., Ham, J. & Ko, S. H. 2015. Highly sensitive and stretchable multidimensional strain sensor with prestrained anisotropic metal nanowire percolation networks. *Nano Letters* 15: 5240–5247.

Kim, T., Park, J., Sohn, J., Cho, D. & Jeon, S. 2016. Bioinspired, highly stretchable, and conductive dry adhesives based on 1D–2D hybrid carbon nanocomposites for all-in-one ECG electrodes. *ACS Nano* 10: 4770–4778.

Klinger, C., Patel, Y. & Postma, H. W. C. 2012. Carbon nanotube solar cells. *PLOS ONE* 7: e37806.

Kong, J., Cassell, A. M. & Dai, H. 1998. Chemical vapor deposition of methane for single-walled carbon nanotubes. *Chemical Physics Letters* 292: 567–574.

Krishnan, A., Dujardin, E., Ebbesen, T. W., Yianilos, P. N. & Treacy, M. M. J. 1998. Young's modulus of single-walled nanotubes. *Physical Review B* 58: 14013–14019.

Kruse, J. E., Lymperakis, L., Eftychis, S., Adikimenakis, A., Doundoulakis, G., Tsagaraki, K. et al. 2016. Selective-area growth of GaN nanowires on SiO_2-masked Si (111) substrates by molecular beam epitaxy. *Journal of Applied Physics* 119: 224305.

Kuang, Q., Lao, C.-S., Li, Z., Liu, Y.-Z., Xie, Z.-X., Zheng, L.-S. & Wang, Z. L. 2008. Enhancing the photon- and gas-sensing properties of a single SnO_2 nanowire based nanodevice by nanoparticle surface functionalization. *The Journal of Physical Chemistry C* 112: 11539–11544.

Kumar, M. 2011. Carbon nanotube synthesis and growth mechanism. In *Carbon Nanotubes—Synthesis, Characterization, Applications*, ed. Yellampalli, S., Ch. 8. Rijeka, Croatia: InTech.

Lee, J., Lee, P., Lee, H., Lee, D., Lee, S. S. & Ko, S. H. 2012. Very long Ag nanowire synthesis and its application in a highly transparent, conductive and flexible metal electrode touch panel. *Nanoscale* 4: 6408–6414.

Lee, J., Lim, M., Yoon, J., Kim, M. S., Choi, B., Kim, D. M., Kim, D. H., Park, I. & Choi, S.-J. 2017a. Transparent, flexible strain sensor based on a solution-processed carbon nanotube network. *ACS Applied Materials & Interfaces* 9: 26279–26285.

Lee, J. U. 2005. Photovoltaic effect in ideal carbon nanotube diodes. *Applied Physics Letters* 87: 073101.

Lee, Y. R., Kwon, H., Lee, D. H. & Lee, B. Y. 2017b. Highly flexible and transparent dielectric elastomer actuators using silver nanowire and carbon nanotube hybrid electrodes. *Soft Matter* 13: 6390–6395.

Li, L., Gu, L., Lou, Z., Fan, Z. & Shen, G. 2017. ZnO quantum dot decorated Zn_2SnO_4 nanowire heterojunction photodetectors with drastic performance enhancement and flexible ultraviolet image sensors. *ACS Nano* 11: 4067–4076.

Lin, J.-Y., Hsueh, Y.-L. & Huang, J.-J. 2014. The concentration effect of capping agent for synthesis of silver nanowire by using the polyol method. *Journal of Solid State Chemistry* 214: 2–6.

Lin, J.-Y., Hsueh, Y.-L., Huang, J.-J. & Wu, J.-R. 2015. Effect of silver nitrate concentration of silver nanowires synthesized using a polyol method and their application as transparent conductive films. *Thin Solid Films* 584: 243–247.

Liu, H., Li, J. & Tan, R. 2017a. Flexible In_2O_3 nanowire transistors on paper substrates. *IEEE Journal of the Electron Devices Society* 5: 141–144.

Liu, N., Gan, L., Liu, Y., Gui, W., Li, W. & Zhang, X. 2017b. Improving pH sensitivity by field-induced charge regulation in flexible biopolymer electrolyte gated oxide transistors. *Applied Surface Science* 419: 206–212.

Liu, Y., Wang, S., Liu, H. & Peng, L.-M. 2017c. Carbon nanotube-based three-dimensional monolithic optoelectronic integrated system. *Nature Communications* 8: 15649.

Liu, Y., Wang, S. & Peng, L.-M. 2016. Toward high-performance carbon nanotube photovoltaic devices. *Advanced Energy Materials* 6: 1600522.

Louis, B., Gulino, G., Vieira, R., Amadou, J., Dintzer, T., Galvagno, S., Centi, G., Ledoux, M. J. & Pham-Huu, C. 2005. High yield synthesis of multi-walled carbon nanotubes by catalytic decomposition of ethane over iron supported on alumina catalyst. *Catalysis Today* 102–103: 23–28.

Lv, T., Yao, Y., Li, N. & Chen, T. 2016. Highly stretchable supercapacitors based on aligned carbon nanotube/molybdenum disulfide composites. *Angewandte Chemie International Edition* 55: 9191–9195.

Lyu, S. C., Liu, B. C., Lee, S. H., Park, C. Y., Kang, H. K., Yang, C. W. & Lee, C. J. 2004. Large-scale synthesis of high-quality single-walled carbon nanotubes by catalytic decomposition of ethylene. *The Journal of Physical Chemistry B* 108: 1613–1616.

Maruyama, S., Kojima, R., Miyauchi, Y., Chiashi, S. & Kohno, M. 2002. Low-temperature synthesis of high-purity single-walled carbon nanotubes from alcohol. *Chemical Physics Letters* 360: 229–234.

Matsuda, Y., Tahir-Kheli, J. & Goddard, W. A. 2010. Definitive band gaps for single-wall carbon nanotubes. *The Journal of Physical Chemistry Letters* 1: 2946–2950.

Matula, R. A. 1979. Electrical resistivity of copper, gold, palladium, and silver. *Journal of Physical and Chemical Reference Data* 8: 1147–1298.

Melzer, K., Bhatt, V. D., Jaworska, E., Mittermeier, R., Maksymiuk, K., Michalska, A. & Lugli, P. 2016. Enzyme assays using sensor arrays based on ion-selective carbon nanotube field-effect transistors. *Biosensors and Bioelectronics* 84: 7–14.

Meng, G., Zhuge, F., Nagashima, K., Nakao, A., Kanai, M., He, Y., Boudot, M., Takahashi, T., Uchida, K. & Yanagida, T. 2016. Nanoscale thermal management of single SnO_2 nanowire: Pico-joule energy consumed molecule sensor. *ACS Sensors* 1: 997–1002.

Meyyappan, M. 2016. Carbon nanotube-based chemical sensors. *Small* 12: 2118–2129.

Michael, K. 2007. Factors affecting synthesis of single wall carbon nanotubes in arc discharge. *Journal of Physics D: Applied Physics* 40: 2388.

O'halloran, A., O'malley, F. & Mchugh, P. 2008. A review on dielectric elastomer actuators, technology, applications, and challenges. *Journal of Applied Physics* 104: 071101.

Pan, Z., Ren, J., Guan, G., Fang, X., Wang, B., Doo, S.-G., Son, I. H., Huang, X. & Peng, H. 2016. Synthesizing nitrogen-doped core–sheath carbon nanotube films for flexible lithium ion batteries. *Advanced Energy Materials* 6: 1600271.

Park, S., Park, B., Nam, S., Yun, S., Park, S. K., Mun, S., Lim, J. M., Ryu, Y., Song, S. H. & Kyung, K.-U. 2017. Electrically tunable binary phase Fresnel lens based on a dielectric elastomer actuator. *Optics Express* 25: 23801–23808.

Pfeifer, S., Park, S.-H. & Bandaru, P. R. 2010. Analysis of electrical percolation thresholds in carbon nanotube networks using the Weibull probability distribution. *Journal of Applied Physics* 108: 024305.

Pratyush Das, K., Heinz, S., Mikael, T. B., Lynne, M. G., Chris, B., John, B., Cedric, D. B. & Heike, R. 2013. Selective area growth of III–V nanowires and their heterostructures on silicon in a nanotube template: Towards monolithic integration of nano-devices. *Nanotechnology* 24: 225304.

Priolo, F., Gregorkiewicz, T., Galli, M. & Krauss, T. F. 2014. Silicon nanostructures for photonics and photovoltaics. *Nature Nanotechnology* 9: 19.

Rakhi, R. B., Chen, W., Cha, D. & Alshareef, H. N. 2012. Substrate dependent self-organization of mesoporous cobalt oxide nanowires with remarkable pseudocapacitance. *Nano Letters* 12: 2559–2567.

Rogers, J. A., Lagally, M. G. & Nuzzo, R. G. 2011. Synthesis, assembly and applications of semiconductor nanomembranes. *Nature* 477: 45–53.

Roslan, M. S., Chaudary, K., Rizvi, S. Z. H., Daud, S., Ali, J. & Munajat, Y. 2016. Arc discharge synthesis of CNTs in hydrogen environment in presence of magnetic field. *Jurnal Teknologi* 78: 257–260.

Ryu, S., Lee, P., Chou, J. B., Xu, R., Zhao, R., Hart, A. J. & Kim, S.-G. 2015. Extremely elastic wearable carbon nanotube fiber strain sensor for monitoring of human motion. *ACS Nano* 9: 5929–5936.

Salvatierra, R. V., Raji, A.-R. O., Lee, S.-K., Ji, Y., Li, L. & Tour, J. M. 2016. Silicon nanowires and lithium cobalt oxide nanowires in graphene nanoribbon papers for full lithium ion battery. *Advanced Energy Materials* 6: 1600918.

Schütt, J., Ibarlucea, B., Illing, R., Zörgiebel, F., Pregl, S., Nozaki, D., Weber, W. M., Mikolajick, T., Baraban, L. & Cuniberti, G. 2016. Compact nanowire sensors probe microdroplets. *Nano Letters* 16: 4991–5000.

Sehrawat, P., Julien, C. & Islam, S. S. 2016. Carbon nanotubes in Li-ion batteries: A review. *Materials Science and Engineering: B* 213: 12–40.

Seichepine, F., Rothe, J., Dudina, A., Hierlemann, A. & Frey, U. 2017. Dielectrophoresis-assisted integration of 1024 carbon nanotube sensors into a CMOS microsystem. *Advanced Materials* 29: 1606852.

Shah, K. A. & Tali, B. A. 2016. Synthesis of carbon nanotubes by catalytic chemical vapour deposition: A review on carbon sources, catalysts and substrates. *Materials Science in Semiconductor Processing* 41: 67–82.

Shehada, N., Cancilla, J. C., Torrecilla, J. S., Pariente, E. S., Brönstrup, G., Christiansen, S. et al. 2016. Silicon nanowire sensors enable diagnosis of patients via exhaled breath. *ACS Nano* 10: 7047–7057.

Shen, G., Chen, P.-C., Ryu, K. & Zhou, C. 2009. Devices and chemical sensing applications of metal oxide nanowires. *Journal of Materials Chemistry* 19: 828–839.

Shi, Z., Lian, Y., Liao, F. H., Zhou, X., Gu, Z., Zhang, Y., Iijima, S., Li, H., Yue, K. T. & Zhang, S.-L. 2000. Large scale synthesis of single-wall carbon nanotubes by arc-discharge method. *Journal of Physics and Chemistry of Solids* 61: 1031–1036.

Shlafman, M., Tabachnik, T., Shtempluk, O., Razin, A., Kochetkov, V. & Yaish, Y. E. 2016. Self aligned hysteresis free carbon nanotube field-effect transistors. *Applied Physics Letters* 108: 163104.

Shulaker, M. M., Hills, G., Patil, N., Wei, H., Chen, H.-Y., Wong, H. S. P. & Mitra, S. 2013. Carbon nanotube computer. *Nature* 501: 526–530.

Silvano, G., Claudio, B., Valter, B., Giorgio, G. & Claudio, N. 2006. Carbon nanotube biocompatibility with cardiac muscle cells. *Nanotechnology* 17: 391.

Smart, S. K., Cassady, A. I., Lu, G. Q. & Martin, D. J. 2006. The biocompatibility of carbon nanotubes. *Carbon* 44: 1034–1047.

Su, M., Zheng, B. & Liu, J. 2000. A scalable CVD method for the synthesis of single-walled carbon nanotubes with high catalyst productivity. *Chemical Physics Letters* 322: 321–326.

Suenaga, K., Wakabayashi, H., Koshino, M., Sato, Y., Urita, K. & Iijima, S. 2007. Imaging active topological defects in carbon nanotubes. *Nat Nano* 2: 358–360.

Sugai, T., Yoshida, H., Shimada, T., Okazaki, T., Shinohara, H. & Bandow, S. 2003. New synthesis of high-quality double-walled carbon nanotubes by high-temperature pulsed arc discharge. *Nano Letters* 3: 769–773.

Sun, D.-M., Timmermans, M. Y., Kaskela, A., Nasibulin, A. G., Kishimoto, S., Mizutani, T., Kauppinen, E. I. & Ohno, Y. 2013. Mouldable all-carbon integrated circuits. *Nature Communications* 4: 2302.

Sun, Y., Yin, Y., Mayers, B. T., Herricks, T. & Xia, Y. 2002. Uniform silver nanowires synthesis by reducing $AgNO_3$ with ethylene glycol in the presence of seeds and poly(vinyl pyrrolidone). *Chemistry of Materials* 14: 4736–4745.

Taube Navaraj, W., García Núñez, C., Shakthivel, D., Vinciguerra, V., Labeau, F., Gregory, D. H. & Dahiya, R. 2017. Nanowire FET based neural element for robotic tactile sensing skin. *Frontiers in Neuroscience* 11: 501.

Tian, Y., Hu, Z., Yang, Y., Wang, X., Chen, X., Xu, H., Wu, Q., Ji, W. & Chen, Y. 2004. In situ TA-MS study of the six-membered-ring-based growth of carbon nanotubes with benzene precursor. *Journal of the American Chemical Society* 126: 1180–1183.

Treacy, M. M. J., Ebbesen, T. W. & Gibson, J. M. 1996. Exceptionally high Young's modulus observed for individual carbon nanotubes. *Nature* 381: 678–680.

Trung, T. Q. & Lee, N.-E. 2016. Flexible and stretchable physical sensor integrated platforms for wearable human-activity monitoring and personal healthcare. *Advanced Materials* 28: 4338–4372.

Wallentin, J., Anttu, N., Asoli, D., Huffman, M., Åberg, I., Magnusson, M. H., Siefer, G., Fuss-Kailuweit, P., Dimroth, F., Witzigmann, B., Xu, H. Q., Samuelson, L., Deppert, K. & Borgström, M. T. 2013. InP nanowire array solar cells achieving 13.8% efficiency by exceeding the ray optics limit. *Science* 339: 1057.

Wan, Q., Dattoli, E. N., Fung, W. Y., Guo, W., Chen, Y., Pan, X. & Lu, W. 2006. High-performance transparent conducting oxide nanowires. *Nano Letters* 6: 2909–2915.

Wang, J., Shea, M. J., Flach, J. T., Mcdonough, T. J., Way, A. J., Zanni, M. T. & Arnold, M. S. 2017a. Role of defects as exciton quenching sites in carbon nanotube photovoltaics. *The Journal of Physical Chemistry C* 121: 8310–8318.

Wang, X., Li, G., Seo, M. H., Lui, G., Hassan, F. M., Feng, K., Xiao, X. & Chen, Z. 2017b. Carbon-coated silicon nanowires on carbon fabric as self-supported electrodes for flexible lithium-ion batteries. *ACS Applied Materials & Interfaces* 9: 9551–9558.

Wei, J., Jia, Y., Shu, Q., Gu, Z., Wang, K., Zhuang, D. et al. 2007. Double-walled carbon nanotube solar cells. *Nano Letters* 7: 2317–2321.

Whitcomb, D. R., Clapp, A. R., Bühlmann, P., Blinn, J. C. & Zhang, J. 2016. New perspectives on silver nanowire formation from dynamic silver ion concentration monitoring and nitric oxide production in the polyol process. *Crystal Growth & Design* 16: 1861–1868.

Wilder, J. W. G., Venema, L. C., Rinzler, A. G., Smalley, R. E. & Dekker, C. 1998. Electronic structure of atomically resolved carbon nanotubes. *Nature* 391: 59–62.

Yan, Y., Miao, J., Yang, Z., Xiao, F.-X., Yang, H. B., Liu, B. & Yang, Y. 2015. Carbon nanotube catalysts: Recent advances in synthesis, characterization and applications. *Chemical Society Reviews* 44: 3295–3346.

Yang, Y., Yang, X., Tan, Y. & Yuan, Q. 2017. Recent progress in flexible and wearable bio-electronics based on nanomaterials. *Nano Research* 10: 1560–1583.

Yao, K., Liang, R. & Zheng, J. P. 2016. Freestanding flexible Si nanoparticles–multiwalled carbon nanotubes composite anodes for Li-ion batteries and their prelithiation by stabilized Li metal powder. *Journal of Electrochemical Energy Conversion and Storage* 13: 011004-011004-6.

Yatom, S., Selinsky, R. S., Koel, B. E. & Raitses, Y. 2017. 'Synthesis-on' and 'synthesis-off' modes of carbon arc operation during synthesis of carbon nanotubes. *Carbon* 125: 336–343.

Yehezkel, S., Auinat, M., Sezin, N., Starosvetsky, D. & Ein-Eli, Y. 2016. Bundled and densified carbon nanotubes (CNT) fabrics as flexible ultra-light weight Li-ion battery anode current collectors. *Journal of Power Sources* 312: 109–115.

Yehezkel, S., Auinat, M., Sezin, N., Starosvetsky, D. & Ein-Eli, Y. 2017. Distinct copper electrodeposited carbon nanotubes (CNT) tissues as anode current collectors in Li-ion battery. *Electrochimica Acta* 229: 404–414.

Yin, Z., Cho, S., You, D.-J., Ahn, Y.-K., Yoo, J. & Kim, Y. 2018. Copper nanowire/multi-walled carbon nanotube composites as all-nanowire flexible electrode for fast-charging/discharging lithium-ion battery. *Nano Research* 11: 769–779.

Yu, J., Lu, W., Pei, S., Gong, K., Wang, L., Meng, L., Huang, Y., Smith, J. P., Booksh, K. S., Li, Q., Byun, J.-H., Oh, Y., Yan, Y. & Chou, T.-W. 2016. Omnidirectionally stretchable high-performance supercapacitor based on isotropic buckled carbon nanotube films. *ACS Nano* 10: 5204–5211.

Yun, J., Lee, M., Jeon, Y., Kim, M., Kim, Y., Lim, D. & Kim, S. 2016. Nanowatt power operation of silicon nanowire NAND logic gates on bendable substrates. *Nano Research* 9: 3656–3662.

Zhang, A., Zheng, G. & Lieber, C. M. 2016a. Nanowire-enabled energy conversion. In *Nanowires: Building Blocks for Nanoscience and Nanotechnology*, 227–254. Cham, Switzerland: Springer International Publishing.

Zhang, A., Zheng, G. & Lieber, C. M. 2016b. Nanowire-enabled energy storage. In *Nanowires: Building Blocks for Nanoscience and Nanotechnology*, 203–225. Cham, Switzerland: Springer International Publishing.

Zhang, A., Zheng, G. & Lieber, C. M. 2016c. Nanowire field-effect transistor sensors. In *Nanowires: Building Blocks for Nanoscience and Nanotechnology*, 255–275. Cham, Switzerland: Springer International Publishing.

Zhang, X., Zhao, J., Dou, J., Tange, M., Xu, W., Mo, L. et al. 2016d. Flexible CMOS-like circuits based on printed P-type and N-type carbon nanotube thin-film transistors. *Small* 12: 5066–5073.

Zhang, Y., Jiao, Y., Liao, M., Wang, B. & Peng, H. 2017. Carbon nanomaterials for flexible lithium ion batteries. *Carbon* 124: 79–88.

Zheng, B., Li, Y. & Liu, J. 2002. CVD synthesis and purification of single-walled carbon nanotubes on aerogel-supported catalyst. *Applied Physics A* 74: 345–348.

3

Flexible 2D Electronics in Sensors and Bioanalytical Applications

Parikshit Sahatiya,
Rinky Sha, and
Sushmee Badhulika

3.1 Introduction

Rapid progress in the synthesis and fundamental understanding of surface phenomena of nanomaterials has enabled their incorporation into sensor architectures [1]. Functional nanomaterials are strong candidates for electronic applications because their reduced dimensions create a ballistic charge transport [2]. The reduced dimensionality also creates structures with exceptionally high surface area to volume ratio, and some materials, such as 2D transition metal dichalcogenides [3], graphene [4], hexagonal–boron nitride (h-BN) [5], and phosphorene [6] are composed almost entirely of surface atoms. These two consequences of reduced size result in a class of materials that has the potential for unsurpassed sensitivity toward changes in its physical and chemical properties.

However, all established classes of high-performance electronics exploit single crystal inorganic materials, such as silicon or gallium arsenide, in forms that are fundamentally rigid and planar [7]. The human body is, by contrast, soft and curvilinear. This mismatch in properties hinders the development of devices capable of intimate, conformal integration with humans, for applications ranging from human-machine interfaces, sensors, electronic skin, and multifunctional sensors for Internet of Things (IoT) [8]. Hence, there is heightened need for not only the flexible materials, but also integrating them on a flexible substrate which would be a step ahead in biointegrated devices. For developing flexible electronics devices, there is need for investigation into materials which are flexible and stretchable. One envisioned solution involves the use of organic electronic materials, whose flexible properties have generated interest in them for potential use in paper-like displays, solar cells, and other types of consumer electronic devices [9,10]. Such materials are not, however, stretchable or capable of wrapping curvilinear surfaces; they also offer only moderate performance, with uncertain reliability and capacity for integration into complex integrated circuits. Functional 2D nanomaterials such as MoS_2, graphene, h-BN, and phosphorene are promising candidates for the development of flexible electronic devices and sensors because of their high mobility, high thermal conductivity, high young's modulus etc.

3.1.1 Review of Status in Research and Development in the Subject

3.1.1.1 2D Materials

For developing flexible electronic devices there is need for investigations into materials which are flexible and stretchable. There are reports on graphene hybrids which have been utilized in various applications such as photodetectors, electrochemical sensors, etc. [11–15]. But such materials configurations are difficult to be integrated at the device level. Metal oxides are costly and have less environmental stability. In the search of functional materials for electronic applications, 2D materials are intensely studied due to their remarkable properties which find potential applications in the field of flexible and wearable electronics [16]. Among them, transition metal dichalcogenides (TMDs) have proven to be noteworthy because of their strong excitonic effects, optical transparency, mechanical flexibility, layer dependent bandgap, tunable optical properties, and high breakdown voltages [17,18]. Molybdenum disulfide (MoS_2) has been the most explored among the TMD family because of its direct bandgap, high electron mobility (up to 480 $cm^2 V^{-1}S^{-1}$), and superior on/off ratio [19], thus making it a better candidate over existing materials such as silicon, organic semiconductors, and oxide semiconductors which possess low carrier mobilities in the field of flexible electronics. Carbon nanomaterials such as two-dimensional graphene have gained much attention for flexible electronics because of their attractive and motivating properties. The carrier mobility of graphene on insulator substrate is reported to be ~100,000 $cm^2V^{-1}s^{-1}$ [20]. Such high mobility values motivate the use of graphene in high speed electronics. The current capacity of graphene has been reported to be 109 cm^{-2}. Thermal conductivity of graphene at room temperature is claimed to be 5,300 $Wm^{-1}K^{-1}$ [21], respectively, with transmittance of nearly 97%. Graphene has outstanding mechanical properties with Young's modulus of 1 TPa and tensile strength of 130 GPa [22]. For the above stated reasons and properties, MoS_2 and graphene are considered to be the most promising materials for next-generation flexible electronics. Apart from MoS_2 and graphene, there are other members of TMDs such as WS_2, WSe_2, and $MoSe_2$ which possess interesting electronic properties and have been utilized in various potential applications [23,24]. Recently, black phosphorous whose single layer is called phosphorene has gained enormous attention due to its charge carrier mobility, unique in plane anisotropic structure and tunable direct bandgap [25]. To fully explore the potential of black phosphorous, hybridization, functionalization, and doping have been utilized for its use in various potential application in field effect transistor, photodetector, photo catalysis, and batteries, etc. The detailed review of black phosphorous in terms of its functionalization and applications is reported by Lei et al. [26].

3.1.1.2 Flexible Substrates

Flexible substrates provide ideal platforms for exploring some of the unique characteristics that arise in metamaterials via mechanical deformation. The use of flexible substrates to demonstrate metamaterials with novel functionalities is gaining increasing attention worldwide. The most commonly used flexible substrates for metamaterials are poly-dimethyl siloxine (PDMS) and polyimide, due to their widespread use in flexible electronics. Other flexible substrates utilized for metamaterial devices include polyethylene naphthalene (PEN) [27], polyethylene terephthalate (PET) [28], polymethylmethacrylate (PMMA) [29], and polystyrene [30]. Polyimide is an ideal choice as substrate for flexible electronics due to its strong adhesion to metal coatings, which provides a high degree of strain delocalization. Polyimide provides an operating range of –269°C to 400°C with very high glass transition which makes it ideal for deposition techniques such as sputtering and E beam evaporation. Its adhesion to photoresist and resistance to corrosive acids used whilst etching is another feature which allows direct patterning of structures onto it [31]. Moreover, it is biocompatible [38] which is of foremost importance for wearable electronics. Also, most of the abovementioned polymer substrates are microfabrication compatible, and the devices can be fabricated using sophisticated cleanroom techniques which offer tremendous applications in the fabrication of reliable flexible electronic devices. Despite the advantages offered by plastic substrates, their inability to withstand high processing temperatures, poor recyclability, and non-biodegradability makes them unsuitable for the development of eco-friendly flexible electronics for IoT applications. However, all flexible substrates are not microfabrication compatible, and hence there is an urgent need to develop lithography-free solution phase processes for the fabrication of devices on flexible substrates. For example, a pencil eraser is extremely stretchable and bendable and can be utilized as a pressure and strain sensor, but a pencil eraser is not microfabrication compatible. Further, cellulose paper is very low cost and biodegradable which can be utilized for disposable sensors. Both pencil eraser and cellulose paper are not microfabrication compatible and hence there is a recent surge of reports utilizing flexible substrates other than plastic substrates for various exciting applications. Such applications not only utilize the properties of an active element, but also utilize the substrate inherent properties for enhanced sensing. A recent report from Sahatiya et al. has demonstrated the direct growth of MoS_2 on different flexible substrates which include cellulose paper, Al foil, Cu foil, polyimide, ceramic paper, and thread using a solution processed hydrothermal method [32]. Figure 3.1 shows the schematic of the fabrication of MoS_2 on different flexible substrates and their corresponding field emission scanning electron microscopy (FESEM) images.

3.2 Electronic System Based on 2D Materials

3.2.1 Field Effect Transistors

Field effect transistors forms the basic building block of an electronic device, and it is indeed important to fabricate FETs based on novel 2D materials which not only aids in exploring new applications, but also helps in understanding the underlying excellent properties of novel 2D materials. The first 2D material, graphene, has been considered and utilized in FETs where the behavior is similar to classical FETs wherein modulating the gate voltage modulates the drain current where the gate electrode is separated by a thin dielectric [33]. But the lack of bandgap in graphene restricts its usage in digital applications, but the high mobility of graphene makes it a strong candidate for radio-frequency (RF) applications. The highest current amplification even at 400 GHz has been demonstrated by the use of graphene field effect transistors (GFETs). Further, the ambipolar characteristics of graphene have been utilized for the development of RF mixers [34]. To overcome the limitations of graphene arising from the fundamental limitation of zero bandgap of graphene, researchers started exploring novel 2D materials whose properties are similar to graphene and also possess permanent bandgap. The search for such materials led to the development of TMDs (MoS_2, $MoSe_2$, WS_2, and WSe_2) and phosphorene [35,36]. The other route followed to overcome the above issue is the fabrication of vertical heterojunctions by stacking 2D materials such as graphene, TMDs, etc. These devices work on different principles which explored vertical current transport which includes tunneling field effect transistors, band to band tunneling resistor, and the transistor whose operation is based on the field effect modulation of Schottky junctions [37,38]. Such FETs exhibit a high ON/OFF ratio typically not achieved by pristine graphene transistors. Due to the permanent bandgap, TMDs are promising candidates for FETs as they have good stability and have been reported with a large I_{on}/I_{off} ratio of 10^4 and the subthreshold swing of 80 mv/decade [39]. The only drawback of the TMDs family is the low mobility wherein the theoretical value reported is 400 $cm^2/V\text{-}s$, whereas the experimental values reported are few tens $cm^2/V\text{-}s$ [40]. Apart from TMDs and graphene, phosphorene has been an excellent candidate for the fabrication of 2D FETs which has reported the mobility value of 286 $cm^2/V\text{-}s$ for a few layer phosphorene, whereas as high as 1000 $cm^2/V\text{-}s$ has been reported for a multilayer phosphorene with 10 nm thickness [25]. Figure 3.2 shows the schematic of the fabricated device with the contacts at different angles and its

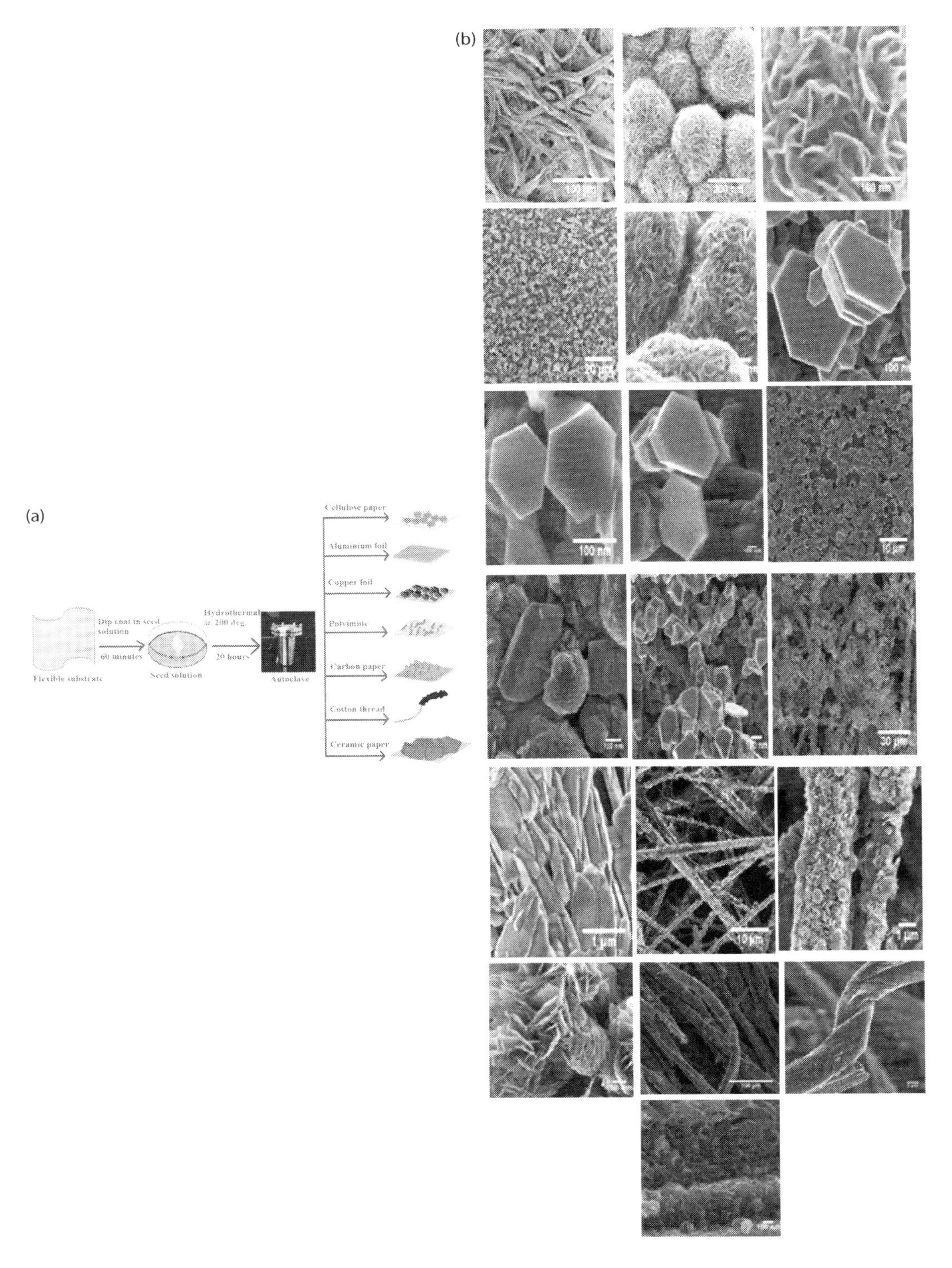

FIGURE 3.1 (a) Schematic of the fabrication procedure for the growth of MoS_2 on different flexible substrates, (b) FESEM images of MoS_2 grown on different substrates, (a–c) cellulose paper, (d, e) Al foil, (f–h) Cu foil, (i–k), polyimide (l, m), ceramic paper, (n–p,) carbon paper, and (q–s) cotton thread showing distinct morphology of MoS_2 on different substrates. (Reprinted from Sahatiya, P. et al., *Flex. Print. Electron.*, 3, 015002, 2018. With permission.)

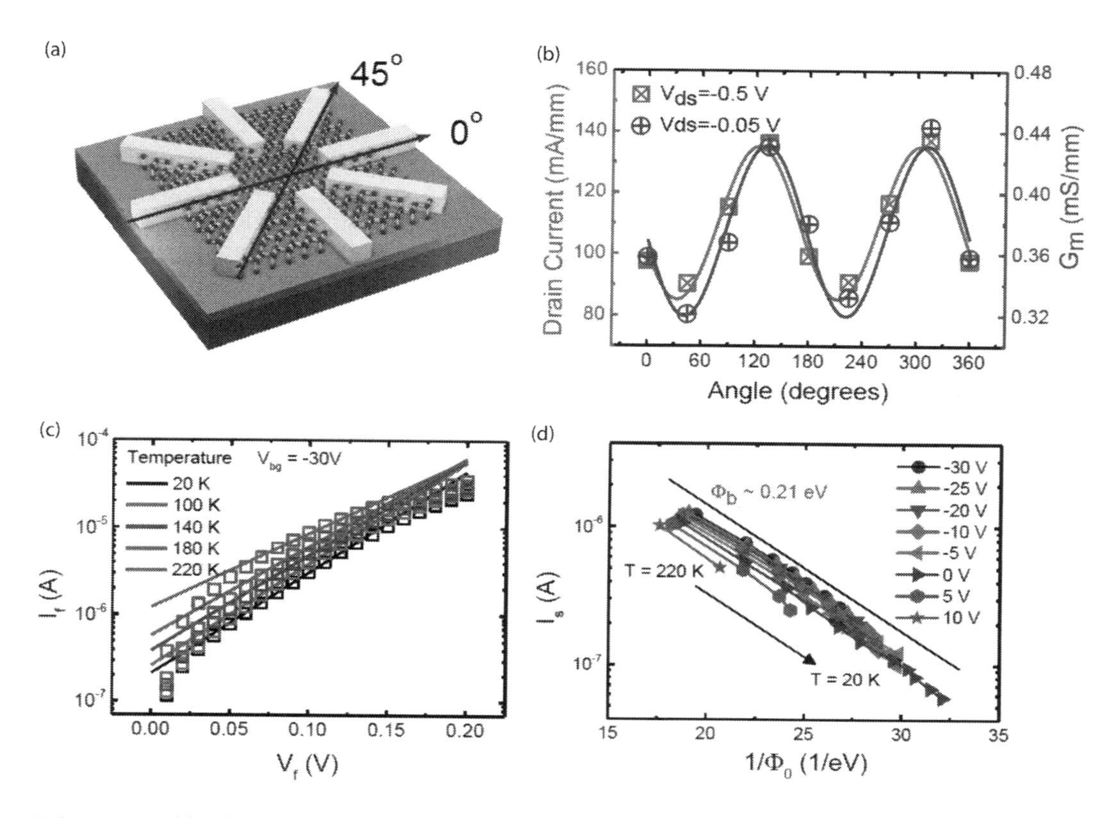

FIGURE 3.2 (a) Schematic of the fabricated device with phosphorene as channel where contacts are assembled in a particular angle, (b) current v/s angle graph showing negligible change in the performance, (c) temperature dependent IV, and (d) calculation of Schottky barrier height. (Reprinted with permission from Liu, H. et al., *ACS Nano*, 8, 4033–4041, 2014. Copyright 2014 American Chemical Society.)

angle dependent IV characteristics. The major drawback of phosphorene is its chemical reactivity with the ambient conditions which restricts its usage in practical applications. Another issue with the fabrication of FETs with graphene/TMDs and phosphorene is the contact resistance which creates a Schottky barrier height and results in the degradation in the performance of the fabricated transistor. Further, recent attempts have been made for the fabrication of FETs on the biodegradable cellulose paper for applications in disposable sensors [41].

3.2.2 Heterostructure Based on 2D Materials

Heterostructure fabricated by interfacing two semiconductors where the electronic band structure at the interface would be modulated due to electrostatics is an important building block for electronic devices which finds potential applications in fabrication of solid state devices, light emitting diodes, lasers, 2D materials possessing unique optical bandgap structures, large specific surface area, and strong light-matter interaction. In the view of above, graphene, TMDs, h-BN, and phosphorene have emerged to be strong candidates for the development of 2D heterostructure. 2D semiconductors exhibit high carrier mobility, excellent bendability, and high on-off ratio that enable for its use in flexible and wearable electronic applications. Further, these materials are composed of strong covalent bonds thereby providing excellence in plane stability [42]. Also, due to the weak vertical van der Waals forces, it becomes easier to isolate the 2D monolayer and restack them with other 2D materials.

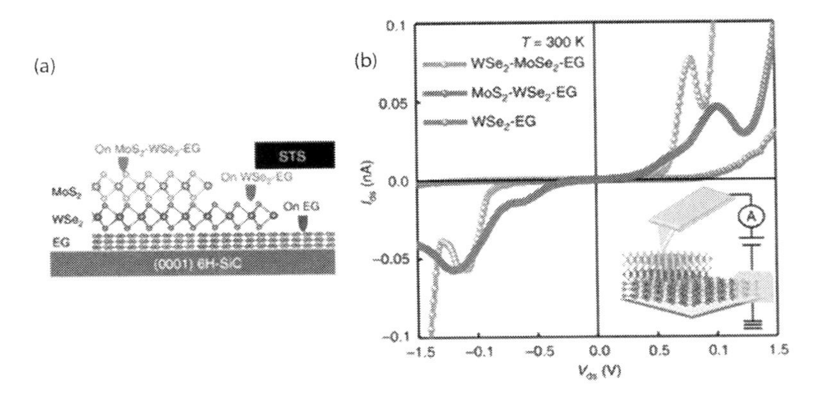

FIGURE 3.3 (a) Schematic of the fabricated device and (b) IV characteristic showing negative differential resistance. (Reprinted with permission from Lin, Y.C. et al., *Nat. Commun.*, 6, 7311, 2015. Copyright 2015 American Chemical Society.)

Apart from the scotch tape exfoliation and chemical exfoliation of a 2D monolayer, the chemical vapor deposition (CVD) process has been utilized for vertical heterostructure. The early experiments were performed on graphene and h-BN since they share similar lattice constants where Yang et al. demonstrated the growth of graphene using a plasma-assisted deposition method on h-BN, and it was observed that the growth of graphene is restricted only on the area of h-BN [43]. Further, even though the lattice constant of MoS_2 is 28% larger than graphene, graphene still can be utilized for the growth of MoS_2 [44] wherein strain develops on graphene, and due to the excellent mechanical properties of graphene, it is able to withstand such strains. Also, the direct growth of MoS_2, WSe_2, and h-BN on graphene through CVD methods has been demonstrated by Lin et al. [45]. Recently, Lin et al. reported the direct growth of MoS_2/WS_2/graphene by an oxide powder vaporization technique, and the electrical characterization revealed sharp negative differential resistance at room temperature resulting from the resonant tunneling charge carriers which is contradicting with the conventional stacked junctions [46]. Figure 3.3 shows the schematic of the device and its electrical characterization revealing negative differential resistance. Such processes have opened new avenues for research. Sahatiya et al. have demonstrated the solution-processed growth of MoS_2 on graphene/cellulose paper where graphene was uniformly coated on cellulose paper followed by the hydrothermal growth of MoS_2 on graphene. Characterization revealed that the growth of MoS_2 affected the crystallinity of graphene [47].

3.3 Applications

3.3.1 Photodetectors

Flexible photodetectors with a broad spectral range starting from the ultra violet (UV) to the near infrared (NIR) find widespread applications in areas such as optoelectronics, sensors, communication, and surveillance. However, the majority of them aim at improving the responsivity in a particular region or wavelength which is achieved by fabricating heterojunctions with 2D materials [48]. The major issue of a photodetector which is the inability to absorb wider regions of the electromagnetic spectrum still remains a challenge mainly due to the lack of synthesizing suitable hybrids which can absorb from UV to NIR. The other issue is improper device fabrication where the placement of metal electrodes plays an important role in collecting the photogenerated carriers. Most photodetectors comprise of the p-n heterojunction, where one of the materials is responsible for absorbance, having metal contacts on p and n type allow for effective separation of photogenerated carriers. The built in electric field at the potential barrier of the heterojunction is responsible for effective separation of photogenerated carriers. But for a broadband photodetector, both the materials of the heterojunctions should participate in the absorbance. In such a case, metal contacts on p and n type will trap either the photogenerated electrons or hole which leads to the failure of the device. Hence, proper device fabrication which decides the metal contact placement plays an important role in the fabrication of the broadband photodetector. To tackle these issues, Sahatiya et al. have developed broadband photodetectors based on MoS_2-ZnS [49], MoS_2-V_2O_5 [50], and MoS_2-CQD [51] wherein it is explained in details that the metal contact placement on MoS_2 leads to the broadband absorption and detection. Discrete distribution is important so that both the materials are exposed to illumination. De Fazio et al. reported the large area of a monolayer graphene/MoS_2 ion gel dielectric-based FET as flexible photodetectors on a flexible PET substrate which reported an excellent responsivity of 45.5 A/W [52]. Figure 3.4 shows the fabricated structure and its responsivity value graphs. A vertical heterostructure of $SnSe_2$/MoS_2 was reported by Zhou et al. using the MoS_2 template method [53]. Recently, to enhance the responsivity of the fabricated photodetector, concepts of piezotronics have been applied wherein one of the materials of the heterojunction is a piezoelectric material. Upon application of external strain, modulation in the depletion region causes the electric field to be increased which helps in the effective separation of the photogenerated charge carriers. Various examples include MoS_2/CuO, etc. [54].

3.3.2 Multifunctional Sensors

Flexible and disposable multifunctional sensors that can sense physical and chemical stimuli are important in the field of flexible and wearable electronics due to the potential applications in sensors, health care, IoT, environmental monitoring, etc. Further, multifunctional sensors responding to different chemical stimuli fabricated using functional nanomaterials and the usage of the same sensor multiple times for different sensing leads to unreliable sensor data. Hence, for the development of multifunctional sensors, there is a need for the development of disposable sensors wherein the user can perform the dedicated sensing and the sensor then can be disposed. The issue is intensified by the lack of suitable techniques for fabricating 2D materials-based disposable sensors which can be integrated to a smartphone with a dedicated application developed for each sensing. Sahatiya et al. have recently demonstrated the growth of MoS_2 on disposable cellulose paper by the solution-processed hydrothermal method. Further, the sensor data were wirelessly transmitted over the Bluetooth to the smartphone where a dedicated Android

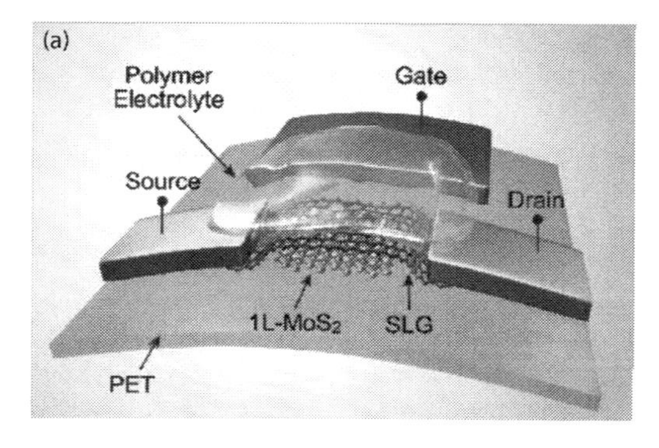

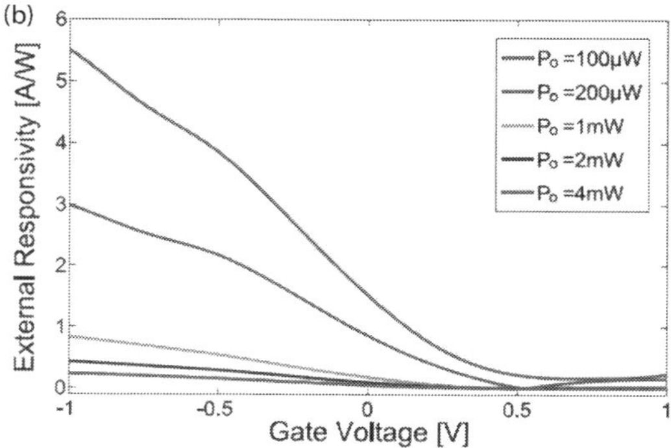

FIGURE 3.4 (a) Digital image of the fabricated device with MoS_2 as channel and ion gel as dielectric and (b) graph of external responsivity with gate voltage at different power. (Reprinted with permission from De Fazio, D. et al., *ACS Nano*, 10, 8252–8262, 2016. Copyright 2016 American Chemical Society.)

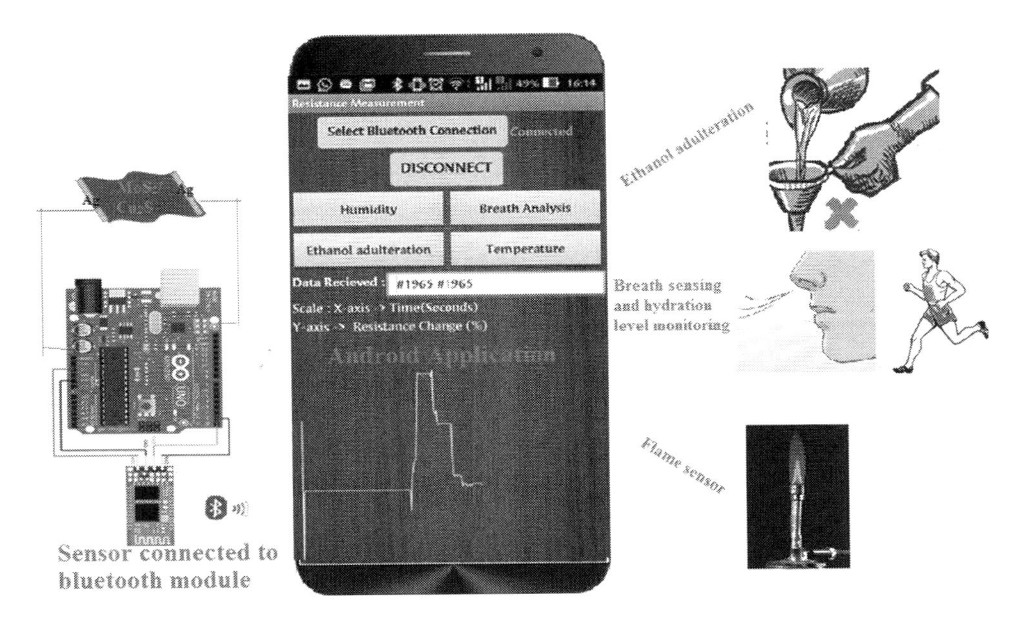

FIGURE 3.5 Multifunctional sensor (ethanol adulteration, humidity monitoring, and flame sensor) fabricated using MoS_2/Cu_2S and wireless integration of the same with smartphone. (Reprinted with permission from Sahatiya, P. et al., *ACS Appl. Mater. Interfaces*, 2018. Copyright 2018 American Chemical Society.)

application was developed thereby achieving user friendly interpretation of data [55]. Lu et al. demonstrated the use of graphene for smartphone-enabled sensing of various volatile organic compounds (VOCs). Authors utilize impedance spectroscopy for distinguishing the response of acetone from other VOCs. Finally, the measurements were carried out to measure the concentration of acetone in the human breath which could be applied for early diagnosis of various diseases [56] (Figure 3.5).

3.3.3 Bioanalytical Applications

Bio-sensing has prime significance in improving the quality of human life in terms of disease diagnosis, therapy, and environmental safety [57–60]. This has led researchers to focus on developing facile, portable, cost-effective, light-weight, implantable, ultra-conformable, and flexible biosensors with high sensitivity, lower limit of detection, and high stability. A biosensor is an analytical device that converts the information about the presence of a chemical species (analyte) to a quantifiable signal. It comprises of two key components: (a) receptor and (b) transducer. The receptor is a biomolecule, which recognizes the analyte, and the transducer converts this bio-recognition event into an analytically useful signal like current, potential, and impedance. Common and extensively used bio-recognition elements in bio-sensing include enzyme, antibody, or oligonucleotide [60,61].

Biomolecules such as proteins, cholesterol, glucose, DNA, etc. play key roles in several disease developments; for instance, an excess level of cholesterol in the human blood causes the risk of numerous diseases including coronary heart disease, whereas Alzheimer's disease and cancer cause DNA damage [62–65]. Henceforth, early detection of biomolecules is of greatest importance in disease diagnosis as well as therapy. Owing to remarkable physico-chemical properties such as large specific surface area, high catalytic ability, tunable band gap, rapid electron transfer rates, and excellent carrier mobility, 2D materials, like MoS_2, WS_2, and graphene have been used widely as a sensing material in the fabrication of various biosensors [66–68].

3.3.4 Detection of Glucose

Glucose is considered as one of the most important energy sources of living cells and metabolic intermediates [69]. The continuous monitoring of glucose in blood is essential for diagnosis and management of diabetes mellitus, a major worldwide public health problem [70]. In glucose biosensors, the glucose oxidase (GOD) enzyme is used as the mediator or recognition element.

Kwak et al. demonstrated a flexible glucose sensor using graphene-based FET wherein PET was employed as a flexible FET substrate and a CVD grown monolayer graphene was used as a channel [71]. A PDMS well was attached on top of the graphene channel to maintain the same active area of 33 mm² for all devices as illustrated in Figure 3.6a. To enhance the electro-catalytic ability of graphene toward glucose detection, graphene was functionalized with 1-pyrenebutanoicacid succinimidyl ester (PSE) linker. For glucose detection, GOD enzyme was immobilized on the surface of graphene which generates gluconic acid and H_2O_2 as shown in the equation (3.1). Direct measurement of H_2O_2 is used for the detection of glucose.

$$\text{D-glucose} + H_2O + O_2 \xrightarrow{\text{GOD}} \text{D-gluconic acid} + H_2O_2. \tag{3.1}$$

Since, graphene exhibits ambipolar characteristics, the graphene-based FET sensor can be operated in both p and n-type regions. In the p type region, V_g is smaller than the Dirac point whilst the n type region is the region to the right of the Dirac point. With the successive addition of glucose, the Dirac point was shifted toward lower values of V_g, to the left as presented in Figure 3.6b and c, thus confirming n-doping property of the graphene-based FET. This flexible FET sensor exhibited excellent linearity, reproducibility, and flexibility toward glucose detection in the range of 3.3–10.9 mM.

You et al. reported the silk fibroin-encapsulated graphene FET-based enzymatic biosensor that utilized silk protein as both device substrate and enzyme immobilization material for glucose detection [72]. The conductance of the graphene channel changed with the increasing glucose concentrations.

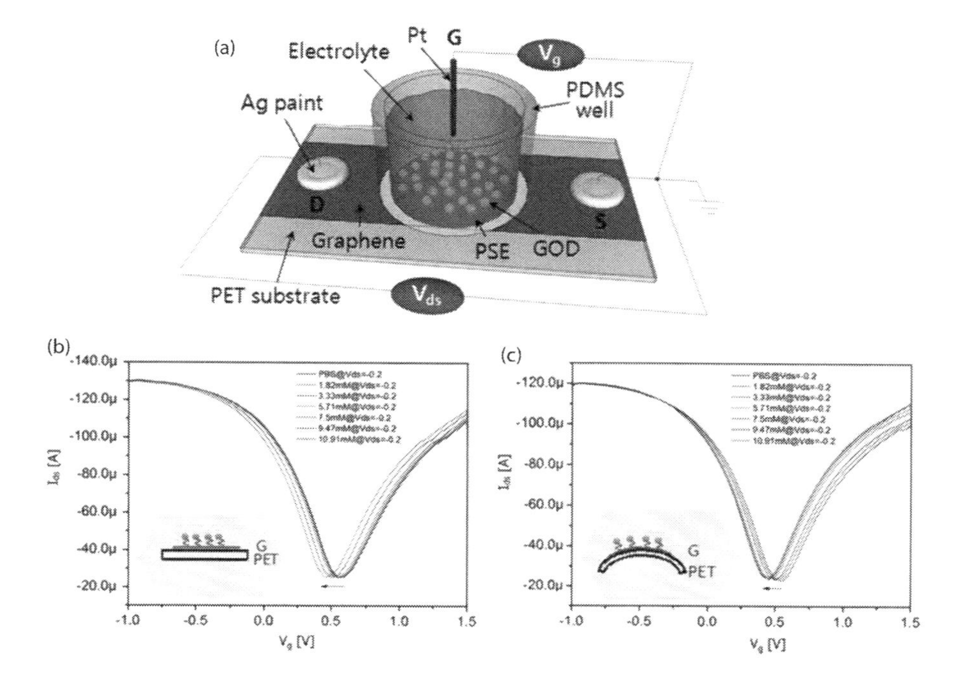

FIGURE 3.6 (a) Flexible glucose sensor using graphene-based FET and charge transfer curves of graphene FET in response to the different glucose concentrations, (b) without, and (c) with bending of PET substrate, respectively. (Reprinted from Sahatiya, P. et al., *2D Mater.*, 4, 025053, 2017. With permission.)

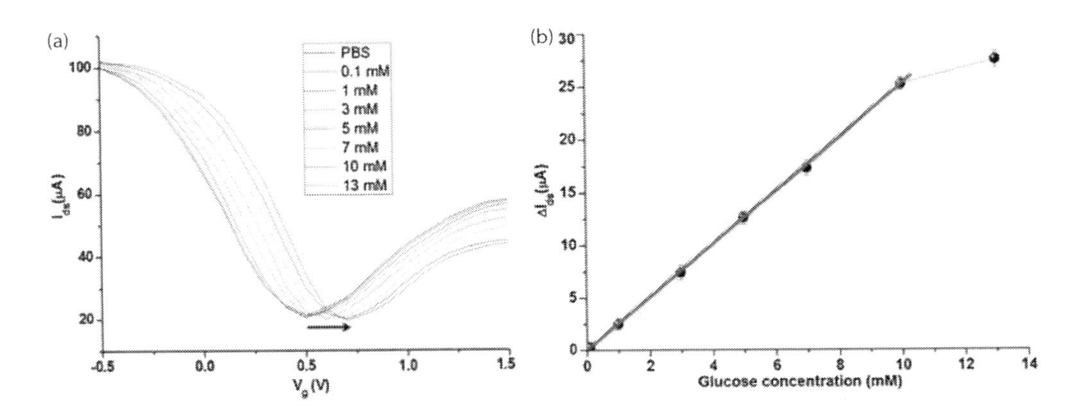

FIGURE 3.7 (a) Charge transfer curves of silk/GO_x functionalized graphene FET toward different concentrations of glucose at $V_{ds} = 100$ mV. (b) The response curve of the silk/GO_x functionalized graphene FET to glucose at $V_g = 0$ V. (Reprinted from Mas-Balleste, R. et al., *Nanoscale*, 3, 20–30, 2011. With permission.)

Figure 3.7a displays the transfer curves at various glucose concentrations. With the increasing glucose concentrations, the Dirac point shifted to the right, indicating the p-doping effect. In Figure 3.7b, the I_{ds} values biased at $V_{ds} = 100$ mV were measured at $V_g = 0$ V for different glucose concentrations in the range of 0–13 mM. I_{ds} increased linearly with increased glucose concentration. This flexible biosensor exhibited good selectivity, linearity with a sensitivity of 2.5 $\mu A{\cdot}mM^{-1}$ in the wide concentration range of 0.1–10 mM. This silk-supported biosensor could also be implantable as the graphene FET device was encapsulated by biocompatible silk fibroin.

Wang et al. developed the non-enzymatic glucose sensor using a molybdenum sulfide (MoS_x)-nickel (II) hydroxide ($Ni(OH)_2$) composite in sequence on a flexible carbon nanotube/polyimide (CNT/PI) composite membrane [73]. The flexible electrochemical sensor showed sensitivity of 0.1451 $\mu A \mu M^{-1}$ in the linear range from 10 to 1600 μM of glucose and limit of detection of 5.4 μM with good selectivity, repeatability, fast response time (<3 s), and long-term stability (2 weeks). The excellent performance toward non-enzymatic detection of glucose was ascribed to the synergistic effect between $Ni(OH)_2$ and MoS_x. Furthermore, this sensor was successfully used for glucose detection in human blood serum samples by the standard addition method with satisfactory recovery.

3.3.5 Detection of Other Biomolecules

Dopamine is an important neurotransmitter and clinically valuable diagnostic indicator. The presence of abnormalities in dopamine levels causes progression of neurological disorders such as Parkinson's and Alzheimer's diseases [74]. Thus, sensitive and selective detection of dopamine is of prime significance for the diagnostics of neurological diseases. Xu et al. reported the fabrication of a flexible, disposable sensor based on a poly(3,4-ethylenedioxythiophene) (PEDOT) modified laser scribed graphene (LSG) for detection of dopamine in the presence of ascorbic acid and uric acid [75]. The LSG electrodes were prepared via direct laser writing on polyimide (PI) sheets whilst PEDOT was electrodeposited on the LSG electrode. The sensor exhibited excellent selectivity, reproducibility, and repeatability with sensitivity of 0.22 ± 0.01 $\mu A \mu M^{-1}$ and a low detection limit of 0.33 μM. This direct laser scribing of PI films offers a commercialization-available approach for efficient fabrication of graphene electrodes.

Sweat lactate is known as one of the foremost biomarkers of pressure ischemia and tissue oxygenation and provides warning for restricted blood flow (tenfold greater than in blood), therefore it can be used to track an individual's performance and exertion level [76]. Wang et al. demonstrated a flexible sweat platform fabricated by depositing Cu buds on freestanding graphene paper carrying monolayer MoS_2 nanocrystals for bio-functional detection of glucose and lactate. This flexible and non-invasive sensor exhibited wide linear ranges from 5 to 1775 μM for glucose and 0.01–18.4 mM for lactate with their corresponding detection limits of 500 nM and 0.1 μM, respectively. The flexible sensor also showed fast response, good selectivity, reproducibility and exceptional flexibility, which enable its usage for monitoring glucose and lactate in human perspiration. This unique strategy of structurally integrating 3D transition metal, 0D transition metal sulfide, and 2D graphene provides a new insight into the design of flexible electrodes for sweat glucose and lactate monitoring and a wider range of applications in biosensing, bioelectronics, and lab-on-a-chip devices.

3.3.6 Detection of Biomarkers

Immunosensors, an important class of biosensors have been extensively used to detect protein biomarkers such as a prostate specific antigen (PSA), heat shock protein 70, (HSP70) etc., based on antigen–antibody interaction with high sensitivity and particularly excellent specificity [77,78]. Xiang et al. reported a graphene FET-based biosensor on a flexible Kapton substrate using 3D inkjet printing for detection of an infectious organism; norovirus target pathogen [79]. The graphene layer was functionalized selectively with coupling molecules (e.g., –COOH, –SH, or $-NH_2$ groups) via passive adsorption. A microfluidic channel was fabricated for the FET to function in flow-through modules. Sensing elements like peptides, proteins, and antibodies can be loaded onto the functionalized graphene and then blocked with bovine serum albumin (BSA) protein. Upon binding of the target molecule, intensity changes of the alternating current (AC) signal are recorded as a function of concentration to changes of AC signal. The value of the voltage gain from source to drain at 10 GHz produced a linear response from 0.07 to 3.70 dB when the concentration of norovirus protein increased from 0.1 to 100 μg/mL. This flexible graphene FET sensor could be used for detection of a variety of disease-causing pathogens.

Majd et al. developed a flexible FET-type aptasensor for detection of ovarian cancer antigen, CA 125 cancer marker using carboxylated multiwalled carbon nanotubes immobilized onto reduced graphene oxide film [80]. PMMA was used as a flexible substrate. The CA 125 biomolecule was bound to the specific CA 125 ssDNA (single-stranded DNA) which transduced a signal in the FET-type aptasensor. From Figure 3.8a, it was observed that the Id of the aptasensor decreased with the increasing of the target molecules concentrations which was due to the depletion of negative charge carriers elicited by the aptamer/CA 125 binding event. Under the experimental condition with pH 7.5, the aptamer/CA 125 binding event induced negative point charges in the liquid-ion gate near the reduced graphene oxide (rGO) layers, thus leading to the depletion of negative charge carriers in layers of rGO channel. The increased

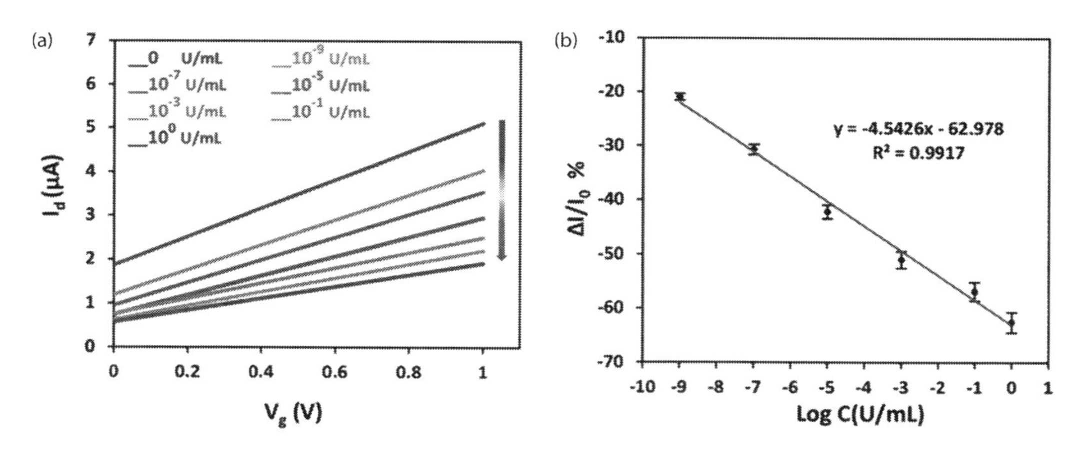

FIGURE 3.8 (a) I_d–V_g responses with increasing CA 125 concentration from 10^{-9} U/mL to 1 U/mL in PBS (pH 7.4, 0.01 M) at $V_d = 0.2$ V and (b) their corresponding calibration curve. (Reprinted from Ovchinnikov, D. et al., *ACS Nano*, 8, 8174–8181, 2014. With permission.)

negative gate voltage by the CA 125 binding event results in the decrease of the Id of n-type rGO. Based on FET sensing results, the decrease in rGO conductivity could be attributed to the gating effect from charged proteins. For the gating effect, charged proteins attached on the surface of rGO are equivalent to a positive or negative potential gating that leads to a change in the rGO carrier density. Figure 3.8b exhibits the calibration curve, and the developed flexible sensor showed a wide linear range for CA 125 (10^{-9} to 1 U/mL) and lower limit of detection of 0.5 nU/mL with good reproducibility, stability, and selectivity. The FET-based aptasensor was also used to detect CA 125 in human serum with a satisfactory result.

2D materials-based flexible biosensors have been used widely for the detection of other important biomarkers and biomolecules like urea, uric acid, hydrogen peroxide, etc. Describing each of them is beyond the scope of this chapter.

3.3.7 Gas and Humidity Sensors

Other most exciting applications of 2D materials and their derivatives are gas and humidity sensors. Nitrogen dioxide (NO_2) gas, a common air pollutant is produced throughout various fossil-fuels combustion in industrial factories, automotive engines, and power plants. NO_2 causes an increase in respiratory symptoms and a reduction in pulmonic function directly in humans and also reacts with water-based chemicals in the atmosphere to form acid rain causing severe environmental destruction [81]. Hence, development of sensitive NO_2 gas sensors is highly desirable for environmental monitoring.

Yun et al. presented a washable and flexible gas sensor composed of a few-layer MoS_2 and rGO-coated cotton-yarn (CY rGO) for NO_2 gas sensing at room temperature [82]. Figure 3.9a illustrates

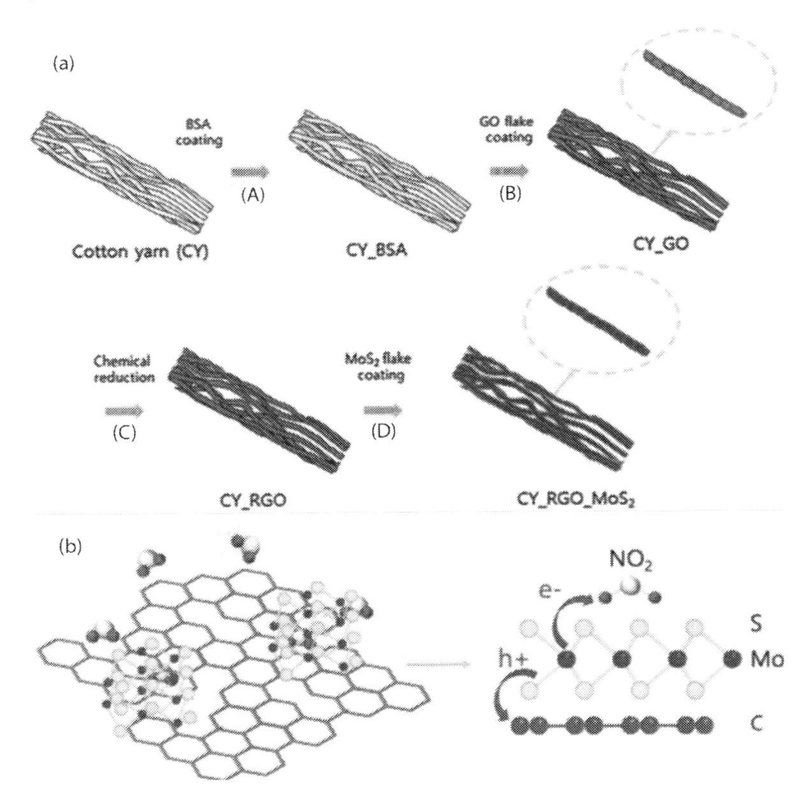

FIGURE 3.9 (a) Detailed fabrication steps for CY rGO-MoS_2 gas sensor and (b) charge transport between MoS_2 and reduced graphene oxide induced by NO_2 sensing. (Reprinted from Lei, W. et al., *Chem. Soc. Rev.*, 46, 3492–3509, 2017. With permission.)

detailed fabrication steps for a CY rGO-MoS_2 gas sensor. The flexible gas sensor response was defined as $\Delta R/R_0(\%) = (R_g - R_0)/R_0 \times 100$, where R_0 and R_g = resistances of the yarn sensor before and after exposure to NO_2, respectively. The only CY RGO and CY rGO-MoS_2 composite sensor exhibited 6% and 28% of response to 0.45 ppm NO_2 at 45% relative humidity. A composites-based flexible sensor showed 4 times more sensitivity to a NO_2 than only rGO-based flexible gas sensor which was due to a large surface area and a synergic effect of rGO with MoS_2. The CY as a flexible substrate which possesses high surface area played an important role in the enhancement of gas sensing properties. Moreover, as shown in Figure 3.9b, the energy levels are favorable for transfer of holes from a MoS_2 valence band to rGO, which could enhance the NO_2 sensing property of rGO by increasing the carrier concentration. This composite gas sensor also showed excellent flexibility and durability toward mechanical and washing stress that can find applications in wearable devices.

Sahatiya et al. demonstrated a flexible, disposable cellulose paper-based MoS_2-Cu_2S hybrid for humidity-sensing application where the sensor was exposed to different relative humidities, and corresponding resistance change was measured in ambient temperature of 27°C [55]. In addition, the measurements were also carried out by measuring the response through an Arduino board and transmitting the data wirelessly via Bluetooth to a smartphone as depicted in Figure 3.10a. As the relative humidity increases, the water molecules get adsorbed on the n-type MoS_2 as well on p-type Cu_2S. Since water molecules are electron donor species, they will increase the charge carrier concentration

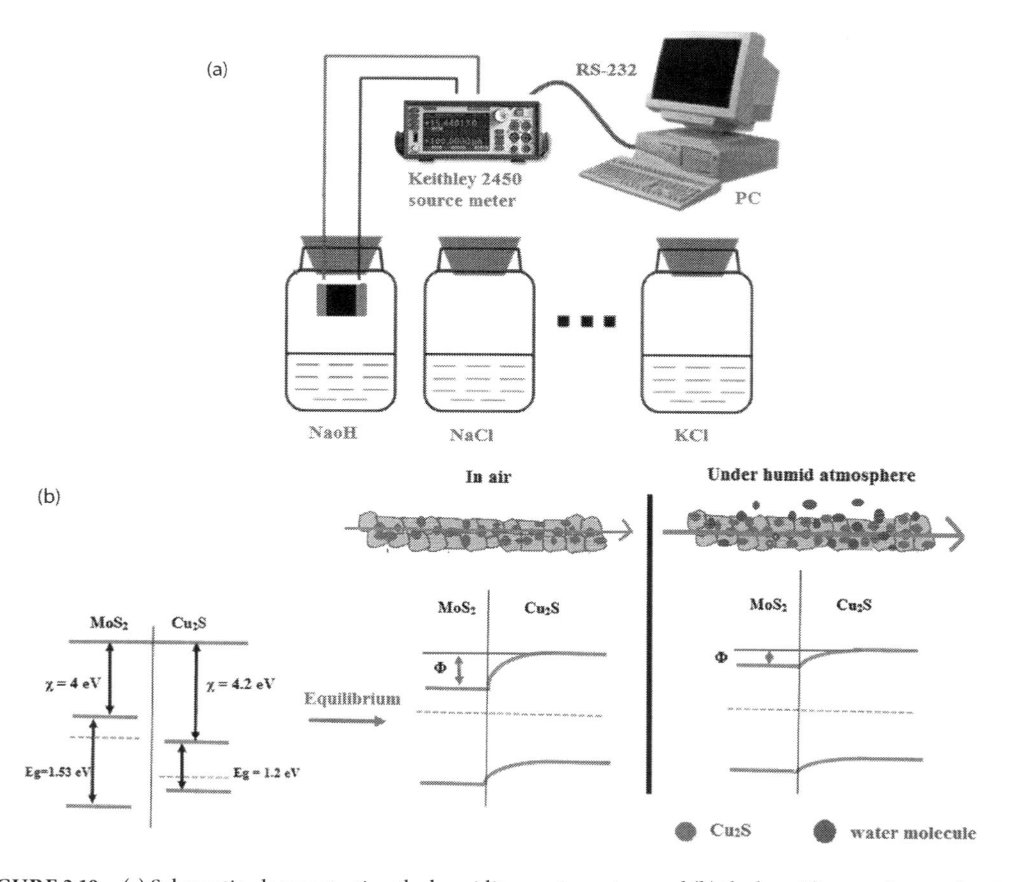

FIGURE 3.10 (a) Schematic, demonstrating the humidity sensing set up and (b) the humidity sensing mechanism of MoS_2/Cu_2S under different humidity levels. (Reprinted from Hwang, H.J. et al., *AIP Adv.*, 8, 015022, 2018. With permission.)

in n-type MoS_2 and will decrease the charge carrier concentration in Cu_2S which results in a decrease in the potential barrier created, thereby increasing the current as depicted in Figure 3.10b.

2D materials-based flexible sensors have been used widely for the detection of other toxic gases such as ammonia. Describing each of them is beyond the scope of this chapter.

3.4 Conclusion and Outlook

An understanding of different 2D materials such as graphene, TMDs, h-BN, and phosphorene has been presented with their electronic properties and the ability to form different heterojunctions and their potential applications in photodetectors, multifunctional sensors, and bio and chemical sensors. Progress in 2D materials offers tremendous opportunity for the researchers to explore new concepts and technologies in electronics, energy, and optoelectronics. Despite considerable advances in the 2D materials research, the progress in the synthesis and fabrication of large and uniform monolayer 2D material is not comparable to that of mechanically exfoliated 2D materials. To develop practical applications, the growth of 2D materials should be defect free with the possibility of doping. With the current ongoing progress in 2D materials, one can expect that the ideal properties of 2D materials which are envisioned now will revolutionize the developments in electronics, energy, and optoelectronics.

References

1. Márquez, F., & Morant, C. Nanomaterials for sensor applications. *Soft Nanoscience Letters*, **2015**, 5(1).
2. Lherbier, A., Blase, X., Niquet, Y. M., Triozon, F., & Roche, S. Charge transport in chemically doped 2D graphene. *Physical Review Letters*, **2008**, 101(3), 036808.
3. Ciarrocchi, A., Avsar, A., Ovchinnikov, D., & Kis, A. Thickness-modulated metal-to-semiconductor transformation in a transition metal dichalcogenide. *Nature Communications*, **2018**, 9(1), 919.
4. Torres, T. Graphene chemistry. *Chemical Society Reviews*, **2017**, 46(15), 4385–4386.
5. Song, L., Ci, L., Lu, H., Sorokin, P. B., Jin, C., Ni, J., & Ajayan, P. M. Large scale growth and characterization of atomic hexagonal boron nitride layers. *Nano Letters*, **2010**, 10(8), 3209–3215.
6. Reich, E. S. Phosphorene excites materials scientists. *Nature*, **2014**, 506(7486), 19.
7. Rogers, J. A., & Huang, Y. A curvy, stretchy future for electronics. *Proceedings of the National Academy of Sciences*, **2009**, 106(27), 10875–10876.
8. Rogers, J. A., Someya, T., & Huang, Y. Materials and mechanics for stretchable electronics. *Science*, **2010**, 327(5973), 1603–1607.
9. Tobjörk, D., & Österbacka, R. Paper electronics. *Advanced Materials*, **2011**, 23(17), 1935–1961.
10. Hayes, R. A., & Feenstra, B. J. Video-speed electronic paper based on electrowetting. *Nature*, **2003**, 425(6956), 383.
11. Sha, R., Puttapati, S. K., Srikanth, V. V., & Badhulika, S. Ultra-sensitive phenol sensor based on overcoming surface fouling of reduced graphene oxide-zinc oxide composite electrode. *Journal of Electroanalytical Chemistry*, **2017**, 785, 26–32.
12. Sha, R., Komori, K., & Badhulika, S. Amperometric pH sensor based on graphene–polyaniline composite. *IEEE Sensors Journal*, **2017**, 17(16), 5038–5043.
13. Sha, R., Puttapati, S. K., Srikanth, V. V., & Badhulika, S. Ultra-sensitive non-enzymatic ethanol sensor based on reduced graphene oxide-zinc oxide composite modified electrode. *IEEE Sensors Journal*, **2017**, 18(5), 1844–1848.
14. Sha, R., & Badhulika, S. Binder free platinum nanoparticles decorated graphene-polyaniline composite film for high performance supercapacitor application. *Electrochimica Acta*, **2017**, 251, 505–512.

15. Sahatiya, P., Jones, S. S., Gomathi, P. T., & Badhulika, S. Flexible substrate based 2D ZnO (n)/graphene (p) rectifying junction as enhanced broadband photodetector using strain modulation. *2D Materials*, **2017**, 4(2), 025053.
16. Mas-Balleste, R., Gomez-Navarro, C., Gomez-Herrero, J., & Zamora, F. 2D materials: To graphene and beyond. *Nanoscale*, **2011**, 3(1), 20–30.
17. Chhowalla, M., Shin, H. S., Eda, G., Li, L. J., Loh, K. P., & Zhang, H. The chemistry of two-dimensional layered transition metal dichalcogenide nanosheets. *Nature Chemistry*, **2013**, 5(4), 263.
18. Manzeli, S., Ovchinnikov, D., Pasquier, D., Yazyev, O. V., & Kis, A. 2D transition metal dichalcogenides. *Nature Reviews Materials*, **2017**, 2(8), 17033.
19. Hong, S., Naqi, M., Jung, U., Liu, N., Kwon, H. J., Grigoropoulos, C. P., & Kim, S. 66-1: Invited paper: High mobility flexible 2D multilayer MoS_2 TFTs on solution-based polyimide substrates. In *SID Symposium Digest of Technical Papers*, **2017**, 48(1), 965–967.
20. Schmitz, M., Engels, S., Banszerus, L., Watanabe, K., Taniguchi, T., Stampfer, C., & Beschoten, B. High mobility dry-transferred CVD bilayer graphene. *Applied Physics Letters*, **2017**, 110(26), 263110.
21. Lee, D., Lee, S., An, B. S., Kim, T. H., Yang, C. W., Suk, J. W., & Baik, S. Dependence of the in-plane thermal conductivity of graphene on grain misorientation. *Chemistry of Materials*, **2017**, 29(24), 10409–10417.
22. Sakhaee-Pour, A. Elastic properties of single-layered graphene sheet. *Solid State Communications*, **2009**, 149(1–2), 91–95.
23. He, J., Hummer, K., & Franchini, C. Stacking effects on the electronic and optical properties of bilayer transition metal dichalcogenides MoS_2, $MoSe_2$, WS_2, and WSe_2. *Physical Review B*, **2014**, 89(7), 075409.
24. Ovchinnikov, D., Allain, A., Huang, Y. S., Dumcenco, D., & Kis, A. Electrical transport properties of single-layer WS_2. *ACS Nano*, **2014**, 8(8), 8174–8181.
25. Liu, H., Neal, A. T., Zhu, Z., Luo, Z., Xu, X., Tománek, D., & Ye, P. D. Phosphorene: An unexplored 2D semiconductor with a high hole mobility. *ACS Nano*, **2014**, 8(4), 4033–4041.
26. Lei, W., Liu, G., Zhang, J., & Liu, M. Black phosphorus nanostructures: Recent advances in hybridization, doping and functionalization. *Chemical Society Reviews*, **2017**, 46(12), 3492–3509.
27. Hwang, H. J., Heo, S., Yoo, W. B., & Lee, B. H. Graphene–ZnO: N barristor on a polyethylene naphthalate substrate. *AIP Advances*, **2018**, 8(1), 015022.
28. Gupta, S. K., Jha, P., Singh, A., Chehimi, M. M., & Aswal, D. K. Flexible organic semiconductor thin films. *Journal of Materials Chemistry C*, **2015**, 3(33), 8468–8479.
29. Zhao, X., Tong, Y., Tang, Q., Tian, H., & Liu, Y. Highly sensitive H_2S sensors based on ultrathin organic single-crystal microplate transistors. *Organic Electronics*, **2016**, 32, 94–99.
30. Kumagai, H., Sato, N., Takeoka, S., Sawada, K., Fujie, T., & Takahashi, K. Optomechanical characterization of freestanding stretchable nanosheet based on polystyrene–polybutadiene–polystyrene copolymer. *Applied Physics Express*, **2016**, 10(1), 011601.
31. Zulfiqar, A., Pfreundt, A., Svendsen, W. E., & Dimaki, M. Fabrication of polyimide based microfluidic channels for biosensor devices. *Journal of Micromechanics and Microengineering*, **2015**, 25(3), 035022.
32. Sahatiya, P., Jones, S. S., & Badhulika, S. Direct, large area growth of few-layered MoS_2 nanostructures on various flexible substrates: Growth kinetics and its effect on photodetection studies. *Flexible and Printed Electronics*, **2018**, 3(1), 015002.
33. Li, X., Wang, X., Zhang, L., Lee, S., & Dai, H. Chemically derived, ultrasmooth graphene nanoribbon semiconductors. *Science*, **2008**, 319(5867), 1229–1232.
34. Parrish, K. N., Ramón, M. E., Banerjee, S. K., & Akinwande, D. A compact model for graphene FETs for linear and non-linear circuits. In *Proceedings of the SISPAD Simulation of Semiconductor Processes and Devices*, **2012**, 75–78.
35. Larentis, S., Fallahazad, B., & Tutuc, E. Field-effect transistors and intrinsic mobility in ultra-thin $MoSe_2$ layers. *Applied Physics Letters*, **2012**, 101(22), 223104.

36. Li, L., Yu, Y., Ye, G. J., Ge, Q., Ou, X., Wu, H., & Zhang, Y. Black phosphorus field-effect transistors. *Nature Nanotechnology*, **2014**, 9(5), 372.
37. Lemaitre, M. G., Donoghue, E. P., McCarthy, M. A., Liu, B., Tongay, S., Gila, B., & Rinzler, A. G. Improved transfer of graphene for gated Schottky-junction, vertical, organic, field-effect transistors. *ACS Nano*, **2012**, 6(10), 9095–9102.
38. Chen, J. R., Odenthal, P. M., Swartz, A. G., Floyd, G. C., Wen, H., Luo, K. Y., & Kawakami, R. K. Control of Schottky barriers in single layer MoS_2 transistors with ferromagnetic contacts. *Nano Letters*, **2013**, 13(7), 3106–3110.
39. Podzorov, V., Gershenson, M. E., Kloc, C., Zeis, R., & Bucher, E. High-mobility field-effect transistors based on transition metal dichalcogenides. *Applied Physics Letters*, **2004**, 84(17), 3301–3303.
40. Yoon, Y., Ganapathi, K., & Salahuddin, S. How good can monolayer MoS_2 transistors be? *Nano Letters*, **2011**, 11(9), 3768–3773.
41. Fortunato, E., Correia, N., Barquinha, P., Pereira, L., Gonçalves, G., & Martins, R. High-performance flexible hybrid field-effect transistors based on cellulose fiber paper. *IEEE Electron Device Letters*, **2008**, 29(9), 988–990.
42. Geim, A. K., & Grigorieva, I. V. Van der Waals heterostructures. *Nature*, **2013**, 499(7459), 419.
43. Yang, W., Chen, G., Shi, Z., Liu, C. C., Zhang, L., Xie, G., & Watanabe, K. Epitaxial growth of single-domain graphene on hexagonal boron nitride. *Nature Materials*, **2013**, 12(9), 792.
44. Shi, Y., Zhou, W., Lu, A. Y., Fang, W., Lee, Y. H., Hsu, A. L., & Idrobo, J. C. van der Waals epitaxy of MoS_2 layers using graphene as growth templates. *Nano Letters*, **2012**, 12(6), 2784–2791.
45. Lin, Y. C., Lu, N., Perea-Lopez, N., Li, J., Lin, Z., Peng, X., & Bresnehan, M. S. Direct synthesis of van der Waals solids. *ACS Nano*, **2014**, 8(4), 3715–3723.
46. Lin, Y. C., Ghosh, R. K., Addou, R., Lu, N., Eichfeld, S. M., Zhu, H., & Wallace, R. M. Atomically thin resonant tunnel diodes built from synthetic van der Waals heterostructures. *Nature Communications*, **2015**, 6, 7311.
47. Sahatiya, P., & Badhulika, S. Strain-modulation-assisted enhanced broadband photodetector based on large-area, flexible, few-layered Gr/MoS_2 on cellulose paper. *Nanotechnology*, **2017**, 28(45), 455204.
48. Yu, S. H., Lee, Y., Jang, S. K., Kang, J., Jeon, J., Lee, C., & Cho, J. H. Dye-sensitized MoS_2 photodetector with enhanced spectral photoresponse. *ACS Nano*, **2014**, 8(8), 8285–8291.
49. Gomathi, P. T., Sahatiya, P., & Badhulika, S. Large-area, flexible broadband photodetector based on ZnS–MoS_2 hybrid on paper substrate. *Advanced Functional Materials*, **2017**, 27(31), 1701611.
50. Sahatiya, P., & Badhulika, S. Discretely distributed 1D V_2O_5 nanowires over 2D MoS_2 nanoflakes for an enhanced broadband flexible photodetector covering the ultraviolet to near infrared region. *Journal of Materials Chemistry C*, **2017**, 5(48), 12728–12736.
51. Sahatiya, P., Jones, S. S., & Badhulika, S. 2D MoS_2–carbon quantum dot hybrid based large area, flexible UV–vis–NIR photodetector on paper substrate. *Applied Materials Today*, **2018**, 10, 106–114.
52. De Fazio, D., Goykhman, I., Yoon, D., Bruna, M., Eiden, A., Milana, S., & Kis, A. High responsivity, large-area graphene/MoS_2 flexible photodetectors. *ACS Nano*, **2016**, 10(9), 8252–8262.
53. Zhou, X., Zhou, N., Li, C., Song, H., Zhang, Q., Hu, X., & Xiong, J. Vertical heterostructures based on $SnSe_2$/MoS_2 for high performance photodetectors. *2D Materials*, **2017**, 4(2), 025048.
54. Sahatiya, P., & Badhulika, S. Fabrication of a solution-processed, highly flexible few layer MoS_2 (n)–CuO (p) piezotronic diode on a paper substrate for an active analog frequency modulator and enhanced broadband photodetector. *Journal of Materials Chemistry C*, **2017**, 5(44), 11436–11447.
55. Sahatiya, P., Kadu, A., Gupta, H., Gomathi, P. T., & Badhulika, S. Flexible, disposable cellulose paper based MoS_2-Cu_2S hybrid for wireless environmental monitoring and multifunctional sensing of chemical stimuli. *ACS Applied Materials & Interfaces*, **2018**. doi:10.1021/acsami.8b00245.

56. Liu, L., Zhang, D., Zhang, Q., Chen, X., Xu, G., Lu, Y., & Liu, Q. Smartphone-based sensing system using ZnO and graphene modified electrodes for VOCs detection. *Biosensors and Bioelectronics*, **2017**, 93, 94–101.
57. Celik, N., Balachandran, W., & Manivannan, N. Graphene-based biosensors: Methods, analysis and future perspectives. *IET Circuits, Devices & Systems*, **2015**, 9(6), 434–445.
58. Sha, R., Badhulika, S., & Mulchandani, A. Graphene-based biosensors and their applications in biomedical and environmental monitoring. In: *Springer Series on Chemical Sensors and Biosensors (Methods and Applications)*. Springer, Berlin, Germany, Chapter 9, **2017**, 1–30.
59. Viswanathan, S., Narayanan, T. N., Aran, K., Fink, K. D., Paredes, J., Ajayan, P. M., & Demirci, U. Graphene–protein field effect biosensors: Glucose sensing. *Materials Today*, **2015**, 18(9), 513–522.
60. Sahatiya, P., & Badhulika, S. Graphene hybrid architectures for chemical sensors. In: Gonçalves G., Marques P., Vila M. (Eds.), *Graphene-Based Materials in Health and Environment*. Carbon Nanostructures. Springer, Cham, Switzerland, **2016**.
61. Yang, Y., Yang, X., Tan, Y., & Yuan, Q. Recent progress in flexible and wearable bio-electronics based on nanomaterials. *Nano Research*, **2017**, 10(5), 1560–1583.
62. Settu, K., Liu, J. T., Chen, C. J., & Tsai, J. Z. Development of carbon–graphene-based aptamer biosensor for EN2 protein detection. *Analytical Biochemistry*, **2017**, 534, 99–107.
63. Lin, X., Ni, Y., & Kokot, S. Electrochemical cholesterol sensor based on cholesterol oxidase and MoS_2-AuNPs modified glassy carbon electrode. *Sensors and Actuators B: Chemical*, **2016**, 233, 100–106.
64. Lin, T., Zhong, L., Song, Z., Guo, L., Wu, H., Guo, Q., & Chen, G. Visual detection of blood glucose based on peroxidase-like activity of WS_2 nanosheets. *Biosensors and Bioelectronics*, **2014**, 62, 302–307.
65. Hu, Y., Li, F., Han, D., & Niu, L. Graphene for DNA biosensing. In *Biocompatible Graphene for Bioanalytical Applications*. Springer, Berlin, Germany, **2015**, 11–33.
66. Liu, Y., Dong, X., & Chen, P. Biological and chemical sensors based on graphene materials. *Chemical Society Reviews*, **2012**, 41(6), 2283–2307.
67. Xu, B., Su, Y., Li, L., Liu, R., & Lv, Y. Thiol-functionalized single-layered MoS_2 nanosheet as a photoluminescence sensing platform via charge transfer for dopamine detection. *Sensors and Actuators B: Chemical*, **2017**, 246, 380–388.
68. Leong, S. X., Mayorga-Martinez, C. C., Chia, X., Luxa, J., Sofer, Z., & Pumera, M. 2H→1T phase change in direct synthesis of WS_2 nanosheets via solution-based electrochemical exfoliation and their catalytic properties. *ACS Applied Materials & Interfaces*, **2017**, 9(31), 26350–26356.
69. Xu, J., Xu, N., Zhang, X., Xu, P., Gao, B., Peng, X., Mooni, S. et al. Phase separation induced rhizobia-like Ni nanoparticles and TiO_2 nanowires composite arrays for enzyme-free glucose sensor. *Sensors and Actuators B: Chemical*, 244, 38–46.
70. Badhulika, S., Paul, R. K., Terse, T., & Mulchandani, A. Nonenzymatic glucose sensor based on platinum nanoflowers decorated multiwalled carbon nanotubes-graphene hybrid electrode. *Electroanalysis*, **2014**, 26(1), 103–108.
71. Kwak, Y. H., Choi, D. S., Kim, Y. N., Kim, H., Yoon, D. H., Ahn, S. S., Yang, J. W., Yang, W. S., & Seo, S. Flexible glucose sensor using CVD-grown graphene-based field effect transistor. *Biosensors and Bioelectronics*, **2012**, 37(1), 82–87.
72. You, X., & Pak, J. J. Graphene-based field effect transistor enzymatic glucose biosensor using silk protein for enzyme immobilization and device substrate. *Sensors and Actuators B: Chemical*, **2014**, 202, 1357–1365.
73. Wang, Q., Zhang, Y., Ye, W., & Wang, C. $Ni(OH)_2/MoS_x$ nanocomposite electrodeposited on a flexible CNT/PI membrane as an electrochemical glucose sensor: The synergistic effect of $Ni(OH)_2$ and MoS_x. *Journal of Solid State Electrochemistry*, **2016**, 20(1), 133–142.

74. Łuczak, T. Preparation and characterization of the dopamine film electrochemically deposited on a gold template and its applications for dopamine sensing in aqueous solution. *Electrochimica Acta*, 53(19), **2008**, 5725–5731.
75. Xu, G., Jarjes, Z. A., Desprez, V., Kilmartin, P. A., & Travas-Sejdic, J. Sensitive, selective, disposable electrochemical dopamine sensor based on PEDOT-modified laser scribed graphene. *Biosensors and Bioelectronics*, **2018**, 107, 184–191.
76. Wang, Z., Dong, S., Gui, M., Asif, M., Wang, W., Wang, F., & Liu, H. Graphene paper supported MoS_2 nanocrystals monolayer with Cu submicron-buds: High-performance flexible platform for sensing in sweat. *Analytical Biochemistry*, **2018**, 543, 82–89.
77. Okuno, J., Maehashi, K., Kerman, K., Takamura, Y., Matsumoto, K., & Tamiya, E. Label-free immunosensor for prostate-specific antigen based on single-walled carbon nanotube array-modified microelectrodes. *Biosensors and Bioelectronics*, **2007**, 22(9–10), 2377–2381.
78. Lin, J., & Ju, H. Electrochemical and chemiluminescent immunosensors for tumor markers. *Biosensors and Bioelectronics*, **2005**, 20(8), 1461–1470.
79. Xiang, L., Wang, Z., Liu, Z., Weigum, S. E., Yu, Q., & Chen, M. Y. Inkjet-printed flexible biosensor based on graphene field effect transistor. *IEEE Sensors Journal*, **2016**, 16(23), 8359–8364.
80. Majd, S. M., & Salimi, A. Ultrasensitive flexible FET-type aptasensor for CA 125 cancer marker detection based on carboxylated multiwalled carbon nanotubes immobilized onto reduced graphene oxide film. *Analytica Chimica Acta*, **2018**, 1000, 273–282.
81. Hyodo, T., Urata, K., Kamada, K., Ueda, T., & Shimizu, Y. Semiconductor-type SnO_2-based NO_2 sensors operated at room temperature under UV-light irradiation. *Sensors and Actuators B: Chemical*, **2017**, 253, 630–640.
82. Qi, H. Y., Mi, W. T., Zhao, H. M., Xue, T., Yang, Y., & Ren, T. L. A large-scale spray casting deposition method of WS_2 films for high-sensitive, flexible and transparent sensor. *Materials Letters*, **2017**, 201, 161–164.

4

Flexible and Stretchable Thin-Film Transistors

Joseph B. Andrews,
Jorge A. Cardenas,
and
Aaron D. Franklin

4.1 Introduction

The design of transistors originates as far back as 1930 with a patent by James Lilienfeld. The patent describes a majority carrier electrical device in which the current is controlled through an externally applied electric field [1]. However, significant challenges in developing capable materials delayed researchers from producing its first embodiment. While bi-polar junction transistors took off, other types of transistors, including metal-oxide-semiconductor field-effect transistors (MOSFETs) and thin-film transistors (TFTs), remained simply an idea. The field then took a large step forward in the 1960s, with the advent of metal-chalcogenide thin films. Paul Weimer, in his seminal paper titled 'The TFT—A New Thin-Film Transistor' experimentally demonstrates the viability of a thin film of cadmium sulfide to act as the channel in a transistor in which the current is controlled by an externally applied gate field [2]. What differentiates Weimer's devices from their contemporary field-effect transistors (primarily silicon-based metal-oxide-semiconductor field-effect transistors) are their thin channel thickness (in which current has no bulk material contribution and instead is primarily driven by majority carrier injection into the channel) and relatively large overall size/area. These initial experiments ignited the TFT field that is still in active development today.

Since the 1960s, the field of TFTs has continued to grow. With that growth, there has been exploration of a large variety of materials that can act as majority carrier semiconducting thin films. Apart from traditional silicon and other bulk semiconductors, most common thin-film semiconductors can be placed into three categories—organics, metal-oxides, and nanomaterials. All three have shown to be viable in TFTs and other thin film-based devices, while each comprise many respective advantages and challenges.

Along with the noteworthy materials research and advancements, there has been an evolving landscape of applications benefitting from TFTs. Throughout the decades since Weimer's work, TFTs have thrived in niche application spaces, specifically low-cost, large-area electronics. For example, metal-oxide TFTs drive many of today's display applications. Of the many aspects of TFTs that make them so useful for these applications, one that has yet to be taken full advantage of is their inherent mechanical flexibility. The flexibility stems from the fact that the channel simply consists of a thin layer of material, making fabrication on most any substrate possible.

In this chapter, we present an overview of the various TFT materials that can be used to fabricate flexible, and even stretchable, transistors. While organic and metal-oxide semiconductor TFTs will be briefly reviewed, our focus will be primarily on carbon nanotube (CNT) thin films, due to their prominence in current research and their significant advantages and potential to move the field forward. We will discuss the operational principles of CNT-TFTs, fabrication methods, CNT thin film deposition techniques, and the complementary materials needed for fully functional, flexible transistors. We will then discuss some of the most prominent recent work and benchmark the CNT-TFTs with competing metal-oxide and organic semiconductor technologies.

4.2 Materials

The three primary materials used in thin-film transistors include organic-based semiconductors, such as pentacene or poly(3-hexylthiophene) (P3HT), metal-oxides, and nanomaterials. Metal chalcogens were the first to jump start the field as early as the 1960s [2]; however, excitement accelerated with the advent of organic semiconductor-based TFTs in the 1990s. Most recently, nanomaterials have generated considerable attention due to their low-cost deposition, environmental robustness, longevity, and performance. While the materials are rather diverse, the TFT structures remain relatively consistent. An example of a flexible TFT structure can be seen in Figure 4.1. The thin-film device relies on a semiconductor, which can be electrostatically controlled by a gate electrode, placed in close proximity, but separated by an insulating dielectric. Conducting contacts must also be made to the semiconducting material to provide carrier injection. The order of the materials (e.g., top or bottom contacts, top or bottom gates) can be modified to suit specific properties and applications.

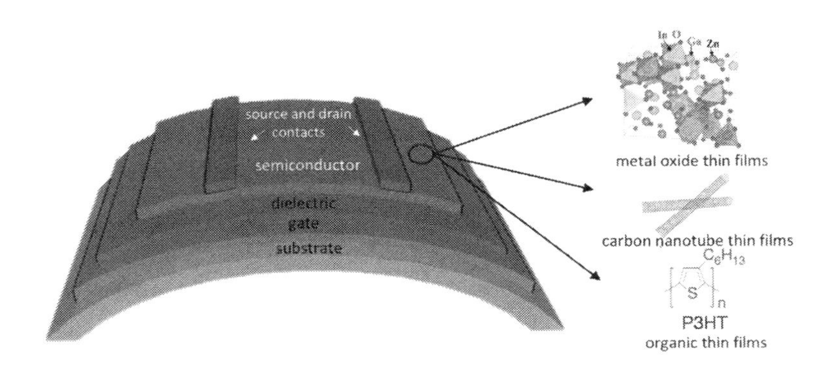

FIGURE 4.1 Generic flexible TFT to illustrate a typical geometry. The semiconductor can be one of a variety of materials, but the three most prevalent are displayed: metal oxides (From T. Kamiya et al., *J. Disp. Technol.*, 5, 468–483, 2009. © 2011 IEEE. With permission), carbon nanotube networks (From Franklin, A.D., *Science*, 349, aab2750, 2015. Reprinted with permission from AAAS), and organic semiconductors (With kind permission from Springer Science+Business Media: *Nature*, ARTICLES A high-mobility electron-transporting polymer for printed transistors, 457, 2009, 679, Kastler, M. et al.) All these materials have shown to maintain semiconducting properties, and a resulting transistor behavior, in thin-film transistors even when flexed and, in some instances, stretched.

4.2.1 Organic TFTs

Organic semiconductors for electronic devices have been previously discussed in Chapter 3, but deserve mention here due to their relevance and influence on all modern thin-film devices. The first demonstration of an organic molecular film being utilized as a semiconductor dates back to 1986, where Tsumura et al. exhibited a 'macromolecular electronic device' composed of a polythiophene thin film [6]. While the hole mobility of the p-type semiconductor was severely limited at $\approx 10^{-5}$ cm^2/(V-s), the idea was novel and generated many new avenues for research. Throughout the 1990s and 2000s, continued efforts to explore and develop organic polymers resulted in increased device performance, manufacturability, and flexibility.

P-type organic semiconductors are most readily available, with some of the most common being pentacene [7–9] and P3HT [10–12]. Realizing n-type semiconductors with similar performance metrics and stability remains a significant challenge. In order to develop complementary logic circuits, which necessitate n- and p-type transistors with similar performance, device geometries often must be asymmetrical as n-type transistors have to be much wider to produce similar current levels to their p-type counterparts [13]. Examples of the most prominent n-type organic semiconductors are copper hexafluorophthalocyanine ($F_{16}CuPc$) [14,15] and poly(benzobis-imidazobenzophenanthroline) (BBL) [16,17]. One exciting aspect of organic semiconductors, and perhaps the most relevant for this chapter, is that they allowed researchers to think creatively about depositing semiconducting films from solution [18,19] and ultimately led to the introduction of transistors into the field of printed electronics [20,21]. Additionally, organic thin films have provided valuable initial benchmarks for what is possible in flexible electronics and displays [22–25].

However, many challenges remain for organic TFTs, such as the substantial trade-off between manufacturing cost and performance [26]. The stability of organic molecules under continued bias stress is also very concerning and leads to over inflated mobility estimates from current laboratory measurements [27]. Overall, while organic semiconductors provide a promising path towards the development of flexible transistors, much work is still needed in improving manufacturing cost, performance, environmental robustness, and longevity.

4.2.2 Metal Oxides

The progress generated as a result of metal-oxides in the field of thin-film transistors easily matches, if not overtakes, that of organic semiconductors. Since Weimer's first paper exhibited the use of cadmium sulfide as a transistor [2], many more materials have been discovered and explored including n-type compounds of SnO_2 [28,29], ZnO [30,31], and the most widely used $InGaZnO_4$ [32–35]. P-type materials have proven to be less common due to the electronic structure and carrier transport mechanisms of metal-oxide semiconductors. As of yet, only a few metal-oxide materials, including SnO_x [36,37] and Cu_xO [38,39], offer hole mobility sufficient for TFT technologies.

N-type metal-oxide thin films offer the highest performing option for TFTs, but that performance comes at a significant fabrication cost. Typically, high performing metal-oxide TFTs must be fabricated using vacuum deposition sputtering techniques [32,40] and often require relatively high temperature anneals (>300°C) [34]. Solution-processed metal-oxide TFTs have been demonstrated as a lower cost option, but typically still require high temperature anneals to increase device performance [40–42].

For applications requiring flexible form factors, metal oxides offer a promising path forward. Using solution processing, metal oxides have been shown to be compatible with flexible substrates. Furthermore, recent reports have demonstrated minimal performance degradation due to flexing [43,44]. Overall, metal-oxide thin-film transistors offer a higher cost, higher performing option compared to organic TFTs, while still maintaining inherent flexibility. The most prominent impact of metal-oxide TFTs has been in the back planes of displays where they upended the use of polysilicon as a lower cost, higher performance alternative. However, their use in displays has not yet taken direct advantage of the mechanical malleability of metal-oxide TFTs; yet, their dominance in display applications provides a strong backbone for flexible display research moving forward.

4.2.3 Carbon Nanotubes

Over the past 20 years, nanomaterials have received tremendous attention and have been acclaimed to solve many challenges for a vast array of electronic devices. While much of the excitement has failed to come to fruition, one area that nanomaterials have provided distinct realizable advantages in is the realm of TFTs [4]. Nanomaterials, when aggregated into a networked thin film, generally maintain a slightly attenuated version of their promising electronic properties. Semiconducting thin-film networks have been demonstrated using CNTs [45–47], various nanowires [48–50], and, most recently, 2-dimensional transition metal dichalcogenides (TMDs) [51,52].

Networked semiconducting carbon nanotubes provide one of the most promising paths forward due to their fantastic electrical and mechanical properties [53,54], low-temperature solution-based processing [47,55–57], and overall compatibility with a variety of substrates and materials. Additionally, nanotubes have been shown to be extremely robust to harsh environments and have exhibited excellent, long-term stability. Lastly, CNT-TFTs maintain their operation when either flexed or stretched [58–63]. Typically, CNT networks are so resilient to strain and flexion that the device breaks down first due to the complementary components (contacts, gate, or dielectric) [60], not the CNT channel, when put under mechanical stress.

Due to the promise and recent significant research attention, the remainder of this chapter will focus on CNT-TFTs. The subsequent sections will detail semiconducting CNT (s-CNT) processing, CNT-TFT fabrication, and the respective operational theory. The performance of various demonstrated CNT-TFTs will then be benchmarked against the competing peer technologies—metal-oxide and organic TFTs.

4.3 Carbon Nanotube Electronic Structure

CNTs obtain their unique electronic properties from their carbon lattice structure. A CNT is a cylindrical shell of carbon atoms bonded together in an sp^2 configuration. One way to visualize a CNT is to first consider a 2D carbon honeycomb lattice (graphene) rolled into a seamless cylinder, as in Figure 4.2a. The diameter and electronic properties of a CNT are dependent upon its chiral vector, which is a circumferential vector associated with the angle in which the tube is 'rolled up.' Effectively, by rolling the sp^2 hybridized graphene, the electronic states of the graphene band structure are quantized into 1D sub-bands that can lead to a CNT exhibiting either semiconducting (if the quantized sub-band does not pass through one of the corners or 'K-points' of the Brillouin zone) or metallic properties (if the sub-band does pass through a corner of the Brillouin zone) [64].

The chirality is defined by a wrapping vector (vector **c** in Figure 4.2a). The wrapping vector can be broken down into two unit vectors, $\mathbf{a}_1$ and $\mathbf{a}_2$, where $\mathbf{c} = \mathbf{n}\mathbf{a}_1 + \mathbf{m}\mathbf{a}_2$ and can be denoted using its two indices (n, m). A tube is considered 'armchair' if $n = m$ and 'zigzag' if $m = 0$, with all other cases being of the 'chiral' type. Armchair, zigzag, and chiral CNTs are all illustrated in Figure 4.2b. 'Armchair' tubes exhibit only metallic properties [65]. Apart from 'armchair' tubes, the electronic nature of CNTs follows this rule: tubes are metallic if $(n - m)$ is equal to a factor of 3, otherwise, the tubes are semiconducting [66]. Along with varying electronic properties, the chiral vector will also determine the CNT diameter. It has been shown that the band gap associated with semiconducting CNTs is directly related to the diameter of the CNT and can be approximated by the following equation [64]:

$$E_G(eV) \approx \frac{0.7}{d_{CNT}(nm)}$$

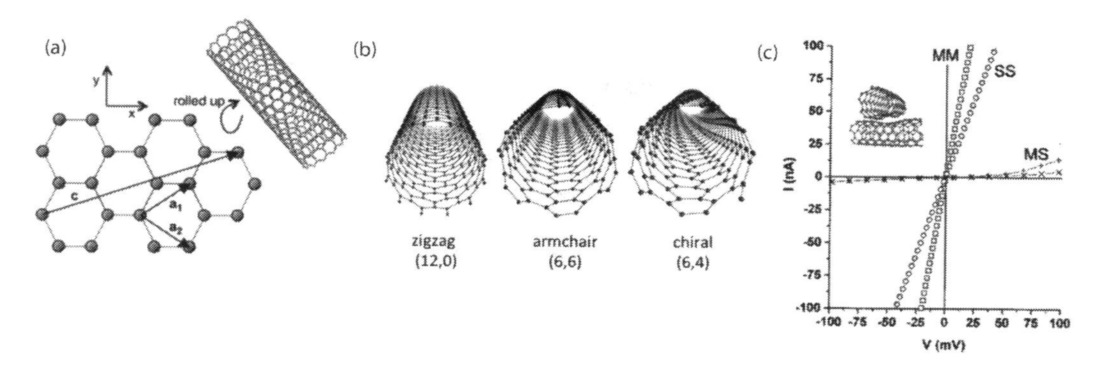

FIGURE 4.2 Structure and electrical properties of carbon nanotubes. (a) Visualization of 2D carbon lattice with a vector illustrating chirality. (b) Single-walled carbon nanotubes of varying chirality, and thus varying diameters. (c) Electrical properties at carbon nanotube junctions. The electrical data are displayed for junctions between two metallic tubes, two semiconducting tubes, and one metallic tube with one semiconducting tube ([a] From Franklin, A.D.: Carbon Nanotube Electronics. *Emerging Nanoelectronic Devices*. 2015. 315–335. Copyright Wiley-VCH Verlag GmbH & Co. KGaA. Reproduced with permission; [b] Reprinted with permission from Charlier, J.C. et al., *Rev. Mod. Phys.*, 79, 677–732, 2007. Copyright 2007 by the American Physical Society; [c] From Fuhrer, M.S. et al., *Science*, 288, 494–497, 2000. Reprinted with permission from AAAS.)

While the electronic structure of a CNT is crucial when understanding the operation of TFTs fabricated from a networked CNT film, another aspect that must be considered is what happens at tube-to-tube junctions. For many thin-film devices, the channel length is greater than any individual CNT's length. Therefore, carriers must percolate through multiple CNT junctions to perpetuate current flow. Initial experiments that analyzed CNT-to-CNT junctions indicated that the junction resistance stems from a tunnel barrier between the two CNTs; however, the tunneling probability is quite high. Metallic-metallic as well as semiconducting-semiconducting CNT junctions exhibit 'Ohmic' behavior with resistance values of 200 and 500 kΩ, respectively. Metallic-semiconducting junctions are more complex—in addition to the tunnel barrier, there exists a Schottky barrier with a height related to the bandgap of the semiconducting CNT. Fortunately, in CNT-TFTs, the majority of the junctions exist between two semiconducting CNTs and sorting methods are used to eliminate the incorporation of metallic CNTs in the film. Results from these initial CNT junction experiments can be found in Figure 4.2c [67].

4.4 Device Theory

The operating principle of a CNT-TFT is very similar to the operating principles that govern single CNT transistors. While commonly referred to as 'p-type' transistor devices, CNTs are intrinsic semiconductors with the Fermi level lying in the middle of their bandgap [64]. This means that carriers in the channel are not induced by the gate field manipulating charge from dopants in the semiconductor, but must instead be injected at the contacts. Schematic diagrams of CNT-TFTs with applied voltages and their respective band diagrams are displayed in Figure 4.3a and b, respectively. The operating mechanism relies on the gate field lowering the energy barrier in the channel to allow for both thermionic emission and tunneling current at the metal-CNT contact interface. Essentially, as the energy band of the CNT network is modulated, the energy barrier at both the source and drain contacts becomes steeper and thinner. As the gradient steepens, the probability for carriers to tunnel through that energy barrier increases, eventually leading to an 'on-state' current [68]. Thus, a CNT-TFT relies

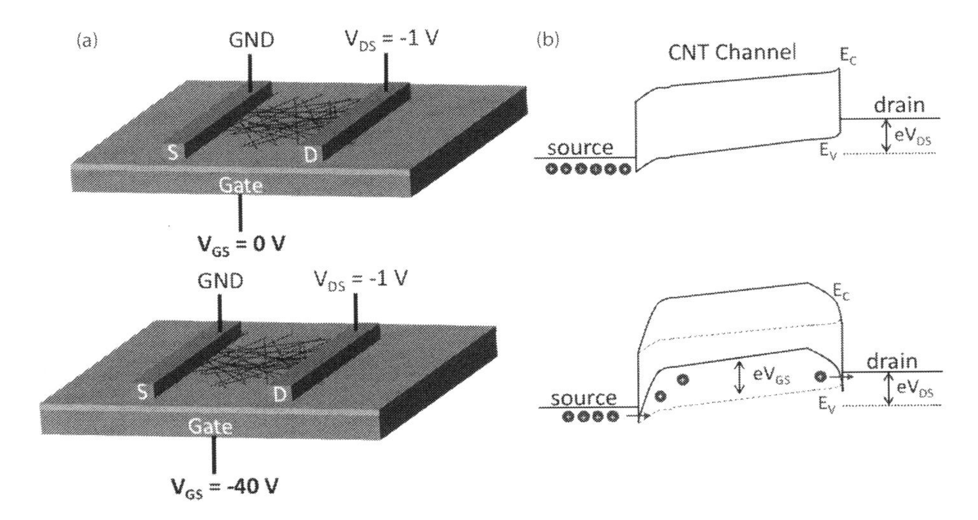

FIGURE 4.3 Theoretical operation of CNT-TFTs depicted with energy band diagrams. (a) Two CNT-TFTs with indicated applied voltages. (b) Corresponding band diagrams to illustrate the Schottky barrier tunneling operation of CNT-TFTs.

on carriers tunneling through a Schottky barrier at the contacts, and the operation and polarity (n-type, p-type, or ambipolar) relies heavily on the contact material (e.g., work function) and applied voltages [69].

While the operating mechanisms of CNT-TFTs can be described similarly to single/parallel nanotube devices, the performance is quite different. This is due to the CNT-CNT junction resistance throughout CNT networks and the large channel dimensions frequently used for TFTs. While many single/parallel tube devices rely on sub-micron channel lengths [70–72], even below 10 nm [73,74], many flexible CNT-TFTs have channel lengths and widths in the 100s of μm or even low mm [60,61,63,75]. While carbon nanotubes themselves present highly conductive pathways, the resistive junctions lead to high sheet resistance values within networked CNT channels. A scanning electron microscopy (SEM) image in Figure 4.4a exemplifies a typical CNT density within a s-CNT film. Additionally, Figure 4.4b shows how

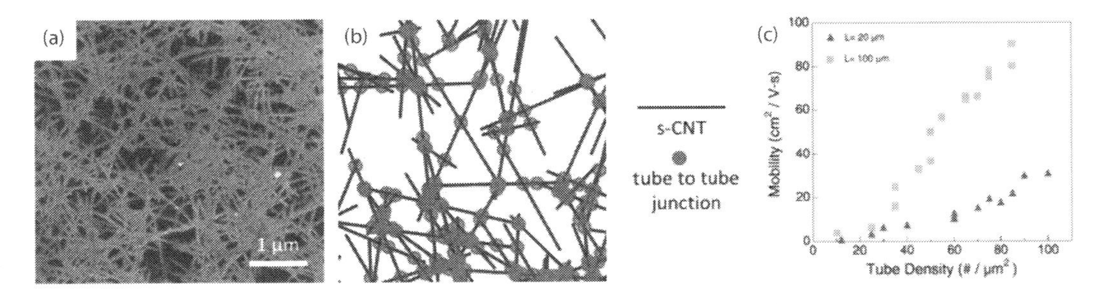

FIGURE 4.4 Electron percolation through CNT networks. (a) SEM of a CNT thin film. (b) Diagram to illustrate the influence of tube-to-tube junctions in s-CNT networks. (c) Mobility vs tube density within a s-CNT network. ([b] Reprinted with permission from Schießl, S.P. et al., *Phys. Rev. Mater.*, 1, 46003, 2017. Copyright 2017 by the American Physical Society; [c] From Rouhi, N. et al.: Fundamental Limits on the Mobility of Nanotube-Based Semiconducting Inks. *Advanced Material.* 2011. 23. 94–99. Copyright Wiley-VCH Verlag GmbH & Co. KGaA. Reproduced with permission.)

CNTs and junctions will ultimately form a pathway for carriers to percolate. Much work has been done to simulate the carrier percolation with respect to different network variables, including CNT density, length, hardness, and others [76–78]. Primarily, the studies corroborate the experimental results introduced below, but also demonstrate the effect of non-rigid, 'wavy' CNTs. Networks of 'wavy' or bent CNTs lead to high resistive networks and higher percolation thresholds [79]. Experimental measurements have also been done to analyze the percolation effects [80–82]. A plot displaying an experimental measurement of the effective mobility of a CNT network in a CNT-TFT, with respect to both tube density and channel length, is shown in Figure 4.4c. Denser networks allow for more junctions to exist in parallel, thus decreasing the sheet resistance, while longer channel lengths are more significantly affected by the tube density.

4.5 Semiconducting Carbon Nanotube Processing

One of the most significant challenges in the early stages of developing CNT-based TFTs was isolating the semiconducting carbon nanotubes for deposition. For transistors, one crucial attribute for many applications is the on/off ratio, which is central for many digital and analogue applications. Semiconducting tubes offer the ability to modulate current flow, but the conductivity of metallic tubes will stay relatively constant regardless of any applied field. Therefore, the presence of metallic tubes will negatively influence the off-current and lead to a transistor, if the description is even still applicable, with little to no on/off ratio. This can be mediated by fabricating TFTs with long channel lengths and low tube density, leading to decent on/off ratios [46]. However, increasing the CNT purity is the best way to fabricate high performing devices with substantial on/off ratios.

CNTs used for thin-film transistors can be synthesized in many ways: chemical vapor deposition (CVD) growth [83,84], arc-discharge (AD) [85–87], and laser ablation (LA) [88,89] are among the most common. CVD offers controlled placement and even a high degree of diameter/chirality control based upon the catalysts and substrates that are used, but often leads to tubes with high defect densities. AD is the most economical option, but offers little control over CNT alignment and chirality. Lastly, LA is not economically favorable, but does allow for control of the CNTs' diameter and electronic type to some extent [90]. Both LA and AD methods produce CNTs without a substrate and therefore they must be deposited in some other manner, which will be discussed in later sections. AD synthesized tubes are the most commonly used in the CNT-TFT space due to overall low-cost production and the recent development of large-scale CNT sorting methods.

Much work was done in the mid-2000s to identify solution-based methods to sort CNTs by their electronic type (semiconducting and metallic). The two primary methods developed, and still used today, include density gradient ultracentrifugation [56,91,92] and selective dispersion using aromatic polymers [93–95]. Density gradient ultracentrifugation works by engineering specific surfactants to attach to CNTs of varying type. The attached surfactants will alter the buoyancy of CNTs and, using a centrifuge, the differential densities cause the varying types of CNTs to separate in solution. One may then use a pipette to selectively separate the desired type of CNTs from the whole solution. Surfactants attaching to differing tubes and the resulting solution after centrifugation can be seen in Figure 4.5a and b, respectively. The selective dispersion through aromatic polymers works in a similar manner, but without the need for centrifugation. Specific polymers produce excellent dispersions of tubes with highly specific chiral species. Therefore, when these polymers are introduced into an unfiltered solution of nanotubes, only the select chirality will be solubilized in the solution. The unwanted types will then aggregate and can be removed from the solution. One of the first polymers used in a demonstration of this method was poly(9,9-dioctylfluorenyl-2,7-diyl) or PFO. PFO can be modified to attach to different chiral species as shown in Figure 4.5c. Both methods lead to nanotubes of sufficient purity stabilized in solution and are therefore advantageous in that solution-based deposition techniques are now scalable and compatible with fabrication on flexible substrates.

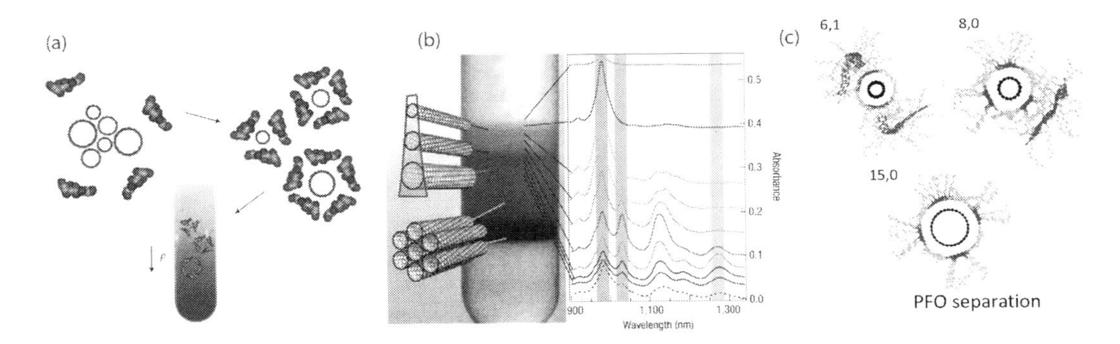

FIGURE 4.5 Figure illustrating density grade ultracentrifugation method of sorting CNTs. (a) Illustrates how different amounts of the polymer surfactant will attach to tubes of varying chirality. After that, ultracentrifugation may be used to separate the CNTs based on their density, ultimately sorting out the different chiralities, as shown in (b). (c) Nanotube chiralities separated using the PFO dispersion technique. ([a,b] With kind permission from Springer Science+Business Media: *Nat. Nanotechnol.*, Sorting carbon nanotubes by electronic structure using density differentiation, 1, 2006, 60, Arnold, M.S.; [c] With kind permission from Springer Science+Business Media: *Nat. Nanotechnol.*, Highly selective dispersion of single-walled carbon nanotubes using aromatic polymers, 2, 2007, 640–646, Nish, A. et al.)

4.6 Deposition Techniques

Once high purity s-CNTs have been properly separated and dispersed in solution, there are many ways to deposit the semiconducting solution onto a flexible substrate. Methods as simple as substrate incubation in s-CNT solution have been shown to produce semiconducting thin films of sufficient electronic properties to enable transistor behavior [57,96,97]. Additionally, solutions of CNTs have been deposited using methods such as spray coating [98] or spin coating [99,100]. While all these methods can produce s-CNT networks on flexible substrates, they offer no dimensionality control (leading to high material waste) and little compatibility with low-cost scalable production.

A subset of solution deposition techniques that has received significant research attention lately is electronics printing. First developed for organic-based TFTs, many of the same printing methods can now be used to deposit s-CNT films with high degrees of control. Three of the most common types of printers for depositing CNT solutions are inkjet printing, aerosol jet printing, and roll-to-roll printing. These printing methods are low-cost, and have varying levels of advantageousness for large-scale manufacturing, with specific pros and cons that deserve attention.

Inkjet printing is the most well-known direct (or additive) deposition technique, as well as the method which has reached prolific application in print media and consumer products. The operation of inkjet printing may vary from printer to printer, but the primary principles remain the same. In essence, ink is first loaded into a syringe. Next, the pressure in that syringe is actuated using a piezoelectric transducer, forming a droplet that is expelled from the syringe [101]. The nozzle of the syringe can be moved using directional motors to effectively write on any substrate. The advantages to inkjet printing include high control over droplet volume and density, accurate spatial positioning, and reasonably high-resolution line widths (typically 50–100 μm). Some disadvantages include high propensity for nozzle clogging, limited number of printable inks, detrimental drying effects, and relatively low-throughput (when compared to template-based printing). The operating mechanism allows for printing of CNTs for TFTs [47,102–104], but the printer is generally constrained due to the tendency of nanotubes to attach to the nozzle and preclude long continuous printing. A schematic of an inkjet deposition head is given in Figure 4.6a.

Aerosol jet printing, developed by Optomec Inc. [105] to open up additive manufacturing (direct deposition) to a larger variety of materials, has also proven to be a viable CNT deposition

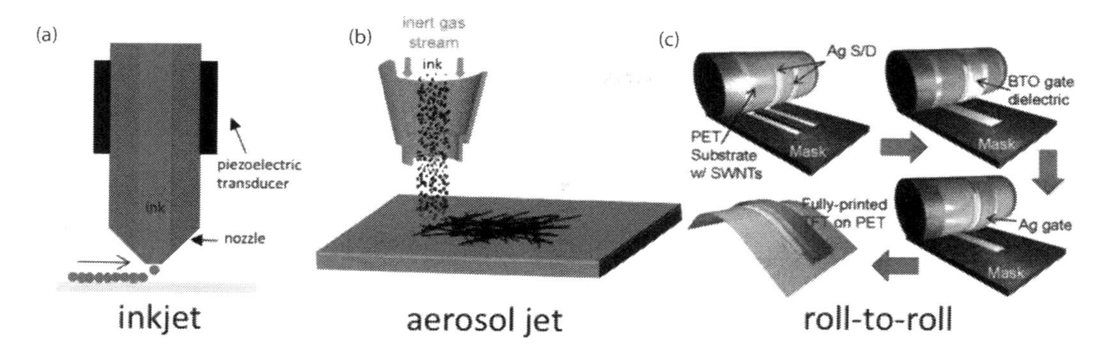

FIGURE 4.6 Depictions of the three most common printing techniques. (a) Inkjet printing of an arbitrary ink. (b) Aerosol jet printing of semiconducting CNTs. (c) A fully printed CNT-TFT can be fabricated using traditional roll-to-roll printing techniques. (Reprinted with permission from Lau, P.H. et al., *Nano Lett.*, 13, 3864–3869, 2013. Copyright 2013 American Chemical Society.)

technique [55,106,107]. The end result is similar to inkjet printed films, but the operation is quite different. First, a functional electronic ink is excited into an aerosol state through ultra-sonication or pneumatic atomization. The aerosol is then carried by an inert carrier gas stream through tubing to a deposition head. At the deposition head, a secondary inert gas stream that flows annularly around the nozzle acts to guide the ink down to the substrate and prevent nozzle clogging. The advantages for aerosol jet printing include high-resolution prints, similar to inkjet printing, in addition to a wider compatibility of printable inks, and a lower predisposition for nozzle clogging. The primary drawback to aerosol jet printing is relatively low throughput and ink overspray. With respect to CNT printing, the use of a sheath gas at the nozzle is uniquely critical. The sheath gas will preclude the nozzle from clogging by preventing CNT agglomeration on the nozzle sidewalls. Additionally, the atomization process of CNT ink provides control over CNT density in the printed film, which can be tuned to fit desired film properties. A figure depicting the operation of the deposition head in a typical aerosol jet printer is shown in Figure 4.6b.

Lastly, template-based roll-to-roll printing techniques have been shown to produce CNT thin-film devices at extremely high throughput [75]. While roll-to-roll printing techniques can be used to print the necessary complementary materials for CNT-TFT devices, there has been no work demonstrating the roll-to-roll printing of semiconducting CNTs. However, due to the compatibility with CNT thin films, and its prevalence in the field, this deposition technique must be mentioned. The method is similar to traditional newspaper printing, in which rolls are used to pick up patterned materials (through metal/hard material stencils) and in turn deposit the materials on the desired substrate. The main methods include knife-over-edge coating, slot-die coating, gravure printing, and screen printing [108]. The primary disadvantage to this technique is the low degree of control. Apart from the pattern designs and ink selection, the only print parameter capable of being adjusted is the roll speed. For research, roll-to-roll printing is especially unfavorable as a new, expensive mask must be manufactured each time a new design is to be printed. Additionally, there is a tight window of ink viscosities compatible with these techniques, which can restrict material selection and ink formulation, and ultimately is why s-CNT inks have not been printed using roll-to-roll techniques. The primary advantages that stem from roll-to-roll printing include throughput and, therefore, the capacity for low-cost large area electronic manufacturing. A device printed using roll-to-roll techniques is illustrated in Figure 4.6c. More information on printing technologies and their relation to flexible electronics can be found in Chapter 9.

It should be noted that the adhesion of CNT films to a surface often necessitates pre-treatment. For example, poly-L-lysine is a monomer that can be used to promote CNT adhesion to smooth surfaces [60,106]. Another molecule that has been used previously is (3-Aminopropyl)triethozysilane

(APTES) [57]. Both molecules provide attachment sites for CNTs and are key for achieving dense and uniform films for all solution-based CNT deposition techniques.

4.7 Key Materials for a Functional CNT-TFT

The various materials necessary for a fully functional TFT include the semiconducting channel for carrier transport, conducting contacts for carrier injection, a gate electrode to apply a gating field, and a dielectric layer to insulate the channel from the gate. A variety of material combinations and compositions have been investigated for CNT-TFT fabrication. Details on contact materials and the gate dielectrics are expounded on below. The gate electrode is of secondary importance, and for simplicity, most devices use the same conducting material used for the contacts. Therefore, conducting gate electrode materials will not be explicitly discussed.

4.7.1 Contact Material

For flexible electronics, many viable contact materials have been identified. The most common of these are metallic nanoparticles, which are used prolifically due to their high conductivity, long-term performance stability, and high compatibility with many of the above-mentioned printing processes [60,75,109,110]. To illustrate the morphology of printed metallic nanoparticles, an optical image and SEM image of Ag nanoparticles are displayed in Figure 4.7a and b, respectively. Along with metallic nanoparticles, metallic CNTs are also favorable due to their compatibility with print processes and inherent flexibility/stretchability. Cao et al. provided a direct comparison between silver and gold nanoparticles and metallic CNTs in terms of the contact resistance and device on-current. Metallic CNTs provided the lowest contact resistance, presumably due to the more conformal contact with the

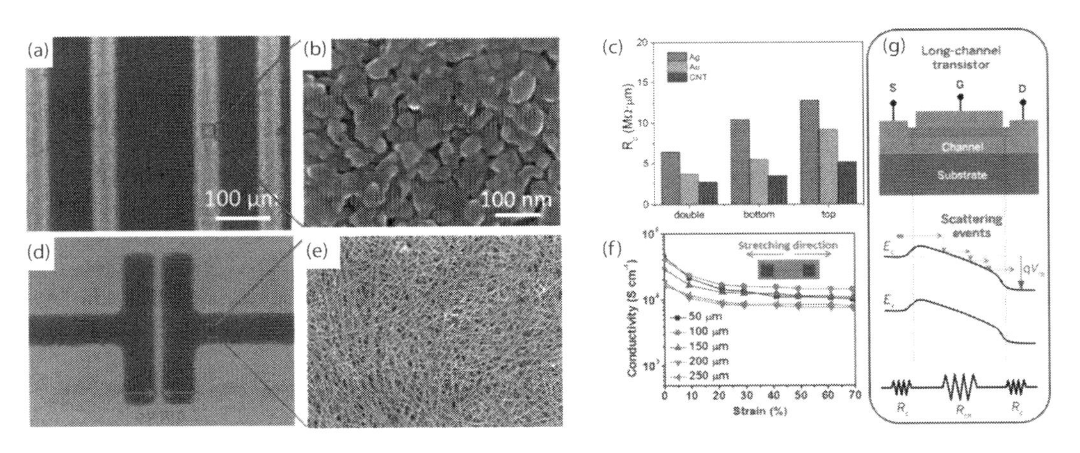

FIGURE 4.7 Printed contacts for CNT-TFTs. (a) Optical and (b) SEM images of aerosol jet printed silver nanoparticle contacts often used in CNT-TFTs. (c) Comparison of CNT-TFT contact resistance among various printed electrode materials and configurations. (d) Optical and (e) SEM images of stretchable, screen printed silver nanowire electrodes. (f) Conductivity versus strain plot of printed silver nanowire films. (g) Schematic diagram and band diagram of a long-channel transistor illustrating the dominant role that the channel plays in overall device performance ([a–c] Reprinted with permission from Cao, C. et al., *ACS Nano*, 10, 5221–5229, 2016. Copyright 2016 American Chemical Society; [d–f] From Liang, J. et al.: A Water-Based Silver-Nanowire Screen-Print Ink for the Fabrication of Stretchable Conductors and Wearable Thin-Film Transistors. *Advanced Materials*. 2016. 28. 5986–5996. Copyright Wiley-VCH Verlag GmbH & Co. KGaA. Reproduced with permission; [g] From Franklin, A.D., *Science*, 349, aab2750, 2015. Reprinted with permission of AAAS.)

thin film of s-CNTs. Gold nanoparticles provided the next lowest, with silver exhibiting the highest contact resistance. A bar graph of the contact resistance results from the work can be seen in Figure 4.7c. The three materials are compared, as well as the contact geometry. Top contact refers to the source and drain being printed on top of the CNT channel, bottom refers to the contacts printed below the channel, and double contact refers to a combination of the two [106]. While metallic CNTs may form a good contact to s-CNT channels, they possess the disadvantage of having high sheet resistance when compared to bulk metals and even metallic nanoparticles. Therefore, metallic CNTs would not be favorable interconnects and another material would be needed to interface the contacts with other components within a circuit.

For stretchability, metallic nanoparticles are not favorable due to a predisposition to delaminate and/or crack after repeated strain cycles. Many stretchable contact materials, including silver nanowires [112] and metallic carbon nanotubes [62,63,113], rely on geometrically enabled stretchability. This occurs when high-aspect ratio nanostructures are deposited in a dense enough network in such a manner that significant strain will not preclude an electrical connection due to large overlap between the nanostructures. This enables the contacts to remain conductive throughout film elongation and allows for a constant supply of carriers to be injected into the s-CNT channel with no discontinuity due to strain. An example of a geometrically enabled stretchable film consisting of silver nanowires can be seen in the SEM images in Figure 4.7d and e. Additionally, the relationship between conductivity and strain in this silver nanowire film is displayed in Figure 4.7f.

In addition to the above-mentioned printed nanomaterials, thin layers of evaporated metals have also been demonstrated as high-performance contacts in flexible electronics [55,114]. However, vacuum processing and high temperature evaporation are not favorable for most flexible substrates or high-throughput production environments. Further, evaporated contacts will not be preferable for large-area or low-cost electronics.

While the contact material for CNT-TFTs does play a role in the device performance, the overall device resistance is dominated by the highly resistive CNT channel. A schematic of the resistive network can be seen in Figure 4.7g. Overall, the contact material must provide a conducting platform to inject carriers into the CNT film that is not compromised by flexing or stretching.

4.7.2 Dielectric

Dielectric materials have proven to be one of the weaker points for printed flexible electronics. Currently, there exists two primary options for printing an insulating, low-leakage dielectric—polymer-based dielectrics and ion-gel dielectrics.

One of the most common dielectrics used for flexible TFTs is a barium titanate/poly(methyl methacrylate) ($BaTiO_3$/PMMA) composite [75,113]. $BaTiO_3$/PMMA has many advantages for printed flexible devices, including a high permittivity constant of 17 and compatibility with roll-to-roll printing techniques and long-term film stability, with demonstrations of performance stability over 1000 bending cycles. Some of its primary disadvantages include lack of thickness control and reproducibility [115] and the toxicity of barium compounds in consumer devices. More recently, a printable polymer-only dielectric consisting of poly(vinylphenol) (PVP) bases has been shown to produce CNT-TFTs with minimal hysteresis and no performance degradation due to stretching [60,61,116,117]. The reproducibility and ease of processing through aerosol or inkjet printing presents promise towards a more integral dielectric for large-area, printed, and flexible devices. However, the overall robustness and longevity of this dielectric, which is a pitfall for many organic electronic materials, has yet to be investigated.

Another candidate for electrostatically isolating CNTs in TFTs is by using an ion gel-based dielectric [55,107,118]. Ion gels create high capacitive networks through the utilization of a nanometer-thick electrical double layer existing at the interface between the metallic gate electrode and electrolyte [119]. However, due to the fact that the double layer relies on the mobility of ions within the gel, the capacitance degrades rather quickly at higher frequencies [75]. Ion gels that have been integrated with s-CNTs show robust performance up to 22 kHz in a 5-stage ring oscillator. However, disadvantages associated with ion-gel dielectrics in CNT-based devices include excessive ambipolar behavior, which can lead to high leakage currents and superfluous power consumption [120].

Apart from printed dielectrics, vacuum processing techniques, such as atomic layer deposition (ALD), have been used to develop ultra-thin dielectrics for flexible CNT-TFTs [114,121,122]. The resulting devices exhibit excellent electrostatic control, due to both the thinness of the dielectric and its low-level of defects associated with the deposition processes. The thinness of the dielectrics allows for operation under flexing, but overall, the vacuum processing and cost of deposition would preclude CVD/ALD-grown dielectrics for low-cost, large-area applications.

4.8 Device Performance and Performance Benchmarks

An overview of device structures, performance, and related cost estimates can be seen in Figure 4.8. Here, we highlight three distinct TFT variations—metal-oxide TFTs, CNT-TFTs, and printed CNT-TFTs. Figure 4.8a–c presents a metal-oxide transistor, consisting of ZnO, that is largely representative

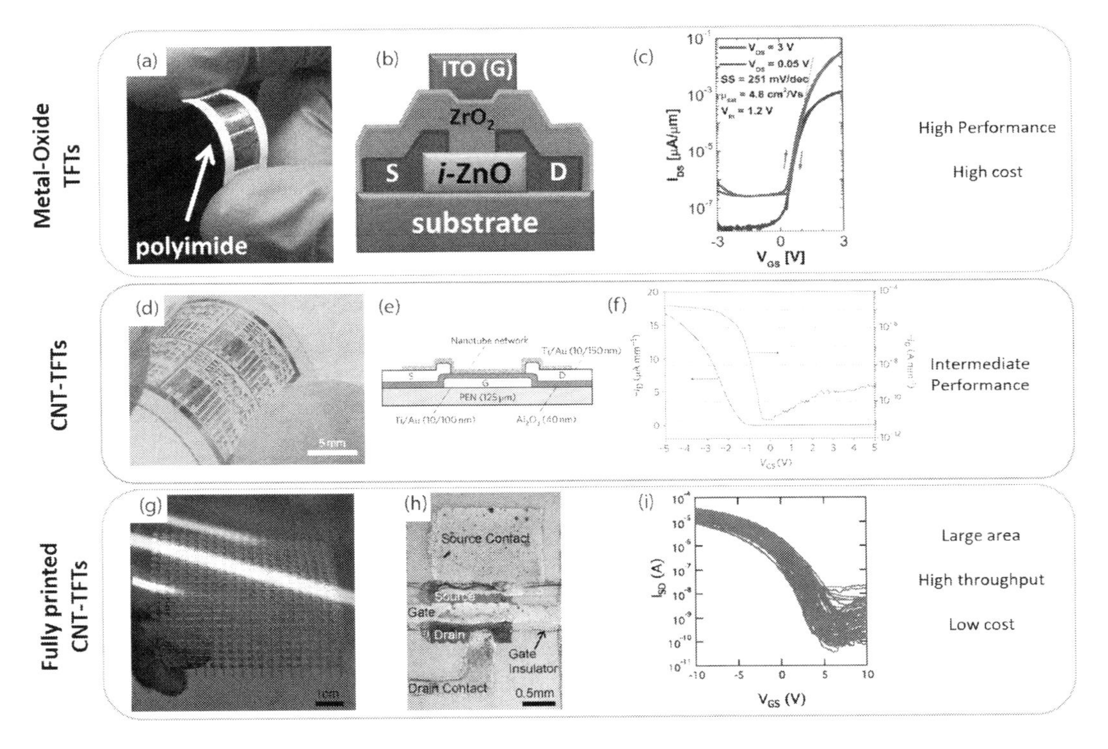

FIGURE 4.8 Key flexible, thin-film devices—design, images, and performance. (a) Optical image of an array of metal-oxide transistors fabricated on a flexible polyimide substrate. (b) Schematic diagram of an i-ZnO TFT and (c) its subthreshold characteristics. (d) Optical image of an CNT-TFT integrated circuit fabricated on polyimide. (e) Schematic diagram of a CNT-TFT and (f) its subthreshold characteristics. (g) Optical image of an array of roll-to-roll printed CNT-TFTs fabricated on a flexible PET substrate. (h) Up-close optical image of an individual printed CNT-TFT and (i) a series of printed CNT-TFT subthreshold characteristics. ([a–c] From Rembert, T. et al.: Room Temperature Oxide Deposition Approach to Fully Transparent, All-Oxide Thin-Film Transistors. *Advanced Materials*. 2015. 27. 6090–6095. Copyright Wiley-VCH Verlag GmbH & Co. KGaA. Reproduced with permission; [d–f] With kind permission from Springer Science+Business Media: *Nat. Nanotechnol.*, Flexible High-Performance Carbon Nanotube Integrated Circuits, 6, 2011, 156–161, Sun, D.M. et al.; [g–i] Reprinted with permission from Lau, P.H. et al., *Nano Lett.*, 13, 3864–3869, 2013. Copyright 2013 American Chemical Society.)

of typical metal-oxide TFT performance and cost. The structure is flexible, but the deposition of ZnO is completed using a cathodic arc deposition, a form of vacuum deposition, which precludes many low-cost applications. However, the deposition is completed at low temperature, which allows for a wider variety of flexible substrates. The performance is typical and representative for low-temperature metal-oxide TFTs, with on/off ratios exceeding 10^4 and an effective electron mobility of 4.8 cm^2/V s [123].

Another more common metal-oxide used for TFTs is $InGaZnO_4$ or IGZO. Similar to ZnO mentioned above, many IGZO based-devices are fabricated using vacuum-based deposition techniques (typically direct current [DC] sputtering), which limit the substrate size and fabrication throughput. However, vacuum processed IGZO films provide the highest performing option for TFTs with typical on/off ratios ranging from 10^5–10^9, mobilities ranging from 10–90 $cm^2/(V\text{-}s)$, and subthreshold swings as low as 100–300 mV/dec, depending on the quality of the dielectric [43,124–126]. Additionally, many metal-oxide films can be deposited from solution through spin-casting, which typically results in lower performing films, though this can be somewhat alleviated by sintering at high temperatures (>300°C), which in turn prevents fabrication on most flexible substrates. There are many emerging low-temperature processing methods with the potential to recover this compromise in performance through combustion synthesis or photochemical activation methods [123,127].

The next set of devices presented in Figure 4.8d–f are representative of CNT-TFTs produced using traditional cleanroom processing methods. In these cases, where cleanroom fabrication is utilized, the CNT thin film is either deposited from solution or transferred to the substrate directly after CVD growth. The source and drain contacts typically consist of lithographically patterned and evaporated palladium or gold. Finally, the gate dielectric consists of a sub-100 nm layer of either Al_2O_3, HfO_2, ZrO_2, or SiO_2 grown using either CVD or ALD. CNT-TFTs are compatible with a wide range of flexible substrates due to low-temperature solution deposition or transfer; however, the most common substrates include polymers such as polyethylene naphthalate (PEN), polyethylene terephthalate (PET), and polyimide (PI). Device performance is often slightly lower than that of metal-oxide-based devices, with on/off ratios ranging from 10^4–10^6, hole mobilities ranging from 5–50 $cm^2/(V\text{-}s)$, and subthreshold swings ranging from 100–300 mV/dec. High device mobilities can be attributed to the high CNT channel densities, evaporated contacts, and high gate voltage [114], while low subthreshold slopes can be attributed to thin and high quality gate dielectrics.

Presented in Figure 4.8g–i is a set of devices representative of CNT-TFTs fabricated using roll-to-roll printing techniques. This class of flexible TFTs has, by far, the lowest fabrication costs when compared against the previous two classes presented in Figure 4.8; however, there is a compromise in performance. All components of these devices, including the s-CNT channel, gate dielectric, gate electrode, and source/drain electrodes are deposited from solution or printed. Common devices exhibit decent performance, with on/off ratios ranging from 10^4 to 10^6, carrier mobilities ranging from 1–30 $cm^2/(V\text{-}s)$ [75], and very large subthreshold swings due to thick polymer-based dielectrics.

In addition to these representative devices, we have performed an extensive literature review to present a table that compares many of the latest, state-of-the-art thin-film transistors, which is shown in Table 4.1. There is a focus, in keeping with the rest of this chapter, on CNT-TFTs, but additional comparisons are made to recent metal-oxide TFTs and even organic TFTs. The table parameters for comparison, along with their descriptions and significance are outlined below:

Polarity—The device polarity indicates the majority carrier type. The TFT is considered n-type if the majority carriers are electrons and p-type if the majority carriers are holes.

Deposition Process—The deposition process alludes to the fabrication method of the semiconducting channel. This process is critical as it sets the parameters for substrate selection, fabrication cost, and overall applicability of the device.

Subthreshold Swing (mV/dec)—The subthreshold swing is a parameter derived from the ability of the gate to electrostatically modulate the channel's potential to allow for current injection. It is measured by the voltage required (in mV) to modulate the current by 1 decade.

TABLE 4.1 Table to Benchmark Highest Performing Thin-Film Transistors across a Variety of Materials, Deposition Techniques, and Substrates

References	Channel Material	Polarity	Channel Deposition Process	SS (mV/dec)	μ (cm^2/(V-s))	Log_{10} [I_{ON}/I_{OFF}]	I_{ON} (μA/mm)	Cost	Flexible	Substrate	Gate Dielectric (Thickness)
[106]	s-CNT	p-type	Aerosol Jet Print	–	6.7	5.5	102	Medium	No	Si	SiO_2 (90 nm)
[60]	s-CNT	p-type	Aerosol Jet Print	–	16.1	4.8	6.7	Low	Yes	PI	PVP/pMSSQ (1–5 μm)
[111]	s-CNT	p-type	Drop Cast	–	35	3	75	Low	Yes	PUA Matrix	polyurethane poly(ethylene oxide) (1 μm)
[114]	s-CNT	p-type	CVD Transfer	–	20	6	17	Medium	Yes	PEN	Al_2O_3 (40 nm)
[75]	s-CNT	p-type	Immersion	–	9	5	29	Low	Yes	PET	$BaTiO_3$/ PMMA (1.5 μm)
[121]	s-CNT	n- and p-type	Drop Cast	162	15	5	183	Medium	Yes	PET	HfO_2 (58 nm)
[110]	s-CNT	p-type	Immersion	–	7.7	4.5	3	Low	Yes	PET	$BaTiO_3$ (5 μm)
[103]	s-CNT	p-type	Inkjet	–	1.6–4.2	4	0.11	Medium	No	Si	SiO_2 (500 nm)
[150]	s-CNT	p-type	Immersion	–	9.8	4.8	0.63	Medium	Yes	PES	pV3D3 (40 nm)
[109]	s-CNT	p-type	Inkjet Print	–	6	4.5	100	Low	Yes	PET	$BaTiO_3$/ PMMA (2.3 μm)
[122]	s-CNT	p-type	Drop Cast	–	2.9	5.8	4.8	Medium	Yes	Paper	Al_2O_3 (70 nm)
[151]	s-CNT	n-type	Immersion	248	14.9	3.6	0.25	Medium	Yes	PET	MgO/Al_2O_3 (40 nm)
[152]	s-CNT	n- and p-type	Drop Cast	140	30	6	60	Medium	No	Si	HfO_2 (50 nm)
[153]	s-CNT	n- and p-type	Immersion	–	6	6	50	Medium	No	Si	SiO_2 (50 nm)
[154]	s-CNT	n- and p-type	Aerosol Jet Print	130	130	4	700	Low	Yes	PI	Ion-gel
[123]	ZnO	n-type	CAD	251	4.5	5	30	Medium–High	Yes	PI	ZrO_2 (20 nm)
[124]	IGZO	n-type	Sputter	180	22.1	5.3	128	High	Yes	PI	HfLaO (40 nm)
[43]	IGZO	n-type	Sputter	330	13	8	208	High	Yes	PEN	SiN_x (200 nm)
[125]	IGZO	n-type	Sputter	130	7.5	9.3	11000	High	Yes	PI	Al_2O_3 (25 nm)

(Continued)

TABLE 4.1 (*Continued*) Table to Benchmark Highest Performing Thin-Film Transistors across a Variety of Materials, Deposition Techniques, and Substrates

References	Channel Material	Polarity	Channel Deposition Process	SS (mV/dec)	μ (cm^2/(V-s))	Log_{10} [I_{ON}/I_{OFF}]	I_{ON} (μA/mm)	Cost	Flexible	Substrate	Gate Dielectric (Thickness)
[126]	IGZO	n-type	Sputter	129	76	5.8	555	High	Yes	PC	$SiO_2/TiO_2/SiO_2$ (80nm)
[155]	IZO	n-type	ALD	290	42.1	9.7	~25000	High	Yes	PI	AlO_x (100 nm)
[156]	ZnO	n-type	Aerosol jet print	77	1.67	5.3	50	Low	Yes	PI	Ion-gel
[157]	In_2O_3	n-type	Spray pyrolysis	900	1.25	4	320	Medium	Yes	PI	Al_2O_3 (25 nm)
[158]	IGZO	n-type	Spin coat	–	84.4	5	55	Low–Medium	Yes	PI	Al_2O_3 (150 nm)
[127]	IGZO	n-type	Spin coat	223	8.58	9	2560	Medium	Yes	PI	$ZrAlO_x$ (35 nm)
[159]	SnO	p-type	Sputter	7630	5.87	3.8	440	High	Yes	PI	HfO_2 (220 nm)
[160]	SnO	p-type	Sputter	4000	3.3	2	6	High	Yes	PI	P(VDF-TrFE) (300 nm)
[161]	In_2O_3/ ZnO	n-type	Spray pyrolysis/spin coat	–	45	7	1000	Medium	No	Si	SiO_2 (400 nm)
[162]	IGZO/ IZO	n-type	Sputtering	–	18	9.3	10600	High	Yes	PEN	SiO_2 (20 nm)
[163]	IGZO	n-type	Sputter/transfer	350	14	7	16.5	High	Yes	PDMS	SiO_2 (100 nm)
[21]	Lisicon® S1200, Merck	p-type	Inkjet print	–	1	6	5.5	Low	Yes	Paralene-C	Lisicon® D207, Merck (360 nm)
[164]	Pentacene	p-type	Thermal evaporation	–	0.1	5	13	High	Yes	Paper	Paralene-C (340 nm)
[51]	WS_2	p-type	Spray cast	–	0.2	2	0.19	Low	Yes	PET	Ion gel

Effective Mobility (μ)—The mobility is related to how easily carriers move through the TFT channel in response to an applied electric field and therefore plays a significant role in determining a device's on-state performance. The mobility is an allusion to a material property used to describe bulk semiconductors. In TFTs, an effective mobility is extracted using the equation shown below:

$$\mu = \frac{L_{CH} g_m}{W_{CH} C_{OX} V_{DS}},$$

where L_{CH} and W_{CH} correspond to the device's physical length and width, respectively, g_m is the device's peak transconductance, C_{OX} is the gate capacitance per unit area, and V_{DS} is the applied drain-source voltage. The gate capacitance is typically calculated using the parallel plate model shown below:

$$C_{OX} = \frac{\varepsilon_0 \varepsilon_{OX}}{t_{OX}}.$$

The effective mobility is an important parameter for comparing the performance between devices as it adjusts for the device dimensions and the drain to source voltage, which can distort the performance arbitrarily.

On/Off Ratio—A crucial aspect of transistor operation is the ratio between the on-state current and the off-state current. The off-state current is defined as the minimum current, and theoretically, the on-current is defined as the current at a specific operating voltage ($V_{DD} = V_{DS} = V_{GS}$). In comparing devices measured using a diverse set of voltage windows, the on-current is extracted from either the author of the study or the highest achieved current.

Cost—The cost is a rough estimate based on the fabrication methods and materials. A 'Low' cost designation indicates that the device was all-solution-processed with no vacuum steps or high-temperature steps above 300°C. A 'Medium' cost designation indicates that the channel was either solution deposited or printed at low-temperature, however, at least one vacuum processing step was used (typically a thin and high-quality gate dielectric grown by CVD or ALD). A 'High' cost designation indicates that multiple vacuum processing steps were used to deposit both the channel and gate dielectric.

Flexibility—This parameter identifies if the TFT was fabricated on a flexible substrate and if the authors of the work attempted to flex the device.

Substrate—A critical component of flexible devices is the substrate selection. We have identified which substrates each report has utilized for the reader's reference.

Gate Dielectric and Thickness—The gate dielectric and thicknesses inform the reader about which materials and parameters are compatible with flexible devices and how they correlate with cost. Additionally, the quality and thickness of the gate dielectric largely determines the subthreshold swing of a device.

4.9 Effects due to Flexing/Stretching

In the ideal case for flexible TFTs, the device performance will not be compromised in any manner due to mechanical strain. Flexing induces strain that is dependent on the radius of curvature and device thickness. Due to the thin channel dimensions of many TFTs, the levels of strain during flexing can be quite low. Metal-oxides are fairly mechanically robust and can withstand moderate amounts of strain [44,128]. The response to various bending curvatures for an IGZO-based device is shown in Figure 4.9a and b [43]. While all parameters remain constant, the strain induced in the devices is quite

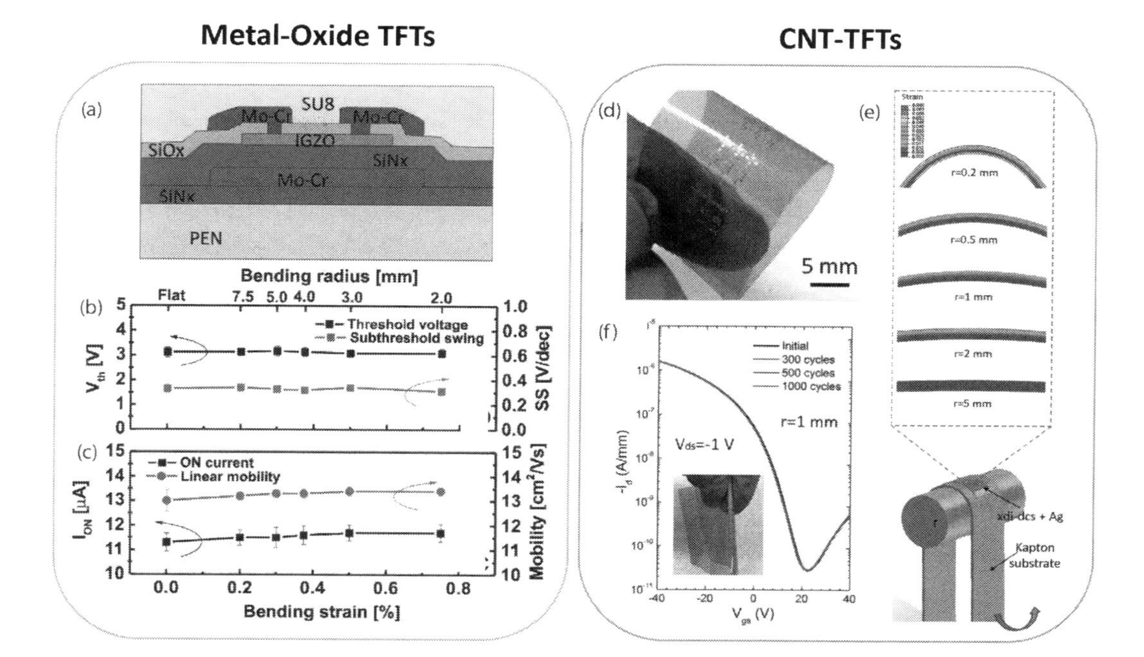

FIGURE 4.9 Strain and bending effects on TFT performance. (a) Schematic diagram of a flexible IGZO TFT fabricated on PEN and (b, c) its electrical response during bending. (d) Optical image of a flexible CNT-TFT fabricated on Kapton along with (e) an illustration of the strain distribution on the substrate at various bending radii. (f) Resulting CNT-TFT performance initially and after 300, 500, and 1000 bending cycles. ([a–c] From A.K. Tripathi, et al. *IEEE Trans. Electron Devices*, 62, 4063–4068, 2015. © 2015 IEEE. With permission; [d–f] From Cao, C. et al.: Completely Printed, Flexible, Stable, and Hysteresis-Free Carbon Nanotube Thin-Film Transistors via Aerosol jet Printing. *Advanced Electronic Materials*. 2017. 3. 1–10. Copyright Wiley-VCH Verlag GmbH & Co. KGaA. Reproduced with permission.)

low (<1%). This same device was shown to breakdown at a bending radius of 0.5 mm, or just 1.5% strain. Therefore, while still remaining highly applicable for flexible electronics where only moderate strain is induced, metal-oxides show little potential for enabling stretchable and foldable electronics.

One of the reasons CNTs are so favorable for flexible electronics, besides their solution process capabilities and their compatibility with a variety of substrates, is the fact that networked CNT films geometrically enable stretching/flexing. Individual CNTs are fairly mechanically robust, with experimentally measured Young's moduli of >1 TPa [129,130], but a thin-film of electrically connected CNTs has the additional property of flexibility due to superfluous electrical connections [131]. As strain is applied, the individual tubes will straighten and slide, but with a network density that is high enough, many conducting pathways will still exist.

Many CNT-TFTs have exhibited little change in properties due to flexing [59,96,114]. One primary example is displayed in Figure 4.9d–f [60]. The fully printed CNT-TFT is bent over a 1 mm radius of curvature over 1000 bending cycles and shows little to no change in electrical properties. At a bending radius of less than 1 mm, the device did break down, but this was due to the generation of cracks and delamination in the Ag nanoparticle gate, and not the CNT thin film.

Additionally, CNT-TFTs have been shown to withstand large amounts of strain in stretching. The challenge with stretchable electronics remains that complementary materials (conducting contacts and insulating gate dielectrics) breakdown prior to the CNT channel. Some devices have utilized the geometric stretchability of metallic CNTs or silver nanowires, coupled with elastomer gate dielectrics, to demonstrate transistor characteristics that are impervious to strain of up to 100% [62,63].

4.10 Applications

Flexible TFTs facilitate many applications in which low-cost, large-area electronics are required, including flexible active sensors [132–134], circuits [75,96,114,135], and display backplanes [57,136–138]. CNT-TFTs specifically have introduced many novel applications due to their inherent flexibility and robust performance. While many applications are still in development, initial experiments and demonstrations have shown the viability of CNT-TFTs to operate in real-world environments.

One application includes using CNT-TFTs as pressure sensors, in which the transconductance of the channel is modulated by environmental pressure. Using flexible CNT-TFTs that are on the size scale of less than 1 square mm, large-area pressure mapping electronics can be manufactured that could revolutionize current environmental pressure monitoring systems. For example, a large-area array of flexible sensors could be printed conformably on the inside of an automobile tire, and owing to the robust properties of CNT-TFTs, maintain function and continuously monitor the pressure differentials throughout the inner surface of the tire. This kind of large-area sensor system could also be useful for various aerospace or medical applications. A photograph of the sensor, along with its respective performance over a pressure range relevant to tires is included in Figure 4.10a [61].

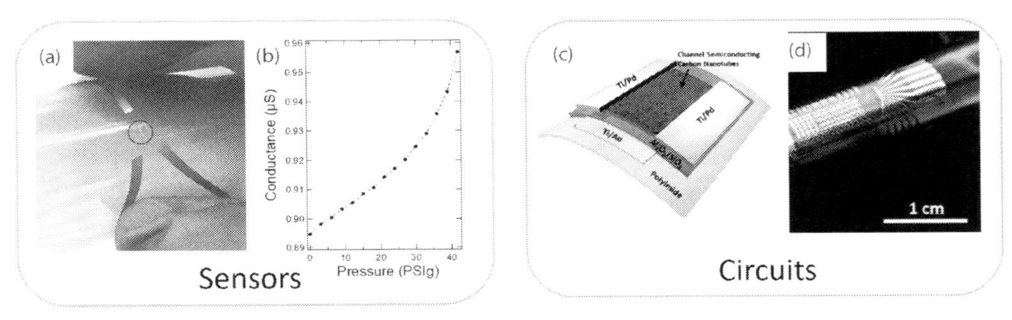

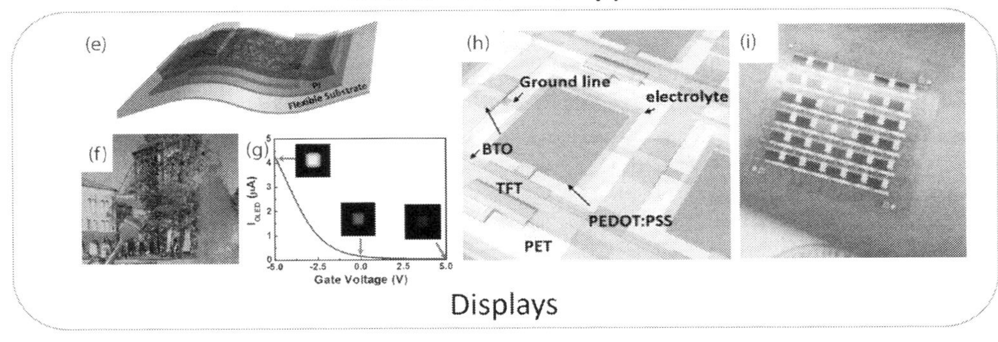

FIGURE 4.10 Applications of flexible CNT-TFTs. (a) A photograph of a fully printed, CNT-TFT used as the active sensor for environmental pressure and (b) the corresponding data throughout the range relevant to a tire. (c) The schematic for a CNT-TFT fabricated on polyimide for flexible circuit applications with a photograph of the circuitry is displayed in (d). (e) A schematic of a fully flexible, CNT-TFT used as the control circuitry for and light emitting diode (LED) display. The circuitry is transparent, as shown in (f) and able to control pixel brightness (g). (h) Displays a CNT-TFT schematic for circuitry used to control the flexible, electrochromic display shown in (i). ([a,b] From J.B. Andrews et al., *IEEE Sensors*, 18, 7875–7880, 2017. © 2017 IEEE. With permission; [c,d] Reprinted with permission from Wang, C. et al., *Nano Lett.*, 12, 1527–1533, 2012. Copyright 2012 American Chemical Society; [e–g] Reprinted with permission from Zhang, J. et al., *ACS Nano*, 6, 7412–7419, 2012. Copyright 2012 American Chemical Society; [h,i] Reprinted with permission from Cao, X. et al., *ACS Nano*, acsnano.6b05368, 2016. Copyright 2016 American Chemical Society.)

Along with physical sensors, CNT-TFTs have been shown to be viable options for biosensors [139–142]. Biosensors necessitate two main components, a biorecognition element and a transduction element. CNT-TFTs have been shown to be viable transduction elements, in which biomolecules act to effectively gate the transistor and modulate the threshold voltage. Little work has been done to fabricate flexible versions of these devices, but current progress is promising for wearable biosensors.

Furthermore, CNT-TFTs have been used to develop flexible logic circuitry capable of completing elementary digital computation. In recent work, CNT-TFTs have been used to fabricate invertors, not-AND (NAND), and not-OR (NOR) gates with effective device mobilities as high as 50 $cm^2/(V\text{-}s)$ and cut-off frequency values of 170 MHz—a value sufficient for wireless communication applications. Figure 4.10c and d display the device schematic and a photograph of the device circuitry [96]. Other flexible CNT-TFT circuitry has been used to facilitate the signal excitation and measurement for a fully flexible cyclic voltammetry tag capable of detecting Wurster's blue (N,N,N′,N′-tetramethyl-p-phenylenediamine [TPMD]), a common redox indicator for oxidase testing [143].

Lastly, low-cost and flexible display backplanes are an exciting application for flexible CNT-TFTs. In order to drive display circuitry, transistors must be used to modulate the signal to effectively turn on and off specific pixels. A flexible transistor capable of controlling the pixel brightness in an LED display is presented in Figure 4.10e–g. Uniquely, the control circuitry is both flexible and transparent. The gate voltage of each transistor is used to modulate the current passing through an LED in order to control brightness [137]. Figure 4.10h and i present CNT-TFT circuitry that is used to drive a wearable electrochromic display. The low-cost and flexible nature of CNT-TFTs enables both conformal contact and large-area fabrication necessary for wearable devices [138].

4.11 Recent Advances in 2D Nanomaterial TFTs

An emerging class of materials with potential for flexible TFTs is 2D nanomaterial thin films. Recent advancements in liquid phase exfoliation and ink formulation engineering have resulted in the development of 2D material thin-film devices. Although 2D thin-film electronics is an emerging field, devices fabricated from 2D nanosheet networks (2DNNs) could potentially yield several advantages over organic, metal-oxide, and potentially CNT-based TFTs. One of these advantages is the wide variety of 2D solutions that can be cheaply produced through liquid phase exfoliation. The possibilities include solutions of conducting, semiconducting (n- and p-type), and insulating nanosheets [52]. Furthermore, McManus et al. have demonstrated an ink formulation method for optimizing water-based solutions of 2D nanosheets for inkjet printing, allowing for the additive and scalable manufacture of 2DNN-based electronics [144–146]. A variety of their water-based inks can be seen in Figure 4.11a. Such inks have been shown to be biocompatible, opening potential use in biomedical applications. In terms of devices, Kelly et al. were the first to demonstrate all-printed TFTs based off of 2DNNs, as shown in Figure 4.11b and c [51]. In this study, the authors investigated numerous 2DNN channel materials and device structures. The resulting devices between references [52] and [51] maintain good interfaces between nanosheet networks without significant remixing (see SEM images in Figure 4.11d). The most notable of the devices reported from Kelly et al. were their spray-printed WS_2 (p-type) and $MoSe_2$ (n-type) TFTs with ionic liquid gates which exhibited on/off ratios of 600 and 100, respectively (as seen in Figure 4.11e). For higher mobility devices, Carey et al. demonstrated the first heterojunctions between printed boron nitride (BN) dielectric layers and graphene channels to produce flexible TFTs with mobilities as high as 91 $cm^2/(V\text{-}s)$ [147].

Moving forward, 2DNN-based thin films face several challenges before they can compete with their organic- and nanomaterial-based thin-film counterparts. First, increasing the network mobility of semiconducting 2DNNs should be paramount. One route for doing so is through junction engineering, where flake thickness is tuned to minimize junction resistance [148]. Another is maximizing the size of each flake in order to minimize the number of junctions throughout the network [149].

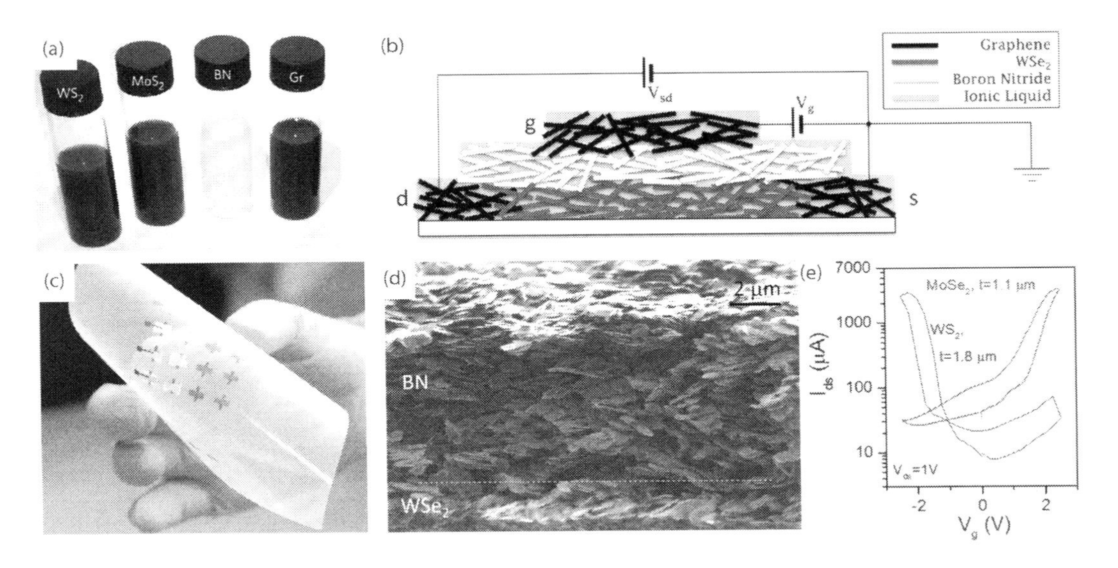

FIGURE 4.11 Recent advancements in liquid phase exfoliation have resulted in 2D TFTs. (a) Optical image of water-based 2D ink dispersions. (b) Schematic diagram of an all-2D TFT and (c) corresponding optical image of an array of 2D TFTs. (d) Cross-sectional SEM image showing BN and WSe_2 layers. (e) Representative subthreshold characteristics of TFTs fabricated from $MoSe_2$ and WSe_2. ([a] With kind permission from Springer Science+Business Media: *Nat. Nanotechnol.*, Water-based and biocompatible 2D crystal inks for all-inkjet-printed heterostructures, 12, 2017, 343–350, McManus, D. et al.; [b–e] From Kelly, A.G. et al., *Science (80-.)*., 356, 69–73, 2017, Reprinted with permission of AAAS.)

Second, the connectivity and interface properties of printed 2DNN films must be continually improved. This is especially critical if 2DNN-based devices are to be used in flexible electronics, as the connectivity of a thin-film network is critical during strain induced by flexing. Finally, the nanoflake concentration in 2D inks must be increased without compromising the printed film's electrical properties. Due to low ink concentrations, current printing methods require multiple print iterations for sufficiently conducting films which is not ideal for high-throughput fabrication environments. Although this list of challenges is not exhaustive, addressing these points will be needed to exploit the potential advantages that 2DNNs have to offer for low-cost flexible electronics.

4.12 Conclusion and Outlook

Research attention on flexible thin-film transistors and the materials of which they consist has accelerated over the past few decades with promise for enabling a new class of electronics. Due to their inherently thin nature and compatibility with solution processing, TFTs remain ideal candidates for facilitating flexible electronic applications. While various materials have been demonstrated to exhibit the properties necessary for a TFT semiconducting channel, the most promising remains networked CNT thin films. In this chapter, we have discussed in detail, various aspects of CNT-TFTs that are critical for flexible devices including operation, fabrication, performance, and application. Looking forward, many obstacles remain before a truly low-cost, large-area, and flexible CNT-TFT technology can be developed. While there has been much progress, challenges associated with device performance, integration, and overall uniformity inhibit widespread commercial penetration. Further development of fabrication and material processing techniques are in need of optimization for an effective CNT-TFT technology. With those challenges in mind, CNT-TFTs present an auspicious path forwards for the future of large-area, flexible devices.

References

1. J. E. Lilienfeld, 'Method and apparatus for controlling electric currents,' US Patent 1,745,175. pp. 1–4, 1930.
2. P. K. Weimer, 'The TFT—A New Thin-Film Transistor,' *Proc. IRE*, vol. 50, no. 6, pp. 1462–1469, 1962.
3. T. Kamiya, K. Nomura, and H. Hosono, 'Origins of high mobility and low operation voltage of amorphous oxide TFTs: Electronic structure, electron transport, defects and doping,' *J. Disp. Technol.*, vol. 5, no. 12, pp. 468–483, 2009.
4. A. D. Franklin, 'Nanomaterials in transistors: From high-performance to thin-film applications,' *Science*, vol. 349, no. 6249, p. aab2750, 2015.
5. M. Kastler et al., 'ARTICLES A high-mobility electron-transporting polymer for printed transistors,' *Nature*, vol. 457, no. 7230, p. 679, 2009.
6. A. Tsumura, H. Koezuka, and T. J. Ando, 'Macromolecular electronic device field-effect transistor with a polythiophne film,' *Appl. Phys. Lett.*, vol. 49, no. 1986, p. 1210, 1986.
7. D. J. J. Gundlach, Y. Y. Y. Lin, T. N. N. Jackson, S. F. F. Nelson, and D. G. G. Schlom, 'Pentacene organic thin-film transistors—Molecular ordering and mobility,' *IEEE Electron Device Lett.*, vol. 18, no. 3, pp. 87–89, 1997.
8. Y. Y. Lin, and D. I. Gundlach, 'Pentacene-based organic thin-film transistors,' *Electron Devices, IEEE*, vol. 44, no. 8, pp. 1325–1331, 1997.
9. H. Klauk, M. Halik, U. Zschieschang, G. Schmid, W. Radlik, and W. Weber, 'High-mobility polymer gate dielectric pentacene thin film transistors,' *J. Appl. Phys.*, vol. 92, no. 9, pp. 5259–5263, 2002.
10. Z. Bao, A. Dodabalapur, and A. J. Lovinger, 'Transistor applications with high mobility soluble and processable regioregular poly (3-hexylthiophene) for thin film field-effect transistor applications with high mobility,' *Appl. Phys. Lett.*, vol. 4108, no. 1996, pp. 21–24, 1996.
11. D. H. Kim et al., 'Enhancement of field-effect mobility due to surface-mediated molecular ordering in regioregular polythiophene thin film transistors,' *Adv. Funct. Mater.*, vol. 15, no. 1, pp. 77–82, 2005.
12. R. Zhang et al., 'Nanostructure dependence of field-effect mobility in regioregular poly(3-hexylthiophene) thin film field effect transistors,' *J. Am. Chem. Soc.*, vol. 128, no. 11, pp. 3480–3481, 2006.
13. H. Klauk, U. Zschieschang, J. Pflaum, and M. Halik, 'Ultralow-power organic complementary circuits,' *Nature*, vol. 445, no. 7129, pp. 745–748, 2007.
14. M. M. Ling, and Z. Bao, 'Copper hexafluorophthalocyanine field-effect transistors with enhanced mobility by soft contact lamination,' *Org. Electron. physics, Mater. Appl.*, vol. 7, no. 6, pp. 568–575, 2006.
15. Z. Bao, A. Lovinger, and A. Dodabalapur, 'Organic field-effect transistors with high mobility based on copper phthalocyanine,' *Appl. Phys. Lett.*, vol. 69, no. 20, pp. 3066–3068, 1996.
16. A. Babel, J. D. Wind, and S. A. Jenekhe, 'Ambipolar charge transport in air-stable polymer blend thin-film transistors,' *Adv. Funct. Mater.*, vol. 14, no. 9, pp. 891–898, 2004.
17. C. R. Newman, C. D. Frisbie, D. A. Da Silva Filho, J. L. Brédas, P. C. Ewbank, and K. R. Mann, 'Introduction to organic thin film transistors and design of n-channel organic semiconductors,' *Chem. Mater.*, vol. 16, no. 23, pp. 4436–4451, 2004.
18. S. K. Park, T. N. Jackson, J. E. Anthony, and D. A. Mourey, 'High mobility solution processed 6,13-bis(triisopropyl-silylethynyl) pentacene organic thin film transistors,' *Appl. Phys. Lett.*, vol. 91, no. 6, pp. 6–9, 2007.
19. A. Afzali, C. D. Dimitrakopoulos, and T. L. Breen, 'High-performance, solution-processed organic thin film transistors from a novel pentacene precursor,' *J. Am. Chem. Soc.*, vol. 124, no. 30, pp. 8812–8813, 2002.

20. S. Chung, S. O. Kim, S. K. Kwon, C. Lee, and Y. Hong, 'All-inkjet-printed organic thin-film transistor inverter on flexible plastic substrate,' *IEEE Electron Device Lett.*, vol. 32, no. 8, pp. 1134–1136, 2011.
21. K. Fukuda et al., 'Fully-printed high-performance organic thin-film transistors and circuitry on one-micron-thick polymer films,' *Nat. Commun.*, vol. 5, pp. 1–8, 2014.
22. G. H. Gelinck et al., 'Flexible active-matrix displays and shift registers based on solution-processed organic transistors,' *Nat. Mater.*, vol. 3, no. 2, pp. 106–110, 2004.
23. C.-H. Wang, C.-Y. Hsieh, and J.-C. Hwang, 'Flexible organic thin-film transistors with silk fibroin as the gate dielectric,' *Adv. Mater.*, vol. 23, no. 14, pp. 1630–1634, 2011.
24. T. Sekitani, U. Zschieschang, H. Klauk, and T. Someya, 'Flexible organic transistors and circuits with extreme bending stability,' *Nat. Mater.*, vol. 9, no. 12, pp. 1015–1022, 2010.
25. C. D. Sheraw et al., 'Organic thin-film transistor-driven polymer-dispersed liquid crystal displays on flexible polymeric substrates,' *Appl. Phys. Lett.*, vol. 80, no. 6, pp. 1088–1090, 2002.
26. C. Reese, M. Roberts, M. M. Ling, and Z. Bao, 'Organic thin film transistors,' *Mater. Today*, vol. 7, no. 8, pp. 20–27, 2004.
27. E. G. Bittle, J. I. Basham, T. N. Jackson, O. D. Jurchescu, and D. J. Gundlach, 'Mobility overestimation due to gated contacts in organic field-effect transistors,' *Nat. Commun.*, vol. 7, pp. 1–7, 2016.
28. A. Aoki and H. Sasakura, 'Tin oxide thin film transistors,' *Jpn. J. Appl. Phys.*, vol. 9, no. 5, pp. 582–582, 1970.
29. R. E. Presley, C. L. Munsee, C. H. Park, D. Hong, J. F. Wager, and D. A. Keszler, 'Tin oxide transparent thin-film transistors,' *J. Phys. D. Appl. Phys.*, vol. 37, no. 20, pp. 2810–2813, 2004.
30. E. Fortunato et al., 'High field-effect mobility zinc oxide thin film transistors produced at room temperature,' *J. Non. Cryst. Solids*, vol. 338–340, no. 1 SPEC. ISS., pp. 806–809, 2004.
31. T. Hirao et al., 'Novel top-gate zinc oxide thin-film transistors (ZnO TFTs) for AMLCDs,' *J. Soc. Inf. Disp.*, vol. 15, no. 1, p. 17, 2007.
32. H. Yabuta et al., 'High-mobility thin-film transistor with amorphous $InGaZnO_4$ channel fabricated by room temperature rf-magnetron sputtering,' *Appl. Phys. Lett.*, vol. 89, no. 11, pp. 10–13, 2006.
33. K. Nomura, H. Ohta, A. Takagi, T. Kamiya, M. Hirano, and H. Hosono, 'Room-temperature fabrication of transparent flexible thin-film transistors using amorphous oxide semiconductors,' *Nature*, vol. 432, no. 7016, pp. 488–492, 2004.
34. M. Kim et al., 'High mobility bottom gate InGaZnO thin film transistors with SiO_x etch stopper,' *Appl. Phys. Lett.*, vol. 90, no. 21, pp. 1–4, 2007.
35. J. K. Jeong, J. H. Jeong, H. W. Yang, J. S. Park, Y. G. Mo, and H. D. Kim, 'High performance thin film transistors with cosputtered amorphous indium gallium zinc oxide channel,' *Appl. Phys. Lett.*, vol. 91, no. 11, pp. 1–4, 2007.
36. Y. Ogo et al., 'Tin monoxide as an s-orbital-based p-type oxide semiconductor: Electronic structures and TFT application,' *Phys. Status Solidi Appl. Mater. Sci.*, vol. 206, no. 9, pp. 2187–2191, 2009.
37. E. Fortunato et al., 'Transparent p-type SnO_x thin film transistors produced by reactive RF magnetron sputtering followed by low temperature annealing,' *Appl. Phys. Lett.*, vol. 97, no. 5, pp. 1–4, 2010.
38. E. Fortunato et al., 'Thin-film transistors based on p-type Cu_2O thin films produced at room temperature,' *Appl. Phys. Lett.*, vol. 96, no. 19, pp. 2–5, 2010.
39. J. Sohn et al., 'Effects of vacuum annealing on the optical and electrical properties of p-type copper-oxide thin-film transistors,' *Semicond. Sci. Technol.*, vol. 28, no. 1, p. 15005, 2013.
40. J. S. Lee, S. Chang, S. M. Koo, and S. Y. Lee, 'High-performance a-IGZO TFT with ZrO_2 gate dielectric fabricated at room temperature,' *IEEE Electron Device Lett.*, vol. 31, no. 3, pp. 225–227, 2010.
41. E. B. Secor, J. Smith, T. J. Marks, and M. C. Hersam, 'High-performance inkjet-printed indium-gallium-zinc-oxide transistors enabled by embedded, chemically stable graphene electrodes,' *ACS Appl. Mater. Interfaces*, p. acsami.6b02730, vol. 8, no. 27, pp. 17428–17434, 2016.

42. D.-H. Lee, S.-Y. Han, G. S. Herman, and C. Chang, 'Inkjet printed high-mobility indium zinc tin oxide thin film transistors,' *J. Mater. Chem.*, vol. 19, no. 20, p. 3135, 2009.
43. A. K. Tripathi, K. Myny, B. Hou, K. Wezenberg, and G. H. Gelinck, 'Electrical characterization of flexible inGaZnO transistors and 8-b transponder chip down to a bending radius of 2 mm,' *IEEE Trans. Electron Devices*, vol. 62, no. 12, pp. 4063–4068, 2015.
44. Y. H. Kim et al., 'Flexible metal-oxide devices made by room-temperature photochemical activation of sol–gel films,' *Nature*, vol. 489, no. 7414, pp. 128–132, 2012.
45. H. E. Unalan, G. Fanchini, A. Kanwal, A. Du Pasquier, and M. Chhowalla, 'Design criteria for transparent single-wall carbon nanotube thin-film transistors,' *Nano Lett.*, vol. 6, no. 4, pp. 677–682, 2006.
46. E. S. Snow, P. M. Campbell, M. G. Ancona, and J. P. Novak, 'High-mobility carbon-nanotube thin-film transistors on a polymeric substrate,' *Appl. Phys. Lett.*, vol. 86, no. 3, pp. 1–3, 2005.
47. P. Beecher et al., 'Ink-jet printing of carbon nanotube thin film transistors,' *J. Appl. Phys.*, vol. 102, no. 4, 2007.
48. X. Duan, C. Niu, V. Sahi, J. Chen, J. W. Parce, S. Empedocles, and J. L. Goldman., 'High-performance thin-film transistors using semiconductor nanowires and nanoribbons,' *Nature*, vol. 425, pp. 274–278, 2003.
49. E. N. Dattoli, Q. Wan, W. Guo, Y. Chen, X. Pan, and W. Lu, 'Fully transparent thin-film transistor devices based on SnO_2 nanowires,' *Nano Lett.*, vol. 7, no. 8, pp. 2463–2469, 2007.
50. H. C. Lin, M. H. Lee, C. J. Su, T. Y. Huang, C. C. Lee, and Y. S. Yang, 'A simple and low-/cost method to fabricate TFTs with poly-Si nanowire channel,' *IEEE Electron Device Lett.*, vol. 26, no. 9, pp. 643–645, 2005.
51. A. G. Kelly et al., 'All-printed thin-film transistors from networks of liquid-exfoliated nanosheets,' *Science (80-.).*, vol. 356, no. 6333, pp. 69–73, 2017.
52. D. McManus et al., 'Water-based and biocompatible 2D crystal inks for all-inkjet-printed heterostructures,' *Nat. Nanotechnol.*, vol. 12, no. 4, pp. 343–350, 2017.
53. T. W. Ebbesen, H. J. Lezec, H. Hiura, J. W. Bennett, H. F. Ghaemi, and T. Thio, 'Electrical conductivity of individual carbon nanotubes,' *Nature*, vol. 382. pp. 54–56, 1996.
54. J. C. Charlier, X. Blase, and S. Roche, 'Electronic and transport properties of nanotubes,' *Rev. Mod. Phys.*, vol. 79, no. 2, pp. 677–732, 2007.
55. M. Ha et al., 'Printed, sub-3V digital circuits on plastic from aqueous carbon nanotube inks,' *ACS Nano*, vol. 4, no. 8, pp. 4388–4395, 2010.
56. M. S. Arnold, A. A. Green, J. F. Hulvat, S. I. Stupp, and M. C. Hersam, 'Sorting carbon nanotubes by electronic structure using density differentiation,' *Nat. Nanotechnol.*, vol. 1, no. 1, p. 60, 2006.
57. C. Wang, J. Zhang, K. Ryu, A. Badmaev, L. G. De Arco, and C. Zhou, 'Wafer-scale fabrication of separated carbon nanotube thin-film transistors for display applications,' *Nano Lett.*, vol. 9, no. 12, pp. 4285–4291, 2009.
58. M. Y. Timmermans, *Carbon Nanotube Thin Film Transistors for Flexible Electronics*. PhD Dissertation published by School of Science in Helinski, Finland, 2013.
59. S. Park, M. Vosguerichian, and Z. Bao, 'A review of fabrication and applications of carbon nanotube film-based flexible electronics,' *Nanoscale*, vol. 5, no. 5, p. 1727, 2013.
60. C. Cao, J. B. Andrews, and A. D. Franklin, 'Completely printed, flexible, stable, and hysteresis-free carbon nanotube thin-film transistors via aerosol jet printing,' *Adv. Electron. Mater.*, vol. 3, no. 5, pp. 1–10, 2017.
61. J. B. Andrews, J. A. Cardenas, J. Mullett, and A. D. Franklin, 'Fully printed and flexible carbon nanotube transistors designed for environmental pressure sensing and aimed at smart tire applications,' *IEEE Sensors*, vol. 18, no. 19, pp. 7875–7880, 2017.
62. A. Chortos et al., 'Highly stretchable transistors using a microcracked organic semiconductor,' *Adv. Mater.*, vol. 26, no. 25, pp. 4253–4259, 2014.
63. A. Chortos et al., 'Mechanically durable and highly stretchable transistors employing carbon nanotube semiconductor and electrodes,' *Adv. Mater.*, vol. 28, no. 22, pp. 4441–4448, 2016.

64. A. D. Franklin, 'Carbon nanotube electronics,' *Emerg. Nanoelectron. Devices*, John Wiley and Sons, West Sussex, UK, pp. 315–335, 2015.
65. L. B. Ebert, 'Science of fullerenes and carbon nanotubes,' *Carbon NY.*, vol. 35, no. 3, pp. 437–438, 1997.
66. J. W. G. Wilder, L. C. Venema, A. G. Rinzler, R. E. Smalley, and C. Dekker, 'Electronic structure of atomically resolved carbon nanotubes,' *Nature*, vol. 391, no. 6662, pp. 59–62, 1998.
67. M. S. Fuhrer et al., 'Crossed nanotube junctions,' *Science (80-.).*, vol. 288, no. 5465, pp. 494–497, 2000.
68. R. Martel, T. Schmidt, H. R. Shea, T. Hertel, and P. Avouris, 'Single- and multi-wall carbon nanotube field-effect transistors,' *Appl. Phys. Lett.*, vol. 73, no. 17, pp. 2447–2449, 1998.
69. V. Derycke, R. Martel, J. Appenzeller, and P. Avouris, 'Controlling doping and carrier injection in carbon nanotube transistors,' *Appl. Phys. Lett.*, vol. 80, no. 15, pp. 2773–2775, 2002.
70. A. Javey, J. Guo, Q. Wang, M. Lundstrom, and H. Dai, 'Ballistic carbon nanotube field-effect transistors,' *Nature*, vol. 424, no. 6949, pp. 654–657, 2003.
71. Q. Cao et al., 'End-bonded contacts for carbon nanotube transistors with low, size-independent resistance,' *Science (80-.).*, vol. 350, no. 6256, pp. 68–72, 2015.
72. A. D. Franklin et al., 'Carbon nanotube complementary wrap-gate transistors,' *Nano Lett.*, vol. 13, no. 6, pp. 2490–2495, 2013.
73. A. D. Franklin et al., 'Sub-10 nm carbon nanotube transistor,' *Nano Lett.*, vol. 12, no. 2, pp. 758–762, 2012.
74. C. Qiu, Z. Zhang, M. Xiao, Y. Yang, D. Zhong, and L.-M. Peng, 'Scaling carbon nanotube complementary transistors to 5-nm gate lengths,' *Science (80-.).*, vol. 355, no. 6322, pp. 271–276, 2017.
75. P. H. Lau et al., 'Fully printed, high performance carbon nanotube thin-film transistors on flexible substrates,' *Nano Lett.*, vol. 13, no. 8, pp. 3864–3869, 2013.
76. L.-P. Simoneau, J. Villeneuve, C. M. Aguirre, R. Martel, P. Desjardins, and A. Rochefort, 'Influence of statistical distributions on the electrical properties of disordered and aligned carbon nanotube networks,' *J. Appl. Phys.*, vol. 114, no. 11, p. 114312, 2013.
77. R. A. Bell, *Conduction in Carbon Nanotube Networks*, Springer, Cham, Switzerland, 2015.
78. S. P. Schießl et al., 'Modeling carrier density dependent charge transport in semiconducting carbon nanotube networks,' *Phys. Rev. Mater.*, vol. 1, no. 4, p. 46003, 2017.
79. L. Simoneau, J. Villeneuve, and A. Rochefort, 'Electron percolation in realistic models of carbon nanotube networks Electron percolation in realistic models of carbon nanotube networks,' *J. Appl. Phys.*, vol. 118, p. 124309, 2015.
80. L. Hu, D. S. Hecht, and G. Grüner, 'Percolation in transparent and conducting carbon nanotube networks,' *Nano Lett.*, vol. 4, no. 12, pp. 2513–2517, 2004.
81. E. Bekyarova et al., *J. Am. Chem. Soc.*, vol. 127, no. 16, pp. 5990–5995, 2005.
82. N. Rouhi, D. Jain, K. Zand, and P. J. Burke, 'Fundamental limits on the mobility of nanotube-based semiconducting inks,' *Adv. Mater.*, vol. 23, no. 1, pp. 94–99, 2011.
83. B. Zheng, C. Lu, G. Gu, A. Makarovski, G. Finkelstein, and J. Liu, 'Efficient CVD growth of single-walled carbon nanotubes on surfaces using carbon monoxide precursor,' *Nano Lett.*, vol. 2, no. 8, pp. 895–898, 2002.
84. C. Kocabas, S. H. Hur, A. Gaur, M. A. Meitl, M. Shim, and J. A. Rogers, 'Guided growth of large-scale, horizontally aligned arrays of single-walled carbon nanotubes and their use in thin-film transistors,' *Small*, vol. 1, no. 11, pp. 1110–1116, 2005.
85. C. Journet et al., 'Large-scale production of single-walled carbon nanotubes by the electric-arc technique,' *Nature*, vol. 388, no. 6644, pp. 756–758, 1997.
86. P. Hou, C. Liu, Y. Tong, S. Xu, M. Liu, and H. Cheng, 'Purification of single-walled carbon nanotubes synthesized by the hydrogen arc-discharge method,' *J. Mater. Res.*, vol. 16, no. 9, pp. 2526–2529, 2001.
87. T. Sugai, 'New synthesis of high-quality double-walled carbon nanotubes by high-temperature pulsed arc discharge,' *Nano Lett.*, vol. 3, pp. 769–773, 2003.

88. C. Kocabas, M. A. Meitl, A. Gaur, M. Shim, and J. A. Rogers, 'Aligned arrays of single-walled carbon nanotubes generated from random networks by orientationally selective laser ablation,' *Nano Lett.*, vol. 4, no. 12, pp. 2421–2426, 2004.
89. M. Kusaba, and Y. Tsunawaki, 'Production of single-wall carbon nanotubes by a XeCl excimer laser ablation,' *Thin Solid Films*, vol. 506–507, pp. 255–258, 2006.
90. R. Das, Z. Shahnavaz, M. E. Ali, M. M. Islam, and S. B. Abd Hamid, 'Can we optimize arc discharge and laser ablation for well-controlled carbon nanotube synthesis?,' *Nanoscale Res. Lett.*, vol. 11, no. 1, 2016.
91. S. Ghosh, S. M. Bachilo, and R. B. Weisman, 'Advanced sorting of single-walled carbon nanotubes by nonlinear density-gradient ultracentrifugation,' *Nat. Nanotechnol.*, vol. 5, no. 6, pp. 443–450, 2010.
92. A. A. Green and M. C. Hersam, 'Nearly single-chirality single-walled carbon nanotubes produced via orthogonal iterative density gradient ultracentrifugation,' *Adv. Mater.*, vol. 23, no. 19, pp. 2185–2190, 2011.
93. A. Nish, J. Y. Hwang, J. Doig, and R. J. Nicholas, 'Highly selective dispersion of single-walled carbon nanotubes using aromatic polymers,' *Nat. Nanotechnol.*, vol. 2, no. 10, pp. 640–646, 2007.
94. J.-Y. Hwang et al., 'Polymer structure and solvent effects on the selective dispersion of single-walled carbon nanotubes,' *J. Am. Chem. Soc.*, vol. 130, no. 11, pp. 3543–3553, 2008.
95. H. W. Lee et al., 'Selective dispersion of high purity semiconducting single-walled carbon nanotubes with regioregular poly(3-alkylthiophene)s,' *Nat. Commun.*, vol. 2, no. 1, pp. 541–548, 2011.
96. C. Wang et al., 'Extremely bendable, high-performance integrated circuits using semiconducting carbon nanotube networks for digital, analog, and radio-frequency applications,' *Nano Lett.*, vol. 12, no. 3, pp. 1527–1533, 2012.
97. L. S. Liyanage et al., 'Wafer-scale fabrication and characterization of thin-film transistors with polythiophene-sorted semiconducting carbon nanotube networks,' *ACS Nano*, vol. 6, no. 1, pp. 451–458, 2012.
98. M. Jeong, K. Lee, E. Choi, A. Kim, and S. B. Lee, 'Spray-coated carbon nanotube thin-film transistors with striped transport channels,' *Nanotechnology*, vol. 23, no. 50, pp. 21–26, 2012.
99. M. C. LeMieux, M. Roberts, S. Barman, W. J. Yong, M. K. Jong, and Z. Bao, 'Self-sorted, aligned nanotube networks for thin-film transistors,' *Science (80-.)*, vol. 321, no. 5885, pp. 101–104, 2008.
100. S. Y. Lee, S. W. Lee, S. M. Kim, W. J. Yu, Y. W. Jo, and Y. H. Lee, 'Scalable complementary logic gates with chemically doped semiconducting carbon nanotube transistors,' *ACS Nano*, vol. 5, no. 3, pp. 2369–2375, 2011.
101. M. Singh, H. M. Haverinen, P. Dhagat, and G. E. Jabbour, 'Inkjet printing-process and its applications,' *Adv. Mater.*, vol. 22, no. 6, pp. 673–685, 2010.
102. J. Vaillancourt et al., 'All ink-jet-printed carbon nanotube thin-film transistor on a polyimide substrate with an ultrahigh operating frequency of over 5 GHz,' *Appl. Phys. Lett.*, vol. 93, no. 24, pp. 1–4, 2008.
103. H. Okimoto et al., 'Tunable carbon nanotube thin-film transistors produced exclusively via inkjet printing,' *Adv. Mater.*, vol. 22, no. 36, pp. 3981–3986, 2010.
104. B. Kim, S. Jang, M. L. Geier, P. L. Prabhumirashi, M. C. Hersam, and A. Dodabalapur, 'High-speed, inkjet-printed carbon nanotube/zinc tin oxide hybrid complementary ring oscillators,' *Nano Lett.*, vol. 14, no. 6, pp. 3683–3687, 2014.
105. Optomec, 'Aerosol Jet 300 Series Systems—Datasheet,' 2015. https://www.optomec.com/wp-content/uploads/2014/04/AJ-300-Systems-Web0417.pdf. (Accessed September, 2019.)
106. C. Cao, J. B. Andrews, A. Kumar, and A. D. Franklin, 'Improving contact interfaces in fully printed carbon nanotube thin-film transistors,' *ACS Nano*, vol. 10, no. 5, pp. 5221–5229, 2016.
107. M. Ha et al., 'Aerosol jet printed, low voltage, electrolyte gated carbon nanotube ring oscillators with sub-5 μs stage delays,' *Nano Lett.*, vol. 13, no. 3, pp. 954–960, 2013.
108. F. C. Krebs, 'Polymer solar cell modules prepared using roll-to-roll methods: Knife-over-edge coating, slot-die coating and screen printing,' *Sol. Energy Mater. Sol. Cells*, vol. 93, no. 4, pp. 465–475, 2009.

109. C. M. Homenick et al., 'Fully printed and encapsulated SWCNT-based thin film transistors via a combination of R2R gravure and inkjet printing,' *ACS Appl. Mater. Interfaces*, vol. 8, no. 41, pp. 27900–27910, 2016.
110. X. Cao et al., 'Screen printing as a scalable and low-cost approach for rigid and flexible thin-film transistors using separated carbon nanotubes,' *ACS Nano*, vol. 8, no. 12, pp. 12769–12776, 2014.
111. J. Liang, K. Tong, and Q. Pei, 'A water-based silver-nanowire screen-print ink for the fabrication of stretchable conductors and wearable thin-film transistors,' *Adv. Mater.*, vol. 28, pp. 5986–5996, 2016.
112. J. Liang et al., 'Intrinsically stretchable and transparent thin-film transistors based on printable silver nanowires, carbon nanotubes and an elastomeric dielectric,' *Nat. Commun.*, vol. 6, p. 7647, 2015.
113. L. Cai, S. Zhang, J. Miao, Z. Yu, and C. Wang, 'Fully printed stretchable thin-film transistors and integrated logic circuits,' *ACS Nano*, vol. 10, no. 12, pp. 11459–11468, 2016.
114. D. M. Sun et al., 'Flexible high-performance carbon nanotube integrated circuits,' *Nat. Nanotechnol.*, vol. 6, no. 3, pp. 156–161, 2011.
115. Z. M. Dang, J. K. Yuan, J. W. Zha, T. Zhou, S. T. Li, and G. H. Hu, 'Fundamentals, processes and applications of high-permittivity polymer-matrix composites,' *Prog. Mater. Sci.*, vol. 57, no. 4, pp. 660–723, 2012.
116. M. W. Lee, M. Y. Lee, J. C. Choi, J. S. Park, and C. K. Song, 'Fine patterning of glycerol-doped PEDOT:PSS on hydrophobic PVP dielectric with ink jet for source and drain electrode of OTFTs,' *Org. Electron. physics, Mater. Appl.*, vol. 11, no. 5, pp. 854–859, 2010.
117. M. E. Roberts, M. C. LeMieux, A. N. Sokolov, and Z. Bao, 'Self-sorted nanotube networks on polymer dielectrics for low-voltage thin-film transistors,' *Nano Lett.*, vol. 9, no. 7, pp. 2526–2531, 2009.
118. H. Li, Y. Tang, W. Guo, H. Liu, L. Zhou, and N. Smolinski, 'Polyfluorinated electrolyte for fully printed carbon nanotube electronics,' *Adv. Funct. Mater.*, vol. 26, no. 38, pp. 6914–6920, 2016.
119. J. H. Cho et al., 'Printable ion-gel gate dielectrics for low-voltage polymer thin-film transistors on plastic,' *Nat. Mater.*, vol. 7, no. 11, pp. 900–906, 2008.
120. L. Cai, S. Zhang, J. Miao, Z. Yu, and C. Wang, 'Fully printed foldable integrated logic gates with tunable performance using semiconducting carbon nanotubes,' *Adv. Funct. Mater.*, vol. 25, no. 35, pp. 5698–5705, 2015.
121. X. Zhang et al., 'Flexible CMOS-like circuits based on printed p-type and n-type carbon nanotube thin-film transistors,' *Small*, vol. 12, no. 36, pp. 5066–5073, 2016.
122. N. Liu, K. N. Yun, H. Y. Yu, J. H. Shim, and C. J. Lee, 'High-performance carbon nanotube thin-film transistors on flexible paper substrates,' *Appl. Phys. Lett.*, vol. 106, no. 10, p. 103106, 2015.
123. T. Rembert, C. Battaglia, A. Anders, and A. Javey, 'Room temperature oxide deposition approach to fully transparent, all-oxide thin-film transistors,' *Adv. Mater.*, vol. 27, no. 40, pp. 6090–6095, 2015.
124. N. C. Su et al., 'Low-voltage-driven flexible InGaZnO thin-film transistor with small subthreshold swing,' *IEEE Electron Device Lett.*, vol. 31, no. 7, pp. 680–682, 2010.
125. N. Münzenrieder, L. Petti, C. Zysset, T. Kinkeldei, G. A. Salvatore, and G. Tröster, 'Flexible self-aligned amorphous InGaZnO thin-film transistors with submicrometer channel length and a transit frequency of 135 MHz,' *IEEE Trans. Electron Devices*, vol. 60, no. 9, pp. 1–6, 2013.
126. H. H. Hsu, C. Y. Chang, and C. H. Cheng, 'A flexible IGZO thin-film transistor with stacked TiO_2-based dielectrics fabricated at room temperature,' *IEEE Electron Device Lett.*, vol. 34, no. 6, pp. 768–770, 2013.
127. J. W. Jo et al., 'Highly stable and imperceptible electronics utilizing photoactivated heterogeneous sol-gel metal-oxide dielectrics and semiconductors,' *Adv. Mater.*, vol. 27, no. 7, pp. 1182–1188, 2015.
128. J. S. Park et al., 'Flexible full color organic light-emitting diode display on polyimide plastic substrate driven by amorphous indium gallium zinc oxide thin-film transistors,' *Appl. Phys. Lett.*, vol. 95, no. 1, pp. 2007–2010, 2009.

129. J.-P. Salvetat, J.-M. Bonard, and N. H. Thomson, 'Mechanical properties of carbon nanotubes,' *Appl. Phys. A*, vol. 69, no. 3, pp. 255–260, 1999.
130. N. Yao, V. Lordi, N. Yao, and V. Lordi, 'Young' s modulus of single-walled carbon nanotubes Young's modulus of single-walled carbon nanotubes,' vol. 1939, no. 1998, 2000.
131. J. A. Rogers, T. Someya, and Y. Huang, 'Materials and mechanics for stretchable electronics,' *Science (80-.)*, vol. 327, no. 5973, pp. 1603–1607, 2010.
132. K. Chen et al., 'Printed carbon nanotube electronics and sensor systems,' *Adv. Mater.*, vol. 28, pp. 4397–4414, 2016.
133. Y. Yamamoto et al., 'Printed multifunctional flexible device with an integrated motion sensor for health care monitoring,' *Sci. Adv.*, vol. 2, no. 11, pp. e1601473–e1601473, 2016.
134. C. Yeom, K. Chen, D. Kiriya, Z. Yu, G. Cho, and A. Javey, 'Large-area compliant tactile sensors using printed carbon nanotube active-matrix backplanes,' *Adv. Mater.*, vol. 27, no. 9, pp. 1561–1566, 2015.
135. W. J. Yu et al., 'Small hysteresis nanocarbon-based integrated circuits on flexible and transparent plastic substrate,' *Nano Lett.*, vol. 11, no. 3, pp. 1344–1350, 2011.
136. A. Schindler, J. Brill, N. Fruehauf, J. P. Novak, and Z. Yaniv, 'Solution-deposited carbon nanotube layers for flexible display applications,' *Phys. E Low-Dimensional Syst. Nanostructures*, vol. 37, no. 1–2, pp. 119–123, 2007.
137. J. Zhang, C. Wang, and C. Zhou, 'Rigid/flexible transparent electronics based on separated carbon nanotube thin-film transistors and their application in display electronics,' *ACS Nano*, vol. 6, no. 8, pp. 7412–7419, 2012.
138. X. Cao et al., 'Fully screen-printed, large-area, and flexible active-matrix electrochromic displays using carbon nanotube thin-film transistors,' *ACS Nano*, p. acsnano.6b05368, vol. 10, no. 11, pp. 9816–9822, 2016.
139. W.-S. Li et al., 'High-quality, highly concentrated semiconducting single-wall carbon nanotubes for use in field effect transistors and biosensors,' *ACS Nano*, vol. 7, no. 8, pp. 6831–6839, 2013.
140. H. R. Byon, and H. C. Choi, 'Network single-walled carbon nanotube-field effect transistors (SWNT-FETs) with increased schottky contact area for highly sensitive biosensor applications,' *J. Am. Chem. Soc.*, vol. 128, no. 7, pp. 2188–2189, 2006.
141. A. Star, E. Tu, J. Niemann, J.-C. P. Gabriel, C. S. Joiner, and C. Valcke, 'Label-free detection of DNA hybridization using carbon nanotube network field-effect transistors,' *Proc. Natl. Acad. Sci.*, vol. 103, no. 4, pp. 921–926, 2006.
142. J. P. Kim, B. Y. Lee, S. Hong, and S. J. Sim, 'Ultrasensitive carbon nanotube-based biosensors using antibody-binding fragments,' *Anal. Biochem.*, vol. 381, no. 2, pp. 193–198, 2008.
143. Y. Jung et al., 'Fully printed flexible and disposable wireless cyclic voltammetry tag,' *Sci. Rep.*, vol. 5, p. 8105, 2015.
144. G. Hu et al., 'Black phosphorus ink formulation for inkjet printing of optoelectronics and photonics,' *Nat. Commun.*, vol. 8, no. 1, 2017.
145. D. Song, A. Mahajan, E. B. Secor, M. C. Hersam, L. F. Francis, and C. D. Frisbie, 'High-resolution transfer printing of graphene lines for fully printed, flexible electronics,' *ACS Nano*, vol. 11, no. 7, pp. 7431–7439, 2017.
146. C. Casiraghi et al., 'Inkjet printed 2D-crystal based strain gauges on paper,' *Carbon NY*, vol. 129, pp. 462–467, 2018.
147. T. Carey et al., 'Fully inkjet-printed two-dimensional material field-effect heterojunctions for wearable and textile electronics,' *Nat. Commun.*, vol. 8, no. 1, p. 1202, 2017.
148. P. N. Nirmalraj, T. Lutz, S. Kumar, G. S. Duesberg, and J. J. Boland, 'Nanoscale mapping of electrical resistivity and connectivity in graphene strips and networks,' *Nano Lett.*, vol. 11, no. 1, pp. 16–22, 2011.
149. I. E. Stewart, M. Jun Kim, and B. J. Wiley, 'Effect of morphology on the electrical resistivity of silver nanostructure films,' *ACS Appl. Mater. Interfaces*, vol. 9, no. 2, pp. 1870–1879, 2017.
150. D. Lee et al., 'Logic circuits composed of flexible carbon nanotube thin-film transistor and ultrathin polymer gate dielectric,' *Sci. Rep.*, vol. 6, pp. 1–7, 2016.

151. G. Li et al., 'Fabrication of air-stable n-type carbon nanotube thin-film transistors on flexible substrates using bilayer dielectrics,' *Nanoscale*, vol. 7, no. 42, pp. 17693–17701, 2015.
152. Q. Xu et al., 'Selective conversion from p-type to n-type of printed bottom-gate carbon nanotube thin-film transistors and application in complementary metal-oxide-semiconductor inverters,' *ACS Appl. Mater. Interfaces*, vol. 9, no. 14, pp. 12750–12758, 2017.
153. J. Zhang, C. Wang, Y. Fu, Y. Che, and C. Zhou, 'Air-stable conversion of separated carbon nanotube thin-film transistors from p-type to n-type using atomic layer deposition of high-κ oxide and its application in CMOS logic circuits,' *ACS Nano*, vol. 5, no. 4, pp. 3284–3292, 2011.
154. M. Ha et al., 'Printed, sub-3V digital circuits on inks,' *ACS Nano*, vol. 4, no. 8, pp. 1–11, 2010.
155. J. Sheng, H.-J. Lee, S. Oh, and J.-S. Park, 'Flexible and high-performance amorphous indium zinc oxide thin-film transistor using low-temperature atomic layer deposition,' *ACS Appl. Mater. Interfaces*, vol. 8, no. 49, pp. 33821–33828, 2016.
156. K. Hong, S. H. Kim, K. H. Lee, and C. D. Frisbie, 'Printed, sub-2V ZnO electrolyte gated transistors and inverters on plastic,' *Adv. Mater.*, vol. 25, no. 25, pp. 3413–3418, 2013.
157. L. Petti et al., 'Low-temperature spray-deposited indium oxide for flexible thin-film transistors and integrated circuits,' *Appl. Phys. Lett.*, vol. 106, no. 9, p. 092105, 2015.
158. Y. S. Rim, H. Chen, Y. Liu, S. H. Bae, H. J. Kim, and Y. Yang, 'Direct light pattern integration of low-temperature solution-processed all-oxide flexible electronics,' *ACS Nano*, vol. 8, no. 9, pp. 9680–9686, 2014.
159. J. A. Caraveo-Frescas, P. K. Nayak, H. A. Al-Jawhari, D. B. Granato, U. Schwingenschlögl, and H. N. Alshareef, 'Record mobility in transparent p-type tin monoxide films and devices by phase engineering,' *ACS Nano*, vol. 7, no. 6, pp. 5160–5167, 2013.
160. J. A. Caraveo-Frescas, M. A. Khan, and H. N. Alshareef, 'Polymer ferroelectric field-effect memory device with SnO channel layer exhibits record hole mobility,' *Sci. Rep.*, vol. 4, pp. 1–7, 2014.
161. H. Faber et al., 'Heterojunction oxide thin-film transistors with unprecedented electron mobility grown from solution,' *Sci. Adv.*, vol. 3, no. 3, p. e1602640, 2017.
162. M. A. Marrs et al., 'Control of threshold voltage and saturation mobility using dual-active-layer device based on amorphous mixed metal—oxide—semiconductor on flexible plastic substrates,' *IEEE Trans. Electron Devices*, vol. 58, no. 10, pp. 3428–3434, 2011.
163. B. K. Sharma et al., 'Load-controlled roll transfer of oxide transistors for stretchable electronics,' *Adv. Funct. Mater.*, vol. 23, no. 16, pp. 2024–2032, 2013.
164. A. T. Zocco, H. You, J. A. Hagen, and A. J. Steckl, 'Pentacene organic thin-film transistors on flexible paper and glass substrates,' *Nanotechnology*, vol. 25, no. 9, 2014.

5

Mechanics of Flexible and Stretchable Electronics

Nouha Alcheikh

5.1 Stress and Strain

To understand the mechanics of devices under deformation, it is first essential to describe specific basic parameters. Stress and strain are the fundamental parameters for the mechanical behavior of material. These parameters have to be independent of the geometry and the dimension of the shape so that they can determine the stiffness and strength of a material. Stress is a measure of the internal forces within a material. The source of the stresses in thin and multilayers film can be internal (residual stress) and external. Then, the total stress within a structure is equal to the residual stress and the applied stress.

Consider an object is loaded with a certain force. How strong the material is stressed depends on the area loaded. If the area is decreased (increased), the stress increases (decreases). Therefore, it is defined as the load versus the area loaded, thus it carries the units of Pa. If the shape is exposed to a direct force $F_{\uparrow}$ (pull or push) perpendicular to the area A or parallel to its axis as shown in Figure 5.1a and b, then it is supposed to be in tension or compression (normal stress). If the force is parallel to the area $F_{\leftrightarrow}$, see Figure 5.1c, the stress is a shear stress. The mathematical equations representing the tension or compression stress (σ), and shear stresses (τ) are provided in Eqs. (5.1) and (5.2).

$$\sigma = \frac{F_{\uparrow}}{A}. \tag{5.1}$$

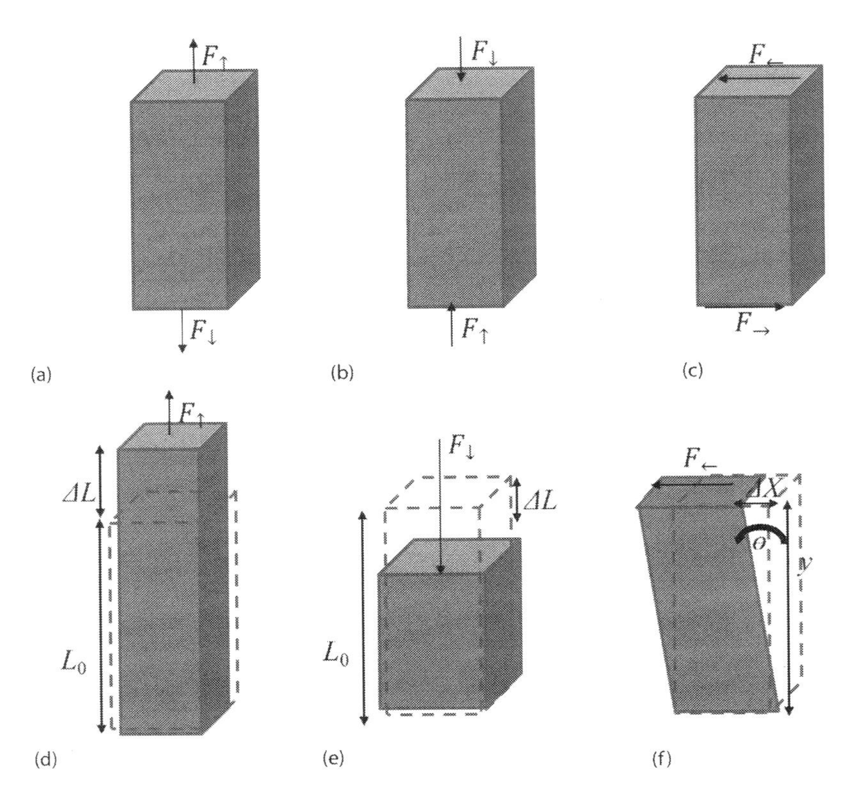

FIGURE 5.1 Types of stress: (a) Normal stress (Tension), (b) Normal stress (Compression), (c) Shear stress; and types of strain: (d) Tensile deformation, (e) Compressive deformation, (f) Shear deformation.

$$\tau = \frac{F\leftarrow}{A}. \tag{5.2}$$

If the shape is exposed to a normal force or a shear force, and hence a stress, the shape will change in length or in angles. Thus, tensile strain is the spatial variation of the displacement with respect to reference dimensions, compressive strain describes a compacted shape, and shear strain describes the change in an initially right angle. To describe changes in length, the tensile or compressive strain (ε) is defined as the difference amount ΔL between the final length and the origin length L_0 (Figure 5.1d and e). The mathematical equation representing the tensile or compressive strain, is provided in Eq. (5.3).

$$\varepsilon = \frac{\Delta L}{L_0}. \tag{5.3}$$

To describe changes in the angles, and for small displacement, Figure 5.1f, the shear strain (γ) is defined as:

$$\gamma = \tan\theta = \frac{\Delta X}{y}. \tag{5.4}$$

5.1.1 Residual Stress

The residual stresses in the bilayer, for example, film/polymer are normally determined by measuring the deflection of the film after the deposition. Jaccodine et al. [1] calculated the residual stresses of the film from the curvature and from the mechanical parameters of the film and the substrate by using a simple equation. The residual stresses have three main origins: intrinsic, thermal, and hygroscopic [2–5]. The intrinsic stresses develop during the deposition process, such as sputtering spin coating, and vapor deposition, which can be low or high or ambient temperature. The intrinsic stresses can be tensile or compressive in inorganic films and generally tensile in organic films, see Table 5.1 [6]. The thermal stresses are diverse from intrinsic stresses in that they are the stresses existing at the deposition temperature. They develop from the mismatched thermal expansion coefficients between two materials. The thermal expansion coefficient $\left(\alpha(K^{-1})\right)$ defined as the changes of the geometry with temperature variations. For example, if a film is deposited onto hot substrate, the intrinsic stress can arise when the substrate is cooled and the temperature of the film and substrate is returned to ambient. Hygroscopic stresses also develop from the mismatched humidity expansion coefficients (CHEs) of the film and substrate. Hygroscopic stresses can be generally tensile in inorganic films and tensile or compressive in organic films, see Table 5.1 [6].

It is important to notice that the distribution of the residual stress depends on the homogeneity of the mismatch strains and upon the presence of stress relaxation mechanisms. Sometimes the distribution throughout the structures can be inhomogeneous (e.g., the formation of oxide layers may be accompanied by the development of inhomogeneous residual stresses). This last can lead to some deformations in the structure, such as cracks or delamination [7]. This failure phenomenon (crack or delamination) is detailed in the following sections.

5.1.2 Stress—Strain Relation

A material response under stress or strain that follows a straight line is said to be linear elastic material. The elasticity implies that the material returns to its original shape when the load is removed without hysteresis. This means full reversibility. Beyond a certain value of strain (yield point), the material starts to deform permanently, this is called plasticity. The plasticity implies that the material will not return to its original shape when the load is removed. Upon more elongation, the stress increases until it reaches a maximum value. At this point, the material enters the necking stage. In the necking stage, the stress increases and the material's cross-section reduces and starts to stretch significantly until failure or fracture or rupture.

Within the elastic limits of materials and for a unidirectional load, the stress can be proportional to strain by the modulus of elasticity or Young's modulus (E) see Eq. (5.5). Usually, for the tensile and compressive deformation, Young's modulus is assumed to be the same. For most materials used in electronics, it has a high value; it varies between 70 GPa for aluminum to 234 GPa for nickel and typical ceramics in the range 69 GPa (window glass) to 1000 GPa (diamond). For the polymers (viscoelastic),

TABLE 5.1 Summary of Intrinsic, Thermal, and Hygroscopic Contributions to Residual Stresses in Organic and Inorganic Films

Stress Component	Origin	Inorganic Films	Organic Films
Intrinsic	Film formation (vapor deposition, solution processing)	− or +	+
Thermal	Difference in coefficient of thermal expansion (CTE) between film and substrate and temperature changes	−	− or +
Hygroscopic	Difference in CHE between film and substrate and humidity changes	+	− or +

Source: Leterrier, Y., Mechanics of curvature and strain in flexible organic electronic devices, in S. Logothetidis (Ed.), *Handbook of Flexible Organic Electronics: Materials, Manufacturing and Applications*, Woodhead Publishing, Oxford, UK, 2014.

it varies from 0.2 GPa (polyethylene) to 5 GPa (polyester) [8]. However, the elastic modulus is one of the basic material properties that describes the mechanical behavior of a device (its resistance to being deformed when a force is applied to it). It reflects the stiffness of the material, the higher the modulus is, the lower is the elastic deformation. Nevertheless, it is important to understand the stiffness of a sample. The stiffness of a material $\left(K\left(\text{N/m}\right)\right)$ indicates its capacity to stretch or flex; a high stiffness means that the material being inflexible, while a very low stiffness material (high compliant material) is easily being flexible or stretched.

$$E = \frac{\sigma}{\varepsilon}. \tag{5.5}$$

If a body is strained by a strain (ε), the body will also exhibit a strain in perpendicular directions. Normally, the tensile strain causes a reduction in dimensions in the transverse direction. The ratio of transverse strain to axial strain is described by *Poisson's ratio* ν. For most materials, Poisson's ratio is lower than 0.35; if its value is 0.5 so the sample volume doesn't change, and in this case the material is incompressible. Another basic material property which describes the mechanical behavior of a device is toughness. Toughness is defined as the ability of a material to resist cracks, and hence to resist failure. It is described as the quantity of energy absorbed at failure per unit volume (J/m^3). Table 5.2 summarizes the definition of the mechanics of materials [9].

TABLE 5.2 Mechanics of Materials: Background Concepts and Definitions

Stress	Force Per Unit Area
Strain	Change in length per unit length
Young's modulus	Constant of proportionality between stress and strain. Units are the same as for stress (i.e., force per unit area), and the most commonly used are pounds/in2 (psi), Pa (pascal), kilopascal (kPa), and megapascal (MPa)
Elasticity	Property of a material to regain its original dimensions (size and shape) at the removal of load or force (e.g., steel is more elastic than rubber)
Plasticity	Property of a material to deform permanently when subjected to external load beyond the elastic limit
Yield point (elastic limit)	Stress at which a material begins to deform permanently (i.e., plastically). Before the yield point, the material will deform elastically (non-permanently) and then return to its original shape when the applied stress is removed. Once the yield point is passed, some fraction of the deformation will be permanent and non-reversible (plastic, permanent deformation)
Ultimate strength	Maximum stress that a material can withstand while being stretched or pulled before necking, the stage during which the specimen's cross-section starts to stretch significantly and finally to fail (fracture)
Stiffness (rigidity)	Resistance to undergoing (elastic) deformation in response to the application of a force, the property of being inflexible and hard to distort. A stiff material has a strong supporting structure and does not deform much when a stress is applied. The stiffness of a material is represented by the ratio between stress and strain (called 'Young's modulus of elasticity,' 'elastic modulus,' or 'modulus of elasticity'). Stiff materials, by definition, have a high modulus of elasticity (i.e., a considerable stress is need for a minor deformation)
Compliance(flexibility)	Reciprocal of stiffness, representing the tolerance of a material to undergo elastic deformation, the property of being flexible and easy to distort. Compliant (flexible) materials, by definition, have a low elastic modulus, and only minor stress is required for a considerable strain. Highly compliant materials are easily stretched or distended

Source: Vegas, M.R. and Del Yerro, J.L.M., *Aesthetic Plast. Surg.*, 37, 922–930, 2013.

5.2 Radius of Curvature of the Film on Substrate

As we explained in Section 5.1.1, in multilayered structures, one of the serious concerns linked to reliability is residual stress caused by the differences in the thermal expansion and humidity coefficients between the multilayer or the deposition film and the substrate. Since the film and the substrate have different mechanical properties, the mechanics of the film-on-substrate structure depend on the Young's modulus and thicknesses of the film (E_f, h_f) and the substrate (E_s, h_s). When $E_f h_f << E_s h_s$, the stress in the film is higher than the stress in the substrate (small), then the structure curves slightly in a spherical cap described by the Stoney equation [10] with the radius of curvature given by [11]:

$$R = \frac{h_s^2}{6(\alpha_f - \alpha_s)\Delta T h_f \dfrac{E_f^*}{E_s^*}} \left[\frac{\left(1 - \dfrac{E_f^* h_f^2}{E_s^* h_s^2}\right) + 4\dfrac{E_f^* h_f}{E_s^* h_s}\left(1 + \dfrac{h_f}{h_s}\right)^2}{1 + \dfrac{h_f}{h_s}} \right]. \tag{5.6}$$

In Eq. (5.6), $E_f^* = E_f/(1-\nu_f)$, $E_s^* = E_s/(1-\nu_s)$, and ΔT, represent, respectively, biaxial strain modulus for the film and substrate and the change in temperature. ν_f and ν_s indicate the Poisson's ratio of the film and the substrate. The radius of curvature R is negative (positive) when the film is under compression (tensile) and the curve being on the convex side (concave). When $E_f h_f \approx E_s h_s$, the strength of the film and the substrate is equal, in this case, instead of forming a spherical cap, it rolls into a cylinder. The mathematical representation of the radius of curvature R is provided by Gleskova et al. [11].

5.3 Films Failure Modes: Cracking and Delamination

The mechanical failure of the multilayer under large deformation is one of the limitations on the flexibility of electronics devices [12]. When a tensile strain is applied to a film/compliant substrate, the main failure mechanisms are parallel channeling cracks that initiate in brittle films on the substrate, and more loading leads to delamination at the film/substrate interface or cracks can channel through both the film and substrate [13–15]. The compressive strain may lead to edge delamination and buckle delamination of thin film [15–17]. These main failure mechanisms are summarized in Figure 5.2.

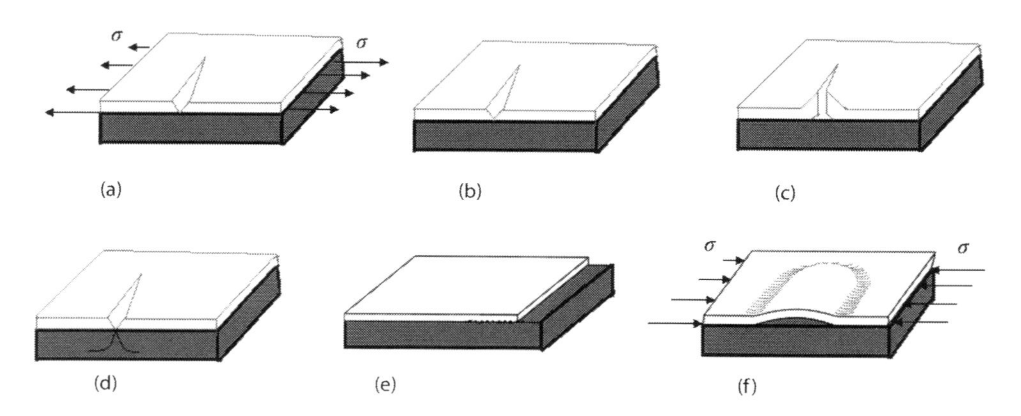

FIGURE 5.2 A schematic showing the main failure mechanisms for film/substrate devices under tensile and compressive. (a) Tensile stress (fully cracked film), (b) Tensile stress (partially cracked film), (c) Tensile stress (cracked and delamination film), (d) Tensile stress (cracked film and failure of the substrate), (e) Compressive stress (edge film delamination) and (f) Compressive stress (buckle film delamination).

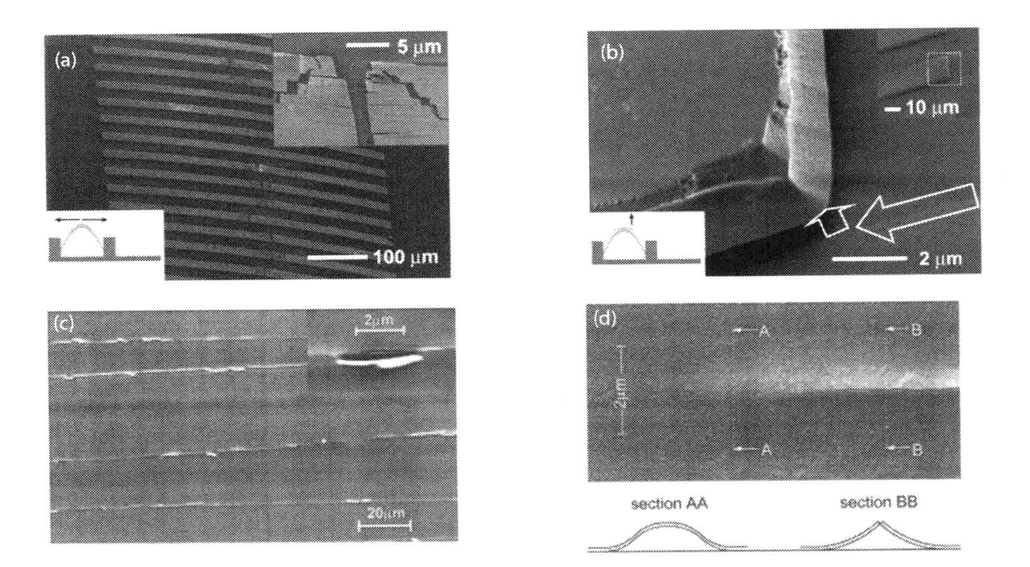

FIGURE 5.3 SEM images of bending in thin Si ribbons/epoxy/polyethylene terephthalate (PET) corresponding to (a) cracking failure and (b) delamination. (From Park, S.I. et al., *Adv. Funct. Mater.*, 18, 2673–2684, 2008.) (c) SEM images showing compression cracks in indium-tin oxide (ITO)/PET and (d) tunneling delamination-buckle-crack. (From Cotterell, B. and Chen, Z., *Int. J. Fract.*, 104, 169–179, 2000.)

Examples of failure modes under tensile loading are shown in Figure 5.3. Cracks usually appear sequentially: as the stress increases, a crack starts from the top surface and propagates perpendicular to the interface. If a particular constituent layer is weaker in cohesive strength than any of the interfaces, the crack will propagate transverse to the interfaces within that layer. If the interface is satisfactorily weak (poor adhesion), the crack deflects into the interface and propagates along it. In addition to channel cracks, delamination is also observed when the stress concentration near the channel root reaches the critical stress, and it can be enough to cause interfacial delamination [6,18].

A number of existing studies have largely originated to analyze the cracking failure under tensile of brittle film attached to substrates [20–23]. Thouless [24,25] investigated the mechanics of film/substrate systems and stress relaxation. Hutchinson and Suo [26] presented the fracture mechanics of thin film/substrate systems and multilayers. However, the mechanics of crack failure will be described for systems in which the cracks are limited to a surface layer. The critical strain (ε_{crit}) is the key parameter to analyze the cracking failure. Below critical strain, no further cracks can generate. For a semi-infinite substrate, the results of Beuth [27] presented the critical tensile strain as:

$$\varepsilon_{crit}^2 = \frac{2\gamma_f}{\pi h_f \bar{E}_f g(\alpha,\beta)}, \tag{5.7}$$

where $\bar{E} = E_f/(1-\nu_f^2)$ is the material plane strain tensile modulus and 'γ_f' is the toughness of the interface, can be calculated as the effective surface energy at critical loading. The 'Dundurs parameters' a and β, are the non-dimensional parameters which define the modulus mismatch of film/substrate [28]. As we show in Eq. (5.7), the critical strain decreases with increasing the thickness and the elastic modulus of the film (h_f, E_f), and the toughness of the film (γ_f). Dundurs work shows that for any problem of a composite body made of two isotropic elastic materials, the plane strain problems a and β, are given by:

$$\alpha = \frac{\bar{E}_f - \bar{E}_s}{\bar{E}_f + \bar{E}_s}, \qquad \beta = \frac{\mu_f(1-2\nu_s) - \mu_s(1-2\nu_f)}{2\mu_f(1-\nu_s) + 2\mu_s(1-2\nu_f)},$$

where $\bar{E} = E_s/(1-\nu_s^2)$ is the material plane strain tensile modulus of the substrate and 'μ_f and μ_s' are the film and the substrate shear modulus. For films having the same properties as substrate, a and β equal zeros. For dissimilar properties, a can vary from –1 to +1 and β usually lies between $\beta = 0$ and $\beta = \alpha/4$ [28,29].

Failure by delamination is often considered as a failure event which can be described by an interfacial low toughness parameter (poor adhesion) which, together with the mismatching mechanical properties of the film and compliant substrate, can lead to increasing the stress at the interfacial until debonding can occur [30]. The delamination and interface crack growth were modeled numerically with the use of different models [31]. Mei et al. [32] studied the effect of interfacial delamination on channel cracking of a brittle thin film on an elastic substrate. A finite element model is used to calculate the energy release rates for both the interfacial delamination and the steady-state channel cracking. A two-dimensional analytical model based on an energy balance is capable to relate measured buckle geometries [33]. For the delamination failure, the critical strain can be written as [32]:

$$\varepsilon_{crit}^2 = \frac{\gamma_d}{h_f \bar{E}_f g_d\left(\frac{d}{h_f}, \alpha, \beta\right)}. \tag{5.8}$$

where 'γ_d' is the toughness of the interface from delamination and d is the channel crack width. g_d is a dimensionless function that can be determined from a two-dimensional plane strain problem. When $d/h_f \to \infty$, the $g_d \to 0.5$, and thus the interfacial crack reaches a steady state with critical strain:

$$\varepsilon_{crit}^2 = \frac{2\gamma_d}{h_f \bar{E}_f}. \tag{5.9}$$

When $d/h_f \to 0$, the $g_d \sim \left(\frac{d}{h_f}\right)^{1-2\lambda}$, [34].

where λ depends on the 'Dundurs parameters' α and β [35]:

$$\cos\lambda\pi = \frac{z(\alpha-\beta)}{(\alpha-\beta)}(1-\lambda)^2 - \frac{\alpha-\beta^2}{1-\beta^2}, \text{ and } \varepsilon_{crit}^2 = \frac{\gamma_d}{h_f \bar{E}_f \left(\frac{d}{h_f}\right)^{1-2\lambda}}. \tag{5.10}$$

These studies principally focused on the fracture of bilayer structures under tension which cannot apply with multilayer structures. Jia et al. [36] studied the critical strains for delamination and channel cracking in a multilayer (oxide-organics-oxide) on a polymer substrate. The internal toughness γ_d can be written as:

$$\gamma_d = f\left(\frac{d}{h_{oxide}}, \frac{E_{organic}}{\bar{E}_{oxide}}, \frac{h_{organic}}{h_{oxide}}\right)\bar{E}_{oxide}\varepsilon^2 h_{oxide}, \tag{5.11}$$

where h_{oxide} and $\bar{E}_{oxide} = E_{oxide}/(1-\nu_{oxide}^2)$ are the thickness and the plane strain modulus of the oxide (E_{oxide} and ν_{oxide} are the Young's modulus and Poisson's ratio of the oxide), and $h_{organic}$ and $E_{organic}$ are the thickness and Young's modulus of the organic film. 'f' is a function that represents the normalized driving force of interfacial delamination, and ε is the applied tensile strain.

If the bilayer film/substrate is under compression loading, the main failure modes are buckle delamination and edge delamination as shown in Figure 5.2. Compressive residual stresses may induce in the film due to the mismatching in the thermal expansion coefficients or intrinsic stresses of film/substrate. Some of the morphologies, such as the 'telephone cord' have been detected in film/substrate devices [26]. An extensive review of buckled delamination has been reported by Giola and Ortiz [37]. For a clamped–clamped plate, the critical strain for buckling is given by:

$$\varepsilon_{crit} = \frac{\pi^2}{12}\left(\frac{h}{b}\right)^2, \tag{5.12}$$

where h and b are the thickness and the width of the plate. The interfacial toughness is the energy released from a steady-state tunneling delamination and buckling. Without cracking, the toughness (γ_b) is given by [16]:

$$\gamma_d = f(\Psi,\alpha,\beta)\frac{\bar{E}_f\varepsilon^2 h_f}{2}\left(1-\frac{\varepsilon_{crit}}{\varepsilon}\right)^2. \tag{5.13}$$

'f' is a function of the mode-mixity angle ψ and 'Dundurs parameters' α and β and ε are the applied compressive strain. As shown in Eq. (5.13), the energy release rates are dependent on the ratio, $\varepsilon_{crit}/\varepsilon$.

With cracking, the toughness (γ_t) is a combination of the energy release rates for delamination (γ_d) and cracking (γ_c), and is given by [16]:

$$\gamma_t = \gamma_d + \gamma_c\left(\frac{h}{2b}\right). \tag{5.14}$$

It should also be noted that numerous devices seem to fail first by buckle delamination initiated away from the edges of the film. To calculate the toughness of edge delamination, the roughness is assumed to be random on the delaminated interface such that once sliding across the interface has occurred on the order of one roughness half-wavelength 'l', the two surfaces become uncorrelated and are thereafter propped open a distance 'R', the amplitude of the roughness [17]. Hence, the interfacial toughness is given by:

$$\gamma_d = f\left(\sigma^*, R^*, \mu_f\right)\gamma_0\left[1+\left(\frac{R}{l}\right)^2\right], \tag{5.15}$$

where σ^*, R^*, and γ_0 are given by:

$$\sigma^* = \frac{\varepsilon h_f\left(1-\nu_f^2\right)}{l},\quad R^* = \frac{R}{l},\ \text{and}\ \gamma_0 = \frac{\varepsilon^2 E_f h_f\left(1-\nu_f^2\right)}{2}.$$

5.4 Mechanics of Flexible—Mode (Bending Deformation)

In a pure bending and for the elastic case, the top surface of the flexure devices shows lengthening while the bottom surface shortens. The top surface has a positive strain (tension), while the bottom surface has a negative strain (compression), and on the surface inside the sheet, the strain is zero (the neutral axis) as shown in Figure 5.4. As we explained in the previous section, the critical strain is the key design parameter for flexible electronics devices because below critical strain, no further cracks can generate. Hence, the total strain within a structure is equal to the residual stress and the applied strain (bending). As we discussed in the previous Section 5.3, the main causes for the residual stresses are the fabrication process. It should be noticed that the side tensile strain may lead to crack failure and delamination, while the opposite side under compressive strain tends to develop buckling delamination failure. The bending

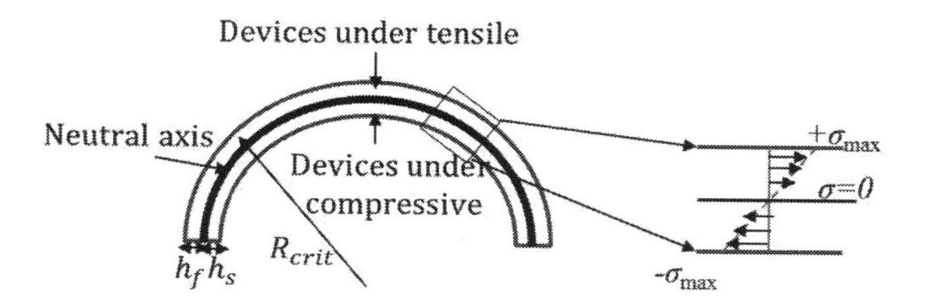

FIGURE 5.4 Schematic of a flexed structure based on a thin film on substrate.

strain outcomes from the applied strain and it is defined on the top surface (ε_{crit}) in the bending direction. In a homogeneous sheet, it equals the distance from the neutral axis divided by critical radius of curvature (R_{crit}) [38].

$$\varepsilon_{crit} = \frac{\left(h_f + h_s\right)}{2R_{crit}}, \tag{5.16}$$

and for $h_f << h_s, \varepsilon_{crit} \approx \frac{h_s}{2R_{crit}}$.

From the Eq. (5.16), it is simple to notice that the critical strain decreases linearly with the thickness of the film and the substrate. At a specific value of radius of the curvature, a thinner film or a thinner substrate shows a lower value of critical strain. For example, at 1 cm of radius of curvature, nanoscale ribbons with thicknesses of 100 nm show 0.0005% of maximum strain, and it slowly increases to 0.1% when attached to a 20 mm of plastic as the substrate while its critical strain is equal to 1% [19,39].

The electronics devices are normally placed on the top surface (under tension) or on the bottom surface (under compression) of a flexed sheet. To reduce the maximum strain and to maximize the flexibility of the devices, the maximum strain should be placed in the neutral axis where the strain does not change under deformation. By adding a fragile material with proper Young's modulus (E_l) and thickness (h_l), the neutral plane is sandwiched between the substrate and the adding layer. When the stiffness of the electronics devices is negligible, the electronics will lie in the neutral surface if:

$$E_s h_s^2 = E_l h_l^2. \tag{5.17}$$

It is clear from Eq. (5.17), for equal mechanics properties of the substrate and the adding layer, ($E_S = E_l$), then equal thickness will ensure that the electronics lie in the neutral axis [38].

5.5 Mechanics of Stretching Mode Deformation

In a flexible mode, the design approaches for high flexible electronics structures rely mainly on thin geometries (micro/nanostructures) and neutral mechanical plane axis where the most critical strain of the material is lower than 2%. The devices with critical strain of the material lower than 2% cannot be stretched. Thus, these concepts cannot be used to protect the devices under stretchable mode. Stretchable electronics can be achieved in two general routes. One relies on developing new material that intrinsically stretched, the other on new materials in design layout. We should notice that the stretchability is the possibility of geometry to be stretched reversibly, with a linear elastic response to applied strain. Under certain applied force, the maximum strain of stretchability ($\varepsilon_{stretchability}$) is the maximum of deformation before cracking, see Eq. (5.3). A complementary approach to extreme deformability is based on organic conductors and semiconductors [40,41]. But, their electrical conductivity is lower than the inorganic semiconductor such as silicon and gallium nitride [42,43]. The main challenge here is to design inorganic materials

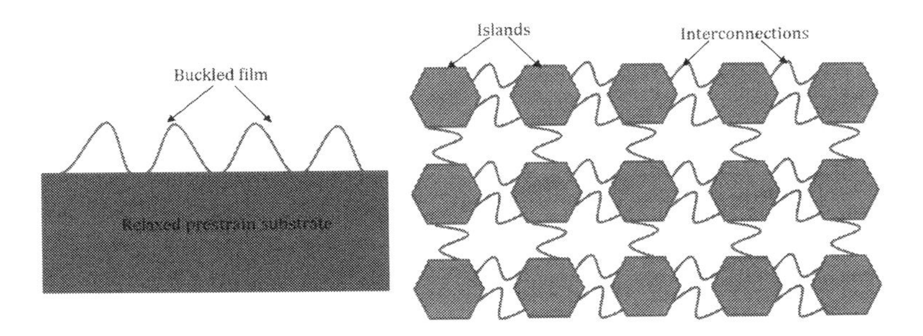

FIGURE 5.5 Schematics of wavy and island-interconnects designs.

for stretchability, when all known inorganic semiconductor materials fracture at strains of the order of 1% (0.3% for the copper). Numerous mechanics approaches have been advanced to achieve devices stretchable by coupling micron size inorganic electronics materials with elastomer- or polymer-based stretchable substrates. They can be classified into: wavy, wrinkled design [44] and island-interconnects design [45–49], Figure 5.5.

5.5.1 Mechanics of Wavy Ribbon Design

In stretchable inorganic electronics devices, the wavy design was the first design introduced by Khang et al. [44]. The strategy of this design is based on coupling of thin hard film inorganic electronics, for example, gold [50] and silicon nano-ribbon [44] on a pre-stretched substrate such as elastomer, and then releasing this pre-strain leads to the formation of wavy, wrinkled configurations. A number of theoretical studies are available to predict the maximum strain of this design and its stretchability limitation. In the regime of small pre-strain (small deformation), a mechanical study has been developed by Song [40] when the thin ribbon and the substrate are modeled as an elastic nonlinear von Karman and as a semi-infinite solid, respectively. The ribbon buckled only if the pre-strain (ε_{pre}) is higher than its critical buckling strain (ε_{crit}). The peak strain (ε_{peak}), which is the sum of the membrane and bending strains, is given by:

$$\varepsilon_{peak} \approx 2\sqrt{\varepsilon_{pre}\varepsilon_{crit}}, \tag{5.18}$$

where the critical strain is small, and it is given by:

$$\varepsilon_{crit} = \frac{1}{4}\left(\frac{3\bar{E}_s}{\bar{E}_f}\right)^{2/3}. \tag{5.19}$$

For the buckled structure under applied strain ($\varepsilon_{applied}$), the peak strain in the ribbon is:

$$\varepsilon_{peak} \approx 2\sqrt{\left(\varepsilon_{pre} - \varepsilon_{applied}\right)\varepsilon_{pre}\varepsilon_{crit}}. \tag{5.20}$$

In the regime of high pre-strain (high deformation), the stretchability is given by the sum of the membrane strain, bending strain, and fracture strain ($\varepsilon_{fracture}$). The peak strain in the ribbon is given by:

$$\varepsilon_{peak} \approx 2\sqrt{\frac{\left(\varepsilon_{pre} - \varepsilon_{applied}\right)\varepsilon_{crit}}{1+\varepsilon_{pre}}}. \tag{5.21}$$

5.5.2 Mechanics of Island-Interconnects Design

The island-interconnects design is suggested to realize high stretchability in all directions. The functional devices are normally placed on the undeformed rigid island (also called active devices), while the electrical current is supported by interconnects (Figure 5.5). Hence, the stretchability of the whole device is established by the elastically deformed interconnects. Structure interconnects designs have developed from straight to arch-shaped, from those strongly bonded or weakly bonded to a pre-stretched substrate [51–53] to fractal/self-similar designs [54], serpentines designs [46], and spiral designs [48,49], Figure 5.6. It is important to notice that, for design approaches that embed to a substrate, the total system stretchability of the system (ε_{system}) is much smaller than the interconnect stretchability ($\varepsilon_{interconnect}$) [55]. In this chapter, the focus of mechanical study of islands-interconnects is on straight and serpentine interconnects.

Bridge (straight-interconnects) design: Song et al. [45] developed an analytic model from the total energy method to understand the mechanical behavior of the islands-straight design. By minimization of total energy in the bridge, the peak strains in the bridge and in the island can be approximated by [40]:

$$\varepsilon_{peak}^{bridge} \approx 2\pi \frac{h_{bridge}}{L_{bridge}^{0}} \sqrt{\frac{\varepsilon_{pre}}{1+\varepsilon_{pre}}}$$

and

$$\varepsilon_{peak}^{island} \approx 2\pi \frac{\left(1-\nu_{island}^{2}\right) E_{bridge}\, h_{bridge}^{3}}{E_{island}\, h_{island}^{2} L_{bridge}^{0}} \sqrt{\frac{\varepsilon_{pre}}{1+\varepsilon_{pre}}}, \tag{5.22}$$

where h_{bridge}, L_{bridge}^{0}, and E_{bridge} are the thickness, length, and Young's modulus of the bridge and h_{island}, E_{island}, and ν_{bridge} are the thickness, Young's modulus, and Poisson's ratio of the substrate. More advanced analytical models have been done by to increase the accuracy in large deformation [56].

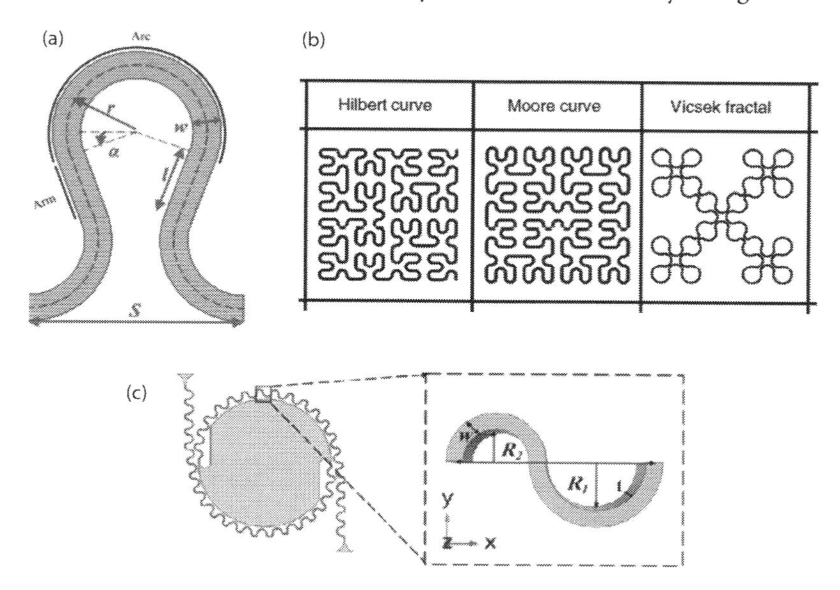

FIGURE 5.6 (a) Definition of the geometric parameters of a serpentine. (From Li, R. et al., *Soft Matter*, 9, 8476–8482, 2013.) (b) Examples of fractal/self-similar designs. (From Fan, J.A. et al., *Nat. Commun.*, 5, 3266, 2014.) (c) Schematic of spiral structure with serpentine arms. (From Rehman, M.U. and Rojas, J.P., *Extreme Mech. Lett.*, 15, 44–50, 2017.)

Bridge (serpentine-interconnects) design: As we noted previously, high stretchable electronics structures rely mainly on the peak strain which depends essentially on the deformation and the length of interconnects. Increases in the lengths can lead to the decreases of the peak strains. Hence, serpentine interconnects are much longer than the straight interconnects, thereby, they can offer higher stretchability. The developing of the analytical models of this design depends on interconnects thickness.

For a thin serpentine thickness (sub-micrometer) and under strain load, the interconnects display a wrinkling deformation (out-of-plane buckling without significant twisting) [53]. It was, for example, found that a silicon consists of sub-micrometer single-crystal elements structured into serpentine geometries and supported by an elastomeric substrate, shows high reversible stretchability based on wrinkling buckling deformation without any twisting (with wavelength smaller than the length of the shortest straight segment [44]). They demonstrate that the peak silicon strain can be written as:

$$\varepsilon_{peak}^{bridge} = \frac{2\varepsilon_{crit}}{(1+\varepsilon_{applied})^{3/2}}\sqrt{\frac{\varepsilon_{pre}-\varepsilon_{applied}}{\varepsilon_{crit}}-1}. \tag{5.23}$$

With a moderate thickness (up to ≈45 μm), the interconnects demonstrate an out-of-plane buckling and twisting deformation for long wavelengths [57]. For weakly-bonded serpentine ribbons, the model shows that an ultimate stretchability is limited by the arc radius α, Figure 5.6a.

$$\varepsilon_{peak}^{bridge} = \frac{\pi+2\alpha-2\cos(\alpha)}{2\cos(\alpha)}. \tag{5.24}$$

For thicker interconnects (>45 μm), the interconnects deformation is purely in-plane bending without any buckling [58]. It was found that the peak strain can be given as:

$$\varepsilon_{peak}^{bridge} = \varepsilon_{crit}\frac{1+12\pi\theta+48\theta^2+6\pi\theta^3}{2\alpha\left[\frac{w}{l^2}+\frac{3w(1+2\alpha)}{l}\right]}, \tag{5.25}$$

where L and w are the height and the width of the serpentine, respectively, and $\theta = r/l$, Figure 5.6a.

5.6 Computational Methods

Computational methods, such as finite-element can be also used to describe the mechanical behavior of the flexible and stretchable structures. Some structures have complex geometries when studying their deformations based on analytical models that are very complicated and limited. Hence, finite element methods can often be employed [61]. Several software programs such as COMSOL, ANSYS, and ABAQUS [48,49,60] are utilized to simulate the performance of the design as its stretchability, the number of cycles, failure through delamination, and fatigue [46,55,59,62].

5.7 Mechanical Characterization Methods

Mechanical testing methods are designed to test the flexibility and the stretching capabilities of devices across the electronics industry to determine and to understand the mechanical properties, response, and the reliability of these devices. Many methods have been developed to test and to measure the performance and the mechanical properties of thin films [63–66], such as (Nano)-indentation, buckling method, (Nano)-scratching, X-ray diffraction, and bending and tensile methods, most of them are shown in Figure 5.7. In the chapter, only some of these techniques (indentation, buckling methods, tensile method, and bending and tensile methods) will be discussed.

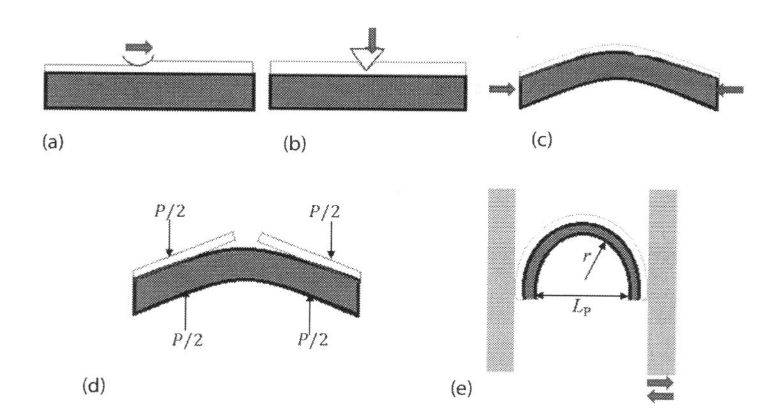

FIGURE 5.7 Five test methods for evaluating the mechanical properties and mechanical failure in films/substrate. (a) Nano-scratching, (b) Nano-identation, (c) Buckling mode, (d) Four-point bending test and (e) The collapsing radius bending test. (a: Bull, S.J. and Berasetegui, E.G., *Tribol. Int.*, 39, 99–114, 2006; c: Chen, J. and Bull, S.J., *J. Phys. D.*, 44, 034001, 2010; e: Grego, S. et al., *J. Soc. Inf. Display*, 13, 575–581, 2005.)

5.7.1 (Nano)-Indentation

(Nano)-indentation is one of the most important techniques that can be used to measure the mechanical properties of a thin film material such as Young's modulus, hardness, and fracture toughness. This technique relies on measuring the imposed force of a sharp diamond tip into the material with the indentation depth (force/displacement curve). Analysis of the load and unload/displacement curve obtained from (Nano)-indentation can provide enough information about the mechanical properties of the material such as Young's modulus (E) and hardness (H) [67]. It should be noticed that, to deduce the mechanical properties of the material, the film should have a minimum thickness 10 times the indenter depth, but much thinner than the substrate. Hainsworth et al. [68] calculated the mechanical properties of the material (E, H) from loading curves fitting for a wide range of materials by using the formulation:

$$P \approx E\left(0.194\sqrt{\frac{E}{H}} + 0.930\sqrt{\frac{H}{H}}\right)^{-2}\delta^2, \tag{5.26}$$

where P and δ are the load and the displacement of the tip in the material.

The main disadvantage of the (Nano)-indentation is, during the loading or unloading test, failure may appear on the thin film [15,66].

5.7.2 Buckling and Tensile Methods

The bucking method is another valuable technique for determining the Young's modulus of thin films, using measured buckle geometry. The key of this method is to introduce a compressive pre-strain in a film/substrate, and then releasing the pre-strain causes a rippled pattern. By measuring the wavelength buckling (λ), the Young's modulus can be deduced [69]. The wavelength buckling can be estimated as [70]:

$$\lambda \approx 2\pi h_f\left[\frac{\left(1-\nu_s^2\right)E_f}{\left(1-\nu_f^2\right)E_s}\right]^{1/3}. \tag{5.27}$$

Tensile tests have also been used to determine the elastic modulus of the film/substrate. The method based on using the laminate plate theory [71]:

$$E_f \approx \frac{\left(1-\nu_f^2\right)}{h_f}\left[E_h + \frac{E_s h_s}{\left(1-\nu_s^2\right)}\right]^{1/3}. \tag{5.28}$$

5.7.3 Bending Methods

The most direct and fastest for calculating the critical strain films for flexible electronics is bending the device to a given radius r. The easier configuration is a rollable device bending around a cylinder of a given radius r. For a small radius, the critical strain can be calculated from Eq. (5.16). Grego et al. [72] developed another method (collapsing radius) for the bending radius in the range of 3–20 mm, Figure 5.7e. In the collapsing radius, the sample is clamped between two parallel plates whose separation distance is controlled by a linear actuator. The bending radius, in a first approximation, is given by half of the distance between the plates (L_p). In such geometry, it is useful to use the parallel plate method [73]. In this method, the sample is mounted between two plates with a large radius, and it is successively squeezed. With the neutral axis at the middle of the sample, the maximum strain is expressed as:

$$\varepsilon_{peak} \approx \frac{C h_{sa}}{\left(L_p - h_{sa}\right)}. \tag{5.29}$$

where h_a is the sample thickness and C is a geometrical constant equal to 1.198 [73]. A detailed discussion on the measurement of the bending-mode strain is given by Harris et al. [70].

5.7.4 Stretching Methods

In a stretching-mode measurement, different types of measurements can be utilized for stretching the samples. The easier one is simple uniaxial or biaxial tension tests (linearly increasing strain) [74]. It is also an important way for studying the critical strain and the number of cycles before failure. Depending on the type of mechanical strain expected (bands and patches, Figure 5.8a and b), Klein et al. [75] developed three discrete techniques for the mechanical testing of stretchable electronic devices, bladder, membrane, and lateral expansion methods. The bladder expansion and membrane expansion methods have the capability for biaxial tensile and normal strain applications, where the lateral expansion method is primarily targeted for uniaxial tensile and normal strain. The first and second methods stressed the sample in biaxial tension using a rubber bladder (Figure 5.8) and an expanding membrane, respectively. The third method uses the lateral expansion of a rubber cylinder to stretch the samples in the direction perpendicular to the compression.

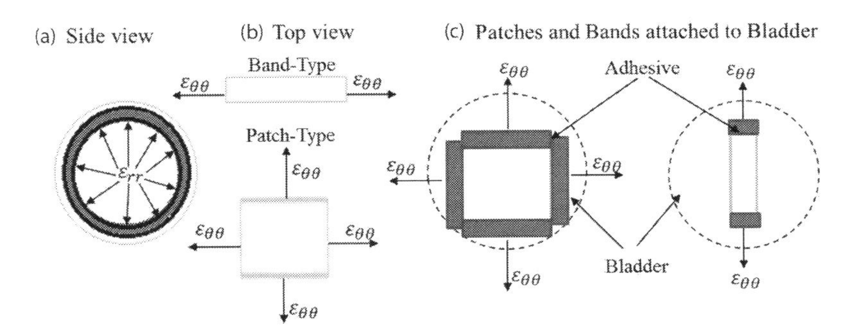

FIGURE 5.8 (a) and (b) Description of band and patches stretchable electronics devices during typical operation. (c) Schematics showing patches and bands attached to bladder (bladder method). (From Klein, S.A. et al., *J. Electron Packaging*, 139, 020905, 2017.)

References

1. Jaccodine, R. J., & Schlegel, W. A. (1966). Measurement of strains at Si-SiO_2 interface. *Journal of Applied Physics*, 37(6), 2429–2434.
2. Ree, M., & Kirby, D. P. (1994). Intrinsic and thermal stress in polyimide thin films. *American Chemical Society*, 482–493.
3. Ledger, A. M., & Bastien, R. C. (1978). Intrinsic and thermal stress modeling for thin-film multilayers. In *Laser Induced Damage in Optical Materials, 1977: Proceedings of a Symposium*, October 4–6, 1977, NBS, Boulder, CO (Vol. 655). NBS: for sale by the Supt. of Docs., US Govt. Print. Off.
4. Hutchinson, J. W. (1996). Stresses and failure modes in thin films and multilayers. Notes for a Dcamm Course. Technical University of Denmark, Lyngby, pp. 1–45.
5. Perera, D. Y., & Nguyen, T. (1996). Hygroscopic stress and failure of coating/metal systems. *Double Liaison*, 43, 66–71.
6. Leterrier, Y. (2014). Mechanics of curvature and strain in flexible organic electronic devices. In S. Logothetidis (Ed.), *Handbook of Flexible Organic Electronics: Materials, Manufacturing and Applications*, Woodhead Publishing, Oxford, UK.
7. Evans, A. G., Crumley, G. B., & Demaray, R. E. (1983). On the mechanical behavior of brittle coatings and layers. *Oxidation of Metals*, 20(5–6), 193–216.
8. Rösler, J., Harders, H., & Baeker, M. (2007). *Mechanical Behaviour of Engineering Materials: Metals, Ceramics, Polymers, and Composites*. Springer Science & Business Media, Berlin, Germany.
9. Vegas, M. R., & Del Yerro, J. L. M. (2013). Stiffness, compliance, resilience, and creep deformation: Understanding implant-soft tissue dynamics in the augmented breast: Fundamentals based on materials science. *Aesthetic Plastic Surgery*, 37(5), 922–930.
10. Janssen, G. C. A. M., Abdalla, M. M., van Keulen, F., Pujada, B. R., & van Venrooy, B. (2009). Developments from polycrystalline steel strips to single crystal silicon wafers. *Thin Solid Films*, 517, 1858–1867.
11. Gleskova, H., Cheng, I. C., Wagner, S., Sturm, J. C., & Suo, Z. (2006). Mechanics of thin-film transistors and solar cells on flexible substrates. *Solar Energy*, 80(6), 687–693.
12. Lacour, S. P., Jones, J., Suo, Z., & Wagner, S. (2004). Design and performance of thin metal film interconnects for skin-like electronic circuits. *IEEE Electron Device Letters*, 25(4), 179–181.
13. Chen, Z., Cotterell, B., & Wang, W. (2002). The fracture of brittle thin films on compliant substrates in flexible displays. *Engineering Fracture Mechanics*, 69(5), 597–603.
14. Ambrico, J. M., & Begley, M. R. (2002). The role of initial flaw size, elastic compliance and plasticity in channel cracking of thin films. *Thin Solid Films*, 419(1), 144–153.
15. Kang, C. W., & Huang, H. (2017). Deformation, failure and removal mechanisms of thin film structures in abrasive machining. *Advances in Manufacturing*, 5(1), 1–19.
16. Cotterell, B., & Chen, Z. (2000). Buckling and cracking of thin films on compliant substrates under compression. *International Journal of Fracture*, 104(2), 169–179.
17. Balint, D. S., & Hutchinson, J. W. (2001). Mode II edge delamination of compressed thin films. *Transactions-American Society of Mechanical Engineers Journal of Applied Mechanics*, 68(5), 725–730.
18. Ma, Q. (1997). A four-point bending technique for studying subcritical crack growth in thin films and at interfaces. *Journal of Materials Research*, 12(3), 840–845.
19. Park, S. I., Ahn, J. H., Feng, X., Wang, S., Huang, Y., & Rogers, J. A. (2008). Theoretical and experimental studies of bending of inorganic electronic materials on plastic substrates. *Advanced Functional Materials*, 18(18), 2673–2684.
20. Evans, A. G., Drory, M. D., & Hu, M. S. (1988). The cracking and decohesion of thin films. *Journal of Materials Research*, 3(5), 1043–1049.
21. Ye, T., Suo, Z., & Evans, A. G. (1992). Thin film cracking and the roles of substrate and interface. *International Journal of Solids and Structures*, 29(21), 2639–2648.

22. Vlassak, J. J. (2003). Channel cracking in thin films on substrates of finite thickness. *International Journal of Fracture*, 119(4), 299–323.
23. Thouless, M. D., Li, Z., Douville, N. J., & Takayama, S. (2011). Periodic cracking of films supported on compliant substrates. *Journal of the Mechanics and Physics of Solids*, 59(9), 1927–1937.
24. Thouless, M. D. (1989). Some mechanics for the adhesion of thin films. *Thin Solid Films*, 181(1–2), 397–406.
25. Thouless, M. D. (1995). Modeling the development and relaxation of stresses in films. *Annual Review of Materials Science*, 25(1), 69–96.
26. Hutchinson, J. W., & Suo, Z. (1991). Mixed mode cracking in layered materials. *Advances in Applied Mechanics*, 29, 63–191.
27. Beuth, J. L. (1992). Cracking of thin bonded films in residual tension. *International Journal of Solids and Structures*, 29(13), 1657–1675.
28. Dundurs, J. (1969). Discussion: 'Edge-bonded dissimilar orthogonal elastic wedges under normal and shear loading.' (Bogy, DB, 1968, ASME J. Appl. Mech., 35, pp. 460–466). *Journal of Applied Mechanics*, 36(3), 650–652.
29. Suga, T., Elssner, G., & Schmauder, S. (1988). Composite parameters and mechanical compatibility of material joints. *Journal of Composite Materials*, 22(10), 917–934.
30. Chen, H., Lu, B. W., Lin, Y., & Feng, X. (2014). Interfacial failure in flexible electronic devices. *IEEE Electron Device Letters*, 35(1), 132–134.
31. Mishnaevsky, L. L., & Gross, D. (2005). Deformation and failure in thin films/substrate systems: Methods of theoretical analysis. *Applied Mechanics Reviews*, 58(5), 338–353.
32. Mei, H., Pang, Y., & Huang, R. (2007). Influence of interfacial delamination on channel cracking of elastic thin films. *International Journal of Fracture*, 148(4), 331–342.
33. Cordill, M. J., Fischer, F. D., Rammerstorfer, F. G., & Dehm, G. (2010). Adhesion energies of Cr thin films on polyimide determined from buckling: Experiment and model. *Acta Materialia*, 58(16), 5520–5531.
34. Ming-Yuan, H., & Hutchinson, J. W. (1989). Crack deflection at an interface between dissimilar elastic materials. *International Journal of Solids and Structures*, 25(9), 1053–1067.
35. Zak, A. R., & Williams, M. L. (1962). Crack point stress singularities at a bi-material interface. *Journal of Applied Mechanics*, 30, 142–143.
36. Jia, Z., Tucker, M. B., & Li, T. (2011). Failure mechanics of organic–inorganic multilayer permeation barriers in flexible electronics. *Composites Science and Technology*, 71(3), 365–372.
37. Gioia, G., & Ortiz, M. (1997). Delamination of compressed thin films. *Advances in Applied Mechanics*, 33(8), 119–192.
38. Suo, Z., Ma, E. Y., Gleskova, H., & Wagner, S. (1999). Mechanics of rollable and foldable film-on-foil electronics. *Applied Physics Letters*, 74(8), 1177–1179.
39. Rogers, J. A., Someya, T., & Huang, Y. (2010). Materials and mechanics for stretchable electronics. *Science*, 327(5973), 1603–1607.
40. Song, J. (2015). Mechanics of stretchable electronics. *Current Opinion in Solid State and Materials Science*, 19(3), 160–170.
41. Lipomi, D. J., & Bao, Z. (2017). Stretchable and ultraflexible organic electronics. *MRS Bulletin*, 42(2), 93–97.
42. Song, J., Jiang, H., Huang, Y., & Rogers, J. A. (2009). Mechanics of stretchable inorganic electronic materials. *Journal of Vacuum Science & Technology A: Vacuum, Surfaces, and Films*, 27(5), 1107–1125.
43. Yu, K. J., Yan, Z., Han, M., & Rogers, J. A. (2017). Inorganic semiconducting materials for flexible and stretchable electronics. *NPJ Flexible Electronics*, 1(1), 4.
44. Khang, D. Y., Jiang, H., Huang, Y., & Rogers, J. A. (2006). A stretchable form of single-crystal silicon for high-performance electronics on rubber substrates. *Science*, 311(5758), 208–212.

45. Song, J., Huang, Y., Xiao, J., Wang, S., Hwang, K. C., Ko, H. C., Kim, D. H., Stoykovich, M. P., & Rogers, J. A. (2009). Mechanics of noncoplanar mesh design for stretchable electronic circuits. *Journal of Applied Physics*, 105(12), 123516.
46. Zhang, Y., Fu, H., Su, Y., Xu, S., Cheng, H., Fan, J. A., Hwang, K. C., Rogers, J. A., & Huang, Y. (2013). Mechanics of ultra-stretchable self-similar serpentine interconnects. *Acta Materialia*, 61(20), 7816–7827.
47. Gray, D. S., Tien, J., & Chen, C. S. (2004) High-conductivity elastomeric electronics (Adv. Mater. 2004, 16, 393). *Advanced Materials*, 16(6), 477–477.
48. Qaiser, N., Khan, S. M., Nour, M., Rehman, M. U., Rojas, J. P., & Hussain, M. M. (2017). Mechanical response of spiral interconnect arrays for highly stretchable electronics. *Applied Physics Letters*, 111(21), 214102.
49. Rojas, J. P., Arevalo, A., Foulds, I. G., & Hussain, M. M. (2014). Design and characterization of ultra-stretchable monolithic silicon fabric. *Applied Physics Letters*, 105(15), 154101.
50. Bowden, N., Brittain, S., Evans, A. G., Hutchinson, J. W., & Whitesides, G. M. (1998). Spontaneous formation of ordered structures in thin films of metals supported on an elastomeric polymer. *Nature*, 393(6681), 146–149.
51. Ko, H. C., Stoykovich, M. P., Song, J. et al. (2008). A hemispherical electronic eye camera based on compressible silicon optoelectronics. *Nature*, 454(7205), 748–753.
52. Lacour, S. P., Jones, J., Wagner, S., Li, T., & Suo, Z. (2005). Stretchable interconnects for elastic electronic surfaces. *Proceedings of the IEEE*, 93(8), 1459–1467.
53. Kim, D. H., Song, J., Choi, W. M., Kim, H. S., Kim, R. H., Liu, Z., Huang, Y. Y., Hwang, K. C., Zhang, Y. W., & Rogers, J. A. (2008). Materials and noncoplanar mesh designs for integrated circuits with linear elastic responses to extreme mechanical deformations. *Proceedings of the National Academy of Sciences*, 105(48), 18675–18680.
54. Fu, H., Xu, S., Xu, R., Jiang, J., Zhang, Y., Rogers, J. A., & Huang, Y. (2015). Lateral buckling and mechanical stretchability of fractal interconnects partially bonded onto an elastomeric substrate. *Applied Physics Letters*, 106(9), 091902.
55. Zhang, Y., Wang, S., Li, X. et al. (2014). Experimental and theoretical studies of serpentine microstructures bonded to prestrained elastomers for stretchable electronics. *Advanced Functional Materials*, 24(14), 2028–2037.
56. Li, R., Li, M., Su, Y., Song, J., & Ni, X. (2013). An analytical mechanics model for the island-bridge structure of stretchable electronics. *Soft Matter*, 9(35), 8476–8482.
57. Yang, S., Ng, E., & Lu, N. (2015). Indium tin oxide (ito) serpentine ribbons on soft substrates stretched beyond 100%. *Extreme Mechanics Letters*, 2, 37–45.
58. Su, Y., Ping, X., Yu, K. J. et al. (2017). In-plane deformation mechanics for highly stretchable electronics. *Advanced Materials*, 29(8) 1604989.
59. Fan, J. A., Yeo, W. H., Su, Y. et al. (2014). Fractal design concepts for stretchable electronics. *Nature Communications*, 5, 3266.
60. Rehman, M. U., & Rojas, J. P. (2017). Optimization of compound serpentine-spiral structure for ultra-stretchable electronics. *Extreme Mechanics Letters*, 15, 44–50.
61. Zhu, J. Z. (2013). *The Finite Element Method: Its Basis and Fundamentals*. Elsevier, Amsterdam, the Netherlands.
62. Xu, W., Lu, T. J., & Wang, F. (2010). Effects of interfacial properties on the ductility of polymer-supported metal films for flexible electronics. *International Journal of Solids and Structures*, 47(14), 1830–1837.
63. Stafford, C. M., Harrison, C., Beers, K. L., Karim, A., Amis, E. J., VanLandingham, M. R., Kim, H. C., Volksen, W., Miller, R. D., & Simonyi, E. E. (2004). A buckling-based metrology for measuring the elastic moduli of polymeric thin films. *Nature Materials*, 3(8), 545–550.
64. Lewis, J. (2006). Material challenge for flexible organic devices. *Materials Today*, 9(4), 38–45.

65. Chang, S. Y., Hsiao, Y. C., & Huang, Y. C. (2008). Preparation and mechanical properties of aluminum-doped zinc oxide transparent conducting films. *Surface and Coatings Technology*, 202(22), 5416–5420.
66. Chen, J., & Bull, S. J. (2010). Approaches to investigate delamination and interfacial toughness in coated systems: An overview. *Journal of Physics D: Applied Physics*, 44(3), 034001.
67. Oliver, W. C., & Pharr, G. M. (1992). An improved technique for determining hardness and elastic modulus using load and displacement sensing indentation experiments. *Journal of Materials Research*, 7(6), 1564–1583.
68. Hainsworth, S. V., Chandler, H. W., & Page, T. F. (1996). Analysis of nanoindentation load-displacement loading curves. *Journal of Materials Research*, 11(8), 1987–1995.
69. Hahm, S. W., Hwang, H. S., Kim, D., & Khang, D. Y. (2009). Buckling-based measurements of mechanical moduli of thin films. *Electronic Materials Letters*, 5(4), 157–168.
70. Harris, K. D., Elias, A. L., & Chung, H. J. (2016). Flexible electronics under strain: a review of mechanical characterization and durability enhancement strategies. *Journal of Materials Science*, 51(6), 2771–2805.
71. Reddy, J. N. (2004). *Mechanics of Laminated Composite Plates and Shells: Theory and Analysis*. CRC Press, Boca Raton, FL.
72. Grego, S., Lewis, J., Vick, E., & Temple, D. (2005). Development and evaluation of bend-testing techniques for flexible-display applications. *Journal of the Society for Information Display*, 13(7), 575–581.
73. Matthewson, M., Kurkjian, C. R., & Gulati, S. T. (1986). Strength measurement of optical fibers by bending. *Journal of the American Ceramic Society*, 69(11), 815–821.
74. Huyghe, B., Rogier, H., Vanfleteren, J., & Axisa, F. (2008). Design and manufacturing of stretchable high-frequency interconnects. *IEEE Transactions on Advanced Packaging*, 31(4), 802–808.
75. Klein, S. A., Aleksov, A., Subramanian, V., Malatkar, P., & Mahajan, R. (2017). Mechanical testing for stretchable electronics. *Journal of Electronic Packaging*, 139(2), 020905.

II

Devices

6

Printed Electronics

Mohammad Vaseem
and Atif Shamim

6.1 Introduction

The traditional manufacturing of electronic devices has relied on printed circuit board (PCB) technology for many years. PCB usually has copper sheets laminated on both sides of a non-conductive substrate. The electronic design (layout pattern) is realized through a mask by etching the copper from the PCB. This can be done in a number of ways, such as mechanical drilling, chemical etching, or through a laser source. As the technology progressed, feature sizes reduced dramatically, and more sophisticated methods, such as photolithography or electron-beam lithography, were adapted. However, it is important to note that traditional PCB etching and other lithography processes are all subtractive methods of fabrication, as is clearly shown in Figure 6.1a. This means that these methods remove or subtract the additional material to create the patterns through the use of expensive masks and photoresists (PR). However, there are two major disadvantages: First, there are substantial material wastages during the process, and second, we need expensive masks and sometimes expensive vacuum processes in a cleanroom environment. It will be beneficial if we can pattern by depositing the material only where we want, without the need of a mask, something which is now known as 'additive manufacturing'.

Printing is an efficient alternative to conventional PCB manufacturing and photolithography processes, as it can deposit materials in a digital fashion at selected places of choice (see Figure 6.1b). Thus, there is less material wastage, which results in huge cost savings. Printing has been fashioned as a low-cost

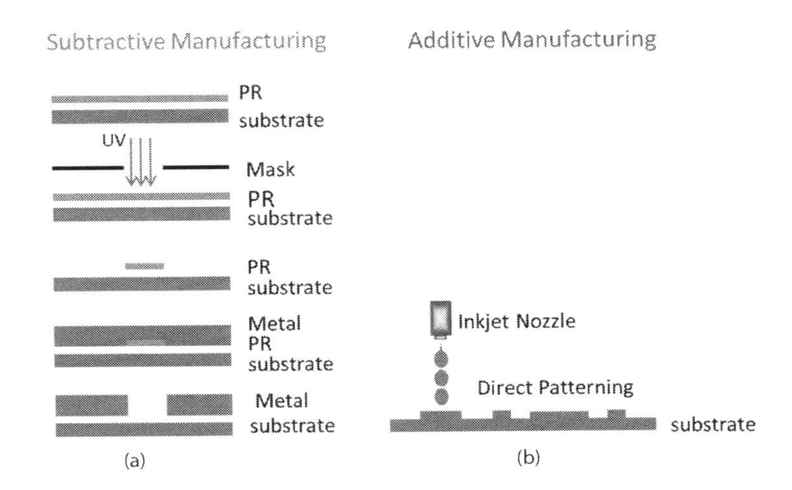

FIGURE 6.1 A schematic depiction of a (a) subtractive and (b) additive manufacturing process.

fabrication technique for electronics, applicable to large-area and flexible electronics manufacturing. Printing differs from other electronics fabrication techniques because it does not require high-vacuum, high-temperature, photolithographic, or etching steps. Although large-scale manufacturing is promising, rapid prototyping is another area in which printing is attractive.

This chapter is focused on printed electronics and its various applications. The chapter provides details of various printing techniques; discusses different types of printers available for material deposition and patterning; provides information on inks, which are the main constituent of printing technology; and discusses various sintering mechanisms for printed electronics. After presenting fundamental aspects of printing, we discuss a wide range of applications, such as displays, sensors, radio-frequency (RF) electronics, wearable devices, and batteries. Finally, future directions for printed electronics are presented.

6.2 Printing Technologies and Types of Printers

The global interest in printed electronics is motivated by the fact that large-area deposition and patterning can be done at a fast pace through printing technologies. In fact, the dream is to have roll-to-roll (R2R) and reel-to-reel printing capabilities for electronics such as newspapers and magazines. In this context, an in-depth examination of these printing techniques and their application to printing-device manufacturing is presented. There are mainly two types of printing technologies available, namely, contact and non-contact printing.

6.2.1 Contact Printing Technology

As the name suggests, contact printing means the inks are transferred from the ink-reservoir to the target substrate through an in-contact patterning mechanism. There are mainly two contact-printing-based techniques available, which are briefly discussed in the following subsections.

6.2.1.1 Flexography Printing

The flexography printing technique is attractive for high-speed prototyping, which requires high-resolution patterns. Flexography consists of three main parts, namely, a large print cylinder, anilox roller, and impression roller, which play an important role in the printing (as shown in Figure 6.2).

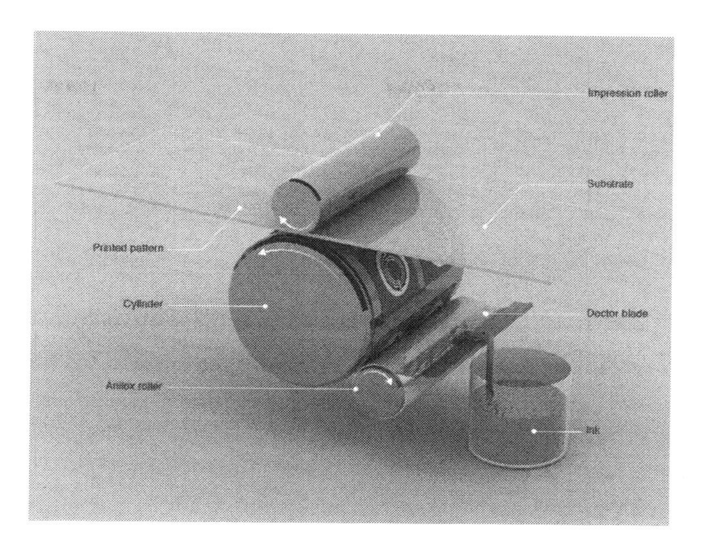

FIGURE 6.2 An illustrative depiction of a flexography printing technique. This figure was produced by Heno Hwang, scientific illustrator at King Abdullah University of Science and Technology (KAUST).

First, the desired patterns are developed by an analogue and digital platemaking process on a rubber or polymer plate (also called print plate), which is then attached to a print cylinder. The ink is fed to the anilox roller from the solution reservoir, which then transfers it to the print cylinder that has the design pattern. The anilox roller has engraved cells that control the quantity of ink to be transferred to the print cylinder. To avoid obtaining a final product with uneven surfaces, it must be ensured that the amount of ink transferred onto the print cylinder is not excessive. This is achieved by using a scraper, called a doctor blade. The print cylinder transfers the ink onto the substrates, which run between the print cylinder and the impression roller. As a result, uniform thin-films are patterned on the substrate. High flexibility and low pressure on the substrate allow printing on rigid and hard substrates. This printer can handle a wide variety of inks, such as aqueous and non-aqueous inks, UV-curable, and e-beam curable inks. Flexographic inks usually have lower viscosity than gravure printer inks, which results in faster drying of the ink and faster production speeds. The resolution of flexography is dependent on how the print plate is made. The printed pattern with a resolution between 50–100 μm could be easily achieved with the proper control of process parameters and substrate surfaces. Recently, micro-contact printing (μCP) has been utilized for patterned elastomeric stamps on the print plate, achieving a 20 μm resolution (Joyce et al. 2014). The Timson T-Flex 508 flexography printer is commonly employed for printing metal traces.

6.2.1.2 Gravure Printing

Gravure printing is another physical-contact–based printing technique, the most promising in terms of resolution and print speed. It transfers functional inks to the substrate by physical contact of the engraved pattern with the substrate. Gravure printing consists of a large cylinder typically made from metals such as copper coated with chromium (as shown in Figure 6.3). The surface of the cylinder is engraved with microcells (which act as a pattern) by using electromechanical means or through a laser. These cells are filled with the ink either through a nozzle dispenser from the top or a through a reservoir beneath the rotating gravure cylinder. Excess ink is then removed from the rotating cylinder using a doctor blade. The main challenge with gravure is overcoming non-idealities of the doctor blade's wiping process. In addition, it is difficult to create inks for which a correct combination of surface energies, ink viscosity, and surface tension is required. Printing velocity, pressure, and cell width/depth ratio are some of the process parameters that affect the consistency of the printing results. Recently, the Vivek

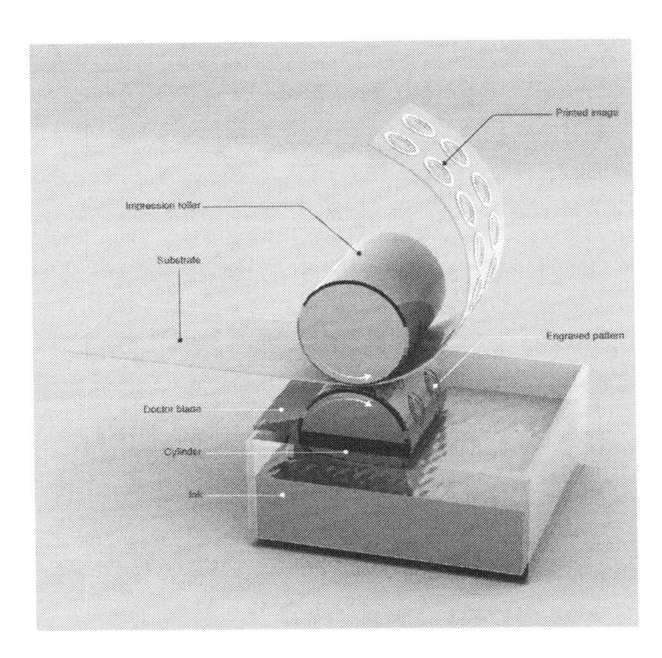

FIGURE 6.3 An illustrative depiction of a gravure printing technique. This figure was produced by Heno Hwang, scientific illustrator at King Abdullah University of Science and Technology (KAUST).

Group (Grau and Subramanian 2016) have demonstrated that by understanding the underlying physics in detail, gravure printing can be pushed into a highly scaled regime, with feature sizes well below 10 μm, whilst still printing at high speeds (on the order of 1 meter per second). Although printing fine features with high ink-transfer volume and speed is advantageous, a higher setup cost than for flexography is the main disadvantage. In addition, an engraved cylinder for every design change adds to complexity and cost. Gravure printers have been successfully employed to print, for example, high-density transistors, circuits, interconnects, photovoltaics, batteries, and Radio-Frequency Identifications (RFIDs). Coatema (Coating Machinery GmbH) provides a wide variety of customized gravure printers on the market.

6.2.2 Non-contact Printing Technology

Screen printing and inkjet printing are the most popular non-contact–based printing technologies. Screen printing is the most popular, and it is a mature technology for printed electronics because it has been used for a long time in the electronics industry. The salient features of these two printing technologies are discussed in the following subsections.

6.2.2.1 Screen Printing

Screen printing is widely used in electronics production and is the typical process for thick-film manufacturing with paste-like materials, which include conductive, resistive, and dielectric materials. A typical screen-printing process is shown in Figure 6.4.

In screen printing, a stencil is produced as the negative image of the pattern (1). A screen frame with stencil is placed on the substrate (2), and then the ink is placed on top of the screen (3). A squeegee (rubber blade) is used to spread the ink across the screen with slight downward force (4&5). The ink is squeezed where the stencil has open space, transferring the pattern of stencil onto the substrate (6). The number of meshes in the screen and ink material with different viscosity plays a major role in high-resolution patterning. Usually, the screen is developed by different sizes of mesh openings and with several materials ranging

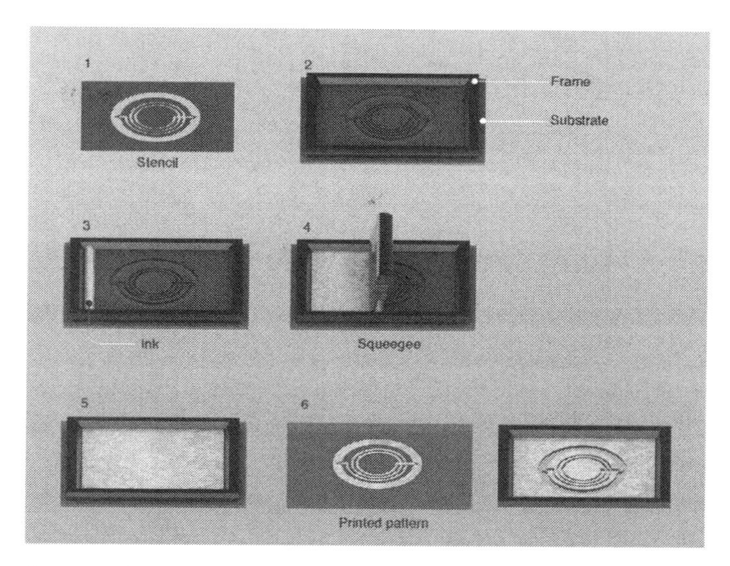

FIGURE 6.4 An illustrative depiction of a screen printing technique. This figure was produced by Heno Hwang, scientific illustrator at King Abdullah University of Science and Technology (KAUST).

from polyester to stainless steel. Screen printing is a well-established process which does not require high capital investment. A wide variety of commercial electronic inks are available that have been designed for the screen-printing process, but the excessive waste of expensive inks during the printing process pose a main disadvantage. In addition, the high wet thickness of the film limits the layer-to-layer printing process. Sometimes ink also dries out on the stencil and deteriorates the mask designs, causing printing quality to suffer. Even though there are several disadvantages, screen printing is widely used for fast prototyping in the lab.

6.2.2.2 Inkjet Printing

Unlike gravure, flexography, and screen-printing techniques, inkjet printing technology is a completely digital process in which one can transfer a design in the form of a digital layout in a computer directly onto the substrate. In addition, low viscosity inks with less material wastage can be used. Dimatix Materials Printers (DMP) are commercially available, with 2800 (lab-scale), 3000 (lab scale with multi-nozzle system), and 5000 (industrial scale) series. In inkjet printing, ink loaded in a reservoir is ejected onto a substrate via a continuous and on-demand printing method (as shown in Figure 6.5). Several actuation mechanisms are used to eject the ink droplet, which include piezo, electrostatic, and thermal. For industrial inkjet printing, piezo-based drop-on-demand technology is the most prevalent for printing electronics. While other types of inkjet printing techniques exist, such as continuous, thermal inkjet, and hydrodynamic, each one of them presents a disadvantage in one way or the other. Thermal inkjet is common for home printers (for black/white and color printing); it is low cost, but has slower throughput. Thermal inkjet has a high-temperature heating element that vaporizes some of the ink, creating a bubble that forces the drop out of the orifice. In continuous inkjet, an ink is needed that can be charged and deflected by a static electric field, which again limits and complicates the type of inks that can be used. In piezoelectric ink-jet printing, a piezoelectric element is used to squeeze individual drops out of a small chamber by changing its shape. When an electric field is applied to a piezo-activated wall of the chamber, the wall's dimensions change according to the applied voltage. Depending on the polarity of the applied voltage, it is either a minute contraction or a minute expansion. In the latter case, an ink drop is pushed out of the nozzle. However, nozzle

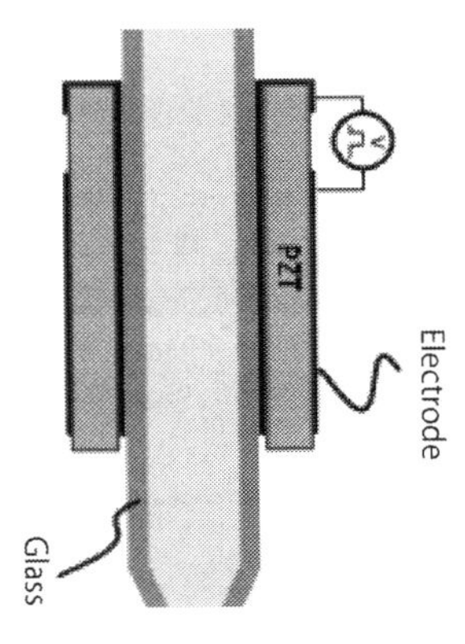

FIGURE 6.5 A schematic depiction of piezo-based drop-on-demand inkjet printing.

clogging and inconsistency in image quality are the main concerns in this technology. Modern piezoelectric inkjet heads can contain thousands of densely spaced nozzles, each jetting 100,000 drops per second. Several kg of material can be deposited each hour with a single printhead (ProJet MJP 5600 from 3D systems).

Table 6.1 summarizes the key features of commercially available printing techniques.

TABLE 6.1 A Comparative Table for Different Printing Techniques

Types of Printer	Printing Resolution (μm)	Printing Speed (m/min)	Ink Viscosity Requirement (cp)	Advantages	Disadvantages
Flexography	20–80	5–180	10–500	• Contact method • R2R compatibility • Lower setup cost • Large area printing	• Prints a 'donut', rather than a dot • Susceptible to 'halos'
Gravure	10–200	8–100	10–1100	• Contact method • R2R compatibility • Able to print on wide variety of substrates • Large-area printing	• Higher setup cost than flexography • Engraved cylinders for every design change
Offset	20–50	0.6–15	2000–5000	• Contact method • R2R compatibility • Large area printing	• Correct combination of surface energies, ink viscosity, and surface tension are required
Screen	30–100	0.6–100	500–5000	• Non-contact method • R2R compatibility • Established technology and major ink for printed electronics • Low-price inks	• Can be expensive due to amount of material wastage
Inkjet	20–100	0.02–5	10–15	• Non-contact method • Digital customization • Amenable to prototyping • Can print thin film and fine features	• Low throughput

6.3 Type of Inks

Inks and writing/printing are not new. These technologies have been practiced for many centuries, for example, to print newspapers and magazines. However, these are color inks, which do not have any electronic functionality. With the advancement of nanotechnology, various studies of electronic nanomaterial have been presented. It has been observed that when electronic materials are prepared in a nanometer regime, their surface-volume ratio increases, and subsequently their melting point is decreased. Due to the low melting temperature of nanoparticles, their functionality starts to compete with its bulk counterpart. Several methods, such as gas-phase synthesis and solution process, have been employed to prepare nanoparticles and their ink-formulation. Currently, most commercial inks are based on screen-printable conductors and dielectrics. Several commercial sources, such as DuPont, Sigma-Aldrich, Sun Chemical Co., Henkel AG & Co., and Creative Materials Inc., are readily available for printable electronic inks. In addition, gravure and flexography printable electronics inks are also commercially available. However, inkjet printable inks are commercially limited. Most of the inkjet inks reported in the literature have not yet entered the market. There are several concerns associated with the inkjet electronics inks, such as storage stability, jetting stability, or loading of materials in ink. An ideal requirement which is suited for inkjet inks is that the nanoparticle size must range from 2 to 1000 nm in diameter, preferably less than 100 nm. During ink formulation, the concentration of nanoparticles in the solvent must be in the range from 10% to 50%. Generally, nanoparticle ink formulations include one or more vehicles (e.g., solvents) in which the nanoparticles are preserved. These include, for example, water, alcohol, and typical organic solvents. The ink formulations also include dispersing agents to enhance the dispersion of the nanoparticles in the vehicles. These dispersants have head groups that can associate with the nanoparticle and a tail group that is compatible with the vehicle (solvent) used in the liquid-phase component mixture of the ink. An example of a dispersing agent is based on amine, Triton, polyvinylpyrrolidone (PVP), polyvinyl alcohol (PVA), and polyethylene glycol (PEG). In some cases, additives can be included in the ink formulations in order to adjust the ink viscosity, surface energy, etc. Generally, nanoparticles are in the form of a solid powder state, so three actions are required when converting a powder into a stable dispersion. These are wetting the surface, breaking up agglomerates, and stabilizing the dispersed particles against flocculation. The balance between the forces of attraction and repulsion will determine whether the particles move to cluster back into agglomerates or remain dispersed. The maintenance of dispersion can be assisted by mechanically breaking up the agglomerates with a ball mill or a similar device. Such a mechanical process is carried out in the presence of a dispersant in order to avoid re-agglomeration once the process has ceased. Compared to nanoparticle ink, metal-organo complex or sol-gel-based ink has recently received significant attention as a potentially lower-cost alternative that is stable at concentrations approaching saturation; neither additional stabilizers nor reducing agents are required. By adjusting the viscosity and surface tension of the solution complex's ink-chemistry, this type of ink could be used for various deposition techniques (e.g., spin-coating, direct ink writing, fine nozzle printing, airbrush spraying, inkjet printing, screen printing, and roll-to-roll processing methods) in order to fabricate dielectric or conductive films. The following subsections will deal with inkjet ink, and its related application will be presented.

6.3.1 Conductive Inks

The field of printed electronics deals with several kinds of conductive inks based on nanoparticles, nanowire, and two-dimensional sheets that include metal, silicon, carbon, and oxides (Lee et al. 2010, 2014, Torrisi et al. 2012, An et al. 2014, Ghosh et al. 2014, Hamedi et al. 2014, Li et al. 2014). However, the printed-electronics market is currently dominated by conductive nanoparticle-based inks. At present, most of the conductive inks available are based on silver nanoparticles, because silver possesses the highest conductivity without any oxidation issue. As a requirement, ink must be stable to aggregation and precipitation to achieve reproducible performance. To meet this requirement, silver nanoparticle ink normally uses organic stabilizers, which unfortunately also act as insulators. Moreover, silver nanoparticle ink with a high solid content is more prone to stability issues, which generally result in clogging of the inkjet nozzles, and it affects the shelf life of the ink. High silver loading is beneficial, but the high temperature removal of organic stabilizers from the nanoparticles can limit the choices of printable substrate.

The fabrication of high-quality and low-cost electronics requires innovative ink formulations that are cheaper and faster than traditional production methods. Inkjet printing with the use of particle-free metal-organic-complexes (MOC) or salts of various metals (sol-gel) is a very attractive low-cost technology for direct metallization. Perelaer et al. (2009a) utilized a silver salt (20 wt% silver) solution in methanol/anisole (ink purchased from TEC-IJ-040, InkTec Co., Ltd, Korea). Their study demonstrated that printed silver tracks on glass have a conductivity of 1.2–2.1 × 10^7 S/m at 150°C. Reactive silver inks, including ammonium hydroxide as a complexing agent and formic acid as a reducing agent, have been reported (Walker and Lewis 2012). Despite several possible drawbacks, this ink possessed the best conductivity at 90°C, which is almost equivalent to that of bulk silver. Dong et al. (2015) synthesized MOC ink through a two-step process. First, silver oxalate was synthesized by silver nitrate and then dissolved in ethylamine as a complexing ligand, with ethyl alcohol and ethylene glycol as solvents, using a low-temperature (0°C) mixing process. The printed patterns on the polyimide (PI) substrate that were cured at 150°C for 30 min showed metalized silver with a conductivity of 1.1 × 10^7 S/m. Recently, Vaseem et al. (2016b) have developed a novel silver-organo-complex (SOC) ink, a silver-ethylamine-ethanolamine-formate-complex-based transparent and stable ink, wherein ethylamine, ethanolamine, and formate species act as in situ complexing solvents and reducing agents (shown in Figure 6.6). As-formulated ink was inkjet printed on a wide range of substrates, including PI, polyethylene terephthalate (PET), polyethylene naphthalate (PEN), and glass. The decomposition of the ink led to uniform surface morphology with excellent adhesion to the substrates. They also reported excellent jetting and storage stability. Inkjet printing with a single layer demonstrated conductivities of ~0.66 × 10^7 and ~0.4 × 10^7 S/m on glass and PEN, respectively. However, with printing numbers of seven to eight layers, the conductivity became saturated at ~2.43 × 10^7 and ~2.12 × 10^7 S/m on glass and PEN substrates, respectively.

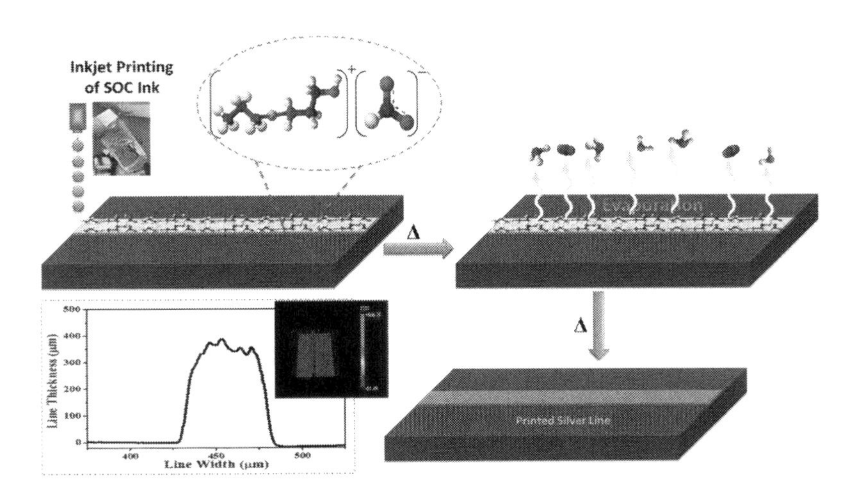

FIGURE 6.6 An illustrative procedure for the printing of silver-ethylamine-ethanolamine-formate-complex-based SOC ink. It also shows the thermal reduction process for printed ink conversion to silver metallic phase. (From Vaseem, M. et al., *ACS Appl. Mater. Interfaces*, 8, 177–186, 2016. With permission.)

Most of the reported silver complex-based inks suffer from several drawbacks, such as storage and jetting stability. The actual volume faction of the metal within the particle free organo-complex or sol-gel inks is comparatively lower than the nanoparticle-based ink. The printed film after sintering is usually thinner and porous and thus has to undergo significant compaction. In the field of printed electronics, like other emerging technologies, new materials and processing methods are required for their ever-improving development and performance.

6.3.2 Dielectric Inks

For fully printed active and passive devices, the development of dielectric ink is essential to provide the required insulation. Dielectrics are used as thin-films to provide proper insulation and prevent leakage currents in devices where high-capacitance is required. Moreover, thin dielectrics facilitate low-voltage operation in transistors. Typically, a large-area printing process capable of producing high-quality dielectric film is desired in a fabrication process. In contrast to conductive inks, dielectric ink preparation is not straightforward. Sol-gel processed dielectric ink is preferred to nanoparticle-based dielectrics due to issues such as high leakage, voids, and cracks in the nanoparticle-based films. Most of the printable dielectrics are based on polymers such as PVP and PIs. However, these polymer dielectrics have low relative dielectric constants in the range of 2.5–4. Among them, PVP is a common dielectric material used in organic thin-film transistor fabrication (Jang et al. 2005, Vornbrock et al. 2010, Tseng 2011). PVP has been shown to successfully inkjet print pinhole-free films. An advantage of PVP polymers for electronic applications is that they can be thermally cross-linked, making them resistant to environmental effects and leaking currents. McKerricher et al. (2015) reported inkjet printing of the PVP dielectric layer, precisely controlling the thickness of the layer. The PVP dielectric ink was assessed by printing metal-insulator-metal (MIM) capacitors using printed layers of silver nanoparticle-based ink in addition to the PVP ink to test the leakage current through the dielectric. The printed PVP film showed good insulating behavior with an applied field of 0.5 MV/cm and a current density of $1.4e^{-7}$ A/cm^2. No breakdown was observed for the printed dielectric film up to 0.9 MV/cm (100 V).

As mentioned above, the dielectric constants for most of the polymers are relatively lower, which is fine for many applications. However, high-k dielectrics are advantageous in transistors because they offer higher capacitance and lower operating voltages. Since the first report of high-mobility amorphous oxide semiconductors in 2004, solution-processed sol-gel-based oxides have quickly gained momentum (Nomura et al. 2004). The drive has been towards thin-film transistors (TFTs) for displays. The main dielectric contenders for these devices have been SiO_2, ZrO_2, HfO_2, Al_2O_3, TiO_2, Y_2O_3, and Ta_2O_5 (Fortunato et al. 2012). Of these sol-gel dielectrics, ZrO_2 has been inkjet printed in a fully inkjet-printed TFT (Jang et al. 2015). This innovative work utilized a sacrificial polymer layer to modify the surface energy before printing ZrO_2, and it serves as a stepping stone for this field. For passive devices, alumina (Al_2O_3) is one of the most commonly used ceramics, since it is abundant, has a relatively high dielectric constant (~9), a decent dielectric temperature coefficient, good chemical stability, and a high bandgap of 8.7 eV. Al_2O_3 in its pure form offers the lowest known dielectric loss, which is important for RF electronics often sensitive to loss. Recently, McKerricher et al. (2017) have shown sol-gel-based Al_2O_3 dielectric ink. In their work, aluminum nitrate is used with the common solvent 2-methoxyethanol (2-MEA) and ethanol. They discovered that the co-solvent system of 2-MEA and ethanol was necessary for inkjet film formation. These sol-gel inks were printed and subsequently annealed/oxidized at 400°C to form the relevant high-k material. They discovered that 50% 2-MEA and 50% ethanol gave the best results, as the higher volatility limited the capillary-driven spreading, which in turn reduced the coffee ring effect (shown in Figure 6.7). In that work, the estimated dielectric constant (Dk) of 6.2 for the inkjet alumina is low (~70%); however, it is similar to spin-coated sol-gel results and other reports in the literature (Avis and Jang 2011, Nayak et al. 2013).

It has been observed from literature that generally it is difficult to realize low-leakage films at lower temperatures (<200°C) due to porosity in the films, and sometimes alumina is not converted well to its dielectric form. Thus, it is desirable for future work in this area to achieve high-quality printed dielectric films with

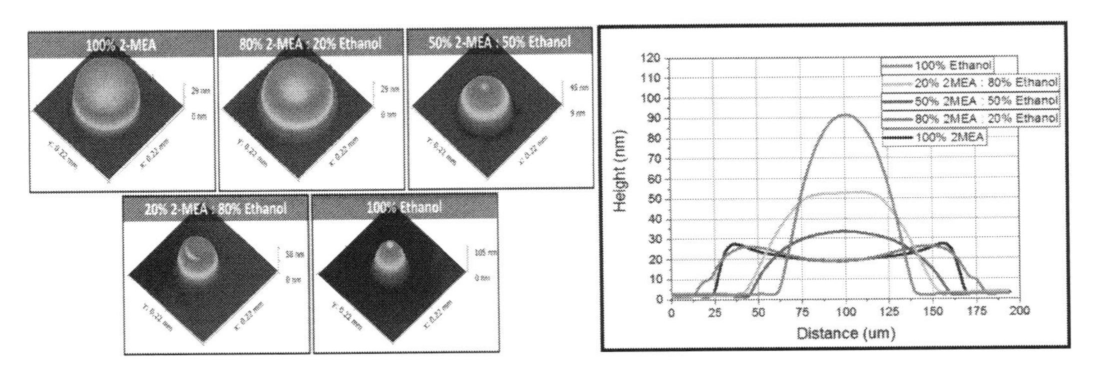

FIGURE 6.7 3D image and profiler of 0.8 M concentration $Al(NO_3)_3$ $9H_2O$ printing of single 10 pl droplet on a PVP-coated gold substrate with 150 seconds of UV ozone treatment (AlO_x droplets imaged after annealing at 400°C). (From McKerricher, G. et al., *Ceram. Int.*, 43, 9846–9853, 2017. With permission.)

low leakage currents, preferably at lower temperatures. In addition, not much investigation has been done regarding the RF properties (particularly loss) of these printed dielectric films. Finally, step coverage, wetting, and pinholes are serious concerns for printed dielectric films and must be addressed in future work.

6.3.3 Semiconductor Inks

Semiconductor materials are critical components for developing active electronic and sensing devices. The deposition of functional semiconductor materials through inkjet printing requires the formulation and development of suitable inks. Since ink formulation determines the quality of printed patterns, ink stability and/or jetting properties are very important for the performance and lifetime of electronic devices, such as displays. Several organic-semiconductor-ink-based printed thin-film transistors have been studied and reported in the literature (Noguchi et al. 2007, Doggart et al. 2009, Vornbrock et al. 2010, Kang et al. 2014, Hyun et al. 2015, Kitsomboonloha et al. 2015, Grau et al. 2016). However, inorganic-semiconductor-material-based ink formulations have been seldom reported. Due to the dissolving capabilities of metal precursors, most of the reported inorganic semiconductor inks are based on the sol-gel process. Dongjo Kim et al. reported an ink based on an n-type zinc tin oxide (ZTO) semiconductor which was prepared by dissolving zinc acetate dehydrate and tin (IV) acetate in 2-ME (Kim et al. 2009a). The concentration of metal precursors was 0.75 M, and the molar ratio of [Sn/(Sn+Zn)] was 0.3. The solubility of the precursor was improved with the ethanolamine-based stabilizing agent. Dasgupta et al. (2012) prepared n-type indium tin oxide (ITO) nanoparticle-based ink for inkjet printing. In their ink-formulation, 10 wt% (1 g) of ITO nanopowder with average particle size of 13 nm (Evonik GmbH) was mixed with 10 mL of deionized water and 5 wt% of commercial stabilizer TEGO 752 W (Evonik gmbH). Through a vigorous mixing and centrifugation process, they were able to develop an ITO-based ink for inkjet printing. Later on, the same group also formulated n-type (In_2O_3) and p-type (Cu_2O) oxide semiconductor nanoparticle-based dispersions (Baby et al. 2015). They utilized the sodium salt of poly(acrylic acid) (PAANa) as a stabilizer. PAANa ligands reduce the tendency of agglomeration due to their adsorption on the surface of nanoparticles. As a result, the ink showed storage stability for a time period of several months. Vaseem et al. (2013) reported the first p-type oxide-based ink-formulations. They have synthesized 5–8 nm CuO quantum dots (QDs) which were formulated as an ink using mixed solvents (ethanol, isopropyl alcohol, ethylene glycol, and de-ionized water). It was observed that the CuO concentration and digitally controlled number of overprinting are important factors for optimizing the uniformity and thickness of printed films with smooth edge definition (shown in Figure 6.8). They have also examined the effect of microwave-assisted annealing

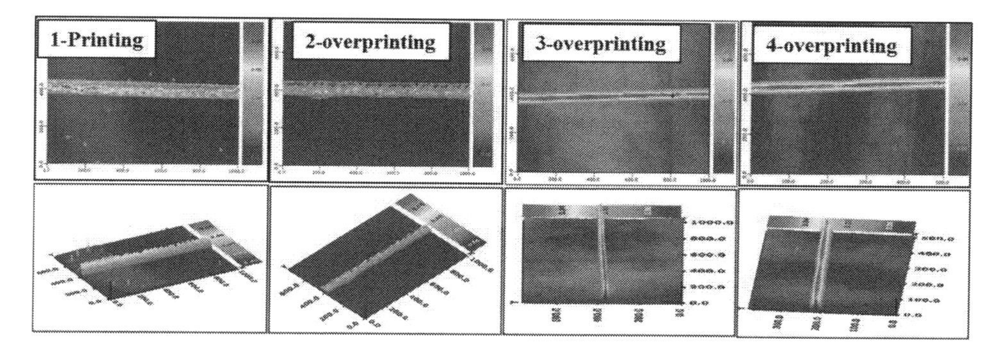

FIGURE 6.8 2D and 3D surface profiler images of inkjet-printed CuO lines with number of over-printing. The ink is printed with drop-spacing of 40 μm.

(MAA) on the structural and electrical properties of the inkjet-printed CuO field effect transistors (FETs) on a standard Si/SiO_2 substrate. The MAA-treated FETs showed a high mobility (i.e., 28.7–31.2 $cm^2 V^{-1} s^{-1}$), which is the best among the p-type inorganic based FETs.

Avis et al. (2014) reported the n-type indium-zinc-tin oxide (IZTO)-based ink formulation. The ink was prepared using indium chloride ($InCl_3$), tin chloride ($SnCl_2$), and zinc chloride ($ZnCl_2$) as the metal precursors of indium, tin, and zinc, respectively. The amount of each precursor was adjusted as 2 mM and mixed in a vial. The combination of acetonitrile (35 vol%) and ethylene glycol (65 vol%) solvents was utilized for the formulation of IZTO ink. Later on, the Vivek Group reported the n-type tin dioxide-based sol-gel ink, whereby the ink was prepared by dissolving 0.001 mol tin chloride dehydrate ($SnCl_2 \cdot 2H_2O$) in ethanol (Jang et al. 2015). Li et al. (2017) came up with a new precursor-based n-type indium gallium oxide (InGaO) semiconductor ink. The InGaO ink was prepared by dissolving 0.19 M indium nitrate [$In(NO_3)_3 \cdot xH_2O$)] and 0.01 M gallium nitrate [$Ga(NO_3)_3 \cdot xH_2O$] into a mixture of 2-ME and ethylene glycol with a volume ratio of 1:1.

In addition to sol-gel and nanoparticle-based semiconductor inks, there are many reports based on 1D semiconducting single-wall carbon nanotube (SWCNT)-based inks. SWNTs have been widely explored as a high-performance semiconductive channel material for TFTs. Many advances have been made towards creating printable SWNT TFTs for high carrier mobility and low operating voltage. However, most of the SWCNT work utilized commercial ink from NanoIntegris, Inc. Most of the inorganic-based semiconductor inks have outperformed their organic counterparts (mostly in terms of higher mobilities); however, they have not been able to penetrate the market, primarily due to the long-term storage stability issue. These reported inks performed well in the lab environment, but their inkjet stability has not been rigorously assessed. The inkjet stability is important to assess the quality of formulated inks. Although ink development for inorganic semiconductors is still in its infancy as compared to their organic counterparts, the progress is rapid. The drive is towards stable, high-mobility inks that can be processed at lower temperatures.

6.3.4 Functional Inks

Fully printed devices that employ dielectric/semiconductor inks have recently been reported. The next generation of fully printed components and systems should have the ability to control their performance, such that they can be tuned or reconfigured when necessary. This requires the development of functional inks that are, for example, magnetic, ferroelectric, or piezoelectric in nature. In RF electronics, tunable or reconfigurable components are becoming important due to the proliferation of new wireless devices, different wireless standards in different parts of the world, and high congestion in the existing bands of wireless communication. Non-printed magnetic materials have been used effectively for tunable and reconfigurable components such as inductors, antennas, and phase shifters. However, there is a paucity of functional inks with magnetic properties, and there are only few reports on magnetic ink printing. Hoseon Lee et al. (2013) demonstrated inkjet printing of commercially available cobalt-based, ferromagnetic nanoparticles (~200 nm) for the miniaturization of flexible printed inductors. These metallic cobalt nanoparticles usually require surface passivation to avoid the oxidation problem. Han Song et al. (2014) utilized an interesting approach to align the cobalt nanoparticle ink with an external magnetic field during printing to enable prototyping and development of novel magnetic and composite materials and components. Murali Bissannagari et al. inkjet-printed NiZn-ferrite films using NiZn-ferrite nanoparticle-based ink, completing its magnetic characterization (Bissannagari and Kim 2015). All the above inks are metallic in nature, but a magnetic ink with dielectric (insulator) properties is required for tunable RF applications. In this regard, Vaseem et al. (2018) reported the development of an iron oxide nanoparticle-based magnetic ink. First, a tunable inductor was fully printed using iron oxide nanoparticle-based magnetic ink. Furthermore, the researchers functionalized iron oxide nanoparticles with oleic acid to make them compatible with a UV-curable SU8 solution. Functionalized iron oxide nanoparticles were then embedded in the SU8 matrix to make a magnetic substrate. The fabricated substrate was characterized for its magnetostatic and microwave properties. A frequency-tunable printed patch antenna was demonstrated using the magnetic and their in-house SOC inks. Such a functional ink is not only highly suitable for tunable and reconfigurable microwave devices, but could also be explored in sensing, biotechnology, and biomedical areas.

6.4 Sintering Mechanisms

Sintering is the process of compacting and forming a solid film of material by heat through interparticle neck formation without reaching melting. How effective is the sintering that governs the conductivity of any printed metal In contrast to sintering in metal, high-temperature annealing or curing governs the electrical properties of semiconductor and dielectric films. Usually, conductive inks are comprised of both a liquid vehicle (water or organic solvent) that determines the basic properties of the ink and a dispersed or dissolved component providing the desired functionality. In nanoparticle-based inks, for reproducible performance, the nanoparticles' dispersion in the ink should be stable without aggregation and/or precipitation (larger-particle sedimentation). Therefore, the addition of a stabilizing agent (to help in dispersion), usually an organic material, is required. However, such nanoparticle (NP)-based inks form patterns composed of conducting metallic NPs capped with organic stabilizers that act as insulators (as shown in Figure 6.9a). Due to the presence of such organics between the particles, the number of percolation paths (or conductive paths, marked with dotted red line in Figure 6.9a) is limited, and the resistivity of the printed pattern is too high to be of practical importance. As the capped organic molecules decompose with heating, nanoparticles start fusing with each other to form the conductive films (as shown in Figure 6.9c with a solid red line).

There are several other important factors, such as the size of nanoparticles that are used to realize the conductors. As the size of the nanoparticle is decreased, its reactivity is increased, and its melting point is decreased. Both of these factors favor the sintering of the nanoparticles into a fused conductor (as shown in Figure 6.9c). For non-nanoparticles-based ink, such as MOC, the ink also follows the same sintering process as nanoparticle-based ink. The only difference in MOC ink is that after printing, the metal ion is reduced to metal and the ligand is expelled at the same time. If the ligands around the metal are still there after decomposition, then the scenario becomes equivalent to the sintering of metal NPs. A sintering temperature such as <300°C is effective in the case of metals due to their soft nature; however, in the case of semiconductors or dielectrics, the curing or annealing temperature is much higher (>400°C). Various sintering techniques are discussed in detail in the following subsection.

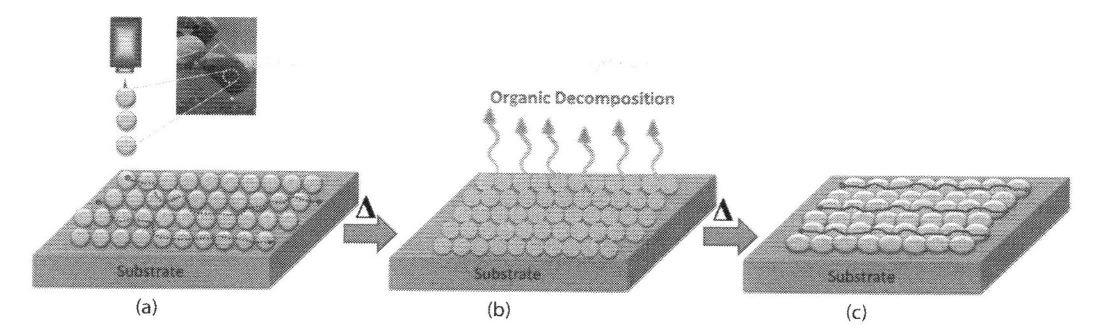

FIGURE 6.9 (a–c) An illustrative procedure for the printing of nanoparticle-based ink and organic decomposition to form the conductive film.

6.4.1 Thermal Sintering

Thermal sintering is usually done in the oven (box oven or furnace) with heating in the range of 150°C–200°C for 10–60 min to achieve conductivity close to bulk metal. After printing nanoparticle-based ink solution on a target surface, the next important step is pre-curing or drying the ink. This step is important to achieve high electrical conductivity in the printed pattern, preferably close to that of the bulk metal for printed metal applications. If the ink solution is printed on a polymer substrate, the thermal pre-curing is performed at temperatures generally less than 200°C and preferably less than 100°C to prevent variations in the substrate, such as changes in elasticity/plasticity, shrinkage, warping, and/or damage to the substrate. The ink solution may be cured in air or other gas environments, such as nitrogen and argon. Curing in inert environments may increase the cost and complexity of the inkjet printing system. The crystallization of printed films with sintering at a low temperature is crucial for the practical use of inks, particularly for flexible electronics. Thermal annealing is a conventional technique, but it is limited by the temperature gradient developed through the surface and by high-temperature treatment over a long period of time. There are several reports of metal inks with thermal sintering temperature of <200°C. Lowering the curing temperature was achieved by properly tailoring the composition of the ink, mainly by using short stabilizers or liquids that have low boiling points and low heats of vaporization (Kamyshny and Magdassi 2014).

6.4.2 Microwave Sintering

It is well known that microwave heating is much faster and uniform as compared to thermal heating. In principle, microwave radiation is absorbed due to coupling with the rotating dipoles or charge carriers (Rao et al. 1999). Microwave-assisted heating is widely used both for the sintering of dielectric materials and in materials synthesis (Rao et al. 1999, Lin et al. 2011, Nadagouda et al. 2011, Yang et al. 2013, Kahrilas et al. 2014). Highly conductive metals like Au, Ag, and Cu can also be sintered by microwave radiation, but they have a very small penetration depth (1–2 μm film thickness) at 2.54 GHz (Kamyshny and Magdassi 2014). Ink-jet-printed conductive films fulfill this requirement; however, much thicker films beyond the penetration depth may not be suitable for microwave sintering. Nevertheless, since Ag is a perfect thermal conductor in comparison to the polymeric substrate, the silver films on a polymeric substrate will be heated uniformly by thermal conductance. The Ulrich S. Schubert Group have reported ~5% conductivity of bulk silver for a printed film thickness of 4.1 μm on a PI substrate at a constant microwave power of 300 W for 4 min (Perelaer et al. 2006). Later, the same group obtained 10%–34% conductivity of bulk silver within 1 second of microwave sintering of the printed silver tracks which were pre-heated at 110°C for 1–2 min (Perelaer et al. 2009a). They have also obtained 40% conductivity in less than 15 seconds by combining photonic and microwave flash treatments (Perelaer et al. 2012). The highest conductivity (60% of bulk) has been obtained by combining low-pressure Ar plasma with microwave flash in less than 10 min on cost-effective PEN foil (Perelaer et al. 2012). Vaseem et al. demonstrated the reduction of CuO nanoparticle-based ink into pure Cu film by microwave plasma (Hahn et al. 2016). The resultant films show the conductivity of 1–2 × 10^7 S/m, which is 2 to 6 times lower than that of bulk Cu. The same group also demonstrated a microwave sintering effect on particle free silver-ethanolamine-formate complex (SEFC)-based ink, which is spin-coated on different substrates. The best conductivity observed was 1.4 × 10^7 S/m for 1.5 min microwave sintering on glass substrate, which is 5 times lower than that of bulk silver (Vaseem et al. 2016a). Jung et al. demonstrated rapid cellulose-mediated microwave sintering for high conductivity Ag patterns on paper (as shown in Figure 6.10). In that work, the silver ink with organic-capped nanoparticles was printed on a paper substrate (step i). Paper absorbs microwaves through a dielectric loss mechanism associated with its complex dielectric constant. Due to the anisotropic dielectric properties of the cellulose fibers, a microwave electric field applied parallel to the paper substrate provides sufficient heating to the substrate (step ii). The heat is transferred to produce Ag patterns (step iii) with a conductivity that is 29%–38% of bulk Ag in a very short period of time (~1 second) at 250°C–300°C (Jung et al. 2016). As mentioned above, microwave sintering is limited to thinner film due to its penetration depth, but it is considered a favorable means of decreasing the sintering time and increasing electrical conductivity for inkjet-printed films on flexible polymeric or paper substrates.

6.4.3 Photonic Sintering

As compared to conventional thermal sintering, photonic sintering irradiates the sample with multiple high-intensity short flashes of light, each with a pulse length in the range of a few micro- to milliseconds (Perelaer et al. 2012). A photonic sintering technique has been developed and commercialized by

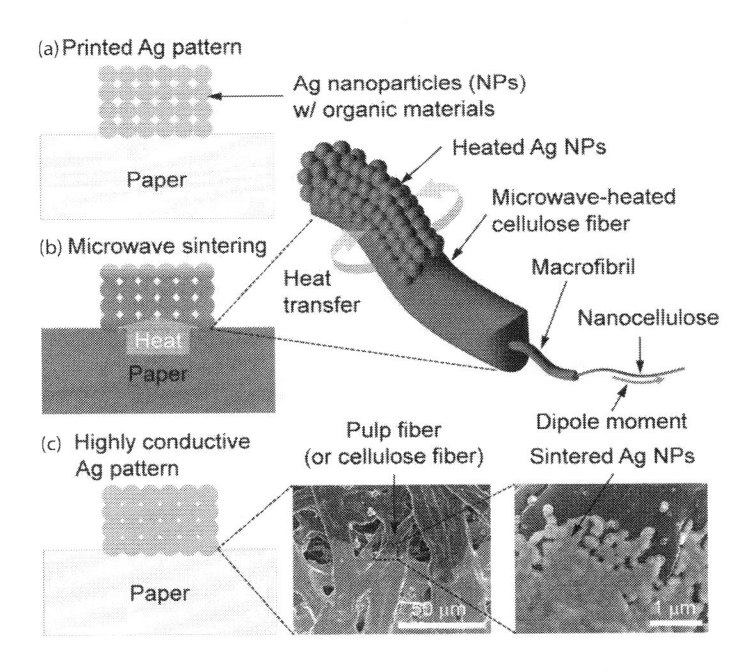

FIGURE 6.10 (a–c) Schematic illustration of the process for creating high-conductivity Ag patterns on paper in a short period of time through rapid and selective microwave heating of cellulose fibers. (From Jung, S. et al., *ACS Appl. Mater. Inter.*, 8, 20301–20308, 2016. With permission.)

NovaCentrix (USA). Photonic sintering uses intense pulsed light (IPL) with a broad spectrum in the visible range. By carefully controlling the intensity of the lamp, the pulse length, and the number of flashes, the temperature inside the printed film can be maintained at a constant value without causing damage to the underlying substrate (Perelaer et al. 2012). When a low amount of energy is applied to the system, no sintering takes place, whereas too much energy damages the substrate since the excess heat dissipates into the polymer foil. NovaCentrix introduced the reduction of copper oxide nanoparticle ink into copper thin-film by the PulseForge 3100 system (Das 2009). With the limitation to a thicker film (<1 μm), they have achieved approximately 3 times lower conductivity than the bulk Cu. With this technique, it is possible to obtain a conductivity only two to four times lower than that of bulk metals for sintered silver and copper printed layers, respectively (Albrecht et al. 2016). With a single printing pass, Secor et al. achieved a conductivity of ~2.5×10^3 S/m for high-solids-concentration graphene ink through photonic sintering (as shown in Figure 6.11). The ink was formulated with graphene flakes (typical thickness of ~2 nm with average particles size of 100 nm) and ethyl cellulose-based polymer (Figure 6.11a and b). To achieve the high electrical conductivity, the polymer is decomposed using photonic annealing, whereby a high-intensity pulsed xenon lamp is used to anneal the printed graphene patterns, as shown in Figure 6.11c. The authors have established graphene ink as a leading candidate for printed, flexible electronics (Secor et al. 2015). The only disadvantage being the high setup cost, this technique is advantageous for R2R processing and offers rapid processing with translational speeds of up to 100 meters per minute (Perelaer et al. 2012).

6.4.4 Laser Sintering

Laser (a coherent light source) can follow the conductive tracks and sinters them selectively, without affecting the substrate (as shown in Figure 6.12a). To perform laser sintering, alignment marks are necessary, which may be printed along with the conductivity test traces. Once printed, a high-power laser (such as carbon dioxide, neodymium:yttrium aluminum garnet [Nd:YAG] layers) is aligned with the printed alignment marks, and then the laser is raster scanned over the traces at several power levels from 0–75 W with a different resolution in dots per inch (dpi). It is important to select a laser with

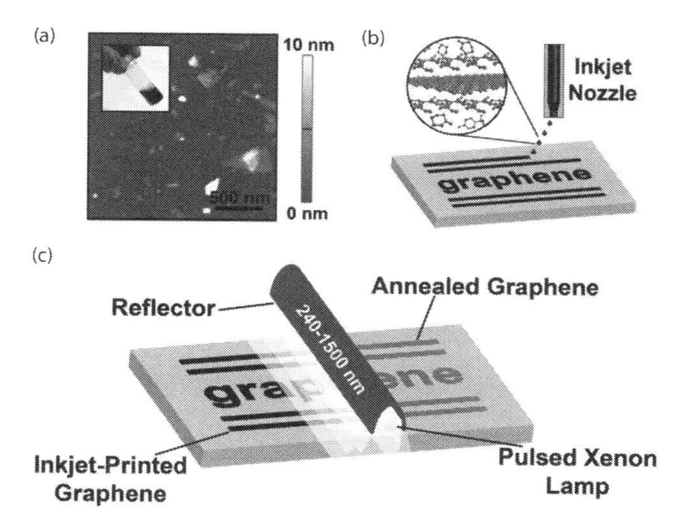

FIGURE 6.11 Inkjet printing and intense pulsed light (IPL) annealing of graphene. (a) Atomic force microscopy image of graphene flakes; inset: image of graphene ink vial. (b) Schematic illustration of inkjet printing of graphene. (c) Schematic illustration of IPL annealing applied to graphene patterns. (From Secor, E.B. et al., *Adv. Mater.*, 27, 6683–6688, 2015. With permission.)

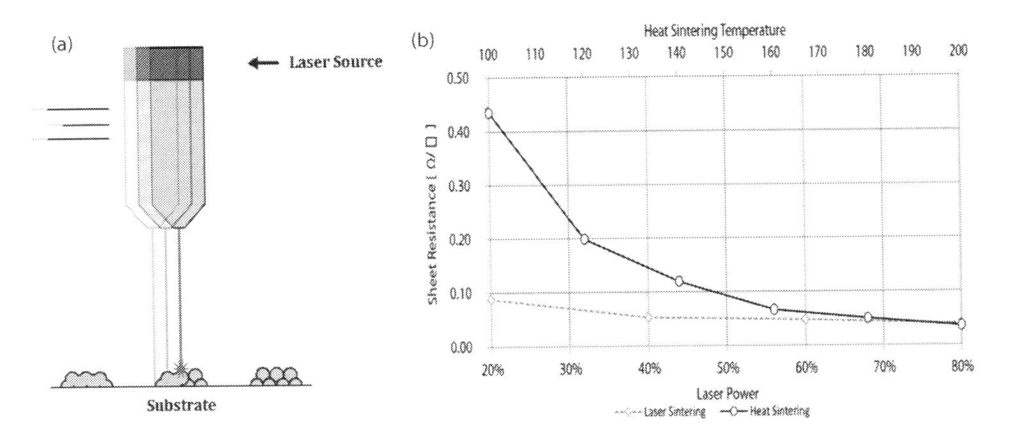

FIGURE 6.12 (a and b) Characterization of silver nanoparticle ink with laser and heat sintering. (From Cook, B.S. and Shamim, A., *IEEE Trans. Antenn. Propag. Eng.*, 60, 4148–4156, 2012. With permission.)

a beam width smaller than the smallest feature being sintered in order to ensure that the substrate is not damaged. Sintering for a 10 × 10 mm area can be performed in several seconds. This sintering method has been tested for silver nanoparticle ink on a paper substrate, which is very sensitive to high temperatures (Cook and Shamim 2012). The authors have inkjet-printed five layers of silver traces on a paper substrate, followed by laser sintering with a beam width of around 100 μm. The resultant sheet resistances have been compared against heat sintering of a trace of the same thickness. It can be seen from the graph (Figure 6.12b) that even at low powers, laser sintering produces sheet resistance similar to that of the trace which has been sintered at 200°C. Furthermore, the authors also show that laser sintering can achieve similar performance to that of heat sintering, but in a very short time (few minutes) as compared to heat sintering (1 hour). This was achieved without any damage to the substrate.

Later on, the Lewis Group introduced laser-assisted direct ink writing (laser-DIW), which combines the printing of concentrated silver nanoparticle inks with focused infrared laser annealing in order to rapidly create high-conductivity, ductile metallic wires and 3D architectures 'on-the-fly' in a one-step, additive process (Skylar-Scotta et al. 2016). They studied the effect of laser intensity, both continuous wave (CW) and high-frequency pulses (100-Hz, 1-ms pulse duration), on the electrical conductivity of laser-DIW silver wires printed on a glass substrate. They reported that by modulating the incident laser intensity over an order of magnitude, the silver resistivity can be varied by more than 3 orders of magnitude. Annealing via a CW laser achieved slightly lower electrical resistivity than that of a pulsed laser of the same peak intensity, reaching a minimum resistivity of 5.4×10^{-6} Ω·cm, compared with that of bulk silver. Unlike bulk thermal annealing methods, laser-DIW enables one to create patterned regions of low to high resistance simply by modulating the local laser intensity during silver ink printing. As compared to photonic sintering, which is better suited for larger patterns, laser sintering is suitable for smaller areas (usually in the scale of cm^2). As compared to thermal and microwave sintering, laser sintering also requires a high setup cost, and stringent alignment is necessary to follow the metal traces.

6.4.5 Room-Temperature Sintering

Another approach is to tailor the ink formulation to sinter at a temperature that is compatible with the polymer substrate. Recently, a new possibility to sinter a metal nanoparticle formulation at room temperature was discovered (Li et al. 2009, Magdassi et al. 2010, Grouchko 2011, Layani 2012, Tang et al. 2012). The idea is to inkjet print a tailored silver ink onto a substrate, which is precoated with a sintering agent, for example, poly(diallyldimethylammoniumchloride) (PDAC). Two opposite charged molecules neutralize each other; as a result, these molecules collide and make a sintered film. In one of the examples of the room-temperature sintering process, a film of concentrated (30 wt%) silver NP dispersion (stabilized with polyacrylic acid, PAA, based polymer) with wet thickness of film was formed, as depicted in the top illustration of Figure 6.13. After drying for 2 min at room temperature, an array of silver NPs was obtained (Figure 6.13a). Then, a solution of polycations (PDAC, 1 wt%) was placed as droplets by inkjet printing on top of the silver array (Figure 6.13b). The sintering agent (PDAC) triggered the neutralization or the removal of the polymer. As a consequence, the small-sized silver nanoparticles came into contact and spontaneously merged into a continuous metal structure (Figure 6.13c), yielding a conductivity of 20% that of bulk silver (Magdassi et al. 2010). Li et al. adopted a reactive inkjet printing approach: Two reactive materials were printed from separate print heads and were combined at the substrate, where a chemical reaction took place; as a product of the reaction, a conductive feature was formed. Conductive copper lines were directly written on paper through the inkjet printing of a copper salt and sodium borohydride ($NaBH_4$) as a reducing agent sequentially from two separate compartments of a multi-color HP cartridge. The obtained conductivity was 1/30 that of bulk metal copper (Li et al. 2009). Room-temperature sintering technology opens a new possibility for printing the conductive patterns on temperature-sensitive substrates for applications in flexible and plastic electronics. However, the conductivity is still lower than heat treatment; thus, the process needs to be tested with other polyelectrolytes and also validated with other types of metal nanoparticles. It will be interesting to see how effective the sintering is if particles size decreases to quantum dots.

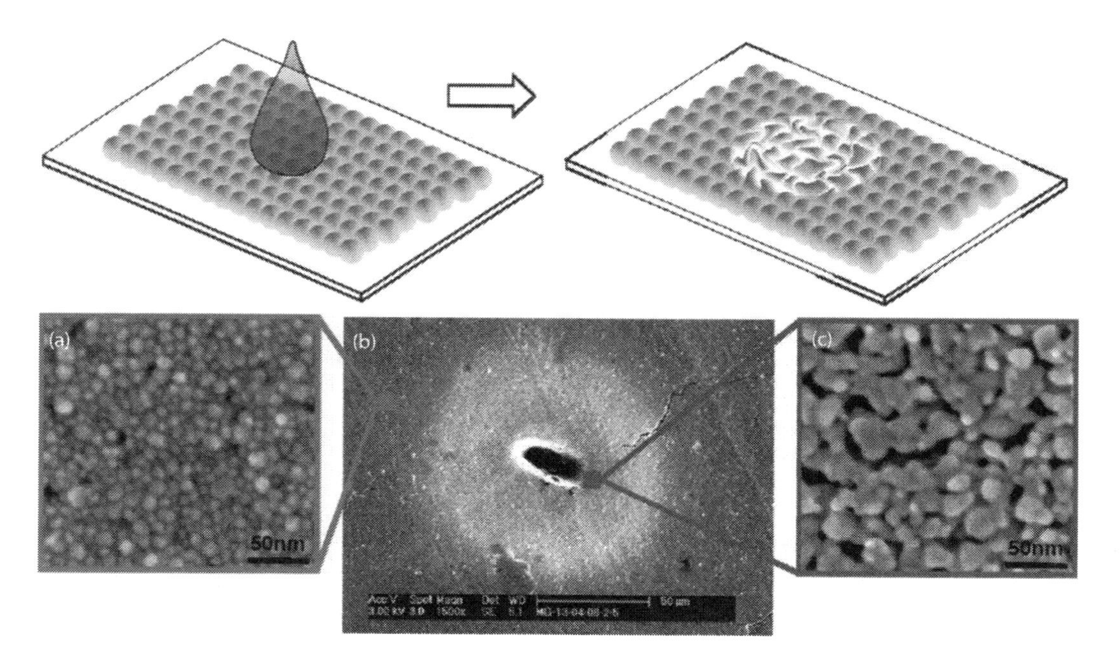

FIGURE 6.13 Top: Schematic illustration showing what happens when a droplet of PDAC solution is printed on silver NP arrays. Bottom: SEM image of a printed drop zone (b) and the magnified images of NP arrays after the contact with PDAC outside (a) and inside (c) the droplet zone. (From Magdassi, S. et al., *ACS Nano*, 4, 1943–1948, 2010. With permission.)

6.5 Printing Applications

There are a variety of applications that have emerged in recent years with the progress in printing techniques. Some of the prominent applications are presented in the following subsections.

6.5.1 Displays

Printing is an attractive choice for display applications due to roll to roll flexibility and lower manufacturing cost. Printing can realize not only the display elements, but also the circuitry involved to drive the display pixel in an active matrix. There are several display elements in consideration, for example, organic light-emitting diodes (OLEDs), liquid crystal display (LCD), and electrophoretic displays. Among them, OLED has received a lot of attention in the recent years. OLED is a light emitting technology made by placing a series of organic thin films between two conductors. When an electrical current is applied, light is emitted. OLEDs can be used for displays and lighting, with possible applications spanning TV sets, computer screens, mobile phones, decorative lighting, and more. It was a big news when Sony launched their 11-inch OLED-TV 'XEL-1' in 2007 (Sony 2007). Currently, Samsung's Galaxy series, equipped with OLED displays, is very popular. Since OLEDs emit light, they do not require a backlight and thus are thinner than LCD displays, more efficient, simpler to make, and they boast a better color contrast. However, in terms of printed electronics, OLED is not yet common, as printing OLED displays is a relatively challenging task for many reasons. A number of layers need to be deposited in the pixels' circuitry. Being able to place the right number of drops of the active materials into the pixels is a challenge, as is developing a process whereby the ink dries to deliver the flat films of the materials into the pixel (OLED ink jet printing: introduction and market status) https://www.oled-info.com/oled-inkjet-printing. There is another constraint based on the power requirement: Usually OLEDs generate their own light and therefore require more energy to operate in the form of a high-operating current. As a result, it seems unlikely that a printed transistors' performance will be able to satisfy the power requirement of such displays. Despite so many constraints, several inkjet-printable polymer light-emitting materials are commercially available, and many companies and research institutions are trying to develop products based on printing technology.

An LCD usually consists of two thin transparent substrates which are separated by a gap of a few micrometers. An LC, primarily of small organic molecules or dye, is inserted into the gap. The inner part of the assembly also consists of electrodes and an alignment layer. The application of an electric field changes the orientation of the LC, and therefore the polarization state of the light passing through the LC layer changes. The electrically controlled change of light polarization in conjunction with properly oriented thin-film polarizer's allows for the control of the transmission at each individual pixel (Alamán et al. 2016). As compared to OLEDs, pixels in LCDs are not self-emitting, thus a back light is required for illumination. In LCD, a color image is generated due to color filters. The color filters of LCD are generally characterized by the dot-matrix pattern that forms the three basic colors of red (R), green (G), and blue (B) for the full-color images. Each of the dots defines one tiny colored region (R, G, or B) through which the white backlight can pass and thus be filtered. In addition to color filters, an absorbing structure (black matrix) is required to prevent light leakage in between these subpixels (RGB). The fabrication of the color filters by inkjet printing supports low-cost and large-area manufacturing. Several groups have provided their efforts in developing inkjet printing technology and deposition strategies for printing color filters. Chang et al. (2004) prepared LCD color filters using UV-curable ink printed onto a glass substrate. The viscosity and surface tension of the inks was adjusted in order to achieve good jetting directionality, uniform drop size, and excellent wetting on the glass substrate. The researchers have observed that a CF_4-plasma-treated substrate showed improved color uniformity of the color filters when compared to O_2-plasma treatment. Koo et al. (2006) demonstrated the preparation and physical characterization of the red and/or green monochrome films on glass substrates by inkjet printing technology with pigment-based colorant resist inks. They have observed that as-synthesized inks with nanoparticle pigments showed improved transmittance and have superior compatibility with the substrate, rib wall, and black matrix in the sub-pixel cells. The organic dyes usually have superior spectral characteristics for color filters, but due to their low thermal stability, they are not widely used. In this regard, Kim et al. (2009) synthesized four perylene- and four phthalocyanine-based high thermal stability dyes and used to inkjet print color filters. They have shown that the transmittance and color properties of the prepared color filters were higher and the thermal stability similar to those of the pigment-dispersed color filters. Inkjet-printed droplets usually exhibit irregular changes of shape during drying, thus a proper control of drying and spreading of dots are important in printing color filters. In this regard, Chen et al. (2010) presented the comprehensive concept and thorough design of inkjet-printed LCD color filters using a microfluidic mechanism. Their experimental results showed that the solid RGB color layers could be self-assembled from the liquid droplets and formed with uniform thickness except for the neighborhood of the sidewalls. It is indicated that the sidewalls, serving as 'physical barriers', had a remarkable effect on confining and self-aligning the droplet flow within the desirable regions. Later on, Chang et al. (2011) also used UV-curable pigment-based ink containing hyperbranched polyurethane-methacrylates (HBPs) polymers. They have made color patterns on a glass substrate with inkjet printing, which exhibited good chemical resistance and high nanoindentation hardness. The black matrix, a grid-like structure that is used in color filters, has also been inkjet printed by Lu et al. They have demonstrated the fabrication with heterostructures to form stripe-arranged and delta-arranged thickness-tunable black matrices. Their self-aligning process with a 10 μm inkjet-printed grid improved the printing resolution and saves more than 75% of the black matrix's photoresist consumption compared to contemporary photolithography (Lu et al. 2014). From the above reports, it is concluded that inkjet printing is a viable technology for printing color filters in display applications. However, non-uniformity in the inkjet-printed colorant film thickness is a main concern, usually leading to undesired visual effects. Thus, ink wettability on the substrate, drop-volume control, and printing positioning are crucial and need to be optimized further for the mass production of industrial displays.

Another display element is the electrophoretic display, often called 'e-paper'. In the simplest implementation of an electrophoretic display, a bi-stable element consisting of a dye-filled cell containing charged colored particles of titanium oxide is used. As soon as an external voltage is applied across the cell, these colored particles move up or down. The polarity of the voltage governs whether the cell appears dark or light. As a result, an electrophoretic display forms images by rearranging charged pigment particles with an applied electric field. There are many approaches to electronic paper, with many companies developing technology in this area. Other technologies being applied to electronic paper include modifications of liquid crystal displays and electrochromic displays. Advantages of electronic paper include low power usage (power is only drawn when the display is updated), flexibility, and better readability than most displays. Comiskey et al. first reported an electrophoretic ink for all printed reflective electronic displays. They have reported the synthesis of an electrophoretic ink based on the microencapsulation of an electrophoretic dispersion. They have observed that the use of a microencapsulated electrophoretic medium solves lifetime issues and permits the fabrication of a bi-stable electronic display solely by means of printing (Comiskey et al. 1998). Andersson et al. (2002) demonstrated an active-matrix paper display entirely based on organic materials. The display was realized on a polyethylene-coated cellulose-based fine paper using ordinary printing techniques. Smart pixels, included in each of the crossings of the matrix, were achieved by combining an electrochemical transistor with an electro-chromic display cell. Later, Daniel et al. (2007) developed jet-printing technology to fabricate active-matrix backplanes for paper-like electrophoretic displays. The described methods are promising for low-cost flexible displays. However, the pixel size is still larger than what can be achieved with conventional processing methods, but it seems suitable for applications such as e-paper. Electrophoretic ink contained in micro-cells was found to be a useful vehicle for testing backplanes and for evaluating the process yield. The sealed electrophoretic medium is promising for flexible displays.

6.5.2 Radio-Frequency Microwave Components (Passives)

Recently, much progress has been made in conductive inks and printing techniques, and almost all the passive components such as antennas, filters (inductors and capacitors), resonators, and interconnects (transmission lines) can be reliably printed. Radio-frequency passives have dimensions that are defined by the frequency of operation, and they are required in all wireless devices from cell phones to frequency modulation (FM) radios. Special attention is given to issues important for RF devices, such as the loss tangent of the dielectric, extreme sensitivity to conductivity of inks, metal thickness at high frequencies due to the skin effect, and surface roughness of printed traces. In the following subsections, printed passive components will be discussed.

6.5.2.1 Antenna

In general, antennas play a critical role in communication systems. They are also typically large in size (of the order of wavelength) and therefore a suitable candidate for printing technologies. Their shape, geometry, and dimensions are controlled by numerous challenging specifications, such as desired radiation pattern, wide bandwidth, miniaturized size, flexibility, and lower cost. The first published inkjet-printed antennas started to appear in the literature around 2007 and were geared towards RFID integration on low-cost flexible substrates to target the inventory control and supply chain management markets (Yang et al. 2007, 2009, Orecchini et al. 2011). Subsequent inkjet-printed antennas have been produced with sensor integration, such as an RFID integrated gas sensor and a fully paper-mounted temperature-sensing RFID module with a printed antenna (Yang et al. 2009). However, the early reported inkjet-printed antennas were relatively low gain and narrow band, typically operating in the ultra-high-frequency (UHF) RFID band. The omnidirectional behavior and low gain of these inkjet-printed antennas restricted them to applications where communication range is not important. On the other hand, several applications with the need of long-range communication require higher gains and directional beam antennas. To demonstrate high-gain antennas on a substrate, a smart choice of antenna is important due to the relatively high loss tangent typically inkjet-printable low-cost substrates exhibit at microwave frequencies. Resonant antennas are highly sensitive to lossy substrates; however, travelling or leaky wave antennas have much less dependence on substrates and tend to have higher directivities. In this regard, the Manos Group, for the first time, demonstrated ultrawideband (UWB) antennas through inkjet printing of conductive inks on commercially available paper sheets. The printed antenna worked for frequencies up to 10 GHz, with antenna efficiency better than 80% throughout the whole band (Shaker et al. 2011). Later on, Cook et al. reported a UWB inkjet-printed 'Vivaldi' antenna on a paper substrate (Cook and Shamim 2012). The fabricated antenna exhibits a significantly higher gain of up to 8 dBi and wide bandwidth from MHz frequencies to 10 GHz. Shah et al. introduced a novel hybrid printing technology that realizes a microstrip patch antenna. The antenna's dielectric materials are built using a 3D printer (using polylactic acid, PLA filament), and a metallic pattern with silver nanoparticle ink is printed on paper using a commercial inkjet printer. The printed patch antenna shows a return loss of 24 dB and a peak gain of 7 dBi at 1.9 GHz (Shah et al. 2016). The Manos Group also utilized an additive inkjet-printing multilayer fabrication process for two high-gain, multidirector Yagi-Uda antennas on a flexible substrate for use within the 24.5 GHz industrial, scientific, and medical (ISM) band. An end fire realized gain of 8 dBi is achieved within the 24.5-GHz ISM band, which presented the highest-gain inkjet-printed antenna in the millimeter-wave regime (Tehrani et al. 2016). McKerricher et al. (2016) demonstrated the first fully 3D multijet printing process with integrated polymer and metal. First, a honeycomb structure is 3D printed using a multijet inkjet-printing process, as shown in Figure 6.14a. The inkjet-printed polymer is smoothed and leveled by a planarizer as it is deposited (Figure 14b), providing surface roughness better than 100 nm root mean square. Both the dielectric polymer and silver conductor are additively deposited by inkjet printing, which makes the process versatile (Figure 6.14c). The silver nanoparticle ink is sintered with a carbon dioxide engraving laser, providing a conductive layer without damaging the dielectric. Finally, the use of a printed honeycomb structure allows tailoring

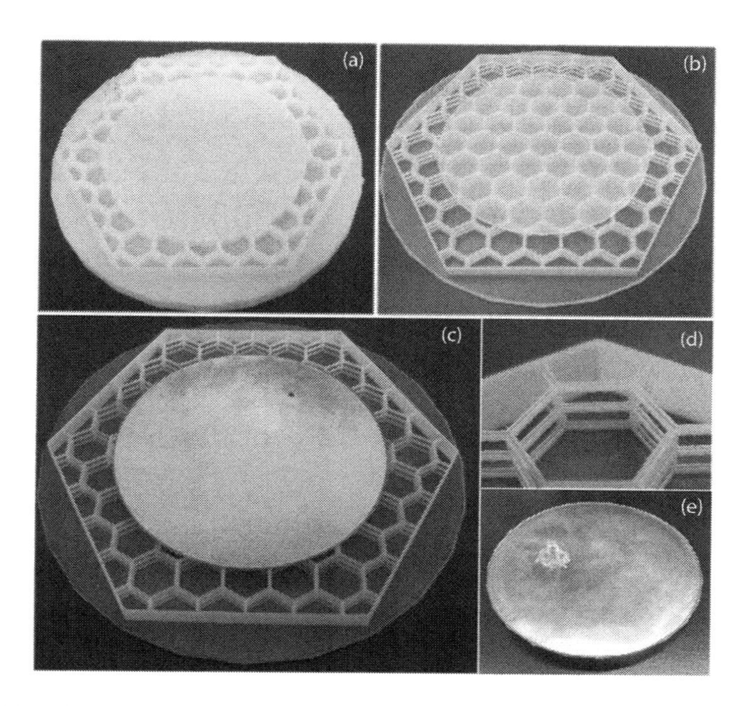

FIGURE 6.14 3D multijet printed structure with (a) wax support material, (b) remaining polymer shell after melting the wax, (c) final antenna with inkjet-printed silver ink, (d) zoomed-in view of honeycomb, and (e) back of antenna with probe feed. (From McKerricher, G. et al., *IEEE Antenn. Wirel. Prop. Lett.*, 15, 544–547, 2016. With permission.)

of the dielectric constant and significantly improves the dielectric loss. Two patch antennas have been fabricated for comparison, one with a solid printed substrate and the other with a honeycomb substrate. The honeycomb substrate antenna has a peak gain of 8 dBi and an efficiency of 81%. For comparison, the solid substrate patch has a gain of 5.8 dBi and a measured efficiency of 67%. Recently, inkjet-printed 3D fractal antennas have been reported by Memon et al. They have printed 3D space-filling Hilbert-curve (HC) fractal-geometry-based antennas on Kodak photo paper, using a home printer. The peak gains for the second- and third-order HC fractal antennas are −8.46 and −1.39 dBi, respectively. The proposed antennas operate in the very-high-frequency (VHF) band and are well suited for military radio communication in the United States of America (Memon et al. 2017).

Recent years have witnessed growing research in the field of flexible electronics. The antenna, among all other components on a flexible substrate, is the most prone to performance degradation because its radiation characteristics are severely affected by substrate deformation. It was therefore critical to study and analyze the flexibility aspect of the antenna. In this regard, Ahmed et al. (2015) presented a novel and conformal multiband antenna using an inkjet-printing process on a Kapton substrate. The antenna covers four wide-frequency bands with measured impedance bandwidths of 54.4%, 14%, 23.5%, and 17.2%, centered at 1.2, 2.0, 2.6, and 3.4 GHz, respectively, and thus enabling it to cover Global System for Mobile (GSM) 900, Global Positioning System (GPS), Universal Mobile Telecommunications System (UMTS), wireless local area network (WLAN), The industrial, scientific and medical (ISM), Bluetooth, long term evolution (LTE) 2300/2500, and Worldwide Interoperability for Microwave Access (WiMAX) standards. The antenna has an omnidirectional radiation pattern with a maximum gain of 2.1 dBi. To characterize the flexibility of the antenna, the fabricated prototype was tested in convex- (Figure 6.15b) and concave-bent (Figure 6.15c) configurations for radii of 78 and 59 mm. The overall performance remained unaffected except for a minor shift of 20 and 60 MHz in S_{11} for concave bending at both radii (as shown in Figure 6.15e). Similarly, Abutarboush et al. (2016) reported an inkjet-printed monopole antenna on a

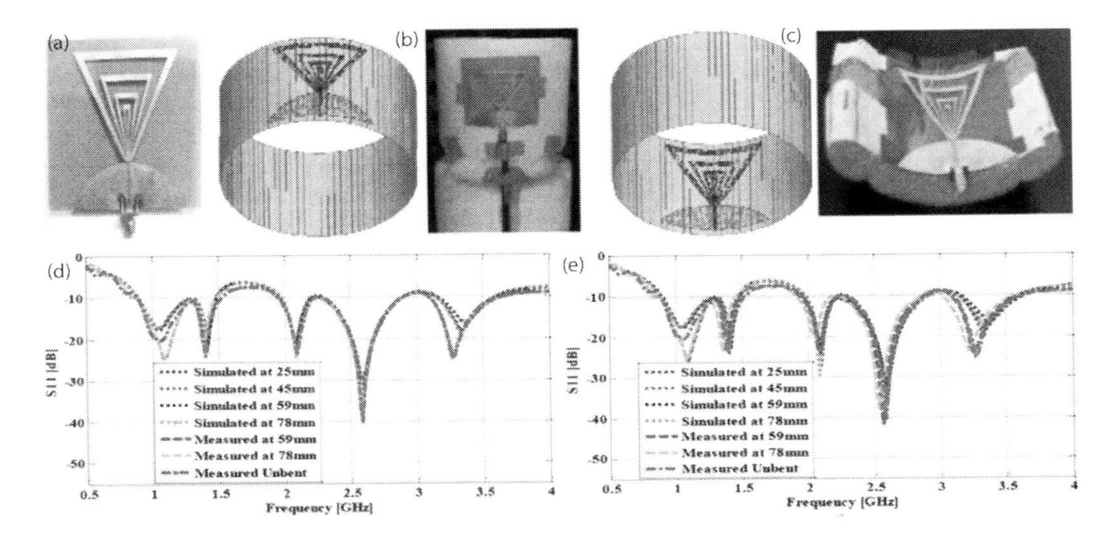

FIGURE 6.15 Antenna in (a) flat position and bending; (b) convex configuration; and (c) concave configuration. Simulated and measured S_{11} of the antenna when unbent and bent on the foam of different radii in (d) convex and (e) concave configurations. (From Ahmed, S. et al., *IEEE Antenn. Wirel. Prop. Lett.*, 14, 1802–1805, 2015. With permission.)

low-cost resin-coated paper substrate. Measured results have shown that the antenna under flat and bent conditions can cover most of the bands for mobile and wireless applications. When the antenna is bent, the bandwidth is increased, but the gain is dropped. Bending the antenna does not change the radiation patterns much. The compact, lightweight, and conformal designs as well as multiband performance in bent configurations prove the suitability of the antenna for future electronic devices.

6.5.2.2 Capacitor, Inductor, and Filter

Capacitors and inductors are fundamental building blocks in circuits. Initially, inkjet printing was successfully utilized to print these components for low-frequency utilization, for example, up to 10 MHz (Ko et al. 2007, Kang et al. 2012, Li et al. 2012). However, inkjet-printed high-performance RF-based capacitors and inductors are limited. Cook et al. demonstrated the first flexible multilayer inkjet-printed capacitors using two custom formulated polymer-based dielectric inks such as SU8 and PVP. The printed microwave capacitors showed values up to 50 pF and self-resonant frequencies of up to 3 GHz, allowing for their use in the 800 MHz RFID and 2.4 GHz bands (Cook et al. 2013). Later, Cook et al. also demonstrated vertically integrated inkjet-printed inductors on a flexible liquid crystal polymer (LCP) with high levels of performance and repeatability. Printed spiral inductors showed an inductance of 10 nH and 25 nH with a maximum quality (Q) factor of over 20 at 1 GHz. (Cook et al. 2014). Garret et al. reported fully inkjet-printed multilayer capacitors and inductors using PVP ink as the dielectric layer and silver nanoparticle ink as the conductor. Inkjet-printing through vias, created with a novel dissolving method, have been used to make RF structures in a multilayer inkjet printing process. Spiral inductors from 10 to 75 nH have been realized with quality factors of ~5. The 10-nH inductor exhibits a self-resonant frequency slightly below 1 GHz. MIM capacitors demonstrated values ranging from 16 to 50 pF with self-resonant frequencies of over 1.5 GHz. It was concluded that the successful implementation of inductors and capacitors in an all inkjet-printed multilayer process with vias is an important step toward fully inkjet-printed large-area and flexible RF systems (McKerricher et al. 2015). Later on, the same group demonstrated 10–35 nH inkjet-printed spiral inductors on a flexible plastic substrate using high conductive particle-free SOC-based ink. The printed inductors showed maximum quality factors of greater than 10 at self-resonant frequencies above 1.5 GHz (Vaseem et al. 2016b). Recently,

McKerricher et al. also demonstrated inkjet-printed thin-film radio-frequency capacitors based on an in-house developed sol-gel-derived alumina dielectric ink. MIM capacitors showed a high capacitance density of >450 pF/mm^2 and quality factors of ~200. The devices have high breakdown voltages of >25 V, with extremely low leakage currents of <2 × 10^{-9} A/cm^2 at 1 MV/cm. The capacitors compare well with similar Al_2O_3 devices fabricated by atomic layer deposition (McKerricher et al. 2017).

Fully inkjet-printed 3D objects with integrated metal provide exciting possibilities for on-demand fabrication of radio-frequency electronics such as inductors, capacitors, and filters. In this regard, McKerricher et al. (2017b) utilized similar SOC ink cured with a low-cost infrared lamp at only 80°C, achieving a high conductivity of 1 × 10^7 S m^{-1}. Higher conductivity at the lower sintering temperature of silver ink made it compatible with 3D-printed objects. By inkjet printing, the infrared-cured silver together with a commercial 3D-inkjet ultraviolet-cured acrylic dielectric, a multilayer process has been demonstrated. By using a smoothing technique, both the conductive ink and dielectric provide surface roughness values of <500 nm. A radio-frequency inductor and capacitor exhibit state-of-the-art quality factors of 8 and 20, respectively, and match well with the electromagnetic simulations. These components (inductor and capacitor) are implemented in a lumped element radio-frequency filter, as shown in Figure 6.16. Figure 6.16a shows a microscopy image of the fabricated filter. There is an inset showing the corresponding placement of the capacitor and inductor for clarity. The capacitor area is further visualized by the cross-sectional SEM image in Figure 6.16b. From the image, the printed silver layers are separated by an 11-μm-thick dielectric layer. The measured frequency response of the filter in Figure 6.16c and d matches well with the simulations with a 3-dB cut-off at 2 GHz. There is a signal rejection of 10 dB at 3 GHz and better than 20 dB at 4 GHz. The printed filter shows an impressive insertion loss of 0.8 dB at 1 GHz, proving the utility of the process for sensitive radio-frequency applications.

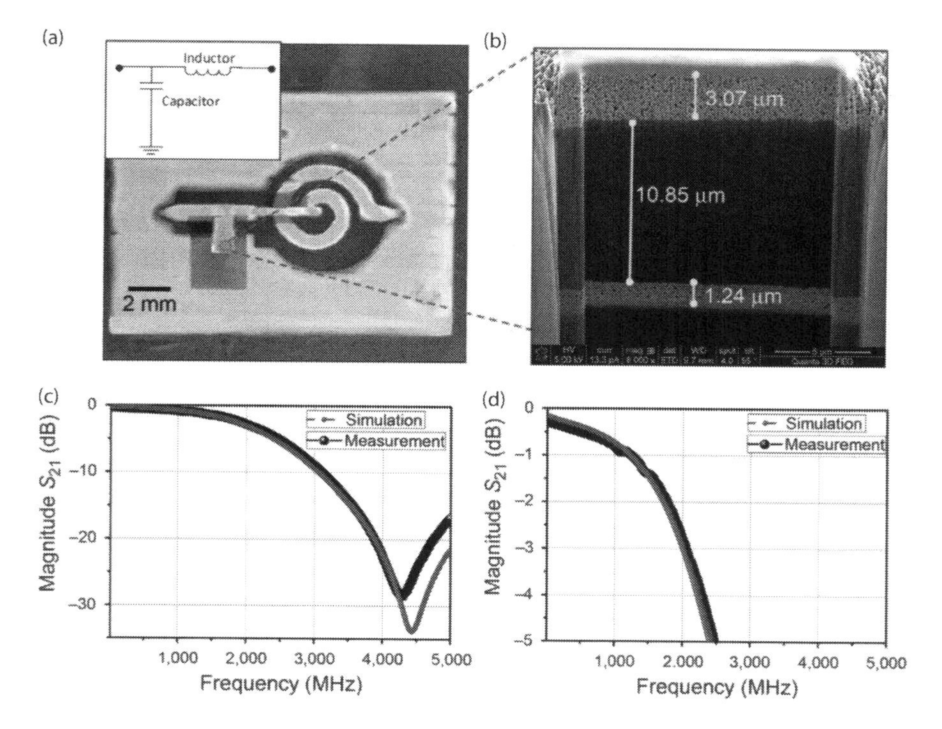

FIGURE 6.16 (a) Microscope image of the printed filter. (b) SEM focused ion beam cross-section image through the capacitor area, showing the thickness of the printed dielectric and the top and bottom electrodes of the filter. (c) Measured and simulated S21 filter response versus frequency. (d) Zoomed in view of the filter response.

6.5.3 RFID Tags

An RFID tag is a device which provides storage and remote reading of data from items equipped with such tags. The RFID tag, typically, consists of a silicon microchip (integrated circuit, IC), mounted on top of an antenna. The antenna is responsible for uni- or bi-directional communication with a reader, as well as for providing power to the tag. RFID devices are generally being deployed in four main communication bands, i.e., the low-frequency range up to 125 kHz (LF band), 13.56 MHz (HF band), 900 MHz (UHF band), and 2.4 GHz (microwave band). RFIDs represent a huge market in applications like identification, logistics, or automation. For such widespread use, tags must cost less than 1–2 cents to be economically viable. Removing ICs can significantly reduce RFID cost. However, the major challenge in designing chipless RFID is data encoding and transmission. Printing antennas on plastic and paper substrates with the use of metal NPs inks is also a promising approach for production of low-cost RFID tags. In this regard, the Manos Group printed a compact RFID tag module with a T-match folded bow-tie structure to match the antenna to the IC on the characterized paper substrate and demonstrated very good overall performance (Yang et al. 2007). The Leena Ukkonen Group demonstrated the design of a novel chipless RFID tag, which has been inkjet printed on a PI substrate. They have combined silver nanoparticle-based ink with a CNT-doped resistive organic ink to create a RFID tag with 'near-transparent' configuration. By adding some transparent elements, they obtained various electromagnetic responses for tags which are visually similar. Their novel coding technique for chipless RFIDs is based on amplitude shift keying of independent resonant peaks. With three scatterers, more than 6 bits can be encoded. The tag is composed of three dual-rhombic loop resonators for a total size in centimeters. It operates within the ultrawideband frequency range of 3–6 GHz (Vena et al. 2013). Later on, the same group also proposed the use of a passive UHF RFID tag integrated into plywood boards for the identification and tracking of individual plywood boards and end products of plywood (Virtanen et al. 2013). The tags are embedded inside plywood by direct inkjet printing the tag antennas on pure birch veneer (a kind of plywood). The measurements showed that the tags printed on veneer and embedded inside 2 mm thick plywood board exhibited theoretical read ranges of 7.9–10.1 meters. Such read ranges obtained meet the demands of the plywood industry and offer reliable identification even in challenging environments.

Jung et al. (2010) demonstrated the first fully printed 1-bit RF tag that contains a rectifier that can provide at least 10 V direct current (DC) from 13.56-MHz-coupled alternating current (AC) from the reader and a ring oscillator that can generate a stable 100-Hz clock signal by the dc power generated from the rectifier. The 1-bit RF tag was completely printed in three steps: (1) antenna, electrodes, gate electrodes, and gate dielectrics were R2R gravure printed on plastic foils at a web speed of 5 m/min (Figure 6.17); (2) the ring oscillator to generate clock signals under 10 V dc was added; and (3) the rectifier to yield 10 V dc from 13.56 MHz AC was added. The printed SWCNT-based transistor showed a mobility of 5.24 $cm^2/V{\cdot}s$ at an on-off ratio of 100, which is good enough to operate 1-bit RF tags at a switching speed of 100 Hz. The 1-bit RF tag in this work has an estimated cost of 0.03 dollar/unit.

6.5.4 Transistors

Though there has been a lot of progress in conductive inks, dielectric and semiconductor inks have not progressed that much. That is why there is a lot of work on printed metallic components such as antennas, interconnects, and electrodes, etc. Fully printed transistors are still in their infancy and therefore most of the printed systems go for a hybrid approach through integration of traditional silicon-based ICs and printed passive components, as will be discussed later in this chapter. Having said that, there is a lot of interest in fully printed transistors for a number of applications such as displays, RFID tags, etc. Mobility, switching speed, on/off current ratio, and low operating voltage are the key performance metrics for printed transistors. A typical transistor consists of a semiconducting channel material, insulating dielectric, conductive metallic source/drain and gate electrodes. Printing different material system layers on top of each other to fabricate a high performance transistor offers its own set

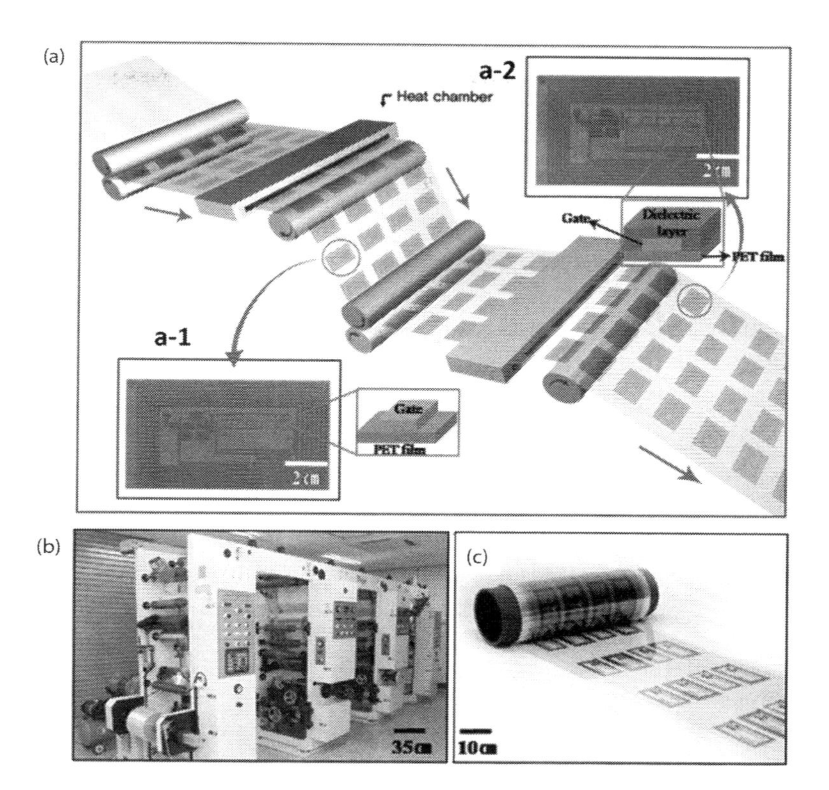

FIGURE 6.17 (a) (a-1) Schematic illustration of R2R gravure-printed antennas, electrodes, and wires after passing through the first printing unit using the silver-based conducting ink. (a-2) Printed dielectric layers on selectively designated spots using high-κ dielectric ink after passing through the second printing unit. (b) Gravure printer used in this work with two printing units. (c) Roll image of the completed R2R gravure-printed 13.56-MHz antenna, electrodes, wires, and dielectrics used as precursor for printing 13.56-MHz-operated 1-bit RF tags. (From Jung, M. et al., *IEEE Trans. Elec. Dev.*, 57, 571–580, 2010. With permission.)

of challenges, i.e., wetting, spreading, and adhesion issues. Conventional lithography-based fabrication is quite mature and thus it is easy to control parameters for realizing very short channel lengths on thin dielectric layers with decent metallic contacts. However, this is not true for fully printed transistors.

In literature, many partially printed transistors have been reported where either Source (S)/Drain (D) is printed or only semiconductor layer is printed (Vaseem et al. 2013, Avis et al. 2014, Baby et al. 2015, Han et al. 2016). However, very few reports are available where fully printed transistors have been demonstrated. The Ali Javey Group demonstrated one of the first high performance fully printed SWCNT-based transistor supported on a flexible substrate (Lau et al. 2013). They used barium titanate (BTO) nanoparticles-based ink as a high k-dielectric and Ag for S/D/Gate (G) electrodes. By using a scalable inverse gravure printing process with high overlay printing registration accuracy of ±10 μm, they were able to achieve a mobility of 9 cm^2/vs with a channel length of 85 μm and a processing temperature of 150°C. Later on, the Vivek Group reported the first fully printed transparent inorganic metal oxide-based transistor devices (Jang et al. 2015). In that work, by employing inkjet printing of thin sol-gel precursor films, they have done a fully printed tin dioxide (SnO_2)-based transistor on glass using printed zirconium oxide (ZrO_2) as dielectric and antimony-doped tin oxide (ATO) for S/D/G electrodes. They have obtained a mobility of 11 cm^2/vs with a conventional channel length of 100 μm and a processing temperature of 500°C. Recently, Junbiao Peng et al. came up with a new precursor-based InGaO semiconductor ink and an indium-doped ITO precursor ink for S/D/G electrodes (Li et al. 2017). They have utilized the solvent etching method and oxygen plasma treatment to prepare surface energy patterns, thus were able to fully inkjet print the transistor. The printed device shows a mobility of 11.7 cm^2/vs with 40 μm channel length and processing temperature in the range of 80°C–500°C. The Vivek Group demonstrated the low temperature processing of a fully inkjet-printed transistor using aqueous indium oxide (InOx) sol-gel-based ink as the semiconductor, sol-gel Al_2O_3 as a dielectric ink and aluminum-doped cadmium oxide (ACO) sol-gel ink for the S/D/G electrodes (Scheideler et al. 2017). The printed transistor device achieved a mobility of 19 cm^2/vs at 250°C with a channel length of 50 μm, however, it may be noted that the dielectric was spin-coated and not printed in this work.

It should be noted that none of the fully printed transistor papers report any kind of RF characterization and subsequently no fully printed RF circuits exist in literature. It is observed that relatively higher mobilities can be achieved with inorganic semiconductors, thus new or different inorganic semiconductor materials must be explored to maximize the transistor mobility. However, one issue is that inorganic semiconductors typically require higher processing temperatures (>200°C). These high temperatures are not compatible with most of the low-cost and flexible plastic substrates. Scaling down to temperatures below 200°C is quite challenging, because the decomposing and converting precursor compounds typically require higher temperatures. Lowering the temperature, especially for the dielectric layer is even more challenging as the leakage and dispersion gets worse for these low temperature processed films. In addition, not much investigation has been done about the RF properties (particularly loss) of these printed dielectric films. It is observed that step coverage, wetting, pinholes, etc., are serious concerns for printed transistors. These issues get magnified for short channel devices which require thinner dielectrics to achieve good RF performance. Thinner dielectrics are inherently more sensitive to the underlying topology of the gate electrode. Further, an ideal channel length for RF transistors should be at least in the submicron range. To achieve narrower channel lengths, new printers and printing processes must be explored which can print at least sub-micron (<1 μm) or even smaller channel lengths. In addition to the above-mentioned materials and printing challenges, overlap capacitance (arises due to poor layer to later registration) is another challenge which needs to be addressed, particularly for the high-frequency operation of fully printed transistors.

6.5.5 Printed Sensor Systems

6.5.5.1 Printed Sensor

A sensor is the vital part of a typical Internet of things (IoT) system that can measure environmental, industrial, and biological parameters. The connectivity mechanism between these sensors and their central controller is also important. There is a plethora of applications using these two enablers in the domain of healthcare, wearables, transportation, smart homes, connected cities, and industries. The staggering number of required sensors and their associated cost has motivated researchers to find ways of making them pervasive and reduce their cost to such an extent that they can potentially become disposable. To that end, RFID technology has emerged as a strong candidate for widespread usage in the above-mentioned scenarios. In the last few years, RFID has transitioned from a mere identification technology to a versatile platform that has been used for tracking, localization, and remote sensing. In the last of these categories, RFID has been demonstrated for wireless gas and humidity sensing (Yang et al. 2009, Andersson et al. 2013, Gao et al. 2013). In addition, many similar inkjet-printed sensors, without RFID functionality, have also been presented on various substrates such as paper polyimide and PET (Sarfraz et al. 2012, Weremczuk et al. 2012, Molina-Lopez et al. 2013, Rivadeneyra et al. 2014).

6.5.5.2 Printed Wireless Sensor System

Advances in wearable and flexible electronics along with the growth of wireless networking have created new paradigms of applications for smart living. An estimated 200 billion devices will be connected to the Internet by the year 2025 in the IoT realm as predicted by Intel corporation (IDC, Intel, United Nations). Recently, Quddious et al. (2016) have demonstrated an inkjet-printed, fully passive sensor capable of sensing humidity and gas. The fabricated sensor prototype is shown in Figure 6.18;

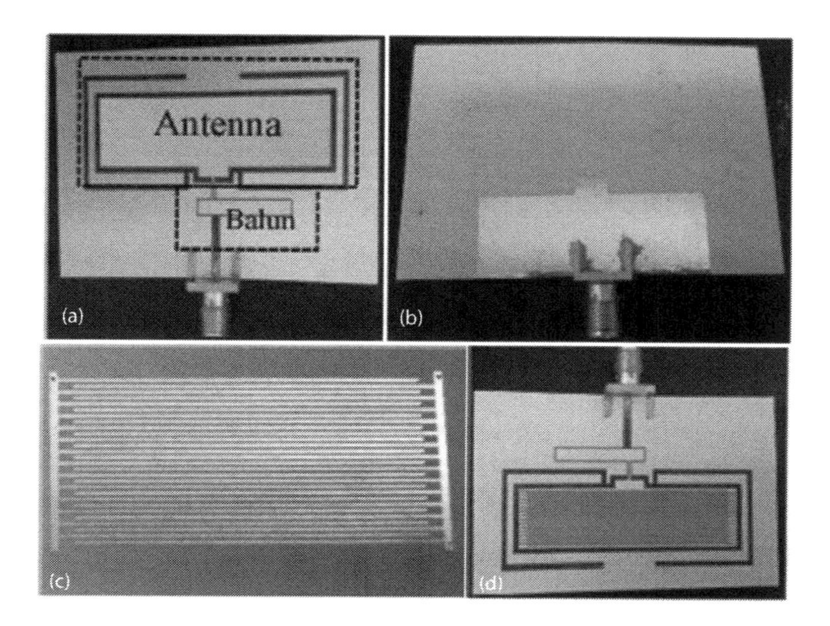

FIGURE 6.18 (a) Top view of antenna; (b) bottom view of antenna; (c) inkjet-printed interdigitated electrode; and (d) complete prototype with antenna, balun, and interdigitated electrode.

the prototype is composed of a printed specialized dipole antenna for wireless sensing (a & b) and a printed interdigitated electrode (c) with a customized printable gas-sensitive ink (d). The interdigitated electrode printed on a paper substrate provides the base conductivity that varies during the sensing process. Aided by the porous nature of the substrate, a change in relative humidity from 18% to 88% decreases the electrode resistance from a few mega-ohms to the kilo-ohm range. For gas sensing, an additional copper-acetate-based customized ink is printed on top of the electrode, which, upon reaction with hydrogen sulphide gas (H_2S), changes both the optical and the electrical properties of the electrode. A fast response time of 3 min is achieved at room temperature for a H_2S concentration of 10 ppm at a relative humidity (RH) of 45%. The passive wireless sensing is enabled through an antenna in which the inner loop takes care of the conductivity changes in the 4–5 GHz band, whereas the outer-dipole arm is used for chipless identification in the 2–3 GHz band.

Farooqui et al. (2017) proposed disposable, compact, dispersible 3D-printed wireless sensor nodes with integrated microelectronics that can be dispersed in the environment and work in conjunction with few fixed nodes for large-area monitoring applications. As a proof of concept, the wireless sensing of temperature, humidity, and H_2S levels are shown, which are important for two critical environmental monitoring situations, namely, forest fires and industrial gas leaks. These inkjet-printed sensors and an antenna have been realized on the walls of a 3D-printed cubic package that encloses the microelectronics developed on a 3D-printed circuit board (as shown in Figure 6.19). Hence, 3D printing and inkjet printing have been uniquely combined in order to realize a low-cost, fully integrated wireless sensor node.

6.5.5.3 Printed Wireless Wearable Sensor System

Farooqui et al. also demonstrated a low-cost inkjet-printed smart bandage for wireless monitoring of chronic wounds (Farooqui and Shamim 2016). They presented an unprecedented low-cost continuous wireless monitoring system, realized through inkjet printing on a standard bandage that can send early warnings for parameters such as irregular bleeding, variations in pH levels, and external pressure at the wound site. In addition to the early warnings, this smart bandage concept can provide long-term wound

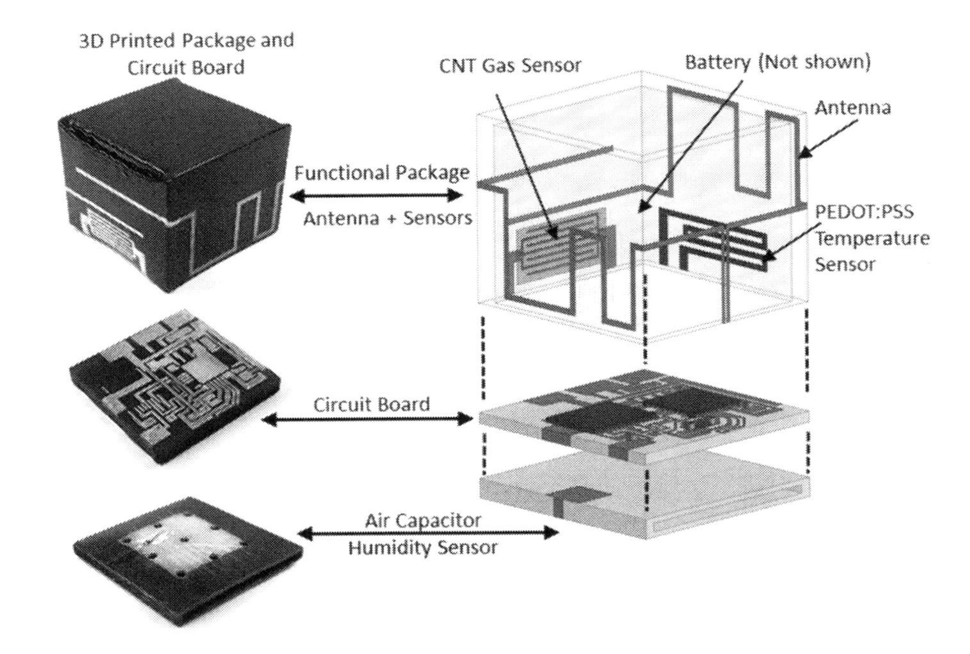

FIGURE 6.19 An illustrative depiction to show the device assembly. The sensor node is comprised of an assembly of a 3D-printed functional package, inkjet-printed circuit board, and inkjet-printed air capacitor. (From Farooqui, M.F. et al., *Adv. Mater. Technol.*, 2, 1700051–1700059, 2017. With permission.)

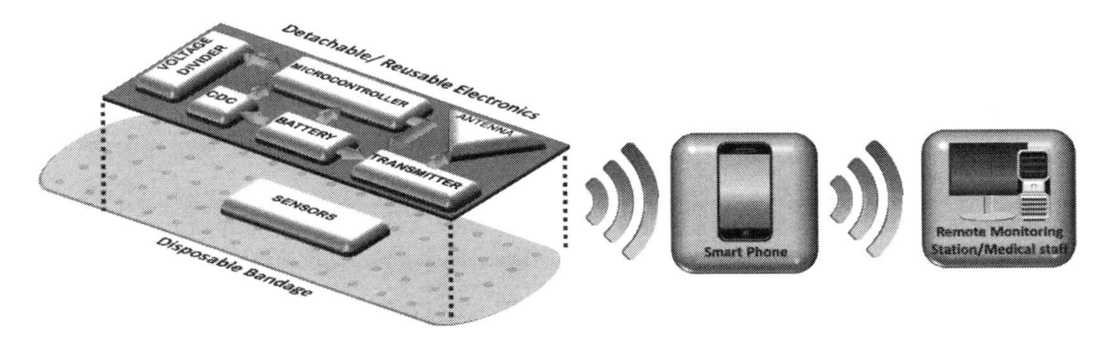

FIGURE 6.20 An illustrative depiction of inkjet-printed smart bandage for the wireless monitoring of chronic wounds.

progression data to the healthcare providers. The smart bandage assembly comprises a disposable and a reusable part (Figure 6.20). The sensors are printed on a disposable bandage strip that can be disposed of after use. The first process involves the fabrication of the bottom sensor's electrode on a paper substrate. Carbon ink is used for the metallization of the bottom electrode. The second process involves the fabrication of both the sensor electronics and the top sensor's electrode on Kapton tape, which is placed on the top of the bandage. Silver nanoparticle ink is used for inkjet printing wireless electronics such as top electrode, circuit board, and antenna. After mounting the components, the detachable sensor electronics as well as the bottom sensor electrode are attached to a disposable bandage strip. The capacitive sensor placed across the bandage both senses bleeding and measures external pressure. For pH measurements, different acid and base solutions are exposed to the electrodes, and resistance change is measured. This work is an important step towards futuristic wearable sensors for remote healthcare applications.

6.5.6 Printed Wearable Tracking System

Localization, that is, the ability to track the position of an object, is another research area which has seen a tremendous boom in the past few years. A diverse range of applications have benefited from advancements in this technology, ranging from healthcare monitoring and generic child, pet, or vehicle tracking to a number of complex location-aware services, such as context-aware marketing and crowd management. Despite the abundance of localization applications, the tracking devices have never been truly realized in E-textiles. Standard PCB-based devices are obtrusive and rigid and hence not suitable for textile-based implementations. An attractive option would be the direct printing of the circuit layout on the textile itself, negating the use of rigid PCB materials. However, the high surface roughness and porosity of textiles prevents the efficient and reliable printing of electronics on textile. In this regard, Krykpayev et al. (2017) manually screen printed an interface layer on the textile, then they inkjet printed a complete localization circuit integrated with an antenna on the textile (Figure 6.21).

6.5.7 Batteries

Batteries are the essential power source for almost all portable electronic devices. A typical battery consists of cathode, anode, and current collector, usually arranged in a closed system. The energy is produced due to an electrochemical reaction which is stored in the form of a chemical solution. As soon as the battery is connected to an external load, the circuit is complete. As a result, an electrical current flows between the anode and the cathode. Traditionally, electronics have been designed around commercial batteries with the shape of prism, cylinder, and coin. However, these batteries are bulky, rigid, and non-flexible, making them unsuitable for powering flexible electronics (Gaikwad et al. 2015). A battery for a flexible electronic device should be thin, bendable, and mechanically compliant. Batteries fabricated using printing processes have the advantages of low cost, flexible form, ease of production, and seamless integration with electronic devices. Over the past years, there has been significant progress towards using printing-based processes to fabricate batteries. The fabrication process for a printed

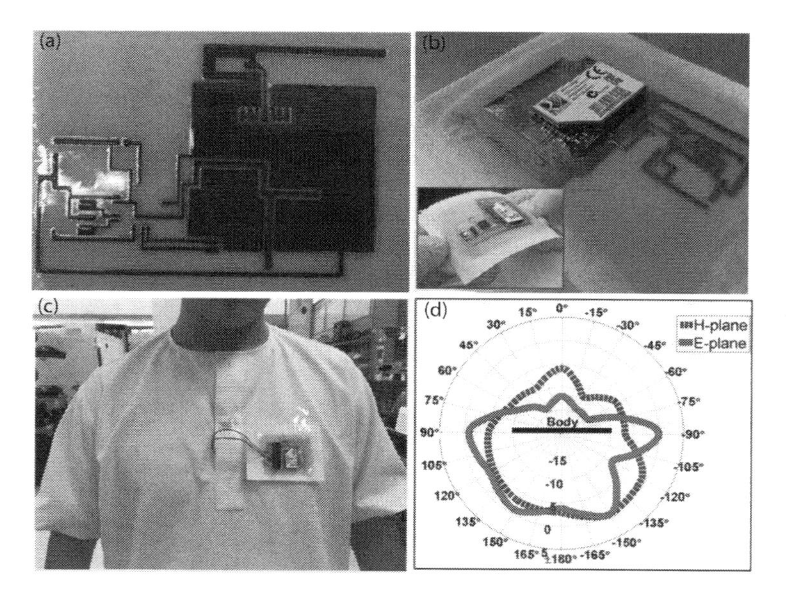

FIGURE 6.21 (a) Circuit layout printed on the interface layer (b) components integration. The inset in (b) shows the flexibility of the device on textile. Radiation pattern measurements on body (c) E-plane (d) curves in E plane. (From Krykpayev, B. et al., *Microelectron. J.*, 65, 40–48, 2017. With permission.)

battery used inks to print active layers, current collectors, and electrolytes. There are several printing techniques, such as blade coating (Hu et al. 2010), dispenser printing (Sun et al. 2013), stencil printing (Gaikwad et al. 2013), spray printing (Singh et al. 2012), inkjet printing (Ho et al. 2009), flexographic printing (Wang et al. 2014), and screen printing (Kang et al. 2014), that are utilized for the fabrication of batteries. Among them, inkjet, screen, and flexographic printing are viable solutions for large-area printing. Ho et al. (2009) fabricated 3D Zn-AgO microbatteries using a custom-built super inkjet (SIJ) printer. Using electrohydrodynamic-actuation-based technology with femtoliter droplet size, they have printed silver electrodes with evenly distributed pillars to increase the areal capacity of the battery. After the two separate electrodes are printed, it was dipped in a KOH/ZnO electrolyte; as a result, Zn was electrodeposited on one electrode, and on the other electrode, silver was oxidized to form a Zn-AgO battery. The authors achieved a higher areal capacity of ~2.4 $mAhcm^{-2}$ in the case of a battery with pillar as compared to a battery with a planar structure (without pillar) (~1.5 $mAhcm^{-2}$). Recently, Wang et al. developed MnO_2 cathode inks for a flexographic-printed zinc-based battery (Wang et al. 2014). In this work, the authors studied the effect of the solvent, binder, and additives on the printability of a MnO_2 ink. The flexographic-printed MnO_2 electrodes were cycled with zinc foils as the counter electrodes. The capacity of the battery after 20 electrochemical cycles was approximately 0.05 $mAhcm^{-2}$. Kang et al. recently demonstrated a flexible thin-film lithium-ion battery by sequentially screen printing the current collectors, active layers, and the polymer gel electrolyte onto the heat-sealable side of an aluminum-laminated pouch (Figure 6.22a and b) (Kang et al. 2014). The areal capacity of the battery was 2.5 $mAhcm^{-2}$ with a capacity retention of 84% after 50 electrochemical cycles.

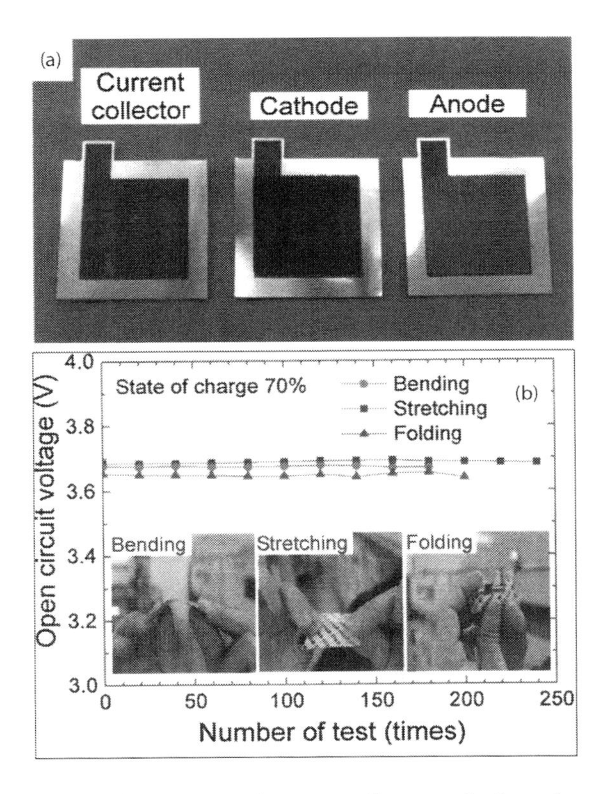

FIGURE 6.22 (a) Photo images of screen-printed current collector, cathode, and anode layer (b) open circuit voltage evolution vs the number of flexibility tests (bending, stretching, and folding) for a pouch-type flexible thin-film lithium-ion battery. The insets in (b) indicate photo images of the bending, stretching, and folding tests. (From Kang, K.Y. et al., *Electrochim. Acta*, 138, 294–301, 2014. With permission.)

The battery was able to maintain its mechanical integrity after bending 200 times to a curvature of 20 mm. It is concluded that the electrochemical capacities of these printed batteries are much lower compared to traditional batteries (100 mAh for silver oxide-based coin cells). Thus, these printed batteries are only suitable where low power is required. There is a need for the development of active materials that have high energy density, which could help improve the areal capacity and volumetric energy density suitable for flexible batteries. The commercialization of flexible batteries will require fully printed battery designs, whereby all layers can be processed on an industrial roll-to-roll manufacturing setup. The smart packaging and models to predict the electro-mechanical performance of flexible batteries could be important in terms of developing future flexible printed batteries.

6.6 Future Direction

It is well established that printing is the best choice for electronics manufacturing and particularly for large-area manufacturing, as it is completely digital, low cost, and high-throughput fabrication of electronics on unconventional flexible substrates such as plastics and papers in a roll to roll (R2R) or sheet to sheet (S2S) manner is possible. In this chapter, we have presented a comprehensive overview of various technologies that have been employed so far for printed devices such as display, RFID tags, RF microwave components, sensors, and batteries. The following subsection provides insight into the future direction of various printing methodologies and techniques.

6.6.1 Inks

It is shown that material development with adjusted chemical properties and optimum processing parameters are the major paradigms for current research on printed electronics. For example, it is desirable to have the highest possible conductivity of the printed conductors, and it should be closer to the bulk metal. Most printed devices have a lower performance than their traditional counterparts precisely because of this lower conductivity. In this way, different sintering mechanisms would play an important role. Room-temperature sintering shows the possibility of lower processing cost, as long as it can yield the right conductivity.

The challenge of semiconductor inks is that there are not many inks that can demonstrate high carrier mobilities. This is because earlier work on semiconductor ink focused on organic-type materials, which have inherently low mobilities, but are suitable for low-speed and low-frequency applications, such as displays or solar cells. Recently, inorganic materials, particularly semiconductor oxide-based inks, have been explored. Although their mobilities are still quite low as compared to conventional silicon, they are considerably higher as compared to the organic inks and can help in printing relatively higher frequency transistors. In addition, there is a limited report of low-loss dielectric printable materials. Another obstacle that these inorganic/dielectric materials typically face is the requirement of high processing temperatures (up to 500°C), which limits their integration on flexible plastic substrates. Therefore, research must be conducted to develop new inks which have higher mobilities, but at the same time can be processed at lower temperatures. In addition, new functional inks, such as magnetic, ferroelectric, or piezoelectric with different curing techniques, would be useful to broaden several applications.

6.6.2 Fully Printed Active High-Frequency Devices

In a RF regime, much work has been done to realize passives such as printed antennas, inductors, capacitors, filters, and resonators. However, there is very limited literature available on high-performance printed actives, such as transistors, due to numerous unresolved challenges in this area. There is some progress regarding R2R fully printed RFID tags that work at 13.56 MHz, but still RFID tags rely on silicon-based transistors with printed antennas and interconnect. There are many challenges, such as overlap capacitance, layer-to-layer registration, shorter channel length, and high mobility, which must be resolved before high-frequency transistors can be completely printed.

6.6.3 Printing Techniques

Another hindrance in the realization of high-performance devices such as transistors is the limitation on the minimum feature size or gap which can be achieved through existing printers. Typical Si-semiconductor technology uses photolithography to pattern high-resolution features on the scale of tens of nanometers. On the other hand, reliable minimum feature sizes in today's printers are of the order of tens of micrometers. It is therefore required to devise new printing techniques/printers which can at least go down to sub-micron feature size reliably for reasonably high-frequency operations. There is a requirement that semiconductors, dielectrics, and conductors can be printed in a multilayer process to fabricate fully printed devices.

6.6.4 Emerging Applications

Overall, fast communication and computations in emerging areas such as the IoT will require cost-effective electronics with high performance. Stretchable and washable inks will support printed wearable electronics. In addition, printing directly on silicon for interconnection will be interesting, possibly replacing fragile wire-bonding in complementary metal-oxide semiconductor (CMOS) technology.

References

Abutarboush, H. F.; Farooqui, M. F. and Shamim, A. 2016. Inkjet-printed wideband antenna on resin-coated paper substrate for curved wireless devices. *IEEE Antenn. Wirel. Prop. Lett.* 15:20–23.

Ahmed, S.; Tahir, F. A.; Shamim, A. and Cheema, H. M. 2015. A compact kapton-based inkjet-printed multiband antenna for flexible wireless devices. *IEEE Antenn. Wirel. Prop. Lett.* 14:1802–1805.

Alamán, J.; Alicante, R.; Peña, J. I. and Sánchez-Somolinos, C. 2016. Inkjet printing of functional materials for optical and photonic applications. *Materials* 9:910–956.

Albrecht, A.; Rivadeneyra, A.; Abdellah, A.; Lugli, P. and Salmerón, J. 2016. Inkjet printing and photonic sintering of silver and copper oxide nanoparticles for ultra-low-cost conductive patterns. *J. Mater. Chem. C* 4:3546–3554.

An, K.; Hong, S.; Han, S.; Lee, H.; Yeo, J. and Ko, S. H. 2014. Selective sintering of metal nanoparticle ink for maskless fabrication of an electrode micropattern using a spatially modulated laser beam by a digital micromirror device. *ACS Appl. Mater. Inter.* 6:2786–2790.

Andersson, H.; Manuilskiy, A.; Gao, J. et al. 2013. Investigation of humidity sensor effect in silver nanoparticle ink sensors printed on paper. *IEEE Sens. J.* 14:623–628.

Andersson, P.; Nilsson, D.; Svensson, P. et al. 2002. Active matrix displays based on all-organic electrochemical smart pixels printed on paper. *Adv. Mater.* 2002, 14:1460–1464.

Avis, C. and Jang, J. 2011. High-performance solution processed oxide TFT with aluminum oxide gate dielectric fabricated by a solgel method. *J. Mater. Chem.* 21:10649.

Avis, C.; Hwang, H. R. and Jang, J. 2014. Effect of channel layer thickness on the performance of indium–zinc–tin oxide thin film transistors manufactured by inkjet printing. *ACS Appl. Mater. Inter.* 6:10941–10945.

Baby, T. T.; Garlapati, S. K.; Dehm, S. et al. 2015. A general route toward complete room temperature processing of printed and high performance oxide electronics. *ACS Nano* 9:3075–3083.

Bissannagari, M. and Kim, J. 2015. Inkjet printing of NiZn-ferrite films and their magnetic properties. *Ceram. Int.* 41:8023–8027.

Chang, C. J.; Chang, S. J.; Wu, F. M.; Hsu, M. W.; Chiu, W. W. and Chen, K. 2004. Effect of compositions and surface treatment on the jetting stability and color uniformity of ink-jet printed color filter. *Jpn. J. Appl. Phys.* 43:8227–8233.

Chang, C. J.; Lin, Y. H. and Tsai, H. Y. 2011. Synthesis and properties of UV-curable hyperbranched polymers for ink-jet printing of color micropatterns on glass. *Thin Solid Films* 519:5243–5248.

Chen, C.-T. Wu, K.-H. Lu, C.-F. and Shieh, F. 2010. An inkjet printed stripe-type color filter of liquid crystal display. *J. Micromech. Microeng.* 20:055004–055014.

Comiskey, B.; Albert, J. D.; Yoshizawa, H. and Jacobson, J. 1998. An electrophoretic ink for all-printed reflective electronic displays. *Nature* 394:253–255.

Cook, B. S.; Cooper, J. R. and Tentzeris, M. M. 2013. Multi-layer RF capacitors on flexible substrates utilizing inkjet printed dielectric polymers. *IEEE Microw. Wirel. Comp. Lett.* 23:353–355.

Cook, B. S.; Mariottit, C.; Cooper, R. et al. 2014. Inkjet-printed, vertically-integrated, high-performance inductors and transformers on flexible LCP substrate. *Microwave Symposium (IMS), 2014 IEEE MTT-S International* 1–4.

Cook, B. S. and Shamim, A. 2012. Inkjet printing of novel wideband and high gain antennas on low-cost paper substrate. *IEEE Trans. Antenn. Propag. Eng.* 60:4148–4156.

Daniel, J.; Arias, A. C.; Wong, W. et al. 2007. Jet-printed active-matrix backplanes and electrophoretic displays. *Japanese J. Appl. Phys.* 46:1363–1369.

Das, R. The game-changer from NovaCentrix: Copper oxide ink, http://www.printedelectronicsworld.com/articles/the-game-changer-from-novacentrix-copper-oxide-ink-00001765.asp, October, 2009.

Doggart, J.; Wu, Y. and Zhu, S. 2009. Inkjet printing narrow electrodes with <50 µm line width and channel length for organic thin-film transistors. *Appl. Phys. Lett.* 94:163503–163506.

Dong, Y.; Li, X.; Liu, S.; Zhu, Q.; Li, J.-G. and Sun, X. 2015. Facile synthesis of high silver content MOD ink by using silver oxalate precursor for inkjet printing applications. *Thin Solid Films* 589:381–387.

Farooqui, M. F. and Shamim, A. 2016. Low cost inkjet printed smart bandage for wireless monitoring of chronic wounds. *Scientific Reports* 6:28949–28961.

Farooqui, M. F.; Karimi, M. A.; Salama, K. N. and Shamim, A. 2017. 3D-Printed disposable wireless sensors with integrated microelectronics for large area environmental monitoring. *Adv. Mater. Technol.* 2:1700051–1700059.

Fortunato, E.; Barquinha, P. and Martins, R. 2012. Oxide semiconductor thin-film transistors: A review of recent advances. *Adv. Mater.* 24:2945–2986.

Gaikwad, A. M.; Arias, A. C. and Steingart, D. A. 2015. Recent progress on printed flexible batteries: Mechanical challenges, printing technologies, and future prospects. *Energy Technol.* 3:305–328.

Gaikwad, A. M.; Steingart, D. A.; Ng, T. N.; Schwartz, D. E. and Whiting, G. L. 2013. A flexible high potential printed battery for powering printed electronics. *Appl. Phys. Lett.* 102:233302–233306.

Gao, J.; Sidén, J.; Nilsson, H.-E. and Gulliksson, M. 2013. Printed humidity sensor with memory functionality for passive RFID Tags. *IEEE Sens. J.* 13:1824–1834.

Ghosh, S.; Yang, R.; Kaumeyer, M. et al. 2014. Fabrication of electrically conductive metal patterns at the surface of polymer films by microplasma-based direct writing. *ACS Appl. Mater. Inter.* 6:3099–3104.

Grau, G. and Subramanian, V. 2016. Fully high-speed gravure printed, low-variability, high-performance organic polymer transistors with sub-5 V operation. *Adv. Electron. Mater.* 2:1500328.

Grau, G.; Cen, J.; Kang, H.; Kitsomboonloha, R.; Scheideler, W. J. and Subramanian, Vivek. 2016. Gravure-printed electronics: Recent progress in tooling development, understanding of printing physics, and realization of printed devices. *Flex. Print. Electron.* 1:023002.

Grouchko, M.; Kamyshny, A.; Mihailescu, C. F.; Anghel, D. F. and Magdassi, S. 2011. Conductive inks with a 'built-in' mechanism that enables sintering at room temperature. *ACS Nano* 5:3354–3359.

Hahn, Y. B.; Vaseem, M. and Hong, A. R. 2016. CuO nanoparticles, ink thereof, and method for preparing Cu thin film by reducing CuO thin film through microwave irradiation. KR101582637B1.

Hamedi, M. M.; Hajian, A.; Fall, A. B. et al. 2014. Highly conducting, strong nanocomposites based on nanocellulose-assisted aqueous dispersions of single-wall carbon nanotubes. *ACS Nano* 8:2467–2476.

Han, Y. H.; Won, J.-Y.; Yoo, H.-S.; Kim, J.-H.; Choi, R. and Jeong, J. K. 2016. High performance metal oxide field-effect transistors with a reverse offset printed Cu source/drain electrode. *ACS Appl. Mater. Inter.* 8:1156–1163.

Ho, C. C.; Murata, K.; Steingart, D. A.; Evans, J. W.; and Wright, P. K. 2009. A super ink jet printed zinc–silver 3D microbattery. *J. Micromech. Microeng.* 19:094013.

Hu, L.; Wu, H.; La Mantia, F.; Yang, Y. and Cui, Y. 2010. Thin, flexible secondary li-Ion paper batteries. *ACS Nano* 4:5843–5848.

Hyun, W. J.; Secor, E. B.; Rojas, G. A.; Hersam, M. C.; Francis, L. F. and Frisbie, C. D. 2015. All-printed, foldable organic thin-film transistors on glassine paper. *Adv. Mater.* 27:7058–7064.

Jang, J.; Kang, H.; Chakravarthula, H. C. N.; Subramanian, V. 2015. Fully inkjet-printed transparent oxide thin film transistors using a fugitive wettability switch. *Adv. Electron. Mater.* 1: 1500086–1500092.

Jang, Y.; Kim, D. H.; Park, Y. D.; Cho, J. H.; Hwang, M. and Cho, K. 2005. Influence of the dielectric constant of a polyvinyl phenol insulator on the field-effect mobility of a pentacene-based thin-film transistor. *Appl. Phys. Lett.*, 87:152105.

Joyce, M. J.; Fleming, P. D. III; Avuthu, S. G.; Emamian, S.; Eshkeiti, A; Atashbar, M. and Donato T. 2014. Contribution of flexo process variables to fine line Ag electrode performance. *Int. J. Eng. Res. Technol.* 3: 1645–1656.

Jung, M.; Kim, J.; Noh, J. et al. 2010. All-printed and roll-to-roll-printable 13.56-MHz-operated 1-bit RF tag on plastic foils. *IEEE Trans. Elec. Dev.* 57:571–580.

Jung, S.; Chun, S. J. and Shon, C.-H. 2016. Rapid cellulose-mediated microwave sintering for high-conductivity Ag patterns on paper. *ACS Appl. Mater. Inter.* 8:20301–20308.

Kahrilas, G. A.; Wally, L. M.; Fredrick, S. J.; Hiskey, M.; Prieto, A. L. and Owens, J. E. 2014. Microwave-assisted green synthesis of silver nanoparticles using orange peel extract. *ACS Sustainable Chem. Eng.* 2:367–376.

Kamyshny, A. and Magdassi, S. 2014. Conductive nanomaterials for printed electronics. *Small* 10:3515–3535.

Kang, B. J.; Lee, C. K. and Oh, J. H. 2012. All-inkjet-printed electrical components and circuit fabrication on a plastic substrate. *Microelectron. Eng.* 97:251–254.

Kang, H.; Kitsomboonloha, R.; Ulmer, K.; Stecker, L.; Grau, G.; Jang, J. and Subramanian, V. 2014. Megahertz-class printed high mobility organic thin-film transistors and inverters on plastic using attoliter-scale high-speed gravure-printed sub-5 μm gate electrodes. *Org. Electron.* 15:3639–3647.

Kang, K.-Y.; Lee, Y.-G.; Shin, D. O.; Kim, J.-C. and Kim, K. M. 2014. Performance improvements of pouch-type flexible thin-film lithium-ion batteries by modifying sequential screen-printing process. *Electrochim. Acta* 138:294–301.

Kim, D.; Jeong, Y.; Song, K.; Park, S. K.; Cao, G. and Moon, J. 2009a. Inkjet-printed zinc tin oxide thin-film transistor. *Langmuir* 25:11149–11154.

Kim, Y. D.; Kim, J. P.; Kwon, O. S.; Cho, I. H. 2009b. The synthesis and application of thermally stable dyes for ink-jet printed LCD color filters. *Dyes Pigments* 81:45–52.

Kitsomboonloha, R.; Kang, H.; Grau, G.; Scheideler, W. and Subramanian, V. 2015. MHz-range fully printed high-performance thin-film transistors by using high-resolution gravure-printed lines. *Adv. Electron. Mater.* 1:1500155–1500161.

Ko, S. H.; Chung, J.; Pan, H.; Grigoropoulos, C. P. and Poulikakos, D. 2007. Fabrication of multilayer passive and active electric components on polymer using inkjet printing and low temperature laser processing. *Sensors Actuat.* 134:161–168.

Koo, H. S.; Chen, M.; Pan, P. C. 2006. LCD-based color filter films fabricated by a pigment-based colorant photo resist inks and printing technology. *Thin Solid Films* 515:896–901.

Krykpayev, B.; Farooqui, M. F.; Bilal, R. M.; Vaseem, M. and Shamim, A. 2017. A wearable tracking device inkjet-printed on textile. *Microelectron. J.* 65:40–48.

Lau, P. H.; Takei, K.; Wang, C. et al. 2013. Fully printed, high performance carbon nanotube thin-film transistors on flexible substrates. *Nano Lett.* 13:3864–3869.

Layani, M.; Grouchko, M.; Shemesh, S. and Magdassi, S. 2012. Conductive patterns on plastic substrates by sequential inkjet printing of silver nanoparticles and electrolyte sintering solutions. *J. Mater. Chem.* 22:14349–14352.

Lee, D.; Paeng, D.; Park, K. and Grigoropoulos, C. P. 2014. Vacuum-free, maskless patterning of Ni electrodes by laser reductive sintering of NiO nanoparticle ink and its application to transparent conductors. *ACS Nano* 8:9807–9814.

Lee, H.; Tentzeris, M. M.; Raj, P. M.; Murali, K. P. and Kawahara, Y. 2013. Inkjet-printed ferromagnetic nanoparticles for miniaturization of flexible printed RF inductors. *Antennas and Propagation Society International Symposium (APSURSI), 2013 IEEE.*

Lee, T. I.; Choi, W. J.; Moon, K. J. et al. 2010. Programmable direct-printing nanowire electronic components. *Nano Lett.* 10:1016–1021.

Li, D. P.; Sutton, D.; Burgess, A.; Graham, D. and Calvert, P. D. 2009. Conductive copper and nickel lines via reactive inkjet printing. *J. Mater. Chem.* 19:3719–3724.

Li, R.-Z.; Hu, A.; Zhang, T.; Oakes, K. D. 2014. Direct writing on paper of foldable capacitive touch pads with silver nanowire inks. *ACS Appl. Mater. Inter.* 6:21721–21729.

Li, Y.; Lan, L.; Sun, S. et al. 2017. All inkjet-printed metal-oxide thin-film transistor array with good stability and uniformity using surface-energy patterns. *ACS Appl. Mater. Inter.* 9:8194–8200.

Li, Y.; Torah, R.; Beeby, S. and Tudor, J. 2012. An all-inkjet printed flexible capacitor for wearable applications. In *Proceedings of Symposium on Design, Test, Integration and Packaging*, MEMS/MOEMS (DTIP), 192–195, IEEE, Cannes, France.

Lin, Y.; Baggett, D. W.; Kim, J.-W.; Siochi, E. J. and Connell, J. W. 2011. Instantaneous formation of metal and metal oxide nanoparticles on carbon nanotubes and graphene via solvent-free microwave heating. *ACS Appl. Mater. Inter.* 3:1652–1664.

Lu, G. S.; You, P. C.; Lin, K. L.; Hong, C. C. and Liou, T. M. 2014. Fabricating high-resolution offset color-filter black matrix by integrating heterostructured substrate with inkjet printing. *J. Micromech. Microeng.* 24:055008–055015.

Magdassi, S.; Grouchko, M.; Berezin, O. and Kamyshny, A. 2010. Triggering the sintering of silver nanoparticles at room temperature. *ACS Nano* 4:1943–1948.

McKerricher, G.; Maller, R.; Vaseem, M.; McLachlan, M. A. and Shamim, A. 2017a. Inkjet-printed thin film radio-frequency capacitors based on sol-gel derived alumina dielectric ink. *Ceram. Int.* 43:9846–9853.

McKerricher, G.; Perez, J. G. and Shamim, A. 2015. Fully inkjet printed RF inductors and capacitors using polymer dielectric and silver conductive ink with through vias. *IEEE Trans. On Elec. Dev.* 62:1002–1009.

McKerricher, G.; Titterington, D. and Shamim, A. 2016. A fully inkjet-printed 3-D honeycomb-inspired patch antenna. *IEEE Antenn. Wirel. Prop. Lett.* 15:544–547.

McKerricher, G.; Vaseem, M. and Shamim, A. 2017b. Fully inkjet-printed microwave passive electronics. *Microsyst. Nanoeng.* 3:16075–16081.

Memon, M. U.; Tentzeris, M. M. and Lim, S. 2017. Inkjet-printed 3D Hilbert-curve fractal antennas for VHF band. *Microw. Opt. Technol. Lett.* 59:1698–1704.

Molina-Lopez, F.; Briand, D. and de Rooij, N. 2013. All additive inkjet printed humidity sensors on plastic substrate. *Sens. Actuators B Chem.* 166:212–222.

Nadagouda, M. N.; Speth, T. F. and Varma, R. S. 2011. Microwave-assisted green synthesis of silver nanostructures. *Acc. Chem. Res.* 44:469–478.

NanoIntegris, http://www.nanointegris.com/en/semiconducting.

Nayak, P. K.; Hedhili, M. N.; Cha, D. and Alshareef, H. N. 2013. High performance In_2O_3 thin film transistors using chemically derived aluminum oxide dielectric. *Appl. Phys. Lett.* 103:33518.

Noguchi, Y.; Sekitani, T. and Someya, T. 2007. Printed shadow masks for organic transistors. *Appl. Phys. Lett.* 91:133502–133504.

Nomura, K.; Ohta, H.; Takagi, A.; Kamiya, T.; Hirano, M. and Hosono, H. 2004. Room temperature fabrication of transparent flexible thin-film transistors using amorphous oxide semiconductors. *Nature* 432:488–492.

OLED ink jet printing: Introduction and market status, https://www.oled-info.com/oled-inkjet-printing.

Orecchini, G.; Alimenti, F.; Palazzari, V.; Rida, A. and Tentzeris, M. M. 2011. Design and fabrication of ultra-low cost radio frequency identification antennas and tags exploiting paper substrates and inkjet printing technology. *IET Microw. Antenn. Propag.* 5:993–1001.

Perelaer, J.; Abbel, R.; Wünscher, S.; Jani, R.; Lammeren, T. and Schubert, U. S. 2012. Roll-to-roll compatible sintering of inkjet printed features by photonic and microwave exposure: From non-conductive ink to 40% bulk silver conductivity in less than 15 s. *Adv. Mater.* 24:2620–2625.

Perelaer, J.; Gans, B.-J. and Schubert, U. S. 2006. Ink-jet printing and microwave sintering of conductive silver tracks. *Adv. Mater.* 18:2101–2104.

Perelaer, J.; Hendriks, C. E.; de Laat, A. W. M. and Schubert, U. S. 2009a. One-step inkjet printing of conductive silver tracks on polymer substrates. *Nanotechnology* 20:165303–165307.

Perelaer, J.; Jani, R.; Grouchko, M.; Kamyshny, A.; Magdassi, S. and Schubert, U. S. 2012. Plasma and microwave flash sintering of a tailored silver nanoparticle ink, yielding 60% bulk conductivity on cost-effective polymer foils. *Adv. Mater.* 24:3993–3998.

Perelaer, J.; Klokkenburg, M.; Hendriks, C. E. and Schubert, U. S. 2009b. Microwave flash sintering of inkjet-printed silver tracks on polymer substrates. *Adv. Mater.* 21:4830–4834.

Quddious, A.; Yang, S.; Khan, M. M.; Tahir, F. A.; Shamim, A.; Salama, K. N. and Cheema, H. M. 2016. Disposable, paper-based, inkjet-printed humidity and H2S gas sensor for passive sensing applications. *Sensors* 16:2073–2085.

Rao, K. J.; Vaidhyanathan, B.; Ganguli, M. and Ramakrishnan, P. A. 1999. Synthesis of inorganic solids using microwaves. *Chem. Mater.* 11:882–895.

Rivadeneyra, A.; Fernández-Salmerón, J.; Banqueri, J.; López-Villanueva, J. A.; Capitan-Vallvey, L. F. and Palma, A. J. 2014. A novel electrode structure compared with interdigitated electrodes as capacitive sensor. *Sens. Actuators B Chem.* 204:552–560.

Sarfraz, J.; Maattanen, A.; Ihalainen, P.; Keppeler, M.; Lindén, M. and Peltonen, J. 2012. Printed copper acetate based H2S sensor on paper substrate. *Sens. Actuators B Chem.* 173:868–873.

Scheideler, W. J.; Kumar, R.; Zeumault, A. R. and Subramanian, V. 2017. Low-temperature-processed printed metal oxide transistors based on pure aqueous inks. *Adv. Funct. Mater.* 27:1606062–1606072.

Secor, E. B.; Ahn, B. Y.; Gao, T. Z.; Lewis, J. A. and Hersam, M. C. 2015. Rapid and versatile photonic annealing of graphene inks for flexible printed electronics. *Adv. Mater.* 27:6683–6688.

Shah, S. I. H.; Lee, D.; Tentzeris, M. M. and Lim, S. 2016. Fabrication of microstrip patch antenna using novel hybrid printing technology. *Microw. Opt. Technol. Lett.* 58:2602–2606.

Shaker, G.; Safavi-Naeini, S.; Sangary, N. and Tentzeris, M. M. 2011. Inkjet printing of ultrawideband (UWB) antennas on paper-based substrate. *IEEE Antenn. Wirel. Prop. Lett.* 10:111–114.

Singh, N.; Galande, C.; Miranda, A.; Mathkar, A.; Gao, W.; Reddy, A. L. M.; Vlad, A. and Ajayan, P. M. 2012. *Paintable Battery. Sci. Rep.* 2:481–485.

Skylar-Scotta, M. A.; Gunasekarana, S. and Lewis, J. A. 2016. Laser-assisted direct ink writing of planar and 3D metal architectures. *PNAS* 113:6137–6142.

Song, H.; Spencer, J.; Jander, A. et al. 2014. Inkjet printing of magnetic materials with aligned anisotropy. *J. Appl. Phys.* 115:17E308.

Sony. 2007. https://www.sony.net/SonyInfo/News/Press/200710/07-1001E/.

Sun, K.; Wei, T.-S.; Ahn, B. Y.; Seo, J. Y.; Dillon, S. J. and Lewis, J. A. 2013. 3D printing of interdigitated Li-Ion microbattery architectures. *Adv. Mater.* 25: 4539–4543.

Tang, Y.; He, W.; Zhou, G.; Wang, S.; Yang, X.; Tao, Z. and Zhou, J. 2012. A new approach causing the patterns fabricated by silver nanoparticles to be conductive without sintering. *Nanotechnology* 23:355304–355309.

Tehrani, B. K.; Cook, B. S. and Tentzeris, M. M. 2016. Inkjet printing of multilayer millimeter-wave Yagi-Uda antennas on flexible substrates. *IEEE Antenn. Wirel. Prop. Lett.* 15:143–146.

Torrisi, F.; Hasan, T.; Wu, W. et al. 2012. Inkjet-printed graphene electronics. *ACS Nano* 6:2992–3006.

Tseng, H.-Y. 2011. *Scaling of Inkjet-Printed Transistors Using Novel Printing Techniques*. Department EECS, University California, Berkeley, CA, Tech. Rep. UCB/EECS-2011-146.

Vaseem, M.; Ghaffar, F. A.; Farooqui, M. F. and Shamim, A. 2018. Iron oxide nanoparticle-based magnetic ink development for fully printed tunable radio-frequency devices. *Adv. Mater. Technol.* 1700242–1700252.

Vaseem, M.; Hong, A.-R.; Kim, R.-T. and Hahn, Y.-B. 2013. Copper oxide quantum dot ink for inkjet-driven digitally controlled high mobility field effect transistors. *J. Mater. Chem. C* 1:2112–2120.

Vaseem, M.; Lee, S.-K.; Kim, J.-G. and Hahn, Y.-B. 2016a. Silver-ethanolamine-formate complex based transparent and stable ink: Electrical assessment with microwave plasma vs thermal sintering. *Chem. Eng. J.* 306:796–805.

Vaseem, M.; McKerricher, G. and Shamim, A. 2016b. Robust design of a particle-free silver-organo-complex ink with high conductivity and inkjet stability for flexible electronics. *ACS Appl. Mater. Inter.* 8:177–186.

Vena, A.; Babar, A. A.; Sydänheimo, L.; Tentzeris, M. M. and Ukkonen, L. 2013. A novel near-transparent ASK-reconfigurable inkjet-printed chipless RFID tag. *IEEE Antennas & Wirel. Prop. Lett.* 12:753–756.

Virtanen, J.; Virkki, J.; Sydänheimo, L.; Tentzeris, M. M. and Ukkonen, L. 2013. Automated identification of plywood using embedded inkjet-printed passive UHF RFID tags. *IEEE Trans. Autom. Sci. Eng.* 10:796–806.

Vornbrock, A. F.; Sung, D.; Kang, H.; Kitsomboonloha, R. and Subramanian, V. 2010. Fully gravure and ink-jet printed high speed pBTTT organic thin film transistors. *Org. Electron.* 11:2037–2044.

Walker, S. B. and Lewis, J. A. 2012. Reactive silver inks for patterning high-conductivity features at mild temperatures. *J. Am. Chem. Soc.* 134:1419–1421.

Wang, Z.; Winslow, R. and Madan, D. 2014. Development of MnO_2 cathode inks for flexographically printed rechargeable zinc-based battery. *J. Power Sources* 268:246–254.

Weremczuk, J.; Tarapata, G. and Jachowicz, R. S. 2012. The ink-jet printing humidity sorption sensor—Modelling, design, technology and characterization. *Meas. Sci. Technol.* 23:014003–014012.

Yang, L.; Rida, A. and Tentzeris, M. M. 2009. Design and development of radio frequency identification (RFID) and RFID-enabled sensors on flexible low cost substrates, *Synthesis Lectures RF/Microw.* 1:1–89.

Yang, L.; Rida, A.; Vyas, R. and Tentzeris, M. M. 2007. RFID tag and rf structures on a paper substrate using inkjet-printing technology. *IEEE Trans. Microw. Theory Tech.* 55:2894–2901.

Yang, L.; Zhang, R.; Staiculescu, D.; Wong, C. and Tentzeris, M. 2009. A novel conformal RFID-enabled module utilizing inkjet-printed antennas and carbon nanotubes for gas-detection applications. *IEEE Antenn. Wirel. Propag. Lett.* 8:653–656.

Yang, Y.; Hu, Y.; Xiong, X. and Qin, Y. 2013. Impact of microwave power on the preparation of silver nanowires via a microwave-assisted method. *RSC Adv.* 3:8431–8436.

7

Ferroic Materials and Devices for Flexible Memory

Saidur R. Bakaul,
Mahnaz Islam, and
Md. Kawsar Alam

7.1 Introduction

An information storage system that complies with mechanical bending to some degree without losing its memory function is described as a flexible memory. With the recent advent of wearable electronics and the Internet of things (IoT), computing has become ubiquitous, which has created a demand for physically flexible memory systems. Accordingly, considerable effort to create different types of flexible memories with high energy efficiency and performance has recently been observed. The growing interest in this field is reflected in the continuous increase in publications on flexible memories per year, as shown in Figure 7.1.

To-date, almost all the mainstream and emerging memories such as ferroic (ferromagnetic and ferroelectric) random access memory (MRAM and FERAM) (de Araujo et al. 1995; Nishimura et al. 2002), flash memory (Bez et al. 2003), resistance random access memory (RRAM) (Rueckes 2000; Tsakalakos and Sands 2000; Won Seo et al. 2010), and phase change random access memory (Chong et al. 2006) have been demonstrated on flexible substrates. However, the performance and energy efficiency of these flexible memories are not as good as their counterparts on rigid substrates. The major underlying reason is the incompatibility of high-performance materials with flexible substrates. Consequently, most of the current memory research focus is on finding routes to integrate the best-performing materials and devices with the flexible platform. In this chapter, we provide an overview and update of these research works. We limit the discussion only to flexible memories based on ferroic materials. Other types of flexible memories have been reviewed recently in Ghoneim and Hussain (2015a) and Han et al. (2013). Our goal is to present an in-depth discussion of the effect of substrate deformation on ferroic materials and memory devices, of new techniques for materials integration, and of the status quo of these devices.

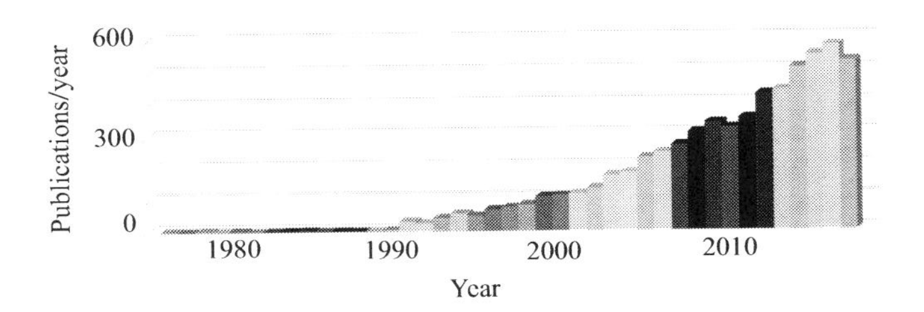

FIGURE 7.1 A bar chart representing the number of publications per year published on flexible memories for the last 3 decades. (Data from ISI Web of Knowledge.)

7.2 Different Types of Ferroic Memories

Ferroic materials exhibit two or more stable states in the absence of an electric, magnetic, or mechanical force. These states are typically energetically degenerate and separated by an energy barrier, which can be overcome by any or a combination of these forces. Based on the conjugate forces, ferroic materials are generally classified in three categories: ferromagnetic (FM), ferroelectric (FE), and ferroelastic (FS). FM and FE memories depend on the properties of electrons, namely, spin and charge, respectively, whereas FS memories rely on the asymmetry in the internal strain of a material and the associated mechanical energy. The common modalities among the three materials types are a hysteretic dependence of the order parameter on the conjugate force and a temperature-dependent phase transition between the para and ferro states. The hysteresis and the tendency of remembering the previously applied stimulus have made these materials candidates to be memory materials.

The hysteresis in FM, FE, and FS materials originates from the spontaneous magnetization, electric polarization, and strain, respectively. A quantum mechanical interaction between electron spins, known as exchange coupling, is the key reason for the FM hysteresis loop. In contrast, the hysteresis in FE and FS materials originates from classical mechanics, namely, the spatial movement of atoms caused by either an electric field or mechanical stress. The atomic movement in a FE material leads to a change in the electrostatic energy environment and to the formation of differently oriented dipoles, for which the orientation can be changed by an electric field. In a FS material, the atomic movement leads to a change in the anisotropic bonding length and structural symmetry, and thereby to switching between different strain states. Significant research on flexible FM and FE memories has been done, whereas a flexible FS memory is yet to be reported.

7.2.1 Ferromagnetic Memory

Giant Magnetoresistance (GMR) and tunnel magnetoresistance (TMR) devices are the most studied FM memory devices. GMR memory cells comprise of an ultrathin non-magnetic layer sandwiched between two FM layers with different coercivity. TMR cells are the advanced version of GMR cells, where the non-magnetic layer is replaced by a thin insulating layer (Figure 7.2a). The FM layer with higher coercivity acts as a spin polarizer and attenuates electrons with spin pointing antiparallel to the material's magnetization direction. The resistance of the cell depends on the relative magnetization direction of the other FM layer (the free layer). Parallel orientation of the magnetization and electron spin direction results in low resistance, and the antiparallel alignment provides a high resistance state. The memory state is toggled between 1 and 0 by changing the magnetization direction of the free layer. The cell is typically selected through a transistor. The write operation is achieved by passing a current through the write line, while the read operation requires measuring the difference in resistivity resulting from parallel or antiparallel spins.

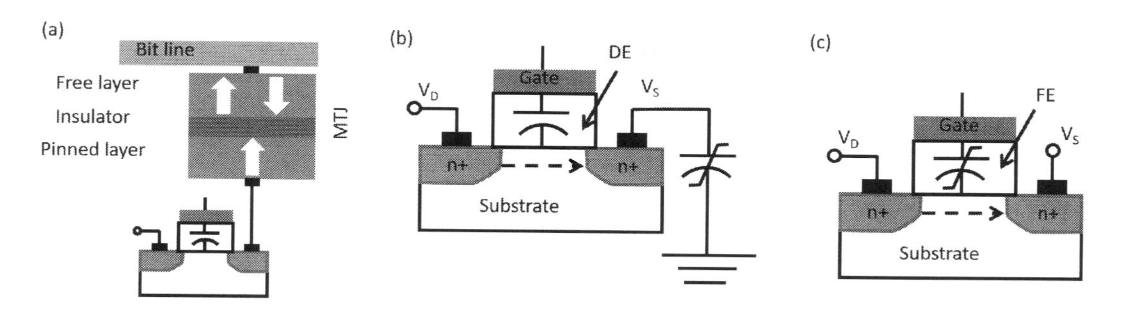

FIGURE 7.2 Schematic representations of different ferroic memory devices. (a) magnetic random access memory, (b) ferroelectric random access memory, and (c) ferroelectric field effect transistor.

Parkin et al. first demonstrated GMR devices on flexible, organic platforms such as mylar, Kapton polyimide, and Ultem polyimide (Parkin et al. 1992; Parkin 1996). A sputter-deposited $Py/Cu/Py/Fe_{46}Mn_{54}$ multilayer on Ultem polyimide showed 300% change in resistance when the free layer was switched between the parallel and antiparallel states. The GMR effect was of similar amplitude to that observed on rigid substrates. This is surprising since the rough surface of the flexible substrates typically reduces the GMR effect (Chen et al. 2008). Flexible GMR-type memory devices have also been demonstrated on other common organic materials such as bathocuproine (Sun et al. 2014) and poly(dimethylsiloxane) (PDMS) membranes (Melzer et al. 2011).

Magnetic tunnel junction (MTJ) is the building block of the already-commercialized magnetic random access memory (MRAM). These devices are more difficult to fabricate than the GMR devices on flexible substrates since the MTJs need very thin (typically less than 2 nm) and defect-free oxide tunnel barriers. A substrate roughness over 2 nm can result in an unwanted short circuit between the two metal electrodes and jeopardize the memory function. Also, these materials often need to be annealed at high temperature, which is typically incompatible with most flexible substrates. Despite these challenges, several groups have successfully developed flexible MTJs using MgO and Al_2O_3 as the tunnel barrier. Barraud et al. (2010) demonstrated the first flexible MTJ using $Co/Al_2O_3/Co$ on a poly(3,4-ethylenedioxythiophene)-poly(styrenesulfonate) (PEDOT-PSS)-buffered organic flexible substrate. Although a moderate level of TMR (20%) was observed, it was comparable to the control devices fabricated on standard rigid Si substrates. In recent years, a number of other groups have shown MgO-based MTJs with TMR up to 300% (Chen et al. 2017; Loong et al. 2016). The exchange bias effect, a crucial ingredient for improving GMR and TMR device performance and size scaling, has also recently been reported on bent polyimide substrates (Vemulkar et al. 2016).

7.2.2 Ferroelectric Memory

Two types of flexible FE memory have been demonstrated: (a) a device consisting of a metal-oxide-semiconductor field effect transistor (MOSFET) along with a FE capacitor, commonly known as FE random access memory (FRAM) (Figure 7.2b) and (b) a MOSFET containing a FE material as the gate capacitor, known as a FEFET (Figure 7.2c). FRAM operation is based on detecting the change of stored electric charge in the FE capacitor. The transistor acts as an access terminal to the capacitor. Data are destructively read out by measuring the switching current of the capacitor. FEFETs offer a non-destructive readout process. Structurally the same as a conventional MOSFET, the insulating gate dielectric in a FEFET is replaced by a FE material. The memory states result from the channel conductance modulation by the bi-stable polarization of the FE material. Data are read out by measuring the channel current, without switching the polarization state. FEFETs offer a higher scaling density compared to FRAMs.

Flexible FE memories have been demonstrated by many groups (Bakaul et al. 2017; Ghoneim et al. 2015; Kim et al. 2014). Typically, inorganic FE-based devices exhibit better performance than organic FE memories due to higher polarization and low leakage current. However, integration of inorganic FE materials with flexible substrates is more challenging due to stringent growth conditions. Inorganic FE memories are mechanically less flexible than the organic FEs, although in many reports both remained unaffected by substrate bending and stretching. $PbZr_xTi_{1-x}O_3$ and polyvinylidene fluoride-trifluoroethylene (PVDF-TrFE) are the two most studied FE materials in the inorganic and organic categories, respectively.

7.3 Effects of Curvature on Ferroic Memory Characteristics

If the substrate is bent, compressive strain is developed inside the curvature, whereas the outer surface experiences tensile strain. A neutral plane with zero uniaxial stress exists in the middle of the substrate. If a film is deposited on one or both sides of the substrate, the neutral plane shifts and the strain landscape is altered. However, the typical functional film thickness is usually several orders smaller than the substrate thickness. Thus, the strain acting in the functional film along the circumference of the bend can be approximated as being the same as the tensile effect exerted on the outer surface of the flexible substrate. The one-dimensional strain acting on the film can be expressed as $\varepsilon = t/2r$, where t is the thickness of the substrate and r is the radius of curvature. This assumption is valid for films that are strongly bonded with the substrate. The strain in the film would be smaller in the case of weak adhesion to substrate such as in transferred materials and van der Waals force-bonded materials.

In the typical applications of wearable electronics, the smallest r is around 1 mm and the tensile strain on a 100-nm thick film is estimated to be 5×10^{-5}. Many of the ferroic materials such as complex oxides and transition metals can easily withstand strain up to 1% before they break down (Faurie et al. 2017). Therefore, although ferroic properties can be affected by substrate bending-induced strain in many different ways, the effects are often negligible, especially when the radius of the curvature is on the millimeter scale (Bakaul et al. 2016; Chen et al. 2017; Lee et al. 2011; Matsumoto et al. 2007; Yoon et al. 2011). However, significant modifications in memory properties can be expected in the case of extreme bending with a radius of curvature below the micron scale.

7.3.1 FM Memory

The properties of the FM material such as the direction of magnetization, domain pattern, and domain wall motion are strongly correlated with the degree of curvature of the FM film. The curvature can induce an effective magnetic anisotropy and an antisymmetric component in the exchange interaction, which is a Dzyaloshinskii-Moriya-like interaction term (Cheong and Mostovoy 2007; Dzyaloshinsky 1958; Moriya 1960; Sheka et al. 2015; Streubel et al. 2016). In a simple way, these effects can be interpreted as a curvature-induced effective magnetic field acting on the magnetization direction in the FM material.

The Dzyaloshinskii-Moriya interaction (DMI) effect can lead to the creation and stabilization of vortex and skyrmion-type curling magnetization patterns. As a result, the magnetic domain pattern and the switching dynamics could be very different in the same device when the substrate is flat or bent. The DMI effect could be both assistive (Bhattacharya et al. 2017) or detrimental (Jang et al. 2015; Sampaio et al. 2016) for current-induced magnetization switching, and the underlying mechanisms are still topics of debate in the scientific community. It may also create a pinning potential for the head-to-head or tail-to-tail domain walls (Yershov et al. 2015) and alter the wall motion speed (Emori et al. 2013; Ryu et al. 2013; Thiaville et al. 2012; Yan et al. 2010).

Besides substrate bending, the high surface roughness of flexible substrates can also create curvature effects in thin FM films deposited on them. The spatially modulated complex dipolar interaction may result in additional perpendicular anisotropy, and its strength and direction depend on the degree of local curvature (Arias and Mills 1999; Gaididei et al. 2014; Tretiakov et al. 2017). Moreover, the peaks and pits in a rough surface can act as defects and pinning center for domain walls, which often leads to increased coercivity of FM materials on flexible substrates (Vemulkar et al. 2016). These effects are correlated to the operational speed and energy efficiency of flexible FM memories.

Another source of magnetic perturbation in flexible FM memory is the strain-induced anisotropy. The relationship between strain and the magnetic anisotropy constant is: $K = 3/2\lambda\varepsilon\gamma$, where ε, λ, and γ are the bending-induced strain on the film, the magnetostriction constant, and Young's modulus, respectively. Inward and outward bending of the film produce compressive and tensile strain on the film, respectively, and accordingly, the sign of the magnetic anisotropy constant can change depending on the nature of the bending. Lee et al. (2015) reported a 1.1% change in the effective anisotropy field in a Co/Pt multilayer on a flexible polyethylene naphthalate substrate coated with a planarizing polymer layer (PENC) substrate at a strain level of 0.3%. A bending-induced tensile strain enhanced the perpendicular magnetic anisotropy, whereas a compressive strain reduced it. Streubel et al. (2014) were able to generate a larger strain by rolling up a 20 nm thick Permalloy (Py) film with a micrometer-scale radius of curvature. The strain-induced anisotropy resulted in a 40% reduction of the coercive field.

The effect of substrate bending on magnetic characteristics can be more substantial in strongly magnetoelastic materials, such as Co, Terfenol-D, Ni, FeAl, FeGa, and $CoFe_2O_4$. The mechanical force exerted by the substrate bending can create a magnetic field in these materials and modify the magnetic susceptibility via the Villari effect. The magnetoelastic energy depends on the angle between the magnetization direction and the substrate bending direction. The energy density is lowest when the stress and magnetization directions are aligned parallel and highest when they are aligned orthogonally. The remnant magnetization (Dai et al. 2013; Tang et al. 2014), coercive field (Dai et al. 2013), squareness of the hysteresis loop (Dai et al. 2012), and the exchange bias effect (Zhang et al. 2013) can be strongly modulated in such FM materials by substrate bending-induced strain.

Wu et al. (2016) demonstrated that a compressive strain can rotate the easy axis from an in-plane to an out-of-plane direction in Fe_3O_4, which is also reflected in a change of domain wall width (Figure 7.3a–c). Interestingly, the original domain patterns are restored when the substrate stress is released. Ota et al. (2016) showed that the change in magnetic anisotropy can be as large as 1.2×10^5 J/m^3 under 2% strain in TbFeCo and Pt/Co/Pt multilayers. Özkaya et al. (2008) utilized the strain-induced magnetic anisotropy to tune the GMR ratio in a Co/Cu/Ni trilayer on a polyimide substrate. Due to the opposite signs of the magneto-elastic coefficients in Co and Ni, the two materials responded differently to mechanical stress. Easy axis rotation and a change in the angle between the magnetization directions in Co and Ni induced by strain were at the root of the observed change in GMR. Strain can also reduce the thickness of the spacer layer, which in turn alters the strength of Ruderman-Kittel-Kasuya-Yosida (RKKY) exchange coupling (Chen et al. 2008). Due to the oscillatory relationship between RKKY coupling and the spacer thickness, both enhancement and decrease in GMR were observed in devices with different spacer thicknesses. However, the estimated change of thickness was less than 1 Å and no direct evidence

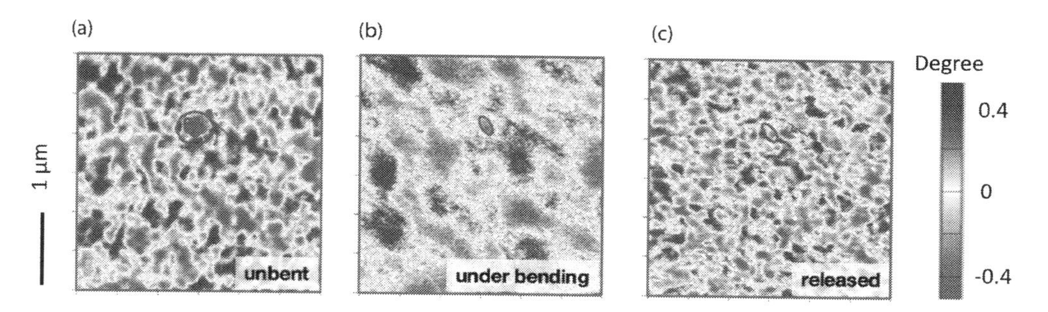

FIGURE 7.3 Magnetic force microscopy images of Fe_3O_4 on a muscovite substrate in (a) unbent, (b) bent (50 μm radius of curvature), and (c) released conditions. (Reprinted with permission from Wu, P.-C. et al., *ACS Appl. Mater. Inter.*, 8, 33794–33801, 2016. Copyright 2016 American Chemical Society.)

of such a subtle change was provided. As such, the underlying mechanism for the strain-dependent GMR change has remained controversial. Loong et al. claimed a 50% enhancement of TMR in a CoFeB/MgO/CoFeB MTJ at a strain level of 0.15%, which was attributed to enhanced electron coherent tunneling, although unambiguous evidence of the mechanism and a quantitative relationship with strain were not provided (Loong et al. 2014, 2016).

7.3.2 FE Memory

In general, inorganic FE materials remain unaffected by substrate bending up to a millimeter-scale radius of curvature (Bakaul et al. 2016; Ghoneim et al. 2015). Studies beyond this limit have not been conducted yet. In contrast to this, Yu et al. (2017) found that the remnant polarization in $Hf_{0.5}Zr_{0.5}O_2$ can be increased by 10% by repeated bending of the substrate (8 mm radius of curvature). This is attributed to an elongation of the c-axis of the tetragonal phase (Hyuk Park et al. 2014) and an increase in the fraction of the FE orthorhombic phase under strain. Compared to the inorganic oxides, the organic FE characteristics are more affected by the substrate bending. Kim et al. reported a 25% reduction of remnant polarization and a 20% increase of coercive voltage in PVDF after 1000 bending cycles with a 4 mm radius of curvature. Kim et al. (2014) fabricated FEFET using PVDF-TrFE as the FE gate and a dicyanomethylene-substituted quinoidal quaterthiophene derivative (QQT(CN)4) as the semiconducting channel (Figure 7.4). The devices exhibited hysteresis in transfer characteristics, although memory retention at zero gate voltage could not be achieved. The on/off ratio was modulated by 3 times upon bending the substrate up to a 500 μm radius of curvature. Although such sharp bending resulted in permanent plastic deformation, the devices remained functional. However, the devices showed a 10 times reduction of the on/off ratio when bending cycles were performed for 1000 times even at a 4 mm radius of curvature.

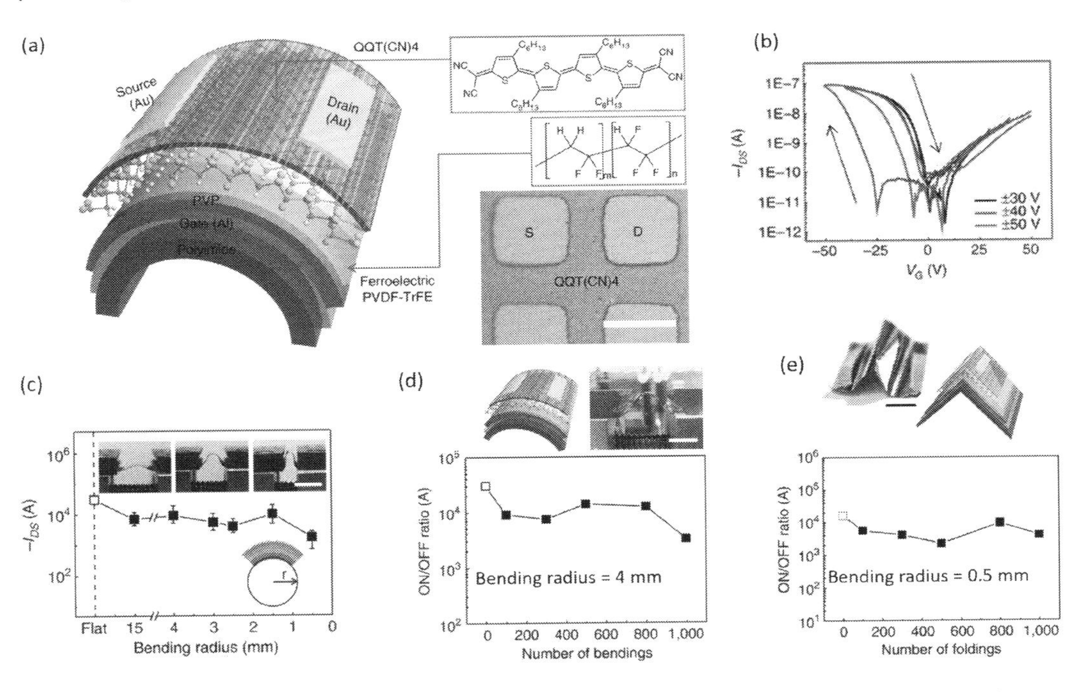

FIGURE 7.4 PVDF-based flexible FEFET and the effect of bending on transport characteristics. (a) Schematic of the device. (b) Transfer characteristics of the as-prepared FEFET at drain voltage = −5 V. (c) Dependence of the channel current on the radius of curvature. (d) and (e) On/off ratio as a function of number of bending cycles for different bending radii. (Reprinted by permission from Macmillan Publishers Ltd. *Nat. Commun.*, Kim, R.H. et al., 2014, Copyright 2014.)

Sharp bending of a substrate can induce a significant strain gradient in the film, developing an effective electric field inside a FE material and perturbing the memory properties. This phenomenon is known as the flexoelectric effect (Kogan 1964). Lu et al. (2012) demonstrated flexoelectric rotation of the polarization in $BaTiO_3$ using the tip of an atomic force microscope. The large flexoelectric coefficient in complex oxide FE materials (Ma and Cross 2001, 2002, 2005, 2006; Zubko et al. 2013) suggests that such an effect may become significant in the case of sharply bent FE memories. Although studied in detail on rigid substrates, the flexoelectric effect in flexible memories has remained an unexplored research area.

7.4 Energy Consumption of Flexible Ferroic Memory Devices

Memory devices consume energy mostly during switching between two states. One of the key challenges in flexible memory research is to reduce the energy consumption without compromising the device performance.

FM memories can be switched by a magnetic field or a current. Field-induced magnetization switching requires more energy than current-induced switching. Current-induced magnetization switching is generally performed by using one of two different mechanisms: spin transfer torque (STT) and spin orbit torque (SOT). In a STT mechanism (Berger 1996, 2001; Slonczewski 1996, 1999), polarized electrons from an FM layer (polarizer) enter the second FM layer and transfer their spin angular momentum to switch the magnetization direction. On the other hand, a SOT is generated from the spin-orbit interaction in heavy metals such as Pt and Ta (Liu et al. 2012; Mihai Miron et al. 2010). Both STT- and SOT-induced magnetization switching on a flexible substrate have been demonstrated. Lee et al. (2015) reported magnetization switching by a SOT in Co/Pt layers on flexible PENC substrates. The switching current density (I_c) was 10^7 A/cm^2, which indicates that for a device of 100×100 nm^2 area, a film thickness of 10 nm, a resistivity of 10^{-6} Ω·m, and a switching time of 1 ns, the energy consumption per bit switching $\left(I_c^2 R t_s\right)$ is 100 fJ. The free layer materials showing perpendicular magnetic anisotropy (PMA) are expected to be switched at a much lower current density than is seen for STT devices (Ikeda et al. 2010; Worledge et al. 2011). Another way to reduce the switching energy is to use materials with a low anisotropy energy. However, this reduces the memory retention time, and therefore a trade-off is needed.

In FE memory, the switching energy can be expressed as $E_{FE} = \frac{1}{2}\Delta P \cdot V_s \cdot A$, where $\Delta \boldsymbol{P}$ is the switchable polarization, V_s is the switching voltage, and A is the active capacitor area. Reduction of $\Delta \boldsymbol{P}$ deteriorates the readability of the two states and the memory window and is therefore not advisable. V_s can be reduced using an ultra-thin film as the switching layer. However, in that case, the leakage current increases and the depolarization effect weakens the FE polarization, leading towards reduced memory retention (Xia et al. 2001).

The coercive field in organic FE materials such as PVDF is generally one order higher than that of an inorganic FE material of the same thickness (Nguyen et al. 2007; Noda et al. 2003), rendering them less suitable for low voltage application. $PbZr_xTi_{1-x}O_3$ (PZT) is one of the best candidates among inorganic FEs due to its low coercive field and leakage current. Bakaul et al. (2017) demonstrated metal/PZT/metal capacitor memory operation on a flexible PEN substrate at 1.5 V with nanosecond switching speed, switchable polarization of 75 μC/cm^2, estimated retention of 10 years, and a hysteresis loop with squareness ratio of 1.6. Attaining such excellent memory properties and low switching voltage in the same device is quite challenging in flexible FE memory and can be obtained only using a single crystal complex oxide. Taking the parameters from Bakaul et al. (2017) and assuming a cell size of 100×100 nm^2, the energy consumption per bit switching can be estimated as 11 fJ, which is one order smaller than the FM energy consumption. Low switching voltages have also been reported in polycrystal PZT and other inorganic FE material devices such as $Hf_xZr_{1-x}O_2$-based memory, albeit only at the expense of deteriorating other memory properties and device performance (Ghoneim et al. 2015; Yu et al. 2017).

A potential pathway to further reduce the energy consumption without compromising memory characteristics could be a negative capacitance (NC) field effect transistor (Salahuddin and Datta 2008).

NC is a non-equilibrium phenomenon that manifests itself as a negative slope in the temporal evolution of voltage across a FE material during polarization state switching. It can be utilized to enhance the equivalent gate capacitance, which may help to turn on a FEFET at an ultra-low voltage. Recently, NC has been observed in single crystal PZT on a flexible PEN substrate (Bakaul et al. 2017) and in the polymer FE material, PVDF, on a rigid substrate (Ku and Shin 2017). Moreover, it has been shown that the channel current (I_D) in a PVDF-based FEFET can be increased by one order of magnitude with a change in gate voltage (V_G) of less than 60 mV (Jo et al. 2015; Salvatore et al. 2012). The sub-60 mV/decade sub-threshold slope in $I_D - V_G$ cannot be obtained in a dielectric-only-gated FET, and it is a signature of the NC effect. It was also shown that substrate bending has a negligible effect on the NC characteristics in PZT (Bakaul et al. 2017). Therefore, an NCFET has the potential to be an energy-efficient, flexible memory device.

7.5 Novel Techniques for Integration of High Performance Ferroic Materials

The key barrier to achieve an energy-efficient and high performance flexible memory is the difficulty in integrating the best-quality functional materials with flexible substrates. The glass-transition temperature (T_g), which is the temperature at which a polymer changes state from glassy to rubbery, is less than 350°C for most of the flexible substrates. Moreover, the coefficient of thermal expansion of these substrates is usually 10–50 ppm/°C, which is one order of magnitude higher than that of silicon. These impose significant challenges for the growth of functional crystalline materials, which often need a high crystallization temperature. Moreover, many of the standard complementary metal-oxide semiconductor (CMOS) device fabrication steps need a processing temperature around 1000°C. Therefore, the decades-old manufacturing knowledge of the silicon industry cannot be directly transferred to the flexible electronics industry. In this section, epitaxial layer transfer (Bakaul et al. 2016, 2017) and van der Waals epitaxy (Kim et al. 2017; Jiang et al. 2017)—two recent materials integration techniques that may solve these problems, are discussed.

7.5.1 Epitaxial Complex Oxide Layer Transfer Technique

A significant number of complex oxides show ferroic order with excellent memory properties (Dawber et al. 2005; Imada et al. 1998). For many of the novel functionalities, it is important to retain the single crystal nature of these oxides when they are finally interfaced with flexible substrates. For example, single crystal, <001>-oriented PZT is considered to be one of the best FE materials due to its very high switchable polarization, low leakage, abrupt polarization switching characteristics, and superior electronic performance (Lallart 2011; Xu et al. 2014). However, integration of single crystalline films of such oxides using direct synthesis has faced significant challenges due to the fundamental crystal chemistry and mechanical incompatibility of dissimilar materials on either side of the interface. Due to this, only polycrystal PZT-based flexible memories have been demonstrated on metal foil (Zuo et al. 2012) and thin silicon (Ghoneim et al. 2015; Lee et al. 2013) substrates so far. Moreover, the PZT films directly grown on these substrates exhibit slow switching and low switchable/remnant polarization due to their polycrystalline or <111>-oriented structure.

The epitaxial layer transfer process is a technique that consists of growing a single crystal oxide on the best matched substrate and then transferring it onto the flexible substrate. Figure 7.5a–c show the steps of integrating single crystal PZT with flexible substrates (Bakaul et al. 2017). Epitaxial PZT is first grown on an $SrTiO_3$ (STO) substrate coated with a 20-nm thick $La_{0.7}Sr_{0.3}MnO_3$ (LSMO) layer, using pulsed laser deposition. Then a layer of poly methyl methacrylate (PMMA) is spin coated on the PZT, and the whole stack is immersed in (KI + HCl + H_2O) solution. This etches only the LSMO and releases the PZT/PMMA bilayer, which is then collected from the solution and transferred onto the flexible substrate using a micromanipulator. Finally, the PMMA is washed away using acetone, leaving the single-crystal

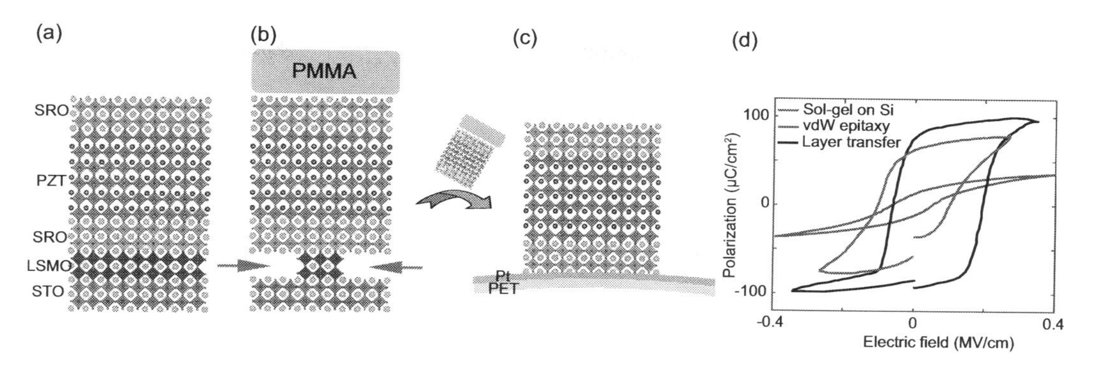

FIGURE 7.5 (a–c) Schematic representation of the complex oxide layer transfer technique. (Adapted from Bakaul, S.R. et al., *Adv. Mater.*, 29, 1605699, 2017. With permission.) (d) A comparison among polarization switching characteristics in $PbZr_{0.2}Ti_{0.8}O_3$ on flexible substrates obtained by sol-gel technique (From Ghoneim, M.T. et al., *Adv. Electron. Mater.*, 1, 1500045, 2015.), vdW epitaxy (From Jiang, J. et al., Science Advances 3(6):e1700121, 2017.), and layer transfer techniques (From Bakaul, S.R. et al.: High Speed Epitaxial Perovskite Memory on Flexible Substrates. *Advanced Materials*. 2017. 29. 1605699. Copyright Wiley-VCH Verlag GmbH & Co. KGaA. Reproduced with permission.)

PZT on the flexible substrate. Since many other complex oxides such as $BaTiO_3$, $CoFe_2O_4$, and $SrRuO_3$ can be epitaxially grown on LSMO and are not affected by the etch solution, they can also be integrated with any flexible substrate using this process.

The epitaxial layer transfer technique allows integration of thin (1 monolayer–100 nm), single crystal ferroelectric films directly onto any flexible substrate. It was shown that the transferred films retain the microstructural, electro-mechanical, and polarization properties of the original films epitaxially grown on lattice-matched oxide substrates and rival the best available properties that can be achieved in such materials. Electronic measurements on capacitor structures containing the transferred oxide layers show excellent characteristics, comparable to those obtained on films grown on single crystal substrates. In terms of all the key metrics for memory operation, such as the remnant polarization, memory retention time, endurance, and switching time, these devices show at least an order of magnitude improvement over any other FE memory devices on flexible substrates that have been reported so far. In addition, the devices can be operated at ~1.5 V which is ~10x lower than the typical voltages required for other demonstrated devices, thereby improving energy efficiency by two orders of magnitude. In fact, the performance of these devices is comparable to those achievable on rigid, lattice-matched substrates.

The concept of the transfer technique dates back to the 1980s when Kongai et al. (1978) separated a device layer from a GaAs substrate in a similar fashion. Following that, industry-standard materials transfer processes such as the smart-cut technique (Moriceau et al. 2012) and laser lift-off (LLO) (Tsakalakos and Sands 2000) have been developed, which can also be used to transfer complex oxide FEs on to flexible substrates (Izuhara et al. 2003; Levy et al. 1998; Park et al. 2006). However, it is important to note that the epitaxial layer transfer technique is significantly different and better than the smart-cut and LLO techniques. The smart cut and LLO techniques can transfer films with thicknesses greater than 100s of nm. The transferred materials usually have significantly large surface roughness (typically >10 nm) due to ion and laser damage and non-uniformity of the ion penetration (Izuhara et al. 2003; Levy et al. 1998). By contrast, the epitaxial layer transfer technique is capable of transferring one single unit cell of PZT (0.4 nm) onto a flexible substrate. This is impossible using a smart-cut technique. Notably, the ability to transfer an ultra-thin film is of critical importance. For example, for a FEFET, a FE gate as thick as 200 nm increases the voltage requirements beyond any reasonable limit. Therefore, the complex oxide layer transfer technique is a fundamental advancement over the conventional smart-cut technique. Recently, Shen et al. (2017) have further improved this technique and demonstrated integration of a centimeter area-scale

FM layer of single crystal spinel ferrite LiFe5O8 (LFO) with a polyimide substrate. Lu et al. (2016) have also demonstrated a similar transfer technique based on a water-soluble $Sr_3Al_2O_6$ (SAO) sacrificial layer. The acid-free solution is an advantage of this technique, although the large lattice constant (15.844 Å) of SAO poses a significant challenge to growing high quality functional layers on top of it.

7.5.2 Van der Waals Epitaxy

Van der Waals (vdW) epitaxy provides another promising route to integrate complex oxides with flexible substrates (Koma 1992). In this technique, epitaxial layers of complex oxides are bonded with the substrate through the weak van der Waals force. The technique is equivalent to growing a material which is almost floating and is free from the strong substrate field and clamping effect. The epilayer grows defect-free and with its own bulk lattice constant right from the first atomic layer. The epilayer can be easily peeled off from the substrate and then transferred onto a flexible substrate. The advantage of this technique over sacrificial layer-based techniques is the absence of any chemical etching and the possibility of contamination. Moreover, this allows direct integration of many single crystal oxides with high temperature flexible substrates such as mica (Jiang et al. 2017; Wu et al. 2016).

While vdW epitaxy helps to grow epilayers free from the influence of the substrate, in many cases, epitaxial registry with the substrate is needed to achieve the intended characteristics. Kim et al. invented a unique fabrication technique called remote epitaxy, which allows the epilayer to 'feel' the lattice registry with the substrate during growth and to be peeled off the substrate easily after growth (Kim et al. 2017). For this, a monolayer of graphene is used between the epilayer and the substrate, which does not fully block the strong potential field of the substrate. Although flexible ferroic memories using this technique are yet to be reported, remote epitaxy certainly opens a new horizon for materials integration in flexible electronics.

Besides functional material transfer techniques, it is also possible to use Si as the original substrate where the devices are first grown, and then transfer the whole assembly onto a flexible substrate. Before transfer, the Si is thinned down to enhance its flexibility. Moreover, if the devices are fabricated on a Si/SiO_2 substrate, the excellent etch selectivity between Si and SiO_2 can be utilized to release the devices, together with the thin SiO_2, from the bulky Si substrate and then transfer and adhere them onto flexible substrates (Loong et al. 2016). Advantage can then be taken of the well-known and reliable silicon electronics manufacturing technologies. Details on different ways of making a Si substrate flexible are provided in Rojas et al. (2013, 2014); Torres Sevilla et al. (2014). Flexible memories with up to 5 mm radius of curvature have been demonstrated by using this technique.

Figure 7.5d shows the polarization (P)—field (E) hysteresis loops of PZT integrated with a flexible substrate using the above-mentioned techniques. The PZT obtained using the epitaxial layer transfer technique shows better characteristics in terms of coherent switching, P-E loop squareness, remnant and switchable polarization, and leakage current. PZT fabricated by vdW epitaxy and by Si-based techniques has a more slanted P-E loop, suggesting the necessity of higher voltage to completely switch the memory state. In addition, the lower switchable polarization would lead to a smaller memory window and reduced retention time. As such, the layer transfer technique provides the best route to fabricate high performance FE memories on flexible substrates, although a lot of engineering complexities and issues need to be overcome before it can be adopted by the industry for a real application.

7.6 Conclusion

Although ferroic memories, particularly TMR and FERAM devices have already been adopted by the computing industry, their flexible versions are lagging far behind and remain topics of academic research. It is still too early to predict the likelihood of ferroic memories being chosen over other contemporary memory technologies, such as flash and optical storage, for a particular application, however, it is clear that ferroic memories have high potential to be implemented in applications where energy-efficiency and high performance are the first-priority design considerations. In order to do this, more research needs to be done to improve techniques for integration of high quality, inorganic materials with flexible substrates, and to understand the physics of local curvature effects on the memory properties.

Acknowledgment

This work is supported by the US Department of Energy, Office of Basic Energy Sciences, Materials Sciences and Engineering Division. A part of the ferroelectric materials characterization was carried out at the Argonne National Laboratory and synthesis was performed at the University of California Berkeley (Office of Naval Research contract no N00014-14-1-0654). The authors acknowledge helpful discussions with A. Petford-Long.

References

Arias, R., and Mills, D. L. 1999. Theory of roughness-induced anisotropy in ferromagnetic films: The dipolar mechanism. *Physical Review B* 59(18):11871–11881.

Bakaul, S. R., Serrao, C. R., Lee, M. et al. 2016. Single crystal functional oxides on silicon. *Nature Communications* 7:10547.

Bakaul, S. R., Serrao, C. R., Lee, O. et al. 2017. High speed epitaxial perovskite memory on flexible substrates. *Advanced Materials* 29(11):1605699.

Barraud, C., Deranlot, C., Seneor, P. et al. 2010. Magnetoresistance in magnetic tunnel junctions grown on flexible organic substrates. *Applied Physics Letters* 96(7):072502.

Berger, L. 1996. Emission of spin waves by a magnetic multilayer traversed by a current. *Physical Review B* 54(13):9353–9358.

Berger, L. 2001. Effect of interfaces on Gilbert damping and ferromagnetic resonance linewidth in magnetic multilayers. *Journal of Applied Physics* 90(9):4632–4638.

Bez, R., Camerlenghi, E., Modelli, A., and Visconti, A. 2003. Introduction to flash memory. *Proceedings of the IEEE* 91(4):489–502.

Bhattacharya, D., Al-Rashid, M. M., and Atulasimha, J. 2017. Energy efficient and fast reversal of a fixed skyrmion two-terminal memory with spin current assisted by voltage controlled magnetic anisotropy. *Nanotechnology* 28(42):425201.

Chen, J.-Y., Lau, Y.-C., Coey, J. M. D., Li, M., and Wang, J.-P. 2017. High performance MgO-barrier magnetic tunnel junctions for flexible and wearable spintronic applications. *Scientific Reports* 7:42001.

Chen, X., Han, X., and Shen, Q.-D. 2017. PVDF-based ferroelectric polymers in modern flexible electronics. *Advanced Electronic Materials* 3(5):1600460.

Chen, Y.-f., Mei, Y., Kaltofen, R. et al. 2008. Towards flexible magnetoelectronics: Buffer-enhanced and mechanically tunable GMR of Co/Cu multilayers on plastic substrates. *Advanced Materials* 20(17):3224–3228.

Cheong, S.-W., and Mostovoy, M. 2007. Multiferroics: A magnetic twist for ferroelectricity. *Nature Materials* 6(1):13–20.

Chong, T. C., Shi, L. P., Zhao, R. et al. 2006. Phase change random access memory cell with superlattice-like structure. *Applied Physics Letters* 88(12):122114.

Dai, G., Zhan, Q., Liu, Y. et al. 2012. Mechanically tunable magnetic properties of Fe81Ga19 films grown on flexible substrates. *Applied Physics Letters* 100(12):122407.

Dai, G., Zhan, Q., Yang, H. et al. 2013. Controllable strain-induced uniaxial anisotropy of Fe81Ga19 films deposited on flexible bowed-substrates. *Journal of Applied Physics* 114(17):173913.

Dawber, M., Rabe, K. M., and Scott, J. F. 2005. Physics of thin-film ferroelectric oxides. *Reviews of Modern Physics* 77(4):1083–1130.

de Araujo, C. A. P., Cuchiaro, J. D., McMillan, L. et al. 1995. Fatigue-free ferroelectric capacitors with platinum electrodes. *Nature* 374(6523):627–629.

Dzyaloshinsky, I. 1958. A thermodynamic theory of 'weak' ferromagnetism of antiferromagnetics. *Journal of Physics and Chemistry of Solids* 4(4):241–255.

Emori, S., Bauer, U., Ahn, S.-M., Martinez, E., and Beach, G. S. D. 2013. Current-driven dynamics of chiral ferromagnetic domain walls. *Nature Materials* 12(7):611–616.

Faurie, D., Zighem, F., Garcia-Sanchez, A., Lupo, P., and Adeyeye, A. O. 2017. Fragmentation and adhesion properties of CoFeB thin films on polyimide substrate. *Applied Physics Letters* 110(9):091904.

Gaididei, Y., Kravchuk, V. P., and Sheka, D. D. 2014. Curvature effects in thin magnetic shells. *Physical Review Letters* 112(25):257203.

Ghoneim, M. T., and Hussain, M. M. 2015a. Review on physically flexible nonvolatile memory for internet of everything electronics. *Electronics* 4(3):424–479.

Ghoneim, M. T., and Hussain, M. M. 2015b. Study of harsh environment operation of flexible ferroelectric memory integrated with PZT and silicon fabric. *Applied Physics Letters* 107(5):052904.

Ghoneim, M. T., Zidan, M. A., Alnassar, M. Y. et al. 2015. Thin PZT-based ferroelectric capacitors on flexible silicon for nonvolatile memory applications. *Advanced Electronic Materials* 1(6):1500045.

Han, S.-T., Zhou, Y., and Roy, V. A. L. 2013. Towards the development of flexible non-volatile memories. *Advanced Materials* 25(38):5425–5449.

Hyuk Park, M., Joon Kim, H., Jin Kim, Y., Moon, T., and Seong Hwang, C. 2014. The effects of crystallographic orientation and strain of thin Hf0.5Zr0.5O2 film on its ferroelectricity. *Applied Physics Letters* 104(7):072901.

Ikeda, S., Miura, K., Yamamoto, H. et al. 2010. A perpendicular-anisotropy CoFeB–MgO magnetic tunnel junction. *Nature Materials* 9(9):721–724.

Imada, M., Fujimori, A., and Tokura, Y. 1998. Metal-insulator transitions. *Reviews of Modern Physics* 70(4):1039–1263.

Izuhara, T., Gheorma, I. L., Osgood, R. M. et al. 2003. Single-crystal barium titanate thin films by ion slicing. *Applied Physics Letters* 82(4):616–618.

Jang, P.-H., Song, K., Lee, S.-J., Lee, S.-W., and Lee, K.-J. 2015. Detrimental effect of interfacial Dzyaloshinskii-Moriya interaction on perpendicular spin-transfer-torque magnetic random access memory. *Applied Physics Letters* 107(20):202401.

Jiang, J., Bitla, Y., Huang, C.-W. et al. 2017. Flexible ferroelectric element based on van der Waals heteroepitaxy. *Science Advances* 3(6):e1700121.

Jo, J., Choi, W. Y., Park, J.-D., Shim, J. W., Yu, H.-Y., and Shin, C. 2015. Negative capacitance in organic/ferroelectric capacitor to implement steep switching MOS devices. *Nano Letters* 15(7):4553–4556.

Kim, R. H., Kim, H. J., Bae, I. et al. 2014. Non-volatile organic memory with sub-millimetre bending radius. *Nature Communications* 5:3583.

Kim, Y., Cruz, S. S., Lee, K. et al. 2017. Remote epitaxy through graphene enables two-dimensional material-based layer transfer. *Nature* 544(7650):340–343.

Kogan, S. M. 1964. Piezoelectric effect during inhomogeneous deformation and acoustic scattering of carriers in crystals. *Soviet Physics-Solid State* 5(10):2069–2070.

Koma, A. 1992. Van der Waals epitaxy—A new epitaxial growth method for a highly lattice-mismatched system. *Thin Solid Films* 216(1):72–76.

Konagai, M., Sugimoto, M., and Takahashi, K. 1978. High efficiency GaAs thin film solar cells by peeled film technology. *Journal of Crystal Growth* 45:277–280.

Ku, H., and Shin, C. 2017. Transient response of negative capacitance in P(VDF0.75-TrFE0.25) organic ferroelectric capacitor. *IEEE Journal of the Electron Devices Society* 5(3):232–236.

Lallart, M. 2011. *Ferroelectrics—Physical Effects, Rijeka, Croatia: InTech.*

Lee, G.-G., Tokumitsu, E., Yoon, S.-M., Fujisaki, Y., Yoon, J.-W., and Ishiwara, H. 2011. The flexible nonvolatile memory devices using oxide semiconductors and ferroelectric polymer poly(vinylidene fluoride-trifluoroethylene). *Applied Physics Letters* 99(1):012901.

Lee, O., You, L., Jang, J., Subramanian, V., and Salahuddin, S. 2015. Flexible spin-orbit torque devices. *Applied Physics Letters* 107(25):252401.

Lee, W., Kahya, O., Toh, C. T., Özyilmaz, B., and Ahn, J.-H. 2013. Flexible graphene–PZT ferroelectric nonvolatile memory. *Nanotechnology* 24(47):475202.

Levy, M., Osgood, R. M., Liu, R. et al. 1998. Fabrication of single-crystal lithium niobate films by crystal ion slicing. *Applied Physics Letters* 73(16):2293–2295.

Liu, L., Pai, C. F., Li, Y., Tseng, H. W., Ralph, D. C., and Buhrman, R. A. 2012. Spin-torque switching with the giant spin hall effect of tantalum. *Science* 336(6081):555–558.

Loong, L. M., Lee, W., Qiu, X. et al. 2016. Flexible MgO barrier magnetic tunnel junctions. *Advanced Materials 28*(25):4983–4990.

Loong, L. M., Qiu, X., Neo, Z. P. et al. 2014. Strain-enhanced tunneling magnetoresistance in MgO magnetic tunnel junctions. *Scientific Reports* 4(1):6505.

Lu, D., Baek, D. J., Hong, S. S., Kourkoutis, L. F., Hikita, Y., and Hwang, Harold, Y. 2016. Synthesis of freestanding single-crystal perovskite films and heterostructures by etching of sacrificial water-soluble layers. *Nature Materials* 15(12):1255–1260.

Lu, H., Bark, C. W., Esque de los Ojos, D. et al. 2012. Mechanical writing of ferroelectric polarization. *Science* 336(6077):59–61.

Ma, W., and Cross, L. E. 2001. Observation of the flexoelectric effect in relaxor Pb(Mgl/3Nb2/3)O3 ceramics. *Applied Physics Letters* 78(19):2920–2921.

Ma, W., and Cross, L. E. 2002. Flexoelectric polarization of barium strontium titanate in the paraelectric state. *Applied Physics Letters* 81(18):3440–3442.

Ma, W., and Cross, L. E. 2005. Flexoelectric effect in ceramic lead zirconate titanate. *Applied Physics Letters* 86(7):072905.

Ma, W., and Cross, L. E. 2006. Flexoelectricity of barium titanate. *Applied Physics Letters* 88(23):232902.

Matsumoto, A., Horie, S., Yamada, H., Matsushige, K., Kuwajima, S., and Ishida, K. 2007. Ferro- and piezoelectric properties of vinylidene fluoride oligomer thin film fabricated on flexible polymer film. *Applied Physics Letters* 90(20):202906.

Melzer, M., Makarov, D., Calvimontes, A. et al. 2011. Stretchable magnetoelectronics. *Nano Letters* 11(6):2522–2526.

Mihai Miron, I., Gaudin, G., Auffret, S. et al. 2010. Current-driven spin torque induced by the Rashba effect in a ferromagnetic metal layer. *Nature Materials* 9(3):230–234.

Moriceau, H., Mazen, F., Braley, C., Rieutord, F., Tauzin, A., and Deguet, C. 2012. Smart Cut™: Review on an attractive process for innovative substrate elaboration. *Nuclear Instruments and Methods in Physics Research Section B: Beam Interactions with Materials and Atoms* 277:84–92.

Moriya, T. 1960. Anisotropic superexchange interaction and weak ferromagnetism. *Physical Review* 120(1):91–98.

Nguyen, C. A., Lee, P. S., Ng, N. et al. 2007. Anomalous polarization switching in organic ferroelectric field effect transistors. *Applied Physics Letters* 91(4):042909.

Nishimura, N., Hirai, T., Koganei, A. et al. 2002. Magnetic tunnel junction device with perpendicular magnetization films for high-density magnetic random access memory. *Journal of Applied Physics* 91(8):5246–5249.

Noda, K., Ishida, K., Kubono, A., Horiuchi, T., Yamada, H., and Matsushige, K. 2003. Remanent polarization of evaporated films of vinylidene fluoride oligomers. *Journal of Applied Physics* 93(5):2866–2870.

Ota, S., Hibino, Y., Bang, D. et al. 2016. Strain-induced reversible modulation of the magnetic anisotropy in perpendicularly magnetized metals deposited on a flexible substrate. *Applied Physics Express* 9(4):043004.

Özkaya, B., Saranu, S. R., Mohanan, S. et al. 2008. Effects of uniaxial stress on the magnetic properties of thin films and GMR sensors prepared on polyimide substrates. *Physica Status Solidi (a)* 205(8):1876–1879.

Park, Y. B., Min, B., Vahala, K. J., and Atwater, H. A. 2006. Integration of single-crystal $LiNbO_3$ thin film on silicon by laser irradiation and ion implantation–induced layer transfer. *Advanced Materials* 18(12):1533–1536.

Parkin, S. S. P. 1996. Flexible giant magnetoresistance sensors. *Applied Physics Letters* 69(20):3092–3094.

Parkin, S. S. P., Roche, K. P., and Suzuki, T. 1992. Giant magnetoresistance in antiferromagnetic Co/Cu multilayers grown on kapton. *Japanese Journal of Applied Physics* 31(Part 2, No. 9A):L1246–L1249.

Rojas, J. P., Ghoneim, M. T., Young, C. D., and Hussain, M. M. 2013. Flexible high-κ/metal gate metal/insulator/metal capacitors on silicon (100) fabric. *IEEE Transactions on Electron Devices* 60(10):3305–3309.

Rojas, J. P., Torres Sevilla, G. A., Ghoneim, M. T. et al. 2014. Transformational silicon electronics. *ACS Nano* 8(2):1468–1474.

Rueckes, T., Kim, K., Joselevich, E., Tseng G. Y., Cheung, C.-L., Lieber C. M. 2000. Carbon nanotube-based nonvolatile random access memory for molecular computing. *Science* 289(5476):94–97.

Ryu, K.-S., Thomas, L., Yang, S.-H., and Parkin, S. 2013. Chiral spin torque at magnetic domain walls. *Nature Nanotechnology* 8(7):527–533.

Salahuddin, S., and Datta, S. 2008. Use of negative capacitance to provide voltage amplification for low power nanoscale devices. *Nano Letters* 8(2):405–410.

Salvatore, G. A., Rusu, A., and Ionescu, A. M. 2012. Experimental confirmation of temperature dependent negative capacitance in ferroelectric field effect transistor. *Applied Physics Letters* 100(16):163504.

Sampaio, J., Khvalkovskiy, A. V., Kuteifan, M. et al. 2016. Disruptive effect of Dzyaloshinskii-Moriya interaction on the magnetic memory cell performance. *Applied Physics Letters* 108(11):112403.

Shen, L., Wu, L., Sheng, Q. et al. 2017. Epitaxial lift-off of centimeter-scaled spinel ferrite oxide thin films for flexible electronics. *Advanced Materials* 29(33):1702411.

Slonczewski, J. C. 1996. Current-driven excitation of magnetic multilayers. *Journal of Magnetism and Magnetic Materials* 159(1–2):L1–L7.

Slonczewski, J. C. 1999. Excitation of spin waves by an electric current. *Journal of Magnetism and Magnetic Materials* 195(2):L261–L268.

Sheka, D. D., Kravchuk, V. P., and Gaididei, Y. 2015. Curvature effects in statics and dynamics of low dimensional magnets, *Journal of Physics A: Mathematical and Theoretical* 48:125202.

Streubel, R., Fischer, P., Kronast, F. et al. 2016. Magnetism in curved geometries. *Journal of Physics D: Applied Physics* 49(36):363001.

Streubel, R., Lee, J., Makarov, D. et al. 2014. Magnetic microstructure of rolled-up single-layer ferromagnetic nanomembranes. *Advanced Materials* 26(2):316–323.

Sun, X., Bedoya-Pinto, A., Llopis, R., Casanova, F., and Hueso, L. E. 2014. Flexible semi-transparent organic spin valve based on bathocuproine. *Applied Physics Letters* 105(8):083302.

Tang, Z., Wang, B., Yang, H. et al. 2014. Magneto-mechanical coupling effect in amorphous Co40Fe40B20 films grown on flexible substrates. *Applied Physics Letters* 105(10):103504.

Thiaville, A., Rohart, S., Jué, É., Cros, V., and Fert, A. 2012. Dynamics of Dzyaloshinskii domain walls in ultrathin magnetic films. *EPL (Europhysics Letters)* 100(5):57002.

Torres Sevilla, G. A., Ghoneim, M. T., Fahad, H., Rojas, J. P., Hussain, A. M., and Hussain, M. M. 2014. Flexible nanoscale high-performance FinFETs. *ACS Nano* 8(10):9850–9856.

Tretiakov, O. A., Morini, M., Vasylkevych, S., and Slastikov, V. 2017. Engineering curvature-induced anisotropy in thin ferromagnetic films. *Physical Review Letters* 119(7):077203.

Tsakalakos, L., and Sands, T. 2000. Epitaxial ferroelectric (Pb,La)(Zr,Ti)O3 thin films on stainless steel by excimer laser liftoff. *Applied Physics Letters* 76(2):227–229.

Vemulkar, T., Mansell, R., Fernández-Pacheco, A., and Cowburn, R. P. 2016. Toward flexible spintronics: Perpendicularly magnetized synthetic antiferromagnetic thin films and nanowires on polyimide substrates. *Advanced Functional Materials* 26(26):4704–4711.

Won Seo, J., Park, J.-W., Lim, K. S., Yang, J.-H., and Kang, S. J. 2008. Transparent resistive random access memory and its characteristics for nonvolatile resistive switching. *Applied Physics Letters* 93(22)3505.

Worledge, D. C., Hu, G., Abraham, D. W. et al. 2011. Spin torque switching of perpendicular Ta|CoFeB|MgO-based magnetic tunnel junctions. *Applied Physics Letters* 98(2):022501.

Wu, P.-C., Chen, P.-F., Do, T. H. et al. 2016. Heteroepitaxy of Fe_3O_4/Muscovite: A new perspective for flexible spintronics. *ACS Applied Materials & Interfaces* 8(49):33794–33801.

Xia, F., Xu, H., Fang, F. et al. 2001. Thickness dependence of ferroelectric polarization switching in poly(vinylidene fluoride–trifluoroethylene) spin cast films. *Applied Physics Letters* 78(8):1122–1124.

Xu, R., Liu, S., Grinberg, I. et al. 2014. Ferroelectric polarization reversal via successive ferroelastic transitions. *Nature Materials* 14(1):79–86.

Yan, M., Kákay, A., Gliga, S., and Hertel, R. 2010. Beating the walker limit with massless domain walls in cylindrical nanowires. *Physical Review Letters* 104(5):057201.

Yershov, K. V., Kravchuk, V. P., Sheka, D. D., and Gaididei, Y. 2015. Curvature-induced domain wall pinning. *Physical Review B* 92(10):104412.

Yoon, S.-M., Yang, S., and Park, S.-H. K. 2011. Flexible nonvolatile memory thin-film transistor using ferroelectric copolymer gate insulator and oxide semiconducting channel. *Journal of The Electrochemical Society* 158(9):H892.

Yu, H., Chung, C.-C., Shewmon, N. et al. 2017. Flexible inorganic ferroelectric thin films for nonvolatile memory devices. *Advanced Functional Materials* 27(21):1700461.

Zhang, X., Zhan, Q., Dai, G. et al. 2013. Effect of mechanical strain on magnetic properties of flexible exchange biased FeGa/IrMn heterostructures. *Applied Physics Letters* 102(2):022412.

Zubko, P., Catalan, G., and Tagantsev, A. K. 2013. Flexoelectric effect in solids. *Annual Review of Materials Research* 43(1):387–421.

Zuo, Z., Chen, B., Zhan, Q.-F. et al. 2012. Preparation and ferroelectric properties of freestanding Pb(Zr, Ti)O3 thin membranes. *Journal of Physics D: Applied Physics* 45(18):185302.

8

Flexible and Stretchable High-Frequency RF Electronics

Juhwan Lee,
Inkyu Lee, and
Zhenqiang Ma

8.1 Introduction

Notwithstanding the difficulties of implementing pioneering experiments, scientists and engineers have always been determined to lay down the solid groundwork for their successors in discovering new inventions. Thanks to such admirable pioneers who have cultivated the silicon industry, the chip is now becoming smaller and faster in a shorter time than ever. The limit is yet to be broken, and the Moore's law still exists. As a result of the development of revolutionary electronics, a new type of device—classified as soft electronics—has evolved over the past few decades, as shown in Figure 8.1.

Firstly, the researchers blazed the trail for the flexible transistor in 1994, which stably performed with mechanical treatments such as bending or twisting (Garnier et al. 1994). Subsequently, soft electronics have triggered the interests of many researchers due to its potential for incorporation in various applications that cannot be attained by conventional rigid electronic devices. This new generation of semiconductor technology has rapidly dominated the field of modern electronics, such as flexible displays (Rogers et al. 2001; Sekitani et al. 2010), smart textiles (Hu et al. 2010; Jang et al. 2014), and medical devices (Gelinck et al. 2004; Kim et al. 2011, 2012a, 2013). While the applications of conventional semiconductor technology were mostly limited to the subject of 'faster' and 'compact,' the new trend of technology began targeting that of 'softer,' to weld the electronics into unusual spaces where it is not possible for devices in conventional rigid forms, such as non-invasive implantable medical devices.

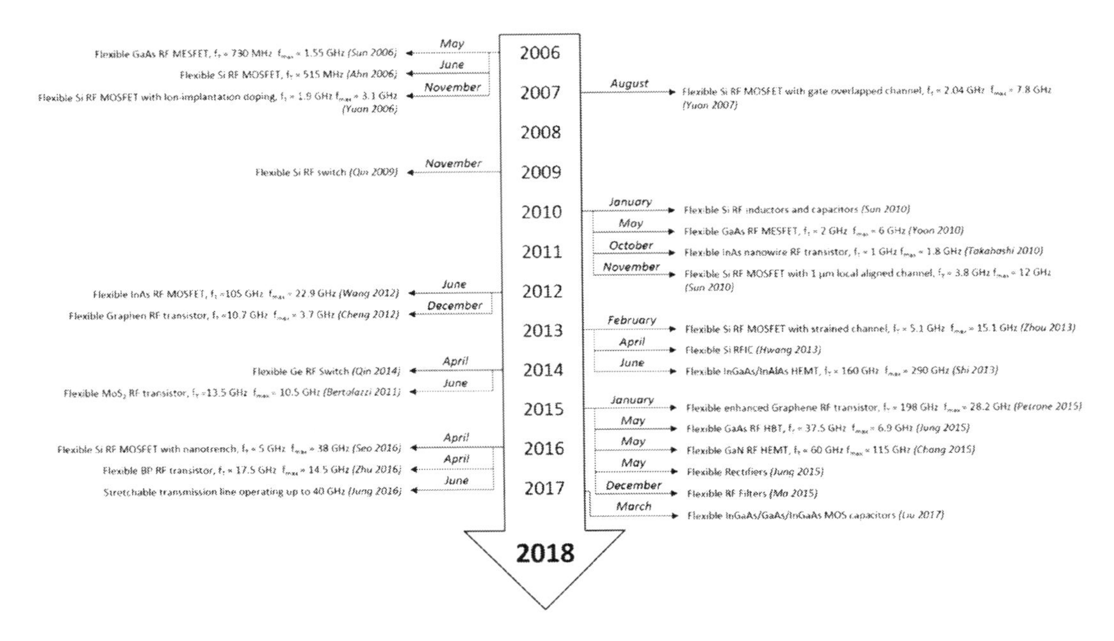

FIGURE 8.1 History of flexible and stretchable high-frequency RF electronics. The material for each device, the RF figure of merits, and a brief description of the device technology are described.

As the form of electronics became anomalous, many components from traditional electronic devices have transitioned to be adapted to the new format of flexible and stretchable electronics. Despite state-of-the-art technology development, some of the fields still have not been heavily investigated. For instance, although the current technology has finally allowed soft electronics to acquire near-field communication capabilities (Samineni et al. 2017), the subject of far-field still remains in the wild frontier. Currently, the applications of conventional electronics are capable of wireless communication for data or for energy via long distance, regardless of the device location. Convenient technologies mostly rely on high-speed radio-frequency (RF) electronics that operate at multi-gigahertz frequencies, which unfortunately have too many restrictions on material and design considerations in the flexible or stretchable form (Jung et al. 2016c; Ma et al. 2015). Surely, flexible and stretchable devices must have conformability, mechanical stability, and fast elastic responses to huge deformations, but all such factors become limitations to RF components, which are highly sensitive to both electrical and mechanical environments.

Building ultrathin RF components requires advanced engineering and careful design procedures, as they must impose specific requirements: (1) possession of high carrier mobility; (2) integrability on foreign substrates; and (3) mechanical flexibility with minimal performance changes (Jung et al. 2016b). On the other hand, the RF passive components must meet the following requirements: (1) fabrication capability of large area; (2) fabrication compatibility with active components on the same substrate; and (3) structural reciprocality or optimum of lumped elements that share relatively uniform dielectric environments. As known, many semiconductor materials have been investigated for almost a century and classified as ideal candidates for certain purposes of electronic application. Some of them are heavily involved in the formation of the active component in microwave circuits which is practically the commercialized RF technology for wireless communication. They have their own singularity and trade-offs, and each material determines the extend-ability of the fabrication process and design architecture. Normally, a microwave device is difficult to build on an unusual substrate because transistors in the microwave circuit require comparably smaller feature sizes than that of devices operating at a lower frequency. Moreover, the fabrication process is exposed to a variety of temperature and pressure conditions that may not be compatible with bendable plastic substrates. Forming thin film transistors from bulk materials, the mechanical stability will be key in maintaining acceptable performances. As a lumped system, passive elements also suffer from the same issues and have more challenges as they occupy a larger area than the active components, which require an advanced dimensional analysis of electrical and mechanical performance variation (Jung et al. 2016a). The above issues are inevitable and must be overcome to merge the traditional RF technology with current flexible and stretchable electronics.

The sections in this chapter aim to present the RF circuit components operating at a high-frequency, including the active and passive elements, followed by RF circuits, and systems in flexible or stretchable form. Depending on the active material, transistors are utilized in specified applications such as amplification, switching signal, etc. Categorizing semiconductor materials of active devices, the sections will cover the most recent technology—including the scheme of device structure and alternative fabrication process to generate the maximum performance—which can be represented by the cut-off frequency (f_T) and the maximum oscillation frequency (f_{max}) of transistors on the flexible substrate. The cut-off frequency is defined as the frequency at which the current gain, the ratio of drain-to-source, and gate-to-source current, are in unity. Generally, the cut-off frequency is a useful parameter to analyze the switching speed of the transistor in an analog or digital circuit, while the maximum oscillation frequency determines the performance of the transistor on a microwave integrated circuit. Nonetheless, both parameters are critical in determining whether the transistors are applicable for the operating frequency of the microwave circuit which should, as a rule of thumb, be approximately 5 times less than that of the transistor.

Consideration of the material selection and the fabrication compatibility for the active components also applies to the passive components and the integrated circuit and system as well. At the circuit level, each component must share a similar environment, including dielectric material, substrate, and encapsulation layer for the enhancement of mechanical stability in bending, twisting, and stretching circumstances. The following sections will introduce various types of passive components, circuits, and systems tuned by modifying their design layout, material, or overall architecture. The combination of design and material considerations can improve the limit of bendability or stretchability of the individual component or the entire circuit. As such, most of the electrodes have become thin and wavy, and were put into the sandwich-structure to improve the mechanical performance on various types of soft substrates, such as polyethylene terephthalate (PET), polyimide (PI), polydimethylsiloxane (PDMS), and silicon rubber (Ecoflex). Design modifications are required after material selection due to the fluctuations of RF gain or loss. For example, the stretchable transmission lines are composed of a single polymer, PI, with metal traces for two different purposes (Jung et al. 2016a). The role of the top and bottom layers is to enhance the mechanical stability of the metal traces, while the mid-layer of PI acts as a dielectric channel where most of the electromagnetic-wave confinements occur. In this case, the selection decision for the polymer within the structure should be followed by concurrent filtering of the electrical and mechanical parameters, such as the dielectric constant and tensile strength. Subsequently, the scheme is expanded to the circuit and system levels.

Besides these key issues, researchers have put tremendous efforts into RF electronics to build them into flexible and stretchable forms. At the end of this chapter, we will explore some of their state-of-the-art flexible and stretchable RF electronic systems and discuss the future work for the realization of flexible and stretchable electronics that can achieve long-distance communication.

8.2 Silicon Transistors

As the flexible electronics industry continues to grow rapidly, requirements for the micro-fabrication system are becoming even more complicated and problematic. In this context, micro-fabrication technology based on Si can be a crucial starting point for developing flexible RF electronics. Since the 1960s, the Si-based micro-fabrication technology has developed at a dashing rate and rooted itself deeply in the semiconductor industry. Now that it has become fully mature, most of the present-day global electronics are fabricated using Si. Thus, it is ideal to utilize a well-established commercial infrastructure of Si-based micro-fabrications in realizing high-performance flexible RF electronics (Rojas et al. 2014; Zhang 2015). In general, fabrication of conventional Si-based transistors is performed on a typical rigid substrate such as the silicon wafer and a self-aligned gate process is utilized to effectively reduce the capacitance of gate-to-source and gate-to-drain (Jung et al. 2016b).

However, when it comes to transforming conventional Si-based micro-fabrication techniques into flexible form, several challenges exist. First, the beneficial self-aligned gate process cannot be used for fabricating flexible Si-nanomembrane (SiNM)-based transistors. Since the fabrication process for flexible SiNM transistors involves one that designates the doping region of the Si layer on the silicon-on-insulator (SOI) wafer before the release of SiNM (Ahn et al. 2007), misalignment between the gate electrode and the channel region can be created, which makes the alignment and patterning process more challenging and consequently decreases the yield of the fabrication process (Sun et al. 2010b).

In addition, the high-temperature process used in the fabrication of conventional Si-based transistors is not available for the fabrication of flexible Si-based RF electronics. In the conventional fabrication process, silicon dioxide (SiO_2) thermally grown at high-temperature is typically used to obtain a high-quality gate insulator. However, the use of such a high-temperature thermal oxidation process cannot be used to fabricate flexible electronics due to the thermal vulnerability of flexible substrates to high-temperature.

In order to address these issues, diverse-deposited oxide films such as SiO, SiO_2, Al_2O_3, and HfO_2, which can be processed at low-temperature (<150°C) with satisfactory quality, have been utilized. Some of the latest breakthroughs in Si-based RF electronics are shown in Figure 8.2. Figure 8.2a illustrates the

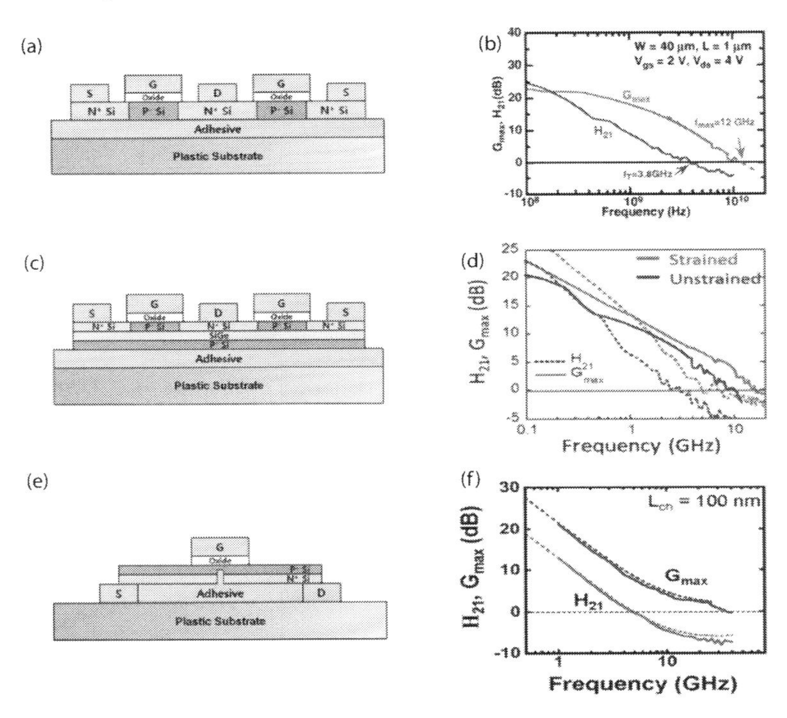

FIGURE 8.2 Si-based high-speed flexible electronics. (a) Structural illustration of a conventional high-speed Si-based MOSFET on a flexible plastic substrate. (b) Measured data for the frequency response characteristics of the Si-based MOSFET shown in (a) (Adapted from Sun et al., *Small* 6, 2553–2557, 2010b.). (c) Structural illustration of a high-speed Si-based MOSFET with strained trilayer. (d) Measured data for the frequency response characteristics of the strained Si-based MOSFET shown in (c). (Adapted from Zhou, H. et al., *Sci. Rep.*, 3, 1291, 2013.) (e) Structural illustration of a high-speed Si-based MOSFET with nanotrench structure. (f) Measured data for the frequency response characteristics of the Si-based nanotrench MOSFET shown in (e). (Adapted from Seo, J.-H. et al., *Sci. Rep.*, 6, 24771, 2016.)

cross-sectional view of traditional high-speed Si-based transistors fabricated on a flexible substrate. In this work, a SiO film was used as the dielectric layer of the flexible metal-oxide-semiconductor (MOSFET) and considerable frequencies (f_T = 3.8 GHz and f_{max} = 12 GHz) were attained, as shown in Figure 8.2b.

8.2.1 Strain on Transistor

In 1996, Fischetti and Laux analyzed the change in carrier mobility in strained Si and found that electron mobility increases under tensile strain due to a strain-induced decrease in the electron effective mass and a reduction in intervalley and interband scattering (Chen et al. 2011; Fischetti and Laux 1996; Yu et al. 2008). Since then, strain engineering on Si-based electronics has been a useful strategy for constructing high-speed electronics systems in enhancing the carrier mobility in Complementary metal-oxide semiconductor (CMOS) devices (Jung et al. 2016b; Lee et al. 2016; Yuan et al. 2006, 2008; Zhou et al. 2013).

For example, beneficial local strain—tensile strain for N-type metal-oxide-semiconductor (nMOS) and compressive strain for P-type metal-oxide-semiconductor (pMOS)—is utilized to improve mobility. Figure 8.2c presents a cross-sectional view of a flexible high-speed transistor with a strained trilayer (Si/SiGe/Si) structure. In this example, the SiGe layer was compressively strained to the Si in-plane lattice constant, due to the lattice mismatch between the SiGe and Si layers, and the outer Si layers were tensile-strained to keep the balance of forces between the Si and SiGe in the freestanding trilayer nanomembrane (NM) during the membrane release process. Because of the tensile strain in the Si channel layer, mobility enhancement of 47.3% was achieved (Rim et al. 2003). Furthermore, the transconductance was also improved from 262 to 386 μS in the strained transistors as a result of the mobility enhancement. Current gain (H_{21}) and power gain (G_{max}) as a function of frequency of unstrained and strained transistors are shown in Figure 8.2d. Compared with the unstrained transistor with f_T = 3.3 GHz and f_{max} = 10.3 GHz, the strained transistor showed highly improved RF properties (f_T = 5.1 GHz and f_{max} = 15.1 GHz).

8.2.2 Nanotrench Transistor

After revealing the secret performance of Si-based transistors in bending, another approach to improving the design of the transistor structure was also achieved to complement the flexible form of transistors. The latest technology of nanoimprint lithography allowed formation of an unusual nanoscale silicon trench on the wafer, which can define the effective channel length of the transistor (Seo et al. 2016). In this case, the channel length is dependent only on the length of the trench so as to allow the larger backside gate where the fabrication of the gate becomes more straightforward. This unique nanotrench transistor design is shown in Figure 8.2e.

The structure overcomes the short-channel effect in MOSFETs with the source-to-drain current via the trench that forces the current flow around the gate to the drain to decrease leakage and prevent a shorted channel. With a 100-nm wide trench, Al_2O_3 acts as the gate insulator. f_T and f_{max} were measured to be 5 and 38 GHz, which show record-breaking speeds compared to that of the transistors in 2006 (Ahn et al. 2006). Moreover, the newly improved design opens up the opportunity of monolithic integration of the active devices by interconnecting the passive devices at the circuit level to ultimately realize flexible RF electronics.

8.3 Compound Semiconductor Transistors

8.3.1 GaAs-Based Transistor

Today, over 80% of modern consumer electronics, such as cell phones and tablets, use gallium arsenide-based (GaAs) microwave devices, chosen for their advantageous qualities, including high-speed wireless communication capability and power handling efficiency (Jung et al. 2015). Thus, GaAs is one of the most widely integrated materials in the compound semiconductor industry (Chang and Kai 1994).

Moreover, GaAs is a thermally stable insulating substrate; the beauty of using GaAs material allows the option to have different types of heterojunctions such as AlGaAs and InGaP, which can be removed or selected (Kuzuhara and Tanaka 2003). Acting as a sacrificial layer, AlGaAs is removable by a chemical etching process. Then, the GaAs membrane in the epitaxial multi-layers can be selected and transferred, and it is eventually utilized for microwave electronic applications in a flexible form (Yoon et al. 2010). The first-ever flexible form of transistors using GaAs was demonstrated by utilizing nanoribbons from bulk wafers while AlGaAs acted as a sacrificial layer. The technique successfully fabricated flexible transistors with outstanding performance (Sun et al. 2006).

Using the same approach, GaAs-based Metal-semiconductor field-effect transistor (MESFET) and heterojunction bipolar transistor (HBT) are also demonstrated in Figure 8.3. An *N*-type GaAs membrane is released from the bulk wafer and transfer-printed onto the flexible substrate and forms a flexible GaAs MESFET shown in Figure 8.3a. The reliable fabrication process achieves a yield of over 99% once the membrane is transferred onto the flexible target substrate (Yoon et al. 2010). As shown in Figure 8.3b, the MESFET recorded f_T and f_{max} of 2 and 6 GHz, respectively, with 3-μm channel length.

On the other hand, since high-speed transistor requires wide-bandgap materials for the emitter region, even AlGaAs and InGaP can serve as candidates for the emitter material. For instance, the HBT can comprise InGaP as an emitter (Jung et al. 2015). The device is then released from its mother substrate and transfer-printed onto the flexible and bio-degradable substrate to enhance its eco-friendly electronic applications. Over one hundred releasable GaAs HBTs were demonstrated on the mother substrate as shown in Figure 8.3c. The transferred HBTs on the flexible substrate are pictured in Figure 8.3d.

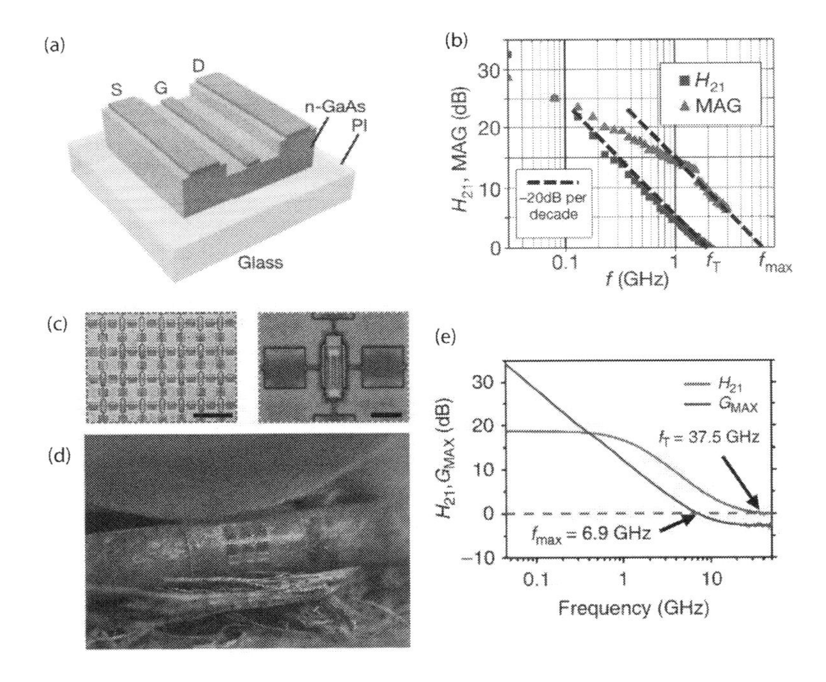

FIGURE 8.3 GaAs-based high-speed flexible electronics. (a) Structural illustration of a GaAs-based MESFET on a PI substrate. (b) Measured data for the frequency response characteristics of the GaAs-based MESFET shown in (a). (Adapted from Yoon, J. et al., *Nature*, 465, 329–333, 2010). (c) Optical microscopy image of an array of releasable GaAs-based HBTs in a dense array (left) and a magnified view of the releasable GaAs-based HBT with photoresist anchors (right). (d) Photography image of the GaAs-based HBTs on the flexible substrate. (e) Measured data for the frequency response characteristics of the GaAs-based HBT shown in (d). (Adapted from Jung, Y.H. et al., *Nat. Commun.*, 6, 7170, 2015.)

f_T and f_{max} of 37.5 and 6.9 GHz, respectively, were recorded in Figure 8.3e on the foreign flexible substrate which is a paper-like bio-degradable material cellulose nanofibers [CNF]. This eco-friendly substrate minimizes electronic waste in consumer electronics.

8.3.2 InAs-Based Transistor

Another compound, indium arsenide (InAs) can promote transistors operating at higher frequencies than others. It is barely utilized within the semiconductor industry because of the costly process for successful epitaxial growth to minimize lattice mismatch. However, it is theoretically feasible to have the fastest speed compared to other compound semiconductor materials such as InAs-based high-electron-mobility transistors (HEMT) (Kim and Del Alamo 2010). The transistor recorded f_T and f_{max} of 644 and 681 GHz, respectively.

Taking this high-speed performance potential of the conventional InAs-based transistor, people gave the unique approach to fabricated extremely high-frequency transistors on the flexible substrate by releasing the InAs layer from the rigid compound semiconductor-on-insulator wafer called 'XOI' (Ko et al. 2010). By etching the sacrificial layer, AlGaSb and InAs patterned as nanoribbons were picked up and transfer-printed onto the target substrate. The remarkably thin structure of the InAs nanoribbon offered excellent gate control and protected the transistor from the short-channel effect.

Furthermore, the flexible InAs-based MOSFET was built using electron-beam lithography, as shown in Figure 8.4a. This fine patterning method allows a small gate length of 75 nm from the T-shaped Al/ZrO_2 gate-stack, and the self-aligned source/drain pads were integrated into the MOSFET structure (Wang et al. 2012). As shown in Figure 8.4b, f_T and f_{max} were achieved at 105 and 22.9 GHz, respectively, on the flexible plastic substrate.

As we have seen in other flexible semiconductor fabrications in the previous section, the releasing and transfer-printing methods are key in flexible electronics fabrication. Utilization of the noble compound material is validated to demonstrate high-speed flexible HEMTs (Shi et al. 2013). This record-breaker shows a stunning performance with f_T and f_{max} at 160 and 290 GHz, respectively.

8.3.3 GaN-Based Transistor

The outstanding properties of gallium nitride (GaN) enable GaN-based HEMTs to have an important role in power electronics in terms of bandgap, electron velocity, gain, breakdown field, and thermal management. Compared to GaAs-based transistors, GaN-based HEMTs have a relatively easier fabrication requirement due to the high-power-per-unit gate width, offering increased input and output impedance that makes the device more suitable to match with the system than that of a GaAs-based transistor.

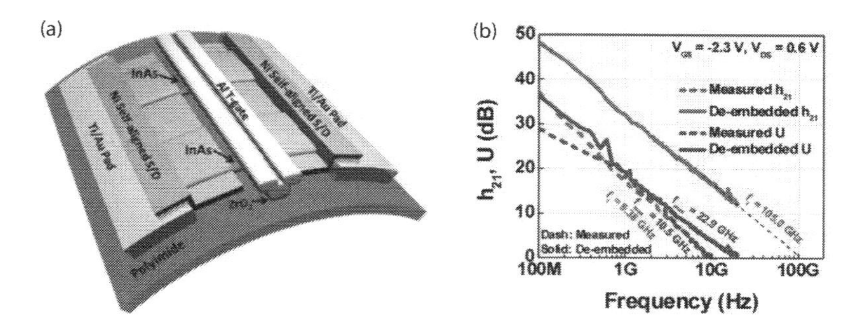

FIGURE 8.4 InAs-based high-speed flexible electronics. (a) Structural illustration of a self-aligned InAs-based MOSFET on a PI substrate. (b) Measured data for the frequency response characteristics of the InAs-based MOSFET shown in (a). (Adapted from Wang, C. et al., *Nano Lett.*, 12, 4140–4145, 2012.)

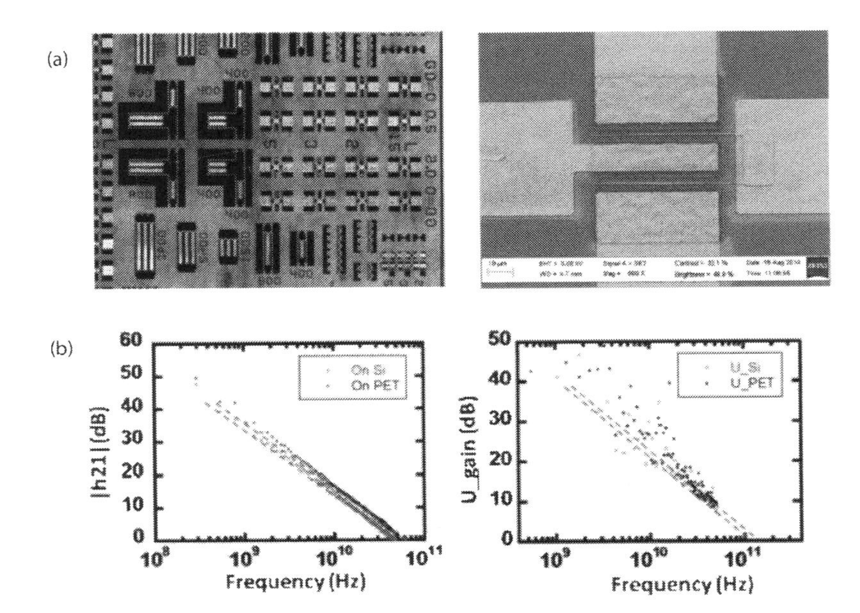

FIGURE 8.5 GaN-based high-speed flexible electronics. (a) Optical microscopy image of the high-speed flexible GaN-based HEMT on a PET substrate (left) and SEM image of a GaN-based HEMT (right). (b) Measured data for the frequency response characteristics of the GaN-based HEMT shown in (a), with current gain plot shown on the left and power gain plot shown on the right. (Adapted from Chang, T.-H. et al., High power fast flexible electronics: Transparent RF AlGaN/GaN HEMTs on plastic substrates, *Microwave Symposium (IMS), 2015 IEEE MTT-S International*, IEEE, 1–4, 2015.)

In this regard, GaN-based HEMTs outperform one order of magnitude higher than other competitors, such as Si- and GaAs-based transistors, in terms of power density (Mishra et al. 2002). Potentially, a GaN-based device can significantly scale down the size, manufacturing costs, and energy consumption. Energy efficiency in electronics means heat dissipation, and high efficiency is crucial in flexible electronics due to a variety of unconventional substrate use which can be easily affected by heat.

The releasing method for a GaN membrane from its mother substrate is quite different from that of Si, GaAs, or InAs. Since GaN film is grown on sapphire or silicon wafers, it is hard to utilize the sacrificial layer to pick up the GaN membrane (Kim et al. 2012a and b; Lee et al. 2006). Thus, high-speed flexible GaN HEMTs were fabricated by an alternate way to transfer-print the active region onto the flexible substrate (Chang et al. 2015; Defrance et al. 2013). The GaN on a silicon wafer was flipped and placed onto a handling substrate, and the backside of the wafer, composed of silicon, was removed through either mechanical lapping or plasma etching. AlGaN/GaN HEMT on a flexible substrate is shown in Figure 8.5a, and the device gave f_T and f_{max} of 60 and 115 GHz, respectively, from $|h_{21}|$ and U-gain curves in Figure 8.5b.

8.4 Low Dimensional Materials in Transistor

8.4.1 1D Materials

A combination of conventional semiconductor technology and the unusual fabrication approach made the rigid semiconductor tuned to state-of-the-art soft electronics in flexible or stretchable form. As the semiconductor industry met its unquestionably advanced level, it has run into the fundamental issue of thin film technology, which is mainly limited from the interfacial lattice mismatch between the semiconductor materials after growth. On this subject, a solution was found by tuning the film into a nanowire that can relax the lattice strain that arises from the mismatch to enable dislocation-free semiconductor growth, for example, GaAs on silicon and germanium (Bakkers et al. 2008; Mårtensson et al. 2004). Nevertheless, nanowire-based flexible transistors could record only f_T and f_{max} of 1 and 1.8 GHz, respectively, which is quite lower than that of a thin-film transistor (TFT) made of the same material (Takahashi et al. 2010).

In terms of flexible RF electronics, not only is it crucial to have good mechanical properties, but also to exhibit exceptional performance. For instance, the carbon nanotube (CNT) is a good candidate for flexible electronics applications due to its mechanical flexibility and remarkable intrinsic mobility of over 100,000 $cm^2 V^{-1}S^{-1}$, which has appealed to high-speed transistor material as proven on the conventional rigid substrate (Cao et al. 2016). However, it has not been intrigued so much for flexible substrates as it is challenging to find a solid technique to implement a pure semiconducting tube.

8.4.2 2D Materials

Additionally, many researchers have put much effort into understanding and taking technological potential of 2D materials, such as graphene, MoS_2, and black phosphorous (BP), as the next generation of high-frequency semiconductor materials.

The discovery of graphene (Novoselov et al. 2004) has gained much popularity among researchers. This relatively raw material, compared to already matured materials in the semiconductor industry, has some flaws in terms of bandgap, carrier mobility, saturation velocity, and current-carrying capacity as a candidate of RF electronics. Nevertheless, its unique electrical and superb mechanical properties are already sufficient to attract interest in implementing flexible and stretchable RF electronics. Graphene has the unique electrical property of long-range ballistic transport, where the electron can efficiently flow without scattering over a considerable distance (Novoselov et al. 2005). Besides the electrical uniqueness, its mechanical hardness that is easily greater than that of diamond also appeals to flexible and stretchable electronic applications (Lee et al. 2008). Due to the electrical imperfection of graphene as a semiconductor material, the synthesizing process, including mechanical exfoliation and chemical vapor deposition (CVD), is crucial in maintaining its maximum performance. Intuitively, a mechanical exfoliation process appears compatible to the flexible electronics fabrication method since graphene can be printed at the end of multiple oxidation and chemical etching processes (Brodie 1859; Hummers and Offeman 1958; Staudenmaier 1899). However, the technique leaves some functional groups on the graphene after reduction of the oxygen-carbon bonding, which can degrade the quality of graphene (Compton and Nguyen 2010).

Demonstrating the mechanical exfoliation process, graphene FET (GFET) was fabricated using solution-based graphene (Sire et al. 2012). It recorded high-speed performances with f_T of 2.2 GHz, but showed a poor f_{max} of 550 MHz which may have resulted from a flaw in the fabrication process that caused anomalous electrostatics in the device that could not saturate the current against the voltage. On the other hand, GFET built by CVD-grown graphene is more attractive for high-speed transistors since it recorded over hundreds of gigahertz on a rigid substrate (Cheng et al. 2012; Wu et al. 2012). It is challenging to fabricate high-speed GFET in flexible form as other transistors as the target substrate may be too pliable or electrically unstable. Having the limitation of the substrate which induces RF loss, efforts were spent on enhancing the performance of GFET on flexible substrates. As shown in Figure 8.6a and b, a considerable high-speed GFET was demonstrated on polyethylene naphthalate (PEN) by electron-beam lithography (Petrone et al. 2012). The fabrication technique can pattern a tiny channel length of 500 nm on a thin layer of graphene, which recorded f_T and f_{max} as 10.7 and 3.7 GHz, respectively. Additionally, the device performed better on a certain amount of strain as it contributed to the removal of trapped charges, which gave a boost to RF properties. Further research could improve f_T up to 39 GHz (Wei et al. 2016), and reducing the gate length also enhanced f_T and f_{max} to 198 and 28.2 GHz, respectively, as shown in Figure 8.6c and d (Petrone et al. 2015).

Similarly, other 2D materials such as MoS_2 and BP, known as transition-metal dichalcogenide (TMD), have been developed in the field of flexible electronics as well. Unlike graphene, these TMDs have a

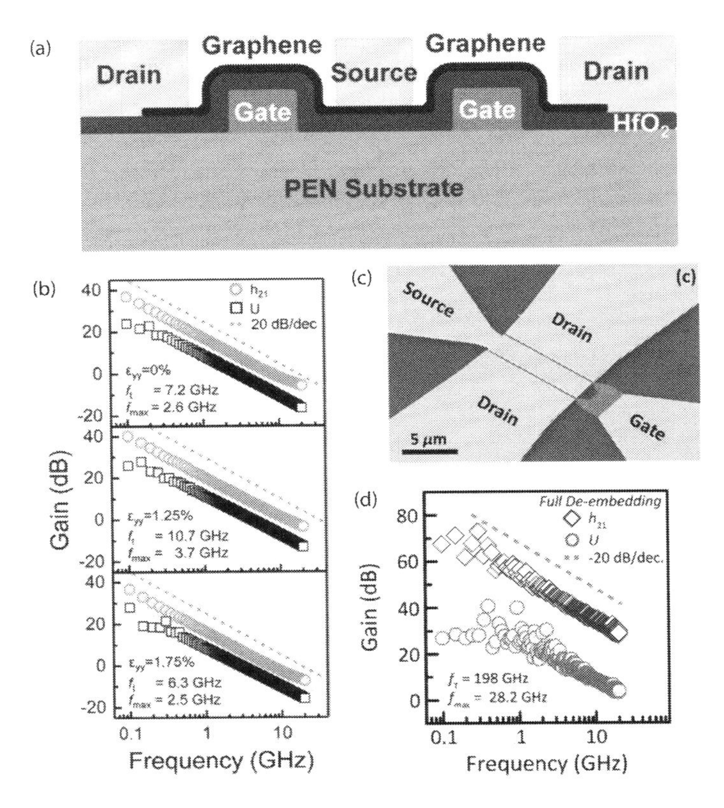

FIGURE 8.6 Graphene-based high-speed flexible electronics. (a) Illustration of a cross-sectional view of a GFET (500-nm gate length) fabricated on a PEN substrate. (b) Measured data for the frequency response characteristics of the flexible GFET shown in (a). Frequency responses are plotted with varying strain values of $\varepsilon_{yy} = 0\%$ (top), $\varepsilon_{yy} = 1.25\%$ (middle), and $\varepsilon_{yy} = 1.75\%$ (bottom). (Adapted from Petrone, N., *Nano Lett.*, 13, 121–125, 2012.) (c) False-colored SEM image of a GFET with 260-nm gate length. (d) Measured data for the frequency response characteristics of the GFET with reduced gate length shown in (c). (Adapted from Petrone, N. et al., *IEEE J. Electron Devices Soc.*, 3, 44–48, 2015.)

bandgap that allows the material to be utilized for RF applications (Splendiani et al. 2010). Composing mono- or multi-layered structures, TMDs can have either a direct bandgap or an indirect bandgap, a unique trait attractive for a variety of applications (Chhowalla et al. 2013).

For example, MoS_2 researchers have extensively improved the limit of Si-based transistors by demonstrating 1-nm gate length to reduce short channel effects (Desai et al. 2016). Additionally, the superior mechanical properties of TMD, similar to that of graphene, made them the ideal candidate material for flexible and stretchable electronics. Thus, many researchers have turned to TMDs to demonstrate high-speed flexible electronics, as shown in Figure 8.7. For example, a flexible MoS_2-based transistor was fabricated by the liquid exfoliation technique, as shown in Figure 8.7a (Cheng et al. 2014). The MoS_2 FET recorded f_T and f_{max} as 13.5 and 10.5 GHz, respectively, as presented in Figure 8.7b. Using the other technique, CVD-growth, a MoS_2-based transistor was demonstrated on a flexible substrate by the transfer-printing method (Figure 8.7c) (Chang et al. 2016). Corresponding frequency of f_T and f_{max} were 5.6 and 3.3 GHz, respectively, which are considered lower in performance than that of MoS_2-based transistors on a rigid substrate (Figure 8.7d). The degraded performance was a result of the disadvantages of unusual substrates as mentioned in the section for graphene.

With the highly sensitive thickness-dependent direct bandgap from 0.3 (bulk) to 2 eV (monolayer) and the high-carrier mobility, BP is currently the best candidate for high-frequency electronic

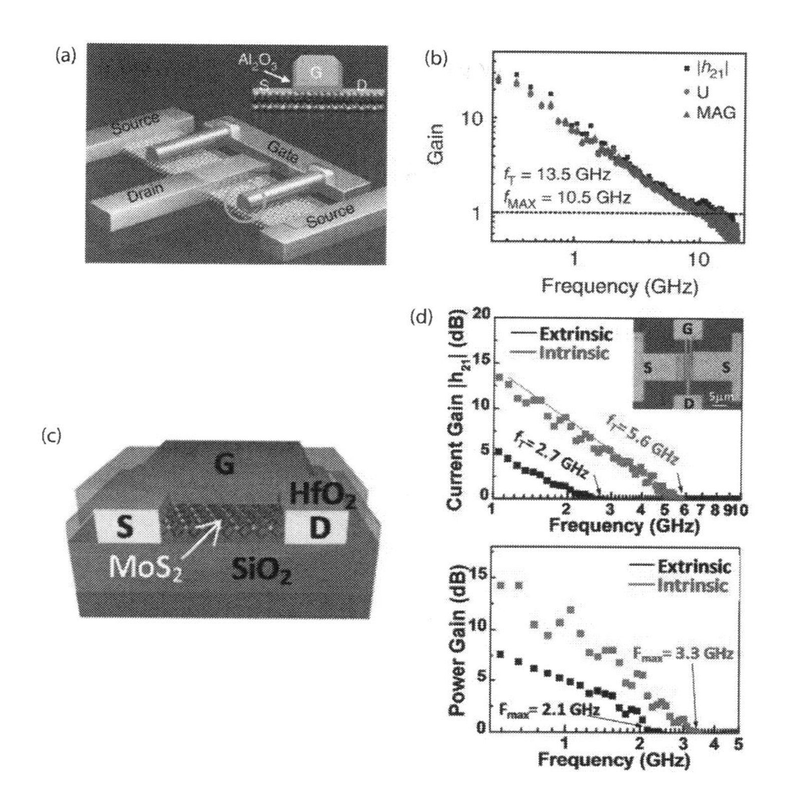

FIGURE 8.7 MoS_2-based high-speed flexible electronics. (a) Structural illustration of a self-aligned MoS_2-based FET fabricated from exfoliated MoS_2 (inset shows the cross-sectional view of the device). (b) Measured data for the frequency response characteristics of the MoS_2-based FET shown in (a). (Adapted from Cheng, R. et al., *Nat. Commun.*, 5, 5143, 2014.) (c) Structural illustration of a MoS_2-based FET based on CVD-grown MoS_2. (d) Measured data for the frequency response characteristics of the MoS_2-based FET shown in (c). (Adapted from Chang, H.Y. et al., *Adv. Mater.*, 28, 1818–1823, 2016.)

applications which require low-power consumption (Castellanos-Gomez et al. 2014; Das et al. 2014; Tran et al. 2014; Zhang et al. 2014). Although challenges still exist in controlling the size and thickness of the film, researchers successfully demonstrated a flexible BP-based transistor by putting it on a flexible plastic substrate, as shown in Figure 8.8a (Zhu et al. 2016). This p-type current transport-based transistor takes advantage of its high-hole mobility, confirming its role in future high-speed flexible electronics. The device recorded f_T and f_{max} of 17.5 and 14.5 GHz, respectively, as shown in Figure 8.8b.

To commercialize the high-speed flexible electronics using the abovementioned 2D materials, there are still numerous requirements to be outlined in terms of easy and cost-effective fabrication, and preparation for a large area of uniformly grown materials on a flexible substrate. Therefore, these 2D material-based transistors have yet to outperform the high-frequency transistors using bulk inorganic semiconductor materials.

8.5 Passive Elements

Integration of basic elements of a microwave circuit is a crucial point when it comes to soft electronics due to the sensitivity of parameters that are affected by the surrounding environment of the circuit. Therefore, the consideration of design must be followed to minimize performance loss on the foreign substrate. The monolithic microwave integrated circuit (MMIC) is composed of both active and passive

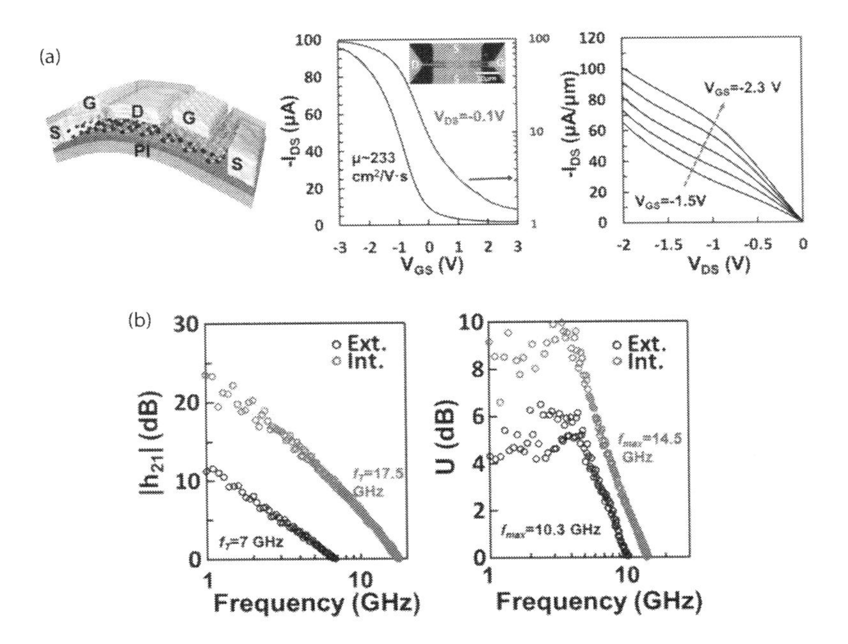

FIGURE 8.8 BP-based high-speed flexible electronics. (a) Structural illustration of a top-gate BP-based TFT on a PI substrate (left), the direct current (DC) transfer curve of the BP-based TFT (middle), and the DC current-voltage characteristics of the BP-based TFT (right). (b) Measured data for the frequency response characteristics of the BP-based TFT shown in (a), with current gain plot shown on the left and power gain plot shown on the right. (Adapted from Zhu, W. et al., *Nano Lett.*, 16, 2301–2306, 2016.)

elements, including inductors, capacitors, transmission lines, attenuators, etc. The conventional fabrication of MMIC is performed concurrently on a substrate, then matching work can be done either before or after manufacture. However, unconventional substrates used in soft electronics are unlikely to be compatible with the originally desired MMICs. The above elements must be rebuilt, manufactured, and matched in an isolated environment for the individual component.

Microwave circuit components are massively affected by enclosing conditions, such as dielectric materials in a lumped circuit, electromagnetic interferences from inner and outer circuits, and mechanical conditions on flexible or stretchable substrates. In the case of an inductor with 1 nH at the operating frequency of 1 GHz on a conventional silicon wafer, it cannot sustain the value on a soft substrate such as plastic and elastomer. The issue unfortunately becomes significant at higher frequencies. As semiconductor technology rapidly advances, the speed of RF electronics must catch up with other circuitries. Thus, the above challenges will appear in wider parts of MMICs.

8.5.1 Inductors and Capacitors

Generic high-frequency electronics are essentially composed of metal-insulator-metal (MIM) capacitors and spiral inductors. Conventionally, a capacitor is employed via a dielectric layer with metal interconnection at a noticeable high temperature. The fabrication process for an inductor requires comparably accurate alignments via the photolithography process due to its reciprocal geometry of the electrodes, which can change the performance of the inductor in a wide range with a small error.

With overcoming these requirements on a flexible substrate, some key elements comprising the capacitor, inductor, and filter were built on plastic film (Sun et al. 2010a). While these first high-speed inductors and capacitors were proven from the mechanical bending test, the performance of self-resonance

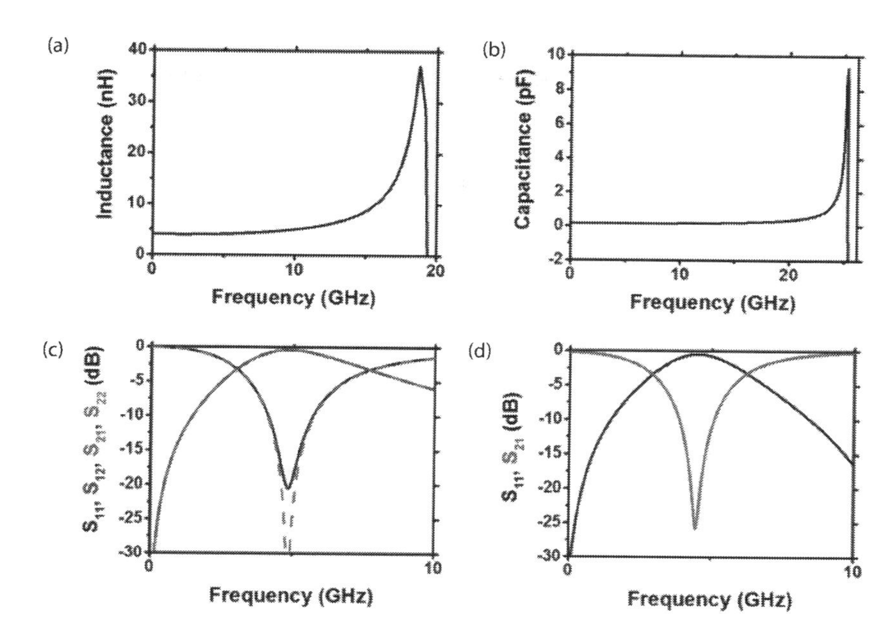

FIGURE 8.9 High-speed flexible passive elements. (a) Measured data for the inductance against frequency of a flexible inductor. (b) Measured data for the capacitance against frequency of a flexible capacitor. (c) Measured data for the S-parameters against frequency for a flexible band-pass filter operating at 4.8-GHz. (d) Measured data for the S-parameters against frequency for a flexible notch filter operating at 4.5-GHz. (Adapted from Ma, Z. et al., Materials and design considerations for fast flexible and stretchable electronics, *Electron Devices Meeting (IEDM), 2015 IEEE International*, IEEE, 19.2.1–19.2.4, 2015.)

frequency, f_{res}, was cut at 8 GHz. After improving the design of the device followed by applying high-k dielectric materials, the performance finally broke the frequency, 12 GHz, which enabled the device at the X-band (Cho et al. 2015). With TiO_2 as a dielectric layer, the self-resonance frequency of the capacitor and inductor produced 25 and 13 GHz, respectively. In Figure 8.9a and b, the enhanced design recorded an even higher operating frequency dropped in the Ku-band (12–18 GHz) with self-resonance frequencies of 25.4 and 19.1 GHz for the capacitor and inductor, respectively (Ma et al. 2015).

Overwhelming the limit, the elements also opened their applications to meet the requirement of the high-frequency operation shown in Figure 8.9c and d that demonstrated the band-pass filter and the notch filter. In the meantime, other noble materials were also utilized to implement flexible microwave passive components (Huang et al. 2015, 2016; Ostfeld et al. 2015).

8.5.2 Stretchable Microwave Transmission Lines

Stretchable and flexible RF electronics technology takes advantage of its tiny size and wireless-communication capability. Thus, most of its applications tend to be oriented towards wearable electronics for both epidermal and implantable bioelectronics (Kim et al. 2011, 2013). The device must be bio-compatible to and suitable for humans as it directly contacts human skin or is implanted inside the human body. Again, these RF electronics are then exposed to abnormal surroundings, and often the issue of RF signal loss that is induced by outer space remains even though an optimized dielectric material is utilized. The challenge becomes severe on the circuit components that must communicate the signal in real time over a distance such as RF transmission lines regardless of the substrate-type.

As an alternate way to resolve the issue, a twisted-pair electrode design was applied to the RF transmission line by constructing a serpentine shape of the overall path (Jung et al. 2016a). As shown in Figure 8.10a,

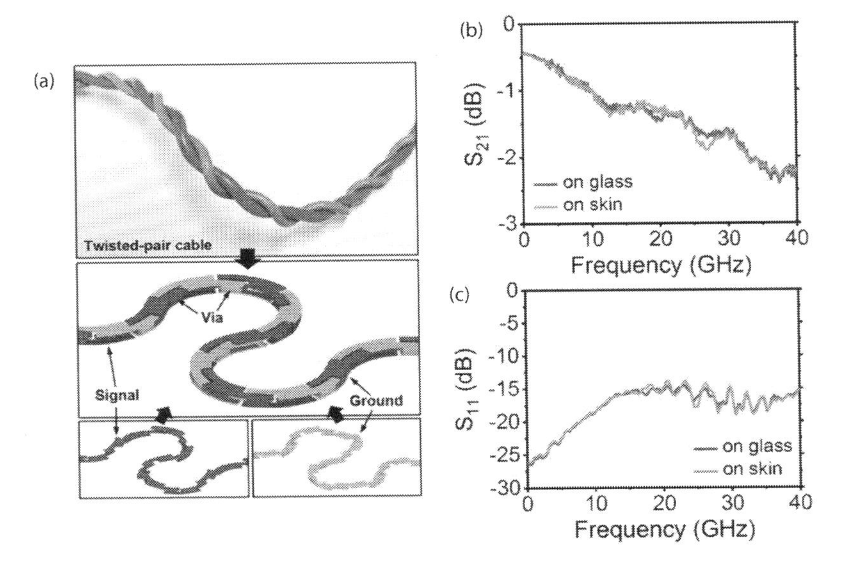

FIGURE 8.10 Stretchable transmission lines operating at microwave frequencies. (a) Simple demonstration of a twisted pair of electrical cables where the microstructure of twisted-pair-based stretchable transmission line is inspired. (b) Measured data for the insertion loss (S_{21}) and the return loss (S_{11}) against the frequency of the stretchable transmission line on the glass substrate and on the animal skin are presented in (c). (Adapted from Jung, Y. H. et al., *Advanced Functional Materials* 26, 4635–4642, 2016a.)

the electromagnetic interference was attenuated by structural confinement of the device itself, and the device could eventually bypass the external noise from the outer environment and effectively transfer the power to the other end. The design minimizes the significant RF signal loss region and freely transfers RF signals, as shown in Figure 8.10b and c.

The internal design modification is also compatible with various serpentine shapes including the self-similar design (Xu et al. 2013; Zhang et al. 2013), fractal design (Fan et al. 2014), nanomesh design (Guo et al. 2014), and bio-inspired design (Jang et al. 2015; Lv et al. 2014). Moreover, the stretchability of the electrode design contributes to conformability of RF electronics onto unusual target surfaces, including human skin, animal tissues, and even commercially available fabrics (Jang et al. 2014). Likewise, this unique design was also applied to stretchable microwave filters working at desired frequency ranges, alluding to the power of structural design improvement on flexible or stretchable high-speed RF electronics.

8.5.3 Flexible Attenuators

In RF electronics, a proper signal level setting is essential to match the instrumentation dynamic range and to avoid circuit overload and damage. Attenuators, simple passive components in RF electronics, are generally used to control the signal power level in various circuit branches or to match pad connecting lines of different impedances (Gallo 2011).

Research on flexible attenuators for constructing flexible microwave systems has been reported by Byun et al. (2015). Figure 8.11a shows the fabricated flexible microwave attenuators using multi-layer graphene. A 40 nm-thick layer of gold was deposited on a PET substrate to form coplanar waveguide (CPW) transmission lines, and a four-layer graphene was used to achieve suitable sheet resistance. Figure 8.11b depicts the measured and simulated S-parameters of the fabricated graphene attenuators. They showed 3 and 6 dB attenuation (S21), which corresponds to the simulation results, and the input/output return losses (S11, S22) were less than −15 dB at the frequencies over 5 GHz. Figure 8.11c is a photograph of

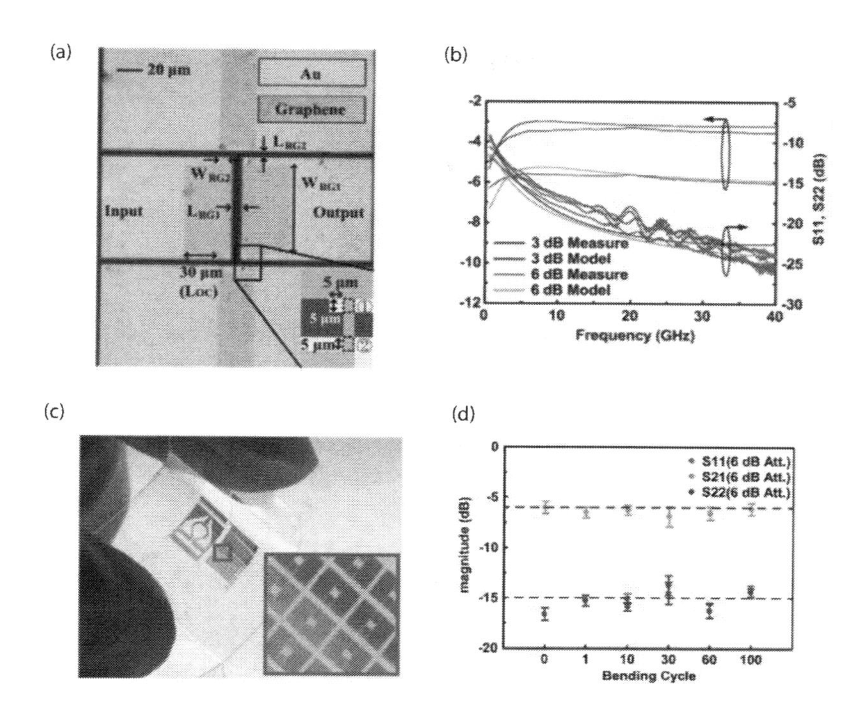

FIGURE 8.11 Flexible graphene-based attenuators. (a) Optical microscopy image of the graphene attenuator. (b) Measured and simulated data for the S-parameter of the 3 and 6 dB graphene attenuators shown in (a). (c) Photograph of an array of devices on a PET substrate. (d) The variations of the attenuation and the return loss at 10 GHz with bending radius of 5 mm. (Adapted from Byun, K. et al., *Nanotechnology*, 26, 055201, 2015.)

flexible graphene attenuator arrays fabricated on a PET substrate. As shown in Figure 8.11d, the attenuation and return loss at 10 GHz were maintained with no significant change even after 100 bending cycles, which demonstrates that the mechanical flexibility and stability of graphene attenuators are suitable for flexible microwave systems.

8.6 RF Circuits and Systems

In order to ultimately realize a high-speed flexible wireless communication device that operates on the order of several GHz, complete integration—not only flexible or stretchable microelectronics and passive components, but also with different kinds of flexible circuit-level devices, such as switches, rectifiers, and amplifiers—is required. As of now, such an integrated flexible wireless communication device where all components including the transistors and their ancillaries are fabricated on a common flexible substrate has not been realized. However, some techniques for flexible circuit-level devices working at low- and high-frequencies have been demonstrated.

8.6.1 Low-Frequency Flexible RF Devices

Research on a frequency doubler using flexible GFETs was reported by Lee et al. (2013). A hexagonal boron nitride (h-BN), an insulating isomorph of graphene with boron and nitrogen arranged in a hexagonal lattice, was used as a gate dielectric and embedded in mechanically flexible PI films. A schematic of the GFET frequency doubler circuit shown in Figure 8.12a and b indicates the output power at the doubled frequency as a function of input power at the fundamental frequency (f_{IN} = 10 MHz).

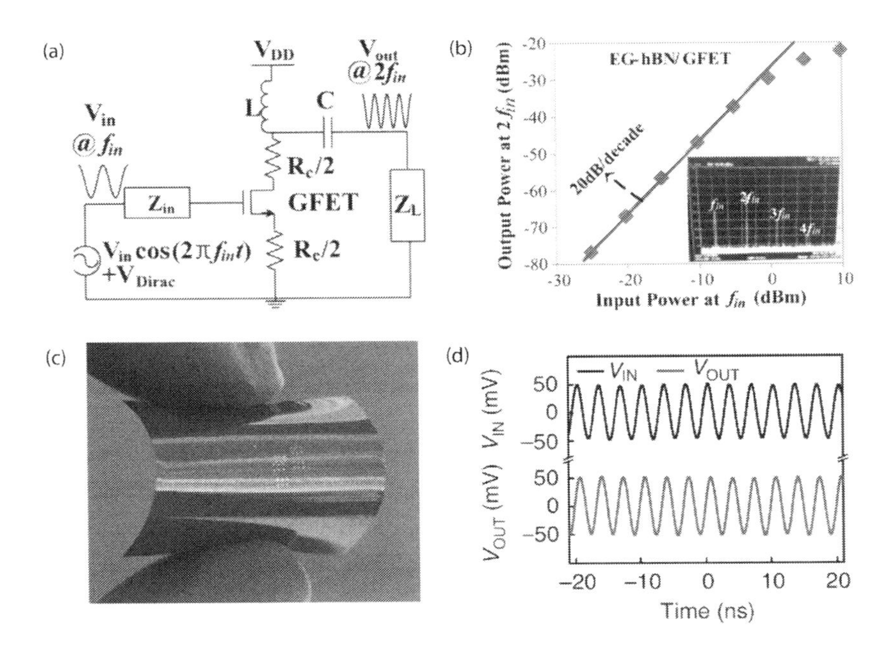

FIGURE 8.12 Low-frequency flexible circuits. (a) Equivalent circuit diagram of a frequency doubler composed of GFET. (b) Measured data for the output power against the input power of the GFET-based frequency doubler (inset shows the spectrum of the output signal). (Adapted from Lee, J. et al., *IEEE Electron Device Lett.*, 34, 172–174, 2013). (c) Photograph of a flexible RF amplifier composed of MoS_2-based FET. (d) Measured data for the input and the output voltages of the flexible RF amplifier composed of MoS_2-based FET. (Adapted from Cheng, R. et al., *Nat. Commun.*, 5, 5143, 2014.)

The graphene frequency doubler has a slope of 20 dB/dec and shows an intrinsically high spectral purity (>90%). Moreover, because of the low interplane thermal conductivity of hexagonal boron nitride and the low impurity interface between graphene and hexagonal boron nitride, a suitable conversion gain (−29.5 dB) and output power (−22.2 dBm) for high-speed electronic applications were attained.

A flexible RF amplifier, constructed by Cheng et al. (2014), was realized by integrating self-aligned multiple MoS_2 transistors. Figure 8.12c and d presents a photograph of MoS_2 circuits on a flexible substrate and the output signal for a 300 MHz sinusoidal input signal, respectively. A voltage gain over a unity was achieved through the flexible RF amplifier, which is desirable for flexible RF applications.

Meanwhile, flexible RF circuits using transistors based on organic semiconductors or amorphous oxide semiconductors are especially suited for applications that require both a low frequency and large area because the materials are relatively cheap and offer a low-carrier mobility. For example, an organic transistor-based flexible wet sensor sheet (FWSS), a transponder with wireless power and data transmission, was developed to detect urination in diapers (Fuketa et al. 2014). The FWSS was composed of flexible organic circuits fabricated on a 12.5 μm-thick PI film and magnetic resonance at 13.56 MHz—instead of conventional electromagnetic induction—was used to increase the distance between the reader and FWSS. A photograph of the FWSS embedded in the cotton of a diaper is presented in Figure 8.13. By measuring the change in resistance between two electrodes and sending the detected signals to the primary reader coil, the presence and absence of liquid were successfully detected.

A flexible thin-film near-field communication (NFC) tag employing amorphous indium-gallium-zinc-oxide (a-IGZO) TFT circuits was created by Myny et al. (2015). The NFC tags were composed of four different a-IGZO circuits and powered by a commercial universal serial bus (USB)-connected reader device at a base carrier frequency of 13.56 MHz. Through the optimized design topologies, the central NFC regulatory standards were remarkably satisfied.

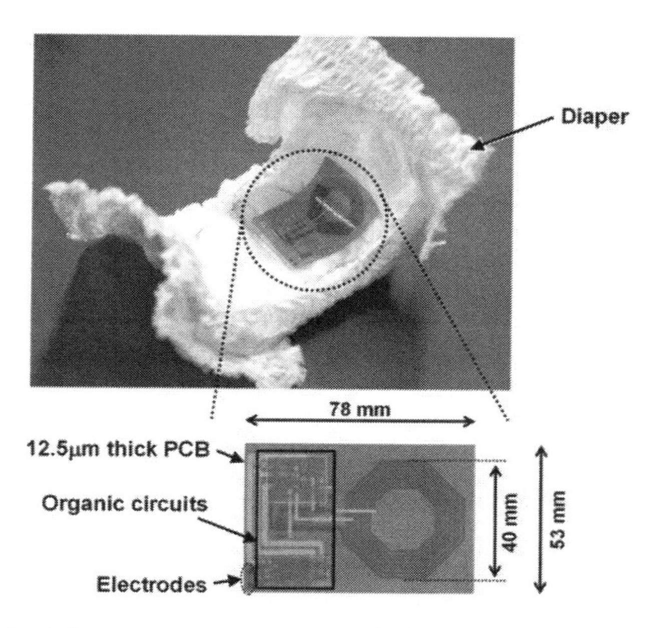

FIGURE 8.13 Flexible wireless wet sensor circuit composed of organic transistors. (Adapted from Fuketa, H. et al., 30.3 Organic-transistor-based 2kV ESD-tolerant flexible wet sensor sheet for biomedical applications with wireless power and data transmission using 13.56 MHz magnetic resonance, *Solid-State Circuits Conference Digest of Technical Papers (ISSCC), 2014 IEEE International*, IEEE, 490–491, 2014.)

8.6.2 High-Frequency Flexible RF Devices

In spite of the difficulties in designing an advanced high-speed flexible wireless communication device, simple flexible RF devices or circuits operating at high frequency were demonstrated. For instance, as shown in Figure 8.14a, a flexible single-pole single-throw (SPST) RF switch based on transferrable single-crystal SiNMs was designed by Qin et al. (2009). In order to form the SPST RF switch, two lateral P-type,intrinsic,N-type (PIN) diodes were connected in a shunt-series configuration. Figure 8.14b and c shows the OFF-state and ON-state RF characteristics of the switch under bending, respectively. By utilizing the shunt-series configuration of the lateral PIN diodes, an improved insertion loss (better than 0.6 dB) and isolation (higher than 22.8 dB up to 5 GHz) were achieved.

Another example is an RF rectifier using GaAs-based Schottky diodes, which are commonly used in high-speed communication systems (Jung et al. 2015). To form a flexible full-bridge rectifier, four microwave GaAs-based Schottky diodes and a MIM capacitor were combined into a simple integrated circuit, as shown in Figure 8.14d. The rectification behavior of RF-to-DC conversion at 5.8 GHz input signal was measured with increasing power level, as shown in Figure 8.14e. From the results, it was demonstrated that the rectifier could produce 2.43-mW DC power with an input signal of 21 dBm, which is ideal for wireless power transferring.

Recently, research on transparent and flexible monopole antennas for wearable glasses applications was conducted by Hong et al. (2016). They utilized a 100-nm-thick transparent indium-zinc-tin-oxide (IZTO)/Ag/IZTO (IAI) multi-layer with a sheet resistance of 4.99 Ω/sq as the conducting electrodes of antennas and ground planes of the wearable glasses. Figure 8.15a and b shows the structure of the fabricated IAI antenna and its optical image embedded on the glasses, respectively. The IAI multi-layer electrode showed not only an optical transparency of over 80% in the visible light range, but also an electrical conductivity of ~2,000,000 S/m, which is enough to be applied to wearable glasses. Figure 8.15c presents the simulated and measured return loss (S_{11}) of the antennas with Ag and IAI

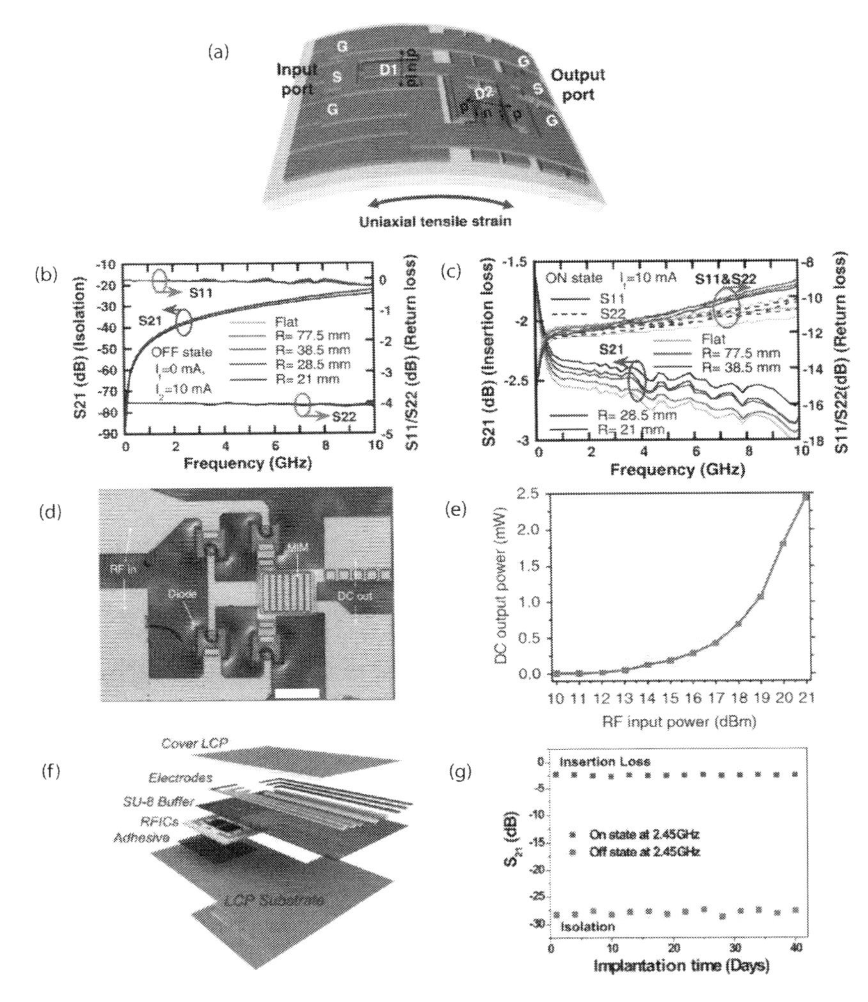

FIGURE 8.14 High-frequency flexible circuits. (a) Structural illustration of a flexible RF switch. Measured data for the frequency response characteristics of the flexible RF switch shown in (a) at the OFF-state and the ON-state are presented in (b) and (c), respectively. (Adapted from Qin, G. et al., *J. Phys. D Appl. Phys.*, 42, 234006, 2009.) (d) Optical microscopy image of a flexible full-bridge rectifier. (e) Measured data for the DC output power against the RF input power of the flexible full-bridge rectifier shown in (d) operating at 5.8 GHz. (Adapted from Jung, Y.H. et al., *Nat. Commun.*, 6, 7170, 2015.) (f) Exploded illustration for the structure of a flexible RFIC. (g) Measured data for the insertion loss under ON- and OFF-states of the RF switch in the flexible RFIC shown in (f) for 6-week duration in rat skin. (Adapted from Hwang, G.-T. et al., *ACS Nano*, 7, 4545–4553, 2013.)

multi-layer films. The fabricated IAI antennas had lower S_{11} values (~−16.1 dB) than those of the Ag antennas (~−22 dB) at the resonance frequency of 2.4–2.5 GHz and showed an average efficiency of 40% and 4-dBi peak gain.

Flexible radio-frequency integrated circuits (RFICs) for medical wireless communication were also developed by Hwang et al. (2013). The RFICs were transferred from an SOI wafer to a flexible liquid-crystal polymer (LCP) substrate and monolithically encapsulated with an LCP layer for packaging for *in vivo* medical applications. Figure 8.14f and g presents the insertion loss and isolation values of the *in vivo* flexible RF switch over a 6-week period. While the isolation was formed near 28 dB, the insertion loss was maintained at ~2.5 dB, which indicates that the performance of flexible

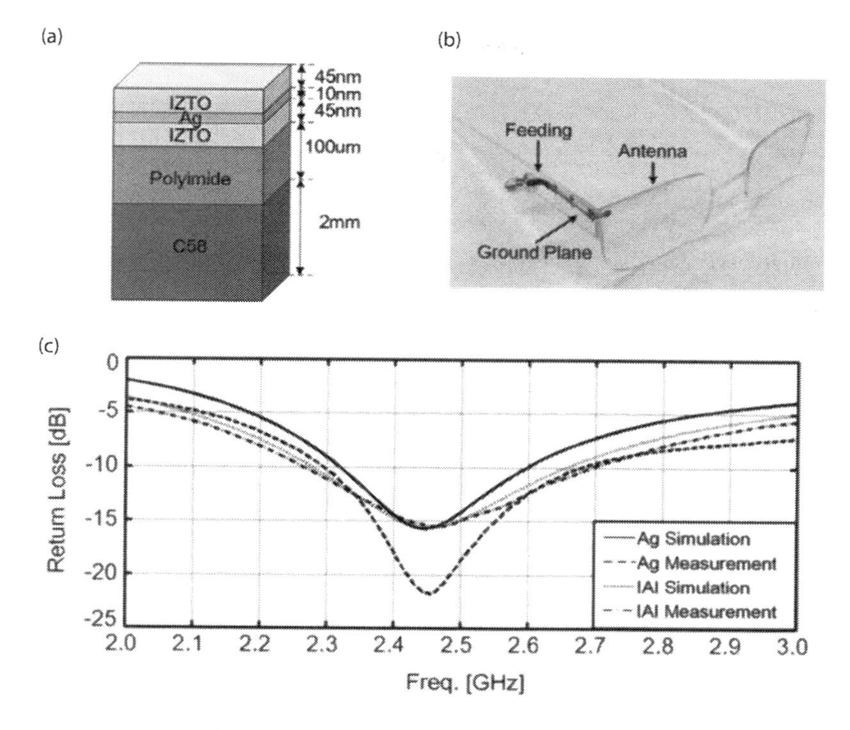

FIGURE 8.15 Transparent and flexible IAI antenna. (a) Structural illustration of a flexible IAI antenna fabricated on a PI substrate and pasted on a C58 lens. (b) Photograph of a flexible IAI antenna integrated with the glasses. (c) Simulated and measured data for the return loss (S_{11}) of the flexible antenna shown in (b). (Adapted from Hong, S. et al., *IEEE Trans. Antennas Propag.*, 64, 2797–2804, 2016.)

RFICs could be maintained under short-term *in vivo* conditions without obvious degradation. These efforts for developing such flexible RF devices operating at high frequency are a significant step towards further advanced high-speed flexible wireless communication devices and systems.

8.7 Conclusion

Consistent efforts are made to address the challenges associated with the designs and fabrication processes in advanced flexible and stretchable RF electronics. First, in the case of RF active components, not only are many different kinds of semiconductors, including Si, GaAs, InAs, and GaN, but also 1D and 2D materials, such as nanowire (NW), CNT, graphene, MoS_2, and BP, have been utilized to improve the electrical characteristics of transistors. These materials have unique electrical and mechanical properties in RF electronic applications, which facilitated fabrication of various active transistors, such as MOSFETs, MESFETs, HBTs, and HEMTs, on a flexible substrate. As far as the fabrication processes are concerned in developing flexible transistors, NM transfer-printing, photolithography, and electron-beam lithography techniques were mainly used to form the inorganic materials on foreign substrates. Unique approaches to patterning the desired feature size of the transistors, improving the electron mobility, and overcoming the short-channel effect were conducted in strained Si-based transistors and nanotrench Si-based transistors. In particular, f_T and f_{max} play a key role in the performance of transistors because they determine the switching speed and power level of RF electronics. The f_T is inversely proportional to the channel length, while the f_{max} is substantially influenced by parasitic capacitances and resistances in a transistor. Therefore, engineering the structure and layout of the device is crucial in achieving a successful RF active element. By reducing the gate length to the sub-micrometer scale and by properly scaling the size of the transistors, remarkable achievements in RF active devices have been made over the past decades.

Meanwhile, the RF performance of integrated circuits on flexible substrates is necessarily degraded as it is significantly affected by the surrounding environment due to redundant noise signals. Thus, minimizing the performance loss on foreign substrates has been a major consideration in realizing passive components in high-speed RF electronics. As part of its efforts, researchers integrated a high-k material, TiO_2, as a dielectric layer to increase the self-resonance frequency of the capacitor and inductor. In addition, they also incorporated a twisted-pair electrode design to the RF transmission line to attenuate the electromagnetic interface from outer space. Furthermore, researchers utilized multi-layered graphene to achieve suitable sheet resistance on a flexible substrate to enhance a flexible attenuator operating at high frequency. In addition to these efforts, a variety of attempts to minimize the RF loss on flexible form by using various designs, such as the fractal design, nanomesh design, and bio-inspired design, and by integrating novel materials have been made.

The selection of substrates became an additional consideration when constructing high-speed flexible RF electronics and systems, due to the direct effect of electrical and mechanical properties of a substrate, such as mechanical stress, thermal conductivity, loss tangent, and dielectric constant, on the performance of the device. For instance, the high thermal conductivity of the substrate prevents the device from the overheating issue, and the low loss tangent induces the RF loss. The substrate that has these characteristics can be chosen for achieving high-performance RF electronics. Unfortunately, common substrates used for flexible and stretchable electronics, such as PI, PET, PEN, and Ecoflex, are usually lacking in the abovementioned properties compared to that of conventional RF electronics (Jung et al. 2017). Although such inadequate properties can be practically improved by depositing dielectric materials on the substrates, it is not a fundamental solution. Consequently, a certain amount of sacrifice has to be made to achieve RF electronics in flexible or stretchable form. Continuous research to offer a more suitable substrate for flexible and stretchable RF electronics are warranted.

Finally, flexible RF devices at the circuit level, including switches, rectifiers, amplifiers, and antennas, operating at low and high frequencies have been demonstrated in recent years. Valuable efforts have been made toward high-performance ICs in reducing the return and insertion loss of RF circuits. A variety of materials such as SiNM, graphene, MoS_2, a-IGZO, and IZTO were employed to fabricate advanced active devices (MOSFETs, PIN diodes, Schottky diodes, etc.) and subsequently combined with flexible or stretchable passive components. Moreover, in some RF circuits, multi-layers were utilized to enhance the electrical characteristics of devices, and the design and layout of circuits were also modified to improve the RF performance of the circuits.

As a significant next step towards complete flexible or stretchable RF electronics, the advanced circuit-integration technology has yet to be developed in order to solder more flexible or stretchable types of elements in an RF circuit. In the meantime, shielding technology to minimize the electromagnetic interference from a noise source and packaging techniques to protect and limit the circuit in a flexible or stretchable environment must be considered. In addition, since only a few frequency bands are approved by the Federal Communications Commission, steps to satisfy or mitigate these regulations are necessary to commercialize flexible or stretchable electronics, such as wearable electronics, smart clothes, medical instruments, etc.

References

Ahn, J.-H., Kim, H.-S., Lee, K. J. et al. 2006. High-speed mechanically flexible single-crystal silicon thin-film transistors on plastic substrates. *IEEE Electron Device Letters* 27: 460–462.

Ahn, J.-H., Kim, H.-S., Menard, E. et al. 2007. Bendable integrated circuits on plastic substrates by use of printed ribbons of single-crystalline silicon. *Applied Physics Letters* 90: 213501.

Bakkers, E., Borgstrom, M. & Verheijen, M. 2008. Epitaxial growth of III–V nanowires on group IV substrates. *MRS Online Proceedings Library Archive* 1068.

Brodie, B. C. 1859. On the atomic weight of graphite. *Philosophical Transactions of the Royal Society of London* 149: 249–259.

Byun, K., Park, Y. J., Ahn, J.-H. & Min, B.-W. 2015. Flexible graphene based microwave attenuators. *Nanotechnology* 26: 055201.

Cao, Y., Brady, G. J., Gui, H. et al. 2016. Radio frequency transistors using aligned semiconducting carbon nanotubes with current-gain cutoff frequency and maximum oscillation frequency simultaneously greater than 70 GHz. *ACS Nano* 10: 6782–6790.

Castellanos-Gomez, A., Vicarelli, L., Prada, E. et al. 2014. Isolation and characterization of few-layer black phosphorus. *2D Materials* 1: 025001.

Chang, C. & Kai, F. 1994. *GaAs High-Speed Devices: Physics, Technology, and Circuit Applications*. New York: John Wiley & Sons.

Chang, H. Y., Yogeesh, M. N., Ghosh, R. et al. 2016. Large-area monolayer MoS2 for flexible low-power RF nanoelectronics in the GHz regime. *Advanced Materials* 28: 1818–1823.

Chang, T.-H., Xiong, K., Park, S. H. et al. 2015. High power fast flexible electronics: Transparent RF AlGaN/GaN HEMTs on plastic substrates. *Microwave Symposium (IMS), 2015 IEEE MTT-S International*. IEEE, pp. 1–4.

Chen, F., Euaruksakul, C., Liu, Z. et al. 2011. Conduction band structure and electron mobility in uniaxially strained Si via externally applied strain in nanomembranes. *Journal of Physics D: Applied Physics* 44: 325107.

Cheng, R., Bai, J., Liao, L. et al. 2012. High-frequency self-aligned graphene transistors with transferred gate stacks. *Proceedings of the National Academy of Sciences* 109: 11588–11592.

Cheng, R., Jiang, S., Chen, Y. et al. 2014. Few-layer molybdenum disulfide transistors and circuits for high-speed flexible electronics. *Nature Communications* 5: 5143.

Chhowalla, M., Shin, H. S., Eda, G. et al. 2013. The chemistry of two-dimensional layered transition metal dichalcogenide nanosheets. *Nature Chemistry* 5: 263–275.

Cho, S. J., Jung, Y. H. & Ma, Z. 2015. X-band compatible flexible microwave inductors and capacitors on plastic substrate. *IEEE Journal of the Electron Devices Society* 3: 435–439.

Compton, O. C. & Nguyen, S. T. 2010. Graphene oxide, highly reduced graphene oxide, and graphene: Versatile building blocks for carbon-based materials. *Small* 6: 711–723.

Das, S., Demarteau, M. & Roelofs, A. 2014. Ambipolar phosphorene field effect transistor. *ACS Nano* 8: 11730–11738.

Defrance, N., Lecourt, F., Douvry, Y. et al. 2013. Fabrication, characterization, and physical analysis of AlGaN/GaN HEMTs on flexible substrates. *IEEE Transactions on Electron Devices* 60: 1054–1059.

Desai, S. B., Madhvapathy, S. R., Sachid, A. B. et al. 2016. MoS2 transistors with 1-nanometer gate lengths. *Science* 354: 99–102.

Fan, J. A., Yeo, W.-H., Su, Y. et al. 2014. Fractal design concepts for stretchable electronics. *Nature Communications* 5: 3266.

Fischetti, M. V. & Laux, S. E. 1996. Band structure, deformation potentials, and carrier mobility in strained Si, Ge, and SiGe alloys. *Journal of Applied Physics* 80: 2234–2252.

Fuketa, H., Yoshioka, K., Yokota, T. et al. 2014. 30.3 Organic-transistor-based 2kV ESD-tolerant flexible wet sensor sheet for biomedical applications with wireless power and data transmission using 13.56 MHz magnetic resonance. *Solid-State Circuits Conference Digest of Technical Papers (ISSCC), 2014 IEEE International*. IEEE, pp. 490–491.

Gallo, A. 2011. Basics of RF electronics. *arXiv preprint arXiv:1112.3226*.

Garnier, F., Hajlaoui, R., Yassar, A. & Srivastava, P. 1994. All-polymer field-effect transistor realized by printing techniques. *Science* 265: 1684–1687.

Gelinck, G. H., Huitema, H. E. A., Van Veenendaal, E. et al. 2004. Flexible active-matrix displays and shift registers based on solution-processed organic transistors. *Nature Materials* 3: 106–110.

Guo, C. F., Sun, T., Liu, Q., Suo, Z. & Ren, Z. 2014. Highly stretchable and transparent nanomesh electrodes made by grain boundary lithography. *Nature Communications* 5: 3121.

Hong, S., Kang, S. H., Kim, Y. & Jung, C. W. 2016. Transparent and flexible antenna for wearable glasses applications. *IEEE Transactions on Antennas and Propagation* 64: 2797–2804.

Hu, L., Pasta, M., La Mantia, F. et al. 2010. Stretchable, porous, and conductive energy textiles. *Nano Letters* 10: 708–714.

Huang, X., Leng, T., Chang, K. H. et al. 2016. Graphene radio frequency and microwave passive components for low cost wearable electronics. *2D Materials* 3: 025021.

Huang, X., Leng, T., Zhu, M. et al. 2015. Highly flexible and conductive printed graphene for wireless wearable communications applications. *Scientific Reports* 5: 18298.

Hummers Jr., W. S. & Offeman, R. E. 1958. Preparation of graphitic oxide. *Journal of the American Chemical Society* 80: 1339–1339.

Hwang, G.-T., Im, D., Lee, S. E. et al. 2013. In vivo silicon-based flexible radio frequency integrated circuits monolithically encapsulated with biocompatible liquid crystal polymers. *ACS Nano* 7: 4545–4553.

Jang, K.-I., Chung, H. U., Xu, S. et al. 2015. Soft network composite materials with deterministic and bio-inspired designs. *Nature Communications* 6: 6566.

Jang, K.-I., Han, S. Y., Xu, S. et al. 2014. Rugged and breathable forms of stretchable electronics with adherent composite substrates for transcutaneous monitoring. *Nature Communications* 5: 4779.

Jung, Y. H., Chang, T.-H., Zhang, H. et al. 2015. High-performance green flexible electronics based on biodegradable cellulose nanofibril paper. *Nature Communications* 6: 7170.

Jung, Y. H., Lee, J., Qiu, Y. et al. 2016a. Stretchable twisted-pair transmission lines for microwave frequency wearable electronics. *Advanced Functional Materials* 26: 4635–4642.

Jung, Y. H., Seo, J.-H., Zhou, W. & Ma, Z. 2016b. High-speed, flexible electronics by use of Si nanomembranes. In J. A. Rogers and J.-H. Ahn (Eds.), *Silicon Nanomembranes: Fundamental Science and Applications*. Weinheim, Germany: John Wiley & Sons, pp. 113–142.

Jung, Y. H., Zhang, H., Cho, S. J. & Ma, Z. 2017. Flexible and stretchable microwave microelectronic devices and circuits. *IEEE Transactions on Electron Devices* 64: 1881–1893.

Jung, Y. H., Zhang, H. & Ma, Z. 2016c. Wireless applications of conformal bioelectronics. In J. A. Rogers, R. Ghaffari, D.-H. Kim (Eds.), *Stretchable Bioelectronics for Medical Devices and Systems*. Basel, Switzerland: Springer, pp. 1–314.

Kim, D.-H. & Del Alamo, J. A. 2010. 30-nm InAs PHEMTs with fT= 644 GHz and fmax= 681 GHz. *IEEE Electron Device Letters* 31: 806.

Kim, D.-H., Ghaffari, R., Lu, N. & Rogers, J. A. 2012a. Flexible and stretchable electronics for biointegrated devices. *Annual Review of Biomedical Engineering* 14: 113–128.

Kim, D.-H., Lu, N., Ma, R. et al. 2011. Epidermal electronics. *Science* 333: 838–843.

Kim, T.-I., Mccall, J. G., Jung, Y. H. et al. 2013. Injectable, cellular-scale optoelectronics with applications for wireless optogenetics. *Science* 340: 211–216.

Kim, T. I., Jung, Y. H., Song, J. et al. 2012b. High-efficiency, microscale GaN light-emitting diodes and their thermal properties on unusual substrates. *Small* 8: 1643–1649.

Ko, H., Takei, K., Kapadia, R. et al. 2010. Ultrathin compound semiconductor on insulator layers for high-performance nanoscale transistors. *Nature* 468: 286–289.

Kuzuhara, M. & Tanaka, S. 2003. GaAs-based high-frequency and high-speed devices. *JSAP International* 7: 4.

Lee, C., Wei, X., Kysar, J. W. & Hone, J. 2008. Measurement of the elastic properties and intrinsic strength of monolayer graphene. *Science* 321: 385–388.

Lee, J., Ha, T.-J., Parrish, K. N. et al. 2013. High-performance current saturating graphene field-effect transistor with hexagonal boron nitride dielectric on flexible polymeric substrates. *IEEE Electron Device Letters* 34: 172–174.

Lee, K. J., Meitl, M. A., Ahn, J.-H. et al. 2006. Bendable GaN high electron mobility transistors on plastic substrates. *Journal of Applied Physics* 100: 124507.

Lee, W., Hwangbo, Y., Kim, J.-H. & Ahn, J.-H. 2016. Mobility enhancement of strained Si transistors by transfer printing on plastic substrates. *NPG Asia Materials* 8: e256.

Lv, C., Yu, H. & Jiang, H. 2014. Archimedean spiral design for extremely stretchable interconnects. *Extreme Mechanics Letters* 1: 29–34.

Ma, Z., Jung, Y. H., Seo, J.-H. et al. 2015. Materials and design considerations for fast flexible and stretchable electronics. *Electron Devices Meeting (IEDM), 2015 IEEE International*. IEEE, pp. 19.2.1–19.2.4.

Mårtensson, T., Svensson, C. P. T., Wacaser, B. A. et al. 2004. Epitaxial III–V nanowires on silicon. *Nano Letters* 4: 1987–1990.

Mishra, U. K., Parikh, P. & Wu, Y.-F. 2002. AlGaN/GaN HEMTs-an overview of device operation and applications. *Proceedings of the IEEE* 90: 1022–1031.

Myny, K., Cobb, B., van der Steen, J.-L. et al. 2015. 16.3 flexible thin-film NFC tags powered by commercial USB reader device at 13.56 MHz. *Solid-State Circuits Conference-(ISSCC), 2015 IEEE International*. IEEE, pp. 1–3.

Novoselov, K., Jiang, D., Schedin, F. et al. 2005. Two-dimensional atomic crystals. *Proceedings of the National Academy of Sciences of the United States of America* 102: 10451–10453.

Novoselov, K. S., Geim, A. K., Morozov, S. V. et al. 2004. Electric field effect in atomically thin carbon films. *Science* 306: 666–669.

Ostfeld, A. E., Deckman, I., Gaikwad, A. M., Lochner, C. M. & Arias, A. C. 2015. Screen printed passive components for flexible power electronics. *Scientific Reports* 5: 15959.

Petrone, N., Meric, I., Chari, T., Shepard, K. L. & Hone, J. 2015. Graphene field-effect transistors for radio-frequency flexible electronics. *IEEE Journal of the Electron Devices Society* 3: 44–48.

Petrone, N., Meric, I., Hone, J. & Shepard, K. L. 2012. Graphene field-effect transistors with gigahertz-frequency power gain on flexible substrates. *Nano Letters* 13: 121–125.

Qin, G., Yuan, H.-C., Celler, G. K., Zhou, W. & Ma, Z. 2009. Flexible microwave PIN diodes and switches employing transferrable single-crystal Si nanomembranes on plastic substrates. *Journal of Physics D: Applied Physics* 42: 234006.

Rim, K., Chan, K., Shi, L. et al. 2003. Fabrication and mobility characteristics of ultra-thin strained Si directly on insulator (SSDOI) MOSFETs. *Electron Devices Meeting, 2003. IEDM'03 Technical Digest. IEEE International*. IEEE, pp. 3.1.1–3.1.4.

Rogers, J. A., Bao, Z., Baldwin, K. et al. 2001. Like electronic displays: Large-area rubber-stamped plastic sheets of electronics and microencapsulated electrophoretic inks. *Proceedings of the National Academy of Sciences* 98: 4835–4840.

Rojas, J. P., Torres Sevilla, G. A., Ghoneim, M. T. et al. 2014. Transformational silicon electronics. *ACS Nano* 8: 1468–1474.

Samineni, V. K., Yoon, J., Crawford, K. E. et al. 2017. Fully implantable, battery-free wireless optoelectronic devices for spinal optogenetics. *Pain* 158: 2108–2116.

Sekitani, T., Zschieschang, U., Klauk, H. & Someya, T. 2010. Flexible organic transistors and circuits with extreme bending stability. *Nature Materials* 9: 1015–1022.

Seo, J.-H., Ling, T., Gong, S. et al. 2016. Fast flexible transistors with a nanotrench structure. *Scientific Reports* 6: 24771.

Shi, J., Wichmann, N., Roelens, Y. & Bollaert, S. 2013. Electrical characterization of In0. 53Ga0. 47As/In0. 52Al0. 48As high electron mobility transistors on plastic flexible substrate under mechanical bending conditions. *Applied Physics Letters* 102: 243503.

Sire, C., Ardiaca, F., Lepilliet, S. et al. 2012. Flexible gigahertz transistors derived from solution-based single-layer graphene. *Nano Letters* 12: 1184–1188.

Splendiani, A., Sun, L., Zhang, Y. et al. 2010. Emerging photoluminescence in monolayer MoS2. *Nano Letters* 10: 1271–1275.

Staudenmaier, L. 1899. Verfahren zur darstellung der graphitsäure. *European Journal of Inorganic Chemistry* 32: 1394–1399.

Sun, L., Qin, G., Huang, H. et al. 2010a. Flexible high-frequency microwave inductors and capacitors integrated on a polyethylene terephthalate substrate. *Applied Physics Letters* 96: 013509.

Sun, L., Qin, G., Seo, J. H. et al. 2010b. 12-GHz thin-film transistors on transferrable silicon nanomembranes for high-performance flexible electronics. *Small* 6: 2553–2557.

Sun, Y., Menard, E., Rogers, J. A. et al. 2006. Gigahertz operation in flexible transistors on plastic substrates. *Applied Physics Letters* 88: 183509.

Takahashi, T., Takei, K., Adabi, E. et al. 2010. Parallel array InAs nanowire transistors for mechanically bendable, ultrahigh frequency electronics. *ACS Nano* 4: 5855–5860.

Tran, V., Soklaski, R., Liang, Y. & Yang, L. 2014. Layer-controlled band gap and anisotropic excitons in few-layer black phosphorus. *Physical Review B* 89: 235319.

Wang, C., Chien, J.-C., Fang, H. et al. 2012. Self-aligned, extremely high frequency III–V metal-oxide-semiconductor field-effect transistors on rigid and flexible substrates. *Nano Letters* 12: 4140–4145.

Wei, W., Pallecchi, E., Haque, S. et al. 2016. Mechanically robust 39 GHz cut-off frequency graphene field effect transistors on flexible substrates. *Nanoscale* 8: 14097–14103.

Wu, Y., Jenkins, K. A., Valdes-Garcia, A. et al. 2012. State-of-the-art graphene high-frequency electronics. *Nano Letters* 12: 3062–3067.

Xu, S., Zhang, Y., Cho, J. et al. 2013. Stretchable batteries with self-similar serpentine interconnects and integrated wireless recharging systems. *Nature Communications* 4: 1543.

Yoon, J., Jo, S., Chun, I. S. et al. 2010. GaAs photovoltaics and optoelectronics using releasable multilayer epitaxial assemblies. *Nature* 465: 329–333.

Yu, D., Zhang, Y. & Liu, F. 2008. First-principles study of electronic properties of biaxially strained silicon: Effects on charge carrier mobility. *Physical Review B* 78: 245204.

Yuan, H.-C., Kelly, M. M., Savage, D. E. et al. 2008. Thermally processed high-mobility MOS thin-film transistors on transferable single-crystal elastically strain-sharing Si/SiGe/Si nanomembranes. *IEEE Transactions on Electron Devices* 55: 810–815.

Yuan, H.-C., Ma, Z., Roberts, M. M., Savage, D. E. & Lagally, M. G. 2006a. High-speed strained-single-crystal-silicon thin-film transistors on flexible polymers. *Journal of Applied Physics* 100: 013708.

Zhang, S., Yang, J., Xu, R. et al. 2014. Extraordinary photoluminescence and strong temperature/angle-dependent Raman responses in few-layer phosphorene. *ACS Nano* 8: 9590–9596.

Zhang, X.-S. 2015. *Micro/Nano Integrated Fabrication Technology and Its Applications in Microenergy Harvesting*. Berlin, Germany: Springer.

Zhang, Y., Xu, S., Fu, H. et al. 2013. Buckling in serpentine microstructures and applications in elastomer-supported ultra-stretchable electronics with high areal coverage. *Soft Matter* 9: 8062–8070.

Zhou, H., Seo, J.-H., Paskiewicz, D. M. et al. 2013. Fast flexible electronics with strained silicon nanomembranes. *Scientific Reports* 3: 1291.

Zhu, W., Park, S., Yogeesh, M. N. et al. 2016. Black phosphorus flexible thin film transistors at gighertz frequencies. *Nano Letters* 16: 2301–2306.

9 Flexible and Stretchable Sensors

Tae Hoon Eom
and Jeong In Han

9.1 Introduction

Recent research into flexible and stretchable sensors has been improved by various applications and utilizations. One significant application is Internet of Things (IOT) technology, which is a main technology of the Fourth Industrial Revolution. IOT refers to recent high-tech industry devices that combine a tremendous amount of data through connecting to the Internet. The development of wearable devices has increased drastically to facilitate the collection of data from the human body. The term 'wearable device' refers to electronic devices that are associated with human body with considerable flexibility. Wearable devices can be classified as accessory type, clothing type, or skin-mounted type, which includes diverse electronic devices and sensors. Accessory type devices were initially developed for portability, while clothing and skin-mounted types can make direct contact with human skin. Thus, the flexible and stretchable properties of wearable devices must be enhanced for clothing and skin-mounted types. Among the diverse range of devices that are considered wearable, sensors play a prominent role by detecting human body motions. Monitoring the human body permits the creation of big data, which can lead to a ubiquitous society and improved healthcare system.

Various physical sensors such as those that monitor temperature, humidity, strain, and pressure, and chemical sensors such as gas and electrochemical sensors have been researched and commercialized briskly for decades. Physical sensors are defined as electronic devices whose electrical features change when a specific physical condition is applied; meanwhile, chemical sensors react to chemical changes. Basically, sensor devices consist of substrate, electrodes, and a sensing layer. However, their structure varies depending on the sensing mechanism such as for the capacitive and resistive types. In addition, piezoelectric and piezoresistive sensing mechanisms are widely used in pressure and strain sensors. Two main types of sensing mechanism and the parameters that evaluate their sensing performance will be discussed for each sensor. Furthermore, various flexible substrates are presented including planar type and cylindrical fiber type.

9.2 Flexible and Stretchable Temperature Sensors

Temperature sensing ability is the most fundamental feature of a wearable device in our daily lives; in particular, measuring the temperature of the human body is essential in human healthcare. Diverse types of temperature sensors have been researched and developed for a wide range of purposes. Traditionally, temperature sensors are fabricated on 2-dimensional rigid substrates such as glass and silica-based films. However, flexibility is a requirement of health monitoring and skin contact, which has induced the development of temperature sensors on flexible substrates. Several types of temperature sensors based on different operation mechanisms and parameters that determine the degree of temperature sensitivity will be discussed in this section.

9.2.1 Temperature Coefficient

The temperature coefficient explains the relative change of certain physical properties with a given amount of temperature change. The basic expression of the temperature coefficient is:

$$\frac{dP}{P} = \alpha dT. \tag{9.1}$$

In this expression, any type of physical property that is influenced by temperature change can be applied. Thus, dP is the variation in physical properties, dT is the temperature change, and α is the temperature coefficient. In principle, resistance, capacitance, and elasticity are the most general physical properties for the temperature coefficient. For a temperature sensor, resistance and capacitance are more likely to be used than elasticity. Among these, the temperature coefficient of resistance (TCR) is most widely used. The TCR signifies the relative resistance change between the initial temperature and the final temperature condition and is stated in the following equation.

$$TCR = \frac{R_2 - R_1}{R_1 (T_2 - T_1)}. \tag{9.2}$$

R_1 is the initial resistance, R_2 is the final resistance, T_1 is the initial temperature, and T_2 is the final temperature. Likewise, the temperature coefficient of capacitance has the same form as Eq. (9.2) after replacing resistance with capacitance.

9.2.2 Thermal Index

The thermal index is also a known criterion with which to measure the ability of a resistive temperature sensor. The thermal index can be derived from the following equations.

$$R = R_0 \exp\left(\frac{E_a}{2kT}\right) = R_0 \exp\left(\frac{B}{T}\right). \tag{9.3}$$

Differentiating Eq. (9.3) gives:

$$\frac{R}{R_0}\frac{dR}{dT} = \exp\left(\frac{B}{T}\right)\cdot -\frac{B}{T^2}. \tag{9.4}$$

Equation (9.2) can be expressed as:

$$TCR = \frac{dR}{R_0 dT}. \tag{9.5}$$

Finally, by integrating Eqs. (9.3) and (9.5), the relationship of between the TCR and the thermal index can be explained as:

$$TCR = -\frac{B}{T^2}. \tag{9.6}$$

$$B = \frac{E_a}{2k}. \tag{9.7}$$

The relationship between the TCR and the thermal index can be derived from Eq. (9.6), where the thermal index was stated as B. Equation (9.3) is the basic definition of the thermal index, where E_a is the activation energy and k is the Boltzmann constant. We can note that the TCR and the thermal index have the inverse sign from Eq. (9.6). Due to their relationship, the thermal index is normally used for temperature sensors that have a negative TCR. Such devices are called thermistors, and we will discuss these in Section 9.2.4. The following sections handle resistive-type temperature sensors and how the TCR and the thermal index are applied in temperature sensing.

9.2.3 Resistive Temperature Detectors

A resistive temperature detector (RTD) is a common type of temperature sensor that reveals resistive behavior in response to temperature changes; i.e., such devices' electrical resistance changes in response to temperature changes. Temperature responsive materials used for a RTD include Pt, Ni, and Cu, which are basically pure metal. Metal alloys such as Nichrome are also applied as the active material in a RTD.

Xiao et al. (2008) reported a RTD fabricated on flexible spin-coated liquid polyimide as depicted in Figure 9.1a. Pt was chosen as a temperature detecting material due to its excellent thermal properties. In addition, Pt showed good compatibility with micro-electro-mechanical systems (MEMS) technology. The polydimethylsiloxane (PDMS) interlayer was used in this work for ease of separation between the sensor device and the silicon carrier. The Pt layer was deposited by a sputtering system by applying various patterns based on a photolithography process. The polyimide sheet and PDMS layer effectively enhanced the flexibility of the RTD. Figure 9.1b shows the basic characterization; the resistance change versus the temperature showed good linearity, which indicated a positive TCR of 2.8×10^{-3}°C^{-1}.

Lichtenwalner et al. (2007) reported an RTD based on Pt and Nichrome thin film; Figure 9.1c shows a real image of the device. These thin films were deposited on a flexible polyimide sheet by a direct current (DC) magnetron sputtering system. Cu conducting lines were worked as interconnected lines that connected the temperature sensing arrays together. The TCR value of the Pt thin film was 2.1×10^{-3}°C^{-1} and that of the Nichrome thin film was 3×10^{-3}°C^{-1}, as shown in Figure 9.1d. The performance of Nichrome thin film was slightly higher than that of the Pt thin film. Moreover, the flexibility of integrated temperature sensing arrays was examined by measuring the strain sensitivity.

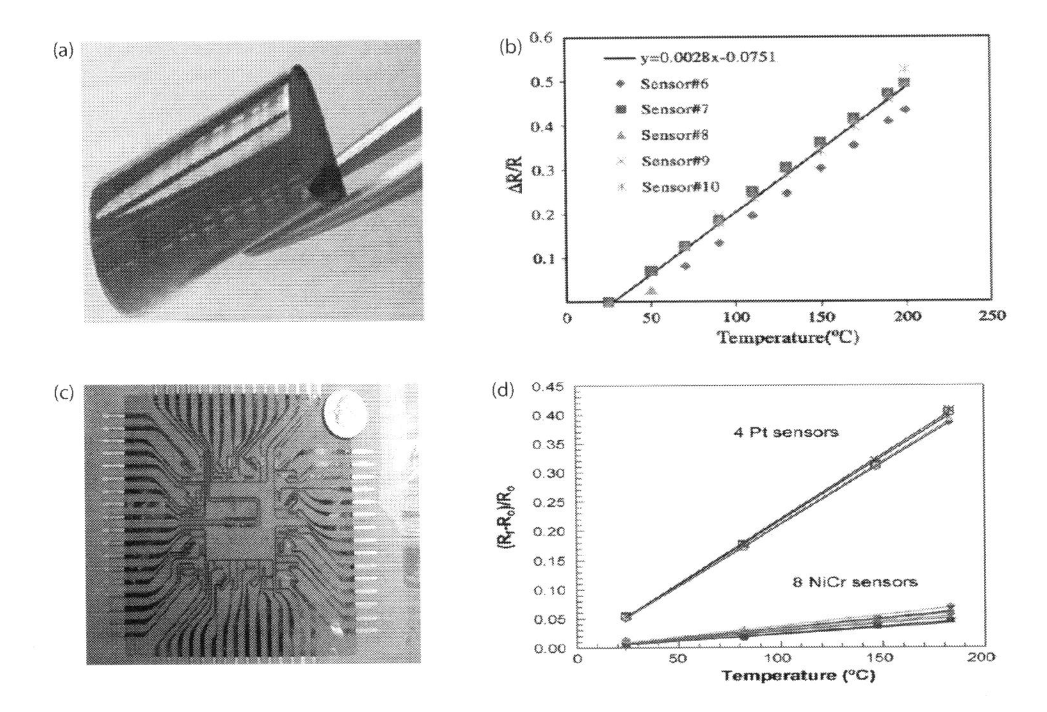

FIGURE 9.1 (a) Flexible RTD based on Pt thin film (b) resistive behavior of flexible RTD with temperature. (Reprinted from *Microelectron. Eng.*, 85, Xiao, S. et al., A novel fabrication process of MEMS devices on polyimide flexible substrates, 452–457, Copyright 2008, with permission from Elsevier.) (c) Flexible RTD array combined with Cu interconnecting lines (d) comparison between Pt-based RTD and NiCr-based RTD. (Reprinted from Sensors and Actuators A: Physical, 135, Lichtenwalner, D. et al., Flexible thin film temperature and strain sensor array utilizing a novel sensing concept, 593–597, Copyright 2007, with permission from Elsevier.)

Instead of planar substrates, various cylindrical materials such as single fiber and woven fibers were selected as substrates for the temperature sensor. Figure 9.2a shows the schematic structure of the polyethylene terephthalate (PET) single fiber-based RTD (Eom and Han 2017b). Ni thin film was uniformly coated on the PET single fiber by radio-frequency (RF) magnetron sputtering, as illustrated in Figure 9.2b. Rotating the PET fiber during the sputtering process resulted in a uniform Ni thin film surface. Figure 9.2c shows the resistive behavior of the Ni thin film on a PET fiber, which showed excellent linearity as the temperature increased; the TCR of the device reached 2.44×10^{-3}°C^{-1}. The flexibility was examined by bending the PET fiber repetitively, and the effect of bending on the TCR was marginal. The TCR could maintain its value after the bending test, which also verified the compatibility of Ni thin film and PET fiber. In addition, Ni–Cr thin film can be a good candidate temperature sensing material rather than Ni thin film (Eom and Han 2017a). By sputtering the Ni and Cr target spontaneously, a Ni–Cr thin film was deposited on a PET fiber as described in Figure 9.2d and e. The Ni–Cr thin film was excellently grown on a PET fiber with good surface state and uniform column length. The concentration of Ni and Cr was determined by controlling the sputtering power of each target, and as the concentration of Cr increased, the TCR of the device was also increased. The TCR of Ni–Cr thin film increased steadily until the [Cr]/[Ni] ratio approached 6. However, when the [Cr]/[Ni] ratio was over 6, the TCR jumped drastically. The maximum TCR was 33×10^{-3}°C^{-1} and Ni–Cr thin film showed an excellent temperature sensing feature, as shown in Figure 9.2f. The alloying effect was apparent in Ni–Cr thin film derived from various scattering. Impurity scattering was the most dominant scattering between various scatterings such as surface scattering, electron scattering, and impurity scattering. The high TCR of thicker Ni–Cr thin film was due to the impurity effect of Ni–Cr thin films.

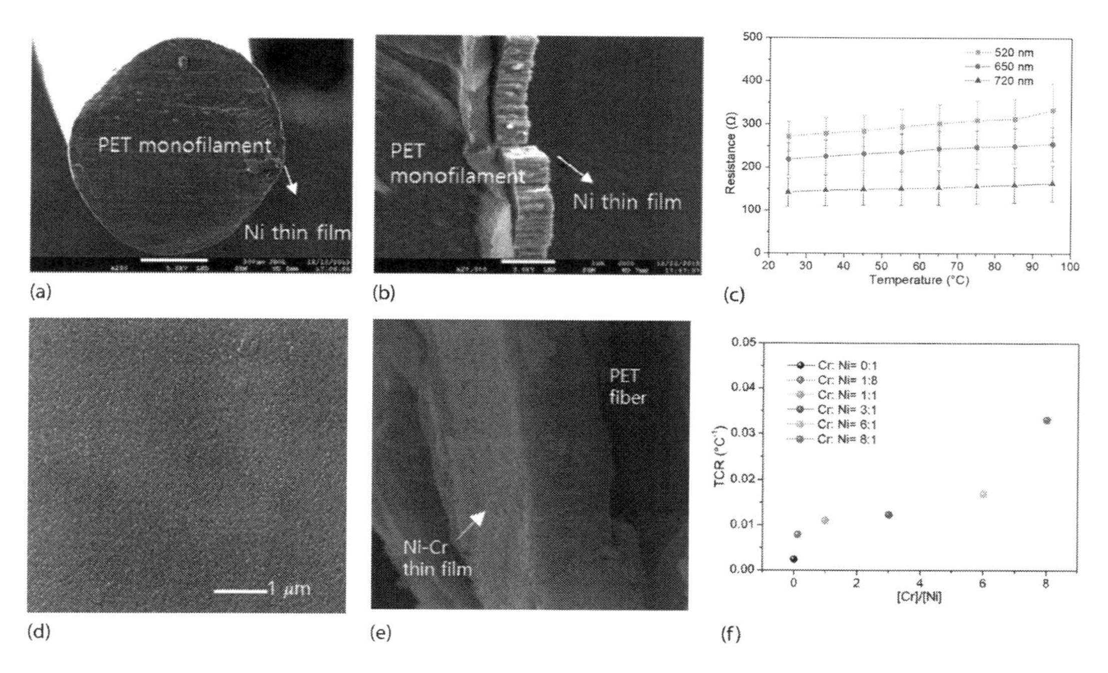

FIGURE 9.2 Cylindrical Ni thin film RTD (a) overall FESEM image of PET single fiber substrate (b) FESEM image of Ni thin film on PET single fiber for RTD (c) R-T curves indicating positive TCR (Reproduced with Eom, 2017, Copyright 2017, with permission from Elsevier.) Cylindrical Ni-Cr RTD (d) FESEM image of surface morphology (e) Ni-Cr thin film (f) relationship between TCR and Ni-Cr ratio. (Reprinted from *Sensor. Actuat. A-Phys.*, 260, Eom, T.H. & Han, J.I., The effect of the nickel and chromium concentration ratio on the temperature coefficient of the resistance of a Ni–Cr thin film-based temperature sensor, 198–205, Copyright 2017, with permission from Elsevier; Reprinted from *Sensor. Actuat. A-Phys.*, 259, Eom, T.H. & Han, J.I., Resistive behavior of Ni thin film on a cylindrical PET monofilament with temperature for wearable computing devices, 96–104, Copyright 2017, with permission from Elsevier; Reprinted from *Appl. Surf. Sci.*, 428, Eom, T.H. & Han, J.I., Single fiber UV detector based on hydrothermally synthesized ZnO nanorods for wearable computing devices, 233–241, Copyright 2018 with permission from Elsevier.)

9.2.4 Thermistors

Thermistors are also commonly used temperature sensors due to their fast response and accuracy. Likewise, the electrical resistance of thermistors is responsive to temperature changes. RTDs normally have a positive TCR, whereas thermistors have a negative TCR. Moreover, semiconductor materials or graphene-based materials are mostly applied in thermistors. Graphene is suitable for various sensors due to its excellent mechanical strength, high carrier mobility, and electrical conductivity. Furthermore, graphene oxide generated via oxidation or graphene nanocomposites is a new potential sensing material.

Yan et al. developed a flexible thermistor using PDMS as a substrate, Ag nanowire as an electrode, and graphene as the temperature sensing material (Yan et al. 2015). PDMS and Ag nanowires have excellent stretching ability and maintain thermal stability after stretching, as shown in Figure 9.3a. They performed several experiments for thermistor performance characterization. The strain was increased from 0% to 50% for the fabricated thermistor device and the TCR was -10.5×10^{-3}°C^{-1} when strain was not applied and was -21.1×10^{-3}°C^{-1} after 50% strain was employed in the device. Figure 9.3b explains the negative TCR of the thermistor as the resistance decreased with the temperature and the effect of strain on sensitivity. The thermal index was increased from 945 K to 1,895 K after applying strain. Moreover, the curves indicate that the temperature sensitivity was maintained when 50% strain was applied.

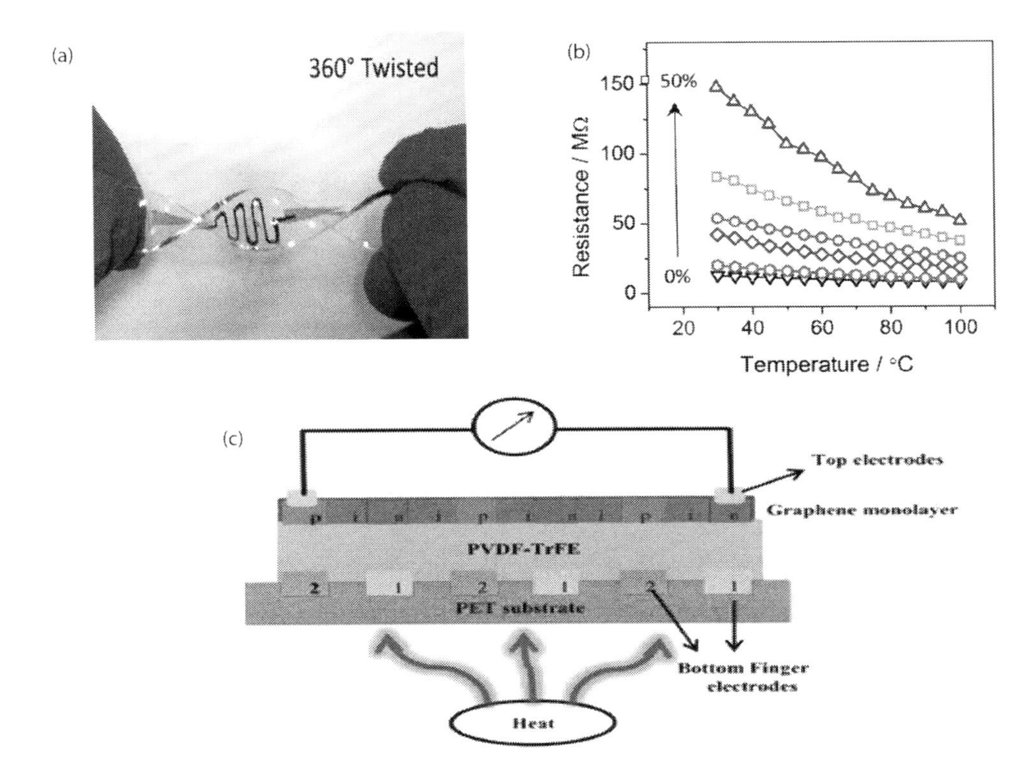

FIGURE 9.3 Graphene-based thermistor (a) 360° twisted flexible graphene thermistor (b) R-T curves indicating negative TCR. (Reproduced with permission Yan, 2015, Copyright 2015, American Chemical Society.) (c) Flexible graphene thermistor with self-power system. (Reprinted from *Nano Energy*, 26, Bendi, R. et al., Self-powered graphene thermistor, 586–594, Copyright 2017, with permission from Elsevier.)

Figure 9.3c displays an enhanced version of a thermistor with an added self-power system (Bendi et al. 2016). A graphene layer was deposited on a Cu substrate by chemical vapor deposition (CVD) and polytetrafluoroethylene (PTFE) was drop-casted on the grown graphene layer. Then, the Cu substrate was removed and replaced by a flexible PET film. Moreover, they applied electrical pooling on the graphene layer that created a p-i-n region and self-power generation capacity. The ratio of final resistance to initial resistance (R/R_0) was lower in the high temperature condition.

9.3 Flexible and Stretchable Humidity Sensors

Humidity measurement is critical in industry because it can affect the business cost of products. Moreover, humidity is essential for recognizing weather changes and other such applications. Polymer-based materials with excellent flexibility are widely used in flexible and stretchable humidity sensors. Capacitive-type, resistive-type, and thermal conductivity-type are the most common sensing mechanisms that have been commercialized in industries. Among these, capacitive-type and resistive-type humidity sensors on a flexible substrate are introduced in this section.

9.3.1 Theory of the Volume Filling of Micropores

The theory of the volume filling of micropores (TVFM) is an essential theory for studying the mechanism of water adsorption in humidity sensing material. In particular, it describes how real gases and vapors behave on microporous surfaces and the adsorption forces of micropores lead to volume filling.

The micropores of a surface become an active site of adsorption between water molecules and surface material. This theory is suitable for studying how humidity sensors operate and is related to other equations such as the Dubinin–Astakhov equation. The Dubinin–Astakhov equation will be further explored in the next section.

9.3.2 Capacitive Humidity Sensors

Capacitance is the ability of an electronic device to store an electric charge. It is fundamentally affected by the area, distance, and dielectric material.

$$C = \varepsilon_0 \varepsilon_r \frac{lw}{d}. \tag{9.8}$$

Besides, humidity or water molecules can change the capacitance value of an electronic device that is applied as a humidity sensing device. Thanks to advantages such as low energy consumption, low cost, and excellent performance, capacitive humidity sensors are widely used in both industry and weather telemetry. Basically, a capacitive humidity sensor consists of a dielectric layer, a sensing layer, and electrodes. Various polymers have been researched to have capacitive behavior that responds to humidity. Among them, polyimide (PI) is one of the most essential materials for capacitive humidity sensors and will be discussed in this section. PI has been researched for decades due to its good mechanical strength, electrochemical stability, and flexibility. In particular, PI can be used as both a humidity sensing material and a substrate for a humidity sensor. In this section, we will further study the diverse uses of PI for humidity sensors and humidity sensing parameters.

Various types of PI can be manufactured such as PI film and PI tape. Yang et al. (2015) reported a flexible capacitive humidity sensor that applied PI film as a substrate. The schematic structure of a capacitive relative humidity sensor on a flexible PI film is depicted in Figure 9.4a. Ag electrodes were interdigitated on PI film through surface modification and an ion-exchange process. This was done by an inkjet printing procedure with PI film that resulted in excellent flexibility. The humidity sensitivity can be evaluated by measuring the capacitance increase, and the capacitance was calculated at five relative humidity points. Then, the humidity sensitivity could be defined as the slope of a capacitance curve. The increase was marginal until the 70% relative humidity condition, after which, the device showed a sudden capacitance increase; this indicates that the humidity sensitivity is larger in a high relative humidity condition. Thus far, we have discussed planar-type PI film-based humidity sensors and confirmed that humidity sensitivity has a different value in different relative humidity regions. The following section explains cylindrical humidity sensors and how water molecules adsorb to the PI sensing layer.

A flexible humidity sensor was fabricated on a cylindrical PET fiber as shown in Figure 9.4b (Jang and Han 2017a). Al thin film was selected for electrodes and was deposited by RF magnetron sputtering, and PI was selected as the humidity sensing material. For capacitive behavior, electrodes were separated into inner and outer electrodes. The inner electrodes (Al) were first deposited on PET fiber followed by coating of the PI layer by the dip-coating method. Finally, the outer electrode (Al) was deposited by micro-patterning. The PI layer appeared between the two outer electrode regions, which was termed the PI exposed area. The surface area of the PI exposed area is the main factor that influences the humidity sensor's capacitance. The basic principle of this humidity sensor can be explained by measuring the capacitance change with the relative humidity. The humidity sensing mechanism was derived from TVFM as mentioned above and adsorption was performed in three steps, one of chemical adsorption and two of physical adsorption. The hydroxyl groups were generated by a combined process between the activated imide groups and the hydrogen ions that were isolated by water molecules. Afterwards, the neighboring hydroxyl groups are associated with a water molecule that forms a hydrogen bonding layer; this step indicates the first physical adsorption as mentioned above. The final step is the second adsorption with additional water molecules. The humidity sensing mechanism was further studied by several

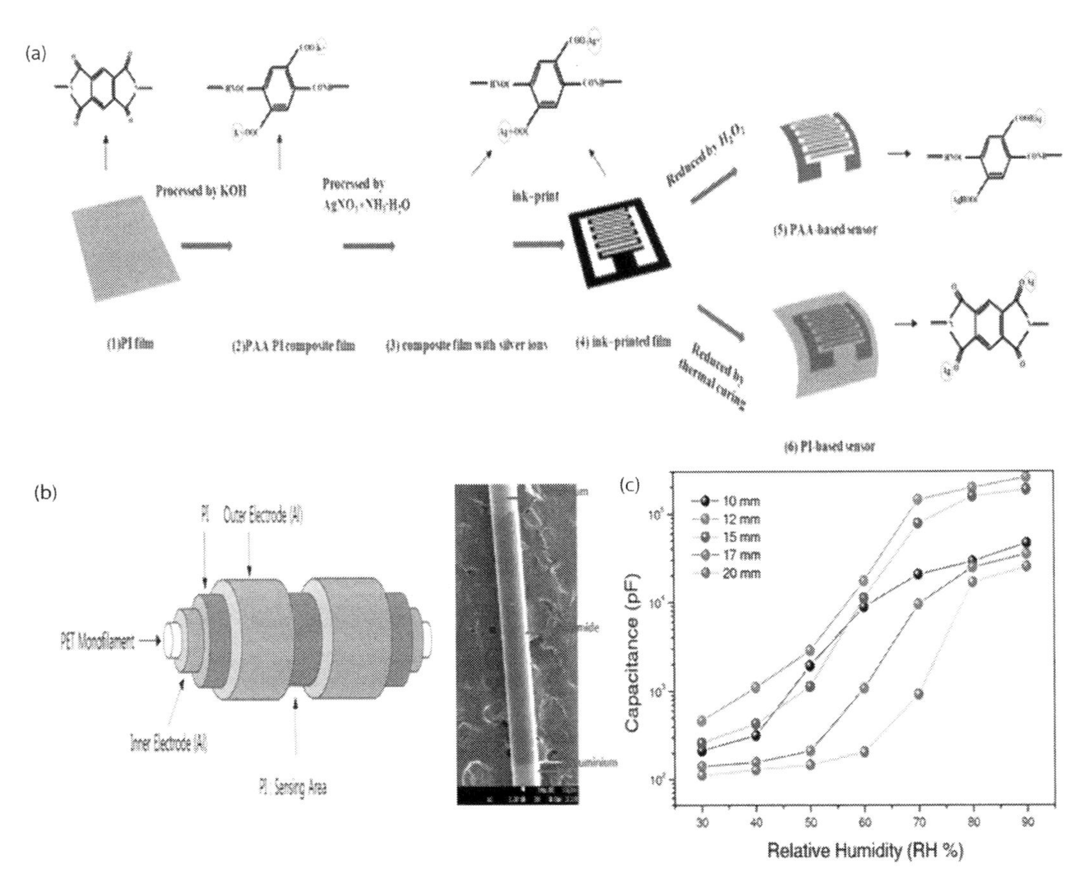

FIGURE 9.4 (a) Fabrication process of capacitive humidity sensor on PI film. (Reprinted from *Sensors and Actuators B: Chemical*, 208, Yang, T. et al., Fabrication of silver interdigitated electrodes on polyimide films via surface modification and ion-exchange technique and its flexible humidity sensor application, 327–333, Copyright 2015, with permission from Elsevier.) Cylindrical capacitive humidity sensor (b) schematic and actual image (c) capacitance change with relative humidity. (From Jang, J. & Han, J.I., *J. Electrochem. Soc.*, 164, B136–B141, 2017; Jang, J.H. & Han, J.I., *Sensor Actuat A-Phys.*, 261, 268–273, 2017.)

experiments and the result was proven by approaching with mathematical methods. As water molecules combined with the PI layer, the capacitance of the humidity sensor was increased in an S-shape, as illustrated in Figure 9.4c. The S-shape curve was formed due to the different capacitance change value in various relative humidity regions. The capacitance jumped between the relative humidity regions at 50%–70%, but the capacitance increased steadily after reaching 70% relative humidity. The capacitance tended to have a higher value when the thickness was higher.

9.3.3 Resistive Humidity Sensors

As we discussed in Section 2.1.1, resistive temperature sensors show resistive behavior in changing temperature conditions; similarly, humidity sensors that have resistive behavior to humidity are called resistive humidity sensors. This type of humidity sensor is commonly used due to its interchangeability and stability. Metal electrodes and polymer- or graphene-based sensing materials are the main structure of resistive humidity sensors. This section handles resistive humidity sensors that consist of flexible substrates.

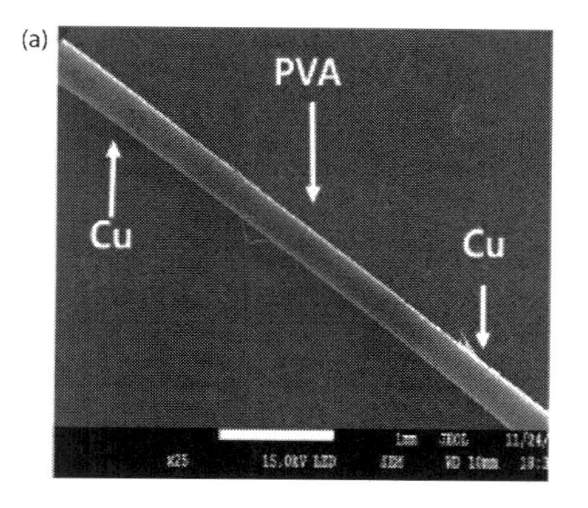

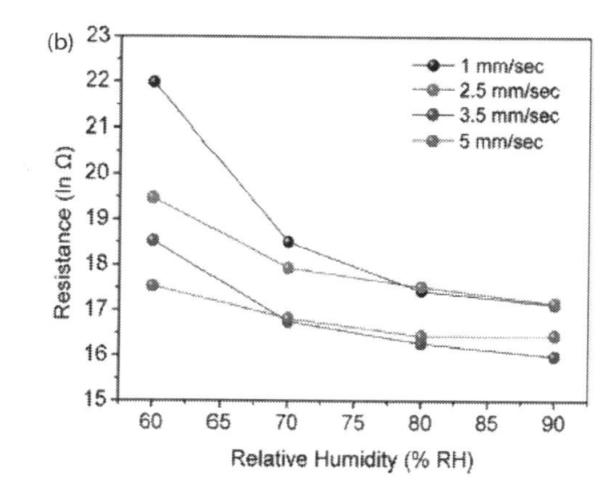

FIGURE 9.5 (a) FESEM image (b) resistance change with relative humidity in different dip coating speed of cylindrical resistive humidity sensor based on PVA sensing layer. (Reprinted from *J. Electrochem. Soc.*, 164, Jang, J. & Han, J.I., High performance cylindrical capacitor as a relative humidity sensor for wearable computing devices, B136–B141, Copyright 2017, with permission from Elsevier; Reprinted from *Sensor Actuat A-Phys.*, 261, Jang, J.H. & Han, J.I., Cylindrical relative humidity sensor based on poly-vinyl alcohol (PVA) for wearable computing devices with enhanced sensitivity, 268–273, Copyright 2017, with permission from Elsevier.)

Jang et al. reported cylindrical resistive humidity sensors fabricated on a PET single fiber, as depicted in Figure 9.5a (Jang and Han 2017b). Compared to the structure of a capacitive humidity sensor, a resistive sensor has a less complicated structure. The humidity sensing mechanism is based on the electrical resistance; thus, only a sensing material and electrodes are required to sense the humidity. Cu thin film was deposited by RF magnetron sputtering at each end of the PET fiber, and a poly vinyl alcohol (PVA) humidity sensing layer was coated by dip-coating between the electrodes. The PVA monomer contains a hydroxyl group, which makes the PVA hydrophilic. In a humid atmosphere, water molecules combine with the hydroxyl group and create hydrogen bonds. As water molecules combine consistently, hydrogen bonds are stacked and Grotthuss chain theory dominates the reaction. The protons are transported through the hydrogen bonds on the PI surface, which results in high electrical conductivity.

Figure 9.5b describes how the resistance changes in accordance with the relative humidity. As the relative humidity increases, the electrical resistance reduces. In addition, the dip-coating speed was varied from 1 to 5 mm/s, which controlled the thickness of the PVA layer. It is prominent that the thickness of the coating layer increases when the dip-coating speed increases. Thus, the figure shows that a thicker PVA layer shows lower resistance. This can be proven by the basic definition of resistance as follows.

$$R = \frac{\rho l}{A} = \frac{l}{\sigma A} \tag{9.9}$$

$$\sigma = \sigma_0 \exp\left(-\frac{E}{kT}\right). \tag{9.10}$$

A is the cross-section area, l is the length, ρ is the resistivity, and σ is the electrical conductivity. The length of the PVA layer is identical, but a larger area is generated with a thicker layer. Modifying Eq. (9.8) to a logarithmic function and integrating the two equations above leads to the derivation of Eq. (9.14).

$$\ln R = \ln l - \ln \sigma - \ln A \tag{9.11}$$

$$\ln R - \ln l = -\ln \sigma - \ln A = -\ln\left\{\sigma_0 \exp\left(-\frac{E}{kT}\right)\right\} - \ln A \tag{9.12}$$

$$\ln\left(\frac{R}{l}\right)=\left(\frac{E}{kT}\right)-\ln\left(A\sigma_0\right) \tag{9.13}$$

$$\ln\left(\frac{R}{l}\right)=\alpha x+\beta. \tag{9.14}$$

This equation indicates that the resistance per length and relative humidity have a linear relationship. In the case of a resistive humidity sensor, α remains negative.

9.4 Flexible and Stretchable UV Sensors

Both the good and bad features of ultraviolet (UV) light are closely related to our everyday lives. UV is defined as electromagnetic radiation of wavelength 10–400 nm, and this is categorized by the wavelength range. Exposure to UV is beneficial to humans and living species for several reasons; however, overexposure leads to harmful conditions in the human body such as skin disease. Therefore, commercial products are divided into UV absorbing devices and UV protection devices. Therefore, the UV sensing capability is the primary property of commercial UV devices. Various materials and methods have been researched to find materials that have UV sensing features. Nitride-based materials and oxide-based materials are the most promising candidates for UV detection materials. GaN, AlGaN, and InN are representative nitride-based materials, while ZnO and TiO_2 are representative oxide-based materials. Among these chemicals, ZnO has revealed the most attractions due to its high exciton binding energy, wide band gap, and high sensitivity to UV exposure. ZnO can be synthesized in plentiful types of shapes such as nanowires, nanorods, nanobelts, nanoflowers, etc. Section 9.4 handles ZnO as a UV sensor and how it is coated on flexible and stretchable substrates. In addition, the operating principle of UV sensors and ZnO growth mechanisms are discussed.

9.4.1 The Basic Mechanism of UV Sensors

Besides the many applications of ZnO such as in solar cells, nanogenerators, and light emitting diodes, we will only discuss UV sensors. As mentioned above, UV is divided into many types: Ultraviolet A, Ultraviolet B, Ultraviolet C, etc.; UV sensors typically look for Ultraviolet A, the wavelength range of which is 315–400 nm. The UV detection process is based on surface adsorbed oxygen molecules that act differently in a dark atmosphere and a UV-illuminated atmosphere. Surface-adsorbed oxygen molecules release free electrons and create a depletion layer; this mechanism is described in the following equations.

$$O_2(g)+e^- \rightarrow O_2^- \tag{9.15}$$

$$O_2(g)+2e^- \rightarrow O_2^{2-}. \tag{9.16}$$

The depletion layer means that ZnO shows poor photoconductivity or photosensitivity. In contrast, when UV is illuminated on the ZnO surface, photoconductivity is drastically increased due to the generation of electron–hole pairs.

$$O_2^- + h^+ \rightarrow O_2(g) \tag{9.17}$$

$$O_2^{2-} + 2h^+ \rightarrow O_2(g). \tag{9.18}$$

The increase of photoconductivity is because the electron–hole pairs transport through the depletion layer and the remaining holes combine with the adsorbed oxygen ions; this reduces the width of the depletion layer and increases the free carrier concentration. The photoconductivity of the ZnO UV sensor depends on the concentration of electrons and holes. Adjusting the synthesis method or precursor solution means that ZnO can be grown in many different shapes as discussed above. Based on these mechanisms, plenty of research studies have reported the improvement of various electrical properties such as the electrical current, response time, and photoelectric properties. The next section discusses the fabrication and characterization of UV sensors based on ZnO nanostructures.

9.4.2 ZnO-Based UV Sensors

The most frequently used ZnO nanostructures for UV sensors are nanowires and nanorods. Similar to the other types of physical sensor we studied, UV sensors are constructed of a substrate, electrode, and sensing layer. Either metal or metal nanowires are used as electrodes, and a substrate was chosen for its flexibility and stretchability. This section describes the preparation of a ZnO sensing layer and how it is coated on a flexible substrate.

The synthesis of the ZnO nanostructure is divided into two steps. The first step is the coating of the seed layer on a substrate via one of various available methods. The second step is the process of growing the ZnO nanostructure. The uniformity and surface roughness of the seed layer contributes significantly to the orientation of the ZnO's growth and shape. RF sputtering is primarily used to deposit metal thin film; however, oxide materials are also sputtering target candidates for thin film deposition. Thus, various coating approaches have been exploited such as RF sputtering, dip coating, spin coating, etc.

A ZnO-based UV sensor on a flexible PET single fiber was depicted in Figure 9.6a (Eom and Han 2018). The PET substrate was cleaned with methanol and acetone to ensure a uniform coating, and Cu electrodes were deposited by RF magnetron sputtering on a PET fiber. Finally, a ZnO seed layer was coated by the dip-coating method, and ZnO nanorods were grown in growth solution at 95°C. Zinc acetate dihydrate, sodium hydroxide, and 2-propanol were applied for the preparation of the seed solution. Zinc nitrate hexahydrate and hexamethylenetetramine (HMTA) were mixed together for the growth solution precursor. These two materials are representative precursors for ZnO growth and synthesis; their chemical reactions are as follows.

$$\mathrm{Zn(NO_3)_2 \cdot 6H_2O \rightarrow Zn^{2+} + 2NO_3^- + 6H_2O} \tag{9.19}$$

$$\mathrm{(CH_2)_6N_4 + 6H_2O \rightarrow 6HCHO + NH_3} \tag{9.20}$$

$$\mathrm{NH_3 + H_2O \leftrightarrow NH_4^+ + OH^-} \tag{9.21}$$

$$\mathrm{2OH^- + Zn^{2+} \rightarrow Zn(OH)_2} \tag{9.22}$$

$$\mathrm{Zn(OH)_2 \rightarrow ZnO + H_2O} \tag{9.23}$$

$\mathrm{Zn(NO_3)_2 \cdot 6H_2O}$ refers to zinc nitrate hexahydrate and $\mathrm{(CH_2)_6N_4}$ refers to HMTA. Zinc nitrate hexahydrate provides zinc ions and H_2O molecules to react with the HMTA; this reaction resources ammonia, which reacts with H_2O molecules, which release hydroxyl ions. The hydroxyl ions react with the zinc ions that were formed in an earlier reaction. Finally, ZnO nanorods are generated and the nuclei of ZnO are created by precipitation when heated to 95°C, and ZnO starts growing from these nuclei.

Many research studies have reported the effects of the molar concentration ratio of seed solution and growth solution on the photoelectric properties. Physical properties such as the aspect ratio and orientation of ZnO nanorods were drastically influenced by the molar concentration ratio. Those with a

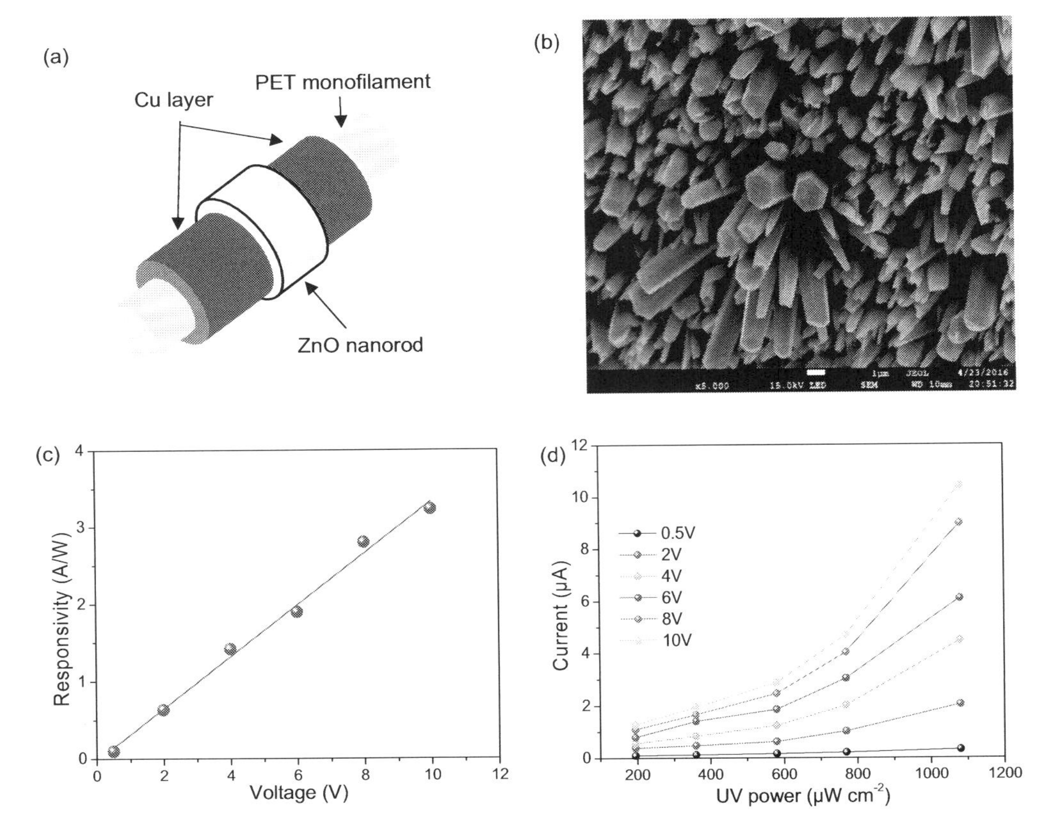

FIGURE 9.6 (a) Structure of PET fiber-based UV sensor (b) FESEM image of ZnO nanorod (c) UV responsivity versus voltage (d) relationship of current and UV power. (Reprinted from *Sensors Actuat B: Chem.*, 239, Gusain, A. et al., Flexible NO gas sensor based on conducting polymer poly [N-9′-heptadecanyl-2, 7-carbazole-alt-5, 5-(4′, 7′-di-2-thienyl-2′, 1′, 3′-benzothiadiazole)](PCDTBT), 734–745, Copyright 2018, with permission from Elsevier.)

higher growth solution ratio tended to have a low aspect ratio and distinct wurzite structure, as shown in Figure 9.6b. The abilities of the UV sensor were evaluated by measuring the amount of electrical current and the photocurrent change. When UV light illuminated ZnO nanorods and a voltage was applied through electrodes, the photocurrent increased. In addition, by measuring the current–time curve, response times such as the rising response time and falling response time could be obtained. During a specific time, the current immediately increased and fell rapidly within a very short time. The time required for the current to reach the maximum value is defined as the rising response time, and the time required for the current to return to its initial value from the maximum is defined as the falling response time. In the same manner to the photocurrent, ZnO nanorods showed a shorter response time when the growth solution ratio was higher.

Besides the photocurrent and response time, the UV sensing ability was also decided by the responsivity. Responsivity is defined as the degree to which the photocurrent increased for a specific initially applied power.

$$R(A/W) = \frac{I_p}{P_{opt}} = \frac{I_{light} - I_{dark}}{P_{opt}}. \tag{9.24}$$

I_p represents the photocurrent, P_{opt} represents the initial optical power, I_{light} represents the current when the UV light is illuminated, and I_{dark} represents the current in dark conditions. The responsivity depends on the photocurrent gain compared to the current in dark conditions. During I–V measurement, the wavelength of the light, exposed area, and illumination power must be identical to compare the responsivity. The responsivity increases for smaller exposed area and higher illumination power, which can be deduced from Eq. (9.23). The relationship between the responsivity and applied voltage is depicted in Figure 9.6c. The responsivity steadily increased in proportion with the applied voltage; when 10 V was applied to the device, the responsivity approached 3 A/W. Additionally, as shown in Figure 9.6d, for identical illumination power, the UV sensor showed high responsivity as the applied voltage increased. Further useful applications of ZnO will be discussed later in this chapter.

9.5 Flexible and Stretchable Strain Sensor

A strain sensor is a type of physical sensor that measures the electrical changes in response to mechanical deformation. The requirement of flexible and stretchable strain sensors has grown dramatically due to its wide applications. Mounting strain sensors to human skin permits the monitoring of diverse human body motions. Thus, the necessity of using biocompatible, flexible, and stretchable materials in strain sensors has risen drastically. Moreover, high sensitivity and durability and low fabrication cost are necessary for human body monitoring. Among the different types of strain sensor, capacitive, resistive, and piezoelectric are the most common. The appropriate material for a strain sensor is determined according to the operating type. Carbon- or graphene-based materials and various semiconductors are the best candidates for strain sensing material, and elastomers such as PDMS and polymers sustain the sensing layer. This section handles how these materials are applied and structured for strain sensing.

9.5.1 Gauge Factor

Whether the operation type of a strain sensor is capacitive or resistive, the gauge factor (GF) is a significant parameter for evaluating the strain sensitivity. The TCR refers to the relative change of electrical resistance compared to the initial resistance, as we discussed in the previous section. Similarly, the GF is the relative change of degree of resistance or capacitance of a strain sensor for a given strain. This can be simplified by the following equation

$$GF = \frac{\Delta C}{C_0 \varepsilon} \text{ or } \frac{\Delta R}{R_0 \varepsilon}. \tag{9.25}$$

The first conventional strain sensor that was developed was based on metal foil, and it showed a GF of around 2. The GF value can be changed by various factors such as the strain sensing material, electrode, operating type, etc. Thus, a tremendous amount of research has been carried out to enhance the GF value of strain sensors. More parameters of strain sensitivity appear later in this section.

9.5.2 Capacitive Strain Sensor

Stretchable elastomer substrates are essential for structuring capacitive strain sensors. Basically, although metals are the fundamental material of electrodes in various electronic devices, metals lose their electrical properties when strain is applied. Researchers have struggled to replace metal electrodes with stretchable materials; a carbon black-based elastomer was the first candidate alternate material, and this was highly favored for decades. However, the resistivity of carbon black was too high when stretched, which made it unsuitable for use as a strain sensor. Recently, other carbon-based materials such as graphene and carbon nanotubes (CNTs) have shown considerable electrical features. The definition of capacitance is derived from this equation.

$$C_0 = \varepsilon_0 \varepsilon_r \frac{l_0 w_0}{d_0}, \tag{9.26}$$

where ε_0 is the vacuum permittivity, ε_r is the relative permittivity, l_0 is the initial length, w_0 is the initial width, and d_0 is the initial thickness. Every parameter in this equation corresponds to the dielectric layer of a capacitor. When strain (ε) is applied to a material, the capacitance increases, which results in a positive GF. The length, width, and thickness change in accordance with the applied strain.

$$l = (1+\varepsilon)l_0 \tag{9.27}$$

$$w = (1 - v_{electrode})w_0 \tag{9.28}$$

$$d = (1 - v_{dielectric})d_0. \tag{9.29}$$

Then, the final capacitance after stretching can be expressed as:

$$C = \varepsilon_0\varepsilon_r \frac{lw}{d} = \varepsilon_0\varepsilon_r \frac{(1+\varepsilon)l_0(1 - v_{electrode})w_0}{(1 - v_{dielectric})d_0}. \tag{9.30}$$

$v_{electrode}$ is the Poisson's ratio of the electrode and $v_{dielectric}$ is the Poisson's ratio of the dielectric layer. Here, Poisson's ratio is the mechanical ratio of the transverse strain to the axial strain of a material when stretched by strain. Poisson's ratio is a unique property of materials, which indicates that selecting an appropriate material for the strain sensor is very important. Assuming that every layer in a capacitor is parallel and the Poisson's ratio of the electrode and dielectric layer are equivalent, Eq. (9.30) can transform into a briefer form.

$$C = (1+\varepsilon)C_0. \tag{9.31}$$

We can see that if the value of a given strain is maintained, the final capacitance and initial capacitance become proportional. In addition, combining Eqs. (9.25) and (9.31) lets us obtain a GF value of 1, which is the ideal value of a capacitor. According to many research studies, this equation is not acceptable for high strain. Several factors such as the surface roughness, disconnection, and tunneling effect also affect the GF.

The most widely used materials for recent capacitive strain sensors are CNT and Ag nanowire. Capacitive strain sensors that are structured by placing a silicon elastomer between two CNT film electrodes are shown in Figure 9.7a (Cai et al. 2013). Silicon elastomer has the role of a dielectric layer that completes the structure of a basic capacitor. CNT films were prepared by the floating catalyst vapor deposition (FCCVD) method and were fitted to a PET frame. Finally, CNT films were transferred to the PDMS layer. As depicted in Figure 9.7b, the capacitance change versus strain had a linear function, while the R^2 value was 0.9999 and the slope value was 0.97. We can easily recognize that the slope of this curve indicates the GF of this strain sensor and detect strain over 200%.

9.5.3 Piezoresistive Strain Sensor

Another useful type of strain sensor is a piezoresistive type that is founded on resistive behavior. The piezoresistive effect indicates the change in electrical resistance when mechanical strain is applied to a material. Thus, piezoresistive strain sensors basically have the same structure as conventional resistive-type sensors. To create a strain sensor, stretchable electrodes and a sensing layer were embedded on a stretchable substrate. In the previous section, a CNT film functioned as a stretchable electrode. Instead of several types of CNTs, a single-walled CNT (SWCNT) showed excellent adequacy for a stretchable strain sensing layer. The schematic diagram of a SWCNT strain sensor is displayed in Figure 9.8a (Yamada et al. 2011). Ti/Au/Ti composite electrodes were deposited at each end of the PDMS substrate that was covered by SWCNT rubber. SWCNT rubber minimized the empty space between the

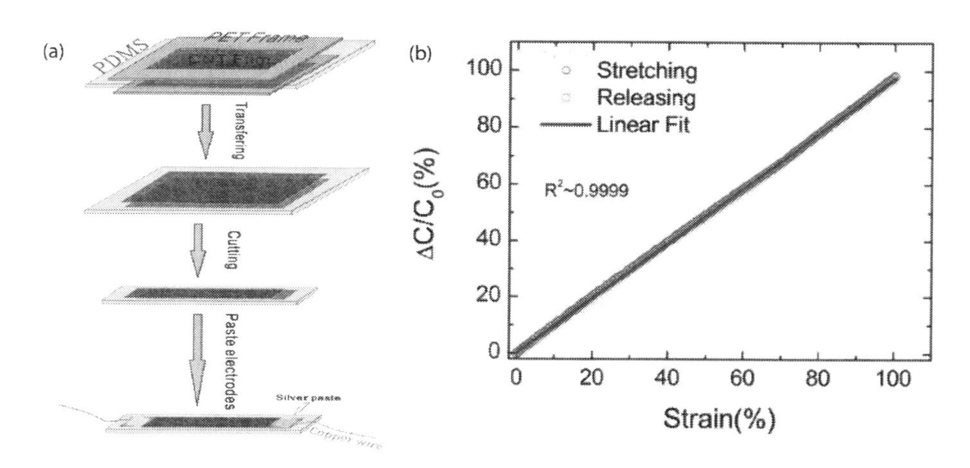

FIGURE 9.7 (a) Fabrication procedure (b) capacitive change with applied strain of stretchable capacitive strain sensor. (Reprinted by permission from Macmillan Publishers Ltd. *Scientific Reports*, Cai, L. et al., Copyright 2013.)

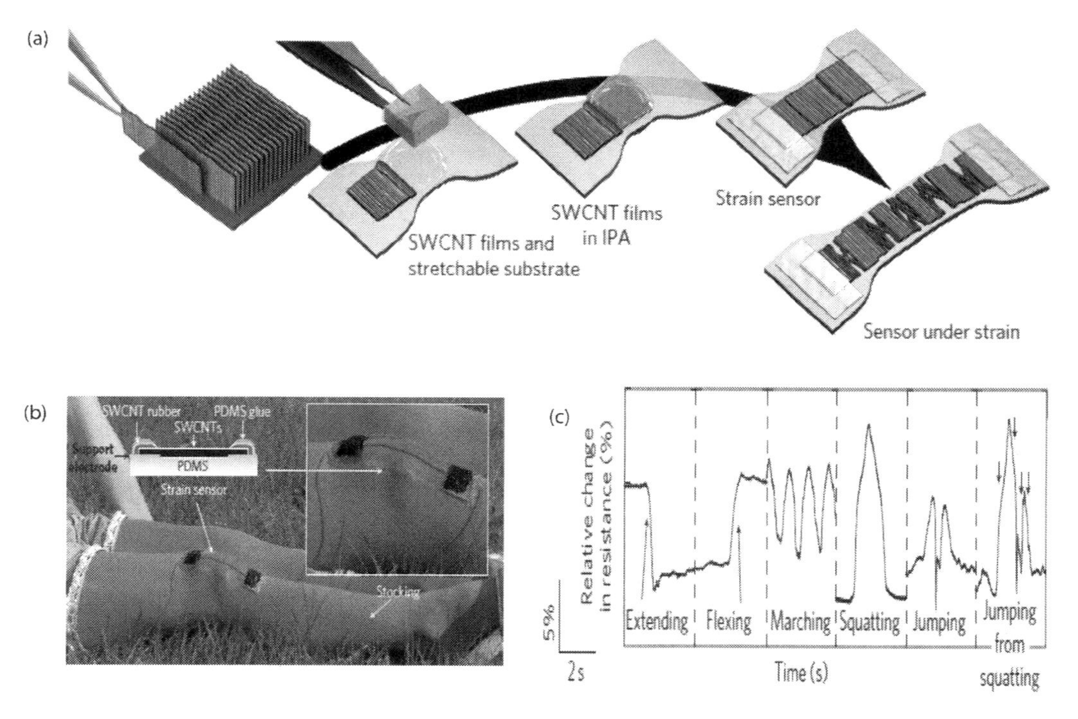

FIGURE 9.8 Stretchable resistive strain sensor based on SWCNT film. (a) Production process (b) practical application of SWCNT strain sensor mounted on human body (c) resistance change according to various human body motions. (Reprinted by permission from Macmillan Publishers Ltd. Yamada, T. et al., *Nat. Nanotechnol.*, Copyright 2011.)

SWCNT and electrodes, which contributes to a better performance. PDMS glue was then adhered to the SWCNT rubber to sustain the whole device. Figure 9.8b and c describes the human body monitoring characterization, which was performed by sticking a strain sensor to the human knee. Various motions were tested such as extending, flexing, jumping, etc. They exhibited distinct resistance changes, which proved the fabricated device's strain sensing capabilities.

9.6 Flexible and Stretchable Pressure Sensor

Recently, pressure sensors have emerged as an essential research area due to their usefulness in health-care and human body monitoring. Pressure is defined as perpendicular force against a material per unit area, and various human body motions can be detected by understanding the basic principles of pressure. The similar sensing principles of strain sensing and pressure sensing mean that the piezoelectric type, piezoresistive type, and capacitive type are the most commonly used in sensing mechanisms. Thus, this section introduces several examples of piezoresistive and capacitive pressure sensors and how pressure sensitivity is measured.

9.6.1 Piezoresistive Pressure Sensor

Figure 9.9a describes an Ag thin film-based piezoresistive pressure sensor whose design was inspired by the natural Mimosa leaf (Su et al. 2015). When a Mimosa leaf is stimulated by an outside attack, it closes itself for protection over a few seconds. An artificial Mimosa leaf was fabricated to follow this phenomenon. The artificially developed Mimosa leaf consists of small piece of Mimosa leaf covered by a PDMS substrate, which was created by repeating the molding process twice. After coating with the PDMS thin film, a titanium thin film and gold thin film were deposited with high uniformity. The role of the gold layer was a conducting layer, while the titanium layer was used to interconnect

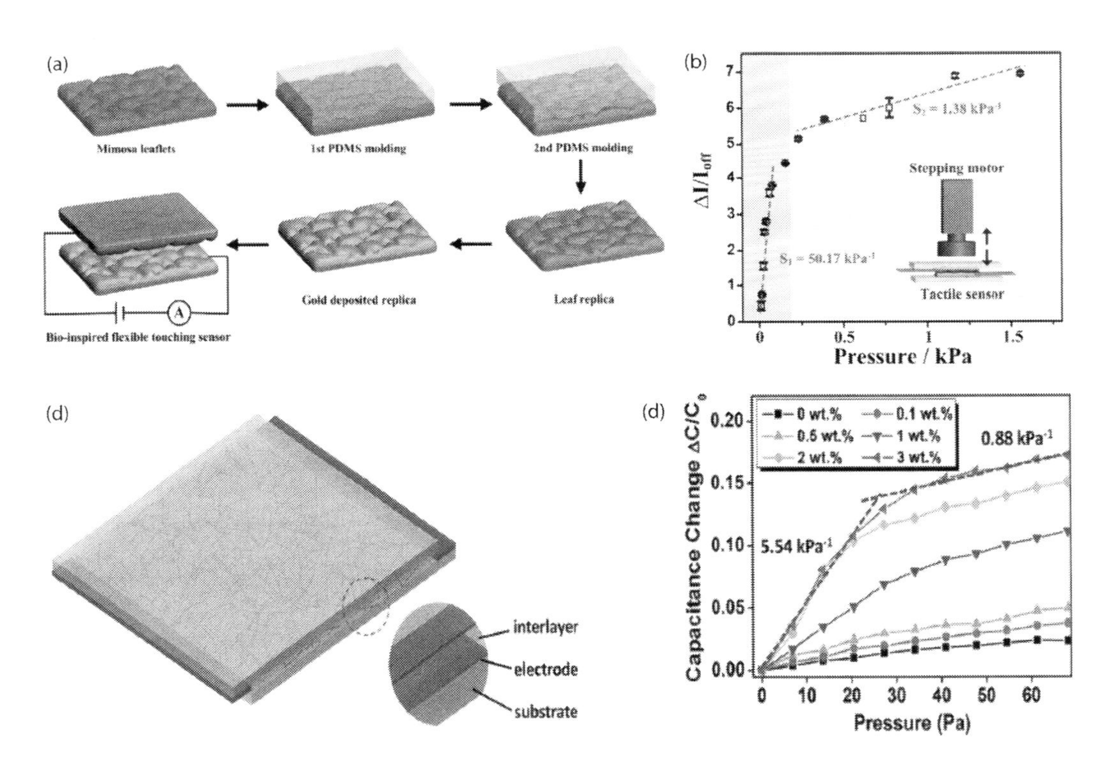

FIGURE 9.9 Flexible piezoresistive pressure sensor on PDMS substrate (a) fabrication inspired by natural Mimosa leaf (b) pressure sensitivity measured by current change. (From Su, B. et al.: Mimosa-inspired design of a flexible pressure sensor with touch sensitivity. *Small*. 11. 1886–1891. 2015. Copyright Wiley-VCH Verlag GmbH & Co. KGaA. Reproduced with permission.) Flexible capacitive pressure sensor (c) basic structure consisting AgNW interlayer (d) pressure sensitivity calculated by capacitive change. (Wang, J. et al., *Nanoscale*, 7, 2926–2932, 2015. Reproduced by permission of The Royal Society of Chemistry.)

the gold layer and PDMS substrate. Two units of the fabricated device faced each other with a wire-shaped source and drain electrodes at each end. The amount of current growth tended to increase as the applied voltage was higher, as can be seen in Figure 9.9b. The pressure sensitivity is defined as follows.

$$S = \frac{I - I_{off}}{I_{off}\,\Delta P}. \tag{9.32}$$

The sensitivity in the low pressure region was 50.17 kPa^{-1}, while the sensitivity was reduced to 1.38 kPa^{-1} in the high pressure region. This phenomenon can be explained by the contact distance between the encountered Ag thin films. As the voltage increased, the distance between them steadily decreased until the distance was saturated. From the saturation point, which is between the purple and yellow regions, deformation proceeded and influenced the pressure sensitivity.

9.6.2 Capacitive Pressure Sensor

The basic mechanism and principle of a capacitive pressure sensor is relevant to that of a capacitive strain sensor. The elastomeric dielectric layer and stretchable electrodes are strongly required to construct a capacitor structure. In the case of a strain sensor, the capacitance change is affected by the length, thickness, and width. However, the thickness is a more important influence than the length, and the thickness is determined in regards to the capacitive pressure sensor because the device is compressed vertically when pressure is applied.

Materials composing the capacitive pressure sensor are like those of the capacitive strain sensor. CNT, Ag nanowire (AgNW), and various ionic conductors are broadly applied for the capacitive pressure sensor. AgNW is a favorable substance for electronic devices due to its transparency and excellent conducting properties. Research has been performed to synthesize AgNW with a high aspect ratio and flexibility by various coating methods. Thanks to these characteristics, AgNW is widely used as a flexible electrode in various electronic devices and pressure sensors. Figure 9.9c shows a capacitive pressure sensor in which AgNW acts as both the electrode and dielectric interlayer (Wang et al. 2015). AgNW and polyurethane (PU) were blended to enhance the dielectric permittivity, and it was demonstrated that metal-based nanoparticles can improve the relative dielectric permittivity of other materials when added to them. AgNW itself was coated on a PET film for an electrode followed by covering the AgNW-PU composite on it. Two assemblies had a sandwich structure as they made contact with the PU layers.

Figure 9.9d shows the capacitance change according to the pressure insertion and the pressure sensitivity, which was calculated by the following equation.

$$S = \frac{C - C_{off}}{C_{off}\,\Delta P}. \tag{9.33}$$

The tendency of capacitance change resembled that of the current change in the piezoresistive pressure sensor in Figure 9.9d. The pressure sensitivity of the low pressure region was 5.54 kPa^{-1} and that of the high pressure region was 0.88 kPa^{-1}. According to Eq. (9.30), the capacitance has a proportional relationship with the relative permittivity. Thus, increasing the AgNW concentration enhances the relative permittivity of the AgNW-PU, which leads to high pressure sensitivity. Moreover, adopting an AgNW-PU composite for the pressure sensing layer resulted in excellent flexibility in the pressure sensor, which was examined by a bending test.

9.7 Flexible and Stretchable Gas Sensor

Various gases such as CO, NO, NO_2, and NH_3 are well known harmful gases for both human health and industry. Skin and respiratory diseases frequently occur upon exposure to such gases. Their hazardous properties make detecting these gases crucial. Gas detection procedures are performed by measuring the electrical properties in different gas concentration atmospheres.

9.7.1 Flexible and Stretchable Ethanol Gas Sensor

We have previously discussed how ZnO is widely used in electronic devices such as UV sensors. ZnO is also widely applied in ethanol gas sensors due to its high surface-to-volume ratio. Zheng et al. developed a flexible ethanol gas sensor based on ZnO nanoparticles, as shown in Figure 9.10a (Zheng et al. 2015). ZnO nanoparticles were drop-casted on a PET substrate and indium tin oxide (ITO) was used as an electrode. The sensing mechanism was a resistive type whose electrical resistance changed in accordance with the ethanol gas concentration. The current increased linearly with the applied voltage when 370 nm UV light was illuminated, which indicated ohmic contact rather than Schottky contact. Similar to the sensing principle of a ZnO UV sensor, excited electron–hole pairs are the main factor that decide the gas sensing ability. The response of the ethanol gas sensor was defined as:

$$\text{Response}(\%) = \frac{R_0}{R_g}. \tag{9.34}$$

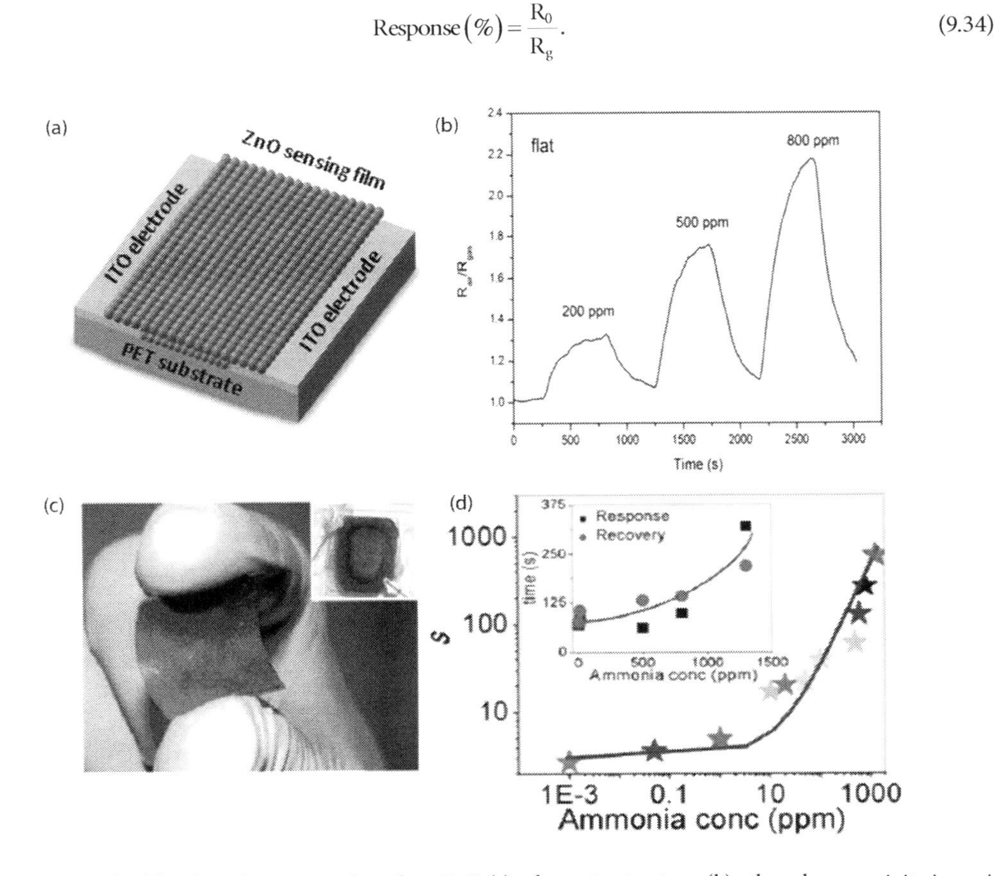

FIGURE 9.10 Flexible ethanol gas sensor based on ZnO (a) schematic structure (b) ethanol responsivity in various ethanol gas concentration. (Reprinted by permission from Macmillan Publishers Ltd. *Scientific Reports*, Zheng, Z.Q. et al., 2015, Copyright 2015.) Flexible ammonia gas sensor (c) actual image (d) ammonia gas sensitivity curve. (Reprinted by permission from Macmillan Publishers Ltd. *Scientific Reports*, Pandey, S. et al., 2013, Copyright 2013.)

R_0 represents the resistance in dark conditions, and R_g represents the resistance during UV illumination. The response of the ethanol gas sensor was increased in a high ethanol concentration atmosphere in Figure 9.10b; the increase in current and response were contributed by the electron–hole pairs that were created when the UV was illuminated. The reaction between the holes and oxygen ions on the ZnO surface reduces the depletion layer width, which exhibits a photocurrent increase. Moreover, oxygen ions such as O_2^- that were explained in Eq. (9.16) react with ethanol molecules according to the following reaction:

$$C_2H_5OH + 3O_2^- \rightarrow 2CO_2 + 3H_2O + 3e^-. \tag{9.35}$$

The electrons produced by this reaction are transferred back to the conduction band of ZnO, and the amount of photocurrent is further increased.

9.7.2 Ammonia Gas Sensor

The existence of ammonia nitrate in explosive chemicals in industry makes the development of a reliable ammonia sensor a vital requirement. Figure 9.10c displays a flexible ammonia gas sensor based on a Guar Gum (GG) and Ag nanoparticle nanocomposite. A GG solution and Ag nitrate were used as precursors for synthesizing the GG/Ag nanoparticle nanocomposite (Pandey et al. 2013). For the characterization of ammonia sensing, the ammonia concentration was in the range 500–1,300 ppm. The ammonia sensing performance was enhanced when the concentration of Ag nanoparticles increased because the GG/Ag nanoparticle interface operated as an active ammonia sensing site. The oxygen molecules were adsorbed on the active site and Ag nanoparticles facilitated this reaction. As shown in Figure 9.10d, Ag nanoparticles destroyed hydrogen bonding, so that O_2^- could be adsorbed on the GG/Ag nanoparticle active site. The ammonia molecules transferred the adsorbed oxygen and electrons to the sensing layer, which resulted in increased current and sensitivity. The sensitivity of the ammonia sensor was defined as follows.

$$S = \frac{\sigma - \sigma_0}{\sigma_0}. \tag{9.36}$$

σ represents the conductivity with no ammonia exposure, and σ_0 represents the conductivity with ammonia exposure. As the ammonia concentration increased, the sensitivity was also increased due to the GG/Ag nanoparticles.

9.7.3 Nitric Oxide Sensor

Air pollution is one of the most significant problems these days; it is caused by various types of gases such as nitric oxide (NO), nitric dioxide (NO_2), and carbon monoxide (CO). NO_x refers to air pollutants including NO and NO_2 and that produce smog and acid rain. NO, which influences the atmosphere dramatically, is primarily emitted from transportation systems; moreover, NO can cause severe conditions in humans such as cell damage and respiratory disease. Thus, detecting NO gas selectively from various gases is very important in industries. A Poly[N-9′-heptadecanyl-2,7-carbazole-alt-5,5-(4′,7′-di-2-thienyl-2′,1′,3′-benzothiadiazole)] (PCDTBT)-based NO gas sensor is introduced in Figure 9.11a (Gusain et al. 2017). PCDTBT is a promising material for diverse applications such as organic solar cells and was spin-coated on a flexible biaxially oriented polyethylene terephthalate (BOPET) substrate. Gold thin film was used as an electrode, and the current was linearly increased by increasing the voltage, which indicated that the sensing mechanism was based on the resistive type. Figure 9.11b shows the response characterization, and other gases such as CO, CH_3, NH_3, H_2S, Cl_2, C_2H_5OH, and NO_2 were composed in the gas sensing chamber to evaluate the selective NO sensitivity. Compared to other gases, the response

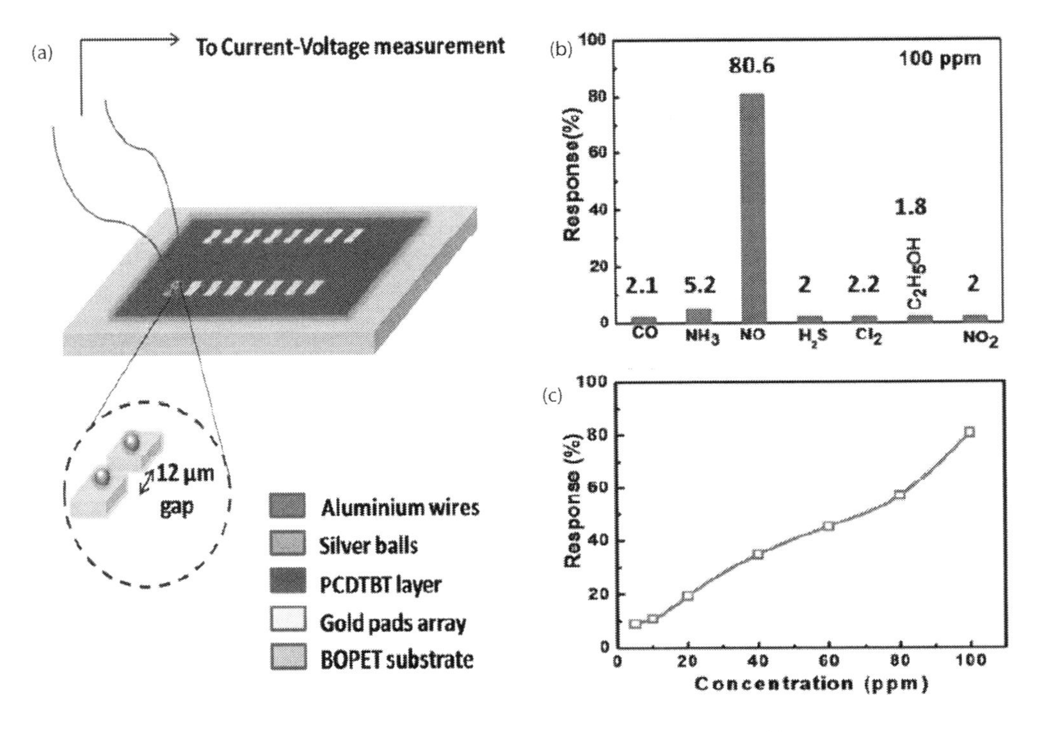

FIGURE 9.11 Flexible NO sensor (a) schematic structure (b) NO gas selectivity (c) NO gas response in various NO concentration. (Reprinted from *Sensor Actuat. B: Chem.*, 239, Gusain, A. et al., Flexible NO gas sensor based on conducting polymer poly [N-9′-heptadecanyl-2, 7-carbazole-alt-5, 5-(4′, 7′-di-2-thienyl-2′, 1′, 3′-benzothiadiazole)] (PCDTBT), 734–745, Copyright 2017, with permission from Elsevier.)

to NO had the highest value of 80.6%, which verified its excellent selectivity. The response of the NO gas sensor was defined as:

$$S(\%) = \frac{I_s - I_{ss}}{I_{ss}} \times 100, \tag{9.37}$$

where I_s represents the current in the gas exposure state, and I_{ss} represents the current before gas exposure. The response showed a linear relationship with the gas concentration whose response increased when the gas concentration was higher in Figure 9.11c.

9.7.4 Nitrogen Dioxide Sensor

NO_2 is also a prominent air pollutant that causes serious respiratory disorders; it is normally generated by the combustion of fossil fuels. Thus, detecting the concentration of NO_2 in the atmosphere has become essential. Materials such as multi-walled carbon nanotubes (MWCNTs) and graphene-based materials such as reduced graphene oxide (RGO) have been actively applied in NO_2 sensing applications. Moreover, a RGO/MWCNT was used spontaneously in a hybrid structure to improve the sensitivity. A RGO/MWCNT-based NO_2 sensor was fabricated on a flexible PI/PET substrate, as depicted in Figure 9.12a (Yaqoob et al. 2016). WO_3 nanoparticles were loaded on a RGO/MWCNT composite as a promoter to enhance the sensitivity. The weak van der Waals force enlarges the space between the

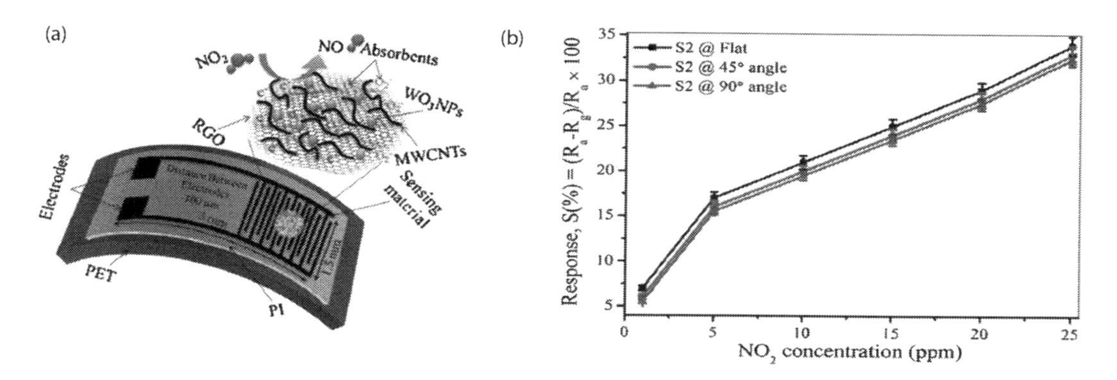

FIGURE 9.12 NO_2 sensor on flexible PI/PET substrate (a) schematic structure (b) response of NO_2 sensor when bended in different angles. (Reprinted from *Sensor Actuat. B: Chem.*, 224, Yaqoob, U. et al., A high-performance flexible NO_2 sensor based on WO_3 NPs decorated on MWCNTs and RGO hybrids on PI/PET substrates, 738–746, Copyright 2016, with permission from Elsevier.)

WO_3 layers and facilitates the adsorption of NO_2 molecules. WO_3 was synthesized by the hydrothermal method, which used $Na_2WO_4{\cdot}2H_2O$ and cetyltrimethylammonium bromide (CTAB) as precursors. Graphene oxide was synthesized by conventional modified Hummer's method. Then, WO_3 nanoparticles were mixed with a RGO/MWCNT composite by sonication. For electrodes, Ag thin film was deposited on a PI tape/Si substrate by RF magnetron sputtering, and the array was transferred to the PET substrate by peeling Si. The response was defined as:

$$\text{Response}(\%) = \frac{R_a - R_g}{R_a} \times 100. \tag{9.38}$$

R_a represents the resistance in air, and R_g represents the resistance when exposed to an NO_2 atmosphere. As shown in Figure 9.12b, the response was increased when the NO_2 concentration increased. This indicates that the resistance was lower in a high NO_2 concentration environment, which resulted in a high current. Moreover, the response ability was maintained even after bending the sensor array.

9.8 Other Types of Sensor

H_2O_2 is a fundamental material in biological applications and industries such as detecting human diseases. Several microorganisms show responses to H_2O_2 by changing color when making contact. The electrochemical approach seems the most adequate method for detecting H_2O_2. Various new materials such as metal nanoparticles, transition metals, and CNTs are being applied as electrochemical electrodes to improve sensing performance. This is achieved by decreasing the overpotential and enhancing the electron transfer kinetics. Figure 9.13a presents copper nanoparticles (Cu NPs) coated on a PET fiber for the non-enzymatic electrochemical detection of H_2O_2 (Kim et al. 2015). Cu NPs were coated on a PET fiber by an electroless plating method at various temperatures. When reactions were carried out at 70°C, Cu NPs were created with 500 nm thickness, which increased when the reaction temperature decreased. The CV curves were obtained to characterize the sensing performance of H_2O_2, as shown in Figure 9.13b. Two anodic peaks were investigated that were the consequence of oxidation, while cathodic peaks were investigated that were caused by reduction. We can note from Figure 9.13c that the peak current was enhanced after Cu NPs were coated on a PET fiber compared to bare PET fiber. The electrochemical sensitivity was 0.387 $mA\mu M^{-1}cm^{-2}$ and the detection limit was 2 μM.

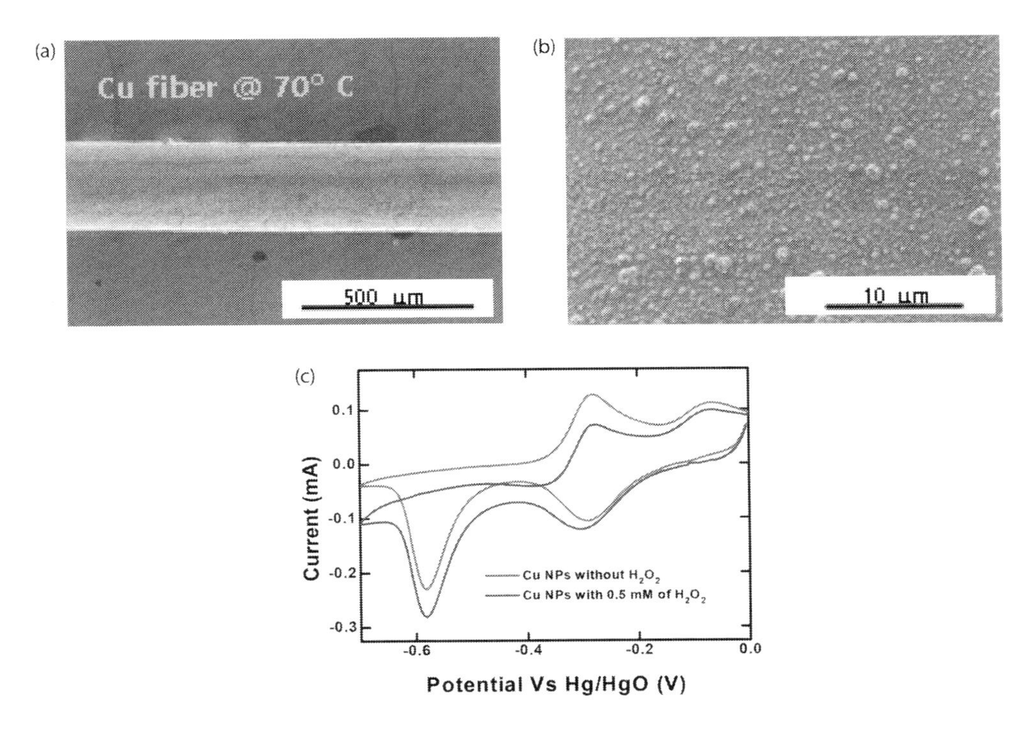

FIGURE 9.13 Cylindrical H_2O_2 sensor (a) overall shape of cylindrical PET fiber (b) Cu NPs coated on PET fiber by electroless plating (c) CY curves indicating electrochemical properties. (Kim, E. et al., *RSC Adv.*, 5, 76729–76732, 2015, Reproduced by permission of The Royal Society of Chemistry.)

References

Bendi, R., Bhavanasi, V., Parida, K., Nguyen, V. C., Sumboja, A., Tsukagoshi, K. & Lee, P. S. 2016. Self-powered graphene thermistor. *Nano Energy*, 26, 586–594.

Cai, L., Song, L., Luan, P., Zhang, Q., Zhang, N., Gao, Q., Zhao, D., Zhang, X., Tu, M. & Yang, F. 2013. Super-stretchable, transparent carbon nanotube-based capacitive strain sensors for human motion detection. *Scientific Reports*, 3, 3048.

Eom, T. H. & Han, J. I. 2017a. The effect of the nickel and chromium concentration ratio on the temperature coefficient of the resistance of a Ni–Cr thin film-based temperature sensor. *Sensors and Actuators A: Physical*, 260, 198–205.

Eom, T. H. & Han, J. I. 2017b. Resistive behavior of Ni thin film on a cylindrical PET monofilament with temperature for wearable computing devices. *Sensors and Actuators A: Physical*, 259, 96–104.

Eom, T. H. & Han, J. I. 2018. Single fiber UV detector based on hydrothermally synthesized ZnO nanorods for wearable computing devices. *Applied Surface Science*, 428, 233–241.

Gusain, A., Joshi, N. J., Varde, P. & Aswal, D. 2017. Flexible NO gas sensor based on conducting polymer poly [N-9′-heptadecanyl-2, 7-carbazole-alt-5, 5-(4′, 7′-di-2-thienyl-2′, 1′, 3′-benzothiadiazole)] (PCDTBT). *Sensors and Actuators B: Chemical*, 239, 734–745.

Jang, J. H. & Han, J. I. 2017a. High performance cylindrical capacitor as a relative humidity sensor for wearable computing devices. *Journal of The Electrochemical Society*, 164, B136–B141.

Jang, J. H. & Han, J. I. 2017b. Cylindrical relative humidity sensor based on poly-vinyl alcohol (PVA) for wearable computing devices with enhanced sensitivity. *Sensors and Actuators A: Physical*, 261, 268–273.

Kim, E., Arul, N. S., Yang, L. & Han, J. I. 2015. Electroless plating of copper nanoparticles on PET fiber for non-enzymatic electrochemical detection of H_2O_2. *RSC Advances*, 5, 76729–76732.

Lichtenwalner, D. J., Hydrick, A. E. & Kingon, A. I. 2007. Flexible thin film temperature and strain sensor array utilizing a novel sensing concept. *Sensors and Actuators A: Physical*, 135, 593–597.

Pandey, S., Goswami, G. K. & Nanda, K. K. 2013. Nanocomposite based flexible ultrasensitive resistive gas sensor for chemical reactions studies. *Scientific Reports*, 3, 2082.

Su, B., Gong, S., Ma, Z., Yap, L. W. & Cheng, W. 2015. Mimosa-inspired design of a flexible pressure sensor with touch sensitivity. *Small*, 11, 1886–1891.

Wang, J., Jiu, J., Nogi, M., Sugahara, T., Nagao, S., Koga, H., He, P. & Suganuma, K. 2015. A highly sensitive and flexible pressure sensor with electrodes and elastomeric interlayer containing silver nanowires. *Nanoscale*, 7, 2926–2932.

Xiao, S., Che, L., Li, X. & Wang, Y. 2008. A novel fabrication process of MEMS devices on polyimide flexible substrates. *Microelectronic Engineering*, 85, 452–457.

Yamada, T., Hayamizu, Y., Yamamoto, Y., Yomogida, Y., Izadi-Najafabadi, A., Futaba, D. N. & Hata, K. 2011. A stretchable carbon nanotube strain sensor for human-motion detection. *Nature Nanotechnology*, 6, 296–301.

Yan, C., Wang, J. & Lee, P. S. 2015. Stretchable graphene thermistor with tunable thermal index. *ACS Nano*, 9, 2130–2137.

Yang, T., Yu, Y., Zhu, L., Wu, X., Wang, X. & Zhang, J. 2015. Fabrication of silver interdigitated electrodes on polyimide films via surface modification and ion-exchange technique and its flexible humidity sensor application. *Sensors and Actuators B: Chemical*, 208, 327–333.

Yaqoob, U., Uddin, A. I. & Chung, G.-S. 2016. A high-performance flexible NO_2 sensor based on WO_3 NPs decorated on MWCNTs and RGO hybrids on PI/PET substrates. *Sensors and Actuators B: Chemical*, 224, 738–746.

Zheng, Z. Q., Yao, J. D., Wang, B. & Yang, G. W. 2015. Light-controlling, flexible and transparent ethanol gas sensor based on ZnO nanoparticles for wearable devices. *Scientific Reports*, 5, 11070.

10

Artificial Skin

Joanna M. Nassar

10.1 Introduction

Skin is a barrier that enables the human body to unknowingly and simultaneously perceive a number of peripheral events such as; distinguishing between smooth and rough textures, the warmth of a human touch against the cold feel of a robotic hand, and the differentiation between shapes such as circles and squares. For decades now, the pursuit to replicate the properties of skin using electronic components has been of profound vitality, with implications to restore the skin sensations for prosthetics and augmenting the capabilities of humanoids. The ability to restore skin sensations to acid and burn victims or amputees could enhance the quality of their life [1]. Providing prosthetic hands with our skin's unique characteristics and complex sense of heat, moisture, touch, force, and texture, calls for further innovations in large-scale multi-sensory skin platforms [2]; electronic skin platforms that can perceive diverse and complex sensations in a simultaneous fashion, but also replicate the mechanical elasticity and durability of human skin by adopting flexibility and stretchability.

The mechanical compliancy and multi-sensory characteristics of skin further stimulated emerging sensors for Internet of Everything (IoE) wearables to mimic the multi-functionality of skin. A smart artificial skin results from increasing sensor density and capabilities for a more informative recognition of external events [3]. For instance, both mechanoreceptors and thermal receptors are essential to identify the wetness of a surface [4]. Artificial skin is key for collecting signals simultaneously, which not only increases the functionality per cost, but also enables a smarter, more personalized, and interactive platform. Consequently, applications where artificial skin has steered into are found in augmented reality gaming and human-machine-interfaces [5,6], large-scale environmental mapping, real-time fitness trackers, and advanced healthcare [7].

The mechanical and sensory requirements dictated by an electronic skin (E-skin) cannot be met using conventional electronics. The development of novel processing techniques and unconventional materials has been vital to realize a skin-inspired platform. In the previous chapters, authors have thoroughly covered the variety of materials, design, and fabrication strategies advanced to achieve mechanically compliant (flexible and/or stretchable) skin-like electronics and sensors. In this chapter, we focus on the integration schemes and material designs adopted to improve sensors characteristics, enabling the development of a multi-sensory artificial skin that mimics the skin's ability to simultaneously sense signals in real-time, such as strain, humidity, pressure, and temperature [8–18]. It is important to note that the diversity of the targeted applications entail different integration approaches for an artificial skin. Whether the E-skin will be used for prosthetics or smart biomedical wearables, requirements for mechanical properties and sensory characteristics may vary, such as the extent of flexibility, need for stretchability, as well as the durability of the platform.

Developing a fully functional and autonomous E-skin doesn't stop at the sensors and actuators mark. Brain machine interfaces coupled with optimized signal processing and transmission methods are essential to deliver the sensory signals into the nervous system for perception and awareness in prosthetics [1]. One perpetual challenge today is the integration of the soft material-based multi-sensory platforms with the rest of the rigid electronic interface and system components. Our vision is to mitigate this matter through the use of flexible and low-cost silicon-based electronics and hybrid heterogeneous integration processes. Using mechanically flexible and/or stretchable silicon can possibly be a solution for a truly high performance yet low-cost and low-power artificial skin.

10.2 Mechanical Properties of Skin

The characteristics and capabilities of a sensor are majorly affected by the mechanical properties of the platform [1]. The low elastic modulus of skin allows it to conform to the surface of objects by altering the distribution of forces and strains at different points [19]. For example, a good grip is achieved by inducing friction, which is in turn realized by increasing contact area with the object [20]; this latter mechanism can be made feasible only if the platform is flexible. Only flexible electronics that can be bent over curved surfaces can provide additional degrees of movement [21–24]. State-of-the-art silicon electronics are stiff and brittle, therefore alternative materials are embraced to simulate the soft mechanical properties of skin. Evidently, organic materials became the first attraction due to their intrinsically flexible and stretchable nature. However, they exhibit limited performance, complex integration strategies, as well as thermal instability, which compelled investigation into 1D materials (e.g., nanowires, nanotubes, and nanorods), 2D material composites, and the use of unconventional substrates such as paper [2]. Although flexibility alone of an E-skin can satisfy many of the requirements desired in robotic applications and smart wearables, stretchable materials with low elastic moduli can further augment the functionality of a prosthetic, making it closer to the feel of real skin. For example, the free movement of joints in the human body is assisted by our naturally stretchable biological skin, capable of withstanding strains up to 75% on average [1]. Making electronics stretchable requires new materials such as intrinsically stretchable organic and hybrid elastomers, or design approaches to transform brittle and inelastic metals and semiconductors into physically deformable materials [22,23,25], providing the desired additional degrees of movement. Biological skin is characterized by a network of touch fibers that governs its soft, yet tough nature. In contrast, elastomers often present a settlement between softness and durability [1]. Elastomers with an elastic modulus comparable to human skin often lack the robustness needed in prosthetics [26]. One common approach to overcome this matter is to adopt a network of a low-modulus materials mixed within a network of a high-modulus materials [1,27]. The elasticity of the resulting composite would be contingent on the applied strain, in such a way to avoid rupture. Moreover, our biological skin displays various designs essential to the perception of force

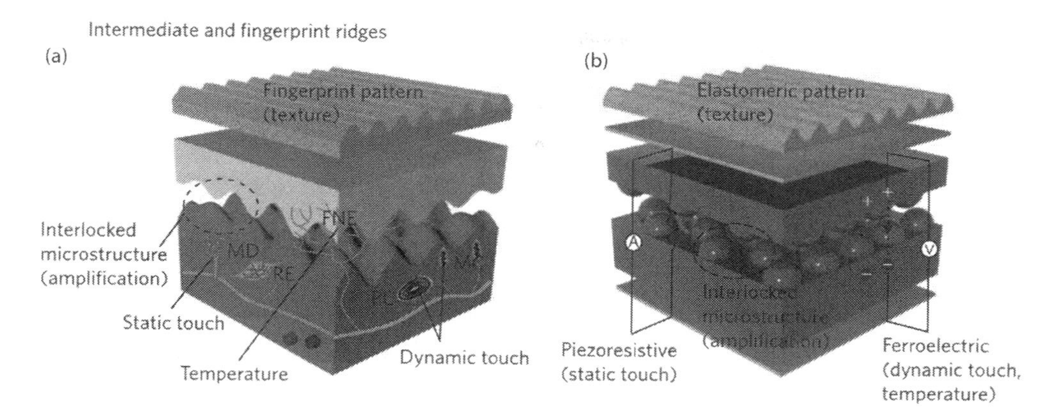

FIGURE 10.1 Bio-inspired micro/nanostructures. (a) Left schematic shows interlocked microstructures and fingerprint patterns found in biological skin. (b) Right schematic reflects the corresponding human-skin-inspired interlocked microdome arrays and texture patterns in E-skin. (From Park, J. et al., *Sci. Adv.*, 1, e1500661, 2015; Figure 1.)

and texture discrimination, such as intermediate ridges and fingerprint ridges found on the surface of our fingers [20,28,29] (Figure 10.1). These are a few of the biomimetic designs that researchers aim to implement into their sensors to improve their performance and get closer to skin sensations [30,31].

10.3 Biomimetic Skin Sensations

10.3.1 Thermal Sensors

Human skin can detect a range of sensations, including moisture, strain, force, and temperature, with the ability to identify temperature changes with a resolution of 0.02°C [1,32]. In electronic skin developments, temperature sensors adopt different operating principles, and are often represented by resistance temperature detectors (RTDs), p-n junction-based sensors, or composite materials that expand under thermal variations. RTDs are very common since they are characterized by a fairly linear slope which is proportional to the temperature coefficient of resistance (TCR); a parameter of sensitivity, whereby an increase in the metal's resistance is directly attributed to increased phonon vibrations at elevated temperatures. The thermal properties of skin were evaluated using flexible RTDs intimately contacting the skin surface, with a detection resolution of 0.014°C [33]; though this sensing mechanism is strain-sensitive and hence poses recurrent limitations for their standard implementation in flexible and stretchable electronics [34]. To overcome this challenge, strain effects could be alleviated using buckling of the metal films or stretchability by design approaches [33,35–37]. *p-n* junction-based temperature sensors are less sensitive to strain and generally exhibit improved sensitivity over TCR sensors [33,38], however, the bottleneck is their high sensitivity to light [39]. Alternatively, highly sensitive thermal sensors can be fabricated using conductive elastomers; hybrid composites made of conductive polymers and fillers [40]. Higher temperatures lead to the expansion of the fillers, causing the conductive particles to move apart from one another, and hence increasing the output resistance. This approach shows improved mechanical stability compared to RTDs, with a temperature sensor resolution of 0.1°C [40]. Still, this method suffers the most from large hysteresis, with additional engineering techniques required to further reduce strain dependence [40].

10.3.2 Static and Dynamic Force Transducers

To mimic the pressure mechanoreceptors, in particular the static force transduction mechanism in the skin, large-scale sensory arrays are needed to measure normal force distributions with sub-mm resolutions [41,42]. Both capacitive and resistive mechanisms can be used to fabricate transducers that are sensitive to static pressure. For capacitive pressure transducers, the generally adopted structure is a parallel plate design, whereby an exerted force will alter the distance or overlap between the two conductive plates, resulting in a capacitance change. It is important that the dielectric of the capacitive sensor is compressive to allow distinction between diverse levels of force. Some examples of materials used are: standard rubber-like polymers (Ecoflex or poly(dimethylsiloxane), PDMS) [43,44], microstructured elastomers [29,31,45], or even simply an air gap [18,46]. The more porous the dielectric is, i.e., filled with air, the more compressible the material is, and consequently allowing for increased sensitivity towards smaller force-induced deformations. When choosing viscoelastic polymers for the dielectric, the sensitivity curve suffers from slight non-linearity; at higher pressures/forces, the response is decreased due to smaller deformations. Also, the recovery of the sensor is slower in comparison to an air-gap-based sensor, where the dielectric quickly returns to its equilibrium shape once the force-induced stress is removed. In a similar manner, microstructuring the rubber polymer allows the dielectric to be more porous and compressible (Figure 10.2a and b). This allows the sensor to deform faster due to the reduced viscoelastic effects, hence leading to improved sensitivity and faster response and recovery times [45] (Figure 10.2c and d). One drawback of capacitive-based pressure transducers is their sensitivity to electromagnetic fields. Although the capacitive mechanism is often implemented for its simplicity, high sensitivity, and linearity [45], capacitive sensors suffer from electromagnetic interference effects [47,48], and the adoption of a shielding material can be challenging to preserve flexibility and stretchability. The alternative is resistive sensors; whereby two mechanisms can be found. The first mechanism is the use of piezoresistive materials. The piezoresistive effect in semiconductors and metals is characterized

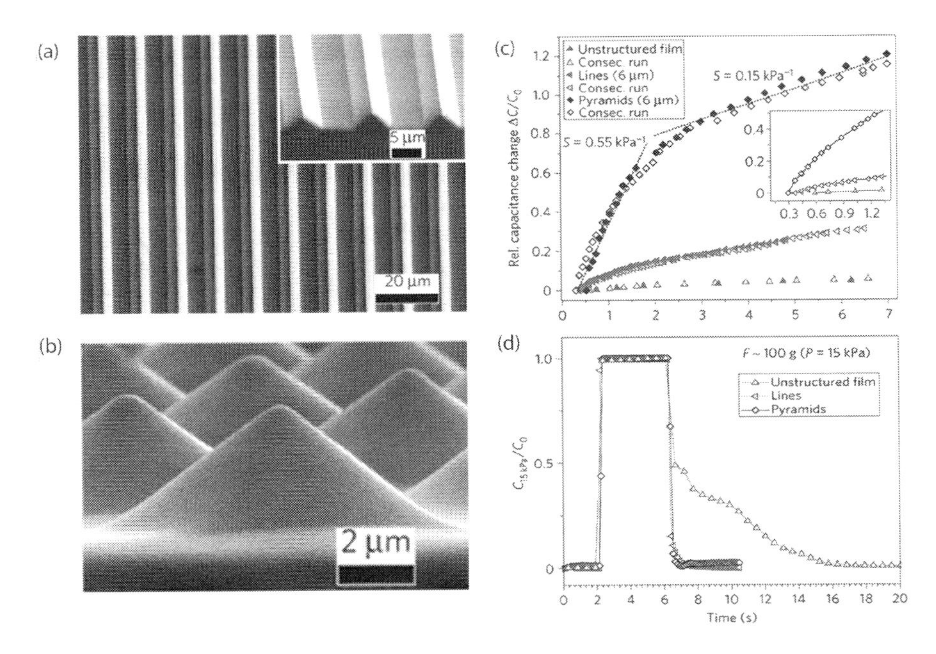

FIGURE 10.2 Biomimetic structural designs. (a and b) SEM images of microstructured PDMS films in the shape of uniform line features and square pyramids. (From Mannsfeld, S. C. et al. *Nat. Mater.*, 9, 859–864, 2010; Figure 1.) (c) Pressure-sensitivity plots showing that the microstructured PDMS sensor films exhibit higher performance much higher (about 30 times higher) pressure sensitivity than the unstructured PDMS films of the same thickness. (d) Real-time response of sensors upon application of external pressure. Both structured and unstructured films exhibit immediate response; however, the structured PDMS-based sensors exhibit much faster recovery (almost immediate), in comparison to a slow relaxation time of more than 10 s for the unstructured sensor. (From Mannsfeld, S. C. et al. *Nat. Mater.*, 9, 859–864, 2010; Figure 2.)

by a change in electrical resistivity under mechanical strain and is orders of magnitude more prominent than geometric effects perceived in the capacitive structures. Depending on the material of choice, when pressure is applied, the band structure is altered in the case of semiconductors, or in the case of conductive elastomers, the distribution of conductive composites (such as carbon nanotubes [CNTs] or nanowires, NWs) in the polymer is altered, leading to a corresponding resistivity change [42,49]. However, piezoresistive materials generally exhibit large hysteresis, have poor pressure sensitivity, and are extremely sensitive to temperature variations [42].

When it comes to dynamic force transduction, piezoelectric and triboelectric mechanisms are used for pressure sensors. Under mechanical deformation, an electric potential or voltage is generated (Figure 10.3) [50]. In piezoelectric materials, the induced strain or mechanical stress changes the magnitude and orientation of dipoles (change of polarization) in the layer, causing an increase in surface charge density on the electrodes. This phenomenon of piezoelectricity is known as the release of electric charge under the application of mechanical stress and occurs in all non-centrosymmetric materials [51]. Because of this characteristic, piezoelectric pressure sensors enable the measurement of slight pressure fluctuations (dynamic pressures) at high static forces, whereas the triboelectric effect is a form of contact electrification. When two distinct materials are rubbed against each other (i.e., inducing frictional contact), electrical charge is built up. The strength and polarity of the electric charges produced will depend on various properties such as mechanical strain, but also temperature and surface roughness. Some of the qualifying materials to induce this effect range are inorganic, such as ZnO [52] and $BaTiO_3$ [53], or organic, such as polyvinylidene fluoride (PVDF) [54,55]. Both piezoelectric and triboelectric-based sensors are highly favorable for skin applications due to their selective sensitivity to dynamic pressures [29,41], but also, they have the advantage of doubling as an energy harvester, producing energy from mechanical stimulation, for low-power and self-powered applications [56].

In general, academic advances show higher static pressure sensitivities compared with the biological skin [1,45,57–59]. The lower boundary for human skin pressure recognition is limited to 1 mN, whereas electronic pressure transducers display sensitivities beyond that of biological skin, with static force measurements below 0.05 mN [60] and 0.08 mN [58]. Future challenges for E-skin development are no longer focused on the sensitivity and performance of the sensors, rather on the integration with the interface circuitry while maintaining the intimate and adaptive characteristics of biological skin.

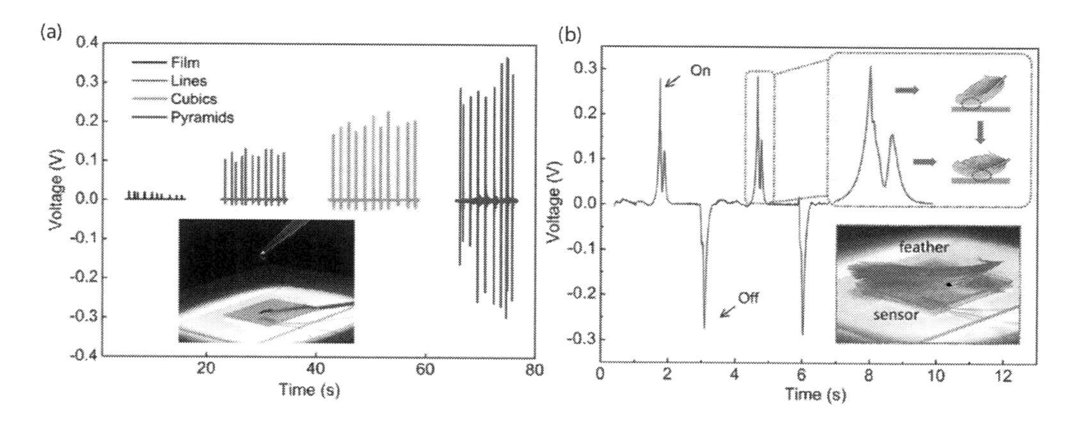

FIGURE 10.3 Dynamic force sensing. (a) Microstructured triboelectric and piezoelectric devices produce increased voltages in response to applied pressure. (b) Dynamic pressure response of a flexible triboelectric sensor induced by a piece of feather. (From Fan, F.-R. et al., *Nano Lett.*, 12, 3109–3114, 2012; Figure 4.)

10.3.3 Strain Sensors

Measuring strain on the surface of the skin can be similarly achieved using the mechanism of static pressure transducers. Strain sensing using the resistive mechanism is described by $R = (\rho L)/A$, where R is the resistance output of the sensor, L is the length of the structure, ρ is the material's resistivity, and A the cross-sectional area of the device. Therefore, strain-induced resistance changes are affected by either a geometrical change in the structure alone, a stress-induced change in the material's resistivity, or both effects in a simultaneous fashion. As described by the Poisson effect, when the sensor structure of length L is stretched, its total length increases by ΔL, whereas its area correspondingly compresses and contracts [61], leading to an increase in the output resistance. But in some cases, stretching alters the band structure of the semiconductor due to induced stress [62], or readjusts the percolation pathways between conductive particles in hybrid elastomers [63], consequently altering the resistivity of the material. Traditional strain gauge designs based on thin film conductors and silicon exhibit high sensitivities, but are limited to small strain variations (<1%). Stretchable designs are hence incorporated to improve the stretchability of these materials without breaking, to enable large-strain measurements [64], but there will be a trade-off with reduced sensitivity to smaller strains. Stretchable capacitors can also be used to measure strain; when the platform is stretched, a change in overlap between the conductive electrodes of a parallel plate structure is translated into a change in capacitance. These structures are desired because of their bimodal characteristic as they are often also sensitive to pressure [44]. Stretchability along with stability of the sensor's electrical characteristics under stretching is of vital importance. Therefore, the choice of material and design of the strain sensor will be imposed by the targeted application, either favoring high responses to small strain variations, but limited stretchability, or preferring a tolerance to higher stretchabilities with a focus on higher strain measurements.

Corresponding techniques to achieve highly stretchable strain sensors include the use of innovative materials such as hybrid conductive elastomers or liquid metals [23]. Hybrid elastomers are characterized by conductivity along with high elasticity. Their soft, transparent, and piezoresistive nature enabled their use in highly stretchable strain, pressure, and acoustic sensors [10,23,65–69]. One of the main advantages of such conductive elastomers is the ability to tune the amount of conductivity versus stretchability depending on their function in the desired application. The high elasticity and conductivity of NW-based elastomer composites rendered them a desired choice for artificial skin developments [65,70,71]. As an example, a self-powered, stretchable, and transparent skin was demonstrated using AgNW/poly(3,4-ethylenedioxythiophene) (PEDOT): polystyrene sulfonate (PSS)/poly (urethane acrylate) (PUA) nanocomposite to build a highly sensitive strain monitoring system [72].

10.3.4 Human Skin-Inspired Strategies for an Artificial Skin

Inspired from the aesthetic and characteristics of biological skin, several engineering strategies have been adopted in artificial skin sensor developments [1]. A skin-like platform suggests a multi-functional and diverse set of sensors that can distinctly and simultaneously perceive the complex range of thermal and mechanical stimuli to resolve sensations such as temperature, vibration, and moisture [73]. One approach to achieve the desired multi-functionality and selectivity to stimuli is through the notion of multi-layer artificial skins, whereby the depth at which the sensors are embedded in the skin can tune their sensitivity and hence selectivity to external stimuli [1]. For instance, a pressure transducer inserted at the interface of the skin's surface is more perceptible to pressure stimuli, whereas the same sensor located deeper in the E-skin is more selective to local stretching of the skin [74]. A similar effect can be achieved by coating the artificial skin with an elastomer, which not only mimics the elasticity and feel of human skin, but also serves in redistributing the forces on the sensor arrays to improve grip for grasping tasks. However, this approach generally leads to increased hysteresis in sensors performance [75].

The unique set of lines and ridges we see at the tip of our fingers do not solely serve as an identification of individuality; their main purpose is to enable human beings to perceive texture by inducing distinctive vibrations on the surface of the skin which permits the differentiation of shear forces, and hence the feel of texture [76]. To mimic this behavior, Park et al. developed a multi-functional flexible E-skin with texture recognition achieved by designing fingerprint-like ridges made out elastomeric patterns. The adopted pressure sensor design is multi-modal and incorporates interlocked microdome arrays to amplify light forces; the piezoresistive effect is used to identify static pressure, whereas dynamic pressure and temperature detection are distinctly identified from the induced ferroelectric effect [29] (Figure 10.4a).

Skin-like interlocked microdome arrays patterned onto the surface of thick PDMS films amplify the signal to small vibrations due to the porosity introduced into the dielectric of the parallel-plate capacitor (Figure 10.4b) [77]. An alternative to these sophisticated engineered materials is a simple kitchen sponge; used as the dielectric of a pressure sensor structure for a low-cost paper skin [18] (Figure 10.4c). Serving a similar function to the microdome arrays, the intrinsic porous structure of the sponge amplifies signals to small vibrations and makes the E-skin ultra-sensitive to light tactile stimuli. Another nature inspired engineering strategy is retrieved from the spider's sensory system. A crack-shaped slit near the leg joints of spiders enables them to sense extremely small variations. Ultrahigh mechanosensitivity can be therefore achieved by purposely engineering well defined nanoscale cracks in strain sensors (Figure 10.4d) [30,78]. Hair-like structures inspired from the whiskers of mammals also display benefits for increased sensitivity to surrounding objects since they are easily deformed in the presence of minimal air flow [79,80].

10.3.5 Self-Healing Property

Skin is our body's largest organ; it is elastic, helps preserve fluid balance, controls body temperature, helps prevent and fight diseases, but it's also self-healing. Skin is constantly subjected to damages from the environment; when the safety of our body is compromised from a wound, it is an essential attribute of skin to act as a barrier against infectants (e.g., germs) and self-heal to close the cut and protect us from external surroundings. The incorporation of a repetitive self-healing capability becomes critical to extend the E-skin's lifetime while maintaining robust functionality, especially in the areas of soft robotics and biomimetic prostheses. It is common to witness insulating self-healing polymers, but the challenging task is the advancement of self-healing materials that incorporate conductive electrical properties, vital for the integration of connectors and sensors in self-healing electronic skins [65]. Self-healing can be either realized through external stimulation (*viz.* heat, light, or solvents) to activate the process or it can be an intrinsic property of the material. Leibler et al. showed the first demonstration of an insulating elastomer that can self-heal at room temperature [81]. The intrinsic self-healing process was accomplished by incorporating hydrogen bonds into the polymer matrix. After which Bao et al. developed the first conductive polymer that can intrinsically self-heal at room temperature, both electrically and mechanically [49]. The self-healing conductor is achieved by embedding nickel nanostructured microparticles in a supramolecular organic polymer (Figure 10.5a). Within few seconds, the skin regains 75% of its mechanical strength and electrical conductivity, and within 30 minutes, the artificial skin is fully restored [49]. A highly stretchable and self-healing actuating elastomer is further demonstrated by incorporating a network of complex crosslinks into the PDMS polymer chains, achieving high dielectric strength and mechanical actuation [82]. Hydrogel material has been recently engineered to incorporate self-healing properties due to its desirable soft and elastic yet tough bonding properties with the human skin [83]. Martin Kaltenbrunner and colleagues demonstrated a self-healing transparent E-skin incorporating ionic hydrogel conductors that can restore function instantly, while the multi-layered structure of the elastomer hybrids and ionic hydrogels can endure biaxial strain beyond 2000% [84] (Figure 10.5b).

White et al. pursued another approach to develop a non-polymeric self-healing conductor. The self-healing liquid metal, eutectic gallium indium (EGaIn), is shown to almost fully recover its electrical properties in less than 1 ms after being damaged [85]. Both EGaIn and Ag ink liquid metals demonstrate intrinsic and rapid mechanical and electrical self-healing properties after rupture [42]. These liquid metal healants show promise due to their instantaneous recovery, however, they are limited to a one

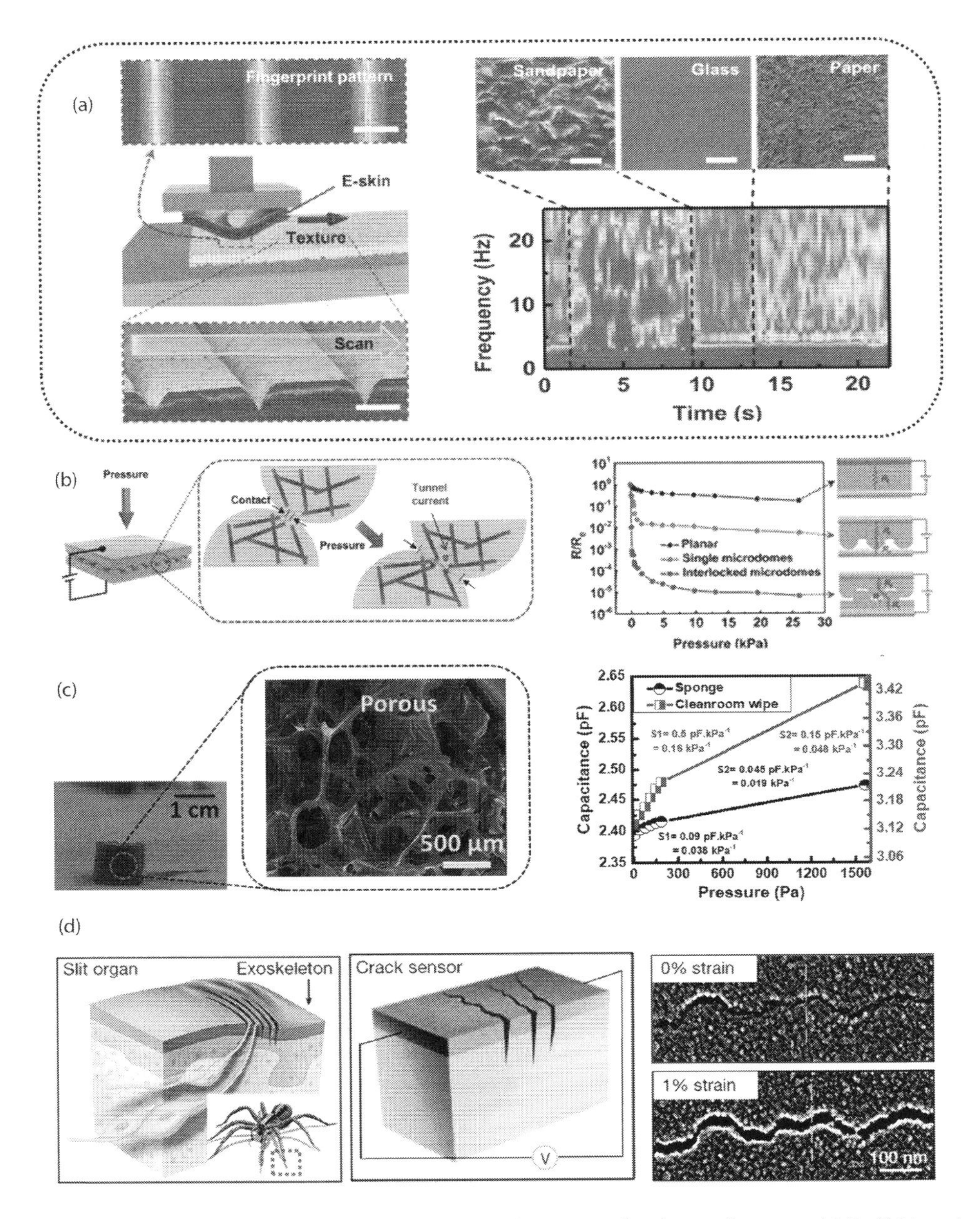

FIGURE 10.4 Human skin-inspired engineering schemes for improved E-skin performance. (a) (Left) Piezoelectric e-skin with periodic fingerprint-like PDMS patterns ($P = 470$ μm, $W = 163$ μm) for texture perception. Scale bar, 200 μm. (Right) Perception and distinction of surfaces (sandpaper, paper, and glass surfaces) with different surface roughness and texture. Scale bar, 200 μm. (From Park, J. et al., *Sci. Adv.*, 1, e1500661, 2015; Figure 6.) (b) Schematic showing the working principle of interlocked microdome-structured pressure sensors. Applied force concentrates stress at the contact spots, deforming the microdomes, causing an increase in the contact area. This leads to increased sensor sensitivity to pressure, as shown by the comparison performance plot on the left. (From Park, J. et al., *ACS Nano*, 8, 4689–4697, 2014; Figure 1 and Figure 2.) (c) Pressure sensor made using a kitchen sponge as the dielectric. SEM image shows the porous structure of the sponge, leading to the improved pressure sensitivity seen in the plot on the right, in comparison to a less porous dielectric material. (From Nassar, J.M. et al., *Adv. Mater. Technol.*, 1, 1600004, 2016; Figure 1, Figure 3, Figure 5.) (d) Schematics show the crack-based strain sensor inspired by the slit organ of spiders. (From Kang, D. et al., *Nature*, 516, 222–226, 2014; Figure 1.)

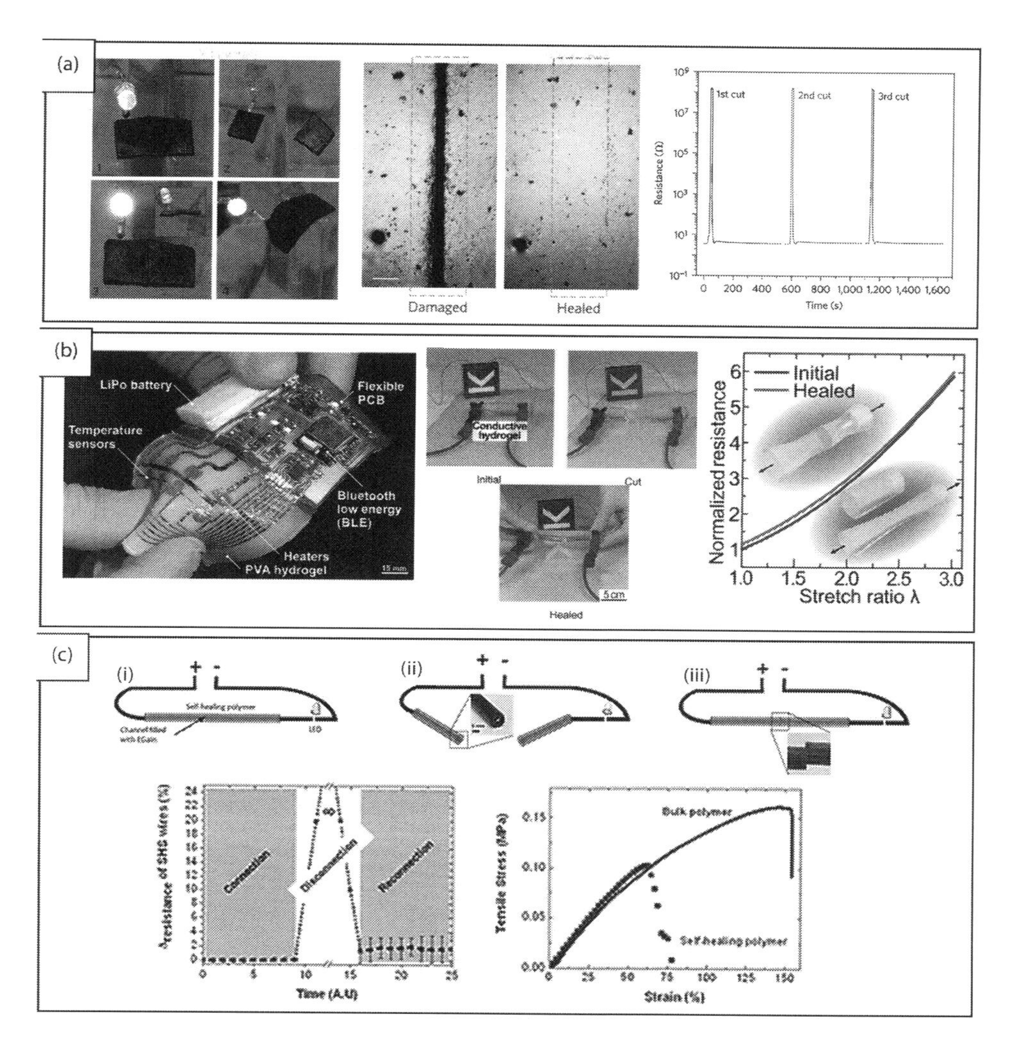

FIGURE 10.5 Self-healing materials and properties for E-skin. (a) (Left) Plastic composite ('supramolecular organic polymer with embedded nickel nanostructured microparticles') capable of repeatedly self-healing at room temperature (middle). (Right) Plot shows that within 30 minutes, the skin fully restores its mechanical strength and conductivity. (From Tee, B.C. et al., *Nat. Nanotechnol.*, 7, 825–832, 2012; Figures 2 & 3.) (b) (Left) Photograph of a self-healing transparent hydrogel-based E-skin integrating stretchable heaters, temperature sensors, and interfaced with battery, control, readout, and Bluetooth low energy (BLE) communication electronics hosted on a flexible PCB board. (Middle) Healing of a conductive hydrogel rod used to light up a light-emitting diode (LED) circuit, with plot (Right) showing both mechanical and electrical properties are restored after complete incision. (From Wirthl, D. et al., *Sci. Adv.*, 3, e1700053, 2017; Figures 1 & 5.) (c) Schematics (i), (ii), and (iii) illustrating the disconnection and reconnection of a simple electronic circuit using an EGaIn self-healing stretchable wire, with corresponding plots reflecting the recovery of the electrical conductivity and mechanical elasticity of the self-healing polymer. (From Palleau, E. et al., *Adv. Mater.*, 25, 1589–1592, 2013; Figure 1.)

time use (non-reproducible healing), whereby the material loses its self-healing properties after the first cycle of healing. Dickey et al. also worked on self-healing stretchable wires to form complex microfluidic networks [86]. The stretchable wires are prepared by injecting EGaIn into microchannels of self-healing polymer. After complete mechanical detachment, the stretchable wires self-heal mechanically and electrically, in a simultaneous fashion (Figure 10.5c) [86].

Findings incorporating both mechanical force perception and repeatable self-healing characteristics show potential to mimic natural skin's functionality. To extend the durability of any electronic interface in wearable applications, it is desired to fully restore both mechanical and electrical damage of the material. Not confined to a skin platform alone, stretchable self-healing materials (insulators, conductors, semiconductors) are further desired in complementary skin systems, such as stretchable displays and energy devices, expected to undergo significant deformation [65]. Further developments are still needed to improve the reliability of the healing process over multiple uses, while maintaining the material's electrical properties under simultaneous stretching and self-healing processes.

10.4 Beyond Human Skin Perceptions

10.4.1 Multi-functional Sensory Platform

Artificial skins also incorporate biological sensors for the recognition of physiological elements, such as humidity. Inspired by the porous nature of human skin that enables transpiration, flexible cellulose paper displays similar characteristics [18]. Although paper is not elastic, its porosity shows to be advantageous for improving the recovery time of integrated humidity sensors (due to a faster evaporation rate) [18,87] and preventing overheating of the electronic skin through a natural venting mechanism. Surpassing the functionalities of human skin, electronic skins that integrate biosensors can identify different physiological elements in sweat, such as Na^+, Ca^{2+}, K^+, Cl^-, and pH, selectively detecting and distinguishing between diverse elements in a fluid [88]. Since this functionality is not intrinsic to human understanding, it is an ambiguity how the brain would identify, interpret, and react to the information sent from sensors carrying sensations not typically recognized by the human body.

A multi-functional platform renders the E-skin smarter. The ability to recognize and distinguish sensations in a simultaneous fashion improves the functionality per cost of the artificial skin. The ability to perceive multiple external sensations from one single sensor structure can relief space restrictions as well as reduce cost. A bimodal capacitive structure integrated into a paper skin [18] uses a low-cost off-the-shelf compressive sponge and aluminum foil, and is shown to have a distinct dual functionality. Traditionally, the structure acts as a highly sensitive tactile/force sensor with a sensitivity of 0.16 kPa^{-1}, highlighted by an increase in the signal in response to external stimuli [18]. When operated at high frequencies above 200 kHz, the sensor doubles as a proximity sensor, where the response is translated into a decrease in the output signal [18,89]. However, selectivity to multiple stimuli without cross-sensitivity is an on-going challenge.

An alternative is to develop 3D stacked artificial skins, superimposing arrays of distinct sensors into one platform that can sense multiple sensations. Each array of sensors reflects one functionality, and each pixel of the array can collectively perceive a range of external stimuli. Kim et al. show a multi-functional smart prosthetic that can measure temperature, pressure, strain, and humidity via the overlap of passive matrices [90] (Figure 10.6a). In this case, sensor density for each layer is not the same, and consequently there is a lack of overlap and direct correlation between the different sensory information perceived per pixel of the E-skin. Someya et al. show a large-scale integration of pressure and temperature sensor arrays for a conformable bionic skin using a net-shaped structure [38,91] (Figure 10.6b and c). However, when it comes to integrating an even larger number of functionalities, the net-shaped structure is limited with reduced spatio-temporal resolution; the pixel area becomes bigger to accommodate more sensors and hence pixel density is reduced. Another design shows the out-of-plane 3D integration of paper-based pressure, temperature, and humidity sensors to build a flexible paper skin [18] (Figure 10.6d). Pixel density for all three layers is matched with a direct overlap. This design is advantageous for a more precise and localized mapping of all the distinct functionalities per pixel.

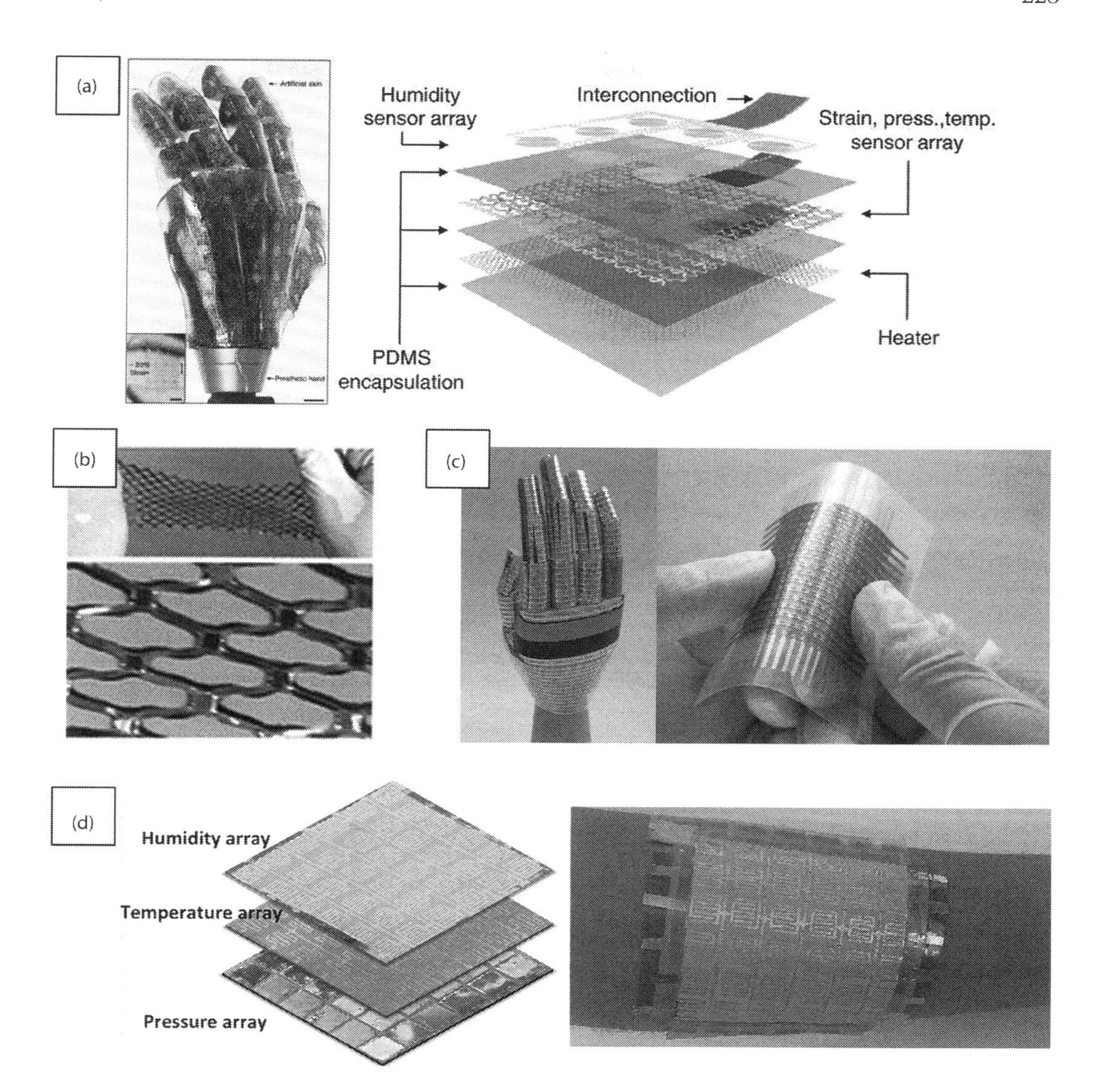

FIGURE 10.6 Multifunctional E-skin. (a) Prosthetic skin based on SiNR electronics, integrating arrays of stretchable sensors and actuators in out-of-plane layers, and covering the entire surface of a prosthetic hand. (From Kim, J. et al., *Nat. Commun.*, 5, 5747, 2014; Figure 1.) (b) Stretchable net-shaped skin structure enabling integrated organic transistors and pressure-sensitive rubber to mechanically extend by 25%. (From Someya, T. et al., *Proc. Natl. Acad. Sci. USA*, 102, 12321–12325, 2005; Figure 1.) (c)Large-scale Bionic skin integrating arrays of conformable flexible pressure and temperature sensors. (From Someya, T., *IEEE Spectrum*, 50, 50–56, 2013; Figure 1.) (d) Schematic of paper skin integrating in 3D arrays of pressure, temperature, and humidity sensors made out of off-the-shelf flexible and recyclable materials. (From Nassar, J.M. et al., *Adv. Mater. Technol.*, 1, 1600004, 2016; Figure 2.)

10.4.2 Transparent Electronic Skin

The semi-transparent or transparent property of flexible and/or stretchable materials (insulators, semiconductors, and conductors) can further extend the field of applications of artificial skins in interactive wearable optoelectronics, such as stretchable displays [92] and biomedical imaging (photodetectors [PDs]) integrated onto prosthetics [42]. As an example, Bao et al. demonstrate the fabrication of a large-scale stretchable and transparent E-skin, integrating 64 pixels of single-walled carbon nanotubes (SWCNTs)/PDMS-based strain and pressure sensors [44] (Figure 10.7a). Transparency is a characteristic of the optimized ratios between the SWCNTs and the intrinsically transparent PDMS. Using this composite material, a strain-insensitive pressure sensor array is demonstrated for reliable functionality under bending conditions [59]. Optical pressure sensors are also developed with high selectivity to humidity and temperature [11], targeting applications in interactive pressure-sensitive and curved displays for robotics, biomedical, and even automobile industries. Recently, a stretchable and bimodal transparent E-skin was developed using a combination of graphene derivatives [93]. The multi-functional E-skin incorporates three sensor matrices laminated on top of one another through a layer-by-layer process and integrates humidity, temperature, and pressure sensing capability (Figure 10.7b). The sensors monitor with high sensitivity and selectivity a variety of daily sensations, such as fi The touch, air flow temperature, and breathing [93]. In a more recent report, graphene tribotronics were introduced for a transparent E-skin [94]. Materials such as

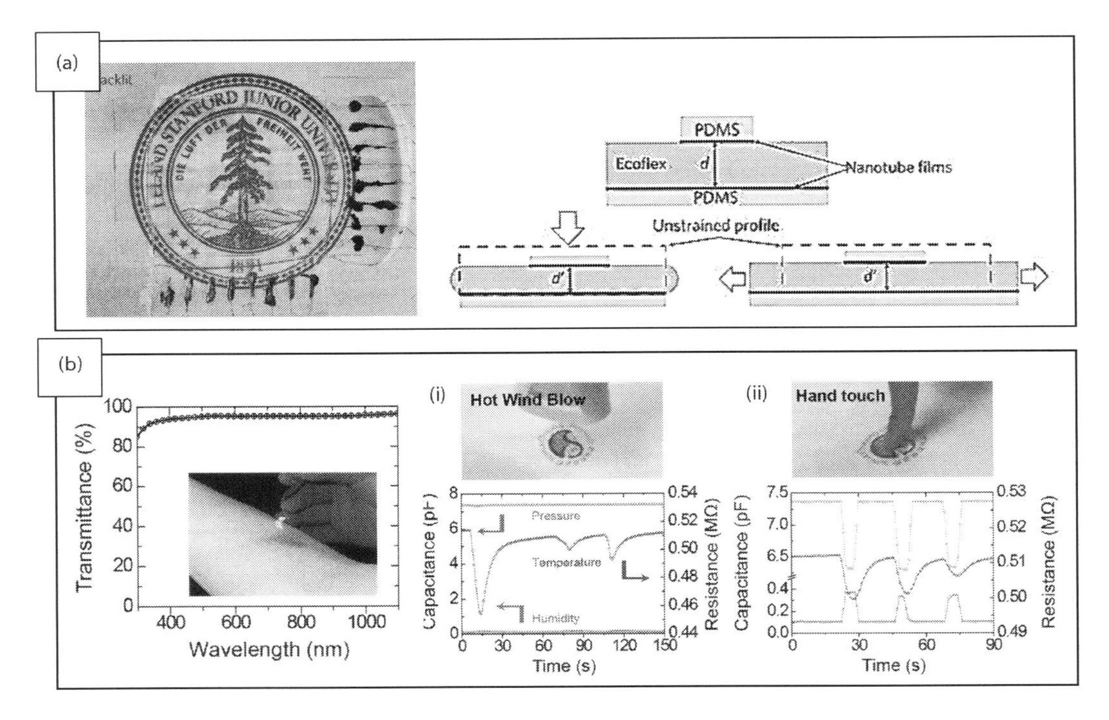

FIGURE 10.7 Transparent E-skin. (a) (Left) 64-pixel array of compressible pressure and strain sensors using SWCNTs/PDMS for transparent electrodes. (Right) Schematics showing a stretchable capacitive pressure sensor, before straining and after being placed under normal pressure and lateral stretching/straining. (From Lipomi, D.J. et al., *Nat. Nanotechnol.*, 6, 788–792, 2011; Figures 3 & 5.) (b) Transparent all-graphene E-skin shows high optical transmittance above 90% for light spectrum range $\lambda = 300$ nm to $\lambda = 1100$ nm (left). E-skin simultaneously monitors temperature, pressure and humidity with high selectivity under two different stimuli: (i) hot wind blowing and (ii) hand touching (right). (From Ho, D.H. et al., *Adv. Mater.*, 28, 2601–2608, 2016; Figures 1 & 4.)

PDMS, indium tin oxide (ITO), graphene, and ionic gels were used to integrate single-electrode-mode triboelectric nanogenerators (S-TENG) and graphene field-effect transistors (FETs) and demonstrate an active matrix of tactile sensors [94].

10.4.3 Responsive E-Skin

Allowing the artificial skin to take action in response to certain external stimuli and sensory perceptions can augment the quality of a prosthetic, and hence of life. As human beings, our biological system as a whole acts as a combination of separate, yet complementary mechanisms and units that work together in a simultaneous fashion, enabling us to perceive the world around us the way that we do. This 'whole' system is a blend of three main elements: (i) detecting external stimuli through our biological receptors/sensors, (ii) which is only perceived and recognized because of the thought process happening in our nervous system (i.e., our brain). This is when we are able to distinguish between different stimuli, give them meaning, categorize them, and feel whether something is hot or cold for instance. (iii) And accordingly, our body reacts and responds through apparent changes on the surface of our biological skin, such as dark spots, wrinkles, inflammation, redness, burns, bumps, etc. As an example, when you dip your finger in boiling water, not only will you feel that the water is hot, but in response to that sensation, your skin will slowly suffer from inflammation, and ultimately a scar. Human skin alters its physical properties depending on external circumstances, whether it's a visual burn or microscopic changes in the pores resulting from dry weather.

Efforts to mimic these functions entail the incorporation of actuation mechanisms within the electronic skin. This can be accomplished through the integration of actuators such as heaters, but also through the use of smart materials that undergo a shape, color, or electrical alteration in reaction to a variation in the environment. External stimuli, such as temperature, light, force, and current, can cause a modification in the electrical, mechanical, or optical properties of these smart materials [95]. This transformation is repeatable and reversible. Shape-memory alloys/polymers (SMAs) undergo a reversible large deformation in response to temperature or stress variations (pseudoelasticity); pH-sensitive polymers undergo a shape change (volume expansion or compression) when the pH of the surrounding medium is altered; chromogenic polymers (electrochromic, photochromic, and thermochromic) change color in response to electrical, optical, or thermal changes; self-healing materials can restore mechanical and electrical properties of the locally induced crack; and dielectric elastomers (DEs) can stretch by up to 500% upon the exertion of an external electric field [95]. As an example, Bao et al. demonstrated stretchable SWNT-based pressure and strain sensors over large surface areas for a visually responsive E-skin [96]. The chameleon-inspired interactive electronic skin uses an electrochromic polymer allowing the platform to change colors in response to distinct applied pressures (Figure 10.8).

Ionic conductors and conductive composite polymers based on 1D materials such as NWs and CNTs are generally preferred when the devices are to be actuated by the movement of ions [2]. Hybrid conductors based on dielectric elastomers, electrostrictive polymers, and liquid crystal elastomers have been engineered with conductive nanocomposites to work as thermal and electric actuators, targeting skin-based therapeutic applications [97,98]. In this case, CNTs and NWs are desired as they display high response to external stresses while maintaining mechanical reliability under bending, twisting, and stretching [23].

10.4.4 Self-Powered E-Skin

A persistent challenge in any electronic device is power. Battery life can be quite challenging, and the need to recharge on a daily basis can be impractical. A common objective shared across the board of mobile electronics, including electronic skins, is the extended supply of power. Increased battery

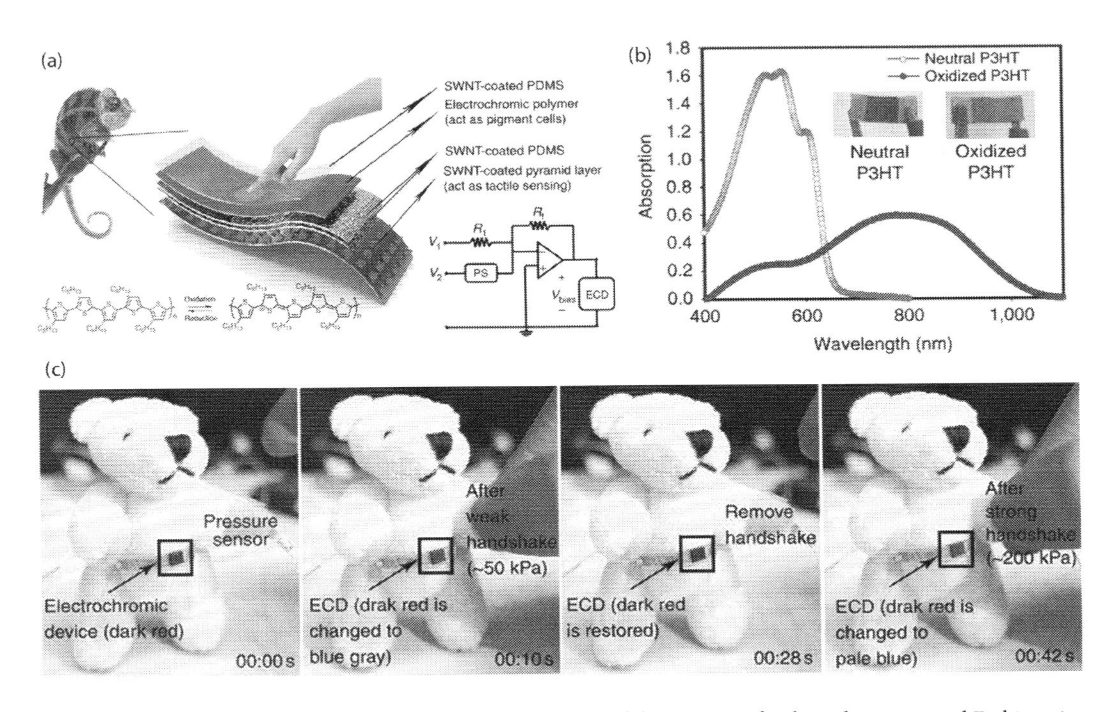

FIGURE 10.8 Responsive interactive E-skin. (a) Illustration of the concept of a chameleon-inspired E-skin using electrochromic polymer poly(3-hexylthiophene-2,5-diyl, P3HT) and SWNT-coated PDMS for pressure sensing. (b) Optical properties of the stretchable polymer-based ECDs, displaying the UV-VIS absorption spectra of the neutral and oxidized states of P3HT. (c) Sequential images of a teddy bear, translating tactile stimuli into visible color changes as a response to various applied pressures. (From Chou, H.-H. et al., *Nat. Commun.*, 6, 8011, 2015; Figures 1, 5, & 7.)

lifetime can be attained by minimizing the device's power consumption, but also by integrating additional energy harvesting techniques such as solar power, thermal power, piezoelectric energy, or even a hybrid of these processes. The energy generated can be then used in conjunction with a battery to recharge the battery on a continuous basis. Generally, the process of collecting data from sensors and transmitting through various communication processes requires microwatts to milliwatts of power [30]. In electronic skin applications, the multi-functional active matrices are going to be more power demanding and not sufficiently addressed with a simple battery integration. Since skin developments aim for large-scale integrations, they are not constrained by surface area, rather by form factor (need for compliancy). Furthermore, an E-skin is constantly exposed to external stimuli from the outside world, and hence provides great potential for the integration of conformal energy storage and harvesting mechanisms from environmental factors such as heat and light, but also from mechanical forces induced from daily physical movements of body parts, such as twisting of the joints on the elbows, knees, ankles, and shoulders (Figure 10.9). In a nutshell, self-powered artificial skins can be used to harvest energy from the mechanical and thermal energy produced by the human body. For example, the human body dissipates heat that can be possibly translated using thermoelectric effect into several watts of stored energy [30].

Much effort has been invested into the development of flexible and stretchable energy generators such as mechanical energy harvesters, solar cells, and supercapacitors [42]. Thermal and mechanical

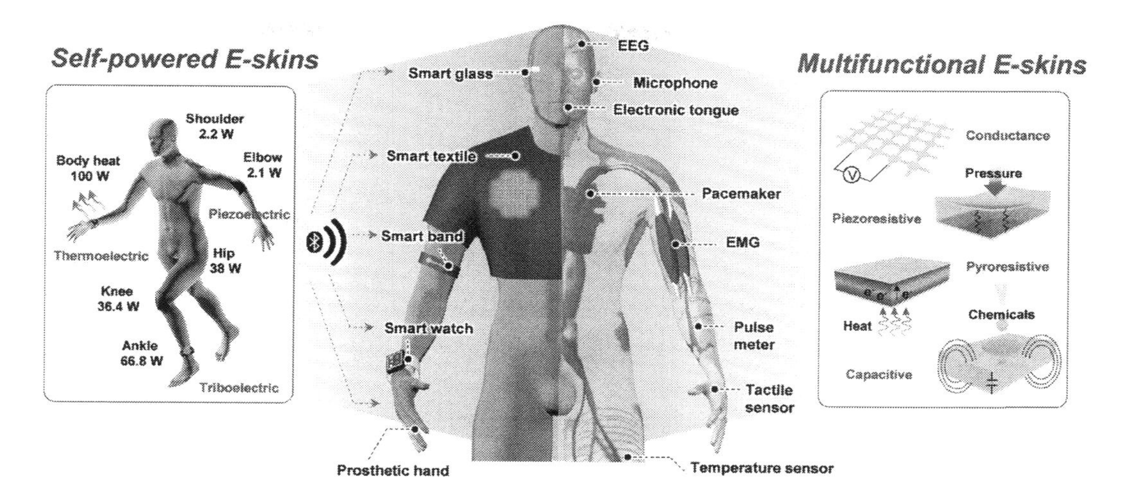

FIGURE 10.9 Self-powered vs. multifunctional E-skins. Concept figure illustrating the possible functionality of self-powered and multifunctional electronic skins. Self-powered E-skins based on piezoelectric, triboelectric, and thermoelectric power harvest mechanical and thermal energy from distinct points of the human body. (From Park, J. et al., *J. Mater. Chem. B*, 4, 2999–3018, 2016; Figure 1.)

energy can be scavenged using different energy harvesting mechanisms that use piezoelectric [30,99], thermoelectric [100–102], pyroelectric [103], and triboelectric [56,104] devices. In a self-powered system, a pressure sensor can use the piezoelectric and triboelectric effect to transform applied mechanical force into an electrical signal. This can be referred to as a bimodal sensor; the device acts as a pressure sensor, but can also act as an energy harvester. However, triboelectric sensors display some drawbacks such as vulnerability to humidity, abrasion resistance, and limitation in detection of biosignals because two counterpart substrates need to effectively interact with one another [105]. An optically transparent and stretchable strain sensor platform integrates a stretchable supercapacitor and a triboelectric nanogenerator for an ultra-low-power and self-powered system that monitors skin strain (Figure 10.10a) [72]. To further improve the high performance of self-powered electronic skin platforms, bio-inspired designs are adopted, integrating interlocked micro and nanostructures. A high-performance E-skin integrating energy harvesting piezoelectric device is presented, whereby the active material takes the structure of interlocked microdomes of piezoelectric PVDF and reduced graphene oxide (rGO) composites [29,30]. Another study implements micropatterned piezoelectric generators for a high-performance self-powered E-skin [106]. Piezoelectric P(VDF-*co*-trifluoroethylene) (P(VDF-TrFE) films are patterned in the form of triangles and pyramids, and the generators display more than five times increase in voltage output compared to thin film non-structured piezoelectric generators [106] (Figure 10.10b). Self-powered pressure sensors that integrate sponge-like mesoporous PVDF structures use the piezoelectric effect to generate electrical energy generation from rainwater and wind flow (Figure 10.10c) [107].

Sunlight is another widely available green source of power. Solar cells with large surface area are desired to maximize light collection, and hence electrical output generated. Rogers et al. demonstrated a stretchable energy harvesting platform based on solar cells. GaAs solar cells are integrated onto rigid islands of the platform and interconnected with one another via stretchable freestanding metal interconnects. When the platform is stretched by ~20%, the devices convert solar power into electrical energy with an efficiency of ~13% [108] (Figure 10.11). However, the high cost accompanied

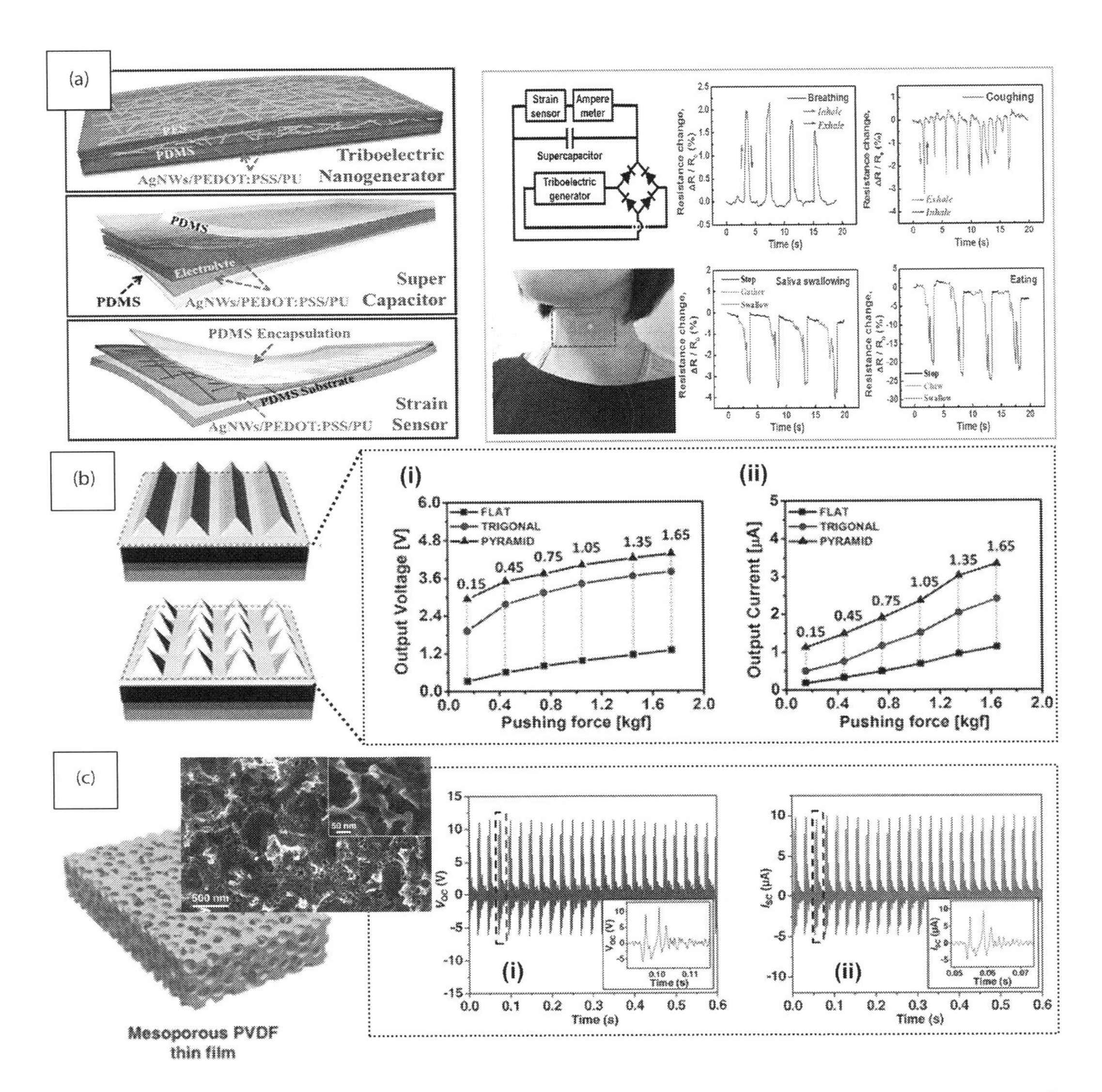

FIGURE 10.10 Self-powered E-skin designs. (a) Self-powered transparent E-skin integrating a strain sensor and a supercapacitor (SC) powered by a triboelectric nanogenerator (TENG). (Right) Schematic description of the TENG, SC, and strain sensor. Monitoring of strain caused by muscle movement for functions of the trachea and esophagus, such as breathing, coughing, swallowing, and eating. (From Hwang, B.-U. et al., *ACS Nano*, 9, 8801–8810, 2015; Figures 2 & 4.) (b) Schematic trigonal line-shaped (top) pyramid-shaped (bottom) P(VDF-TrFE)-based PNGs. (i) Output voltage and (ii) current response, comparing sensor sensitivity from the flat, trigonal line-shaped, and pyramid-shaped P(VDF-TrFE)-based PNGs in response to various applied normal pressures. (From Lee, J.H. et al., *Adv. Funct. Mater.*, 25, 3203–3209, 2015; Figures 1 & 4.) (c) Structure of mesoporous PVDF thin films. (i) Voltage and (ii) current output of the PVDF NG under continuous surface oscillation. Insets show the output curve features during one cycle of surface oscillation. (From Mao, Y. et al., *Adv. Energy Mater.*, 4, 1301624, 2014; Figures 1 & 2.)

with the GaAs limits its application and practicality in large-area E-skins. To consider stretchable solar cells as an economical option for energy harvesting, the efficiency per cost per area of the device needs to be high, and although many attempts have been made to use stretchable organic photovoltaics (OPVs) [109], they display limited efficiencies [110], that ultimately render them expensive for large-scale implementation in E-skins.

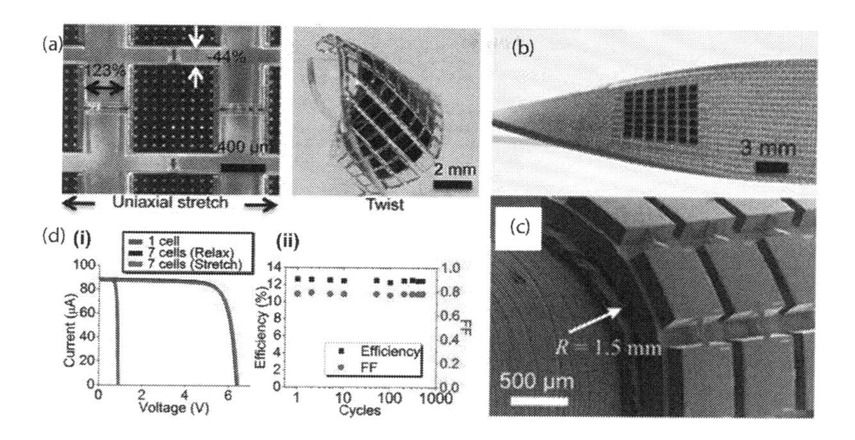

FIGURE 10.11 Solar power energy harvesters. (a) GaAs solar cells—Optical microscope images of stretchable GaAs photovoltaic module under 20% uniaxial strain and twisting. (b) Digital photograph of the solar cell array. (c) SEM image of the module wrapped on a cylinder with a radius of 1.5 mm. (d) (i) Current–voltage characteristics of a single microcell in the array (blue) and of seven microcells with series interconnections in relaxed (black) and stretched (red) states. (ii) Solar cell efficiency and fill factor (FF) under mechanical cycling tests, revealing no degradation in performance of the module under more than 500 cycles. (From Lee, J. et al., *Adv. Mater.*, 23, 986–991, 2011; Figures 2 & 4.)

10.5 System-Level Integration for a Compliant E-Skin

10.5.1 Hybrid Integration of Active Electronics on Stretchable Substrates

In order to process and communicate tactile data, as well as receive corresponding feedback all on one skin-like platform, the stretchable electronic skin design needs to be integrated alongside the embedded interface electronic components. As we have seen in the previous sections, advances in electronic skins have inspired the development of smartly engineered stretchable soft materials and conductive hybrid elastomers [10,42,100,111–113] in the goal of augmenting skin capabilities. However, a recurrent challenge is the accompanied rigid and bulky electronics and connectors to achieve a completely autonomous skin system. Some approaches adopted the development of active components using intrinsically compliant materials such as organics and conductive elastomers. Flexible and/or stretchable thin film transistors (TFTs) and FETs have been demonstrated using indium gallium zinc oxide (IGZO) [114], graphene-NW conductive hybrid elastomers [115], and organic-based materials [38,46,54]. However, the fabrication process of these approaches is limited by integration and scalability, and the output device performance is not yet sufficient to process and manage the large amount of data required for multi-sensory E-skin platforms and biomimetic signals [36]. Low-power and high-performance electronics are necessary to meet the demands of high-speed electronics and large data management accompanied with a large-scale and multi-functional E-skin matrix. Silicon electronics display a high yield advantage and unmatched performance per cost, making them the core of today's digital world. Therefore, a preferable approach would be the hybrid integration of rigid state-of-the-art active components with stretchable E-skin sensors. The stretchable platform would adopt spring-like interconnects coupling together discrete islands as a stress-relief design [116–119] (Figure 10.12a). Bare dies of Si interface components and processing units (e.g., analog to digital converters, metal-oxide-semiconductor field-effect transistors (MOSFETs), etc.) are thinned down and tightly bonded onto the islands through a flip-chip mechanism [120] (Figure 10.12b). The serpentine metal connectors render the platform stretchable while maintaining mechanical reliability and stability under deformation. The embedded unpackaged dies are then encapsulated with a soft packaging, adopting a core/shell encapsulation [120,121] (Figure 10.12c), rendering the

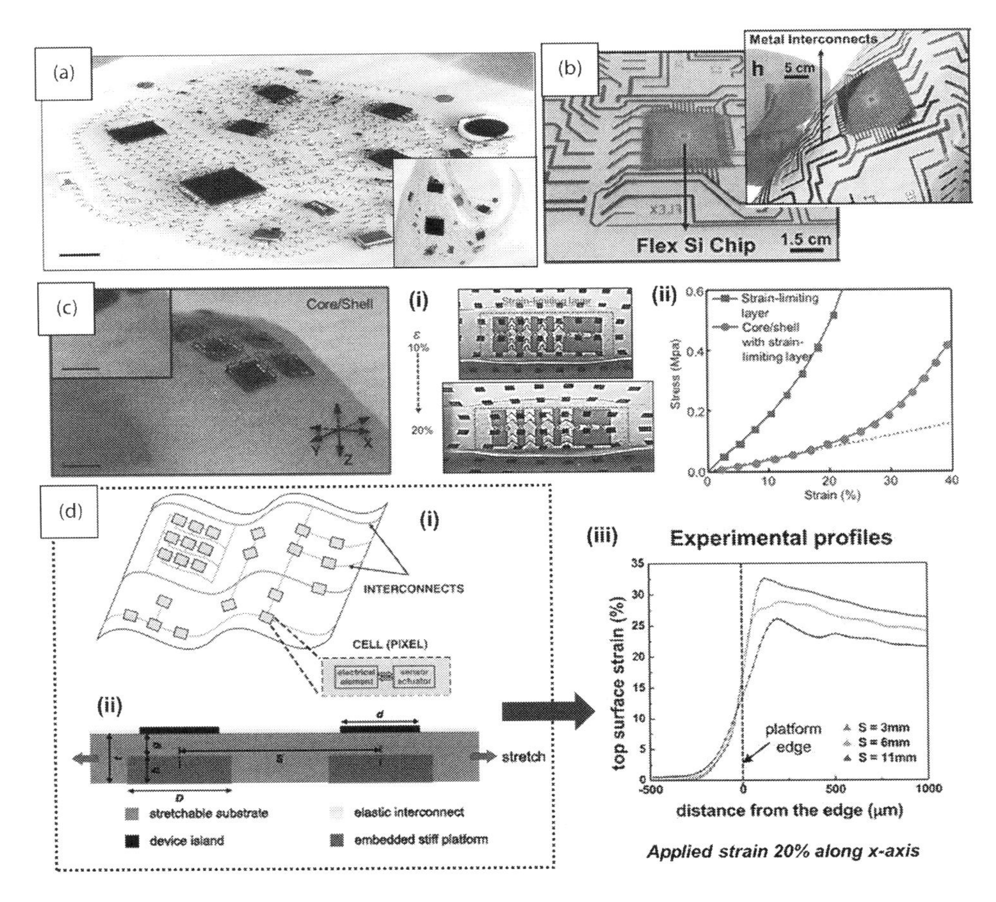

FIGURE 10.12 Design strategies for flexible/stretchable E-skin system integration. (a) 3D network of helical coils as electrical interconnects for stretchable electronic systems, integrating ~250 3D helices, ~500 bonding sites, ~50 component chips. (From Jang, K.-I. et al., *Nat. Commun.*, 8, 15894, 2017; Figure 2). (b) Digital images of a bonded 61 pin flexible silicon chip on a thin flexible Kapton sheet using sputtered conformal metal interconnects. (From Sevilla, G.A.T. et al., *Adv. Mater. Technol.*, 2, 1600175, 2017; Figure 4.) (c) Optical images of the stretchable system using a core/shell elastomer package (scale bar, 1 cm), (i) displaying a strain-limiting layer at 20% strain. (ii) Experimental results for the strain–stress response of a strain-limiting layer only and of the core/shell structure with the strain-limiting layer. (From Lee, C.H. et al., *Adv. Funct. Mater.*, 25, 3698–3704, 2015; Figures 4 and 5.) (d) (i) Concept schematic of a conformable electronic system on stretchable substrate. The cells are interconnected with flexible and stretchable metallization. (From Wagner, S. et al., *Physica E*, 25, 326–334, 2004; Figure 1.) (ii) Cross-sectional view of the alterative design, whereby rigid ICs are embedded within the stretchable substrate via cavities. Electronic devices are then manufactured directly onto the flat, elastomer surface, above the rigid ICs, and interconnected with stretchable metal/wires. (iii) Strain profiles of the later structure taken along the *x*-axis from the center of the SU8 platform under 20% applied strain. (From Romeo, A. et al., *Appl. Phys. Lett.*, 102, 131904, 2013; Figures 1 & 3.)

electronics resistant to moisture and wear-tear cycles when worn on the body. This thickness of the core/shell (made out of Silbione and Ecoflex) encapsulation is controlled in such a way the shear and normal stress is reduced at the core under 20% strain [121] (Figure 10.12c). Stresses are concentrated at the perimeter of the core/shell encapsulation, but no effect on the central portion that contains the electronics. Stephanie Lacour Group also worked on a similar process whereby stiff active surface mount device (SMD) components or bare dies (such as MOSFETs) are embedded and bonded within soft stretchable substrates to achieve completely comfortable electronic skin systems [122,123] (Figure 10.12d), in which the islands containing the electronics are isolated from strain up to 20% stretching.

10.5.2 Heterogeneous and CMOS-Compatible Flexible Silicon E-Skin

Although the approach of hybrid integration of active state-of-the-art components onto soft polymeric substrates is promising, a lot still needs to be done to achieve the robustness and efficiency required. One of the main limitation currently faced is the delamination of the SMD components or bare dies from the stretchable platform, for strains above 20%. Furthermore, integrating separate ICs (even when thinned down) with the E-skin onto a host polymeric platform with stable and efficient metallization interconnects between the islands and the ICs is a challenging task that is still under development [120]. To mitigate this issue, one possible approach is the monolithic integration of artificial skin onto the same flexible/stretchable inorganic Si platform hosting the circuit electronics [43]. Advances in transforming physically rigid silicon electronics into mechanically flexible and even stretchable platforms have been demonstrated using low-cost flexing and design approaches [21,37,124] (Figure 10.13). Mechanically flexible and semi-transparent sub-20 nm Si

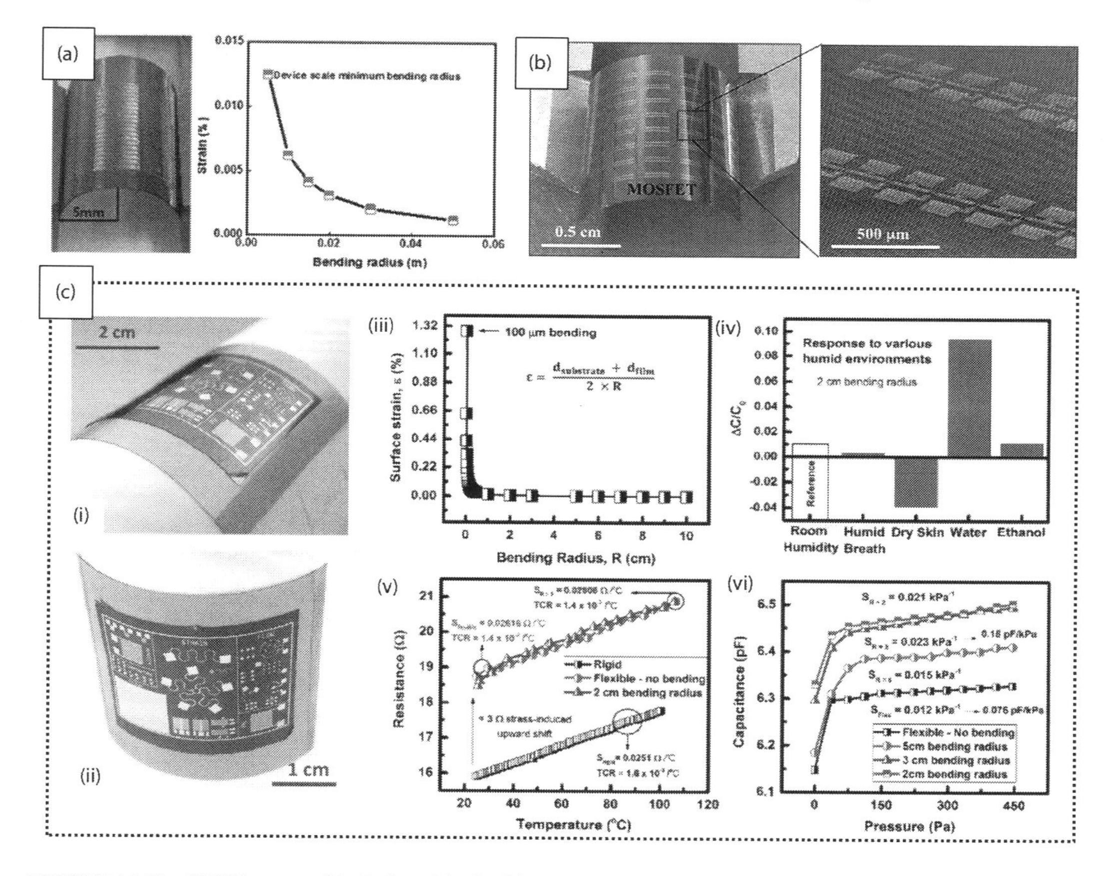

FIGURE 10.13 CMOS compatible E-skin. (a) Flexible FinFET silicon fabric at minimum bending radius of 5 mm, with nominal strain calculation for different bending radius (50, 30, 20, 15, 10, 5 mm), showing minimal strain induced. (From Sevilla, G.A.T. et al., *Adv. Mater.*, 26, 2794–2799, 2014; Figure 4.) (b) Flexible semi-transparent Si (100) MOSFETS using low-cost trench-protect-release mechanism. (From Rojas, J.P. et al., *ACS Nano*, 8, 1468–1474, 2014; Figure 3.) (c) (i) and (ii) Photographs of flexible and bendable Si (100) E-skin with multisensing capabilities. (iii) Bending-induced strain percentage with respect to different bending radii, calculated for a flexible a 20 μm thick Si (100) substrate. (iv) Differential humidity response to different percentage of relative humidity levels. (v) Temperature and (vi) pressure sensitivity plots before flexing, after flexing, before bending, and under tensile bending conditions. (From Nassar, J.M. et al., A CMOS-compatible large-scale monolithic integration of heterogeneous multi-sensors on flexible silicon for IoT applications, in *Electron Devices Meeting (IEDM), 2016 IEEE International*, 2016, 18.6.1–18.6.4; Figures 1, 3, 5, & 6.)

FinFETs [125] (Figure 10.13a), flexible and low-cost Si (100) MOSFETs [126] (Figure 10.13b), and high-κ/metal gate complementary metal-oxide semiconductor (CMOS) circuit elements (such as inverters) integrated on a flexible Si (100) substrate [120,124] have shown great promise with electrical stability under various mechanical deformations, opening doors to highly scalable and high-performance multi-sensory devices. E-skin sensors generally incorporate soft elastomeric materials, and hence a heterogeneous and CMOS-compatible approach was recently developed to enable the on-chip integration of E-skin sensors (strain, humidity, temperature, and pressure) onto a flexible Si (100) platform [43] (Figure 10.13c). A flexible E-skin built using CMOS compatible approaches can eradicate integration and interconnects challenges, increase yield and scalability, and guarantee a truly high speed and efficient management of the continuous data flow required for biomimetic applications and Internet of Everything wearables.

10.6 Applications of Artificial Skin

10.6.1 Robotics and Prosthetics

An artificial skin is a hub of multiple sensors that can selectively and simultaneously map external stimuli in real-time. As the name suggests, an E-skin is mainly used as a second skin for robotics and prosthetics [10,89,113], as it enables a more realistic feel and an augmented user interaction. Patients with prostheses covered with an E-skin will be able to feel and act in response to surrounding events in a more intimate fashion; a smart prosthetic hand will be capable to simultaneously resolve the heat, moisture, and force emanating from another person's touch [10,89], as well as perform complex and delicate tasks such as hand shaking and holding a cup; accredited to the heightened senses delivered by the E-skin.

Specifically for robotics and prosthetics applications, a multi-functional artificial skin must be capable of distinguishing between a variety of mechanical and environmental stimuli with high sensitivity, accuracy, and selectivity; enabling distinctive perceptions for dynamic, static, and shear stresses, the recognition of surface texture/roughness, as well as various environmental changes such as moisture, heat, and light. The major challenge is to achieve high selectivity, high mechanical reliability, and without crosstalk between multiple signals [127]. A stretchable prosthetic skin based on silicon nanoribbons (SiNRs) has been developed by Kim et al. to simultaneously and selectively identify temperature, strain, pressure, and humidity from the surroundings (Figure 10.14a) [64]. Electrical nerve stimulation is roughly demonstrated via multi-electrode arrays (MEAs); this becomes significant to communicate information collected from the prosthetic sensors to the brain for further processing. In a prosthetic, distinct sensors can play a collective role in determining a certain sensation or feel. For instance, humidity sensors alone provide us with a sense of moisture, but thermal actuators along with force sensors, together can deliver a more realistic warm sense of touch [30]. An example of an electronic skin implemented in robotic applications is demonstrated by Gao et al., taking use of rGO to achieve highly sensitive piezoresistive strain sensors [128]. Distinct robotic movements are distinguished in real-time by selectively perceiving multiple deformations such as strain, normal force, bending, and torsion (Figure 10.14b). Texture perception also plays an important role in robotics and prosthetics. Park et al. show a microstructured ferroelectric skin capable of discriminating between static/dynamic pressure and temperature stimuli for selective discrimination of various surface textures [29]. Lipomi and colleagues further demonstrated the ability to discriminate surface chemistry by touch [129]. They use a skin-inspired piezoresistive and conductive fluoroelastomer to build strain sensors embedded into a wearable glove. The piezoresistive strain sensors integrated along with accelerometers allow distinct tactile feedback for different textures, capable of discriminating between hydrophobic and hydrophilic surfaces, enabled through induced force vibrations and velocity changes [129].

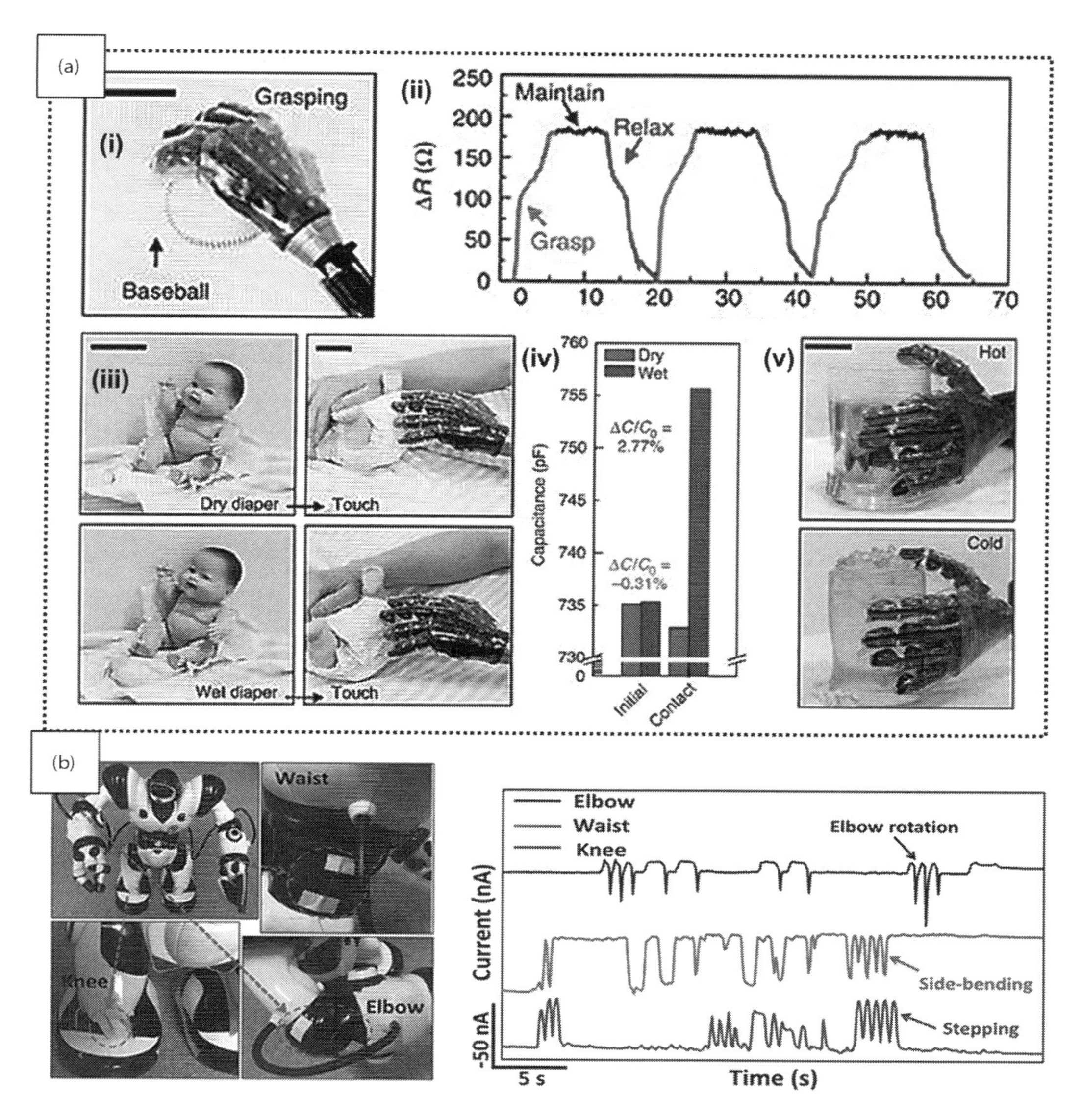

FIGURE 10.14 E-skin in prosthetics and robotic applications. (a) SiNR-based E-skin in various situations of daily lives. (i) An image of the prosthetic hand grasping a baseball, and (ii) corresponding plot showing temporal resistance change of the pressure sensors, in response to dynamic motions of the prosthetic hand, such as grasping and relaxing. (iii) Images of prosthetic hand touching the dry and wet diaper of a baby doll. (iv) Corresponding bar plot of the humidity sensor's response in before and after touching the dry (red) and wet (blue) diaper. (v) Prosthetic hand wrapped with SiNR E-skin, sensitive to temperature recognition when touching a hot versus ice cold drink. (From Kim, J. et al., *Nat. Commun.*, 5, 5747, 2014; Figure 5.) (b) Images of a robot wearing fiber strain sensors at movable joints (elbow, waist, and knee). The sensors are capable of detecting the robot's dance movements to 'Gangnam Style,' reflected in the response curves of plot in real-time. (From Cheng, Y. et al., *Adv. Mater.*, 27, 7365–7371, 2015; Figure 5.)

10.6.2 Biomedical Devices

Wearable health monitors are getting increased popularity for future personalized advanced healthcare. Flexible sensors available in electronic skins are used to continuously and simultaneously track vital signals as means of prevention and non-invasive medical diagnosis. Advanced healthcare and wellness applications implicate the integration of wearable and seamless sensors to monitor a person's physiological and biological signs; such as body temperature, heartbeat variability, respiration rate, and blood pressure. For example, arterial pulse pressure reflects real-time heart rate and arterial blood pressure variations. The extracted information can be then studied in a timely fashion to link any sudden unexpected vital variations to potential cardiovascular health problems, such as hypertension, diabetes, and arteriosclerosis [87]. Assessment of breathing variability and respiration rate are also crucial to indicate the healthy function of the lungs and sleep quality of the user [130,131]. Physical movement and motion of the user can be also used to correlate the physiological signs to the user's physical health and activity.

Today, emerging flexible and stretchable sensors are implemented in biomedical devices to ensure continuous monitoring of vital signs in a non-invasive way [33,57,132–136]. Strain and pressure sensors seamlessly placed on the surface of the body are used to collect information such as heart rate, blood pressure, respiration rhythm, physical activity, as well as skin strain and hydration. Materials such as graphene [137], NWs [138], and CNTs [139] have been adopted to produce stretchable pressure and strain sensors to monitor distinct body vitals. A stretchable and biocompatible E-skin developed by Rogers et al. can also be interfaced on biological organs such as the heart and brain, as a means to guide surgical procedures and improve signal output from disease monitoring [140,141]. Used on the surface of the skin, the piezoelectric compliant modulus E-skin can identify skin lesion, as well as blood pressure when in contact with an artery around the wrist or neck (Figure 10.15a) [141]. An example of a self-powered flexible piezoelectric pulse sensor based on lead zirconate titanate (PZT) thin film is used for real-time healthcare monitoring system [105]. The epidermal piezoelectric sensor enables complete conformal attachment on a rugged skin to respond to the faint and fast human pulses. The self-powered pulse sensor is demonstrated to enable *in vivo* measurement of radial/carotid pulse signals in near-surface arteries, as well as monitor breathing patterns (Figure 10.15b). For a more complete healthcare monitoring system, the BioStamp (MC10™) is a skin-mounted biosensing patch that integrates multi-modal sensing capability with wireless communication, in a compact design of 2.7 mm in thickness. Vital signs from the surface of the human body can be monitored from the integrated tri-axis accelerometer, gyroscope, and temperature sensor. Interface system electronics such as Bluetooth low-energy (BLE) and low-energy battery are connected to the sensors via the stretchable island/serpentine design [121]. Various medical waveforms can be extracted such as secure service gateway (SSG), electrocardiogram (ECG), and photoplethysmogram (PPG), important for assessing medical human dynamic metrics.

The majority of these E-skin developments can be quite expensive, rendering these health monitors unaffordable for the mainstream population. One key challenge is to deliver advanced, yet simple healthcare monitors in the most low-cost and accessible manner. A step towards accessibility is demonstrated through an ultra-low-cost paper health monitor using a flexible paper-based artificial skin (paper skin) wrapped around the wrist [87]. The non-functionalized platform integrates pressure, temperature, and humidity sensors, and can resolve with high sensitivity and selectivity various body vitals such as body temperature, heart rate, arterial stiffness, blood pressure, and skin hydration [87] (Figure 10.15c). A step forward towards personalized biomedical devices is represented by a non-invasive glucose monitor patch, for a skin-like wearable diabetes treatment (Figure 10.15d) [142]. The patch integrates multi-sensors that analyze humidity, pH, glucose, and temperature levels. The wearable E-skin glucose monitor also integrates a drug delivery mechanism *via* microneedles and thermal actuators for accurate transdermal drug delivery control.

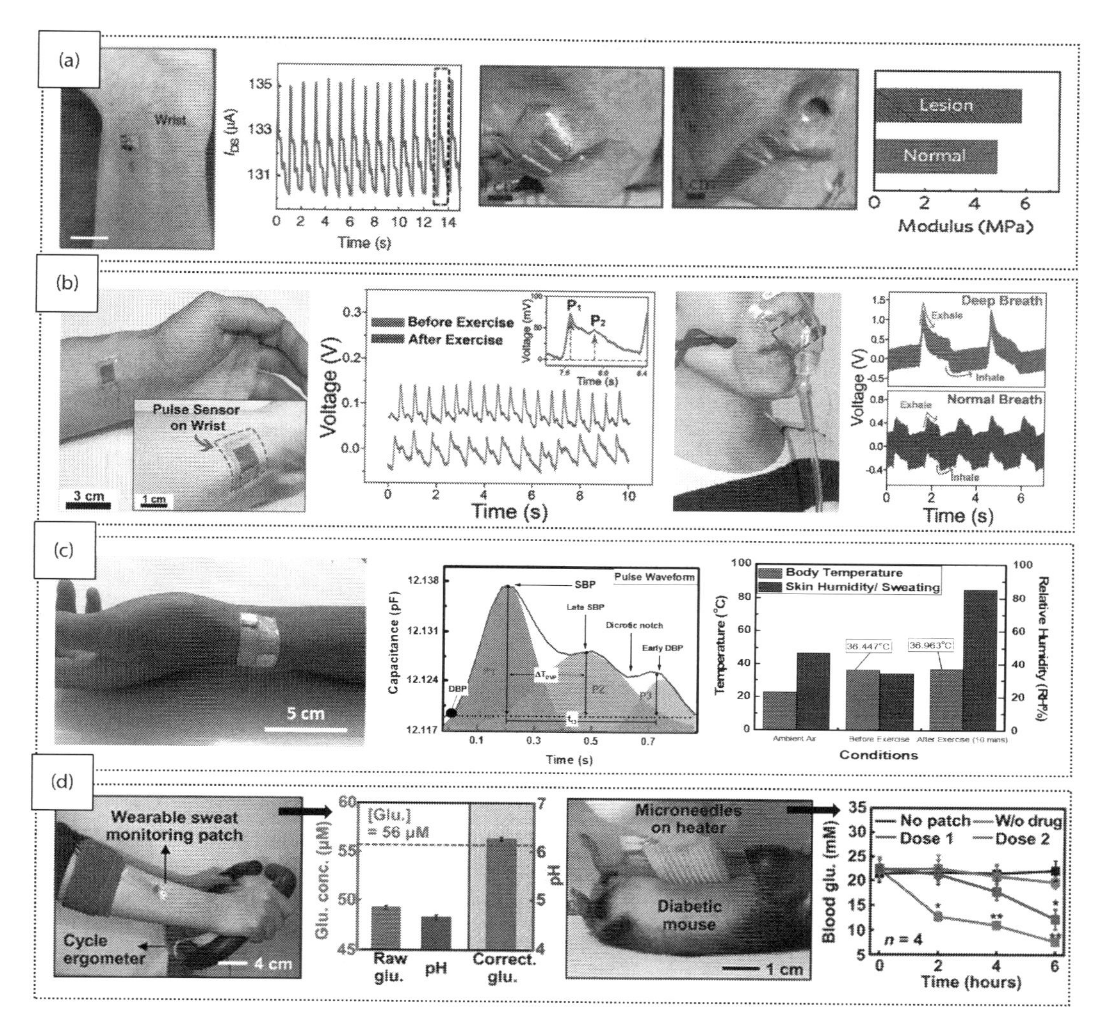

FIGURE 10.15 E-skin for advanced healthcare and biomedical devices. (a) Piezoelectric compliant E-skin measures blood pressure from the wrist through the detection of pressure pulse waveform; Detection of skin lesion from near the nose as shown in the images of skin with lesion (left) and under normal conditions (right). (From Dagdeviren, C. et al., *Nat. Mater.*, 14, 728–736, 2015; Figure 4.) (b) Photograph of a piezoelectric self-powered pulse sensor conformally attached on human wrist using a biocompatible liquid bandage, with corresponding radial artery pulse signals reflecting different heart rates and before and after physical exercise (left); Inset indicates a magnified view of one pulse after exercise, highlighting peaks of pulse pressure (P_1) and late systolic augmentation (P_2); (Right) Photograph of medical mask integrated with a flexible pressure sensor for monitoring periodic human respiration modes, as reflected in the corresponding plot. (From Park, D.Y. et al., *Adv. Mater.*, 29, 1702308, 2017; Figure 3.) (c) Image of a wearable paper-based health monitoring system 'Paper skin' for real-time body vitals detection; (Middle) Plot showing the detection of the radial artery pulse waveform as the Paper Skin is placed around the wrist, with high distinction of the P1, P2, and P3 peaks from which blood pressure and artery information can be extracted; (Right) Plot showing the simultaneous and selective detection of body temperature and skin hydration from the surface of the skin around the wrist. (From Nassar, J.M. et al., *Adv. Mater. Technol.*, 2, 1600228, 2017; Figures 2 and 4.) (d) Digital image of the wearable sweat monitoring patch placed on the subject's arm as he is on a cycle ergometer, with corresponding plot showing the detection of raw sweat glucose, pH levels, and corrected sweat glucose level using the averaged pH; (Right) optical photograph of a transdermal drug delivery device placed inside a diabetic mouse; corresponding plot reflects the detection of blood glucose levels of the diabetic mice for the treated groups (microneedles with the drugs) and control groups (without the patch, microneedle without the drugs). (From Lee, H. et al., *Sci. Adv.*, 3, e1601314, 2017; Figure 4.)

10.7 Restoring Skin Sensations in Neuroprosthetics

10.7.1 Translating Biomimetic Data

To enable amputees to feel their surroundings through a smart prosthetic, electrical data extracted from the sensors of the E-skin must be communicated to the brain, processed, and then translated into information that our human body can comprehend. Sensors signals are accordingly amplified and transmitted to different parts of the nervous system in order to stimulate distinct functions [64]. The process of transducing stimuli from the artificial skin sensors to the peripheral nerves of the brain entails sequential steps of: (i) signal collection from the transducers, (ii) signal encoding, (iii) signal transmission, and finally (iv) neural interfacing [1] (Figure 10.16a). Electrical stimulation was initially adopted to mimic action potentials, with pulse-like waveforms that enable safe stimulation of the tissue. But to directly interface an electronic skin with the nervous system, electrical output from the sensors have to be comprehensible

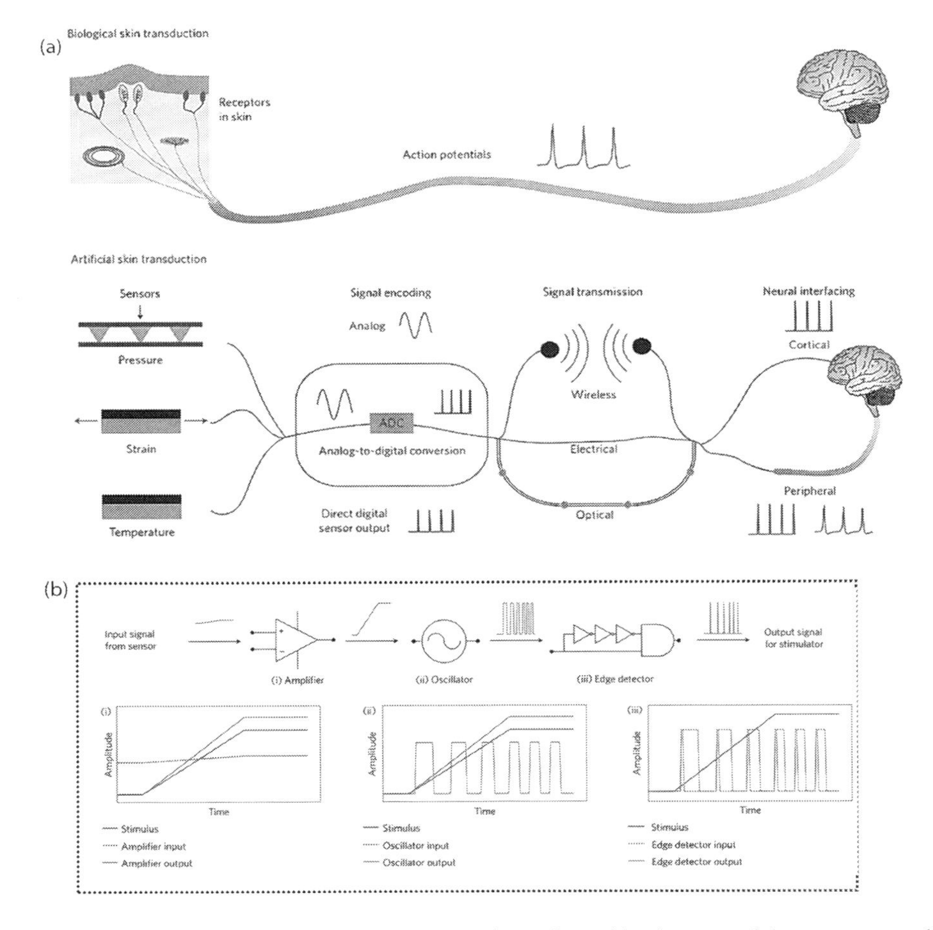

FIGURE 10.16 Translating biomimetic data using neural interfaces. (a) Schematic of the steps required for the transduction of sensory stimuli from biological (top) or artificial (bottom) receptors in the brain, from electrical signals collected from sensors mimicking action potentials, to transmission, and finally neural interfacing; the illustrated steps are key steps that need to be addressed to enable sensations in prosthetic limbs. (From Chortos, A. et al., *Nat. Mater.*, 15, 937, 2016; Figure 1.) (b) A schematic for the conversion of the input signal into electrical spikes, requiring an amplifier (i), an oscillator (ii), and an edge detector (iii). Below plots reflect the input and output signals from each stage of the signal processing circuit. (From Chortos, A. et al., *Nat. Mater.*, 15, 937, 2016; Figure 4.)

to the nervous system; this requires on-board signal modification through the integration of flexible readout interfaces to enable data sampling and transmission [1].

Readout electronics are as important as the development of the sensory arrays for the artificial skin; they form an integral part for collecting data and making sense out of them. Sensations experienced by our brain are tuned by the magnitude of stimulation received from the sensors; therefore, amplifiers are needed to alter the intensity of the stimulation from the sensor array [1]. The amplified analogue signals collected from the E-skin sensors are then converted into digitized frequency signals via a ring oscillator (Figure 10.16b). Nevertheless, to maintain a large-scale compliant skin approach, the complementary interface circuits need to be also designed in such a way to comply with the flexibility and stretchability of the skin. System elements for signal readout and addressing such as CMOS transistors, analog to digital (A/D) converters, amplifiers, and oscillators need to be thin and flexible enough, without compromising their sensitivity, efficiency, and mechanical reliability. Accordingly, flexible and/or stretchable electrical components have been developed, such as oxide and CNT-based devices [114,143], silicon nanomembrane ICs [144], and high-performance CMOS circuit elements on low-cost flexible silicon (100) [120,124,145]. Bao and colleagues adopted newly engineered polymers and organic materials to develop the corresponding mechanically soft skin electronics [146–148]. Stretchable amplifiers and transistors can be developed using stretchable engineered and electroconductive PEDOT:PSS-based hydrogels integrated with 'Utah' microelectrode arrays, for stable, mechanically soft, and biocompatible *in vivo* implantable electrode neural interfaces. With current and on-going advances in flexible and stretchable electronics, there is potential in assembling a conformal neural modulator that can record, convert, and transmit E-skin sensations into the nervous system [1].

10.7.2 Neural Interfacing for Restoring Skin Perception

To deliver the described biomimetic signals to the nervous system, a stable and cell-specific interface is crucial; however, this idea is rather problematic as we have limited understanding of how neural coding works in order to translate electronic information into perception. Still, efforts in brain-machine interfaces are being developed using different strategies and focusing on a conformable form factor to increase signal to noise ratio. Stephanie Lacour and colleagues aim at overcoming the biological mismatch between traditional rigid MEAs and neural tissues by developing flexible and stretchable MEAs on plastic. The neural interface platform is designed to monitor and stimulate *in vitro* and *in vivo* extracellular neuronal activity (Figure 10.17a) [149]. A/D converters and amplification electronics are also integrated within the stretchable platform to ensure a comfortable and compliant brain interfacing platform. Bao et al. are also working on digital tactile (DiTacts) systems, integrating artificial mechanoreceptors that combine pressure sensors with ring oscillators, whereby stimulation is communicated to the brain via optogenetic stimulation to translate touch perception (Figure 10.17b) [150].

The three main methods adopted to interface sensors to the central and peripheral nervous system implement electrical [151–156], optical [157–159], and magnetic stimulations [160,161]. In electrical stimulations, low impedance of MEAs is essential to ensure effective charge insertion into the peripheral nerves [162]. Surface MEAs are used to directly stimulate residual human skin reinnervated by peripheral nerves. For example, texture discrimination can be enabled by interfacing the peripheral nerves using a tactile sensor integrated with intrafascicular electrodes [163]. Platinum nanowires (PtNWs) alongside ceria nanoparticles can be used to achieve stretchable MEAs that can be prepared by growing [64] (Figure 10.18a). Peripheral nerve stimulations for pressure sensing and other electrophysiological parameters were recorded from the ventral posterolateral nucleus (VPL) of a rat's brain [64] (Figure 10.18a). Rogers et al. show high-density microelectrocorticography (ECoG) electrode arrays integrated with flexible Si nanomembranes (SiNMs) transistors for active multiplexing capability and local amplification, enabling the *in vivo* mapping of cat's brain auditory cortex's activity with higher spatial resolution (Figure 10.18b) [164].

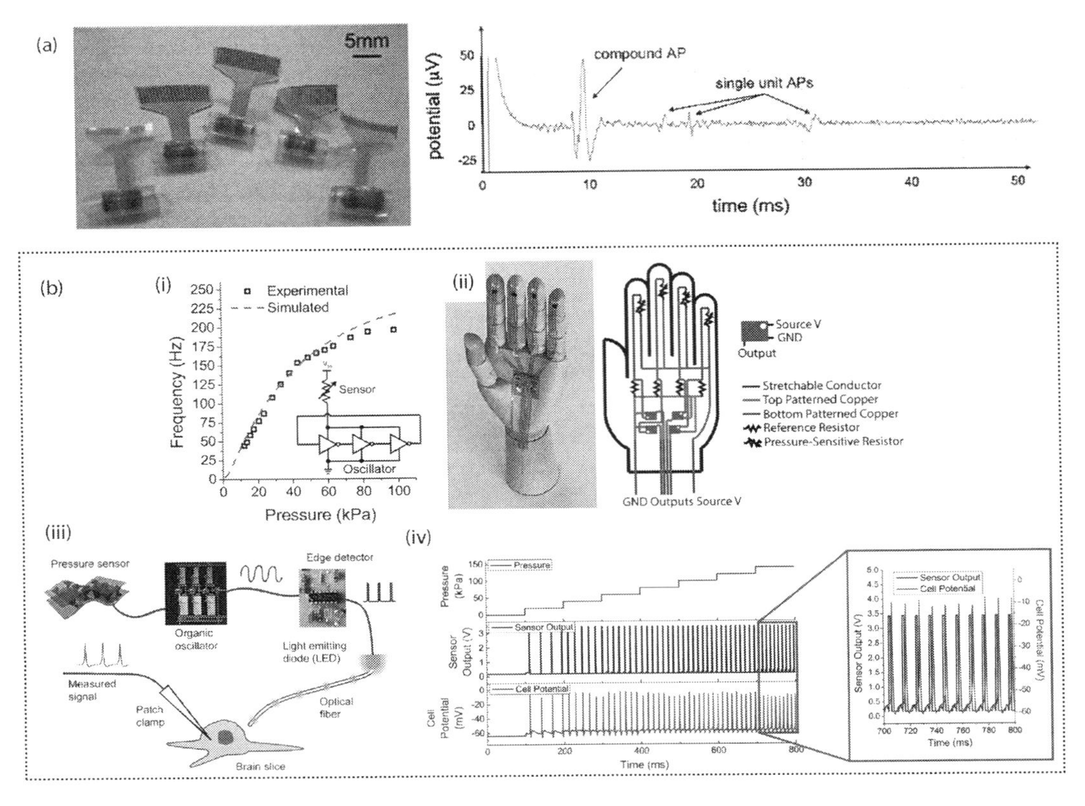

FIGURE 10.17 Neural interface designs. (a) (Left) Flexible micro-electrode arrays (MEA) accommodating ~190 micro-channel; (Right) plot of *in vivo* nerve activity recorded by a rolled MEA at 12 weeks post-implantation, whereby the sensory signals were generated by stimulating the rat's toe with needle electrodes. (From Lacour, S.P. et al., *Med. Biol. Eng. Comput.*, 48, 945–954, 2010; Figure 5.) (b) (i) Plot of frequency output as a function of pressure for the DiTact sensor; (ii) image of DiTact sensors attached on a model hand, integrating pressure-sensitive tactile elements, an organic ring oscillator, and stretchable interconnects. (From Tee, B.C.-K. et al., *Science*, 350, 313–316, 2015; Figure 2.) (iii) Setup of the optoelectronic stimulation system for pressure-dependent neuron stimulation; (iv) correlation between pressure, sensor output, and cell potential signals. The expanded section shows that the neurons' action potentials closely follow the pulses from the LED. (From Tee, B.C.-K. et al., *Science*, 350, 313–316, 2015; Figure 3.)

Alternatively, optical stimulations (i.e., optogenetics) are less invasive and are currently the preferable route for future developments, whereby light pulses penetrate the skin and excite the corresponding neurons for responsive stimulation. *In situ* pressure senses were interpreted into biomimetic optical signals by interfacing fiber optics and skin-inspired organic digital mechanoreceptors to the somatosensory neurons in the brain (Figure 10.17b) [150]. Rogers et al. also worked on compliant implantable electronics for stimulating, inhibiting, and monitoring neural dynamics in the deep brain through optogenetics [165,166]. Tethered fiber optics can be still quite invasive, and are affected by animal movement, leading to data artifacts. Therefore, wireless technology was adopted, demonstrating the development of fully stretchable, implantable, and wireless (antenna frequency 2.3 GHz) cellular-scale light-emitting diodes (LEDs) for insertion into the deep brain for optogenetic studies of neural processes in freely moving mice [165,166]. Figure 10.18c shows the flexible battery-free near-field wireless optoelectronic system used for subdermal implants in various optogenetics applications such as brain stimulation or even spinal opto-stimulations [167,168]. As for magnetic stimulation, coils are employed to generate magnetic fields. The field lines penetrate the skin barrier in a non-invasive manner and induce an electric current for neurons stimulation [1].

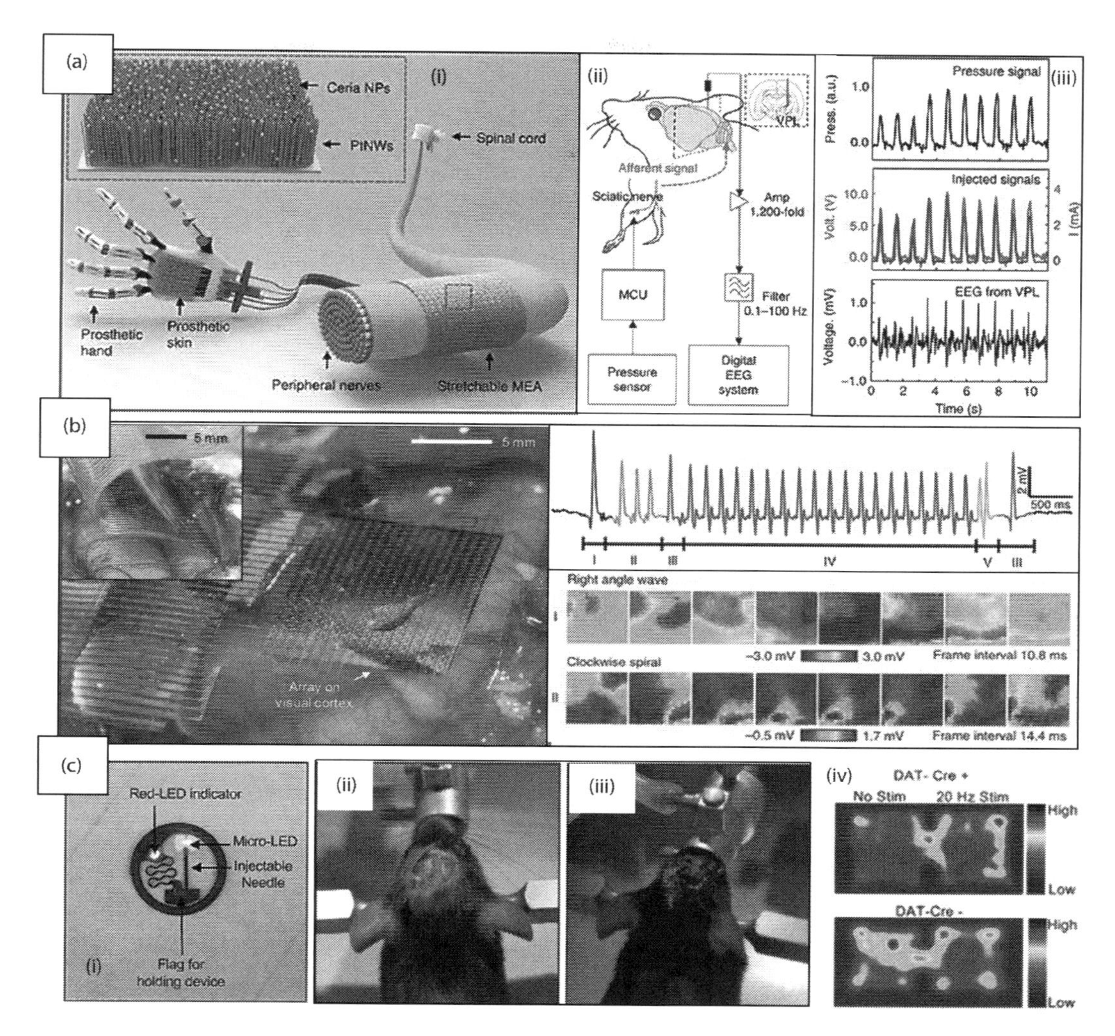

FIGURE 10.18 Emerging technologies for *in vivo* nervous system stimulation. (a) (i) An illustration displaying an approach for interconnecting the PtNWs/ceria NPs-based prosthetic skin to the peripheral nervous fibers using a stretchable MEA; (ii) schematic drawing of the experimental setup for peripheral nerve stimulations in the brain of a rat, for pressure signals and electrophysiological signal recordings from the ventral posterolateral nucleus (VPL) of the thalamus in the right hemisphere; (iii) plots displaying that measured signals from the pressure sensors (top) of the E-skin simultaneously applied voltage to nerves (middle), with corresponding EEG responses recorded from the VPL of the thalamus in the right hemisphere (bottom). (From Kim, J. et al., *Nat. Commun.*, 5, 5747, 2014; Figure 6.) (b) (Left) Digital photo of a compliant and high density active electrode array placed on the visual cortex. (From Viventi, J. et al., *Nat. Neurosci.*, 14, 1599–1605, 2011; Figure 2.) (Right) Data plot from electrographic seizures in feline neocortex, displaying the μECoG signal from one channel of the electrode array during a short electrographic seizure; corresponding movie frames (bottom) show varied spatial-temporal μECoG voltage patterns from all 360 electrodes during the labeled time intervals. (From Viventi, J. et al., *Nat. Neurosci.*, 14, 1599–1605, 2011; Figure 5.) (c) (i) Digital image of the flexible and wireless optoelectronic implantable device for optogenetic stimulation. (ii) and (iii) Images of implanted device before suturing skin and after suturing the skin of the rat. (From Shin, G. et al., *Neuron*, 93, 509–521.e3, 2017; Figure 4.) (iv) Heatmaps showing real-time mouse behavior in response to a 20 Hz photostimulation, for mice expressing ChR2 in the NAc of DAT-Cre+ and DAT-Cre–. (From Shin, G. et al., *Neuron*, 93, 509–521.e3, 2017; Figure 6.)

Comparing the available techniques, optical-based stimulations are desired because they are cell-specific, display high stimulation efficiency, and they avoid artifacts typically found in electrical stimulations. However, the method is still non-trivial for translation into clinical trials on human [169]. Electrical-based stimulation techniques are currently being tested on human subjects. However, because of the high skin resistance, they require high voltages for actuation; both electrical and magnetic stimulations also lack the required specificity and display undesirable low stimulation efficiencies and resolution [1]. Therefore, to restore natural skin perceptions and achieve an effective skin-inspired brain-machine interface for E-skins, the ongoing challenge is the fabrication of a more advanced skin-inspired bioelectronic interface, capable of mitigating the difficulties of interfacing electronic skin platforms with a network of neurons with sufficient cell-specificity and high spatiotemporal resolution.

10.8 Concluding Remarks

The development of an artificial skin is quite a challenging and multi-dimensional task that requires complex advances in a diversity of fields, such as flexible and/or stretchable transducers, biocompatible and biodegradable materials, compliant integrated interface circuit components, stable and efficient stretchable, and soft brain-machine neural interfaces; which in turn necessitates the discovery and engineering of a new set of materials, device architectures, and integration strategies. Correspondingly, progress in multi-functional and compliant artificial skins shows promising results for getting a step closer towards mimicking natural human skin function and sensations. Many tend to focus on the traditional thinking of solely realizing highly sensitive sensors, coupled with fast response and recovery times. However, for large-scale multi-functional electronic skin developments, additional parameters that remain critical for their practical implementation in robotics, prosthetics, and diagnostic devices, are; selectivity to stimuli, wide dynamic ranges, high spatiotemporal resolution *viz.* increased sensors density, increased mechanical robustness and electrical durability, compatible integration with state-of-the-art interface and processing electronics, and accuracy in data collection and translation. Other desirable properties currently being implemented include self-healing, transparency, and self-powering. However, questions are still raised when it comes to the practical functionality of these elementary energy harvesters in such large-scale and complex compliant systems. Restoring sensations and comprehension of the external world through a prosthetic goes beyond data collection from sensors. The human brain needs to understand and interpret the meaning of the random electrical signals gathered by the sensors; only then can the artificial skin be responsive. Current attempts in stable neural interfaces between the E-skin and the nervous system are restricted by the number of input channels that the neural interfaces provide, and the limited understanding of neural coding; however, at this point, any successful transduction can considerably improve the function of a prosthetic. A closing topic to recall is the ultimate integration for an autonomous E-skin system. Given the high speed and complex tasks required by a neural-interfaced prosthetic or E-skin in a wearable multi-functional monitoring system, silicon-based electronics will be of absolute necessity for data management. In this case, a hybrid heterogeneous approach is envisioned whereby heterogeneous materials for soft skin-like transducers could be integrated in a stable and robust way to the state-of-the-art low-cost silicon ICs. Si CMOS technology is by far the most reliable, comprehensive, high speed, and efficient technology that exists. Transforming CMOS electronics into flexible and stretchable ones and interfacing them with soft artificial skin platforms could be the leading technology for efficient and safe brain machine interfacing and biomimetic data translation.

References

1. A. Chortos, J. Liu, and Z. Bao, "Pursuing prosthetic electronic skin," *Nature Materials*, vol. 15, p. 937, 2016.
2. B. T. Nghiem, I. C. Sando, R. B. Gillespie, B. L. McLaughlin, G. J. Gerling, N. B. Langhals, et al., "Providing a sense of touch to prosthetic hands," *Plastic and Reconstructive Surgery*, vol. 135, pp. 1652–1663, 2015.

3. R. S. Johansson, and Å. B. Vallbo, "Tactile sensibility in the human hand: Relative and absolute densities of four types of mechanoreceptive units in glabrous skin," *The Journal of Physiology*, vol. 286, pp. 283–300, 1979.
4. D. Filingeri, and R. Ackerley, "The biology of skin wetness perception and its implications in manual function and for reproducing complex somatosensory signals in neuroprosthetics," *Journal of Neurophysiology*, vol. 117, pp. 1761–1775, 2017.
5. R. C. L. Suen, K. T. Chang, M. P.-H. Wan, W. Y. Chua, and Y. C. Ng, "Mobile and sensor integration for increased interactivity and expandability in mobile gaming and virtual instruments," in *Proceedings of the 2015 Annual Symposium on Computer-Human Interaction in Play*, 2015, pp. 703–708.
6. D. Kogias, E. Michailidis, S. Potirakis, and S. Vassiliadis, "Communication protocols for vital signs sensors used for the monitoring of athletes," in V. Koncar (Ed.), *Smart Textiles and Their Applications*, pp. 127–144. Woodhead Publishing, Cambridge, UK, 2016.
7. W.-P. Hu, B. Zhang, J. Zhang, W.-L. Luo, Y. Guo, S.-J. Chen, et al., "Ag/alginate nanofiber membrane for flexible electronic skin," *Nanotechnology*, vol. 28, p. 445502, 2017.
8. L. Gao, Y. Zhang, V. Malyarchuk, L. Jia, K.-I. Jang, R. C. Webb, et al., "Epidermal photonic devices for quantitative imaging of temperature and thermal transport characteristics of the skin," *Nature Communications*, vol. 5, p. 4938, 2014.
9. X. Huang, Y. Liu, H. Cheng, W. J. Shin, J. A. Fan, Z. Liu, et al., "Materials and designs for wireless epidermal sensors of hydration and strain," *Advanced Functional Materials*, vol. 24, pp. 3846–3854, 2014.
10. S. Park, H. Kim, M. Vosgueritchian, S. Cheon, H. Kim, J. H. Koo, et al., "Stretchable energy-harvesting tactile electronic skin capable of differentiating multiple mechanical stimuli modes," *Advanced Materials*, vol. 26, pp. 7324–7332, 2014.
11. M. Ramuz, B. C. K. Tee, J. B. H. Tok, and Z. Bao, "Transparent, optical, pressure-sensitive artificial skin for large-area stretchable electronics," *Advanced Materials*, vol. 24, pp. 3223–3227, 2012.
12. M. Segev-Bar, A. Landman, M. Nir-Shapira, G. Shuster, and H. Haick, "Tunable touch sensor and combined sensing platform: Toward nanoparticle-based electronic skin," *ACS Applied Materials & Interfaces*, vol. 5, pp. 5531–5541, 2013.
13. T. Sekitani, and T. Someya, "Stretchable, large-area organic electronics," *Advanced Materials*, vol. 22, pp. 2228–2246, 2010.
14. T. Takahashi, K. Takei, A. G. Gillies, R. S. Fearing, and A. Javey, "Carbon nanotube active-matrix backplanes for conformal electronics and sensors," *Nano Letters*, vol. 11, pp. 5408–5413, 2011.
15. K. Takei, T. Takahashi, J. C. Ho, H. Ko, A. G. Gillies, P. W. Leu, et al., "Nanowire active-matrix circuitry for low-voltage macroscale artificial skin," *Nature Materials*, vol. 9, p. 821, 2010.
16. N. T. Tien, S. Jeon, D. I. Kim, T. Q. Trung, M. Jang, B. U. Hwang, et al., "A flexible bimodal sensor array for simultaneous sensing of pressure and temperature," *Advanced Materials*, vol. 26, pp. 796–804, 2014.
17. C. Wang, D. Hwang, Z. Yu, K. Takei, J. Park, T. Chen, et al., "User-interactive electronic skin for instantaneous pressure visualization," *Nature Materials*, vol. 12, p. 899, 2013.
18. J. M. Nassar, M. D. Cordero, A. T. Kutbee, M. A. Karimi, G. A. T. Sevilla, A. M. Hussain, et al., "Paper skin multisensory platform for simultaneous environmental monitoring," *Advanced Materials Technologies*, vol. 1, p. 1600004, 2016.
19. P. Jenmalm, I. Birznieks, A. W. Goodwin, and R. S. Johansson, "Influence of object shape on responses of human tactile afferents under conditions characteristic of manipulation," *European Journal of Neuroscience*, vol. 18, pp. 164–176, 2003.
20. M. J. Adams, S. A. Johnson, P. Lefèvre, V. Lévesque, V. Hayward, T. André, et al., "Finger pad friction and its role in grip and touch," *Journal of The Royal Society Interface*, vol. 10, p. 20120467, 2013.
21. S. F. Shaikh, M. T. Ghoneim, G. A. T. Sevilla, J. M. Nassar, A. M. Hussain, and M. M. Hussain, "Freeform compliant CMOS electronic systems for internet of everything applications," *IEEE Transactions on Electron Devices*, vol. 64, pp. 1894–1905, 2017.

22. A. C. Cavazos Sepulveda, M. Diaz Cordero, A. Carreño, J. Nassar, and M. M. Hussain, "Stretchable and foldable silicon-based electronics," *Applied Physics Letters*, vol. 110, p. 134103, 2017.
23. J. M. Nassar, J. P. Rojas, A. M. Hussain, and M. M. Hussain, "From stretchable to reconfigurable inorganic electronics," *Extreme Mechanics Letters*, vol. 9, pp. 245–268, 2016.
24. Y.-Y. Noh, X. Guo, M. M. Hussain, Z. Ma, D. Akinwande, M. Caironi, et al., "Special issue on flexible electronics foreword," ed: IEEE-INST Electrical Electronics Engineers Inc, Piscataway, NJ, 2017.
25. J. P. Rojas, D. Singh, D. Conchouso, A. Arevalo, I. G. Foulds, and M. M. Hussain, "Stretchable helical architecture inorganic-organic hetero thermoelectric generator," *Nano Energy*, vol. 30, pp. 691–699, 2016.
26. F. De Boissieu, C. Godin, B. Guilhamat, D. David, C. Serviere, and D. Baudois, "Tactile texture recognition with a 3-axial force MEMS integrated artificial finger," in *Robotics: Science and Systems*, Seattle, WA, 2009, pp. 49–56.
27. K.-I. Jang, H. U. Chung, S. Xu, C. H. Lee, H. Luan, J. Jeong, et al., "Soft network composite materials with deterministic and bio-inspired designs," *Nature Communications*, vol. 6, p. 6566, 2015.
28. J. Scheibert, S. Leurent, A. Prevost, and G. Debrégeas, "The role of fingerprints in the coding of tactile information probed with a biomimetic sensor," *Science*, vol. 323, pp. 1503–1506, 2009.
29. J. Park, M. Kim, Y. Lee, H. S. Lee, and H. Ko, "Fingertip skin–inspired microstructured ferroelectric skins discriminate static/dynamic pressure and temperature stimuli," *Science Advances*, vol. 1, p. e1500661, 2015.
30. J. Park, Y. Lee, M. Ha, S. Cho, and H. Ko, "Micro/nanostructured surfaces for self-powered and multifunctional electronic skins," *Journal of Materials Chemistry B*, vol. 4, pp. 2999–3018, 2016.
31. Y. Zhang, Y. Hu, T. Zhao, P. Zhu, R. Sun, and C. Wong, "Highly sensitive flexible pressure sensor based on microstructured PDMS for wearable electronics application," in *Electronic Packaging Technology (ICEPT), 2017 18th International Conference on*, 2017, pp. 853–856.
32. R. Dykes, "Coding of steady and transient temperatures by cutaneous 'cold' fibers serving the hand of monkeys," *Brain Research*, vol. 98, pp. 485–500, 1975.
33. R. C. Webb, A. P. Bonifas, A. Behnaz, Y. Zhang, K. J. Yu, H. Cheng, et al., "Ultrathin conformal devices for precise and continuous thermal characterization of human skin," *Nature Materials*, vol. 12, pp. 938–944, 2013.
34. J. A. Rogers, T. Someya, and Y. Huang, "Materials and mechanics for stretchable electronics," *Science*, vol. 327, pp. 1603–1607, 2010.
35. M. Kaltenbrunner, T. Sekitani, J. Reeder, T. Yokota, K. Kuribara, T. Tokuhara, et al., "An ultra-lightweight design for imperceptible plastic electronics," *Nature*, vol. 499, pp. 458–463, 2013.
36. A. M. Hussain, and M. M. Hussain, "CMOS-technology-enabled flexible and stretchable electronics for internet of everything applications," *Advanced Materials*, vol. 28, pp. 4219–4249, 2016.
37. J. P. Rojas, A. Arevalo, I. Foulds, and M. M. Hussain, "Design and characterization of ultra-stretchable monolithic silicon fabric," *Applied Physics Letters*, vol. 105, p. 154101, 2014.
38. T. Someya, Y. Kato, T. Sekitani, S. Iba, Y. Noguchi, Y. Murase, et al., "Conformable, flexible, large-area networks of pressure and thermal sensors with organic transistor active matrixes," *Proceedings of the National Academy of Sciences of the United States of America*, vol. 102, pp. 12321–12325, 2005.
39. T. Yokota, Y. Inoue, Y. Terakawa, J. Reeder, M. Kaltenbrunner, T. Ware, et al., "Ultraflexible, large-area, physiological temperature sensors for multipoint measurements," *Proceedings of the National Academy of Sciences*, vol. 112, pp. 14533–14538, 2015.
40. T. Q. Trung, S. Ramasundaram, B. U. Hwang, and N. E. Lee, "An all-elastomeric transparent and stretchable temperature sensor for body-attachable wearable electronics," *Advanced Materials*, vol. 28, pp. 502–509, 2016.
41. R. S. Dahiya, G. Metta, M. Valle, and G. Sandini, "Tactile sensing—from humans to humanoids," *IEEE Transactions on Robotics*, vol. 26, pp. 1–20, 2010.

42. M. L. Hammock, A. Chortos, B. C. K. Tee, J. B. H. Tok, and Z. Bao, "25th anniversary article: The evolution of electronic skin (e-skin): A brief history, design considerations, and recent progress," *Advanced Materials*, vol. 25, pp. 5997–6038, 2013.
43. J. M. Nassar, G. A. T. Sevilla, S. J. Velling, M. D. Cordero, and M. M. Hussain, "A CMOS-compatible large-scale monolithic integration of heterogeneous multi-sensors on flexible silicon for IoT applications," in *Electron Devices Meeting (IEDM), 2016 IEEE International*, 2016, pp. 18.6.1–18.6.4.
44. D. J. Lipomi, M. Vosgueritchian, B. C. Tee, S. L. Hellstrom, J. A. Lee, C. H. Fox, et al., "Skin-like pressure and strain sensors based on transparent elastic films of carbon nanotubes," *Nature Nanotechnology*, vol. 6, pp. 788–792, 2011.
45. S. C. Mannsfeld, B. C. Tee, R. M. Stoltenberg, C. V. H. Chen, S. Barman, B. V. Muir, et al., "Highly sensitive flexible pressure sensors with microstructured rubber dielectric layers," *Nature Materials*, vol. 9, pp. 859–864, 2010.
46. Y. Zang, F. Zhang, D. Huang, X. Gao, C.-a. Di, and D. Zhu, "Flexible suspended gate organic thin-film transistors for ultra-sensitive pressure detection," *Nature Communications*, vol. 6, p. 6269, 2015.
47. J. Meyer, P. Lukowicz, and G. Troster, "Textile pressure sensor for muscle activity and motion detection," in *Wearable Computers, 2006 10th IEEE International Symposium on*, 2006, pp. 69–72.
48. L. Zhang, G. Hou, Z. Wu, and V. Shanov, "Recent advances in graphene-based pressure sensors," *Nano Life*, vol. 6, p. 1642005, 2016.
49. B. C. Tee, C. Wang, R. Allen, and Z. Bao, "An electrically and mechanically self-healing composite with pressure-and flexion-sensitive properties for electronic skin applications," *Nature Nanotechnology*, vol. 7, pp. 825–832, 2012.
50. F.-R. Fan, L. Lin, G. Zhu, W. Wu, R. Zhang, and Z. L. Wang, "Transparent triboelectric nanogenerators and self-powered pressure sensors based on micropatterned plastic films," *Nano Letters*, vol. 12, pp. 3109–3114, 2012.
51. R. Whatmore, "Piezoelectric and pyroelectric materials and their applications," in L. S. Miller and J. B. Mullin (Eds.), *Electronic Materials: From Silicon to Organics*, p. 283. Springer, Boston, MA, 1991.
52. J. Chun, K. Y. Lee, C. Y. Kang, M. W. Kim, S. W. Kim, and J. M. Baik, "Embossed hollow hemisphere-based piezoelectric nanogenerator and highly responsive pressure sensor," *Advanced Functional Materials*, vol. 24, pp. 2038–2043, 2014.
53. C. Dagdeviren, Y. Su, P. Joe, R. Yona, Y. Liu, Y.-S. Kim, et al., "Conformable amplified lead zirconate titanate sensors with enhanced piezoelectric response for cutaneous pressure monitoring," *Nature Communications*, vol. 5, p. 4496, 2014.
54. A. Spanu, L. Pinna, F. Viola, L. Seminara, M. Valle, A. Bonfiglio, et al., "A high-sensitivity tactile sensor based on piezoelectric polymer PVDF coupled to an ultra-low voltage organic transistor," *Organic Electronics*, vol. 36, pp. 57–60, 2016.
55. T. Sharma, K. Aroom, S. Naik, B. Gill, and J. X. Zhang, "Flexible thin-film PVDF-TrFE based pressure sensor for smart catheter applications," *Annals of Biomedical Engineering*, vol. 41, pp. 744–751, 2013.
56. Z. L. Wang, J. Chen, and L. Lin, "Progress in triboelectric nanogenerators as a new energy technology and self-powered sensors," *Energy & Environmental Science*, vol. 8, pp. 2250–2282, 2015.
57. C. L. Choong, M. B. Shim, B. S. Lee, S. Jeon, D. S. Ko, T. H. Kang, et al., "Highly stretchable resistive pressure sensors using a conductive elastomeric composite on a micropyramid array," *Advanced Materials*, vol. 26, pp. 3451–3458, 2014.
58. L. Pan, A. Chortos, G. Yu, Y. Wang, S. Isaacson, R. Allen, et al., "An ultra-sensitive resistive pressure sensor based on hollow-sphere microstructure induced elasticity in conducting polymer film," *Nature Communications*, vol. 5, p. 3002, 2014.
59. S. Lee, A. Reuveny, J. Reeder, S. Lee, H. Jin, Q. Liu, et al., "A transparent bending-insensitive pressure sensor," *Nature Nanotechnology*, vol. 11, pp. 472–478, 2016.

60. C. M. Boutry, A. Nguyen, Q. O. Lawal, A. Chortos, S. Rondeau-Gagné, and Z. Bao, "A sensitive and biodegradable pressure sensor array for cardiovascular monitoring," *Advanced Materials*, vol. 27, pp. 6954–6961, 2015.
61. J.-B. Chossat, Y.-L. Park, R. J. Wood, and V. Duchaine, "A soft strain sensor based on ionic and metal liquids," *IEEE Sensors Journal*, vol. 13, pp. 3405–3414, 2013.
62. Y. Kanda, "Piezoresistance effect of silicon," *Sensors and Actuators A: Physical*, vol. 28, pp. 83–91, 1991.
63. N. Hu, Y. Karube, C. Yan, Z. Masuda, and H. Fukunaga, "Tunneling effect in a polymer/carbon nanotube nanocomposite strain sensor," *Acta Materialia*, vol. 56, pp. 2929–2936, 2008.
64. J. Kim, M. Lee, H. J. Shim, R. Ghaffari, H. R. Cho, D. Son, et al., "Stretchable silicon nanoribbon electronics for skin prosthesis," *Nature Communications*, vol. 5, p. 5747, 2014.
65. S. J. Benight, C. Wang, J. B. Tok, and Z. Bao, "Stretchable and self-healing polymers and devices for electronic skin," *Progress in Polymer Science*, vol. 38, pp. 1961–1977, 2013.
66. M. Amjadi, A. Pichitpajongkit, S. Lee, S. Ryu, and I. Park, "Highly stretchable and sensitive strain sensor based on silver nanowire–elastomer nanocomposite," *ACS Nano*, vol. 8, pp. 5154–5163, 2014.
67. J. Park, I. You, S. Shin, and U. Jeong, "Material approaches to stretchable strain sensors," *ChemPhysChem*, vol. 16, pp. 1155–1163, 2015.
68. S. Zhang, H. Zhang, G. Yao, F. Liao, M. Gao, Z. Huang, et al., "Highly stretchable, sensitive, and flexible strain sensors based on silver nanoparticles/carbon nanotubes composites," *Journal of Alloys and Compounds*, vol. 652, pp. 48–54, 2015.
69. N. Annabi, S. R. Shin, A. Tamayol, M. Miscuglio, M. A. Bakooshli, A. Assmann, et al., "Highly elastic and conductive human-based protein hybrid hydrogels," *Advanced Materials*, vol. 28, pp. 40–49, 2016.
70. S. Yao, and Y. Zhu, "Wearable multifunctional sensors using printed stretchable conductors made of silver nanowires," *Nanoscale*, vol. 6, pp. 2345–2352, 2014.
71. S. Lee, S. Shin, S. Lee, J. Seo, J. Lee, S. Son, et al., "Ag nanowire reinforced highly stretchable conductive fibers for wearable electronics," *Advanced Functional Materials*, vol. 25, pp. 3114–3121, 2015.
72. B.-U. Hwang, J.-H. Lee, T. Q. Trung, E. Roh, D.-I. Kim, S.-W. Kim, et al., "Transparent stretchable self-powered patchable sensor platform with ultrasensitive recognition of human activities," *ACS Nano*, vol. 9, pp. 8801–8810, 2015.
73. I. Graz, M. Krause, S. Bauer-Gogonea, S. Bauer, S. P. Lacour, B. Ploss, et al., "Flexible active-matrix cells with selectively poled bifunctional polymer-ceramic nanocomposite for pressure and temperature sensing skin," *Journal of Applied Physics*, vol. 106, p. 034503, 2009.
74. Y.-L. Park, C. Majidi, R. Kramer, P. Bérard, and R. J. Wood, "Hyperelastic pressure sensing with a liquid-embedded elastomer," *Journal of Micromechanics and Microengineering*, vol. 20, p. 125029, 2010.
75. M. Shimojo, "Spatial filtering characteristic of elastic cover for tactile sensor," in *Robotics and Automation, 1994. Proceedings, 1994 IEEE International Conference on*, 1994, pp. 287–292.
76. Y. Jung, D.-G. Lee, J. Park, H. Ko, and H. Lim, "Piezoresistive tactile sensor discriminating multi-directional forces," *Sensors*, vol. 15, pp. 25463–25473, 2015.
77. J. Park, Y. Lee, J. Hong, M. Ha, Y.-D. Jung, H. Lim, et al., "Giant tunneling piezoresistance of composite elastomers with interlocked microdome arrays for ultrasensitive and multimodal electronic skins," *ACS Nano*, vol. 8, pp. 4689–4697, 2014.
78. D. Kang, P. V. Pikhitsa, Y. W. Choi, C. Lee, S. S. Shin, L. Piao, et al., "Ultrasensitive mechanical crack-based sensor inspired by the spider sensory system," *Nature*, vol. 516, pp. 222–226, 2014.
79. K. Takei, Z. Yu, M. Zheng, H. Ota, T. Takahashi, and A. Javey, "Highly sensitive electronic whiskers based on patterned carbon nanotube and silver nanoparticle composite films," *Proceedings of the National Academy of Sciences*, vol. 111, pp. 1703–1707, 2014.

80. R. A. Grant, B. Mitchinson, C. W. Fox, and T. J. Prescott, "Active touch sensing in the rat: Anticipatory and regulatory control of whisker movements during surface exploration," *Journal of Neurophysiology*, vol. 101, pp. 862–874, 2009.
81. P. Cordier, F. Tournilhac, C. Soulié-Ziakovic, and L. Leibler, "Self-healing and thermoreversible rubber from supramolecular assembly," *Nature*, vol. 451, pp. 977–980, 2008.
82. C.-H. Li, C. Wang, C. Keplinger, J.-L. Zuo, L. Jin, Y. Sun, et al., "A highly stretchable autonomous self-healing elastomer," *Nature Chemistry*, vol. 8, pp. 618–624, 2016.
83. Y. S. Zhang, and A. Khademhosseini, "Advances in engineering hydrogels," *Science*, vol. 356, p. eaaf3627, 2017.
84. D. Wirthl, R. Pichler, M. Drack, G. Kettlguber, R. Moser, R. Gerstmayr, et al., "Instant tough bonding of hydrogels for soft machines and electronics," *Science Advances*, vol. 3, p. e1700053, 2017.
85. B. J. Blaiszik, S. L. Kramer, M. E. Grady, D. A. McIlroy, J. S. Moore, N. R. Sottos, et al., "Autonomic restoration of electrical conductivity," *Advanced Materials*, vol. 24, pp. 398–401, 2012.
86. E. Palleau, S. Reece, S. C. Desai, M. E. Smith, and M. D. Dickey, "Self-healing stretchable wires for reconfigurable circuit wiring and 3D microfluidics," *Advanced Materials*, vol. 25, pp. 1589–1592, 2013.
87. J. M. Nassar, K. Mishra, K. Lau, A. A. Aguirre-Pablo, and M. M. Hussain, "Recyclable nonfunctionalized paper-based ultralow-cost wearable health monitoring system," *Advanced Materials Technologies*, vol. 2, p. 1600228, 2017.
88. W. Gao, H. Y. Nyein, Z. Shahpar, L.-C. Tai, E. Wu, M. Bariya, et al., "Wearable sweat biosensors," in *Electron Devices Meeting (IEDM), 2016 IEEE International*, 2016, pp. 6.6.1–6.6.4.
89. J. M. Nassar, M. C. Diaz, and M. M. Hussain, "Affordable dual-sensing proximity sensor for touchless interactive systems," in *Device Research Conference (DRC), 2016 74th Annual*, 2016, pp. 1–2.
90. J. Kim, M. Lee, H. J. Shim, R. Ghaffari, H. R. Cho, D. Son, et al., "Stretchable silicon nanoribbon electronics for skin prosthesis," *Nature Communications*, vol. 5, p. 5747, 2014.
91. T. Someya, "Building bionic skin," *IEEE Spectrum*, vol. 50, pp. 50–56, 2013.
92. S. Choi, J. Park, W. Hyun, J. Kim, J. Kim, Y. B. Lee, et al., "Stretchable heater using ligand-exchanged silver nanowire nanocomposite for wearable articular thermotherapy," *ACS Nano*, vol. 9, pp. 6626–6633, 2015.
93. D. H. Ho, Q. Sun, S. Y. Kim, J. T. Han, D. H. Kim, and J. H. Cho, "Stretchable and multimodal all graphene electronic skin," *Advanced Materials*, vol. 28, pp. 2601–2608, 2016.
94. U. Khan, T. H. Kim, H. Ryu, W. Seung, and S. W. Kim, "Graphene tribotronics for electronic skin and touch screen applications," *Advanced Materials*, vol. 29, p. 1603544, 2017.
95. J. M. Jani, M. Leary, A. Subic, and M. A. Gibson, "A review of shape memory alloy research, applications and opportunities," *Materials & Design*, vol. 56, pp. 1078–1113, 2014.
96. H.-H. Chou, A. Nguyen, A. Chortos, J. W. To, C. Lu, J. Mei, et al., "A chameleon-inspired stretchable electronic skin with interactive colour changing controlled by tactile sensing," *Nature Communications*, vol. 6, p. 8011, 2015.
97. J. H. Choi, J. Ahn, J. B. Kim, Y. C. Kim, J. Y. Lee, and I. K. Oh, "An electroactive, tunable, and frequency selective surface utilizing highly stretchable dielectric elastomer actuators," *Small*, vol. 12, pp. 1840–1846, 2016.
98. S. Rosset, and H. R. Shea, "Flexible and stretchable electrodes for dielectric elastomer actuators," *Applied Physics A*, vol. 110, pp. 281–307, 2013.
99. C. Bowen, H. Kim, P. Weaver, and S. Dunn, "Piezoelectric and ferroelectric materials and structures for energy harvesting applications," *Energy & Environmental Science*, vol. 7, pp. 25–44, 2014.
100. X. Wang, Y. Gu, Z. Xiong, Z. Cui, and T. Zhang, "Silk-molded flexible, ultrasensitive, and highly stable electronic skin for monitoring human physiological signals," *Advanced Materials*, vol. 26, pp. 1336–1342, 2014.
101. M. M. Hussain, J. P. Rojas, D. N. Singh, G. A. T. Sevilla, H. M. Fahad, and S. B. Inayat, "Flexible and stretchable thermoelectric generators," in *Meeting Abstracts*, 2017, pp. 1172–1172.

102. J. P. Rojas, D. Singh, S. B. Inayat, G. A. T. Sevilla, H. M. Fahad, and M. M. Hussain, "Micro and nano-engineering enabled new generation of thermoelectric generator devices and applications," *ECS Journal of Solid State Science and Technology*, vol. 6, pp. N3036–N3044, 2017.
103. C. R. Bowen, J. Taylor, E. LeBoulbar, D. Zabek, A. Chauhan, and R. Vaish, "Pyroelectric materials and devices for energy harvesting applications," *Energy & Environmental Science*, vol. 7, pp. 3836–3856, 2014.
104. Z. L. Wang, "Triboelectric nanogenerators as new energy technology for self-powered systems and as active mechanical and chemical sensors," *ACS Nano*, vol. 7, pp. 9533–9557, 2013.
105. D. Y. Park, D. J. Joe, D. H. Kim, H. Park, J. H. Han, C. K. Jeong, et al., "Self-powered real-time arterial pulse monitoring using ultrathin epidermal piezoelectric sensors," *Advanced Materials*, vol. 29, p. 1702308, 2017.
106. J. H. Lee, H. J. Yoon, T. Y. Kim, M. K. Gupta, J. H. Lee, W. Seung, et al., "Micropatterned P (VDF-TrFE) film-based piezoelectric nanogenerators for highly sensitive self-powered pressure sensors," *Advanced Functional Materials*, vol. 25, pp. 3203–3209, 2015.
107. Y. Mao, P. Zhao, G. McConohy, H. Yang, Y. Tong, and X. Wang, "Sponge-like piezoelectric polymer films for scalable and integratable nanogenerators and self-powered electronic systems," *Advanced Energy Materials*, vol. 4, p. 1301624, 2014.
108. J. Lee, J. Wu, M. Shi, J. Yoon, S. I. Park, M. Li, et al., "Stretchable GaAs photovoltaics with designs that enable high areal coverage," *Advanced Materials*, vol. 23, pp. 986–991, 2011.
109. D. J. Lipomi, B. C. K. Tee, M. Vosgueritchian, and Z. Bao, "Stretchable organic solar cells," *Advanced Materials*, vol. 23, pp. 1771–1775, 2011.
110. O. A. Abdulrazzaq, V. Saini, S. Bourdo, E. Dervishi, and A. S. Biris, "Organic solar cells: A review of materials, limitations, and possibilities for improvement," *Particulate Science and Technology*, vol. 31, pp. 427–442, 2013.
111. S. Harada, K. Kanao, Y. Yamamoto, T. Arie, S. Akita, and K. Takei, "Fully printed flexible fingerprint-like three-axis tactile and slip force and temperature sensors for artificial skin," *ACS Nano*, vol. 8, pp. 12851–12857, 2014.
112. G. Schwartz, B. C. Tee, J. Mei, A. L. Appleton, D. H. Kim, H. Wang, et al., "Flexible polymer transistors with high pressure sensitivity for application in electronic skin and health monitoring," *Nature Communications*, vol. 4, p. 1859, 2013.
113. J. Y. Sun, C. Keplinger, G. M. Whitesides, and Z. Suo, "Ionic skin," *Advanced Materials*, vol. 26, pp. 7608–7614, 2014.
114. H. Chen, Y. Cao, J. Zhang, and C. Zhou, "Large-scale complementary macroelectronics using hybrid integration of carbon nanotubes and IGZO thin-film transistors," *Nature Communications*, vol. 5, p. 4097, 2014.
115. J. Kim, M. S. Lee, S. Jeon, M. Kim, S. Kim, K. Kim, et al., "Highly transparent and stretchable field-effect transistor sensors using graphene–nanowire hybrid nanostructures," *Advanced Materials*, vol. 27, pp. 3292–3297, 2015.
116. Y. Zhang, H. Fu, S. Xu, J. A. Fan, K.-C. Hwang, J. Jiang, et al., "A hierarchical computational model for stretchable interconnects with fractal-inspired designs," *Journal of the Mechanics and Physics of Solids*, vol. 72, pp. 115–130, 2014.
117. X. Hu, P. Krull, B. de Graff, K. Dowling, J. A. Rogers, and W. J. Arora, "Stretchable inorganic-semiconductor electronic systems," *Advanced Materials*, vol. 23, pp. 2933–2936, 2011.
118. N. Qaiser, S. Khan, M. Nour, M. Rehman, J. Rojas, and M. M. Hussain, "Mechanical response of spiral interconnect arrays for highly stretchable electronics," *Applied Physics Letters*, vol. 111, p. 214102, 2017.
119. K.-I. Jang, K. Li, H. U. Chung, S. Xu, H. N. Jung, Y. Yang, et al., "Self-assembled three dimensional network designs for soft electronics," *Nature Communications*, vol. 8, p. 15894, 2017.
120. G. A. T. Sevilla, M. D. Cordero, J. M. Nassar, A. N. Hanna, A. T. Kutbee, A. Arevalo, et al., "Decal electronics: Printable packaged with 3D printing high-performance flexible CMOS electronic systems," *Advanced Materials Technologies*, vol. 2, p. 1600175, 2017.

121. C. H. Lee, Y. Ma, K. I. Jang, A. Banks, T. Pan, X. Feng, et al., "Soft core/shell packages for stretchable electronics," *Advanced Functional Materials*, vol. 25, pp. 3698–3704, 2015.
122. A. Romeo, Q. Liu, Z. Suo, and S. P. Lacour, "Elastomeric substrates with embedded stiff platforms for stretchable electronics," *Applied Physics Letters*, vol. 102, p. 131904, 2013.
123. S. Wagner, S. P. Lacour, J. Jones, I. H. Pai-hui, J. C. Sturm, T. Li, et al., "Electronic skin: Architecture and components," *Physica E: Low-dimensional Systems and Nanostructures*, vol. 25, pp. 326–334, 2004.
124. G. Torres Sevilla, A. Almuslem, A. Gumus, A. M. Hussain, M. Cruz, and M. M. Hussain, "High performance high-κ/metal gate complementary metal oxide semiconductor circuit element on flexible silicon," *Applied Physics Letters*, vol. 108, p. 094102, 2016.
125. G. A. T. Sevilla, J. P. Rojas, H. M. Fahad, A. M. Hussain, R. Ghanem, C. E. Smith, et al., "Flexible and transparent silicon-on-polymer based sub-20 nm non-planar 3D FinFET for brain-architecture inspired computation," *Advanced Materials*, vol. 26, pp. 2794–2799, 2014.
126. J. P. Rojas, G. A. Torres Sevilla, M. T. Ghoneim, S. B. Inayat, S. M. Ahmed, A. M. Hussain, et al., "Transformational silicon electronics," *ACS Nano*, vol. 8, pp. 1468–1474, 2014.
127. C. Bartolozzi, L. Natale, F. Nori, and G. Metta, "Robots with a sense of touch," *Nature Materials*, vol. 15, pp. 921–925, 2016.
128. Y. Cheng, R. Wang, J. Sun, and L. Gao, "A stretchable and highly sensitive graphene-based fiber for sensing tensile strain, bending, and torsion," *Advanced Materials*, vol. 27, pp. 7365–7371, 2015.
129. C. W. Carpenter, C. Dhong, N. B. Root, D. Rodriquez, E. E. Abdo, K. Skelil, et al., "Human ability to discriminate surface chemistry by touch," *Materials Horizons*, vol. 5, pp. 70–77, 2017.
130. N. Douglas, D. White, C. K. Pickett, J. Weil, and C. Zwillich, "Respiration during sleep in normal man," *Thorax*, vol. 37, pp. 840–844, 1982.
131. R. Parkes, "Rate of respiration: The forgotten vital sign: Racheal Parkes explains why emergency department nurses should document the respiratory rates of all patients, irrespective of their presenting complaints," *Emergency Nurse*, vol. 19, pp. 12–17, 2011.
132. K. Takei, W. Honda, S. Harada, T. Arie, and S. Akita, "Toward flexible and wearable human-interactive health-monitoring devices," *Advanced Healthcare Materials*, vol. 4, pp. 487–500, 2015.
133. Y. Shu, C. Li, Z. Wang, W. Mi, Y. Li, and T.-L. Ren, "A Pressure sensing system for heart rate monitoring with polymer-based pressure sensors and an anti-interference post processing circuit," *Sensors*, vol. 15, pp. 3224–3235, 2015.
134. S. C. Mukhopadhyay, "Wearable sensors for human activity monitoring: A review," *Sensors Journal, IEEE*, vol. 15, pp. 1321–1330, 2015.
135. M. M. Rodgers, V. M. Pai, and R. S. Conroy, "Recent advances in wearable sensors for health monitoring," *Sensors Journal, IEEE*, vol. 15, pp. 3119–3126, 2015.
136. Y. Zang, F. Zhang, C.-a. Di, and D. Zhu, "Advances of flexible pressure sensors toward artificial intelligence and health care applications," *Materials Horizons*, vol. 2, pp. 140–156, 2015.
137. Y. Wang, L. Wang, T. Yang, X. Li, X. Zang, M. Zhu, et al., "Wearable and highly sensitive graphene strain sensors for human motion monitoring," *Advanced Functional Materials*, vol. 24, pp. 4666–4670, 2014.
138. S. Gong, W. Schwalb, Y. Wang, Y. Chen, Y. Tang, J. Si, et al., "A wearable and highly sensitive pressure sensor with ultrathin gold nanowires," *Nature Communications*, vol. 5, p. 3132, 2014.
139. A. Chortos, and Z. Bao, "Skin-inspired electronic devices," *Materials Today*, vol. 17, pp. 321–331, 2014.
140. D.-H. Kim, N. Lu, R. Ma, Y.-S. Kim, R.-H. Kim, S. Wang, et al., "Epidermal electronics," *Science*, vol. 333, pp. 838–843, 2011.
141. C. Dagdeviren, Y. Shi, P. Joe, R. Ghaffari, G. Balooch, K. Usgaonkar, et al., "Conformal piezoelectric systems for clinical and experimental characterization of soft tissue biomechanics," *Nature Materials*, vol. 14, pp. 728–736, 2015.
142. H. Lee, C. Song, Y. S. Hong, M. S. Kim, H. R. Cho, T. Kang, et al., "Wearable/disposable sweat-based glucose monitoring device with multistage transdermal drug delivery module," *Science Advances*, vol. 3, p. e1601314, 2017.

143. J. Viventi, D.-H. Kim, J. D. Moss, Y.-S. Kim, J. A. Blanco, N. Annetta, et al., "A conformal, bio-interfaced class of silicon electronics for mapping cardiac electrophysiology," *Science Translational Medicine*, vol. 2, pp. 24ra22–24ra22, 2010.
144. D.-H. Kim, J.-H. Ahn, W. M. Choi, H.-S. Kim, T.-H. Kim, J. Song, et al., "Stretchable and foldable silicon integrated circuits," *Science*, vol. 320, pp. 507–511, 2008.
145. M. M. Hussain, J. P. Rojas, G. T. Sevilla, A. Hussain, M. T. Ghoneim, A. Hanna, et al., "Free form CMOS electronics: Physically flexible and stretchable," in *Electron Devices Meeting (IEDM), 2015 IEEE International*, 2015, pp. 19.4. 1–19.4. 4.
146. Y. Wang, C. Zhu, R. Pfattner, H. Yan, L. Jin, S. Chen, et al., "A highly stretchable, transparent, and conductive polymer," *Science Advances*, vol. 3, p. e1602076, 2017.
147. T. Someya, Z. Bao, and G. G. Malliaras, "The rise of plastic bioelectronics," *Nature*, vol. 540, pp. 379–385, 2016.
148. R. T. Hassarati, H. Marcal, L. John, R. Foster, and R. A. Green, "Biofunctionalization of conductive hydrogel coatings to support olfactory ensheathing cells at implantable electrode interfaces," *Journal of Biomedical Materials Research Part B: Applied Biomaterials*, vol. 104, pp. 712–722, 2016.
149. S. P. Lacour, S. Benmerah, E. Tarte, J. FitzGerald, J. Serra, S. McMahon, et al., "Flexible and stretchable micro-electrodes for in vitro and in vivo neural interfaces," *Medical & Biological Engineering & Computing*, vol. 48, pp. 945–954, 2010.
150. B. C.-K. Tee, A. Chortos, A. Berndt, A. K. Nguyen, A. Tom, A. McGuire, et al., "A skin-inspired organic digital mechanoreceptor," *Science*, vol. 350, pp. 313–316, 2015.
151. G. A. Tabot, J. F. Dammann, J. A. Berg, F. V. Tenore, J. L. Boback, R. J. Vogelstein, et al., "Restoring the sense of touch with a prosthetic hand through a brain interface," *Proceedings of the National Academy of Sciences*, vol. 110, pp. 18279–18284, 2013.
152. S. J. Bensmaia, and L. E. Miller, "Restoring sensorimotor function through intracortical interfaces: Progress and looming challenges," *Nature Reviews Neuroscience*, vol. 15, pp. 313–325, 2014.
153. R. van den Brand, J. Heutschi, Q. Barraud, J. DiGiovanna, K. Bartholdi, M. Huerlimann, et al., "Restoring voluntary control of locomotion after paralyzing spinal cord injury," *Science*, vol. 336, pp. 1182–1185, 2012.
154. T. A. Kuiken, P. D. Marasco, B. A. Lock, R. N. Harden, and J. P. Dewald, "Redirection of cutaneous sensation from the hand to the chest skin of human amputees with targeted reinnervation," *Proceedings of the National Academy of Sciences*, vol. 104, pp. 20061–20066, 2007.
155. D. J. Tyler, and D. M. Durand, "Functionally selective peripheral nerve stimulation with a flat interface nerve electrode," *IEEE Transactions on Neural Systems and Rehabilitation Engineering*, vol. 10, pp. 294–303, 2002.
156. S. Raspopovic, M. Capogrosso, F. M. Petrini, M. Bonizzato, J. Rigosa, G. Di Pino, et al., "Restoring natural sensory feedback in real-time bidirectional hand prostheses," *Science Translational Medicine*, vol. 6, pp. 222ra19–222ra19, 2014.
157. D. Huber, L. Petreanu, N. Ghitani, S. Ranade, T. Hromádka, Z. Mainen, et al., "Sparse optical microstimulation in barrel cortex drives learned behaviour in freely moving mice," *Nature*, vol. 450, 2007.
158. T.-i. Kim, J. G. McCall, Y. H. Jung, X. Huang, E. R. Siuda, Y. Li, et al., "Injectable, cellular-scale optoelectronics with applications for wireless optogenetics," *Science*, vol. 340, pp. 211–216, 2013.
159. C. Towne, K. L. Montgomery, S. M. Iyer, K. Deisseroth, and S. L. Delp, "Optogenetic control of targeted peripheral axons in freely moving animals," *PloS One*, vol. 8, p. e72691, 2013.
160. R. Chen, G. Romero, M. G. Christiansen, A. Mohr, and P. Anikeeva, "Wireless magnetothermal deep brain stimulation," *Science*, vol. 347, pp. 1477–1480, 2015.
161. R. S. Johansson, and J. R. Flanagan, "Coding and use of tactile signals from the fingertips in object manipulation tasks," *Nature Reviews Neuroscience*, vol. 10, pp. 345–359, 2009.
162. M. E. Spira, and A. Hai, "Multi-electrode array technologies for neuroscience and cardiology," *Nature Nanotechnology*, vol. 8, pp. 83–94, 2013.

163. C. M. Oddo, S. Raspopovic, F. Artoni, A. Mazzoni, G. Spigler, F. Petrini, et al., "Intraneural stimulation elicits discrimination of textural features by artificial fingertip in intact and amputee humans," *Elife*, vol. 5, p. e09148, 2016.
164. J. Viventi, D.-H. Kim, L. Vigeland, E. S. Frechette, J. A. Blanco, Y.-S. Kim, et al., "Flexible, foldable, actively multiplexed, high-density electrode array for mapping brain activity in vivo," *Nature Neuroscience*, vol. 14, pp. 1599–1605, 2011.
165. J. A. Rogers, "Recent advances in flexible/stretchable optoelectronics: From next-generation displays to skin-mounted wearables," in *CLEO: Science and Innovations*, 2017, p. SW1K. 1.
166. S. I. Park, G. Shin, J. G. McCall, R. Al-Hasani, A. Norris, L. Xia, et al., "Stretchable multichannel antennas in soft wireless optoelectronic implants for optogenetics," *Proceedings of the National Academy of Sciences*, vol. 113, pp. E8169–E8177, 2016.
167. G. Shin, A. M. Gomez, R. Al-Hasani, Y. R. Jeong, J. Kim, Z. Xie, et al., "Flexible near-field wireless optoelectronics as subdermal implants for broad applications in optogenetics," *Neuron*, vol. 93, pp. 509–521. e3, 2017.
168. V. K. Samineni, J. Yoon, K. E. Crawford, Y. R. Jeong, K. C. McKenzie, G. Shin, et al., "Fully implantable, battery-free wireless optoelectronic devices for spinal optogenetics," *Pain*, vol. 158, pp. 2108–2116, 2017.
169. B. Y. Chow, and E. S. Boyden, "Optogenetics and translational medicine," *Science Translational Medicine*, vol. 5, pp. 177ps5–177ps5, 2013.

11

Flexible and Stretchable Actuators

Nadeem Qaiser

11.1 Introduction

Actuators typically convert various forms of input energy into output energy. The input energy might be in the form of optical, thermal, pneumatic, and electrical, while the output energy is usually mechanical. Figure 11.1 exhibits the schematic of a typical actuator in action. These devices execute a physical motion/displacement or induce a force in response to a controlled input signal, which in turn changes its environment. Regarding motions, actuators might provide the linear, rotary, and bending motions, which depend on the design of device. The sizes of actuator devices range from macro to micro/nano scale (10^{-9} to 10^{-6} m). These actuators provide motion over small distances, i.e., of the orders of μm to mm or small forces of the order of pN (piconewton) to mN (millinewton).

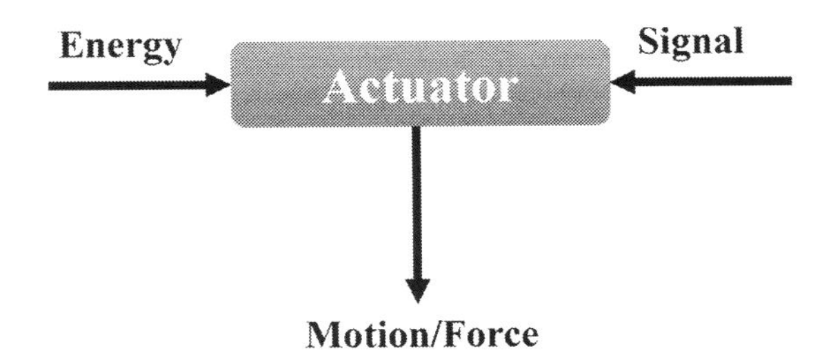

FIGURE 11.1 The schematic of an actuator, which converts input energy into motion or force.

11.1.1 Methods of Actuation

The method of actuation of any actuator dictates the type of that particular actuator. There are several kinds of actuators, which utilize the various physical principles including thermal, chemical, piezoelectric, optical, and others to get the required actuation. For instance, if electrical signals convert the electrical energy into mechanical deformations, these devices are called the electrical actuators. Electrical actuators utilize the clean form of energy. Similarly, when fluid and compressed air deliver the required actuation, these devices are called the hydraulic and pneumatic actuators, respectively. A hydraulic actuator induces the large forces, but has limited acceleration, whereas pneumatic devices have fast response time. Shape memory alloys (SMAs) utilize the thermal energy, provided by Joule heating, to acquire the output strain. SMAs are lightweight, cheaper, and contain the high power density. Each of these actuators have different performance in terms of output, i.e., force. Figure 11.2 shows the performance comparison between different classes of actuators [1].

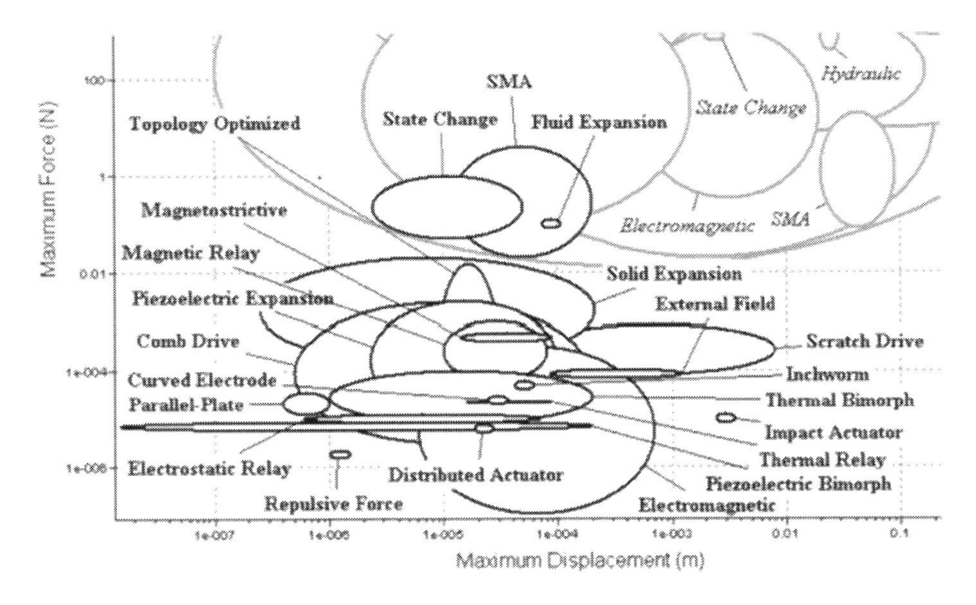

FIGURE 11.2 Performance comparison of different microactuators. (Printed from Bell, D.J. et al., *J. Micromech. Microeng.*, 15, S153–S164, 2005. With permission.)

11.1.2 Characteristics of an Actuator

The followings features of the actuators are essential to characterize and evaluate their performance.

1. **Force/torque:** The maximum force or torque that an actuator can yield
2. **Stroke:** The maximum possible displacement that an actuator can cover
3. **Linearity:** The extent to which the force and stroke remain as linear
4. **Response time:** The time by which an actuator responses as a reaction to the input signal
5. **Efficiency:** The ratio between the input and output energy
6. **Stiffness:** The rate at which the output force diminishes with respect to stroke
7. **Drift:** The unintended shift in output parameter even when the input energy is steady
8. **Bandwidth:** The range of frequencies at which the actuator provides reliable and rated output.

Although the above-mentioned characteristics are crucial for the evaluation of actuators, some additional parameters are equally important and thus need to be considered. These parameters, as shown in Figure 11.3, include the higher precision, power density, reliability, low cost, and capability to execute motions in multi degree of freedoms (DOFs).

11.1.3 Stretchable and Flexible Actuators

Contrary to conventional rigid actuators, the applications in current advanced technologies such as haptics, medical devices, artificial muscles, Micro-Electro-Mechanical Systems (MEMS), and robotics require the use of stretchable, flexible, and soft actuators. Moreover, current wearable electronics-based smart textile and biomimetic should be adoptable [2,3], and conform to human skin. Soft actuators are not only flexible, but also demonstrate high power to density ratio, low weight, less pollution, and noise free operation. Soft actuators are also able to realize complex and multiple DOFs output motions with simple control inputs [4–6]. Furthermore, the integration of different components, i.e., actuators, sensors, energy harvesting, and storage devices for flexible, wearable health monitoring system, and resultant surgical tools,

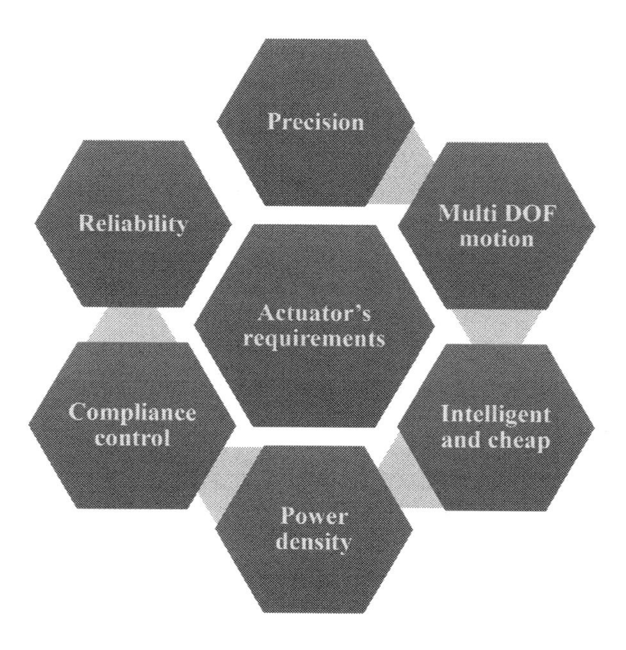

FIGURE 11.3 The general requirements of an optimum actuator.

require the actuators to be flexible. The key idea is to employ a material, which has elastic moduli comparable to that of human skin or muscle tissue, i.e., in the range of 10^4 to 10^9 Pa. Compliant materials with such low moduli readily adopt themselves to different tasks.

Therefore, the following sections will explore the different kinds of actuators, their actuation mechanism, fabrication techniques, persistent challenges, and current applications.

11.2 Actuators Based on Flexible Materials

11.2.1 Electroactive Polymers

In 1980, Roentgen fixed one end of a rubber strip, applied an electric field on the mass attached to the other end of the strip, and tested the first electroactive polymer (EAP) experimentally [7]. As a result, the elongation occurred along the rubber strip. However, the strain was too low until the polymers with significant strain emerged in the 1990s. EAPs consist of a wide range of polymers, which are potential candidates to act as active materials for stretchable and flexible actuators. The main principle of their working includes the capability of changing shape and size when subjected to electrical signals. EAP-based actuators consist of two main classes such as electrochemical/ionic polymers and electric-field activated polymers.

When electrical signals result into mass movement of ions or electrically charged species, these are called ionic EAPs. On the other hand, if they change shape when electric fields act directly on charges within a polymer, they are called field-activated EAPs. In other words, a Coulomb interaction (electrostatic fields) stimulates these actuators. Ubiquitous applications of flexible actuators, i.e., in biomedical, soft robotics, and tactile displays require the large strains as output and thus make EAPs a suitable choice since EAPs yield large strains with high fracture strength [8]. Among others, low power density and high flexibility due to the compliant nature of polymers are some of the major advantages of EAPs. Each of the EAPs classifications has its own merits and demerits. Ionic EAPs experience slow response and low actuation force. Nevertheless, ionic EAPs require the low voltage (~3 V) and thus are highly suitable for smart circuitry systems since they are capable to operate through a small and/or flexible battery. On the other hand, electronic EAPs are not able to show any activation at lower volts and demand a very high voltage (~several kV). Therefore, these are expensive and bulky. Still, they have a fast response, high actuation forces, and operate at room temperature for longer periods.

The overall classification of EAPs is shown in Figure 11.4 and details will be discussed in later sections of this chapter. Ionic EAP-based actuators are further classified into three categories, which include conducting polymers (CPs), ionic polymer metal composites (IPMCs), and carbon nanotubes (CNTs). It is worthy to mention that an ionic EAP-based actuator consists of three main components, i.e., (i) EAP electrode, (ii) electrolyte (gel type and/or ionic liquid type), and (iii) counter electrode. They utilize a redox reaction (electrochemical charging/discharging) to induce the actuation. Electric-field-driven actuators are comprised of dielectric elastomers, electrostrictive polymers, Maxwell-stress-based, and piezoelectric. The working principle of these EAP-based actuators is important to understand and thus is described in the subsequent sections.

11.2.1.1 Conducting Polymers

In 1990, Baughman exhibited the first of the CPs as an actuator that showed the large strain as a response to electrochemical doping and dedoping [9]. Typically, CPs actuators are comprised of two layers of electrodes and a solid-state electrolyte. The most common types of CPs are polyaniline (PANi), polypyrrole (PPy), polythiophene (PTh), and their derivatives. As mentioned-above, CPs usually need lower voltages (up to ~3 V), provide strains ranging from 1%–40%, and induce relatively higher stresses (~35 MPa).

CPs are electronically conducting organic materials, which show changes in size as a result of transportation of solvent ions between a polymer matrix and electrolyte as shown in Figure 11.5 [10]. During doping or oxidation, an external circuit removes the π-electrons, and thus the key reason of expansion

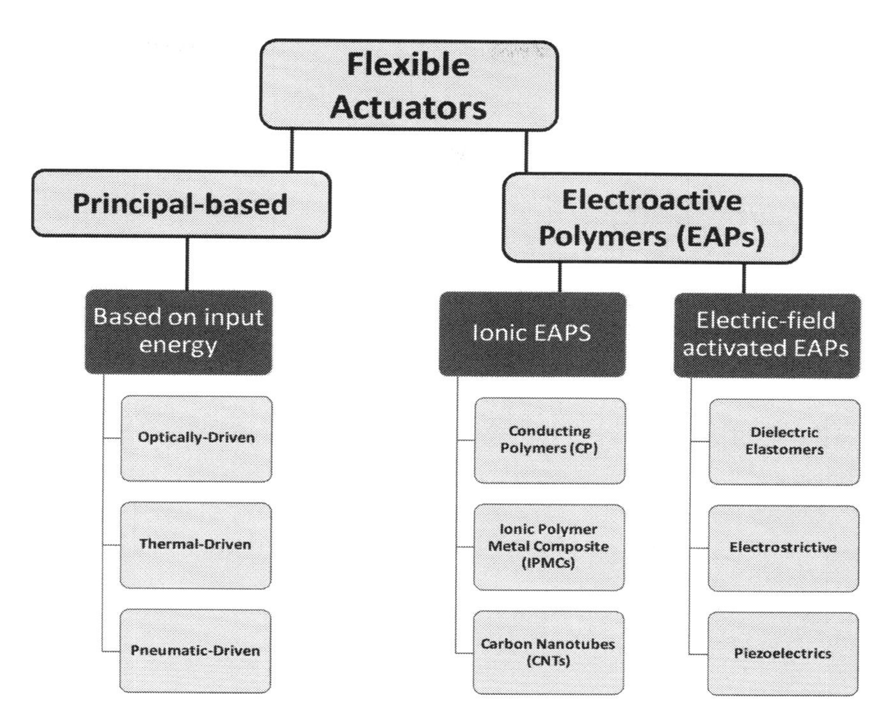

FIGURE 11.4 Types of flexible actuators based on different physical principles.

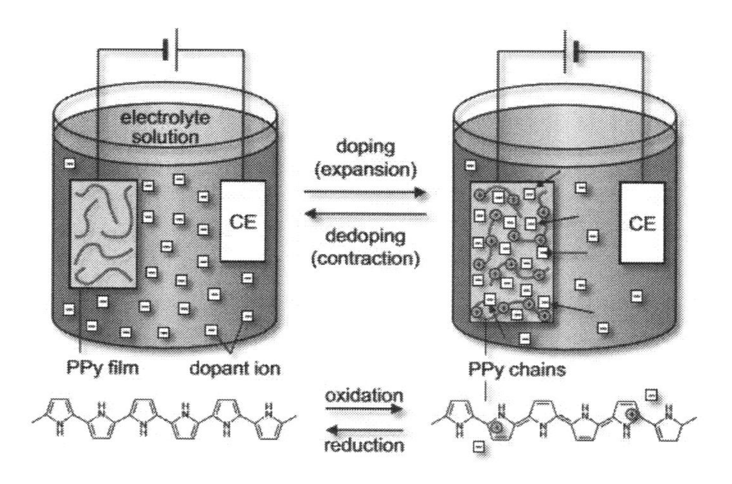

FIGURE 11.5 Electrochemical volume changes of polypyrrole film based on reversible doping and dedoping. (Printed from Okuzaki, H., Progress and current status of materials and properties of soft actuators, in *Soft Actuators*, Springer, Tokyo, Japan, pp. 3–18, 2014. With permission.)

(by the volume of total inserted cations) of a CPs actuator is intercalation of small cations from the surrounding electrolyte solution into the polymer matrix. Likewise, during dedoping (or reduction), these small cations move out of the polymer that results in the contraction of a CPs actuator. PPy is one of the most commonly explored CPs. The reasons include that they are easy to fabricate by the electrochemical deposition method and have a large stable output strain. The value of maximum strain depends on the size and total number of intercalated cations. Free-standing PPy-based actuators have limited stretchability and are prone to failure in wearable devices. Therefore, to provide stretchability, a composite of CPs and a soft substrate is commonly used.

Zheng et al. reported a simple way to prepare a PPy-based stretchable actuator, which was capable to attain effective displacements and pressure. They used a gold-coated and corrugated surface of rubber as a substrate, which provided the stretchability up to 30% with a small change in electrical resistance [11]. Conventional organic electrolytes are not suitable for the operation of CPs actuators in the air. To enhance the performance and cyclic life, several alternative approaches exist. Lu et al. tested for the first time the use of ionic liquids (ILs) as an electrolyte and reported a comparable performance to that of organic electrolyte-based actuators [12]. The reason for efficacious use of ILs could be attributed to its low vapor pressure, wide potential windows, higher ionic conductivity, and non-volatile nature. As a consequence, ILs as electrolytes became an emerging field that made the CPs actuators capable to operate in the air and even with better performance in terms of strain and cyclic life [13–15]. A household-assistive robot based on a multilayer CPs actuator was proposed, which was comprised of an ILs gel electrolyte [16]. Similarly, PANi exhibited the increased strain of 6.7% by the use of high salt concentration in an electrolyte [17]. CPs actuators have large strains and better results, however, performance still needs to be improved since they provide efficiency up to ~1%.

11.2.1.2 Ionic Polymer Metal Composite

In 1992, Oguro et al. showed the bending of Nafion-based IPMC for the first time by using the platinum (Pt) electrodes [18]. IPMCs are one of the promising materials for EAPs actuators, especially when used as artificial-muscle since they are biocompatible and have high flexibility that provides the required conformation for the human body. The schematic diagram of the actuation mechanism for IPMCs actuators is shown in Figure 11.6 [8]. Ions in IPMCs are cross-linked and are not able to move in a dry state, however, in the presence of water, they are free to move. Pressure and water concentration gradients in the ionic polymer are the main reasons for the bending of IPMCs. The IPMCs actuator contains a gel or water-swollen ion exchange membrane, i.e., Nafion or Flemion, both surfaces of which are attached to thin metal electrodes, i.e., Pt, Pd, or carbon electrodes and thus are called polymer-metal composites.

When there is a potential difference between the electrodes (under an applied voltage), the fast migration of a counter cation from the anode to cathode side occurs, as shown in Figure 11.6c. This migration also causes the movement of water molecules via hydration and a pumping effect. Consequently, IPMCs swell near the cathode and experience a fast anode bending due to a gradient. The large ionic current provides a large motion. A back relaxation motion follows this anode bending since the pressure results into the water diffusion out of the cation-rich areas. Several theoretical models have been reported to reveal the underlying mechanism of actuation in IPMCs actuators [19–22]. The major benefits of IPMCs actuators include the large strain, high flexibility, relatively fast response, and excellent formability. Therefore, IPMCs actuators were prepared for various application including underwater microrobot [23].

For optimum performance of IPMCs actuators, the appropriate plating of thin metal electrodes with an ionic polymer membrane is an essential process. Therefore, the good adhesion between membrane and electrodes and large interfacial area for electrochemical activity play a significant role for higher efficiency. Chemical plating is the most commonly used method to attach the ionic polymer membrane with Pt or gold electrodes. Furthermore, the size of the anion and its ionic mobility can affect the performance of the IPMCs actuator. Similar to CPs actuators, ILs can replace the conventional water solvent since they permit the actuator to operate in air for a longer period. Various types of ILs are compatible with an IPMCs actuator [24]. The fluctuation in the bending curvature of an IPMC actuator containing ILs was found to be lower than 21% during 180 minutes of operation [25]. Currently, high conductivity electrodes such as carbon nanomaterials also showed promising outcomes [26–28]. One of the examples, where different 1D, 2D, and 3D graphene/CNT-based electrodes were used, is shown in Figure 11.7 [26].

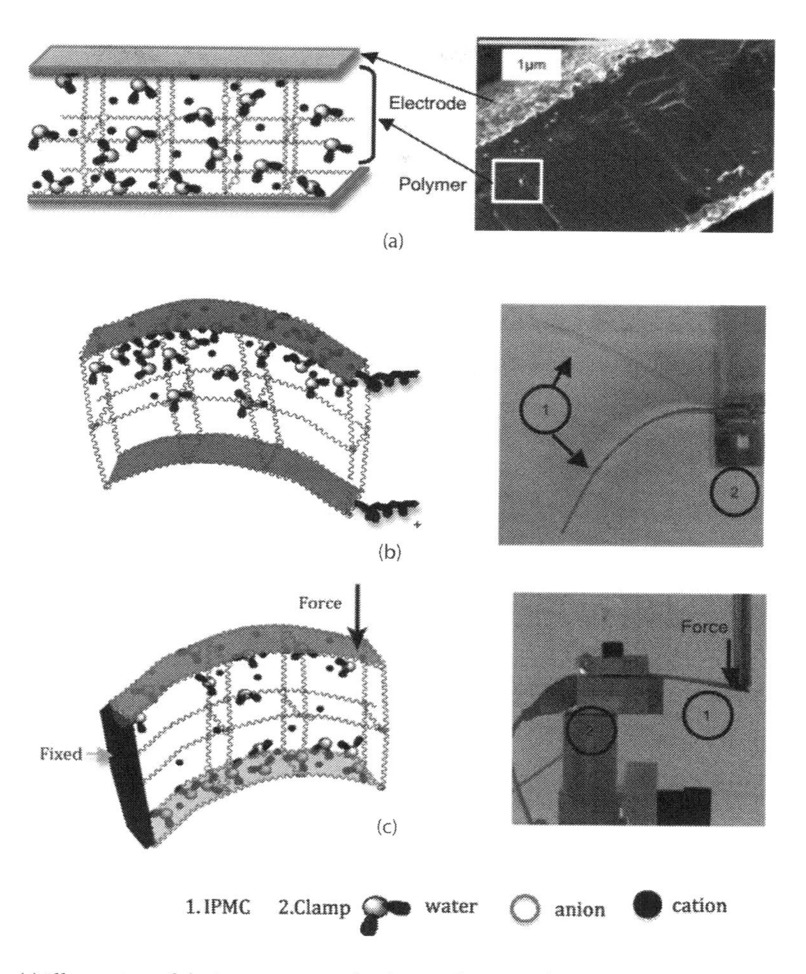

FIGURE 11.6 (a) Illustration of the ions, water, and polymer chain inside an IPMC, along with an SEM image of the cross-section when no external input is exerted, (b) the electromechanical effect in an IPMC causing the IPMC to bend, with a photograph showing the IPMC bending under electric field, and (c) the mechanoelectrical effect due to charge motion on mechanical bending. (Printed from Tiwari, R. and Garcia, E., *Smart Mater. Struct.*, 20, 83001, 2011. With permission.)

Lastly, the intrinsic elastic modulus of IPMCs also plays a significant role for the final strain. The value of modulus changes from 100 to 500 MPa, which actually depends on the contents of the water. The range of frequency and output force is usually up to ~100 Hz and in ~mN, respectively. Composites of different materials are another technique to enhance the performance of these IPMCs actuators. For instance, a composite of polyvinylidene fluoride (PVDF) with bacterial cellulose nanowhiskers (BCNW) significantly improved the output displacement because of the effect of ion migration and doping when subject to voltage [29].

The measurement of electrochemical characteristics for IPMCs is conducted through various methods and techniques. Among others, voltammetry, AC impedance, and measurement of capacitance are important parameters, which assist to understand the actuation mechanism and performance of IPMCs-based actuators. Similarly, various characterization techniques including FTIR spectroscopy and small-angle X-ray scattering are readily used. Although, IPMCs show relatively better output, the required level of optimum performance still needs to be reached and thus has originated the extensive research in this area.

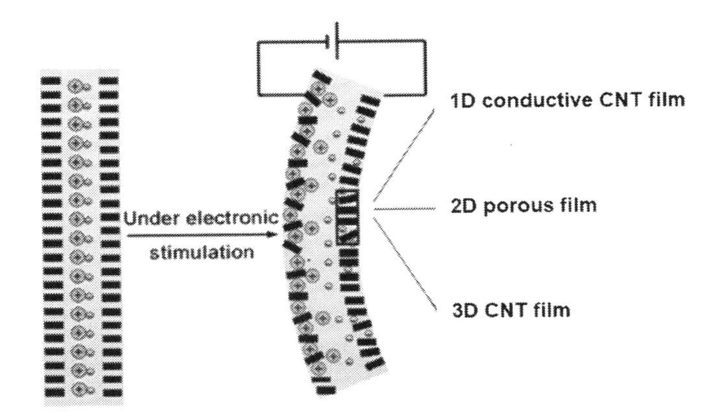

FIGURE 11.7 Different carbon-based electrode materials for IPMC actuators. (Printed from Kong, L. and Chen, W., *Adv. Mater.*, 26, 1025–1043, 2014. With permission.)

11.2.1.3 Carbon Nanotubes

A single-walled CNT (SWCNT) is comprised of a single layer of graphene, which is rolled into the shape of a tube, as shown in Figure 11.8a [30]. The size of this tube is typically in the nanometer range. In 1999, Baughman et al. performed the first study on the electrochemical response of SWCNT under low magnitude of applied voltage [31]. They demonstrated that SWCNT generated higher stresses than natural muscle. The inherent structure of CNTs provides excellent electrical, mechanical, and thermal properties. For instance, the tensile modulus of SWCNT is ~640 GPa which is not much lower than that of a diamond. Due to certain mechanical properties and actuation mechanisms, i.e., ionic transportation; CNTs actuators may also classified as ionic EAPs actuators. The actuation mechanism of CNTs actuator was named as 'charge injection' in which the covalently bond direction of CNTs experience change in dimensions due to charge injection. However, the electrostatic forces are repulsive in nature since like-charges are injected into the nanotubes.

Figure 11.8b demonstrates that an applied voltage injects charges into two CNTs in solution (red shows positive ions, yellow shows negative ions) [32]. This is compensated by the amount of ions at the CNT-electrolyte interface (double layer). The charge injection effect is originated from the double

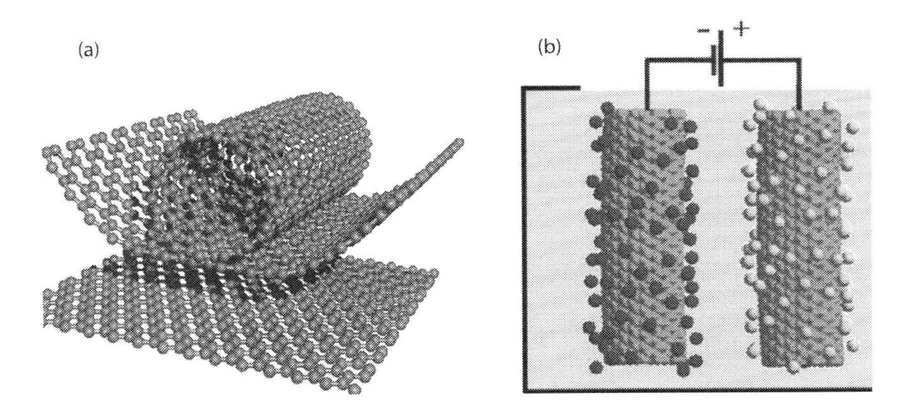

FIGURE 11.8 (a) A graphene sheet rolled into a nanotube. (Printed from Endo, M. et al., *Philos. Trans. A. Math. Phys. Eng. Sci.*, 362, 2223–2238, 2004. With permission.) (b) An applied potential injects charges in the two nanotube electrodes in solution. (Printed from Mirfakhrai, T. et al., *Mater. Today*, 10, 30–38, 2007. With permission.)

layer electrostatic effect. These repulsive forces counteract with stiff carbon-carbon bonds and make the nanotubes to expand through a quantum mechanical effect. Additionally, due to the stiff nature of these bonds, the output strains are lower than 2% [32]. It is clear from the actuation mechanism that by maximizing the surface area, the stored charge would increase because of a double-layer effect.

11.2.2 Field-Activated Electroactive Polymer-Based Actuators

Field-activated polymers belong to a second class of EAPs. FEAPs actuators are comprised of insulating polymers, and their main actuation mechanism includes the Coulombic interaction and dipole formations in the polymers, when subject to external electric signals. The mechanical deformation or motion within the actuator depends on various mechanisms of polarization. FEAPs actuators are highly efficient, yield relatively higher strains (>100%), and respond quickly. Therefore, these characteristics make FEAPS a strong candidate for the current medical needs such as artificial muscles. Following are the details of different types of the FEAPs actuators.

11.2.2.1 Dielectric Elastomers

A dielectric elastomer (DE) actuator is one of the most commonly used type of FEAPs actuators which have great benefits such as light in weight, fast response time, and easily adoptable to various applications including soft robotics, tunable optics, and artificial muscle. These DEs actuators are the rubber-like polymers and sandwiched between two compliant electrodes. A coating of graphite powder or gold on both surfaces of thin elastomer film usually fulfill the purpose of compliant electrodes. However, due to limited fabrication techniques, new materials for compliant electrodes such as CNTs and ionic gels have been introduced. One of the interesting and desirable features of DEs actuators includes the yield of large strain and high electrochemical efficiency.

It is important to understand the underlying mechanism of DEs actuators. When these compliant electrodes are subjected to input voltage, both of the electrodes experience the accumulation of opposite electrostatic charges [34]. According to Coulomb's law, these charges will attract each other with an electrostatic force, which induces the effective mechanical pressure on the elastomer film. In other words, compressive Maxwell stress (σ) is induced, which acts normal to the film surface. The induced Maxwell stress and the relation of electric field (E) may be expressed as [35]:

$$\sigma = \varepsilon_0 \varepsilon E^2 = \varepsilon_0 \varepsilon \left(\frac{V}{z} \right)^2, \tag{11.1}$$

where, ε_0, ε, V, and z are relative dielectric constants of free space and elastomeric polymer, applied voltage, and thickness of polymer film, respectively. The value of the relative dielectric constant of free space is 8.85×10^{-12} F/m. We can conclude from the equation (11.1) that the magnitude of this compressive force on the polymer film depends on the magnitude of applied electric field and relative dielectric constants. Likewise, the z should be as small as possible and material with high dielectric constant is beneficial. Additionally, as dictated by Young's modulus law, the elastic modulus of a polymer would dictate the magnitude of output strain.

Because of the generation of Maxwell stress, electrodes provide the lateral expansion in the area of polymers and compression in their thickness (as shown in Figure 11.9) [33]. However, the values of the stroke are low for a single film, therefore, a stack of elastomer films is used to attain the high stroke. Similarly, by using different configurations of films or elastomer film, out-of-plane motion (bending or folding) might be achieved. Elastomers used in DEs actuators include different kinds of materials derived from chemically cross-linked homopolymers, however, among these materials, silicon rubber, polyurethanes (PUs), and acrylic elastomers deliver optimum performance in terms of high output strains. These homopolymers should have low stiffness, low viscosity, low conductivity, and high dielectric constant. For instance, acrylic elastomers have a stiffness of 1 to 3 MPa and a dielectric

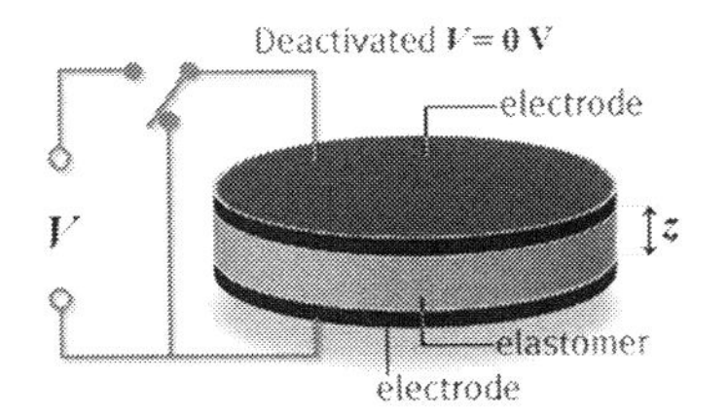

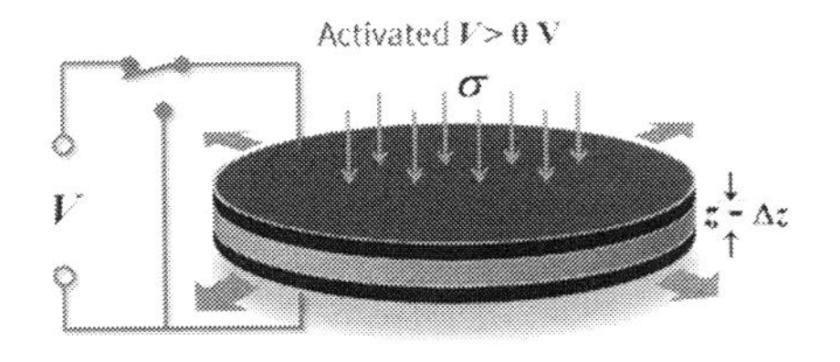

FIGURE 11.9 Structure and working principle of operation for soft dielectric EAP. (Printed from Giousouf, M. and Kovacs, G., *Smart Mater. Struct.*, 22, 104010, 2013. With permission.)

constant of 4.8, which in turn induces high strains (~ up to 158%) as an output. Furthermore, one of the benefits of Si rubber is the temperature range for their operation, i.e., −50°C to over 250°C.

Giousouf et al. have reported a DE actuator, wherein they introduced a pneumatic valve in addition to applied volts of 4 kV [33]. The purpose of the pneumatic valve was to provide the additional force, which acts as a pre-strain on the elastomer film. The only drawback is that they need a very high voltage to actuate the elastomers, i.e., usually ~>1 kV or 150 MV/m of volts.

11.2.2.2 Electrostrictive Polymers

An electrostrictive polymer (EP) actuator is another class of FEAPs, which induce actuation by pulling or pushing of electric dipoles when subject to high voltage. The presence of non-uniform distribution of space charges may induce the piezo-like response in an insulating material, and that material may be called as electrets. The push or pull generates the deformations or displacements within EPs actuator. In other words, the relation between strain and electric polarization can be written as [37]:

$$S_{electrostriction} = -Q\varepsilon_0^2 \left(\varepsilon_r - 1\right)^2 E^2. \quad (11.2)$$

It is clear from equation (11.2) that strain depends on the electrostrictive coefficient (Q), relative permittivities, and electric field. These rearrangements of charged side groups may occur in crystalline, amorphous, solid, or liquid phases. However, the electrostrictive strain is not significant in most of the substances and needs some modification technique to enhance its response for an applied electric field at room temperature.

EPs actuators are comprised of ferroelectric polymers (ferromagnets) with the introduction of imperfection on purpose. These imperfections lead to disruption of the inherent long-range order, which in turn generates the strain under applied voltage. The dipoles of ferromagnets are able to align themselves in response to an applied electric field. Huang et al. made use of defects modification in P(VDF-poly(vinylidene fluoride-trifluoroethylene) [TrFE]), to gain a strain of 10% during the phase transformation process between the ferroelectric and paraelectric [38]. The best known performance of an actuator based on EPs is made of P(VDF-TrFE) material. The actuation mechanism of an EPs-based actuator is shown in Figure 11.10 [36], which reveals that the chain elongates along the chain length and contracts along the perpendicular direction of the chain direction [36]. Likewise, other EPs such as graft copolymer and liquid crystal elastomers operate on the similar mechanism, i.e., utilize the pushing/pulling of dipoles and induce strains. For graft elastomers, in order to reach up to 2.5% output strain, an electric field of 10 MV/m is needed [32]. Similarly, for the liquid crystal elastomer, 4% strain was generated at 133 Hz, when an electric amplitude of 1.5 MV/m was applied [39]. Non-uniform local field or the presence of space charges enhances the electrostriction response. Nevertheless, the advancements of these EPs-based actuators are still at an early stage.

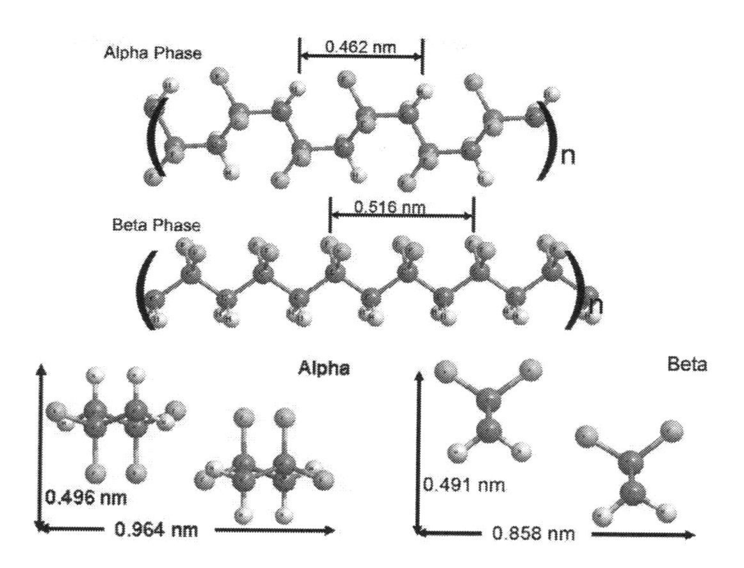

FIGURE 11.10 Alpha and beta phases in PVDF looking along the chains (top) and perpendicular to the chains (bottom). A similar alpha-to-beta phase change is induced in P(VDF-TrFE) relaxor ferroelectrics that begin in the non-polar alpha phase and switch to polar beta phase upon the application of an electric field. This results in dimensional changes (expansion along the chain length and contraction perpendicular to the chain direction). (Printed from Bar-Cohen, Y., *Electroactive Polymer (EAP) Actuators as Artificial Muscles: Reality, Potential, and Challenges*, SPIE Press, Bellingham, WA, 2004. With permission.)

11.2.2.3 Piezoelectrics

Piezoelectricity is defined as the total amount of electric charge which accumulates in dielectric material in response to applied pressure. Piezoelectrics (PEs) are the dielectrics which generate electric charge when subject to mechanical stress/strain, also called direct effect. However, for the case of PEs actuators, the inverse is also possible, i.e., if we apply voltage on the PE; it would generate the mechanical deformation in the materials, and is thus called the inverse PE effect. In solids, dipole moments lead to dipole density or polarization. In a piezoelectric material, there exist a region called the Weiss domain, wherein nearby dipoles tend to be aligned. These randomly oriented domains may be aligned by an applied electric field. In a piezoelectric material, polarization (P) alters due to the rearrangement of molecular dipoles moments, when subject to mechanical stress. Constitutive relation for a PE actuator is given below:

$$D_i = e_{ij}^{\sigma} E_j + d_{im} \sigma_m , \tag{11.3}$$

where D is electric displacement, E is applied electric field, σ_m is stress, e_{ij}^{σ} is dielectric permittivity, and d_{im} is piezoelectric coefficient.

Among others, PVDF and polylacticacid (PLLA) are the most commonly used piezo materials for PEs actuators. In fact, PVDF belongs to the ferroelectric polymer group, i.e., they exhibit permanent polarization in the presence of an applied electric field [34]. This alignment or re-orientation provides the actuation in PVDF-based actuators. In 1969, Kawai reported the piezoelectricity of PVDF for the first time. A direct current (DC) voltage is applied on PVDF film, which generates the residual polarization called the polling process. A typical actuation mechanism for a PEs actuator is comprised of phase transformation from ferroelectric to paraelectric, which in turn induces a large strain or change in lattice constants, as shown in Figure 11.11 [40]. Generally, the output strain along the thickness of a PVDF-based actuator reaches up to ~4.5%. For PVDF film of thickness of 100 μm or higher, it is difficult

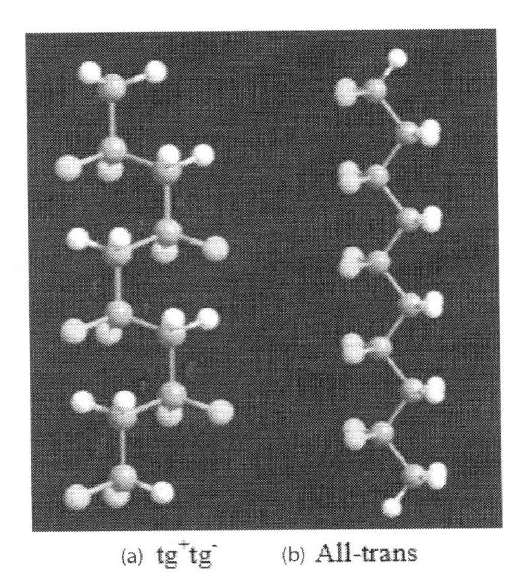

FIGURE 11.11 Schematic description of two conformations of PVDF: (a) tg+ tg− (b) all trans; yellow sphere: fluorine atom, white sphere: hydrogen atom, grey sphere: carbon atom. (Printed from Lovinger, A.J., *Science*, 220, 1115–1121, 1983. With permission.)

to initiate the polling process at room temperature. Therefore, a thermal-assisted polling, where a voltage of several kV is applied at a temperature higher than room temperature, is carried out. Alternatively, a uniaxial pre-stretch is utilized, which supports the process of orientation of chain molecules and thus induces a more effective piezoelectric effect.

Other materials such as PLLA are also being used in this area. Although, covalent bonding in chain molecules of PLLA makes its atoms to displace at a small displacement, a study shows that a PLLA fiber-based actuator also successfully demonstrated the shear piezoelectricity when subjected to an electric field [41,42]. To enhance the piezoelectric effect, various methods have been employed. A copolymer of PVDF, i.e., P(VDF-TrFE) also has shown the promising results of enhanced piezoelectricity. For instance, Choi et al. reported a vertically aligned core-shell structure of P(VDF-TrFE) on the substrate of polyurethane acrylate (PUA), which showed the enhancement of piezoelectricity by 1.85 times [43]. The reason for this enhancement was attributed to a strain confinement effect provided by a particular designed core-shell structure. Porous PVDF structures are another way to enhance the piezoelectricity since the pores provide the strain confinement effect and thus increase the piezoelectric effect [44]. Likewise, when organometallic filler (with high dielectric constant) is added in a PVDF-TrFE copolymer, it increases the dielectric constant of a composite [34].

11.3 Actuators Based on Actuation Principle

Other than applying an electrical field, there are few other strategies which provide the stimulation for actuators. In other words, actuator materials deform as a response to alternative forms of input physical energy such as light (optical), heat (thermal), and fluid (pneumatic). It is interesting to note that the current applications in biomimetic, wearable textile, soft robotics, medical surgeries, and display use the combination of various soft material and/or amalgamation of different physical principles, i.e., thermal-electrical in single system are used. The following section of the chapter covers the materials, which are actuated by different physical principles such as thermal, optical, and pneumatic, etc.

11.3.1 Optically Driven Actuators

The wavelength of light is less than 1 μm, which may irradiate the local region with a great control of resolution and timings. Therefore, it is a promising technique to stimulate the materials, which respond to light energy and are called photoresponsive materials. The actuation mechanism of these optically driven actuators (ODAs) is comprised of interaction of light and photosensitive material, which uses the energy of photons to displace the material. The energy of light is utilized by different properties of light, i.e., wavelength and intensity. For instance, an Si-based cantilever exhibited the displacement due to a change in electrostatic force by the energy of photons [46]. ODAs have fast response, and noiseless operation, but induce small output force/displacements. There are several materials used in ODAs, however, lanthanum-doped lead zirconate titanate (PLZT) ceramics, liquid crystal elastomers, chalcogenide glasses (ChG), SWCNT, hydrogels functionalized with spiropyran, and a fibrous network are the most commonly used materials. In addition to light-solid material interaction, other actuation principles, which include the light-fluid and laser-material interaction are also being used.

There are various photoactuation mechanisms which consist of heating by light, pyroelectric, and piezoelectric effects. A hydrogel composed of a polymer that responds to heat (thermoresponsive) and a photochromic chromophore showed an expansion-shrinkage when irradiated with blue light [47]. These mechanisms attribute the change in shape or size of ODAs, called photostrain, to photon absorption, radiation, or their combinations. Likewise, in ordered materials such as liquid crystal elastomer, the phase change by photon energy can induce the photostrain up to ~>100%. Liquid crystal elastomers are widely used as artificial muscle due to their compliant nature and excellent formability. Photoresponsive polymers with the advantage of non-contact actuation have been explored [48,49]. Among others, azobenzene, spiropyran, and diarylethene are also commonly used as polymer ODAs and have shown promising results [50–52]. Liquid crystal network film exhibited bending, when irradiated with linearly polarized light [53]. ODAs have great potential in areas such as medical, biomimetic, and artificial muscles.

11.3.2 Thermal Driven Actuators

Thermoresponsive materials respond, i.e., expand or contract when exposed to heat. The volume of most of the materials increases when their temperature elevates and vice versa. The expansion of these materials can be utilized as force to displace the materials. There might be other ways to heat up the materials; which include the direct thermal radiation, laser-induced heat, photo-thermal effects, and electrothermal effects [54]. Thermal driven actuators (TDAs) show relatively large output force AC compared to that of EAPs. Amjadi et al. demonstrated a powerful actuation of the multiresponsive paper actuator, which they stimulated by increasing the temperature also. They used a bilayer that contains a copy paper and PPy film. Since the PPy has a large thermal expansion coefficient, the actuator showed a bending of about 360° upon heating [55]. Likewise, Zhang et al. reported a programmable composite of CNT-hydrogel polymer, which demonstrated the actuation when heated up to ~48°. Same actuators were also acted as ODAs, when subjected to ultrafast near-infrared light [56]. Likewise, Tokudome et al. studied a hydrogel actuation on bilayer materials, where they did mimic the wrinkles in nature and that these wrinkles exhibited large deformations [57].

The idea of utilizing the thermoresponsive material's expansion to perform some useful activity or task (i.e. actuator, generator) is bringing forth many interesting applications. Recently, Prof. M. M. Hussain and colleagues have explored the thermos-electro-generator (TEG) effect and demonstrated the interesting applications. For instance, they demonstrated that thermally expandable polymeric material and flexible Si could destroy the solid-state electronics in less than 10 sec, thus making the destructible electronics for a security system [58]. Another class of material, which use thermal energy to remember and regain its original state, is called shape memory alloys (SMAs), the details of which are described in a subsequent section.

11.3.2.1 Shape Memory Alloys

SMAs have the ability to remember their initial position and can recover their original position or shape when subjected to a thermal process. The name of SMAs depicts that they have memory of their initial position and recover from a large strain in a later process. The resistance of the SMAs region increases, which experience bending during phase transition. The most commonly known material of SMAs is comprised of nickle-titanium alloy, Cu-Zn, Au-Cd, Fe-Mn-Se, and others. The cyclic life of SMAs, i.e., nitinol, is comprised of millions of cycles, and the reason could be attributed to superelasticity of SMAs. The working principle of SMAs actuators consists of transition of one form of crystalline structure to another, i.e., monoclinic (martensitic phase at low temperature) to a cubic crystal (austenite phase at higher temperature) form. The temperature-dependent elastic modulus of a SMAs actuator is usually from ~1 to ~100 MPa and can provide output deformation up to ~800% when subjected to a temperature change of 100°C. Typical applications of SMAs actuators in robotics are, i.e., fish robots, walker robots, flower robots, and a biomimetic hand [59–64].

Another strategy to acquire the 3D complex geometries includes the foldable 2D multifunctional components. Thin SMA sheets were used as unit cells for an actuator and mounted to form the different 3D assemblies such as linear, rotational, and surface actuators [45]. Figure 11.12a shows the unit cell and array configuration of 24 connected unit cells in a series. The array might be mounted in different trajectories such as linear, rotational, and surface, as shown in Figure 11.12b [45]. Similarly, different modes of deformations such as bending and/or twisting are achievable using SMAs wires embedded in a polydimethylsiloxane (PDMS) matrix and then actuating them selectively [65]. The major benefits of SMAs actuators include the light weight, flexible, high cyclic life, and force to weight ratio. For instance, an Flexinol wire-based SMA actuator with a 510 μm diameter exerted a force of 35 N, during tensile deformation [66].

11.3.3 Pneumatic Driven Actuators

Following the trend of trying out the different physical principles, studies have demonstrated the actuators, which experience large displacements or bending when subjected to compressed air, and thus are called pneumatic driven actuators (PDAs). Soft PDAs are made out of compliant or soft material such a

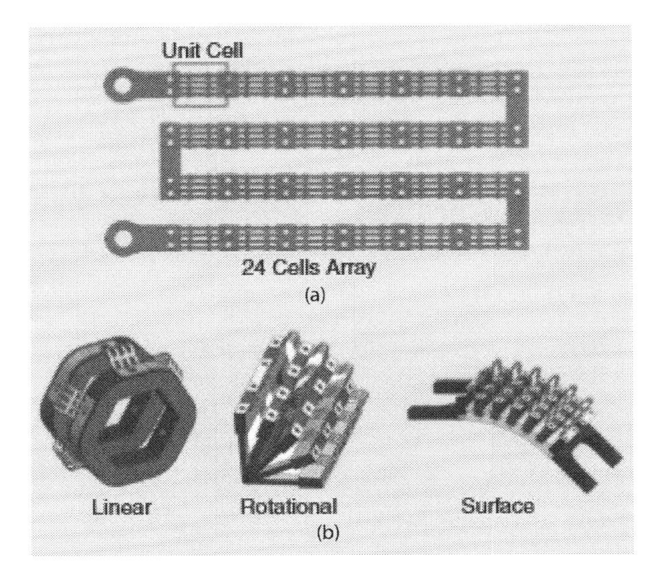

FIGURE 11.12 (a) SMA pattern that contains 24 unit cells electrically in a series. (b) A unit cell array can be mounted in different support structures to produce different trajectories. (Printed from Torres-Jara, E. et al., *IEEE Robot. Autom. Mag.*, 17, 78–87, 2010. With permission.)

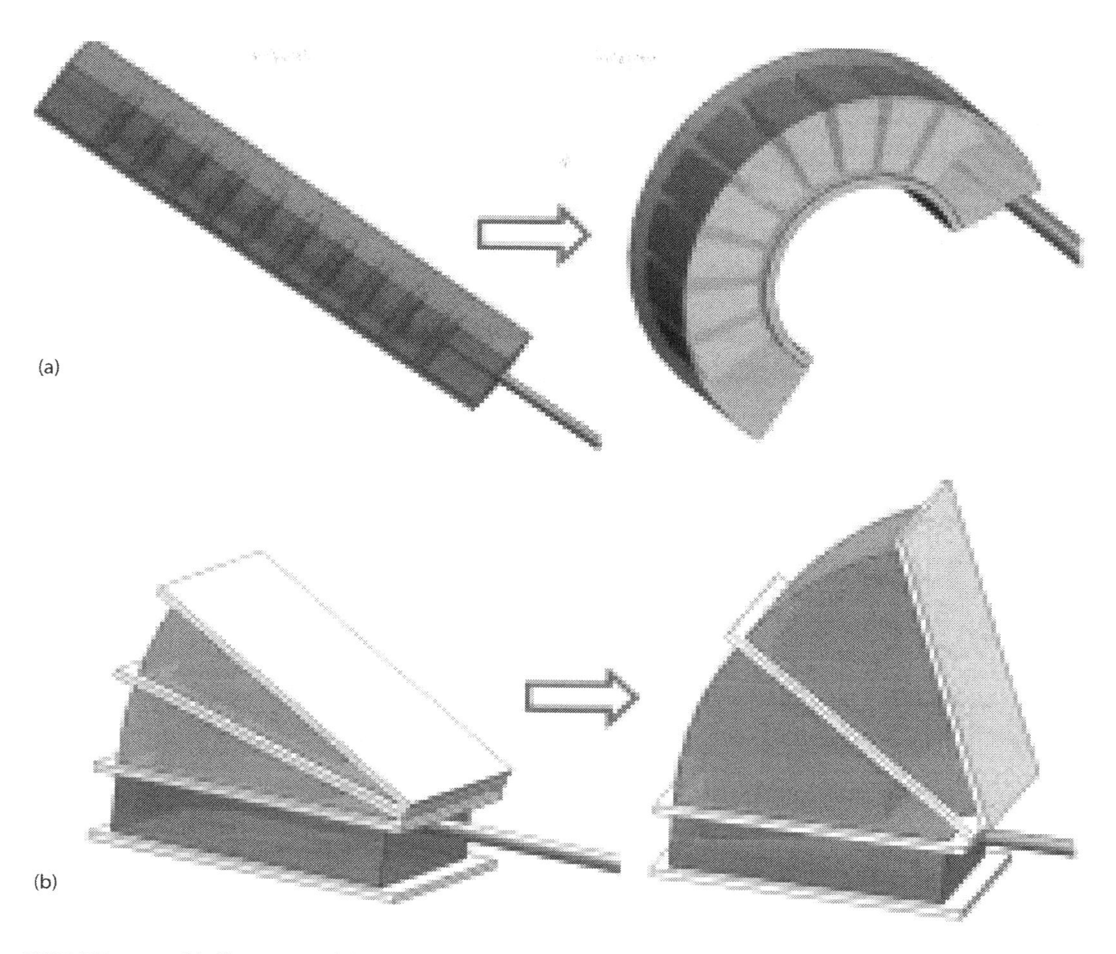

FIGURE 11.13 (a) Illustration of the two actuators configuration. (a) Bending PDA before and after actuation, and (b) the rotary PDA before and after actuation. (Printed from Sun, Y. et al., Characterization of silicone rubber based soft pneumatic actuators, in *IEEE/RSJ International Conference on Intelligent Robots and Systems*, 4446–4453, 2013. With permission.)

silicon rubber. However, there are limited techniques, which are used to characterize the PDAs actuators, i.e., measure the response time and output force or displacements. Few reports have illustrated a PDA actuator, which shows bending and rotary motions (as shown in Figure 11.13), and also shown a method to characterize the silicon rubber-based PDA actuator [67,68]. A rotary PDA actuator, which can rotate, based on elastomer material and actuated by negative pressure, has also successfully demonstrated its operation [69].

It is believed that using the soft actuators on hard materials may susceptibly damage the soft actuators in the long run. Therefore, studies have illustrated ways to recover or repair these damages. For instance, Terrya et al. developed a self-healing PDA actuator, where they used a Diels-Alder (DA) polymer that provides a self-healing capability in the system [70]. Likewise, a study showed that a robust PDA actuator based on an elastomer Ecoflex, when subjected to low input compressed air pressure of 40–50 kPa, delivered the double output force (~14 N) as compared to existing PDAs actuators. They did show the linear and rotary motions (deformations up to ~150%, and bending angles up to ~150°) and modulated the mechanical performance by varying the input pressure [71].

11.4 Fabrication of Flexible Actuators

Soft actuators technology was not possible until the invention of unique and exciting fabrication techniques especially used for manufacturing of high-performance actuators at a small scale, i.e., micro/nano sizes. Fabrication of flexible and stretchable actuators comprises a variety of fabrication methods and tools. This section, nonetheless, covers a summary of these techniques and a respective type of actuator. Ionic polymers might be fabricated by using a casting process. For instance, thin films of Nafion, a Teflon-based polymer, can be prepared by casting a solution and evaporating that solution by hot molding. The electrodes are one of the crucial components of soft actuators. In order to get a high performance out of these flexible actuators, the rational fabrication of electrodes is also an essential footstep. The electrodes of IPMCs were patterned by using the photolithography technique [72,73]. An improved method for the same was reported by using the selective plasma treatment. Similarly, metal layers were directly deposited on a soft Nafion material by using the lithography technique [74,75].

In order to get multiple DOFs and complex motion-based actuators, the above-mentioned methods are not a good choice since the fabricated actuators can exhibit only bending motions. Chen et al. incorporated the multiple fabrication steps in a series, i.e., plasma etching, ion exchange, vapor deposition, and electroless plating, which gave the artificial pectoral fins as a final product. These fins were capable to move in complex paths or, in other words, multiple DOFs deformations were carried out [76]. Other techniques include laser machining [77]. Very recently, Yan et al. fabricated the EAP with a monolithically integrated Au electrode, by using the supersonic cluster beam implantation (SCBI) method. One of the main advantages of the SCBI technique consists of no change observed in mechanical properties of the soft polymer [28]. Similarly, CNT-based EAP was fabricated, where Nafion was first soaked in IL overnight, and then sandwiched by using hot-pressing [78,79].

Numerous nanofabrication techniques are used to fabricate the FEAPs actuators. For instance, spin-cast technology is employed to prepare the thin film of P(VDF-TrFE) [80]. The sizes of thin film range from micro to nano scale. It is well known that properties and functionalities of thin films depend on the thickness of the film itself. Therefore, material properties such as elastic modulus, tensile, and fracture strengths depend on the size of the feature/material. Moreover, the interaction between the thin film of soft material and substrate also depends on the thickness of the film. Therefore, the spin-cast technique is useful to perform the comparative analysis of different thicknesses of P(VDF-TrFE) films. Zhang et al. have prepared the thin film of P(VDF-TrFE) by using solution spin-coating on a Pt-coated Si wafer [81]. They found a critical thickness of film, ~100 nm, below which the crystalline process is significantly slowed down, which results in the low crystallinity of films, which in turn effects the magnitudes of dielectric constant and polarizations in the films of the actuator [81].

Similarly, another strategy, called the Langmuir-Blodgett (L-B) technique, has been widely used to fabricate the PVDF-based polymers. By using this process, the film thickness can be controlled up to nanometer range, i.e., 1 nm. Since the scaling down of PVDF-based polymer film to nm size provides the ultra-thin interface between the PVDF film and a substrate, thus it exhibits the 100% crystalline phase. As a result, it makes it easier to analyze the possible cross-link from bulk PVDF to 2D ultra-thin film. Moreover, they found that for ultra-thin 2D films, the transition temperature needed for first-order ferroelectric phase transition is comparable to bulk value [82]. Similarly, the films of crystalline P(VDF-TrFE) with 15 nm thickness were prepared by the Langmuir-Blodgett technique [83]. A microfabrication process such as direct pattern transfer was also used to fabricate the polymer sample by photoetching. The X-ray (1–16 keV) from a synchrotron storage ring was used in that photoetching process [84]. Nanofibers of PVDF and P(VDF-TrFE) were fabricated by using the electrospinning process [85]. Crystalline nanotubes and nanowires (diameter from ~55 to ~360 nm) of soft copolymers of PVDF were prepared by using the wetting of polymers, i.e., template technique [86]. Nanoimprint lithography is another technique which was used to control the polymer crystallization at nanoscale and fabricated the nanopattern of crystalline PVDF-based actuator [87].

DEs are prepared by using cast, spinning, or coating methods. In order to get the complex geometries, a dip coating technique was used [88]. Spin coating is the most commonly used method to fabricate uniform and good quality DEs actuators. Silicone and acrylic elastomers are common types of DEs. A silicon elastomer is prepared by dissolving the Si fluid and diluted hardener into tetrahydrofuran (THF), and then polishing the mixture on Teflon plate [89].

ODAs, TDAs, and PDAs are also fabricated by using a variety of techniques. For instance, Tang et al. fabricated a liquid-crystalline actuator by casting azobenzene liquid-crystalline polymer on a polyethylene substrate [50]. Likewise, SMAs actuators were patterned by using the diode laser of a 10 μm resolution [45]. Drop-cast was used for uniformly dispersing the Ag nanowires on paper, and then the product was used as a multifunctional paper actuator [55]. There are numerous techniques which can be used to fabricate the different kinds of soft actuators, however, the details of each method is out of the scope of the chapter, and thus the reader might look up to the reference section for further details.

11.4.1 3D Printing of Soft Actuators

Manufacturing the soft structures via a single step technique (rapid prototyping) such as 3D printing is also an emerging field. 3D printing/spray forming would make the manufacturing of soft actuators relatively faster, cheaper, modular, and easy to control. 3D printers are comprised of various types, which depend on the nature of the application. Photosensitive polymer curing is one of the distinctive examples. In photosensitive-based polymer curing, light energy heats up the polymer until its melting point, and then it solidifies to form the required final design. The material used for this purpose includes acrylonitrile butadiene styrene (ABS), nylon, Si, etc. Similarly, for sinter-based printer, Al-, Ti-, and for bio-based printer hydrogels are common materials, respectively. It is worthy to mention that 3D printing has been utilized to fabricate almost every class of actuators; however, this chapter will illustrate only few of these.

There are two ways to incorporate the 3D printing technology; of particular interest is printing of sub-mm size structures. The first approach is comprised of printing of auxiliary components of the actuator system, i.e., linkages and external flexible parts. For instance, tentacle structure, which exhibited two-way motions, and steerable wormlike components with 3D motions were printed by using SMA [93,94]. The second approach includes printing of a fully functional actuator as a whole system. Fused deposition modeling (FDM) printing utilizes the extrusion of thermoplastic material. For instance, Yang et al. fabricated a shape memory polymer-based simple gripper by using the FDM [95]. EAP-based actuators such as DE and IPMC were also developed [96]. A few ODAs were also fabricated by using 3D printers [97,98]. Inkjet printing was used to manufacture a PVDF-based actuator [99]. Similarly, high-performance DEs actuators have been 3D printed successfully. Very recently, Aslan et al. have used the 3D printing process to fabricate a DEs (Si rubber matrix)-based flexible actuator, which is capable to illustrate a very high strain up to 900%, and even proves to be 3 times stronger than natural muscle [100]. In conclusion, 3D printing technology has great potential to fabricate the high-performance flexible and stretchable actuators.

11.5 Applications and Contemporary Research Trends

For tangible applications, flexible and stretchable actuators are typically integrated with other components (such as sensors, battery, feedback system) of wearable textile, bio-inspired and bio-integrated devices, soft robotics, and other areas of current advanced medical technologies. The efficient flexible and stretchable network remains a challenge, since the high-performance soft actuators are difficult to control especially when device integration is considered. This section will discuss the various areas where these flexible actuators are currently being used.

11.5.1 Actuators as Soft Robotics

The main trends of soft robotics are comprised of replicating the various complex motions of living creatures or programming the bio-inspired displacements. Soft robotics might play an essential role to assist the humans to perform different complex activities, especially where a human has limited access such as in a harsh environment, space exploration missions, and others. Since ionic EAPs require only ~2–3 volts, therefore, these EAP might be very handy to study the underwater resources. In order to design these EAP-based flexible robots, a long time pursuit is to mimic the soft-bodied animals, called biomimetic or bio-inspired robotics. Swimming motions (i.e., forward, backward, up/downward) were replicated by fabricating the mantra ray fish and squids, as shown in Figure 11.14a [90,103,104]. Similarly, snake-like robots and octopuses were studied using IPMCs, as shown in Figure 11.14b [91,105,106]. Currently, high-voltage run soft robotics have also attracted the attention because of their fast response and large output strain. For instance, Li. et al. have demonstrated a fast moving fish, which was made of commercial Si elastomer and stimulated by 10 kV for the highest speed (~13.5 cm/sec), as shown in Figure 11.14c and d. Additionally, with only one single battery (~450 mAh), aquatic robotic-fish was able to swim for about 3 hours continuously. Their electronic fish was also capable to sustain its operation in high temperature such as up to ~74.2°C [92].

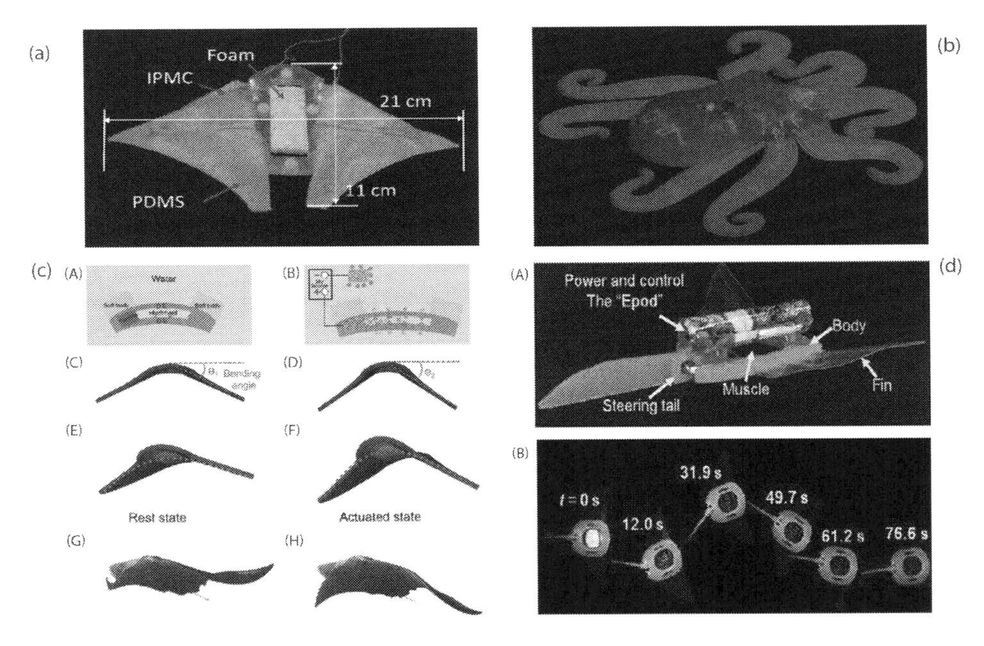

FIGURE 11.14 (a) Robotic manta ray body. (Printed from Chen, Z. et al., *Int. J. Smart Nano Mater.*, 3, 296–308, 2012. With permission.) (b) An octobot with fluorescently dyed fugitive inks (red, not auto-evacuated) and hyperelastic actuator layers (blue) fabricated by molding and EMB3D printing. (Printed from Wehner, M. et al., *Nature*, 536, 451–455, 2016. With permission.) (c) Front view of the actuating mechanisms. (A) In water, the soft body (silicone body) and the muscle laminates (two DE membranes and one hydrogel film) are deformed by the shrinking of the pre-stretched DE membranes with a bending curvature. (B) When a high voltage (HV) is applied to the muscle laminates, the electric field drives the ions in both the surrounding water and the hydrogel. Positive and negative charges accumulate on both sides of the DE membranes, inducing Maxwell stress and relaxing the DE membranes. The bending of the electro-ionic fish decreases. The surrounding water functions as the electric ground. (C) Front view (FEA) of the robotic fish in the rest state with a large bending angle θ_1. (D) Front view (FEA) for the actuated state of the robotic fish with a small bending angle θ_2. (E) Tilted view of FEA simulation for the rest state of the robotic fish. (F) Tilted view of FEA simulation for the actuated state of the robotic fish. Red dashed curves indicate the variation of bending. (G) Snapshot (similar tilted view) of a swimming manta ray. The body and fins of the manta ray buckle down with a large bending angle and (H) a small bending angle. (Printed from Li, T. et al., *Sci. Adv.*, 3, e1602045, 2017. With permission.) and (d) (A) tilted view of the fish showing the onboard system for power and remote control. (B) Live snapshots of the swimming of the robotic fish under remote control (voltage of 8 kV and 5 Hz). (Printed from Li, T. et al., *Sci. Adv.*, 3, e1602045, 2017. With permission.)

11.5.2 Medical Application

Flexible and stretchable actuators are rapidly replacing the conventional bulky and hard robots, which we used in the medical area. The main reason includes that these flexible actuators are biocompatible, biodegradable, harmless to the human body, adoptable, and perform complex motions as of the artificial tissues and muscles. Therefore, flexible actuators are paving the way to enhance the performance of medical related tasks such as a drug delivery system and implantation of artificial human organs. Very recently, Miriyev et al. have reported a breakthrough in soft material-based actuator's performance [100]. They utilized a composite of elastic polymer matrix and ethanol, which was dispersed in the form of micro-bubbles. In order to get a high strain output (~900%) they utilized the rapid expansions, which they obtained during reversible liquid-vapor phase transitions. They implemented the printed soft actuator in different areas, as shown in Figure 11.15a.

Applications such as smart wearable textile/robots to perform various neuromuscular tasks have also attracted the attention of relevant researchers. For instance, a soft, foldable, wearable, and light weight robot for the shoulder has been designed, which is useful for rehabilitation purposes [101]. The fabricated shoulder design was capable to support the upper shoulder through shoulder abduction, i.e., abduction actuators (ABA) along with horizontal extension/flexion actuators (HEFA), as shown in Figure 11.15b.

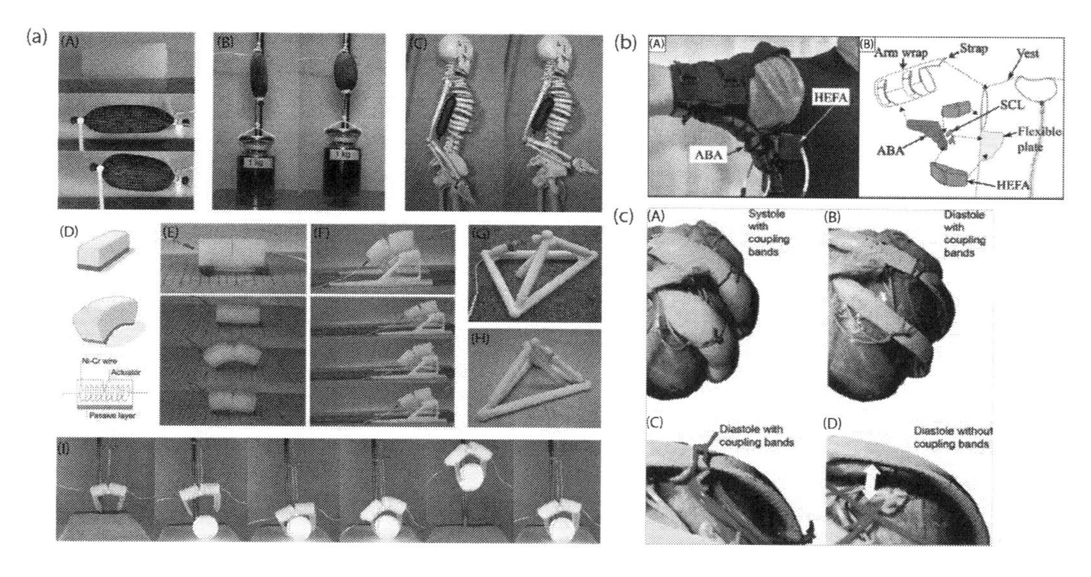

FIGURE 11.15 (a) Implementation of the soft composite material as an actuator. (A) McKibben-type artificial muscle (soft composite material inside braided mesh sleeving) shows displacement of about 25%. (B) 13 g artificial muscle lifts the weight of 1 kg. (C) Soft artificial muscle implemented as a bicep lifting skeleton's arm to 90° position at the elbow [A–C: actuation powered at 45 W (30 V, 1.5 A)]. (D) Design of the bi-morph bending actuator. (E) All-soft two-leg 'worm' and its locomotion powered at 8 W (8 V, 1 A). (F) The sleigh robot and its locomotion powered at 8 W (8 V, 1 A). (G) Tetrahedral robot evolved and 3D-printed in 2000 with embedded electrical motor. (H) The same robot with the soft composite material as an actuator embedded instead of the electrical motor. (I) Soft gripper lifting an egg (sequence from left to right; 8 V, 1 A). (Printed from Miriyev, A. et al., *Nat. Commun.*, 8, 596, 2017. With permission.) (b) (A) The effect of inflation pressure on the shoulder using the abduction actuators (ABA) with the horizontal extension/flexion actuator (HEFA) also visible, (B) exploded view of the final robot including the soft cruciate linkage (SCL). (Printed from O'Neill, C.T. et al., A soft wearable robot for the shoulder: Design, characterization, and preliminary testing, in *2017 International Conference on Rehabilitation Robotics (ICORR)*, pp. 1672–1678, 2007. With permission.) (c) Images from the *in vivo* study: (A) Left ventricle (LV) and soft robotic device with coupling bands in systole, (B) LV and soft actuators with coupling bands in diastole, (C) close-up of a soft actuator and coupling band in diastole, and (D) close-up of a soft actuator with coupling band released and arrow indicating dead zone. (Printed from Payne, C.J. et al., *Soft Robot.*, 4, 241–250, 2017. With permission.)

To connect the ABA to the textile-based vest, they replicated the cruciate ligaments as soft four-bar cruciate linkages (SCL). Figure 11.15b also shows the exploded view of an integrated device. Their pneumatic-based actuators were aimed to exhibit the acceptable assistive forces or shoulder support to the people, who have disability or are impaired.

Other applications of flexible actuators are composed of drug delivery systems, atherectomy, and hand/finger and eyeball prosthesis devices. Apart from these applications, a very interesting application is to grow artificial muscles such as a cardiac muscle assist device, which are useful to assist the patients with failing heart problems. These devices are usually wrapped around the heart and provide the contraction motion to the heart when assistance is needed. These devices, for instance EAPs actuators, contain the self-charging systems and sensory circuits, which sense the contraction pattern of the heart and trigger the actuation to contract the muscle when needed [107]. Very recently, Payne et al. performed an *in vivo* study with an implantable flexible robot, which was wrapped around the ventricles of the heart and programmed to assist the heart by contraction/expansion, as shown in Figure 11.15c. To aid the refilling during diastolic phases, elastic elements were integrated into the actuator. They used thermoplastic elastomer, dipped rubber, and elasticated rubber bands to fabricate different parts of the system. They implemented the McKibben-based actuator, a device that has an inflatable bladder, which provides the pressure to surrounded mesh. The exerted pressure on mesh would allow the mesh to expand/contract linearly [102].

11.6 Summary and Limitations

This chapter has discussed the actuation mechanism of flexible actuators, their materials, and fabrication methods. However, there are a few limitations, which need more attention. For instance, the limitation of a DEs-based actuator is creep and non-linearities. Applications of IPMCs are limited since the response is non-linear and the output force is low. Existing CNT assemblies have a very high Young's modulus (typically 1 GPa for nanotube sheets or 80 GPa for partially aligned nanotube fibers), however, failure strengths are up to 1800 MPa for fibers, that is much lower than that of the inherent properties of the individual nanotubes. For optical driven actuators, conversion efficiency and fatigue resistance are major limitations. Likewise, during operation of a thermal driven actuator, temperature might elevate. Therefore, they might not be suitable for smart textiles, i.e., employing directly on human body.

Despite the limitations, these flexible actuators have shown a great potential and paved a way to implement the technology in various fields including medical devices, drug-delivery systems, rehabilitation, mimicking the human and animal muscles, artificial organs, and soft robotics. In the future, integration of flexible sensory circuits, self-powered battery, and actuators might provide the real-time feedback to the system. Thus, the feedback system will assist the actuators in real-time to attain the required efficiency for different applications.

References

1. D. J. Bell, T. J. Lu, N. A. Fleck, and S. M. Spearing, "MEMS actuators and sensors: Observations on their performance and selection for purpose," *J. Micromech. Microeng.*, vol. 15, no. 7, pp. S153–S164, 2005.
2. K. C. Galloway et al., "Soft robotic grippers for biological sampling on deep reefs," *Soft Robot.*, vol. 3, no. 1, pp. 23–33, 2016.
3. N. W. Bartlett et al., "Soft robotics. A 3D-printed, functionally graded soft robot powered by combustion," *Science*, vol. 349, no. 6244, pp. 161–165, 2015.
4. D. Rus and M. T. Tolley, "Design, fabrication and control of soft robots," *Nature*, vol. 521, no. 7553, pp. 467–475, 2015.

5. S. Wakimoto, K. Suzumori, and K. Ogura, "Miniature pneumatic curling rubber actuator generating bidirectional motion with one air-supply tube," *Adv. Robot.*, vol. 25, no. 9–10, pp. 1311–1330, 2011.
6. C. Majidi, "Soft robotics: A perspective—Current trends and prospects for the future," *Soft Robot.*, vol. 1, no. 1, pp. 5–11, 2014.
7. W. C. Roentgen, "About the changes in shape and volume of dielectrics caused by electricity," *Annu. Phys. Chem. Ser.*, vol. 11, pp. 771–786, 1980.
8. R. Tiwari and E. Garcia, "The state of understanding of ionic polymer metal composite architecture: A review," *Smart Mater. Struct.*, vol. 20, no. 8, p. 83001, 2011.
9. R. H. Baughman, L. W. Shacklette, R. L. Elsenbaumer, E. Plichta, and C. Becht, "Conducting polymer electromechanical actuators," in *Conjugated Polymeric Materials: Opportunities in Electronics, Optoelectronics, and Molecular Electronics*, Dordrecht, the Netherlands: Springer, 1990, pp. 559–582.
10. H. Okuzaki, "Progress and current status of materials and properties of soft actuators," in *Soft Actuators*, Tokyo, Japan: Springer, pp. 3–18, 2014.
11. W. Zheng et al., "Polypyrrole stretchable actuators," *J. Polym. Sci. Part B*, vol. 51, no. 1, pp. 57–63, 2013.
12. W. Lu et al., "Use of ionic liquids for pi-conjugated polymer electrochemical devices," *Science*, vol. 297, no. 5583, pp. 983–987, 2002.
13. C. Plesse, F. Vidal, H. Randriamahazaka, D. Teyssié, and C. Chevrot, "Synthesis and characterization of conducting interpenetrating polymer networks for new actuators," in *Polymer*, vol. 46, no. 18, pp. 7771–7778, 2005.
14. D. Zhou et al., "Solid state actuators based on polypyrrole and polymer-in-ionic liquid electrolytes," in *Electrochim. Acta*, vol. 48, no. 14–16, pp. 2355–2359, 2003.
15. J. Ding et al., "Use of ionic liquids as electrolytes in electromechanical actuator systems based on inherently conducting polymers," *Chem. Mater.*, 15(12), 2392–2398, 2003.
16. K. Ikushima, S. John, K. Yokoyama, and S. Nagamitsu, "A practical multilayered conducting polymer actuator with scalable work output," *Smart Mater. Struct.*, vol. 18, no. 9, p. 95022, 2009.
17. W. Takashima, M. Nakashima, S. S. Pandey, and K. Kaneto, "Enhanced electrochemomechanical activity of polyaniline films towards high pH region: Contribution of Donnan effect," *Electrochim. Acta*, vol. 49, no. 24, pp. 4239–4244, 2004.
18. K. Oguro, "Bending of an ion-conducting polymer film-electrode composite by an electric stimulus at low voltage," *Trans. J. Micromach. Soc.*, vol. 5, pp. 27–30, 1992.
19. P. G. de Gennes, K. Okumura, M. Shahinpoor, and K. J. Kim, "Mechanoelectric effects in ionic gels," *Europhys. Lett.*, vol. 50, pp. 513–518, 2007.
20. T. Yamaue, H. Mukai, K. Asaka, and M. Doi, "Electrostress diffusion coupling model for polyelectrolyte gels," *Macromolecules*, vol. 38, no. 4, pp. 1349–1356, 2005.
21. T. Wallmersperger, D. J. Leo, and C. S. Kothera, "Transport modeling in ionomeric polymer transducers and its relationship to electromechanical coupling," *J. Appl. Phys.*, vol. 101, no. 2, p. 24912, 2007.
22. A. Onuki and K. Kawasaki, *Dynamics and Patterns in Complex Fluids*, vol. 53. Berlin, Germany: Springer, 1989.
23. S. Guo, T. Fukuda, and K. Asaka, "A new type of fish-like underwater microrobot," *IEEE/ASME Trans. Mech.*, vol. 8, no. 1, pp. 136–141, 2003.
24. J. Wang, C. Xu, M. Taya, and Y. Kuga, "A Flemion-based actuator with ionic liquid as solvent," *Smart Mater. Struct.*, vol. 16, no. 2, pp. S214–S219, 2007.
25. K. Kikuchi and S. Tsuchitani, "Nafion® -based polymer actuators with ionic liquids as solvent incorporated at room temperature," *J. Appl. Phys.*, vol. 106, no. 5, p. 53519, 2009.
26. L. Kong and W. Chen, "Carbon nanotube and graphene-based bioinspired electrochemical actuators," *Adv. Mater.*, vol. 26, no. 7, pp. 1025–1043, 2014.
27. H. Rasouli, L. Naji, and M. G. Hosseini, "Electrochemical and electromechanical behavior of Nafion-based soft actuators with PPy/CB/MWCNT nanocomposite electrodes," *RSC Adv.*, vol. 7, no. 6, pp. 3190–3203, 2017.

28. Y. Yan et al., "Electroactive ionic soft actuators with monolithically integrated gold nanocomposite electrodes," *Adv. Mater.*, vol. 29, no. 23, p. 1606109, 2017.
29. S.-S. Kim and C.-D. Kee, "Electro-active polymer actuator based on PVDF with bacterial cellulose nano-whiskers (BCNW) via electrospinning method," *Int. J. Pr. Eng. Man*, vol. 15, no. 2, pp. 315–321, 2014.
30. M. Endo, T. Hayashi, Y. A. Kim, M. Terrones, and M. S. Dresselhaus, "Applications of carbon nanotubes in the twenty-first century.," *Philos. Trans. A. Math. Phys. Eng. Sci.*, vol. 362, no. 1823, pp. 2223–2238, 2004.
31. R. H. Baughman et al., "Carbon nanotube actuators," *Science*, vol. 284, no. 5418, pp. 1340–1344, 1999.
32. T. Mirfakhrai, J. D. W. Madden, and R. H. Baughman, "Polymer artificial muscles," *Mater. Today*, vol. 10, no. 4, pp. 30–38, 2007.
33. M. Giousouf, and G. Kovacs, "Dielectric elastomer actuators used for pneumatic valve technology," *Smart Mater. Struct.*, vol. 22, no. 10, p. 104010, 2013.
34. R. Shankar, T. K. Ghosh, and R. J. Spontak, "Dielectric elastomers as next-generation polymeric actuators," *Soft Matter*, vol. 3, no. 9, p. 1116, 2007.
35. R. Pelrine, R. Kornbluh, Q. Pei, and J. Joseph, "High-speed electrically actuated elastomers with strain greater than 100%," *Science*, vol. 287, no. 5454, pp. 836–839, 2000.
36. Y. Bar-Cohen, *Electroactive Polymer (EAP) Actuators as Artificial Muscles: Reality, Potential, and Challenges*. Bellingham, WA: SPIE Press, 2004.
37. I. Krakovský, T. Romijn, and A. P. de Boer, "A few remarks on the electrostriction of elastomers," http://oasc12039.247realmedia.com/RealMedia/ads/click_lx.ads/www.aip.org/pt/adcenter/pdfcover_test/L-37/386502181/x01/AIP-PT/JAP_ArticleDL_092017/scilight717-1640 × 440.gif/434f71374e315a556e61414141774c75?x, 1998.
38. C. Huang et al., "Poly(vinylidene fluoride-trifluoroethylene) based high performance electroactive polymers," *IEEE Trans. Dielectr. Electr. Insul.*, vol. 11, no. 2, pp. 299–311, 2004.
39. W. Lehmann et al., "Giant lateral electrostriction in ferroelectric liquid-crystalline elastomers," *Nature*, vol. 410, no. 6827, pp. 447–450, 2001.
40. A. J. Lovinger, "Ferroelectric polymers," *Science*, vol. 220, no. 4602, pp. 1115–1121, 1983.
41. M. Sawano, K. Tahara, Y. Orita, M. Nakayama, and Y. Tajitsu, "New design of actuator using shear piezoelectricity of a chiral polymer, and prototype device," *Polym. Int.*, vol. 59, no. 3, pp. 365–370, 2010.
42. S. Ito et al., "Sensing using piezoelectric chiral polymer fiber," *Jpn. J. Appl. Phys.*, vol. 51, no. 9S1, p. 09LD16, 2012.
43. Y.-Y. Choi et al., "Vertically aligned P(VDF-TrFE) core-shell structures on flexible pillar arrays," *Sci. Rep.*, vol. 5, no. 1, p. 10728, 2015.
44. S. Cha et al., "Porous PVDF as effective sonic wave driven nanogenerators," *Nano Lett.*, vol. 11, no. 12, pp. 5142–5147, 2011.
45. E. Torres-Jara, K. Gilpin, J. Karges, R. J. Wood, and D. Rus, "Compliant modular shape memory alloy actuators," *IEEE Robot. Autom. Mag.*, vol. 17, no. 4, pp. 78–87, 2010.
46. M. Tabib-Azar, "Optically controlled silicon microactuators," *Nanotechnology*, vol. 1, no. 1, pp. 81–92, 1990.
47. K. Sumaru, K. Ohi, T. Takagi, T. Kanamori, and T. Shinbo, "Photoresponsive properties of Poly(N-isopropylacrylamide) hydrogel partly modified with spirobenzopyran," *Langmuir*, vol. 22, no. 9, pp. 4353–4356, 2006.
48. J. Deng et al., "Tunable photothermal actuators based on a pre-programmed aligned nanostructure," *J. Am. Chem. Soc.*, vol. 138, no. 1, pp. 225–230, 2016.
49. M. Kondo, Y. Yu, and T. Ikeda, "How does the initial alignment of mesogens affect the photoinduced bending behavior of liquid-crystalline elastomers?" *Angew. Chemie Int. Ed.*, vol. 45, no. 9, pp. 1378–1382, 2006.

50. R. Tang, Z. Liu, D. Xu, J. Liu, L. Yu, and H. Yu, "Optical pendulum generator based on photomechanical liquid-crystalline actuators," *ACS Appl. Mater. Interfaces*, vol. 7, no. 16, pp. 8393–8397, 2015.
51. B. Ziółkowski, L. Florea, J. Theobald, F. Benito-Lopez, and D. Diamond, "Self-protonating spiropyran-co-NIPAM-co-acrylic acid hydrogel photoactuators," *Soft Matter*, vol. 9, no. 36, p. 8754, 2013.
52. F. Terao, M. Morimoto, and M. Irie, "Light-driven molecular-crystal actuators: Rapid and reversible bending of rodlike mixed crystals of diarylethene derivatives," *Angew. Chemie Int. Ed.*, vol. 51, no. 4, pp. 901–904, 2012.
53. Y. Yu, M. Nakano, and T. Ikeda, "Directed bending of a polymer film by light," *Nature*, vol. 425, no. 6954, p. 145, 2003.
54. Y. Hu, G. Wang, X. Tao, and W. Chen, "Low-voltage-driven sustainable weightlifting actuator based on polymer-nanotube composite," *Macromol. Chem. Phys.*, vol. 212, no. 15, pp. 1671–1676, 2011.
55. M. Amjadi, and M. Sitti, "High-performance multiresponsive paper actuators," *ACS Nano*, vol. 10, no. 11, pp. 10202–10210, 2016.
56. X. Zhang et al., "Optically- and thermally-responsive programmable materials based on carbon nanotube-hydrogel polymer composites," *Nano Lett.*, vol. 11, no. 8, pp. 3239–3244, 2011.
57. Y. Tokudome, H. Kuniwaki, K. Suzuki, D. Carboni, G. Poologasundarampillai, and M. Takahashi, "Thermoresponsive wrinkles on hydrogels for soft actuators," *Adv. Mater. Interfaces*, vol. 3, no. 12, p. 1500802, 2016.
58. A. Gumus et al., "Expandable polymer enabled wirelessly destructible high-performance solid state electronics," *Adv. Mater. Technol.*, vol. 2, no. 5, p. 1600264, 2017.
59. H. L. Huang, S.-H. Park, J.-O. Park, and C.-H. Yun, "Development of stem structure for flower robot using SMA actuators," in *2007 IEEE International Conference on Robotics and Biomimetics (ROBIO)*, pp. 1580–1585, 2007.
60. K.-J. Cho, E. Hawkes, C. Quinn, and R. J. Wood, "Design, fabrication and analysis of a body-caudal fin propulsion system for a microrobotic fish," in *2008 IEEE International Conference on Robotics and Automation*, pp. 706–711, 2008.
61. Y. Sugiyama, and S. Hirai, "Crawling and jumping of deformable soft robot," in *2004 IEEE/RSJ International Conference on Intelligent Robots and Systems (IROS)*, vol. 4, pp. 3276–3281, 2004.
62. C. Menon, and M. Sitti, "Biologically inspired adhesion based surface climbing robots," in *Proceedings of the 2005 IEEE International Conference on Robotics and Automation*, pp. 2715–2720, 2005.
63. M. Carlo, and S. Metin, "A biomimetic climbing robot based on the gecko," *J. Bionic Eng.*, vol. 3, no. 3, pp. 115–125, 2006.
64. M. Nishida, K. Tanaka, and H. O. Wang, "Development and control of a micro biped walking robot using shape memory alloys," in *Proceedings 2006 IEEE International Conference on Robotics and Automation*, pp. 1604–1609, 2006.
65. H. Rodrigue, W. Wang, B. Bhandari, M.-W. Han, and S.-H. Ahn, "SMA-based smart soft composite structure capable of multiple modes of actuation," *Compos. Part B Eng.*, vol. 82, pp. 152–158, 2015.
66. A. Villoslada, A. Flores, D. Copaci, D. Blanco, and L. Moreno, "High-displacement flexible shape memory alloy actuator for soft wearable robots," *Rob. Auton. Syst.*, vol. 73, pp. 91–101, 2015.
67. Y. Sun, Y. S. Song, and J. Paik, "Characterization of silicone rubber based soft pneumatic actuators," in *IEEE/RSJ International Conference on Intelligent Robots and Systems*, pp. 4446–4453, 2013.
68. C.-P. Chou and B. Hannaford, "Measurement and modeling of McKibben pneumatic artificial muscles," *IEEE Trans. Robot. Autom.*, vol. 12, no. 1, pp. 90–102, 1996.
69. A. Ainla, M. S. Verma, D. Yang, and G. M. Whitesides, "Soft, rotating pneumatic actuator," *Soft Robot.*, vol. 4, pp. 297–304, 2017.

70. S. Terryn, G. Mathijssen, J. Brancart, D. Lefeber, G. Van Assche, and B. Vanderborght, "Development of a self-healing soft pneumatic actuator: A first concept," *Bioinspir. Biomim.*, vol. 10, no. 4, p. 46007, 2015.
71. G. Agarwal, N. Besuchet, B. Audergon, and J. Paik, "Stretchable materials for robust soft actuators towards assistive wearable devices," *Sci. Rep.*, vol. 6, no. 1, p. 34224, 2016.
72. I.-K. Oh, J.-H. Jeon, and Y.-G. Lee, "Mutiple electrode patterning of ionic polymer metal composite actuators," *Smart Structures and Materials 2006: Electroactive Polymer Actuators and Devices (EAPAD)*, vol. 6168, p. 616828, 2006.
73. K. Kikuchi, S. Tsuchitani, M. Miwa, and K. Asaka, "Formation of patterned electrode in ionic polymer-metal composite using dry film photoresist," *IEEJ Trans. Electr. Electron. Eng.*, vol. 3, no. 4, pp. 452–454, 2008.
74. J. W. L. Zhou, H.-Y. Chan, T. K. H. To, K. W. C. Lai, and W. J. Li, "Polymer MEMS actuators for underwater micromanipulation," *IEEE/ASME Trans. Mech.*, vol. 9, no. 2, pp. 334–342, 2004.
75. G.-H. Feng, and R.-H. Chen, "Improved cost-effective fabrication of arbitrarily shaped μIPMC transducers," *J. Micromech. Microeng.*, vol. 18, no. 1, p. 15016, 2008.
76. Z. Chen, and X. Tan, "Monolithic fabrication of ionic polymer–metal composite actuators capable of complex deformation," *Sensors Actuat. A Phys.*, vol. 157, no. 2, pp. 246–257, 2010.
77. Y. Nakabo, T. Mukai, and K. Asaka, "Kinematic modeling and visual sensing of multi-DOF robot manipulator with patterned artificial muscle," in *Proceedings of the 2005 IEEE International Conference on Robotics and Automation*, pp. 4315–4320, 2005.
78. I.-W. P. Chen, M.-C. Yang, C.-H. Yang, D.-X. Zhong, M.-C. Hsu, and Y. Chen, "Newton output blocking force under low-voltage stimulation for carbon nanotube–electroactive polymer composite artificial muscles," *ACS Appl. Mater. Interfaces*, vol. 9, no. 6, pp. 5550–5555, 2017.
79. J. Kim, S.-H. Bae, M. Kotal, T. Stalbaum, K. J. Kim, and I.-K. Oh, "Soft but powerful artificial muscles based on 3D graphene-CNT-Ni heteronanostructures," *Small*, vol. 13, no. 31, p. 1701314, 2017.
80. K. Kimura, and H. Ohigashi, "Ferroelectric properties of poly(vinylidenefluoride-trifluoroethylene) copolymer thin films," *Appl. Phys. Lett.*, vol. 43, no. 9, pp. 834–836, 1983.
81. Q. M. Zhang, H. Xu, F. Fang, Z.-Y. Cheng, F. Xia, and H. You, "Critical thickness of crystallization and discontinuous change in ferroelectric behavior with thickness in ferroelectric polymer thin films," *J. Appl. Phys.*, vol. 89, no. 5, pp. 2613–2616, 2001.
82. A. V. Bune et al., "Two-dimensional ferroelectric films," *Nature*, vol. 391, no. 6670, pp. 874–877, 1998.
83. A. V. Bune et al., "Piezoelectric and pyroelectric properties of ferroelectric Langmuir–Blodgett polymer films," http://oasc12039.247realmedia.com/RealMedia/ads/click_lx.ads/www.aip.org/pt/adcenter/pdfcover_test/L-37/386502181/x01/AIP-PT/JAP_ArticleDL_092017/scilight717-1640 × 440.gif/434f71374e315a556e61414141774c75?x, 1999.
84. M. Manohara, E. Morikawa, J. Choi, and P. T. Sprunger, "Transfer by direct photo etching of poly(vinylidene flouride) using X-rays," *J. Microelectromech. Syst.*, vol. 8, no. 4, pp. 417–422, 1999.
85. D. Li, and Y. Xia, "Electrospinning of nanofibers: Reinventing the wheel?" *Adv. Mater.*, vol. 16, no. 14, pp. 1151–1170, 2004.
86. S. T. Lau, R. K. Zheng, H. L. W. Chan, and C. L. Choy, "Preparation and characterization of poly(vinylidene fluoride–trifluoroethylene) copolymer nanowires and nanotubes," *Mater. Lett.*, vol. 60, no. 19, pp. 2357–2361, 2006.
87. Z. Hu, G. Baralia, V. Bayot, and J.-F. Gohy, and A. M. Jonas, "Nanoscale control of polymer crystallization by nanoimprint lithography," *Nano Lett.*, vol. 5, pp. 1738–7143, 2005.
88. R. Pelrine, R. Kornbluh, and S. Chiba, "Artificial muscle for small robots and other micromechanical devices," *IEEJ Trans. Sensors Micromachines*, vol. 122, no. 2, pp. 97–102, 2002.
89. S. Michel, X. Q. Zhang, M. Wissler, C. Löwe, and G. Kovacs, "A comparison between silicone and acrylic elastomers as dielectric materials in electroactive polymer actuators," *Polym. Int.*, vol. 59, no. 3, pp. 391–399, 2009.

90. Z. Chen, T. I. Um, and H. Bart-Smith, "Bio-inspired robotic manta ray powered by ionic polymer–metal composite artificial muscles," *Int. J. Smart Nano Mater.*, vol. 3, no. 4, pp. 296–308, 2012.
91. M. Wehner et al., "An integrated design and fabrication strategy for entirely soft, autonomous robots," *Nature*, vol. 536, no. 7617, pp. 451–455, 2016.
92. T. Li et al., "Fast-moving soft electronic fish," *Sci. Adv.*, vol. 3, no. 4, p. e1602045, 2017.
93. P. Walters and D. McGoran, "Digital fabrication of 'smart' structures and mechanisms-creative applications in art and design."
94. T. Umedachi and B. A. Trimmer, "Design of a 3D-printed soft robot with posture and steering control," in *IEEE International Conference on Robotics and Automation (ICRA)*, pp. 2874–2879, 2014.
95. Y. Yang, Y. Chen, Y. Wei, and Y. Li, "3D printing of shape memory polymer for functional part fabrication," *Int. J. Adv. Manuf. Technol.*, vol. 84, no. 9–12, pp. 2079–2095, 2016.
96. J. D. Carrico, N. W. Traeden, M. Aureli, and K. K. Leang, "Fused filament 3D printing of ionic polymer-metal composites (IPMCs)," *Smart Mater. Struct.*, vol. 24, no. 12, p. 125021, 2015.
97. O. Ivanova, A. Elliott, T. Campbell, and C. B. Williams, "Unclonable security features for additive manufacturing," *Addit. Manuf.*, vol. 1–4, pp. 24–31, 2014.
98. Y. Liu, J. K. Boyles, J. Genzer, and M. D. Dickey, "Self-folding of polymer sheets using local light absorption," *Soft Matter*, vol. 8, no. 6, pp. 1764–1769, 2012.
99. O. Pabst et al., "Inkjet printed micropump actuator based on piezoelectric polymers: Device performance and morphology studies," *Org. Electron.*, vol. 15, no. 11, pp. 3306–3315, 2014.
100. A. Miriyev, K. Stack, and H. Lipson, "Soft material for soft actuators," *Nat. Commun.*, vol. 8, no. 1, p. 596, 2017.
101. C. T. O'Neill, N. S. Phipps, L. Cappello, S. Paganoni, and C. J. Walsh, "A soft wearable robot for the shoulder: Design, characterization, and preliminary testing," in *2017 International Conference on Rehabilitation Robotics (ICORR)*, pp. 1672–1678, 2007.
102. C. J. Payne et al., "An implantable extracardiac soft robotic device for the failing heart: Mechanical coupling and synchronization," *Soft Robot.*, vol. 4, pp. 241–250, 2017.
103. A. Azuma, *The Biokinetics of Flying and Swimming, Second Edition*. Reston,VA: American Institute of Aeronautics and Astronautics, 2006.
104. M. Sfakiotakis, D. M. Lane, and J. B. C. Davies, "Review of fish swimming modes for aquatic locomotion," *IEEE J. Ocean. Eng.*, vol. 24, no. 2, pp. 237–252, 1999.
105. M. Yamakita, N. Kamamichi, T. Kozuki, K. Asaka, and Z.-W. Luo, "A snake-like swimming robot using IPMC actuator and verification of doping effect," in *IEEE/RSJ International Conference on Intelligent Robots and Systems*, pp. 2035–2040, 2005.
106. C. Laschi, M. Cianchetti, B. Mazzolai, L. Margheri, M. Follador, and P. Dario, "Soft robot arm inspired by the octopus," *Adv. Robot.*, vol. 26, no. 7, pp. 709–727, 2012.
107. M. M. Mower, and J. L. Roberts, "Augmentation of electrical conduction and contractility by biphasic cardiac pacing," 0217774 Al, 2006.

12

Flexible and Stretchable Photovoltaics and Its Energy Harvesters

Devendra Singh

12.1 Introduction and Background

In recent years, the enormous change in landscape of researchers in energy-based research and technologies in the market can be clearly observed. In the recent decades, clean, affordable, and domestic renewable energy and energy harvesters has been one of the key demands addressed towards the robust and diverse electronics applications. A renewable energy sector is always in demand due to the need for a sustainable and clean resource of energy and to save our environment from exhaustive usage of carbon, oils, etc. Also, to keep up with the demand and supply ratio, and to deal with cost factor, sincere attempts are being made in various energy sectors. Efficient and sustainable renewable energy resources are the current challenge with tedious development for the energy scavenging systems which can capture even nominal power of energy from light, vibration, or thermal sources, etc. To fulfill the need of consumer electronics, plenty of innovative development is in progress in flexible and stretchable conformal electronics industry. Whereas, the compatible energy harvesting devices are still lacking the breakthrough for the flexible electronics industry. With the help of advanced materials and technology, these photovoltaic-based energy harvesting devices can possibly replace the current energy storage systems like batteries, supercapacitors, hydrogen storage, etc.

12.1.1 Introduction of Flexible and Stretchable Photovoltaics and Its Energy Harvesters

With a fast growing and increasing demand of energy, solar energy offers a wide variety of abundant energy solutions through power conversion by buildings/windows to automobiles and consumer electronics etc. [1–3]. Conventional methods of energy harvesting through photovoltaic (PV) on rooftops, solar farms, street lights, etc. have been further revolutionized with innovative materials perspective and design of the modules [4–6]. Recently, PV technology has emerged as a fast-growing renewable source of energy and has tremendous potential for alternatives of traditional batteries and inverters [6]. To maintain the economic growth and as per the expert's prediction, the world would need around 30 terawatts of energy resources by the year 2050. Due to the non-ideal distribution of the earth's surface and many body surfaces on earth, PV technologies are more concerned with the constant capturing of maximum sun radiation throughout its energy spectrum. Although, the solar PV energy varies through availability of solar radiation at different locations, reliable large-scale and cost-effective solutions are still of huge interest towards sustainable energy future. Recently, Ripalda et al. demonstrated how the latest technology such as machine learning techniques can reduce the yearly spectral sets by three orders of magnitude to sets of a few characteristic spectra and use the resulting proxy spectra to find the optimal solar cell designs maximizing the yearly energy production [7]. To overcome the current energy demand and supply problem, PVs are a favorably clean and ideal power source for numerous potential and feasible reasons. To serve the low cost, reliable, efficient, and sustainable electronic devices, integrated solar PV modules are a great candidate for energy harvesting resources. Efficient PV device materials with high performance and their conformal integration of design architecture still are of deep interest towards relevant industry research for flexible and large area consumer electronics [8–10]. Current PV technology materials, methods, and devices are of huge interest to produce flexible (specifically stretchable) and sustainable durable module-based solar energy devices. To fabricate and market the robust solar panels, key components such as transparent and flexible conducting substrates (semiconductors and reflective electrodes) are coherently required with optimum optoelectronic performance along with their mechanical compatible and durability [11]. For the next generation PV devices, flexible and stretchable solar cells are of significant importance to fulfill the need of low cost and light weight modules with easy processing and compatible adaptability. The emerging field of flexible and stretchable electronics has recently been explored by various research groups [12–15] towards advanced semiconductors-based integrated devices. Scientific contributions have been demonstrated and are significantly in progress which can integrate the versatile electronic as well as mechanical properties of different categories of well-known structural materials.

As we are familiar, solar cell (also known as photovoltaic cell) devices directly convert the photon energy of light into electrical energy through the well-known photovoltaic effect (first observed in 1839 by a French physicist, Edmund Becquerel). These solar cells do not utilize the chemical reactions and do not require fuel to produce the electric power (unlike batteries or fuel cells) and also do not have any moving parts (unlike conventional electric generators). The PV effect, is the basis of the conversion of light to electricity in solar cells, occurs when a light enters a PV cell and imparts enough energy to some charge carriers (electrons) to free them. A built-in-potential barrier in the solar cell acts on these free electrons to produce a voltage, which can be used to drive an observed current through a circuit [16].

In this current chapter, we mainly focus on the advances in flexible and stretchable photovoltaics systems with a brief introduction of the basics of solar cells. Here, flexible and stretchable PV systems are described based on various types/generations of solar cells, including different advanced materials and technologies involved, with potential energy harvesting applications.

12.1.2 Working Principle of Solar Cells and Its Fundamental Operational Components

Towards functioning of a solar cell device (which directly converts sunlight into electricity), entered light on the cell produces both a current and a voltage (to generate electric power). The operation of a PV cell follows briefly three steps: (1) the absorption of light, generating either electron-hole pairs (or excitons), (2) the separation of various types of charge carriers (electron and/or holes), and (3) the extraction of those charge carriers to an external circuit. Although a variety of relevant materials and processes can usually satisfy the requirements for PV energy conversion, but in usual practice, nearly all PV energy conversion uses mostly semiconductor-based materials in the form of a p-n junction. For a detailed

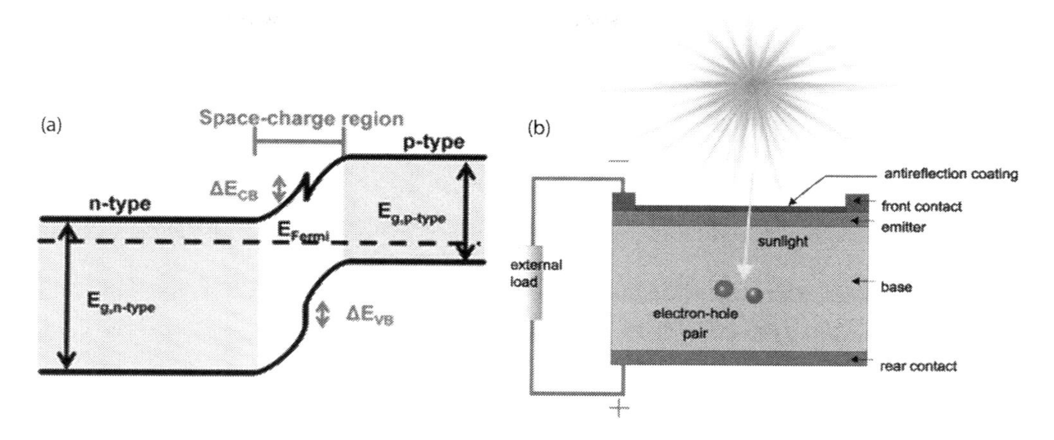

FIGURE 12.1 Generic sketches of (a) the band diagram for a p-n heterojunction in an inorganic solar cell. (From Bakke, J.R. et al., *Nanoscale*, 3, 3482, 2011.) and (b) Schematic illustration of cross section of the photovoltaic cell. (From http://www.pveducation.org/pvcdrom/solar-cell-structure.)

description for the working principle of a junction solar cell (shown in Figure 12.1), the above principle steps can be elaborated in the following fashion; *first step*: photon absorption by a valence band electron occurs if the photon energy amounts to at least the bandgap energy. The excited electron is then in the conduction band of the semiconductor material and free to move, as is the hole in the valence band. In the case of the exciton, the electron and hole are bound until separation occurs, *second step*: because of the electric field at the junction, photo-excited charge carriers (electrons and holes) flow in opposite directions close to the p-n-junction: either electrons flow towards n-layer or holes towards p-layer, depending on where the excitation due to photons occurred, and *third step*: due to charge flow across the junction, there occurs a charge imbalance between both semiconductor regions. Excited electrons (in the n-layer) have a tendency to escape to justify the charge imbalance. A current flow is produced, by providing an external load circuit, which will continue as long as photons strike the PV cell. Thus, the charge carrier then reaches the electrodes through semiconductors (electrons through n-type and holes through p-type), depending on the majority charge carriers. The electrodes are made out of either metal or transparent conductive oxide. These junction PV cells can be classified into different types according to the photoactive semiconductor materials classified into II–VI, III–V, etc. materials [17,18].

In general, standard sunlight conditions on a clear day at the equator at midday are assumed to give an irradiance of 1,000 watts of solar energy per square meter (1 kW/m^2), and this is generally referred to as a 'full sun' condition, i.e., full irradiance. In order to produce more electrical power from a PV solar cell, either increase the photovoltaic effect, the energy of photons, or produce a different type of cell that is more efficient at converting the solar energy into electricity. Figure 12.1 shows the schematic band diagram of an inorganic p-n junction solar cell and a generic current–voltage (I–V) plot of a photovoltaic device in operation. Whereas, p-type material acts as the absorber of incident light, and when an n-type semiconductor is placed in contact with it, a built-in voltage potential gradient, depletion region (space-charge region), is created. This gradient serves as a bias that assists in electrons flowing from the p-type material to the n-type material. When a photon generates an electron-hole pair in the absorber, the electron travels along the conduction band through the space-charge region and through the n-type semiconductor to a front contact, while the hole travels through the neutral region of the p-type material to the back contact [19].

A schematic and basic structure for thin-film PV cells are shown in Figure 12.1b along with the fundamental components of a PV cell, i.e., an active layer (base), consisting of one or more semiconducting materials that absorb light (to generate a photocurrent), and two contacts (front and rear) that collect the current, also show the photons absorbed in the active layer generate electron-hole (e$^-$-h) pairs.

As discussed earlier, an electric field must always be functioning to collect the e^--h carriers as an electric current across the device, commonly by using multiple doping types in the active layer to create a p-n junction, or multiple materials to create a heterojunction, and/or by employing contacts (with two different work functions). This built-in field causes photoelectrons and holes to flow in the opposite directions, until they are collected at their respective contacts, resulting in a net current. The rear (back) contact is typically a layer of a reflective metal, while the front contact must transmit as much light as possible (anti-reflective coating, commonly used as transparent conductive materials) into the active layer.

Figure 12.2 shows the typical PV solar cell I–V characteristics and P-V curves based on the cell model and (b) I–V curves showing the V_{OC} and J_{SC} and where shunt and series resistances affect device performance. When the solar cell acts as a diode, the output current increases exponentially as soon as the forward bias voltage increases, for which a typical schematic of current—voltage characteristic, whereas resulting output power of the solar cell versus applied voltage. For possible minimum losses, series resistance (due to the resistances of the active layer, contacts, and interfaces) should be made as small as possible, whereas shunt resistance (represents leakage pathways, e.g., due to non-uniform thickness of the active layer) should be made as large as possible.

Under short circuit conditions (zero applied voltage to the solar cell), a relatively large photocurrent can be extracted, known as the short-circuit current (I_{sc}). This current depends on the number of photons absorbed in the active layer, which in turn depends on the spectrum and irradiance of the light source as well as the bandgap and thickness of the active layer. The I_{sc} is roughly proportional to the irradiance for a given spectrum and also efficiency of the device is often compared in terms of the short-circuit current density (J_{sc}), which is the current per unit area. At the open circuit, zero net current shows equal and opposite behavior of the dark current and the photocurrent (this particular voltage is known as open-circuit voltage, V_{oc}) [20].

As indicated in Figure 12.2b, typical photovoltaic solar cell I–V characteristics and P-V curves based on the cell model, each solar cell has a maximum power point, point in its I–V curve at which the output power is maximized, where the power, current, and voltage at this point are P_{max}, I_{max}, and V_{max}, respectively, and the power conversion efficiency of the cell is defined as the ratio of P_{max} to the total power incident (light) on the solar cell. To evaluate solar cells, fill factor (FF) is another common figure of merit, FF describes the quality and idealness of a solar cell, equal to the ratio of P_{max} to the

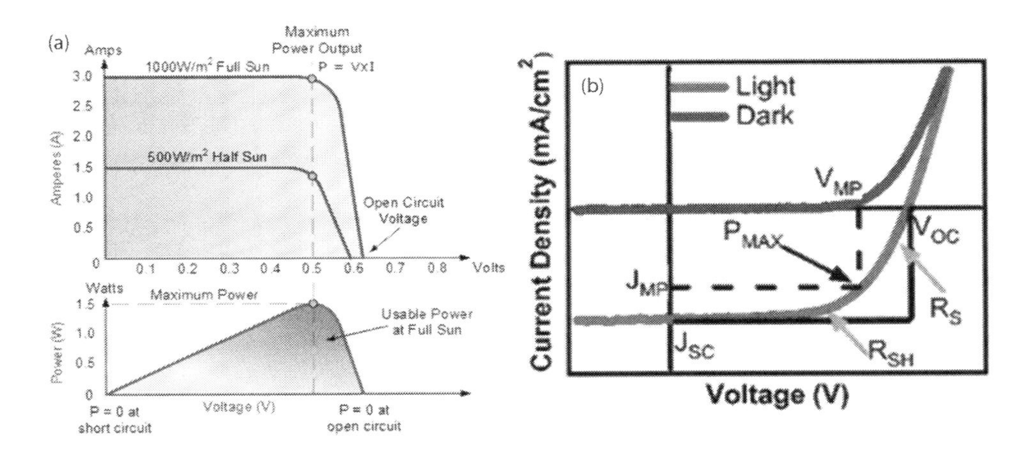

FIGURE 12.2 (a) The typical photovoltaic solar cell current-voltage (I–V) characteristics and power-voltage (P–V) curves based on the cell model. (From http://www.alternative-energy-tutorials.com/solar-power/photovoltaics.html) (b) I–V curves showing the V_{OC} and J_{SC} and where shunt and series resistances affect device performance. (From Bakke, J.R. et al., *Nanoscale*, 3, 3482, 2011.)

theoretical maximum power, $I_{sc}\ V_{oc}$. In practice, losses due to recombination, series resistance, and shunt resistance lead to a reduced FF. The efficiency and the fill factor of the solar cell can be derived as following:

$$\text{Efficiency}, \acute{\eta} = P_{max} / P_{in}, \text{ where } P_{max} = V_{oc} \cdot I_{sc} \cdot FF$$

or

$$FF = P_{max} / P_{theoretical} = I_{MPP} \times V_{MPP} / I_{sc} \times V_{oc}.$$

Optimizing the materials properties, thicknesses, and manufacturing processes for the different layers of the respective solar cell enables a cell with high J_{sc}, V_{oc}, FF, and therefore a high-power conversion efficiency to be achieved.

PV advanced generation has grown up as key energy contributors with about 1.7% of world electricity provided by solar cells, for promising flexible and efficient photovoltaic devices, such as multi-junction (which holds high efficiency records), ultra-thin, intermediate band, and hot-carrier solar cells, and on printable solar cell materials such as colloidal quantum dots for cost-effective solutions [22]. Apart from solutions for the complexity of stacks and cost factor, current technology trends provide simpler alternatives of highest efficiency devices. With new sustainable materials like organic photovoltaic (OPV), dye-sensitized solar cell (DSC), and hybrid perovskite, which are abundant in nature show fast progression and scope for the alternative resources of solar technology. The advantage of these materials can be ease of preparation at relatively low temperatures as well as under atmospheric pressure. For the flexible and hybrid requirement of PV technologies, the above kinds of materials can be printed on a variety of substrates including conformal and integrated platforms. Including the variability of the solar source, challenges with these materials include long-term stability impact, reliable studies of advanced modules, and possible environmental concerns due to involved toxic elements.

12.2 Types and Generations of Solar Cells

Due to the clean and greenest forms of renewable abundant sources, solar energy is still the highest priority along with hydroelectric and wind power, with the tremendous capacity of 100 GW solar power generation globally [23]. Researchers are interested in alternative PV technologies to replace traditional silicon-based panels with higher efficiency, flexibility, customized shape, low cost, and transparency. Due to the light weight potential of flexible solar cells (lack of heavy glass sheets, metal frames), which would also significantly reduce the transportation as well as deployment costs, infinite efforts have been reported very recently. As of today, PV technology has seen various developmental stages, categorized in four generations of solar cells, according to the futuristic need in terms of efficiency, cost, and life-time etc. Figure 12.3 shows the systematic classification of solar cells based on the primary active material.

Traditionally, Si-based crystalline films-based solar cells (wafer-based) were named into the first-generation photovoltaics (FGPV) technology which were the first highly efficient, but high cost invention towards renewable solar power production. The silicon technology is further categorized into sub-groups of single crystalline Si solar cells with efficiency ~18% (more than 90% module production globally), and polycrystalline Si solar cells reached the worldwide production up to 48%, yet are less efficient ~14%. Further, to reduce the cost burden and bulk behavior of fourth-generation solar cells (FGSC), *second-generation solar cells* (SGSC) were developed by introducing thin-film technology. The SGSC were improved by including amorphous/polycrystalline Si, copper-indium gallium selenide (CIGS), and cadmium telluride (CdTe) materials, but the overall efficiency was less compared to FGSC, whereas active volume aspects were critically developed, the cost per watt delivery was improved. Interestingly, the light absorbing layers of Si wafer-based solar cells have a much higher thickness of

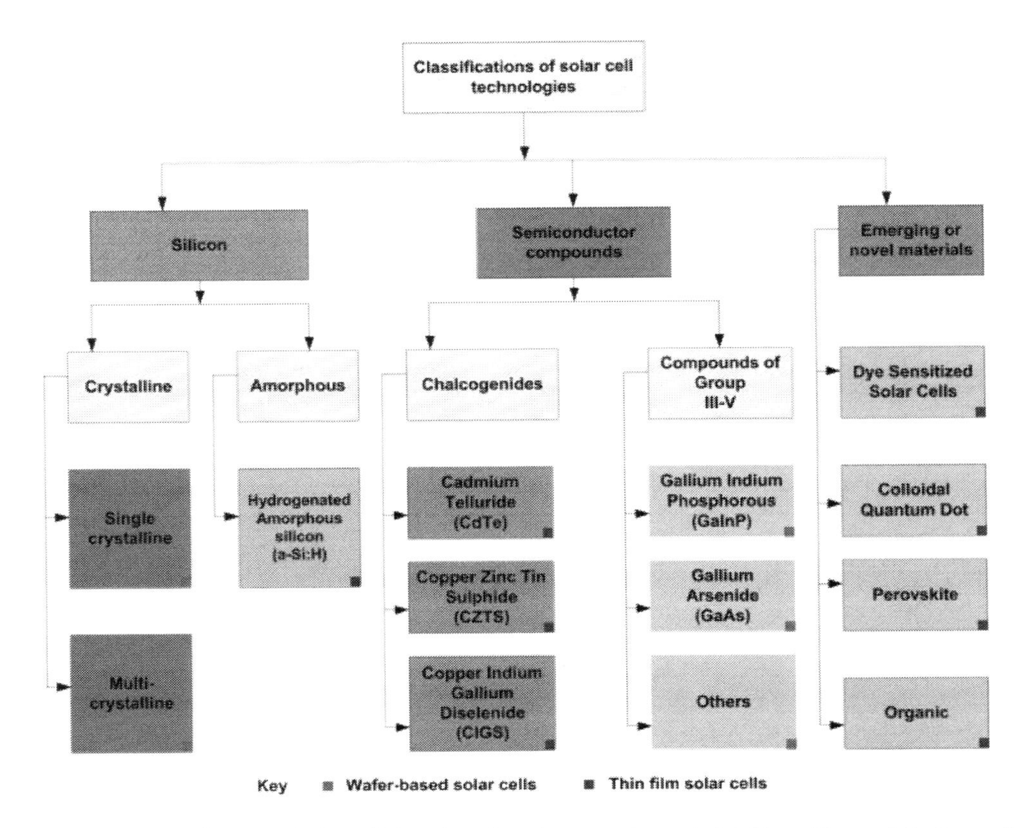

FIGURE 12.3 Classification of solar cells based on the primary active material. (From Verma, S.S., *Renew. Sust. Energ. Rev.*, 80, 1321–1344, 2017.)

up to 350 μm thick, whereas thin-film-based PV cells have very thin light absorbing layers with the thickness of the order of 1 μm. Although, CdTe has a direct optimum bandgap (~1.45 eV) with high absorption coefficient and efficiency up to 11% [24], cadmium is regarded as a heavy toxic metal, thus their recycling can be highly expensive and damaging too to our environment, which cause issues with this CdTe technology [2,24].

Compared to the CdTe thin-film solar cell, CIGS hold a higher efficiency ~10%–12%. The processing of CIGS can be done by the sputtering, evaporation, electrochemical coating technique, and electron beam deposition techniques [25] with the compatibility of various substrates from glass plate, polymers substrates, steel, aluminum, etc. The advantages of CIGS thin-film solar cells include its high durability without a considerable degradation. The integration of thin-film-based solar cells leads to significant manufacturing cost reduction compared to crystalline Si technology and, since the CdTe and CIGS modules share common structural elements and similar manufacturing cost per unit area, the module efficiency will eventually be the discriminating factor that determines the cost per watt. In fact, these two thin-film technologies have a common device/module structure: substrate, base electrode, absorber, junction layer, top electrode, and patterning steps for monolithic integration, as shown in Figure 12.4 where cross sections of scanning electron micrographs provide true physical perspectives of the structures [24].

For comparison, external quantum efficiency of commercial PV modules based on various generation solar cell materials (c-Si, a-Si, GaAs, mc-Si, CdS/CdTe) at short wavelengths has been shown in Figure 12.5 [26]. Precisely, these discussed PV technologies use the toxic heavy metal cadmium (CdSe, CdS, CdTe), whereas CIGS PV cells use indium, a metal in relatively short supply [27]. Although

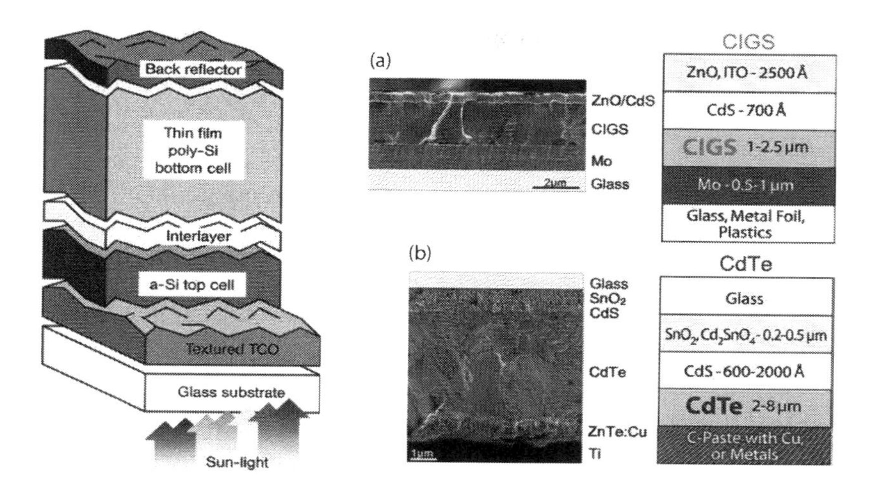

FIGURE 12.4 (a) CIGS and (b) CdTe device structures. (From Pagliaro, M. et al., *Flexible Solar Cells,* WILEY-VCH Verlag GmbH & Co, 2008; Yamamoto, K. et al., *Solar Energy*, 77, 939, 2004.)

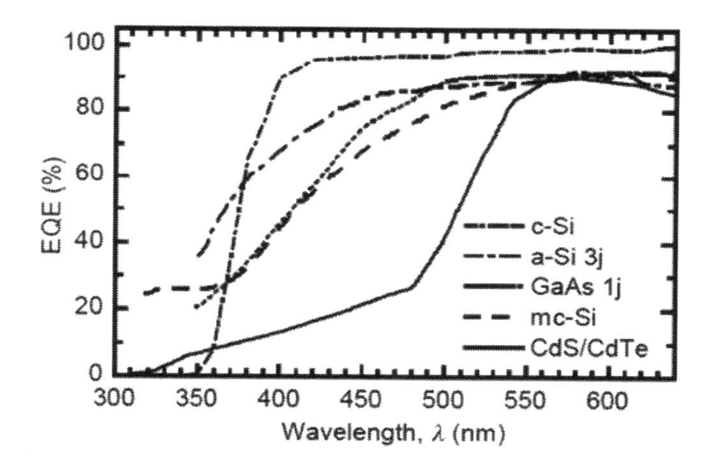

FIGURE 12.5 External quantum efficiency of different types of commercially produced PV modules. (From Richards, B.S. and McIntosh, K.R., Enhancing the Efficiency of Production of CdS/CdTe PV Modules by Overcoming Poor Spectral Response at Short Wavelengths Via Luminescence Down-Shifting, Photovoltaic Energy Conversion, *Conference Record of the 2006 IEEE 4th World Conference*, 213, 2006.)

indium-based transparent electrodes (specifically indium tin oxide (ITO)) are the most frequently used commercial product in the industry, limitation of indium and its recycling is a big challenge. As of today, only 15% of all ITO goes into actual products; the rest of the unused ITO is being scrapped and not recycled [2]. These challenges further pushed the PV technologies into *third generation solar cells* (TGSC) for highly efficient as well as cost-effective modules by introducing quantum dots, tandem/stacked multi-layers of inorganics based on III–V materials (e.g., GaAs/Ge/GaInP2, etc.) [18,28]. Fabrication technology was improved in terms of large surface area to deliver the customized solar cells. Many alternatives, from polycrystalline silicon to CdTe and CIGS alloy cells have been explored to replace Si by designing nanomaterials-based PV devices. Due to low cost, light weight, flexibility, and high optical absorption of solar spectrum, organic (specifically dye sensitized solar cells, DSSCs, which use organic dyes as light absorbers) materials have been introduced (since the 1990s) in TGSC's, along with photo-electrochemical cells, Gratzel cell, dye-sensitized hybrid solar cells, and organic solar cells [2]. Recently, this organic-based PV technology has crossed the efficiency of ~10%, which also needs the following crucial materials distribution; polymer used for light absorption and hole transport, another material was used for electron transport, and to a lesser degree also for light absorption, e.g., fullerene [29]. To enhance energy output, charge dissociation, charge transport, and flexibility of the device, the *fourth-generation solar cells* (FGSCs) have been developed by integrating the stability of novel inorganic nanostructures with low cost thin-film PVs. For FGSCs, the concept of bulk heterojunction has been found the most effective polymer solar cells for efficient solar harvesting devices. Now, for the further advancement of the next generation of solar cells, confined dimensions of the nanomaterials has been in rapid progress (such as 2D, 1D, and quantum dot materials) to obtain inexpensive, highly efficient, transparent, and flexible hybrid PV technology [30]. The recent trend of PV research shows the integration of the above aspects of hybrid organic-inorganic materials-based solar market which have shown recent progress of up to 20% of efficiency on a cell level [31]. These highly efficient so-called perovskite solar cells are able to combine the process of light absorption as well as electron and hole transport just like other typical inorganic thin-film technologies, with a similar fabrication and characterization process compared to inorganic thin-film solar cells. Apart from the materials and structure aspects, for efficient light trapping, surface textures can be modifying for contact layers, with a lower refractive index compared to absorber layer, that leads to scattering of light which further increases the path lengths of weakly absorbed light in the absorber layer. Although the Si-based PVs have dominated the world requirement with more than 90% demand before 2010, thin-film-based solar cells have grown their market up to 25% and leading further with rapid progress due to their low cost, light weight, and the ability to fabricate on flexible platforms (plastic, stainless steels) [32,33]. The leading companies in the a-Si thin-film photovoltaics (TFPV) field have undergone tremendous and rapid expansion globally from an annual production capacity ~30 MW–300 MW by 2010, by applying their PV products to flat surface land, roof, and building-facade applications. As an overview of the existing PV technologies, cell efficiency of various solar cells devices has been summarized in Figure 12.6 [34], the reference temperature is 25°C and the area is the cell total area, or the area defined by an aperture. The summarized cell efficiency results are provided within different categories (with 26 sub-categories) of semiconductors: (1) multi-junction cells, (2) single-junction GaAs cells, (3) crystalline Si cells, (4) thin-film technologies, and (5) emerging photovoltaics.

Among typically classified solar cells, such as Si-based solar cells (limitations of brittleness due to metal oxides and electrodes), DSSCs, organic solar cells (low efficiency), quantum dot solar cells, and perovskite solar cells (fast growing), to increase the cell efficiency many approaches are developed including novel materials, processing techniques, and light harvesting device structures [35]. To avoid the brittle nature of metal oxides-based electrodes (ITO, fluorine-doped tin oxide [FTO], doped-ZnO, etc.), many alternatives have been implemented such as carbon-based materials (e.g., carbon nanotubes (CNTs) and graphene), conductive coated flexible substrates, and the hybrid approach of inorganic-organic materials on flexible substrates [36]. Thus, to conceptualize and fabricate integrated and hybrid PV energy harvesting and storage devices, the understanding of the involved materials and processes needs to be explored multi-dimensionally to out-perform key properties like flexibility, stretchability, bendability, stacking, etc. for potential device applications. Also, to manufacture the next generation of flexible and efficient photovoltaic modules, the material used for contact electrodes including window and buffer/interface layers with optimum composition and processing conditions plays a significant role in the overall device performance and its stability. Indeed, for efficient and flexible solar systems, indium-free ultra-thin absorbers along with the choice of contacts, buffers, and the interfaces need redesigning to further decrease losses, specifically losses associated with interface recombination [37].

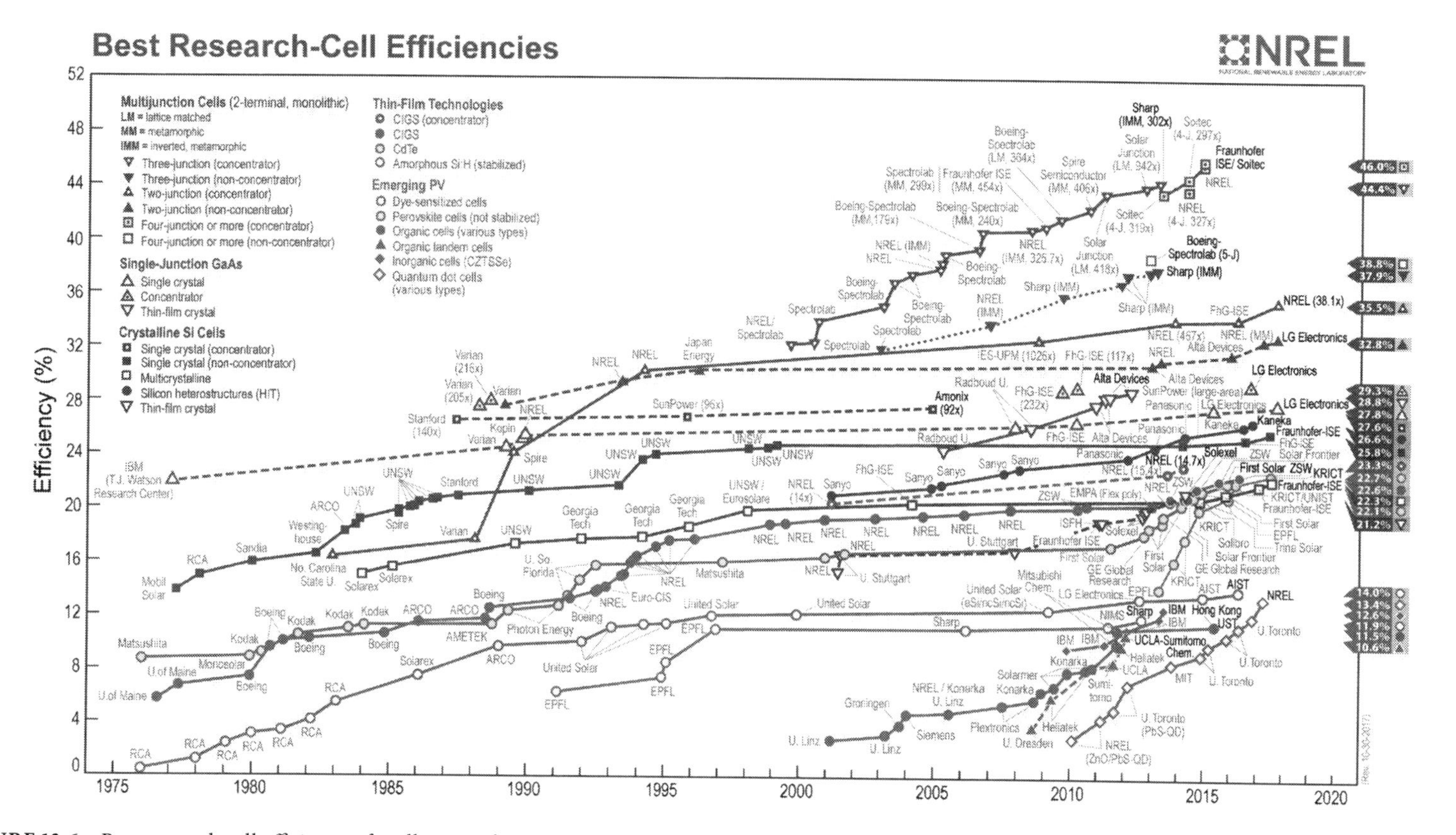

FIGURE 12.6 Best research cell efficiencies for all types of existing PV technology. Data from National Renewable Energy Laboratory (NREL). (From Best Research Cell Efficiencies for all types of existing PV technology.)

12.3 Advanced Materials and Technologies for Flexible and Stretchable Solar Cells

Advanced electronic materials and precise methods of manufacturing that produce flexible, stretchable, and fracture-proof devices would indeed revolutionize consumer electronics, specifically biomedical devices. As we discussed, thin-film semiconductors-based flexible PVs are nearing commercialization, but stretchable electronic materials-based devices are in their infancy, which need advances in the materials, fabrication methods, and devices for stretchable PV modules [38].

Towards the demanding researches for the applications based on flexible as well as stretchable electronics, integration of polymer substrates with organic-inorganic hybrid systems, semiconducting materials with compatible molecular structures are of keen interest. Several strategies have been explored to impart intrinsic elastic/plastic properties along with the electronic properties of semiconducting materials. From the intrinsic mechanical behavior of transparent electrodes to the active layers of interest to pursue stretchable electronic devices preferably with complaint substrate integrating with flexible packaging technology. Towards scalable production of high-quality epitaxial films on flexible, lightweight, and inexpensive substrates for optoelectronic devices, Dutta et al. have demonstrated a method to produce epi-Ge films on metal foils using a continuous roll-to-roll plasma-enhanced chemical vapor deposition (PECVD) process [39]. The Ge templates were subsequently used for a proof-of-concept single-junction flexible GaAs device development with an 11.5% power conversion efficiency (which is a record for GaAs solar cells) directly deposited on alternative substrates, which led to the possible development of high-efficiency and low cost GaAs solar cells on inexpensive flexible metal substrates with further improvement in defect density, grain-boundary passivation, and device efficiency for commercial success [39]. Figure 12.7 (a) shows a schematic of the device architecture of the flexible GaAs solar cell on metal foils, (b) shows the photos of a metal foil rolled in a spool used as a starting substrate for roll-to-roll depositions, epitaxial single-crystalline-like Ge layer deposited by roll-to-roll CVD, and sputter deposition on the metal foil, a photograph of a flexible single-junction solar GaAs solar cell on metal foil substrates, respectively, and (c) shows the illuminated J-V plots of GaAs solar cells fabricated on CVD and sputtered Ge templates. Very recently, to trade-off the qualities of rigid Si-wafers, Rabab et al. [40] have demonstrated a complementary metal-oxide semiconductor-based integration strategy where corrugation architecture enables ultra-flexible (shown in Figure 12.7d–7e) and low cost solar cell modules from bulk monocrystalline large-scale Si solar wafers with a 17% power conversion efficiency. Figure 12.7d and e shows a schematic representation of ultra-flexible c-Si solar cells on a 5-inch wafer scale, and the weight and total power (mW) measured of the devices.

The first *organic solar cells* with enough current output were based on an active bilayer made of donor and acceptor materials and were invented in the mid-1980s, achieved a power conversion efficiency of about 1%. Light is usually absorbed mainly in the so-called donor material, a hole-conducting small molecule, or conjugated polymer. Organic PV cells can be constructed by various approaches, including single layer, bilayer hetero junction, as well as bulk hetero junction cells. Single layer cells consist of a metal, organic, metal (crystals of highly conjugated and polycyclic molecules) sandwich structure and are the oldest and simplest examples, whereas, bilayer heterojunction PV cells consist as a sandwich of anode, donor, acceptor, and cathode. The electrodes are chosen based on their respective matches with the HOMO (highest occupied molecular orbital) of the donor and the LUMO (lowest unoccupied molecular orbital) of the acceptor [1].

The advanced *bulk heterojunction solar cell* concept was introduced (early 1990s), taking into account the low exciton diffusion length in disordered organic semiconductors and the required thickness for the sufficient light absorption. Bulk heterojunction cells are similar in fashion as they also use separate donor and acceptor molecules. Most conjugated polymers in their un-doped state are electron donors when photo-excited, whereas, the bulk heterojunction is usually based on blends of polymer donors and highly soluble fullerene-derivative acceptors [9,41]. Further, optimization of process methods and materials has been developed to achieve higher power conversion efficiency (e.g., 5.8% for polythiopene-fullerene cells). However, disordered organic photovoltaic cells have better potential in

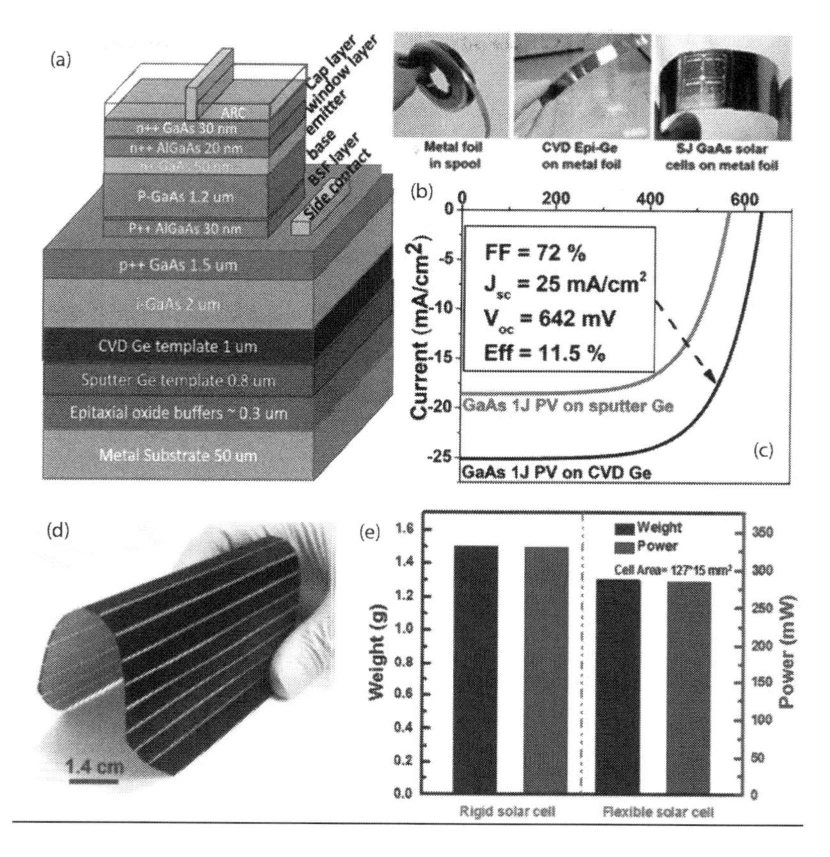

FIGURE 12.7 (a) Schematic of the complete device architecture of the flexible GaAs solar cell made on metal foils. (b) Photograph of metal foil rolled in a spool used as starting substrate for roll-to-roll depositions, epitaxial single-crystalline-like Ge layer deposited by roll-to-roll CVD and sputter deposition on the metal foil, and a photograph of a flexible single-junction solar GaAs solar cell on metal foil substrates, respectively. (c) Illuminated J–V plots of GaAs solar cells fabricated on CVD and sputtered Ge templates. (From Dutta, P. et al., *Energy Environ. Sci.*, 12, 756–766, 2019.) (d) Schematic progression of the deep reactive ion-etching (DRIE) and corrugation technique process with stepwise representation of the corrugation flow followed by the finally achieved ultra-flexible c-Si solar cells on a 5-inch wafer scale. Electrical performance of the flexible c-Si solar cells compared to the rigid ones. (e) Weight and total power (mW) measured of devices of an area = 127 × 15 mm². (From Bahabry, R.R. et al., *Adv. Energy Mater.*, 8, 1702221–1702235, 2018.)

view of manufacturing by roll-to-roll printing for low cost, flexible, and light weight panels/modules. To cover the broader range of the solar spectrum (including UV-visible and IR) are solution-processed multi-junction solar cells integrating different absorption ranges of existing constituent materials. One example is the tandem solar cell, made from the connection in a series of two sub-cells with complementary absorption ranges, high efficiencies achieved for organic tandem solar cells were recently reported [42,43]. An efficient PV cell based on an organic semiconductor would be less expensive and more easily manufactured compared to that of Si-based PV cells, whereas their integration would make the flexible and light weight challenges easier. Since this decade, inkjet printing technology has been intensively explored by the respective research community to execute the large area effective and efficient PV panels, specifically organic-based solar cells. Towards the flexible PV generation of thin-film-based solar panels, Figure 12.8 visualizes the various examples with flexible perovskite solar cells and tandem solar cell on plastic.

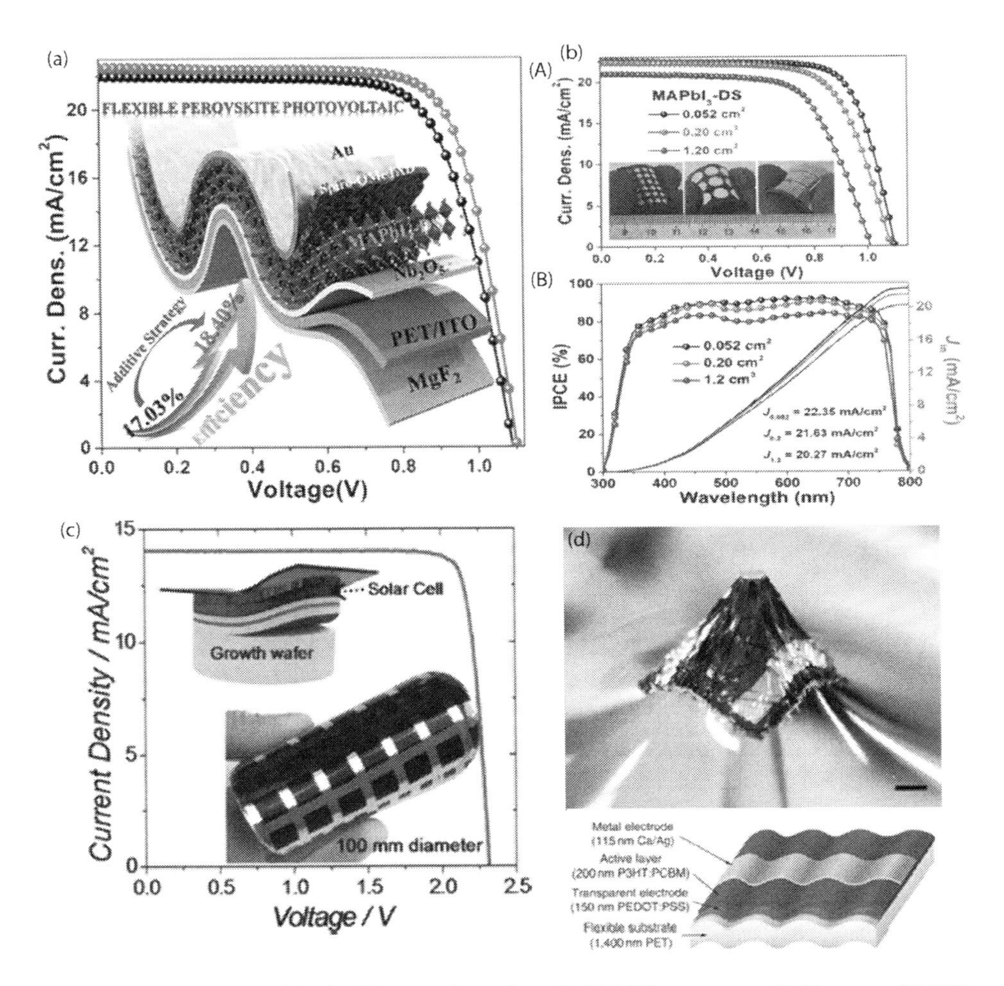

FIGURE 12.8 (a) Illustration of the flexible perovskite solar cells (F-PSC) structure and J-V curves of F-PSCs under reverse scan directions and (b) J-V curves and induced photon to current efficiency (IPCE) of different active areas for the methylammonium lead triiodide (MAPbI3–DS) flexible devices, and the inset shows the picture of F-PSCs with various active areas. (From Feng, J. et al., *Adv. Mater.*, 30, 1801418–1801427, 2018.). (c) Photograph of the final 100 mm-diameter flexible InGaP/(In)GaAs tandem solar cell on plastic. (From Kaltenbrunner, M. et al., *Nat. Commun.*, 3, 1772–1779, 2012.) (d) Scheme of sub-2 μm-thick ultra-light and flexible organic solar cells, device attached to the elastomeric support, under three-dimensional deformation by pressure from a 1.5 mm-diameter plastic tube. Scale bar 500 μm. (From Shahrjerdi, D. et al., *Adv. Energy Mater*, 3, 566–571, 2013.)

The roll-to-roll printing process is significantly less expensive with ease of processing compared to the multi-step assembly of traditional solar cells. The polymer devices are conformal and compatible with various substrates (also do not need additional patterning), whereas inkjet printing enables manufacturing of solar cells with multiple color options as well as patterns for product applications that require lower power. Also, to reduce the cost burden, techniques like polymer-based solution brush-painting and spin-coating have been carried out to build bulk hetero-junction (BHJ) solar cells with enhanced performance of the plastic solar cells. The non-uniformity of thickness for other printing processes (doctor blading, slot extrusion, etc.), which are more compatible with the roll-to-roll system compared to spin-coating, show variations in efficiency. The low fill factor and poor efficiency may be attributed to the formation of large aggregates during the brush painting, which may limit the efficient charge separation and transport phenomena.

In general, polymer-based solar cells are approaching power conversion efficiencies of greater than 5%, still one fifth of the 15% average efficiency of existing commercial Si-based modules [47]. To gain the maximum output efficiency, current ideas to develop an ordered nanostructure with controlled dimensions include, small bandgap polymers with absorption edges as low as 1 eV, absorption coefficients larger than 10^5 cm^{-1}, and charge carrier mobilities higher than 10^4 cm^2 V^{-1} s^{-1}, along with minimizing energy loss at the donor–acceptor interface by tuning energy levels [48]. Recently, Cui et al. have reported that, for improved OPV cells, a chlorinated non-fullerene acceptor which exhibits an extended optical absorption and meanwhile displays a higher voltage than its fluorinated counterpart in the devices ascribes to the reduced non-radiative energy loss (0.206 eV). A high efficiency of 16.5% is achieved due to the simultaneously improved short-circuit current density and open-circuit voltage, which shows that finely tuning the OPV materials to reduce the bandgap-voltage offset has great potential for boosting the efficiency. With these kinds of results for high efficiency and compatibility, industries are digging into an organic PV-based flexible product which can efficiently power the portable devices and charge power banks [49].

Towards hybridizing organic and inorganic solar cells, dye-based materials have also been developed to create a better solar cell with cover crystals of semiconductor titanium dioxide with a layer of chlorophyll. However, the electrons were reluctant to move through the layer of pigment, so the efficiency of the first solar cells sensitized in this way was about 0.01%. Invented in the early 1990s, DSSC (also known as Gratzel cell) entered with the first commercial modules based on this versatile organic-inorganic approach [50]. With plastic solar cells, DSSCs share the low weight, flexibility, and the low cost of production due to roll-to-roll manufacturing of thin-film modules. Yet, their typical 7% efficiency in commercial modules is about twice the efficiency of polymeric modules; whereas their good performance in diffuse light conditions is a feature they have in common with inorganic thin-film solar modules. Finally, dye cells work well in a wide range of lighting conditions and orientation, and they are less sensitive to partial shadowing and low-level illumination. All these attributes make them particularly well suited for architectonic applications. In general, products envisaged by the producers cover a broad range from colorful decorative elements to electric power-producing glass tiles for facade integration in buildings. DSSCs are based on absorption and charge transfer from separate species which is contrary to conventional solar cells (in which the same semiconductor absorbs light and works as a charge carrier). In addition to being a highly effective semiconductor, TiO_2 is also abundant, low cost, non-toxic, and bio-compatible [50]. DSSCs are very tolerant to the effects of impurities because both light absorption and charge separation occur near the interface between two materials. The relative impurity tolerance and simplicity allow for easy, inexpensive scale-up to non-vacuum- and low-temperature-based high-volume manufacturing via continuous processes including screen-printing, spraying, pressing, or roll-to-roll production.

Furthermore, in contrast to Si-based modules, the performance of dye PV modules actually increases with temperature, as a result, they outperform amorphous Si-based modules, despite their lower efficiency. According to the latest reports, stable and highly efficient modules are certainly within reach, with the energy-payback period being significantly shorter than other PV technologies. More remarkably in the past decade, the Gratzel group has reported on a 15% tandem cell consisting of a DSC top cell and a Cu(In, Ga)Se bottom cell [51]. By stacking the two cells, an efficiency of 15% is obtained, which is roughly twice the value of each cell individually. Due to its elegant colored and transparent aspect, the DSSC technology is particularly well suited for being integrated into buildings and allows for building-integrated windows, walls, and roofs of varying color and transparency that will simultaneously generate electricity, even in diffuse light or at relatively low light levels for a cost-effective PV integration.

In context to flexible DSSCs, apart from the cost-effective and scalable fabrication methods, electrode materials (photo-electrode and counter-electrode) are the most challenging aspects including the conformal substrate, whereas sensitizers and electrolytes also play a significant role for future devices. During the operation of DSSCs, redox mediators in the electrolyte regenerate the oxidized sensitizers, recover itself at the counter-electrode, and thus serve as a medium for the transporting charge carriers. Flexible types with solid-state hole conductors show much lower performance than glass-based DSCs and even quasi-solid-state flexible dye-sensitized solar cells (FDSCs) [52]. The option of high transparency in the near infrared region also reflects the use of DSC as a top cell in tandem solar cells devices. Towards flexible and with a potential approach, recently, Yun et al. have fabricated textile-based DSSC devices showing high flexibility and high performance under 4-mm radius of curvature over thousands of deformation cycles, reflecting a huge range of applications, including transparent, stretchable, wearable devices [53].

In recent years due to a huge demand of PVs, which has led to a silicon feedstock shortage, cells based on a different material system and production technology such as the strain-balanced quantum well solar cell (QWSC) have been explored [54]. The first generation of these new quantum well photovoltaic cells operate at a 27% efficiency, which is approximately twice the efficiency of the current Si-based PV cells and close to the single junction cell efficiency record of 27.8%. Overall, these third-generation cells based on GaAs and other III–V semiconductor materials have advantages over commonly used Si-based cells for flat panel PV. When the strain balanced QWSC is incorporated in a tandem cell, the wider spectral range leads to higher cell efficiency. Interestingly, *nanostructured solar cells* (NSC) have come up as a new and promising possibility to reduce the cost of PV cells and modules for bulk power generation as well as to improve the cell conversion efficiency. Different nanostructured materials, such as metallic or semiconducting wires, nanorods, nanotubes, etc. can be synthesized/deposited/mounted on conductive glass by various process methods to overcome the efficiency issues of conventional silicon-based cells [55]. On one side, production costs of solar cells can be reduced with less expensive semiconducting materials, but also, the solar cell may exploit the reduced loss of energy due to the shorter distance of energy transportation in the respective solar cell. For example, producing cheaper and flexible solar cells composed of group III–V nanowires (such as gallium arsenide, indium gallium phosphide, aluminum gallium arsenide, and gallium arsenide phosphide) could achieve a conversion efficiency of 20% and, in the longer term, 40% [31,55]. Whereas, the length of the wire maximizes absorption, but also their nanoscale width (thus the optimum aspect ratio) permits a much freer movement and collection of electrons. Although nanomaterials and nanostructures have promising potential by using the intrinsic advantages associated with these materials, including efficient photon management, rapid charge transfer, and short charge collection distances, recently, a rule for nanostructured solar cells by concurrently engineering the optical and electrical design is proposed for further exceeding the theoretical limit of solar cell efficiency [56]. As discussed earlier, fabricating efficient and transparent OPVs presents a few optimization challenges, alternatives to ITO-based transparent top electrodes include metal nanowires, conductive polymers, thin metal layers, and graphene. Recently, Song et al. demonstrated graphene as both the anode and cathode for low bandgap polymer devices by developing a room temperature dry-transfer technique to transfer graphene onto organic layers. They combined highly transparent electrodes with organic compounds that absorb primarily in the UV and near infra-red (NIR) regimes, to achieve high optical transmittance (~61%) across the visible spectrum and power conversion efficiency of 2.8%–4.1%. Furthermore, they showed that devices can be fabricated on a variety of flexible substrates including plastic and paper, and that devices with graphene electrodes are more resilient to bending than those with ITO-based electrodes [57].

Another advanced PV technology has been evolved as *graphene-based solar cells*, in which graphene can be used as ultra-thin transparent conductive film for window electrodes in cells, where graphene can most commonly be obtained from exfoliated graphite oxide, followed by thermal reduction [32]. Excellent properties including high conductivity, good transparency (in both the visible and near IR regions), tunable wettability, and high chemical and thermal stabilities, stand graphene-based electrodes more suitable for advanced PV cells. Graphene has emerged as a promising alternative electrode due to its simple processing, it enables inexpensive and large-scale industrial manufacturing, for advanced optoelectronic devices. With the ability of processing in the form of free-standing graphene layers with tuned bandgap capacity, it has opened the route to a new generation of carbon-based PV cells. By the right selection of respective substrates, graphene-based PV cells could eliminate the issues of indium scarcity that may affect development of thin-film photovoltaics. Apparently, graphene/silicon (Gr/Si) Schottky junction solar cells have also drawn much interest as an alternative low cost, easy fabrication structure in photovoltaic devices [58].

Recently developed *perovskite solar cells* (PSCs) are a very promising thin-film PV technology due to their high output power conversion efficiencies (>18% efficiency), which combine a light-absorbing material such as mesoporous TiO_2 for electron collection. Unlike conventional DSSCs, PSCs are solid-state devices and typically utilize organo-metal lead tri-halide perovskites as the absorbing layer along with an organic hole transporting layer, with the recent advantages in low-temperature processing compliant with inexpensive, flexible plastic substrates [59]. The promise of organic and perovskite solar cells lies in the ability to fabricate them on a large scale, on flexible substrates, using high-speed and low-energy manufacturing processes and low cost abundant materials. The power conversion efficiency of organic-inorganic hybrid perovskite solar cells has been boosted to be comparable with that of commercial silicon solar cells. Although the fabrication cost of solution-processing PSCs can be lower than that of silicon solar cells, PSCs are still facing the big challenge of instability in air due to the presence of organic components in the perovskites [60]. To obtain stable PSCs, fully replacing the A-site organic cations with pure inorganic cations is one of the promising methods. The power conversion efficiency of PSCs has remarkably increased from 3.8% to 23.7%, but due to poor stability, it creates a huge barrier in its commercialization.

For a hybrid approach of efficient and flexible solar cells, Kang et al. demonstrated ultralight and PSCs with orthogonal Ag nanowires-based transparent electrodes fabricated on thick polyethylene-terephthalate (PET) foils [61]. The resultant PSCs with orthogonal AgNW transparent electrodes exhibit substantially improved device performance, achieving a power conversion efficiency (PCE) of 15.18%, over PSCs with random AgNW network electrodes (10.43% PCE). Moreover, ultralight and flexible PSCs with the orthogonal AgNW electrodes exhibit an excellent power-per-weight of 29.4 W gl, which is the highest value reported for a light weight solar cell device.

Recently, to provide a continuous source of energy in wearable devices, textile-compatible photovoltaics have been shown keen interest towards better energy harvesting (on the order of milliwatts) keeping important aspects of environmental stability, sufficient energy efficiency, and mechanical robustness. Jinno et al. reported ultra-flexible organic photovoltaics coated on both sides with elastomer that simultaneously realize stretchability and stability in water whilst maintaining a high efficiency of 7.9% [62]. The efficiency of double-side-coated devices decreases only by 5.4% after immersion in water for 120 min. Furthermore, the efficiency of the devices remains at 80% of the initial value even after 52% mechanical compression for 20 cycles with 100 min of water exposure. In a similar approach of wearable devices, textile DSSCs woven using PV yarns have been demonstrated by Liu et al., but there are challenges in their implementation arising from the mechanical forces in the weaving process, evaporation of the liquid electrolyte, and partially shaded cells area, which all reduce the performance of the cell. Liu et al. proposed a novel fabrication process for a monolithic-structured printed solid-state DSSC on glass-fiber textile (contain multiple layers of electrodes and active materials) using all solution-based processes, these PV textile devices have shown a peak efficiency of 0.4%, which is potentially suitable for the low cost integration of PV devices onto high temperature textiles [63].

Towards high-efficiency polycrystalline thin-film-based solar cells, Kranz et al. have demonstrated the promising way to enhance the efficiency of CIGS solar cells is by combining them with perovskite solar cells in tandem devices [64]. They presented a process methodology for the fabrication of NIR-transparent perovskite solar cells, which enables power conversion efficiencies up to 12.1% combined with an average sub-bandgap transmission of 71% for photons with a wavelength between 800 and 1000 nm. This combination of a NIR-transparent perovskite top cell with a CIGS bottom cell enabled a tandem device with 19.5% efficiency, which is the highest reported efficiency for a polycrystalline thin-film tandem solar cell, a further enhancement up to 27% for the future developments of perovskite/CIGS tandem devices are also discussed.

In summary, the recent PV technologies rely on the following most important aspects of successful device design: materials, device physics, and manufacturing technologies, and among all the field, organic-inorganic photovoltaics is seemingly shown tremendous and successful interests by the research and industrial community. High-performance and conformable solar cells are of interest for many applications, ranging from the power source for consumer and wearable electronics, to building integrated PV and large-scale power generation. Following the cost-effective technology and large-scale commercialization, the advantages like system integration and reel-to-reel large-scale manufacturing issues might open the route towards improving efficiency by integrating the abovementioned three crucial aspects.

12.4 Flexible Photovoltaic-Based Energy Harvesting Systems

Towards the PV-based energy harvesting devices, over the past several years, flexible electronic materials and printing-based fabrication techniques have emerged as a potentially transformative set of technologies for photovoltaics, energy storage, and other electronics industries. As mentioned earlier, thin-film and emerging technologies in PV offer advantages for conformal, light weight, and flexible power over the rigid silicon panels that dominate the present market. One important advantage is high specific power (the power-to-weight ratio). Reese et al. categorized various strategies to understand the different aspects of PV markets for consumers. They mainly examined the cost-production experience curves of Si, CdTe, and CIGS PV, whereas they assessed the critical role of the substrate, packaging, and interconnects and provide a quantitative assessment of pathways to maximize specific power. With all requisite components included, along with requirements for safety and reliability, they proposed four general strategies which can advance high-specific-power PV [65]. First, heavier, flexible, but high-temperature-compatible materials (that is, metal foils) offer more immediate deployment in flexible markets, their complex packaging sets a weight limit and ultimately adds material costs that exceed those of rigid products. The second strategy utilizes materials like flexible glass, which allows reduced complexity and hence weight, while still allowing the high-temperature growth used on rigid products.

The third strategy requires a shift to low-temperature processes that achieve high efficiency when paired with light weight plastics/polymers substrates. In general, the challenge of this approach is to improve efficiency and/or stability. The fourth strategy lifts-off or thins freestanding cells to obtain high efficiencies that may be integrated into any package.

The use of flexible electronic materials can reduce weight, improve portability, and simplify PV system installation, in addition to enabling entirely new electronics applications such as wearable sensors and smart labels. Additive printing and coating techniques allow materials to be deposited over large areas at high speeds and low temperatures, enabling customizable electronic systems on plastic substrates with low cost and low embodied energy. The successes of these technologies inspire visions of printed, flexible, integrated electronic systems such as that illustrated in Figure 12.9 in which a flexible PV module and battery layers are integrated with printed power management electronics and printed load devices [5].

As we discussed, the majority of industry research based on flexible and solution-processed devices has focused on the development and characterization of new materials, device architectures, and manufacturing processes. The constraints (cost, performance, mechanical properties, degree of integration) must drive system-level design choices such as system sizing and selection of power management electronics, as well as device-level choices such as the materials and manufacturing processes used for the photovoltaic and energy storage devices. To achieve the thin and flexible form-based energy storage devices, flexible batteries (dominantly lithium-ion) have also been developed and integrated into higher-voltage multi-cell energy storage modules that could be combined with solar cells. Unlike batteries, the voltage of a supercapacitor is zero when the capacitor is fully discharged, as a result, if a supercapacitor connected to a PV module is allowed to discharge completely, the operating point of the PV module will be far away from the maximum power point, and the efficiency of charging will be low until the voltage has increased again. The efficiency can be improved by not allowing the capacitor to discharge completely, thus maintaining the PV module closer to its maximum power point; however, this reduces the usable energy storage capacity of the supercapacitor. Currently, the benefits of integrating a flexible battery or supercapacitor with the PV module, either by layering the components or as a

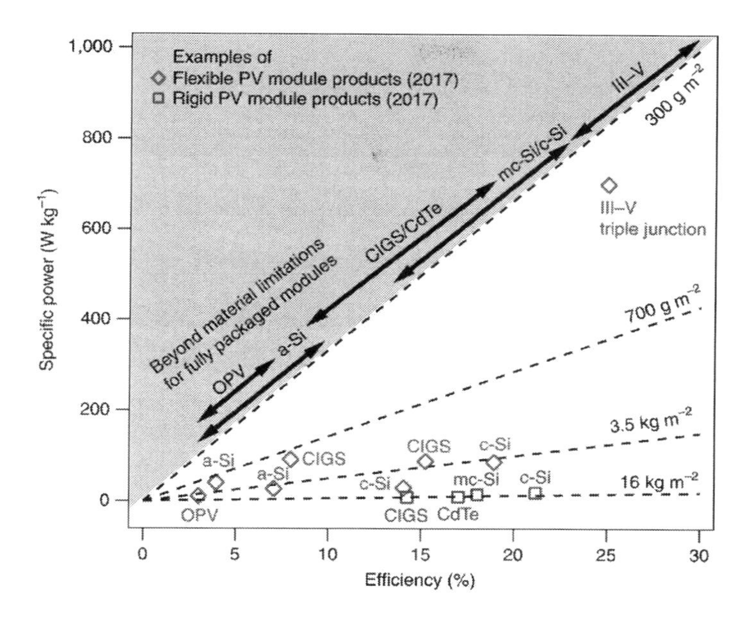

FIGURE 12.9 Specific power as a function of AM1.5G module efficiency. Data points represent current commercial products including amorphous Si (a-Si), CdTe, CIGS, III–V triple junction, multi-crystalline Si (mc-Si), OPV, and single crystal Si (c-Si). The highest available specific power is from a III–V module (>700 W kg^{-1}), with a one-Sun efficiency of 25% compared to the III–V mini-module record of 34.5%. Thin-film modules can have efficiencies >15% and specific powers of around 90 W kg^{-1} compared to a CIGS module record of 19.2%. (From Reese, M.O. et al., *Nat. Energy*, 3, 1002–1012, 2018.)

single photo-rechargeable device, apply primarily to consumer products, portable systems, and indoor energy harvesting applications, as batteries tend to be sensitive to extreme temperatures and have lifespans shorter than the often-cited 20-year PV module lifespan.

When integrating the components of power management electronics, it is necessary to include protecting batteries from over-charging or over-discharging, ensuring maximum power is extracted from the PV module even as illumination or load conditions change, and converting from the output power characteristics of the PV system to the requirements of the load [68]. Characteristics such as amount of power, type of energy storage device, and variability of illumination conditions determine which types of power management electronics are needed. Usually, PV systems may also include inverters, which convert from DC to AC power. AC power is also of interest for wireless sensors because it can be transmitted wirelessly through coupled inductors, capacitors, or antennas, avoiding the need for physical connection between energy harvesting and load devices [68]. Also, diodes are used in PV systems as blocking diodes, bypass diodes, and components of DC-DC converters. Similarly, to the passive components, RF energy harvesting applications have spurred the development of a variety of flexible thin-film diodes. While these recent reports of printed passive components and diodes show great promise for applications in power electronics, thin-film transistors (TFTs) present a greater challenge for such applications. A few recent reports have shown integrated TFT-based power electronics for photovoltaic energy harvesting applications. In one example, Fuketa et al. developed a flexible voltage regulating circuit based on complementary organic TFTs [69]. Meister et al. designed an integrated system consisting of a thin-film OPV module, a thin-film NiMH battery, and indium gallium zinc oxide (IGZO) TFT-based power management electronics, in a layered structure as shown in Figure 12.10 of [67]. Both a 6 V/14.4 mAh system and a 24 V/5.5 mAh system were demonstrated; the dimensions of each were selected to allow a full charge in 4 hours under full sun. A diode-connected TFT (gate and drain

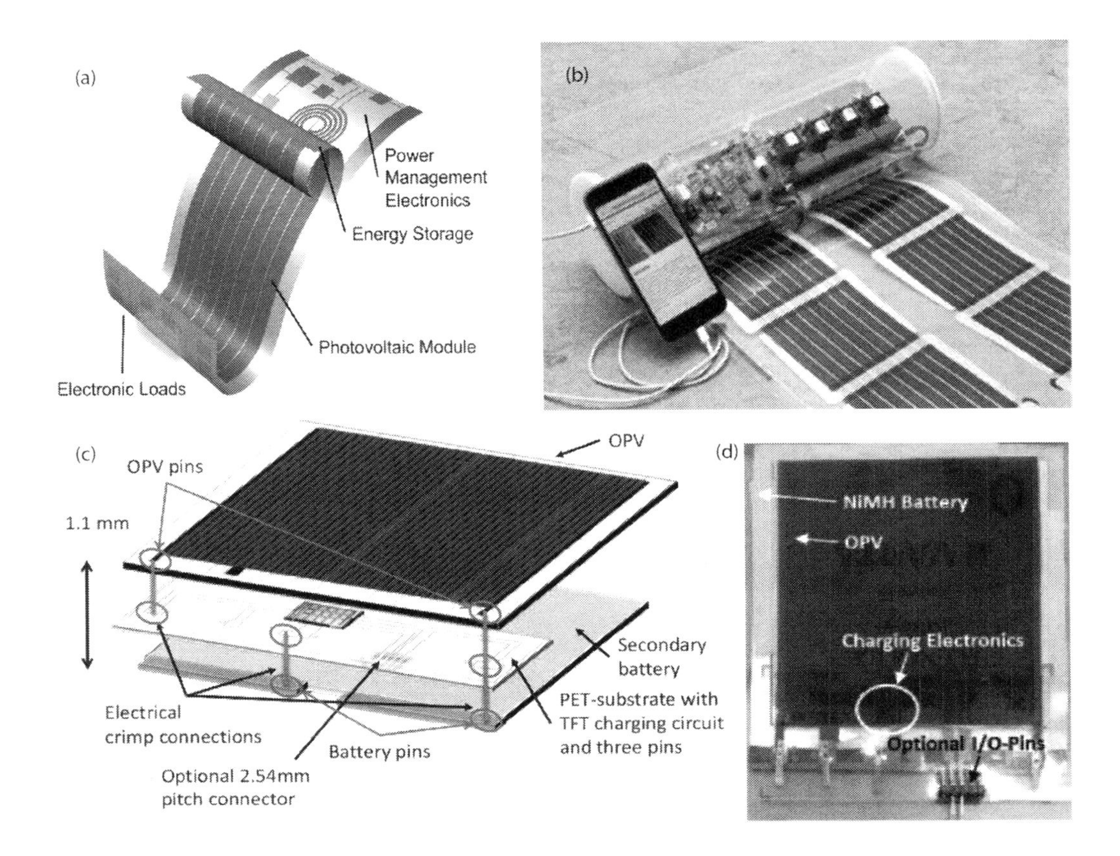

FIGURE 12.10 (a) Concept illustration of a fully printed and flexible photovoltaic system integrating solar module, energy storage, power management, and loads. (From Ostfeld, A. E. and Arias, A. C. *Flex. Print. Electron.*, 2, 013001–013024, 2017.) (b) Photographs of a rollable solar charger, in which a flexible OPV module can be wrapped around a cylinder containing a rigid battery and power management electronics. (From Garcia-Valverde, R. et al., *Sol. Energy Mater. Sol. Cells*, 144, 48–54, 2016.) (c,d) Schematic diagram and photograph of solar energy harvesting and storage system based on OPV module, NiMH battery, and IGZO TFT-based power management electronics. (From Meister, T. et al., Bendable energy-harvesting module with organic photovoltaic, rechargeable battery, and a-IGZO TFT charging electronics 2015 *European Conf. on Circuit Theory and Design (ECCTD)*, IEEE, Trondheim, Norway, 1–4, 2015.)

connected together) was used as a blocking diode. Since the PV open-circuit voltage dropped below the battery maximum voltage under low light conditions, charge pump circuits were designed to boost the voltage and enable low-light charging. The switching signals for the charge pumps were provided by IGZO TFT ring oscillators [69]. Recently reported review work demonstrates possibly all in one integration of energy harvesting and storage devices, which includes photovoltaic, piezoelectric, triboelectric, pyroelectric, and thermoelectric effects, and is of high interest due to the possibility of replacing batteries or at least extending the lifetime of a battery [70]. The development of advanced integrated technologies for the direct conversion of mechanical, thermal, and solar energy into electrochemical energy have high potential towards sustainable and maintenance-free operation of micro-/nano-systems and mobile/portable electronics, additionally to the self-powered wearable electronics by integration of wearable energy harvester/storage devices.

Towards the high energy demand for the applications based on flexible as well as stretchable hybrid electronics, integration of organic-inorganic systems are of keen interest for energy harvesting applications. Industries are exploring alternatives such as an organic PV-based flexible product which can efficiently power the portable devices. Various printing and coating methods allow PV devices with conformal large areas working at high speeds and low cost. These technologies inspire visions of printed, flexible, stretchable, and an integrated electronic system in which flexible organic-inorganic PV module and battery layers are integrated with printed power management system.

12.5 Summary

To achieve efficient and cost-effective flexible and stretchable photovoltaic systems, printing-based fabrication methods with integration of the components on a desired substrate (based on the design of systems with flexible and innovative form factors) are advanced approaches overcoming the conventional process and methods. The PV system design and selection of advanced materials and manufacturing techniques and integrated with the hybrid power electronics are the key aspects for the futuristic scope of commercialization in cheaper prices. Non-ITO-based indium free organics or nanostructure materials-based PV cells/modules built on conformal plastic substrates seem the good alternatives for solution processed flexible and/or wearable (e.g., to charge Li-ion batteries in thin-film solar systems) energy harvesters. Thus, high-performing PV-based energy systems are of great importance for Internet of Things (IoT) applications such as wearable healthcare and/or sensors. Integrated for fully compatible with stand-alone flexible and stretchable platforms.

In summary, out of various potential energy sources available, the use of PVs power management electronics is one of the best alternate options for supplying and harvesting power (at optimum sunlight irradiance and ambient temperature) to various portable and wearable devices. The latest trend emphasizes the importance of considering all components of an electronic system (both power sources and loads) preferably in a single design process. The ideal choices of load duty cycle, battery architecture, energy harvester characteristics, and power electronics must be compatible, for given amount of light exposure. Overall, integrated energy harvesting and storage systems in flexible thin-film form factors can support the vision of ubiquitous, flexible, and wireless robust electronics to be realized.

References

1. A. Luque and S. Hegedus (Eds.) *Handbook of Photovoltaic Science and Engineering,* John Wiley & Sons (2002).
2. M. Pagliaro, G. Palmisano and R. Ciriminna, *Flexible Solar Cells,* WILEY-VCH Verlag GmbH & Co. (2008).
3. A. Shah, P. Torres, R. Tscharner, N. Wyrsch and H. Keppner, "Photovoltaic technology: The case for thin-film solar cells", *Science* 285 (1999) 692–698.
4. B. J. Tok and Z. Bao, "Recent advances in flexible and stretchable electronics, sensors and power sources", *Sci China Chem* 55 (2012) 718–725.
5. A. E. Ostfeld and A. C. Arias, "Flexible photovoltaic power systems: Integration opportunities, challenges and advances", *Flex. Print. Electron* 2 (2017) 013001–013024.
6. L. Gao, L. Chao, M. Hou, J. Liang, Y. Chen, H. Yu and W. Huang, "Flexible, transparent nanocellulose paper-based perovskite solar cells", *NPJ Flexible Electron* 3 (2019) 4–8.
7. J. M. Ripalda, J. Buencuerpo and I. García, "Solar cell designs by maximizing energy production based on machine learning clustering of spectral variations", *Nat Comm* 9 (2018) 5126–5134.
8. M. O. Reese, S. Glynn, M. D. Kempe, D. L. McGott, M. S. Dabney, T. M. Barnes, S. Booth, D. Feldman and N. M. Haegel, "Increasing markets and decreasing package weight for high-specific-power photovoltaics", *Nat. Energy* 3 (2018) 1002–1012.
9. C. Waldauf, G. Dennler, P. Schilinsky and C. J. Brabec, Chapter 12: Bulk heterojunction solar cells for large-area PV fabrication on flexible substrates", *Flexible Electronics: Materials and Applications, Electronic Materials: Science & Technology,* Springer (2009).

10. D. J. Lipomi and Z. Bao, "Stretchable, elastic materials and devices for solar energy conversion", *Energy Environ Sci* 4 (2011) 3314–3328.
11. K. J. Yu, Z. Yan, M. Han and J. A. Rogers, "Inorganic semiconducting materials for flexible and stretchable electronics", *Npj Flexible Electron* 4 (2017) 1–14.
12. D. McCoul, W. Hu, M. Gao, V. Mehta and Q. Pei, "Recent advances in stretchable and transparent electronic materials", *Adv Electron Mater* 2 (2016) 1500407–1500468.
13. D. J. Lipomi, B. C. Tee, M. Vosgueritchian and Z. Bao, "Stretchable organic solar cells", *Adv Mater.* 23 (2011) 1771–1775.
14. J. A. Rogers, T. Someya and Y. Huang, "Materials and mechanics for stretchable electronics", *Science*, 327 (2010) 1603–1607.
15. Basic Photovoltaic Principles and Methods SERI/SP-290–1448 Solar Information Module 6213 (1982).
16. M. A. Green, *Solar Cells: Operating Principles, Technology and System Applications*, Prentice-Hall, NJ (1982).
17. Peter Wurfel, "*Physics of Solar Cells: From Principles to New Concepts*", Wiley-VCH Verlag-GMBH & Co (2005).
18. J. A. Luceño-Sánchez, A. M. Díez-Pascual and R. P. Capilla, "Materials for photovoltaics: State of art and recent developments", *Int J Mol Sci* 20 (2019) 976–1018.
19. J. R. Bakke, K. L. Pickrahn, T. P. Brennana and S. F. Bent "Nanoengineering and interfacial engineering of photovoltaics by atomic layer deposition", *Nanoscale* 3 (2011) 3482–3508.
20. http://www.pveducation.org/pvcdrom/solar-cell-structure.
21. http://www.alternative-energy-tutorials.com/solar-power/photovoltaics.html.
22. S. Almosni, A. Delamarre, Z. Jehl, D. Suchet, L. Cojocaru, M. Giteau, B. Behaghel, A. Julian, C. Ibrahim, L. Tatry, H. Wang et al., "Material challenges for solar cells in the twenty-first century: Directions in emerging technologies", *Sci Technol Adv Mater* 19 (2018) 336–369.
23. T. I. Mohammed, S. C. L. Koh, I. M. Reaney, A. Acquaye, G. Schileo, K. B. Mustapha, R. Greenough, "Perovskite solar cells: An integrated hybrid lifecycle assessment and review in comparison with other photovoltaic technologies", *Renew Sust Energ Rev* 80 (2017) 1321–1344.
24. K. Yamamoto, A. Nakajima, M. Yoshimi, T. Sawada, S. Fukuda, T. Suezaki, M. Ichikawa, Y. Koi, M. Goto, T. Meguro, T. Matsuda et al., "A high efficiency thin film silicon solar cell and module," *Solar Energy*, 77 (2004) 939–949.
25. W. N. Shafarman, B. M. Basol, J. S. Britt, R. B. Hall and R.E. Rocheleau, Semiconductor processing and manufacturing, *Prog Photovoltaics: Res Appl* 5 (1997) 359.
26. B. S. Richards and K. R. McIntosh, Enhancing the Efficiency of Production of CdS/CdTe PV Modules by Overcoming Poor Spectral Response at Short Wavelengths Via Luminescence Down-Shifting, Photovoltaic Energy Conversion, *Conference Record of the 2006 IEEE 4th World Conference* (2006) 213–216.
27. E. Gilioli, C. Albonetti, F. Bissoli, M. Bronzoni, P. Ciccarelli, S. Rampino and R. Verucchi, CIGS-based flexible solar cells. In: Tolio T., Copani G., Terkaj W. (Eds.), *Factories of the Future*, Springer, (2019) 365–382.
28. Y. Kao, H. Chou, S. Hsu, A. Lin, C. Lin, Z. Shih, C. Chang, H. Hong and R. Horng, "Performance comparison of III–V//Si and III–V//InGaAs multi-junction solar cells fabricated by the combination of mechanical stacking and wire bonding", *Sci Rep* 9 (2019) 4308–4319.
29. S. N. F. Mohd-Nasir, M. Y. Sulaiman, N. Ahmad-Ludin, M. A. Ibrahim, K. Sopian, and M. A. Mat-Teridi, "Review of polymer, dye-sensitized, and hybrid solar cells", *Int J Photoenergy* 370160 (2014) 1–12.
30. T. Kim, J. Kim, T. Kang, C. Lee, H. Kang, M. Shin, C. Wang, B. Ma, U. Jeong, T. Kim and B. Kim, "Flexible, highly efficient all-polymer solar cells", *Nat Commun* 6 (2015) 8547–8554.
31. Y. Wu, X. Yan, X. Zhang and X. Ren, "Photovoltaic performance of a nanowire/quantum dot hybrid nanostructure array solar cell," *Nanoscale Res Lett* 13 (2018) 62–69.

32. S. Das, D. Pandey, J. Thomas and T. Roy, "The role of graphene and other 2D materials in solar photovoltaics", *Adv Mater* 31 (2019) 1802722–1802731.
33. B. Liu, L. Bai, T. Li, C. Wei, B. Li, Q. Huang, D. Zhang, G. Wang, Y. Zhao and X. Zhang, "High efficiency and high open-circuit voltage quadruple-junction silicon thin film solar cells for future electronic applications", *Energy Environ Sci* 2017, 10, 1134–1141.
34. Best Research Cell Efficiencies for all types of existing PV technology. Data from National Renewable Energy Laboratory (NREL) (2018).
35. J.-H. Lee, J. Kim, T. Y. Kim, M. S. Al Hossain, S.-W. Kim and J. H. Kim. "All-in-one energy harvesting and storage devices" *J Mater Chem A*, 4, (2016) 7983–7999.
36. D. Singh, R. Tao and G. Lubineau, "A synergetic layered inorganic–organic hybrid film for conductive, flexible, and transparent electrodes", *Npj Flexible Electron* 3 (2019) 10–18.
37. A. Reinders, P. Verlinden, W. van Sark and A. Freundlich, *Chapter: Contacts, Buffers, Substrates and Interfaces; Photovoltaic Solar Energy: From Fundamentals to Applications*, First Edition. (2017) John Wiley & Sons, Ltd.
38. Q. Lin, H. Huang, Y. Jing, H. Fu, P. Chang, D. Li, Y. Yao and Z. Fan, "Flexible photovoltaic technologies", *J Mater Chem C*, 2 (2014), 1233–1247.
39. P. Dutta, M. Rathi, D. Khatiwada, S. Sun, Y. Yao, B. Yu, S. Reed, M. Kacharia, J. Martinez, A. P. Litvinchuk, Z. Pasala et al., "Flexible GaAs solar cells on roll-to-roll processed epitaxial Ge films on metal foils: A route towards low-cost and high-performance III–V photovoltaics", *Energy Environ Sci* 12 (2019) 756–766.
40. R. R. Bahabry, A. T. Kutbee, S. M. Khan, A. C. Sepulveda, I. Wicaksono, M. Nour, N. Wehbe, A. S. Almislem, M. T. Ghoneim, G. A. T. Sevilla, A. Syed et al., "Corrugation architecture enabled ultraflexible wafer-scale high-efficiency monocrystalline silicon solar cell", *Adv Energy Mater* 8 (2018) 1702221–1702235.
41. Y. He and Y. Li, "Fullerene derivative acceptors for high performance polymer solar cells", *Phys Chem Chem Phys* 13 (2011) 1970–1983.
42. Z. Shi, Y. Bai, X. Chen, R. Zeng and Z. Tan, "Tandem structure: A breakthrough in power conversion efficiency for highly efficient polymer solar cells", *Sustain Energy Fuels*, 3 (2019) 910–934.
43. T. Ameri, N. Li and C. J. Brabec, "Highly efficient organic tandem solar cells: A follow up review", *Energy Environ Sci* 6 (2013) 2390–2413.
44. J. Feng, X. Zhu, Z. Yang, X. Zhang, J. Niu, Z. Wang, S. Zuo, S. Priya, S. Liu and D. Yang, "Record efficiency stable flexible perovskite solar cell using effective additive assistant strategy", *Adv Mater* 30 (2018) 1801418–1801427.
45. M. Kaltenbrunner, M. S. White, E. D. Głowacki, T. Sekitani, T. Someya, N. Serdar Sariciftci and S. Bauer, "Ultrathin and lightweight organic solar cells with high flexibility" *Nat Commun* 3 (2012) 1772–1779.
46. D. Shahrjerdi, S. W. Bedell, C. Bayram, C. C. Lubguban, K. Fogel, P. Lauro, J. A. Ott, M. Hopstaken, M. Gaynessmand and D. Sadana, "Ultralight high-efficiency flexible InGaP/(In)GaAs tandem solar cells on plastic", *Adv Energy Mater* 3 (2013) 566–571.
47. X. Gu, Y. Zhou, K. Gu, T. Kurosawa, Y. Guo, Y. Li, H. Lin, B. C. Schroeder, H. Yan, F. Molina-Lopez, C. J. Tassone, C. W. Stefan et al., "Roll-to-roll printed large-area all-polymer solar cells with 5% efficiency based on a low crystallinity conjugated polymer blend", *Adv Energy Mater* 7 (2017) 1602742–1602753.
48. A. C. Mayer, S. R. Scully, B. E. Hardin, M. W. Rowell and M. D. McGehee, "Polymer-based solar cells", *Mater Today* 10 (2007) 28–33.
49. Y. Cui, H. Yao, J. Zhang, T. Zhang, Y. Wang, L. Hong, K. Xian, B. Xu, S. Zhang, J. Peng, Z. Wei, F. Gao and J. Hou, "Over 16% efficiency organic photovoltaic cells enabled by a chlorinated acceptor with increased open-circuit voltages", *Nature Comm* 10 (2019) 2515–2523.

50. K. Sharma, V. Sharma and S. S. Sharma, "Dye-sensitized solar cells: Fundamentals and current status", *Nanoscale Res Lett* 13 (2018) 381–427.
51. P. Liska, K. R. Thampi and M. Grätzel, "Nanocrystalline dye-sensitized solar cell/copper indium gallium selenide thin-film tandem showing greater than 15% conversion efficiency", *Appl Phys Lett* 88 (2006) 203103.
52. B.-M. Kim, H.-G. Han, D.-H. Roh, J. Park, K. M. Kim, U.-Y. Kim, T.-H. Kwon, "Book chapter: Flexible dye-sensitized solar cells", *Flexible Energy Conversion and Storage Devices*, Wiley-VCH Verlag GmbH & Co (2018) 239–281.
53. M. J. Yun, S. I. Cha, S. H. Seo and D. Y. Lee, "Highly flexible dye-sensitized solar cells produced by sewing textile electrodes on cloth", *Scientific Reports* 4 (2014) 5322–5328.
54. I. Syed and S. M. Bedair, "Quantum Well Solar Cells: Principles, Recent Progress, and Potential", *IEEE J Photovolt* 9 (2019) 402–424.
55. M. Yu, Y.-Z. Long, B. Sunb and Z. Fan, "Recent advances in solar cells based on one-dimensional nanostructure arrays", *Nanoscale* 4 (2012) 2783–2796.
56. H.-P. Wang and J.-H. He, "Toward highly efficient nanostructured solar cells using concurrent electrical and optical design", *Adv. Energy Mater* 7 (2017) 1602385–1602413.
57. Y. Song, S. Chang, S. Gradecak and J. Kong, "Visibly-transparent organic solar cells on flexible substrates with all-graphene electrodes", *Adv Energy Mater* (2016) 1600847–1600854.
58. M. F. Bhopal, D. W. Lee, A. Rehman and S. H. Lee, "Past and future of graphene/silicon heterojunction solar cells: A review", *J Mater Chem C*, 5 (2017) 10701–10714.
59. J. H. Heo, D. S. Lee, D. H. Shin and S. H. Im, "Recent advancements in and perspectives on flexible hybrid perovskite solar cells", *J. Mater. Chem. A*, 7 (2019) 888–900.
60. Q. Tai, K-C. Tan and F. Yan, "Recent progress of inorganic perovskite solar cells", *Energy Environ Sci* 12 (2019) 2375–2405.
61. S. Kang, J. Jeong, S. Cho, Y. J. Yoon, S. Park, S. Lim, J. Y. Kim and H. Ko, "Ultrathin, lightweight and flexible perovskite solar cells with an excellent power-per-weight performance", *J Mater Chem A*, 7 (2019) 1107–1114.
62. H. Jinno, K. Fukuda, X. Xu, S. Park, Y. Suzuki, M. Koizumi, T. Yokota, I. Osaka, K. Takimiya and T. Someya, "Stretchable and waterproof elastomer-coated organic photovoltaics for washable electronic textile applications", *Nature Energy* 2 (2017) 780–785.
63. J. Liu, Y. Li, S. Yong, S. Arumugam and S. Beeby, "Flexible printed monolithic-structured solid-state dye sensitized solar cells on woven glass fibre textile for wearable energy harvesting applications", *Scientific Reports* 9 (2019) 1362–1373.
64. L. Kranz, A. Abate, T. Feurer, F. Fu, E. Avancini, J. Löckinger, P. Reinhard, S. M. Zakeeruddin, M. Grätzel, S. Buecheler and A. N. Tiwari, "High-efficiency polycrystalline thin film tandem solar cells", *J Phys Chem Lett* 6 (2015) 2676–2681.
65. M. O. Reese, S. Glynn, M. D. Kempe, D. L. McGott, M. S. Dabney, T. M. Barnes, S. Booth, D. Feldman and N. M. Haegel, "Increasing markets and decreasing package weight for high-specific-power photovoltaics", *Nature Energy* 3 (2018) 1002–1012.
66. R. Garcia-Valverde, J. A. Villarejo, M. Hösel, M. V. Madsen, R. R. Søndergaard, M. Jørgensen and F. C. Krebs, Scalable single point power extraction for compact mobile and stand-alone solar harvesting power sources based on fully printed organic photovoltaic modules and efficient high voltage DC/DC conversion, *Sol Energy Mater Sol Cells* 144 (2016) 48–54.
67. T. Meister et al., Bendable energy-harvesting module with organic photovoltaic, rechargeable battery, and a-IGZO TFT charging electronics 2015 European Conf. on Circuit Theory and Design (ECCTD), Trondheim, Norway: IEEE. 2015, 1–4.
68. A. E. Ostfeld, *PhD Thesis _ Printed and Flexible Systems for Solar Energy Harvesting*, UC Berkeley (2016).

69. H. Fuketa, M. Hamamatsu, T. Yokota, W. Yukita, T. T. Someya, T. Sekitani, M. Takamiya, T. T. Someya, and T. Sakurai, Energy-autonomous fever alarm armband integrating fully flexible solar cells, piezoelectric speaker, temperature detector, and 12V organic complementary FET circuits in 2015. *IEEE International Solid-State Circuits Conference - (ISSCC) Digest of Technical Papers*. IEEE, San Francisco, CA, 2015, 1–3.
70. J.-H. Lee et al. "All-in-one energy harvesting and storage devices", *J. Mater. Chem. A*, 4 (2016) 7983–7999.

13

Flexible and Stretchable Energy Storage

Arwa Kutbee

13.1 Introduction

The Internet of Everything (IoE) is the most awaited titan in this era of today's web infrastructure which spans over a wide spectrum of businesses: industry, infrastructure, home, automotive, and personal wearable electronics to tackle pressing issues including: product management, smart cities, security, entertainment, and healthcare with more than 50 billion devices connected to the Internet and a shear projected economic growth value of $14.4 trillion by 2020. Recent advances in personal IoE are the new wave of wearable/implantable body sensory networks (WIBSNs) which rely heavily on flexible sensors, actuators, and integrated circuits (ICs) to monitor, sense, compute, and communicate daily environmental and human physiologies (e.g., the human's breath, brain, heart, sweat, movement, etc.) for readily available data to doctors, databases, and hospitals. The intuitive WIBSN notion [1] harnesses the benefits of replacing rigid traditional bulky electronics with smart flexible platforms [2] embedded into garments or patches to be worn or implanted into the human body; overcoming the physical boundaries between people, machine, time, place, service, and network by bringing life into the non-living.

Newly developed powering technologies have been a main key enabler in the rise of wearable and implantable IoE applications, especially with the notable growth of monitoring, diagnosing, and treating personal and medical devices that are in dire need of a remote energy supply [3–5]. Consequently, direct power grid connection or localized energy sourcing (e.g., battery) might not be the only two powering solutions, as opposed to status quo rigid and bulky healthcare devices. This comes from the fact that many IoE-based applications are fundamentally human-integrated techniques which conform on irregular body contours, asymmetric surfaces (e.g., body organs and joints), and soft tissues. Hence, size and form restrictions on powering solutions to fit tight movable spaces for the operational convenience to the wearer or the sheer infeasibility of wire connection in implantable scenarios become an inevitable reality. To solve this, several conformal, biocompatible rechargeable and non-rechargeable (primary) batteries have successfully been developed to substitute the conventional rigid form factor (e.g., coin cell, prismatic, and cylindrical) of batteries [6,7].

This new generation of mechanically compliant batteries is particularly beneficial in consumer applications such as smart watches or phones. However, for other wearable and implantable applications, a fully functional operation over several months or years is required. The dilemma comes from three main issues associated with batteries: first, long-lasting batteries are essentially larger in size, thus, sacrificing the main advantage of a free-form factor in favor of an increased battery volume. Second, high cost and risk factors are associated with replacement procedures of primary batteries after they are depleted (e.g., surgical procedures for implantable peacemakers). Finally, rechargeable batteries require continued charging capabilities that are not always present in implantable applications. To address these, energy harvesting technologies have been used to extract and generate power to prolong the battery life or even function as a sole power supply. Various ambient energy available from sunlight, radio-frequency waves, or from living subject's body [8,9] (e.g., body motion, joints friction, and periodic heart movement) can extend the lifetime of wearable and implantable devices. Consequently, a holistic approach to the IoE energy predicament might not be the ideal solution due to the wide range of implantable and wearable devices. Thus, an optimum energy solution is based upon product longevity or lifespan, availability of environmental extractable energy, safety, system power demand, size, and form factor.

Research efforts on device-level considerations explore bendable batteries [7,10], supercapacitors [11], and storage devices in general [6,12]. Additionally, harvesting devices [9,13] including solar cells [14–16], piezoelectric [17,18], and triboelectric nanogenerators [19,20] were studied in great detail by emphasizing the development of the material, structural [21], device fabrication, and design optimizations for bendable flexible [6], stretchable [21,22], and textile [23] platforms. The realization of fully compliant system-level demonstrations for energy and harvesting platforms with efficient power management are very important in the next generation IoE systems. There are three main architectures of energy storage and harvesting devices. These main scenarios present future powering technology for wearable/implantable IoE applications depending on their storage, harvesting, and power management variations.

1. *Battery-powered bendable architecture*: These types of devices rely on the concept of (store-use) notion in which a single battery is used to provide the necessary energy needed by the system. This type of powering solution with both primary and rechargeable variations offers a continuous operation lasting from hours, days, and even years for consumer electronics such as patches, smart watches, and phones.
2. *Harvester-powered bendable architecture*: IoE devices that rely on nearby harvestable energy to direct usage in powering the electronic system, henceforth are based on the (harvest-use) concept. As an example, implantable cardiac/brain healthcare devices that utilize the constant movement of biological energy and natural body movement.
3. *Self-powered bendable architecture*: The third group of architectures is the epitome of both arrangements which are called self-powered devices. They rely on the concept of (harvest-store-use) which means that these devices harvest the energy and store it for the on-demand future usages. It combines a rechargeable battery and an energy harvester to extract the ambient energy into usefully stored electrical energy (e.g., electronic-skin).

Thus, we begin with understanding the main design criteria of different wearable and implantable platforms and efficient power utilization of the power-hungry subsystem. We provide a summary of the main basic power (battery-powered, harvester-powered, and self-powered) architectures. In each architecture, we provide introductory background on the main building blocks including: mechanically compliant batteries, their operation principle, battery parameters, followed by the recent advances in lithium-ion battery, and the usage of thin film battery technology.

13.2 Design Criteria for Self-powered IoE Systems

Opportunities in energy storage and harvesting technologies in wearable/implantable IoE devices are expected to combine the best of the two worlds: mechanical compliance and optimized power utilization to the subsystem components. Thus, bringing a new dimension to rigid and industrial-grade

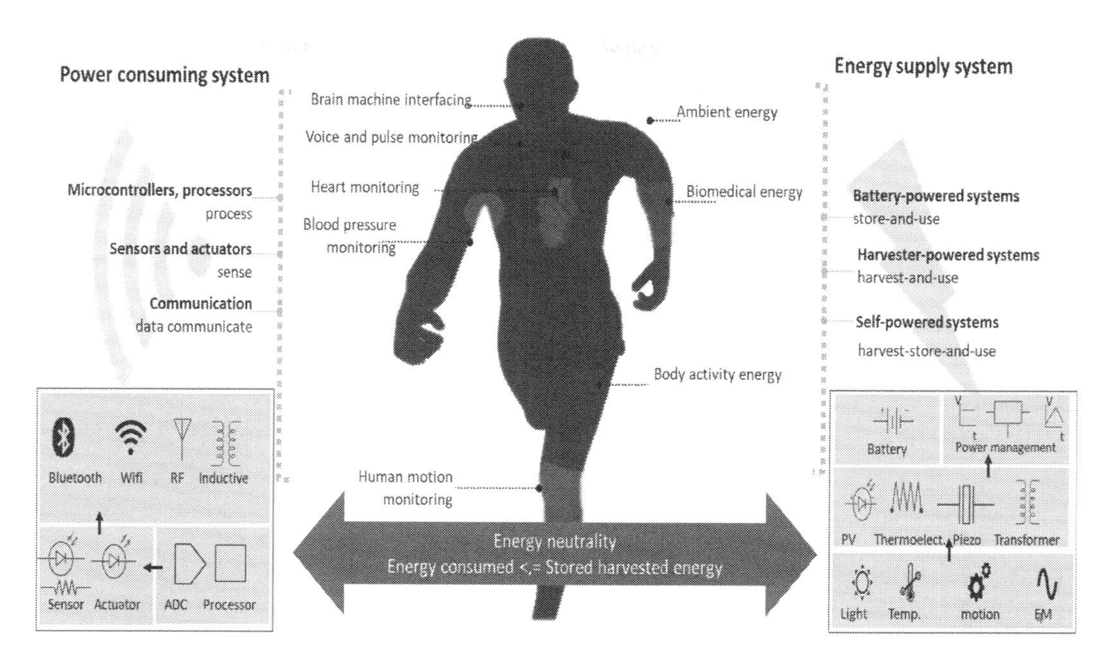

FIGURE 13.1 The main components of an IoE system.

technologies. Figure 13.1 shows a block description of a complete self-powered (type 3) wearable and implantable device which consists of mainly two systems: the power consuming system and the power supply system. The power consuming electronics include the sensing subsystem to acquire data, the processing subsystem for local data processing, and a wireless communication subsystem. On the other hand, the power supply system may consist of an energy storage and/or harvesting unit as well as a power management module. In the general case, the ambient energy (e.g., indoor light, solar light, temperature gradient, motion, and electromotive energy) is harvested and regulated through a power management circuit and stored in an energy storage unit (e.g., battery, fuel cell). Since energy harvesting technologies often exhibit unpredictable behavior and efficiency discrepancies due to variations in environmental conditions (e.g., light intensity variation, frequency variability), the power management circuit is necessary to regulate the voltage over time fluctuations and alter the harvester's current-source-like behavior to a stable voltage-source behavior. Different energy storage, harvesting, and management variations depend on the IoE application intended as shown in Table 13.1.

Recent research efforts in the past several years have been dedicated towards flexible, inexpensive, lightweight, and safe energy storage and harvesting devices for (*in vivo* and *in vitro*) soft surfaces and intimate conformal integration to the human interface. Despite the big development of new materials, design, and fabrication processes for emerging flexible, foldable, stretchable, and textile single element storage and harvesting devices, a main gap lies in the approach to manage all of these elements together with a fully flexible power management module. Case studies of different IoE applications such

TABLE 13.1 Powering Architectures of IoE Applications

	Battery Powered	Harvester-Powered	Self-Powered
Energy storage	X		X
Power management		X	X
Energy harvester		X	X

as real-time health monitoring devices or neural mapping applications require a large sampling rate with an approximated data generation of (1.8 MB/h) in the former and (4TB/h) in the latter case [24]. This places a big burden on the energy efficient module that communicates and processes these data. Thus, power management holds key decisions to efficient conversion of all powering architectures.

We will overview the main challenging considerations of the power consumption by the sub-digital components to achieve low power performance and increase the lifetime of the wearable/implantable devices. We believe that the design of power optimizing hardware electronics and their intersection with newly developed free-form platform is one of the immediate and urgent research directions that need to be addressed towards a fully compliant energy storage and harvesting IoE platforms.

13.2.1 Platform Considerations

Many human-integrated WIBSN devices are envisioned to be cordless or self-powered to prevail for months or years. They must sustain large mechanical strains due to the asymmetric nature of the human body and the wide range of the dynamic motion of the human muscular entity including both in-plane (stretching, contracting) or out-of-plane deformations (flexing, bending, swelling, shrinking, and cramping). Additionally, power demand for each unique therapeutic or monitoring IoE product must match with its distinctive environmental constraints and functionality. Therefore, bendable platforms can be divided into four main sections depending on the mechanical deformations they undergo and the materials they utilize as shown in Figure 13.2.

Inherently flexible (e.g., polyethylene terephthalate [PET], polydimethylsiloxane [PDMS], and polyimide [PI]) elastomers have been widely utilized, as these rubbers or elastomeric materials exhibit a high degree of viscoelasticity or reversible deformation. They can be ink-jetted , nanoimprinted, or lithographically patterned for the fabrication of different flexible and stretchable devices. In order to achieve higher electrical conduction of such rubbers, different hybrid composites of 2D and 1D materials such as CNT and graphene and molybdenum disulfide are embedded in the rubber matrix to use them as active materials in battery electrodes [25] and supercapacitors [26–29].

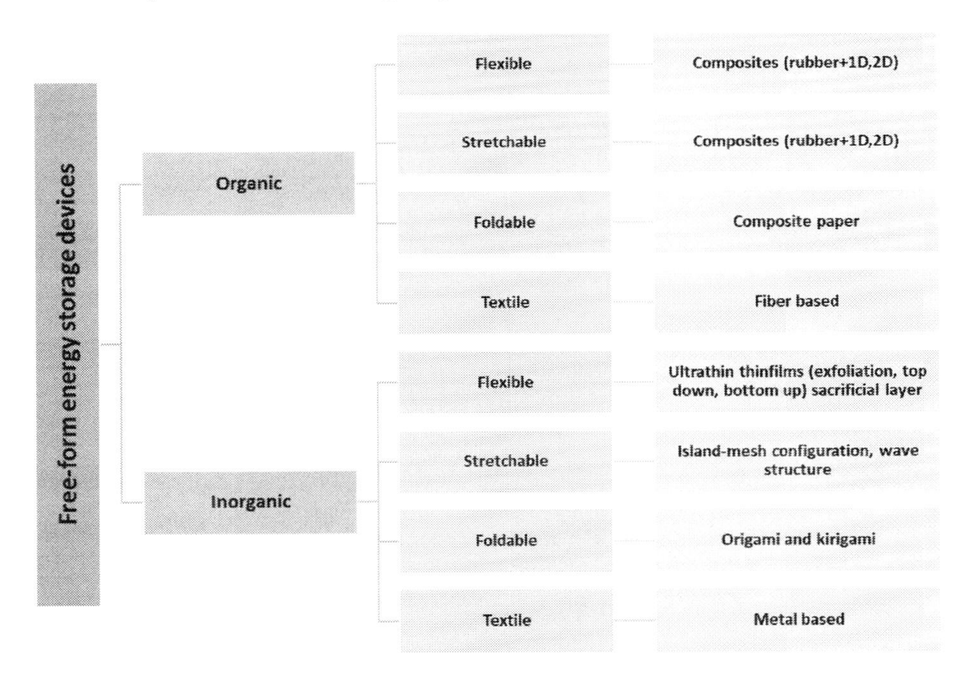

FIGURE 13.2 The different types of free-form energy storage devices.

On the other hand, synthetic sponge-like substrates made of cellulose and polyester are an interesting type of the flexible substrates due to the inherent porosity which can be a main advantage in devices requiring large surface area electrodes to increase surface reactivity of the electrochemical reaction between the electrolyte and electrodes in batteries and supercapacitor applications. Active polymeric materials such as polyvinylidene fluoride [30] can also be synthesized to achieve sponge-like architecture which is also beneficial to increase the harvested electrical energy from surface oscillations by mechanical means in piezoelectric nanogenerator applications. Still a main challenge of plastic-based substrates is the limitation on the maximum fabrication processes below the glass transitions temperature to maintain the integrity of plastics without any inelastic deformations.

Compared to flexible plastic substrates that are relatively cheap (PET = 2 cents dm^{-2}, PI = 30 cents dm^{-2}), the substantial lower price of paper (0.1 c cents dm^{-2}) makes an attractive option for low cost and mass production [31]. Remarkably, commercial paper materials used as household items (wiping, packaging, etc.) can be also utilized in the fabrication of a complex sensory platform. Research efforts have shown the commercially manufactured (Post-it note) based on non-uniform randomly ordered cellulose fibres and their utilization in advanced complex environmental [32] and health monitoring [33] applications. Another form of paper based on CNT-coated paper can be used in flexible [34], miura-folding origami [35] (folded patterns), and kirigami [36] (cut and folded patterns). As opposed to the previously discussed synthetic methods, flexible inorganic rigid thin films which utilize vacuum deposition and complementary metal-oxide semiconductor (CMOS) microfabrication technology are more of a generic approach with no limitations or restrictions on material combinations and their final destination in flexible, foldable, and stretchable substrates [37]. For example, conventional metal oxides, zirconate titanate (PZT), and thin films can be used easily to fabricate batteries [38] and piezoelectric triboelectric nanogenerators on the same common platform and even with other circuit components such as transistors [39], logic memories, and devices [40]. Despite those advantages, a major hindrance in the thin film approach is that it requires a rigid substrate single crystalline (100) Si substrate of a total thickness of (0.5 mm). According to the equation [41] $\varepsilon = t/2r \times 100\%$, the radius of curvature (r) which a material experiences under mechanical strain ε is controlled by the material's thickness (t). Therefore, several approaches to remove this substrate using etching [42], exfoliation [43], laser ablation [44], and transfer printing [45] have allowed the realization of flexible energy storage and harvesting devices. Also packaging strategies for such substrates to push the thin film into a region with minimal induced stress on the active material of the device (~1%, or less) is achieved by the usage of the mechanical neutral plane concept to estimate and the correct packaging of the thin film device in the polymer matrix. Stretchable materials and structures [40], as is the case with flexible materials, can utilize 1D and 2D materials embedded or articulated in stretchable structures: including buckled (planer and coplanar) waves and island-mesh geometries (2D sinusoidal serpentine patterns). These materials or structures have allowed devices to go even beyond the limited flexible regime with an induced strain of $\gg$1%.

Textile platforms are a three-dimensional projection of stretchable structures that enable a spring-like behavior, thus, it abides perfectly to the simple Hook's law. Each stretchable fiber can be interwoven into a full textile. They exhibit a porous structure which allows the high mass loading via dipping of active materials in energy storage devices [46]. Such fibres are also used to get wire-shaped storage and conversion devices from inorganic materials [47]. A summary of recent research efforts are summarized in Figure 13.3.

Research efforts call for finding ways to minimize the circuit power consumption to maintain the energy neutrality of WIBSN systems, long-time serviceability, and operation times. Estimating the energy consumption or power requirements in these devices can be a daunting task as it is very much application-reliant. For the sensing module, the type, geometry, and system complexity of sensors can affect significantly the amount of power consumed.

For instance, capacitive-based sensors are expected to require a less amount of power (1–10 μW) than resistive-based sensors (>10 μW). On the other hand, an increased number of sensors will increase the complexity of the read-out-circuitry and its processing power. Hence, one would expect that WIBSN

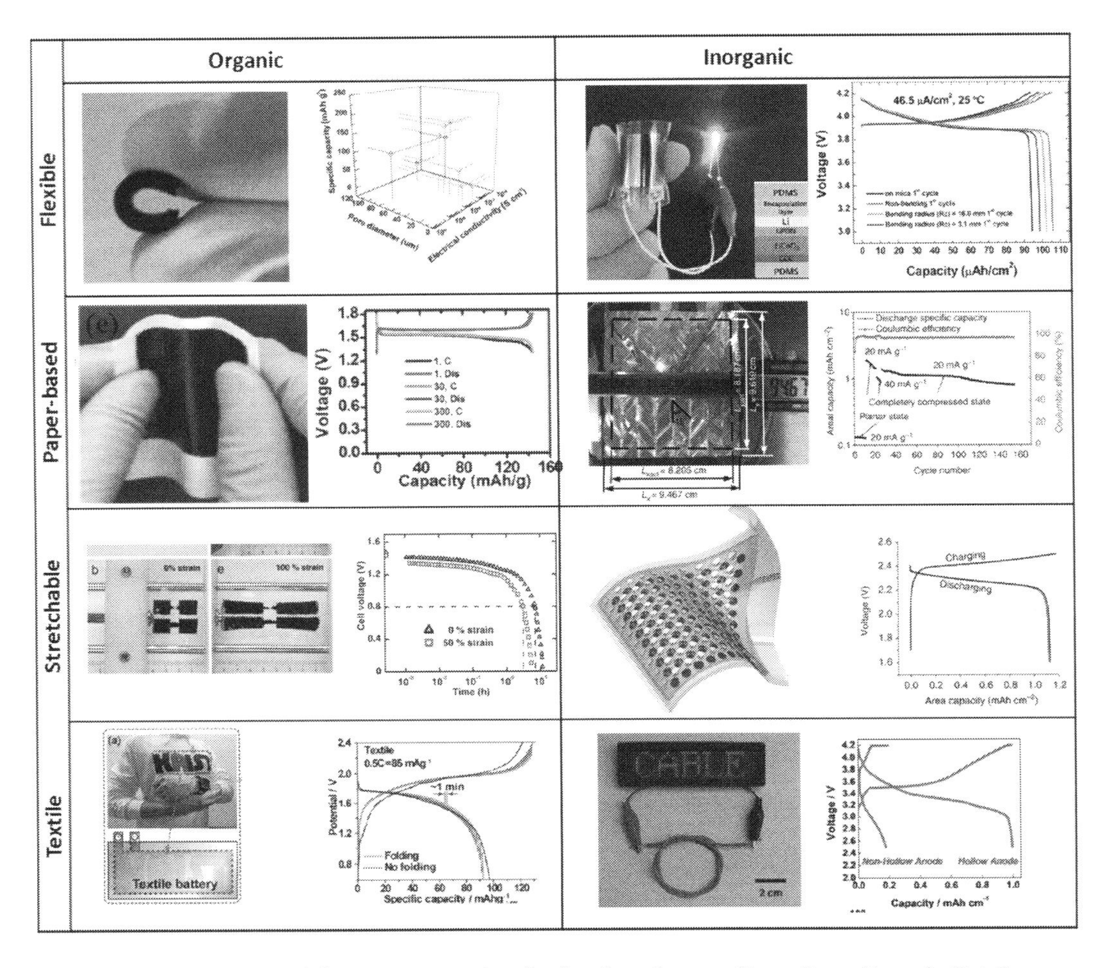

FIGURE 13.3 Summary of the main approaches for free-form battery. (From Song, Z. et al., *Nat Commun*, 5, 3140, 2014; Koo, M. et al., *Nano Letters*, 12, 4810–4816, 2012; Xu, S. et al., *Nat Commun*, 4, 1543, 2013; Kwon, Y.H. et al., *Adv Mater*, 24, 5192–5197, 2012; Lee, Y.-H. et al., *Nano Lett*, 13, 5753–5761, 2013; Hu, L., *ACS Nano*, 4, 5843–5848, 2010; Lee, H. et al., *Adv Energy Mater*, 2, 976–982, 2012; Kaltenbrunner, M. et al., *Adv Mater*, 22, 2065–2067, 2010.)

applications such as E-skin [5,48] or brain machine interfacing with hundreds of sensor arrays to have larger energy demand due to the increased energy demand for higher order polynomials that are performed by the readout circuitry to correlate the sensed-data into real-data (1–50 mW). On the other hand microprocessors based on low-cost CMOS technology have enabled a wide range of free-form flexible transistors. The main consuming part in these devices is the frequency of the switching activity and strategies for reducing the clock frequency could help in power reduction. In the communication subsystem, flexible-based communications are mainly research focused on near field communication via inductive coupling or short distance radio frequency (RF). Challenges remain to the inherent issue with coil design to increase inductance and quality factor and tune the transmission power to achieve larger distances: Among all the subsystems, communication wireless networks (e.g., Wi-Fi, Bluetooth, and Zigbee) will need a few milliwatts for a 10–100 seconds of operation, thus they consume more energy on average than the processing and sensing subsystems.

13.2.2 Power Management Considerations

Depending on the amount of energy harvested compared to the energy consumption, duty cycling and power management techniques can be used to adjust the storage or utilize the harvester powering option. These techniques are mainly used to regulate and reduce the power consumption of the digital subsystem. This is achieved by making the system alternate between sleep and wakeup modes to save energy (active mode of the system in which it senses and transmits data).

$$E_{consumed} = E_{sleep} + DE_{active}.$$

The sleep mode or standby status (E_{sleep}) is often normally longer than the wakeup or active status E_{active}. Thus, the rate of duty cycling (D) is directly linear to the sampling frequency, data transmission, and processing. The decision on autonomous energy utilization during the duty cycles is different when considering (battery-powered architecture) compared to (harvester and self-powered architectures). This deterministic difference in power management strategies is based on the accounts of the dynamics of energy consuming system in the case of battery-powered while more focus on the dynamics of the powering system itself is also accounted for in harvester-powered and self-powered architectures.

Recent research efforts in wearable and implantable IoE devices still may not reach the same level of maturity in terms of power management techniques as in the case with wireless sensor networks (WSNW) such as adaptive duty cycling and harvesting-aware power management [53], due to limitations on the material and large scale integration strategies which are still missing in these approaches.

13.3 Battery Powered System

13.3.1 Principle of Operation

Electrochemical energy storage is a concept discovered by Galvani in the seventeenth century. An electrochemical cell involves the transduction, storage, and transportation of electrical energy from the simple chemical reaction between two substances. The electrical energy can be stored as a chemical energy (charging) and converted back to electrical energy when needed (discharging). The relatively cheap chemicals and their large reaction energy $\left(1\ eV = 96.4\ \frac{\text{kJ}}{\text{mol}}\right)$ make electrochemical cells (e.g., supercapacitors, batteries, fuel cells) very attractive in real life applications. As can be seen from the Ragone plot (Figure 13.4), batteries exhibit high energy densities (large energy storage) and low power density

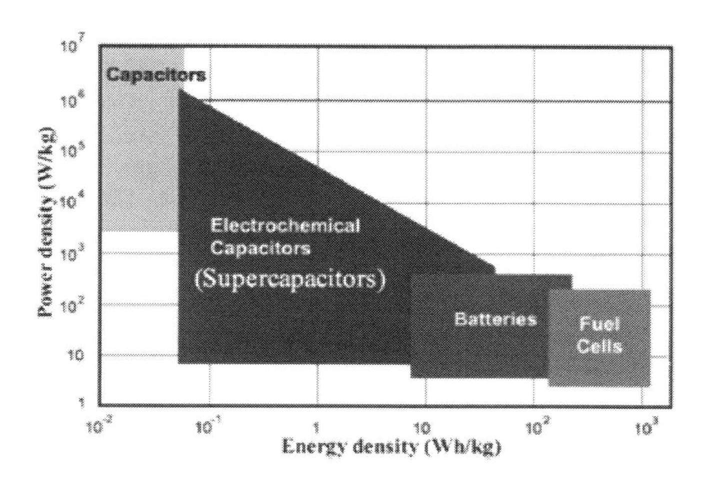

FIGURE 13.4 Ragone plot of different electrochemical energy storage solutions.

(low charging and discharging times). Research efforts in the past few decades have aimed to push the battery limits towards high energy and power densities. In our discussion, we refer to a battery as the smallest packaged form an electrochemical cell can take, while a battery module consists of several cells connected in either in-series or in-parallel.

A battery consists of an anode (where the oxidation half-reaction takes place), electrolyte, and cathode (where reduction half-reaction takes place). Depending on the usage of an external loading circuit, a galvanic cell uses the spontaneous redox reaction to generate electricity with a faradic current from the anode to the cathode. An electrolytic cell uses electricity to force the non-spontaneous chemical reaction to take place. With this in mind, we can divide the battery types into two main types: primary batteries are non-rechargeable batteries that can only work as a galvanic cell. Primary batteries (similar to batteries in cars) utilize conversion electrodes (e.g., FeF_3) which undergo a chemical reaction in which its structure is consumed. On the other hand, rechargeable batteries or secondary batteries can be used as both galvanic and electrolytic cells. They utilize intercalation electrodes (graphite, $LiCoO_2$, $LiFePO_4$) that function as a host structure for ionic species (e.g., lithium ions).

13.3.2 Rechargeable Battery Parameters

Figure 13.5 shows a generic case scenario of a rechargeable battery operating principle. The species being stored (S_z) in a host lattice (A: anode, C: cathode) undergoes the overall macroscopic redox reaction

$$S_zA + C \leftrightarrow A + ZS^+ + C.$$

During battery discharge, an oxidation reaction in the left-hand-side electrode (anode) $\left(S_zA \rightarrow ZS^+ + e + A\right)$, where ionic species ($ZS^+$) are released from the host and e's are extracted by a resistive load. The electrolytes dissolve and transfer ZS^+ ions to the right hand side (RHS) electrode while insulating electron flow. In the right-hand-side electrode, a reduction reaction $\left(ZS^+ + e + C \rightarrow S_zC\right)$ occurs at the (cathode). In case of charging, an external power supply forces the electrons to transfer in the opposite direction, causing an interchange in half cell reaction as the left hand side (LHS) electrode becomes a cathode and the LHS electrode becomes an anode.

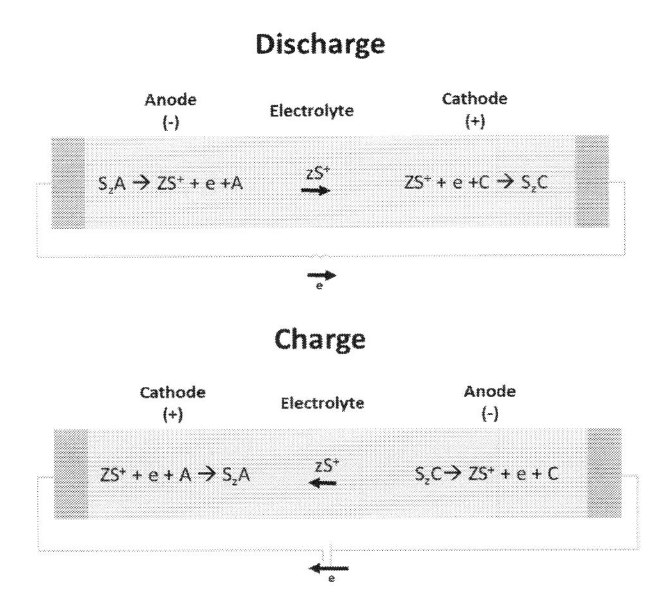

FIGURE 13.5 The operation principle of a rechargeable battery.

13.3.2.1 Potential

The potential difference between half cells is correlated to the electromotive force driving the cell to reach equilibrium at standard conditions: $\Delta G^o = -ZFE^o$, where: F: the Faraday constant, Z: number of electrons, and E: cell potential.

The voltage can be measured as the difference in the free energy of ionic species between the anode and cathode.

$$\Delta E = E_{RHS} - E_{LHS}.$$

Each half cell potential can be deduced from reduction reactions that are obtained with respect to a standard hydrogen electrode (SHE). Nevertheless, potential values in real case examples deviate from the standard value and follow potential plateaus with respect to variations of ionic concentration. The potential dependence on ionic concentration (more accurately activity), temperature of the oxidized and reduced species is generally governed by the Nernst equation.

In other words, the electrode potential is a measure of the extent in which the concentration of the species in a half cell differ from their equilibrium values, thus, if the change in the concentration during charging and discharging can exceed the solubility limits of a specific electrode phase. If the ionic concentrations exceeds the solubility limits, it can have a significant effect on the structural properties of the battery materials and produce unwanted phase changes which influence the overall cell potential.

13.3.2.2 Capacity

The capacity of the battery is defined as the time integral of the current flow out of the battery from the beginning of the current flow ($t = 0$) to a time when it reaches a specified cut-off voltage:

$$Capacity = \int_{t_0}^{t_f} I(t)dt.$$

The capacity is specified in terms of ampere-hours and should be expressed in terms of a load current flow ($t = 0$) to a time when it reaches a specified cut-off voltage.

Similar to the battery potential, the capacity is a material-dependent property. The amount of charge Q that can be stored in an electrode material depends on the batteries volume V and the solubility limit C_T of each electrode as follows:

$$Q_{A,C} = ZFVC_T^{A,C}.$$

13.3.3 Lithium-Ion Battery Technologies

Different battery chemistry will provide different energy and power density specifications. Lithium ion technology is attractive because of its high energy density as shown in Figure 13.6. This comes from the fact that lithium is the most electropositive element (high voltage battery) and a light weight element (highest specific energy). Lithium ions are small for fast diffusion (high power).

13.3.4 Thin Film Lithium-Ion Battery (LIB)

The advantage of thin film battery technology is that it allows to save space and increases the energy density as can be seen in Figure 13.7. Compared to wet battery cells which utilize liquid electrolytes that need an extra membrane to separate the cathode and anode and avoid any leakage, on the contrary, a

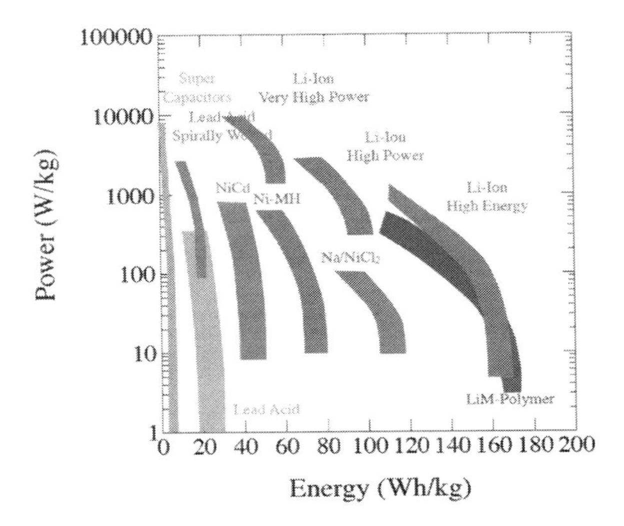

FIGURE 13.6 Ragone plot for different battery technologies.

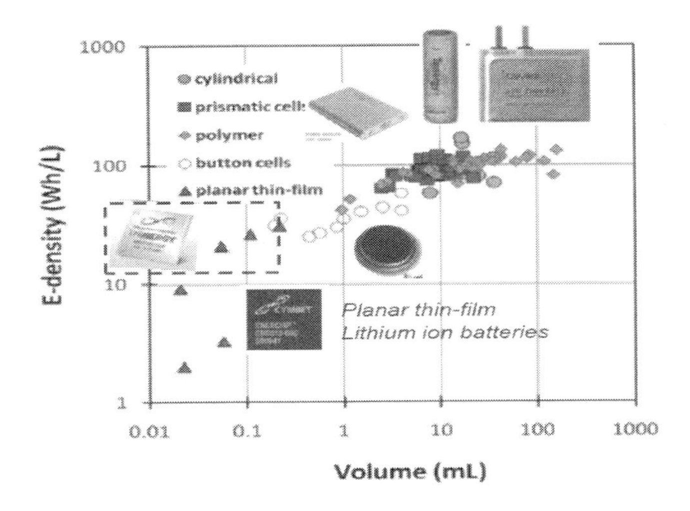

FIGURE 13.7 Comparison between different form factors for a LIB.

solid electrolyte in a thin film battery is a structural component, which allows shorter ion travel distance between the electrodes. Another important advantage in utilizing all solid state materials is that there is very little risk in thermal runaway and, thus, there is no need for extra packaging and bulky power safety management and a cooling system.

References

1. Y. K. Kim, H. Wang, and M. S. Mahmud, "9—Wearable body sensor network for health care applications A2 - Koncar, Vladan," in *Smart Textiles and their Applications*, Oxford: Woodhead Publishing, 2016, pp. 161–184.
2. A. Nathan, A. Ahnood, M. T. Cole, S. Lee, Y. Suzuki, P. Hiralal, et al., "Flexible electronics: The next ubiquitous platform," *Proceedings of the IEEE*, vol. 100, pp. 1486–1517, 2012.

3. W. Gao, S. Emaminejad, H. Y. Y. Nyein, S. Challa, K. Chen, A. Peck, et al., "Fully integrated wearable sensor arrays for multiplexed in situ perspiration analysis," *Nature*, vol. 529, pp. 509–514, 2016.
4. J. R. Corea, A. M. Flynn, B. Lechêne, G. Scott, G. D. Reed, P. J. Shin, et al., "Screen-printed flexible MRI receive coils," *Nature Communications*, vol. 7, p. 10839, 2016.
5. C. Wang, D. Hwang, Z. Yu, K. Takei, J. Park, T. Chen, et al., "User-interactive electronic skin for instantaneous pressure visualization," *Nature Materials*, vol. 12, pp. 899–904, 2013.
6. X. Wang, X. Lu, B. Liu, D. Chen, Y. Tong, and G. Shen, "Flexible energy-storage devices: Design consideration and recent progress," *Advanced Materials*, vol. 26, pp. 4763–4782, 2014.
7. G. Zhou, F. Li, and H.-M. Cheng, "Progress in flexible lithium batteries and future prospects," *Energy & Environmental Science*, vol. 7, pp. 1307–1338, 2014.
8. A. Proto, M. Penhaker, S. Conforto, and M. Schmid, "Nanogenerators for human body energy harvesting," *Trends in Biotechnology*, vol. 35, pp. 610–624, 2017.
9. C. Dagdeviren, Z. Li, and Z. L. Wang, "Energy harvesting from the animal/human body for self-powered electronics," *Annual Review of Biomedical Engineering*, vol. 19, pp. 85–108, 2017.
10. A. M. Gaikwad, A. C. Arias, and D. A. Steingart, "Recent progress on printed flexible batteries: Mechanical challenges, printing technologies, and future prospects," *Energy Technology*, vol. 3, pp. 305–328, 2015.
11. D. P. Dubal, J. G. Kim, Y. Kim, R. Holze, C. D. Lokhande, and W. B. Kim, "Supercapacitors based on flexible substrates: An overview," *Energy Technology*, vol. 2, pp. 325–341, 2014.
12. J. D. MacKenzie and C. Ho, "Perspectives on energy storage for flexible electronic systems," *Proceedings of the IEEE*, vol. 103, pp. 535–553, 2015.
13. S. Bauer, S. Bauer-Gogonea, I. Graz, M. Kaltenbrunner, C. Keplinger, and R. Schwödiauer, "25th Anniversary article: A soft future: From robots and sensor skin to energy harvesters," *Advanced Materials*, vol. 26, pp. 149–162, 2014.
14. Q. Lin, H. Huang, Y. Jing, H. Fu, P. Chang, D. Li, et al., "Flexible photovoltaic technologies," *Journal of Materials Chemistry C*, vol. 2, pp. 1233–1247, 2014.
15. C. H. Lee, D. R. Kim, and X. Zheng, "Transfer printing methods for flexible thin film solar cells: Basic concepts and working principles," *ACS Nano*, vol. 8, pp. 8746–8756, 2014.
16. D. J. Lipomi and Z. Bao, "Stretchable, elastic materials and devices for solar energy conversion," *Energy & Environmental Science*, vol. 4, pp. 3314–3328, 2011.
17. C. Dagdeviren, P. Joe, O. L. Tuzman, K.-I. Park, K. J. Lee, Y. Shi, et al., "Recent progress in flexible and stretchable piezoelectric devices for mechanical energy harvesting, sensing and actuation," *Extreme Mechanics Letters*, vol. 9, pp. 269–281, 2016.
18. F. R. Fan, W. Tang, and Z. L. Wang, "Flexible nanogenerators for energy harvesting and self-powered electronics," *Advanced Materials*, vol. 28, pp. 4283–4305, 2016.
19. Z. L. Wang, "Triboelectric nanogenerators as new energy technology and self-powered sensors—Principles, problems and perspectives," *Faraday Discussions*, vol. 176, pp. 447–458, 2014.
20. Z. L. Wang, J. Chen, and L. Lin, "Progress in triboelectric nanogenerators as a new energy technology and self-powered sensors," *Energy & Environmental Science*, vol. 8, pp. 2250–2282, 2015.
21. K. Xie and B. Wei, "Materials and structures for stretchable energy storage and conversion devices," *Advanced Materials*, vol. 26, pp. 3592–3617, 2014.
22. C. Yan and P. S. Lee, "Stretchable energy storage and conversion devices," *Small*, vol. 10, pp. 3443–3460, 2014.
23. K. Jost, G. Dion, and Y. Gogotsi, "Textile energy storage in perspective," *Journal of Materials Chemistry A*, vol. 2, pp. 10776–10787, 2014.
24. S. F. Shaikh, M. T. Ghoneim, G. A. T. Sevilla, J. M. Nassar, A. M. Hussain, and M. M. Hussain, "Freeform compliant CMOS electronic systems for internet of everything applications," *IEEE Transactions on Electron Devices*, vol. 64, pp. 1894–1905, 2017.
25. Q.-H. Wu, C. Wang, and J.-G. Ren, "Sn and SnO_2-graphene composites as anode materials for lithium-ion batteries," *Ionics*, vol. 19, pp. 1875–1882, 2013.

26. M. F. El-Kady and R. B. Kaner, "Scalable fabrication of high-power graphene micro-supercapacitors for flexible and on-chip energy storage," *Nature Communications*, vol. 4, p. 1475, 2013.
27. D. Kim, G. Shin, Y. J. Kang, W. Kim, and J. S. Ha, "Fabrication of a stretchable solid-state micro-supercapacitor array," *ACS Nano*, vol. 7, pp. 7975–7982, 2013.
28. G. Xiong, C. Meng, R. G. Reifenberger, P. P. Irazoqui, and T. S. Fisher, "Graphitic petal electrodes for all-solid-state flexible supercapacitors," *Advanced Energy Materials*, vol. 4, p. 1300515, 2014.
29. M. Li, Z. Tang, M. Leng, and J. Xue, "Flexible solid-state supercapacitor based on graphene-based hybrid films," *Advanced Functional Materials*, vol. 24, pp. 7495–7502, 2014.
30. Y. Mao, P. Zhao, G. McConohy, H. Yang, Y. Tong, and X. Wang, "Sponge-like piezoelectric polymer films for scalable and integratable nanogenerators and self-powered electronic systems," *Advanced Energy Materials*, vol. 4, pp. 1301624–n/a, 2014.
31. D. Tobjörk and R. Österbacka, "Paper electronics," *Advanced Materials*, vol. 23, pp. 1935–1961, 2011.
32. J. M. Nassar, M. D. Cordero, A. T. Kutbee, M. A. Karimi, G. A. T. Sevilla, A. M. Hussain, et al., "Paper skin multisensory platform for simultaneous environmental monitoring," *Advanced Materials Technologies*, vol. 1, pp. 1600004–n/a, 2016.
33. J. M. Nassar, K. Mishra, K. Lau, A. A. Aguirre-Pablo, and M. M. Hussain, "Recyclable nonfunctionalized paper-based ultralow-cost wearable health monitoring system," *Advanced Materials Technologies*, vol. 2, pp. 1600228–n/a, 2017.
34. N. Aliahmad, M. Agarwal, S. Shrestha, and K. Varahramyan, "Paper-based lithium-ion batteries using carbon nanotube-coated wood microfibers," *Nanotechnology, IEEE Transactions on*, vol. 12, pp. 408–412, 2013.
35. Z. Song, T. Ma, R. Tang, Q. Cheng, X. Wang, D. Krishnaraju, et al., "Origami lithium-ion batteries," *Nature Communications*, vol. 5, p. 3140, 2014.
36. Z. Song, X. Wang, C. Lv, Y. An, M. Liang, T. Ma, et al., "Kirigami-based stretchable lithium-ion batteries," *Scientific Reports*, vol. 5, p. 10988, 2015.
37. J. P. Rojas, G. A. Torres Sevilla, N. Alfaraj, M. T. Ghoneim, A. T. Kutbee, A. Sridharan, et al., "Nonplanar nanoscale fin field effect transistors on textile, paper, wood, stone, and vinyl via soft material-enabled double-transfer printing," *ACS Nano*, vol. 9, pp. 5255–5263, 2015.
38. A. T. Kutbee, M. T. Ghoneim, S. M. Ahmad, and M. M. Hussain, "Free-form flexible lithium-ion microbattery," *IEEE Transactions on Nanotechnology*, vol. 15, pp. 402–408, 2016.
39. G. A. T. Sevilla, J. P. Rojas, H. M. Fahad, A. M. Hussain, R. Ghanem, C. E. Smith, et al., "Field-effect transistors: Flexible and transparent silicon-on-polymer based sub-20 nm non-planar 3D FinFET for brain-architecture inspired computation (Adv. Mater. 18/2014)," *Advanced Materials*, vol. 26, pp. 2765–2765, 2014.
40. J. M. Nassar, J. P. Rojas, A. M. Hussain, and M. M. Hussain, "From stretchable to reconfigurable inorganic electronics," *Extreme Mechanics Letters*, vol. 9, Part 1, pp. 245–268, 2016.
41. Z. Suo, E. Y. Ma, H. Gleskova, and S. Wagner, "Mechanics of rollable and foldable film-on-foil electronics," *Applied Physics Letters*, vol. 74, pp. 1177–1179, 1999.
42. A. M. Hussain, S. F. Shaikh, and M. M. Hussain, "Design criteria for XeF2 enabled deterministic transformation of bulk silicon (100) into flexible silicon layer," *AIP Advances*, vol. 6, p. 075010, 2016.
43. M. Koo, K.-I. Park, S. H. Lee, M. Suh, D. Y. Jeon, J. W. Choi, et al., "Bendable inorganic thin-film battery for fully flexible electronic systems," *Nano Letters*, vol. 12, pp. 4810–4816, 2012.
44. Y. H. Do, W. S. Jung, M. G. Kang, C. Y. Kang, and S. J. Yoon, "Preparation on transparent flexible piezoelectric energy harvester based on PZT films by laser lift-off process," *Sensors and Actuators A: Physical*, vol. 200, pp. 51–55, 2013.
45. S. Xu, Y. Zhang, J. Cho, J. Lee, X. Huang, L. Jia, et al., "Stretchable batteries with self-similar serpentine interconnects and integrated wireless recharging systems," *Nature Communications*, vol. 4, p. 1543, 2013.

46. L. Hu, M. Pasta, F. L. Mantia, L. Cui, S. Jeong, H. D. Deshazer, et al., "Stretchable, porous, and conductive energy textiles," *Nano Letters*, vol. 10, pp. 708–714, 2010.
47. Y. H. Kwon, S.-W. Woo, H.-R. Jung, H. K. Yu, K. Kim, B. H. Oh, et al., "Cable-type flexible lithium ion battery based on hollow multi-helix electrodes," *Advanced Materials*, vol. 24, pp. 5192–5197, 2012.
48. A. Chortos, J. Liu, and Z. Bao, "Pursuing prosthetic electronic skin," *Nature Materials*, vol. 15, pp. 937–950, 2016.
49. Y.-H. Lee, J.-S. Kim, J. Noh, I. Lee, H. J. Kim, S. Choi, et al., "Wearable textile battery rechargeable by solar energy," *Nano Letters*, vol. 13, pp. 5753–5761, 2013.
50. L. Hu, H. Wu, F. La Mantia, Y. Yang, and Y. Cui, "Thin, flexible secondary Li-ion paper batteries," *ACS Nano*, vol. 4, pp. 5843–5848, 2010.
51. H. Lee, J.-K. Yoo, J.-H. Park, J. H. Kim, K. Kang, and Y. S. Jung, "A Stretchable polymer–carbon nanotube composite electrode for flexible lithium-ion batteries: Porosity engineering by controlled phase separation," *Advanced Energy Materials*, vol. 2, pp. 976–982, 2012.
52. M. Kaltenbrunner, G. Kettlgruber, C. Siket, R. Schwödiauer, and S. Bauer, "Arrays of ultracompliant electrochemical dry gel cells for stretchable electronics," *Advanced Materials*, vol. 22, pp. 2065–2067, 2010.
53. J. Hsu, S. Zahedi, A. Kansal, M. Srivastava, and V. Raghunathan, "Adaptive duty cycling for energy harvesting systems," presented at the *Proceedings of the 2006 international symposium on Low power electronics and design*, Tegernsee, Bavaria, Germany, 2006.

14

3D Printed Flexible and Stretchable Electronics

Galo Torres Sevilla

14.1 Introduction

Over the last few years, concepts such as the Internet of Things (IoT) and Internet of Everything (IoE) have gathered increased attention from researchers around the world due to the virtually unlimited number of applications in areas such as biotechnology, medical devices, wearables, and implantable electronics [1–3]. However, the constant growth of new applications brings with it the need for new forms of electronics where performance and power consumption are not the most important factors, but where the introduction of new form factors are an absolute necessity in order to make such applications possible. For this reason, several approaches have been made in order to fill the gap between traditional mechanically rigid high-performance electronics and the need for flexible and lightweight devices [4–6]. Among these approaches, the use of organic electronics with inherent mechanical flexibility and stretchability is one of the main topics of research in the field of flexible electronics [7]. Also, different methods have been developed in order to transform high-performance inorganic rigid electronics into free-form (flexible and stretchable) electronics [8,9]. Among them, transfer printing of thin films from inorganic donor substrates and the creation of nanoribbons, nanomembranes, and nanowires have produced the highest performances by combining the excellent electrical properties of high-quality inorganic semiconductors with the extreme flexibility of organic substrates. Although all these methods have been able to produce excellent devices ranging from single MOS (metal oxide semiconductor) devices to complete systems capable of performing computation tasks, there is still a lack of fast, inexpensive, and reliable prototyping necessary for emerging IoT and IoE applications.

For this reason, new approaches in the area of printed electronics (inkjet printing, 3D printing, transfer printing, nanoimprinting, etc.) have opened new horizons towards the creation of truly low-cost high-performance electronics [10,11]. Among printed electronics techniques, 3D printing has become one of the most widely investigated due to the capability of producing high-quality devices with a wide variety of materials and extremely reduced costs and times [12–14].

3D printing (a member of additive manufacturing—AM) consists of sequential printing of various layers of materials (polymers, metals, semiconductors) on top of each other to create complex structures with different functionalities. One of the most significant features of 3D printing consists in the extremely high prototyping speeds and low production costs due to its additive nature. With the use of 3D printing, one-of-a-kind complex prototypes can be made in just a few hours while the cost of the first item is the same as the cost of the last, making 3D printing technologies one of the most competitive manufacturing methods in biomedical, implantable, and wearable applications. For example, by using 3D printing, it may be possible to monitor the potential of an implantable device on soft 3D printed tissue, just as the one found in the human body, in just a few hours, significantly cutting production and testing times and costs. Another important feature of 3D printing is that the initial cost of set-up is minimal, which allows for a high degree of customization without the need of expensive tools. For these reasons, 3D printed technologies have become widely used in a wide range of devices where fabrication, testing, and characterization of single prototypes are of imperative importance to assess their potential for different applications. In the future, 3D printing is considered to become one of the most heavily used technologies in a wide variety of fields ranging from mass manufactured complex medical devices, where performance and reliability are of critical importance, to design-it-yourself (DIY) systems where each individual can print their own simple systems with reduced costs of materials and tools with price tags below $1000 US dollars [15]. Recently, 3D printing has also been investigated as a possible source to repair and replace defective living tissue in organs such as heart, kidneys, eyeballs, and skin [16–18]. In addition, with a certain amount of tool customization, it has the potential to create entire artificial organs, which could perform the same biological functions as non-functioning organs such as pancreas in diabetic patients or hearts in people who suffer of acute cardiovascular disorders [19]. In this sense, this might become a significant advancement towards the development of new methods that could alleviate the constantly increasing shortage of donor organs for organ transplant. This is due to the fact that one of the most important considerations in organ transplant involves finding a correct tissue match. However, if organs could be 3D printed using artificial tissue grown from the patients own cells and embedded with electronics that monitor and actuate to perform the same functions as their natural counterparts, this issue could potentially disappear and would open new possibilities in the area of personalized medicine. Due to all the advantages of 3D printing, it is only logical that several research groups from around the world take on the task of creating several materials and process variations to contribute in the development of 3D printing for applications where fast, reliable, and inexpensive prototyping is required. In this chapter, we introduce some of the most sophisticated 3D printing technologies and materials for future flexible IoT and IoE applications. Here, we also introduce several examples of different flexible devices and systems that have been achieved either partially or fully with the use of 3D printing technologies.

14.2 3D Printing Technologies for Flexible and Stretchable Electronics

Currently, there are several different methods to produce high-quality 3D printed devices. Among them, stereolithography (SLA), digital light processing (DLP), fused deposition modeling (FDM), selective laser sintering (SLS), selective laser melting (SLM), laminated object manufacturing (LOM), and binder jetting (BJ) are the ones that have attracted the most attention due to their unique properties and advantages in the area of additive manufacturing. In this section, these 3D printing technologies are clearly described and assessed according to their advantages and disadvantages for rapid prototyping of flexible electronics systems.

14.2.1 Stereolithography

Stereolithography consists in the sequential exposure of a photochemical polymer (photopolymer) [20]. This way, each layer of the final 3D object is exposed by a UV laser which crosslinks the photopolymer to convert it into a non-soluble material. Commonly, a computer assisted design (CAD) software guides the UV laser to draw a pre-programmed design or shape onto the printing surface. In stereolithography, two different methods can be used for printing; the first one (most commonly used) consists in printing each layer of the 3D object while an elevator apparatus attached to the printing surface descends a distance equal to the thickness of a single layer of the design (Figure 14.1a). On the other hand, the second approach (bottom-up) consists in the sequential printing of the layers of the 3D object starting from the bottom surface while the elevator apparatus ascends an equal distance equal to the thickness of a single layer of the final design. For this reason, in order to perform bottom-up printing a transparent bottom is required in the tank that holds the liquid-state photopolymer (Figure 14.1b).

Some of the advantages of SLA include the extremely fast prototyping speeds. For example, in contrary to common subtractive manufacturing methods, a complete complex design can be made in less than 2 days. For this reason, SLA is excellent equipment for demonstration purposes. Another advantage of SLA is the smooth surfaces that can be obtained with this method when comparing to other additive manufacturing technologies. Also, the smooth surfaces of the final product can be translated into a high level of design detail, achieving 3D objects with intricate surfaces with an excellent accuracy. Furthermore, due to the printing nature of SLA, the final surfaces of the objects can also be of extremely high quality. For this reason, SLA is the preferred method for printing medical devices that must also be water resistant. Additionally, as in every additive manufacturing technology, complex designs and intricate geometrical configurations can be achieved using SLA. Moreover, with SLA, there are several finish options to choose from, allowing the prototype to be very close in aesthetics to the final product. Finally, SLA can print in a wide variety of materials ranging from flexible to rigid.

In terms of disadvantages, although SLA can produce very high quality prototypes, the fragility and necessity of post curing of the 3D printed objects are still issues that have not been solved for this printing method. For this reason, even if the printed object is of great quality in terms of aesthetics and functionality, SLA printed devices are not compatible with mechanical testing. Also, stereolithography machines are expensive and are not easy to justify if used only for prototyping processes. For this reason, most companies rely on outsourced printing of their prototypes. Finally, one of the most important drawbacks in SLA is that even though they allow fast prototyping speeds, mass production cannot be achieved due to its unit production mechanism.

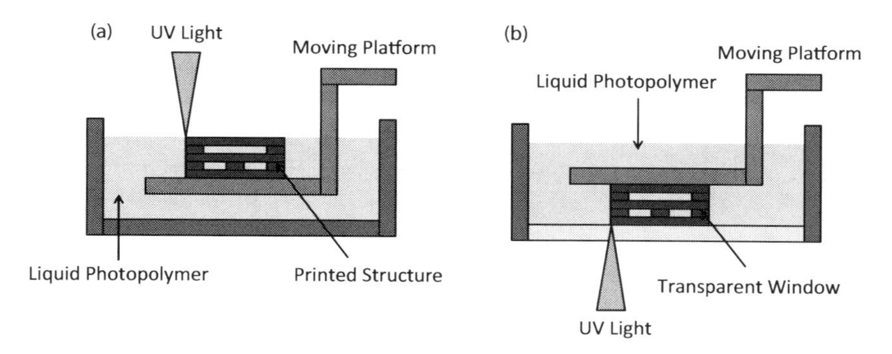

FIGURE 14.1 Schematic illustration of stereolithography system. (a) Top-down approach and (b) bottom-up approach.

14.2.2 Digital Light Processing

Just as SLA, DLP makes use of a light source in order to expose and crosslink a photopolymer that becomes insoluble upon exposure (Figure 14.2) [21]. One of the major differences when comparing to SLA is that DLP only requires a conventional light source instead of an expensive UV laser. For this reason, DLP tools can be obtained at much lower costs. Also, DLP can produce prototypes faster than SLA due to the pass exposure length of each layer. In other words, DLP can expose the entire surface of the 3D printed layer in a single pass instead of exposing each section of the layer by a raster mechanism. In DLP, once the 3D model is sent to the printer, a pool (vat) of photopolymer is exposed to light from a DLP projector under safelight conditions. Once the complete 3D model is printed, the excess of liquid resin is drained from the vat and the solidified object is left on the printing surface.

Advantages of DLP printing include the same points previously mentioned as advantages for SLA printing. However, in terms of cost, DLP printing further reduces the initial investment for process setup due to the use of regular arc light sources instead UV lasers. Also, DLP can produce prototypes with extremely good resolution with much less raw material and by simply using a shallow vat of resin, reducing the amount of material waste and hence the production costs. In terms of disadvantages, just as SLA, DLP requires post-production curing in order to obtain the final 3D printed prototype. Also, in some cases, freestanding assemblies require several layers of support structures in order to allow correct placing of the final printed layers, these support structures have to be manually removed from the printed object and can cause damage if not removed carefully.

14.2.3 Fused Deposition Modeling

FDM is a 3D printing technology that was developed in the early 1980s [22]. Historically, FDM has received increased attention due to its extremely low production costs and fast prototyping speeds. This technology has also received other names such as fused filament fabrication (FFF). By making use of FDM, one can print not only functional prototypes, but also concept models and final consumer products. FDM consists on a simple extrusion of a polymeric material through a heated nozzle that melts the polymer and prints it on top of the printing table. Just as in the case of SLA and DLP technologies, FMD prints each layer with the help of an elevator apparatus that controls the separation between consecutive layers

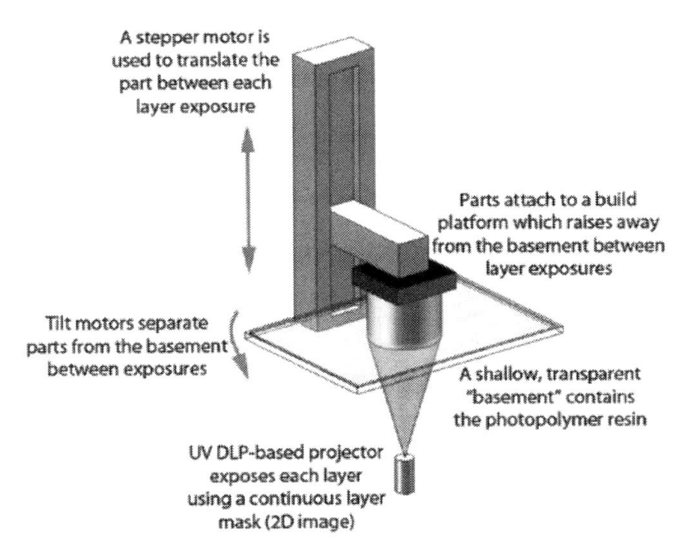

FIGURE 14.2 Continuous digital light processing (cDLP) additive manufacturing system. (© IOP Publishing. Reproduced from Wallace, J. et al., *Biofabrication*, 6, 015003, 2014. With permission.)

in the Z-axis. However, in contrast to DLM, FDM must raster the entire surface of the printed layer before it can move to the next layer. For this reason, FDM uses CAD software, which cuts the entire design into a mesh and calculates the extruder position for each portion. In order to print freestanding structures in FDM, the printer may calculate the position of a special material called support structure that can be dissolved or easily removed after the printing task is completed (Figure 14.3). One of the main advantages of FDM is that depending on the number of nozzles present in the printing machine, one can use several types of materials in the same object. For example, in the case of 3D printed electronics, the designer may choose to print some areas of the design with conductive material that could represent the interconnection of the electronic components that comprise the final systems, while some other areas may be printed with dielectric polymers which can contribute as the packaging structures of the devices. In this sense, FDM printing provides not only the advantage of producing functional prototypes in a single print, but also multifunctional systems. Another advantage of FDM is that the final structures are usually printed with commercial grade thermoplastics. For this reason, printed objects usually have excellent inherent mechanical, thermal, and chemical qualities. Recently, FDM has been widely spread in a variety of industries such as automobile companies like BMW and Hyundai or food companies such as Nestle and Dial. Finally, due to the large spread of this printing technology, companies can choose from a wide variety of models and costs. For example, a professional grade printer may range in tens of thousands of US dollars, while low-end machines can be purchased nowadays for a few hundreds of US dollars. In terms of disadvantages, in FDM, as in many other 3D printing technologies, final objects can show visible layer-lines depending on the design. For this reason, prototypes have to usually be cleaned or polished using hand sanding in order to get an even surface.

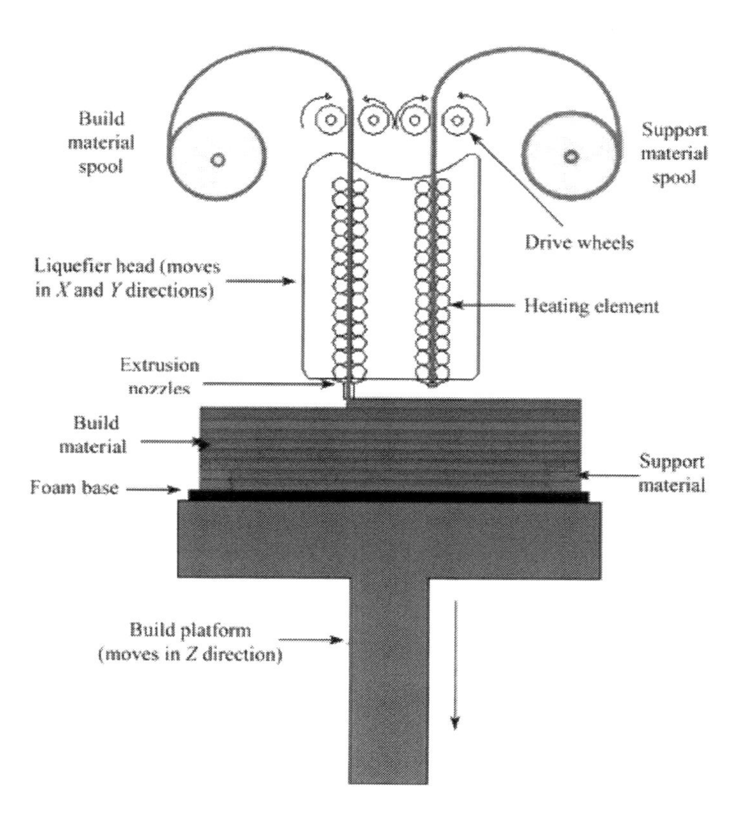

FIGURE 14.3 Schematic illustration of fused deposition modeling system. Here, each layer of the 3D printed structure is printed by extruders that melt the polymer and deposit it on the printing surface.

14.2.4 Selective Laser Sintering

SLS is a 3D printing technology that was developed in the 1980s at Texas University [23]. SLS can be considered to be very similar to SLA in terms of the printing methodology. However, there are some differences that separate SLS from SLA, especially in terms of the composition of the materials that are commonly used for printing. For example, in SLA, the resin of photochemical polymer is normally in a liquid state. On the other hand, for SLS, the material used for printing is usually a powder that is held on a vat adjacent to the printing surface. As in most 3D printing technologies, SLS starts by creating a 3D model in a CAD software, which then transforms the entire object into different layers that can be printed one at a time. One of the main advantages of SLS over SLA is that it does not require the use of support structures for freestanding designs. This is due to the fact that the printed structure is always surrounded by non-sintered powder (Figure 14.4). Another advantage is that in SLS, the powder can be made of a wide variety of materials such as nylon, ceramics, and even metals such as aluminum, steel, or silver. Due to the wide diversity of materials, SLS is an extremely powerful printing method for customized prototypes.

In terms of disadvantages, SLS is one of the most expensive printing methods due to the high power laser source that is required to sinter the powder. For this reason, SLS has not been adopted yet as one of the leading printing mechanisms for fast prototyping of flexible and stretchable 3D printing electronics.

14.2.5 Selective Laser Melting

SLM is a 3D printing technology developed in the mid 1990s in Germany at Fraunhofer Institute [24]. As in SLS, SLM makes use of a high power laser to melt and weld the printing material that is usually in a powder form (Figure 14.5). One of the key features of SLM over SLS is that in SLM, the powder is completely melted to form the 3D object, while in SLS, the powder is only partially melted (sintered). For this reason, SLM can produce much higher quality objects with almost no voids in the design. Due to the high cost of SLM, it is mostly used in high-end industries such as aerospace industries where quality of the final products are more important than the production costs. For this reason, nowadays it is very difficult to find research in the area of flexible 3D printed electronics that make use of SLM as their main printing mechanism.

14.2.6 Laminated Object Manufacturing

LOM consists in a rapid prototyping technology where each layer of the final 3D object is fused or laminated one on top of the other by using both heat and pressure [25]. Also, by using a CAD-assisted blade or laser, in LOM, each layer is cut to the desired shape. Once the layer is laminated and cut, the printing

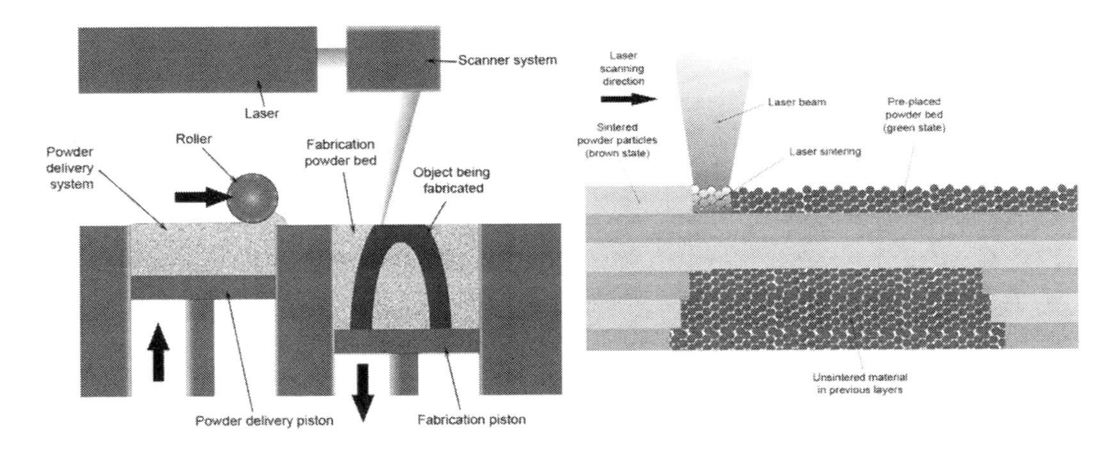

FIGURE 14.4 Schematic illustration of a selective laser sintering system.

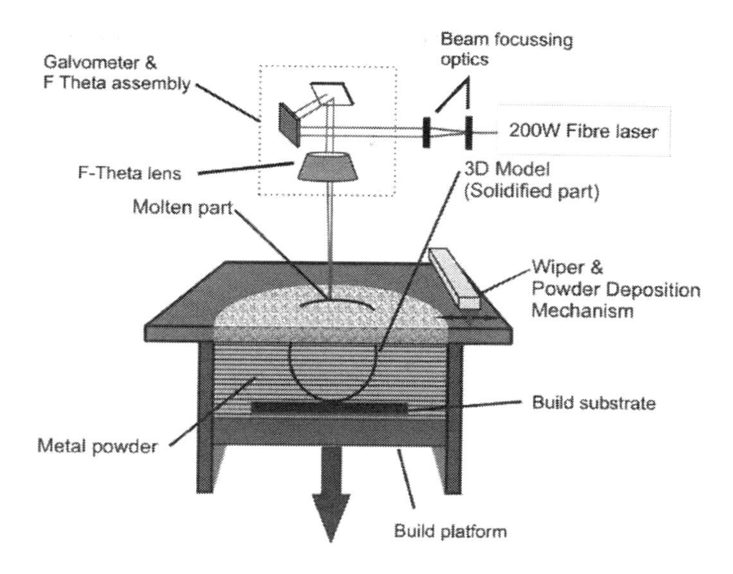

FIGURE 14.5 Schematic illustration of a selective laser melting system.

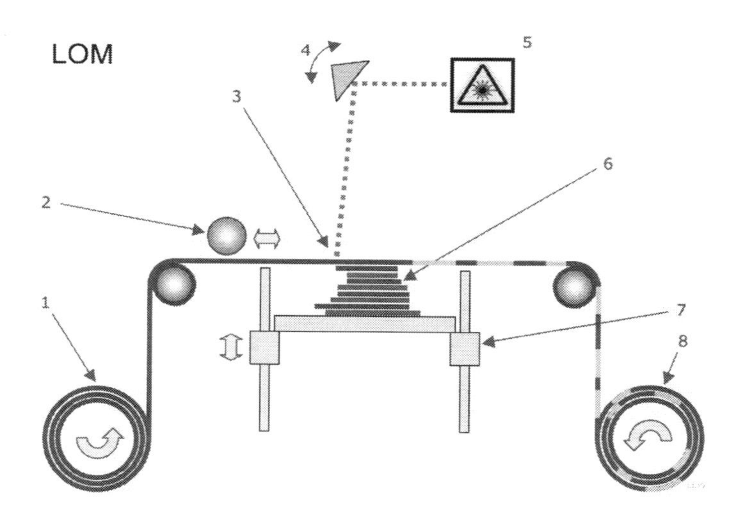

FIGURE 14.6 Schematic illustration of a laminated object manufacturing system. Here, each layer is continuously bonded to the structure by using a lamination system and cutting tools.

table descends by a distance equal to the thickness of each layer (Figure 14.6). Next, the next lamination sheet is pulled onto the printing surface and the lamination and cutting processes are repeated. Although LOM is not one of the most popular 3D printing techniques, it still remains as one of the fastest in terms of producing rough prototypes. Also, since it uses extremely low cost materials, it is mainly used in prototyping of 3D models by architects and artists.

14.2.7 Binder Jetting

BJ is a 3D printing technology that was first developed at the Massachusetts Institute of Technology (MIT) and has also been named as inkjet 3D printing and drop on powder [26]. This technology makes

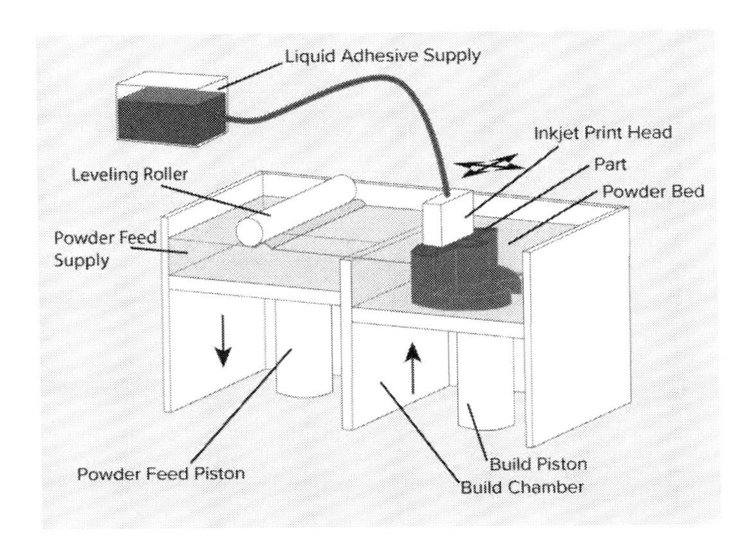

FIGURE 14.7 Schematic illustration of binder jetting system. Here, the first material usually consists of a powdered material, which constitutes the structural material of the 3D printed object. The second material usually consists on a bonding agent that "welds" the powder to crate the final structure.

use of two different materials to print the final 3D object. The first material usually consists of a powdered material, which constitutes the structural material of the 3D printed object. The second material usually consists on a bonding agent that "welds" the powder to crate the final structure (Figure 14.7). Just as in every additive manufacturing technology, in BJ, an elevator that holds the printing table is usually lowered by a distance equal to the thickness of each layer after the previous layer has been fully printed. Among the advantages of BJ 3D printing, we can find the wide variety of materials that can be printed using this technology (ceramics, metals, and polymers). Also, BJ can produce a fully colored design by just adding regular pigments to the binding agent. However, BJ presents several disadvantages in terms of resolution and intricate surfaces. For this reason, BJ printing is not commonly used in the area of rapid prototyping for flexible and stretchable electronics, but instead only to create models that can be used to assess the functionality of the product at low costs.

14.3 Materials for 3D Printed Flexible and Stretchable Electronics

One of the main advantages of 3D printing in the area of flexible and stretchable electronics is that no matter which material needs to be used, if made thin enough, the final structure can become flexible and stretchable. For these reasons, materials used in the area of 3D printed flexible electronics can be divided into four different groups:

1. Polymers
2. Electronic materials
3. Ceramics
4. Heterogeneous materials (composites).

14.3.1 Polymers for 3D Printing of Flexible and Stretchable Electronics

Polymeric materials with low melting points are some of the most widely used materials in the area of 3D printed electronics due to the low cost, chemical strength, and low processing costs. Also, polymers offer unparalleled advantages in terms of processing flexibility (i.e., can be processed in with many different additive manufacturing technologies) [27]. However, low mechanical strength and limited functionality of final prototypes are still some problems that need to be solved. For this reason, several different approaches have been tried to develop combinations of various materials to achieve the desired mechanical and functional prototypes. Among the most promising developments, we can find the use of polymers reinforced with different nanocomposites such as particles, fibres, and other nanomaterials. Excellent review papers have recently been published depicting an in-depth analysis of the advantages of polymer reinforcement [28].

Due to the low added cost and ease of mixing, particle reinforcement is one of the most heavily used methodologies to improve the mechanical and functional properties of 3D printed polymer materials. For example, particles can be easily mixed in liquid form for SLA and DLP printing or in powder form for SLS. Also, they can be combined into polymeric filaments for FDM. The addition of particles into polymers has also helped to reduce some of the difficulties in the printing process. For example, one obstacle for FDM printing is the distortion of the final parts due to the thermal expansion of the polymeric materials. However, it has been found that embedding metal particles can help mitigate this problem [28]. Another problem when printing with polymers is the anisotropy of the final structures that can cause low mechanical strengths in the perpendicular direction of the 3D printed structure.

Just as in the case of particle-reinforced polymers, fiber-based reinforcements can also enhance the properties of polymeric materials to produce mechanically stronger 3D objects. Until now, fiber-reinforced polymers were mostly used in FDM-based 3D printing. In this case, the polymer pallets and the fibres are mixed in a blender first before being delivered to the extruder in the form of filaments. Due to the complexity into making smooth layers in powder-based 3D printing techniques, fibres cannot be used as reinforcement techniques.

Finally, nanomaterials such as carbon nanotube, graphene, graphite, and several kinds of ceramics and metals often exhibit unparalleled advantages in terms of mechanical, electrical, and thermal properties [29]. For this reason, recently, several different approaches have been made in order to introduce nanomaterials as reinforcement for polymers. Here, homogeneous dispersion of the nanoparticles into the polymers is essential in order to achieve the desired characteristics of the composite. However, agglomeration of nanoparticles can usually damage the polymeric composite and cause non-uniformity across the materials. For this reason, chemical treatment of the base materials is usually required before embedding nanoparticles.

In addition to polymer reinforcement, several different approaches have recently been made to produce high quality polymer composites for 4D printing, where the 4th dimension is considered as time. Here, a polymeric material consisting of two or three photopolymers in specific microstructures and ratios are used to create functional electronic prototypes with tunable characteristics and that can change their geometry under different external stimuli. This allows the fabricated devices to accommodate different kinds of sensors, actuators, and electronics processing devices that can be used on demand depending of the application.

14.3.2 Electronic Materials

Over the last few years, several different research groups from around the world have focused their attention into the creation of electronic materials for additive manufacturing technologies. All these approaches have been made with the single objective of creating a completely functional 3D printed device that can perform the same or similar functions as their mass-produced counterparts. Until now, additive manufacturing techniques allow printing of simple passive devices such as resistors, capacitors, and inductors in a single step and without the need of post processing. This has allowed printing of several kinds of sensors and actuators such as capacitive-based touch and proximity sensors [30], resistive-based temperature sensors, and fully printed resistive based heaters [14], as well as several different components for communication devices such as tunable antennas. Nonetheless, printing of active components still requires further development in terms of the active materials and the correct tuning of their electrical properties. For this reason, there have been promising advancements in the research of active materials for 3D printed electronics. For example, recently, Kong et al. [31] demonstrated a fully printed prototype of quantum dot-based (QD) light emitting diodes (LEDs) where the active, conductive, and encapsulating layers are fully 3D printed to create a functional prototype. Here, the quantum dots, which comprise the active layer of the LEDs, are based on CdSe/ZnS semiconductors. This shows that semiconductors can also be 3D printed while maintaining their semiconducting characteristics. Although printing of single devices might seem trivial, it represents an excellent development in the area of fully printed electronic devices due to the need of semiconductors as the base of most active electronic components.

14.3.3 Ceramic Materials

In additive manufacturing, in contrast to polymeric and metallic materials that can be fused and printed by applying heat to their respective melting point, certain materials such as ceramics and concrete are particularly difficult to print due to their extremely high melting temperatures and problems that can arise during fusing them in a powder state. However, ceramic materials are of specific interest to researchers around the world in the area of 3D printing due to their ability to produce 3D models without cracks and that are resistant to harsh environments [32]. For this reason, research in 3D printing of ceramic materials has previously been described by adding several different steps to the printing mechanism as is the case of incorporating colloidal processing techniques or post densification steps after the printing process has been completed. Here, it is important to note that depending of the printing process, the choice of adding different steps to the printing process can be done depending on the application and functionality of the final prototype. For example, in the case of printing a single structure with multiple ceramic materials, colloidal processing may be a better solution than direct printing of ceramics due to the advantage of obtaining more reliable prototypes. However, if rapid prototyping and optical investigation of the final design are the only requirements, direct printing would be a better choice due to the advantage of obtaining the prototype in shorter times. Finally, it is to be noted that ceramic materials are difficult to print and incorporate into flexible and stretchable electronics due to their mechanical properties. Nonetheless, they are described here since in the future, if they become readily available for 3D printing processes, they could become an excellent supporting structure for flexible and stretchable electronics that could be used in harsh environments and industrial applications.

14.3.4 Heterogeneous Materials (Composites)

Composites materials can be considered as combinations of polymeric materials and additives that can help improve the printability and properties of such polymers. However, composite materials are relatively new and only few demonstrations have been made in the area of additive manufacturing. Recently, Simon et al. [14] demonstrated a thermoplastic-based material with embedded conductive carbon black (CB) as an additive to improve the functionality of the final device. Here, the final structure consisted in a flexible sensor that can measure mechanical strain and capacitive changes through changes in the conductance of the sensor. One of the most important points to consider when creating composite materials is the loading of the additive since excessive weight percentage can reduce the printability of the material due to nozzle blockage, while if loading is not sufficient, the conductivity of the composite would not be enough to create a reliable sensor. Another example of the use of composite materials as a means to print high quality 3D functional electronic sensors can be found in a recent report by Kim et al. [33]. Here, an efficient piezoelectric nanoparticle polymer with embedded barium titanate (BTO) nanoparticles was used to create competent mechanical energy harvesters. Although composite materials have not yet been demonstrated in applications that can be commercially available, they open new opportunities in the area of additive manufacturing technologies due to their tunable properties and low costs. For this reason, in the future, it is expected that most functional 3D printed flexible and stretchable electronics would be made by a combination of different kinds of materials that can be tailored to achieve electrical, mechanical, and functional properties close to the ones currently found in commercially available consumer electronics.

14.4 Flexible and Stretchable 3D Printed Devices

3D printed electronics can be divided into two major groups, fully 3D printed electronics, where all the components of the system or individual devices are printed using one or more 3D printing methods, and hybrid 3D printed electronics, where traditional electronics are combined with 3D printing in order to produce fast prototypes of electronic systems. Here, we describe some of the most promising examples of fully printed and hybrid 3D printed electronics.

14.4.1 Fully 3D Printed Flexible and Stretchable Electronics

Recently, Kong et al. demonstrated the use of 3D printing to produce a fully printed array of quantum dot LEDs (QDLEDs) [31]. Here, the LEDs consisted in a combination of inorganic emissive nanoparticles and different layers of conductive and dielectric polymers that constitute the flexible 3D printed electrical interconnects and the substrate. Figure 14.8 shows a schematic of the 3D printed structure and the different layers that were required in order to produce a fully functional 3D printed QDLED. To fabricate these devices, the authors first studied the characteristics of the hole transport layer consisting on poly (3,4-ethylenedioxythiophene) polystyrene sulfonate (PDOT:PSS), which is directly adjacent to the QD

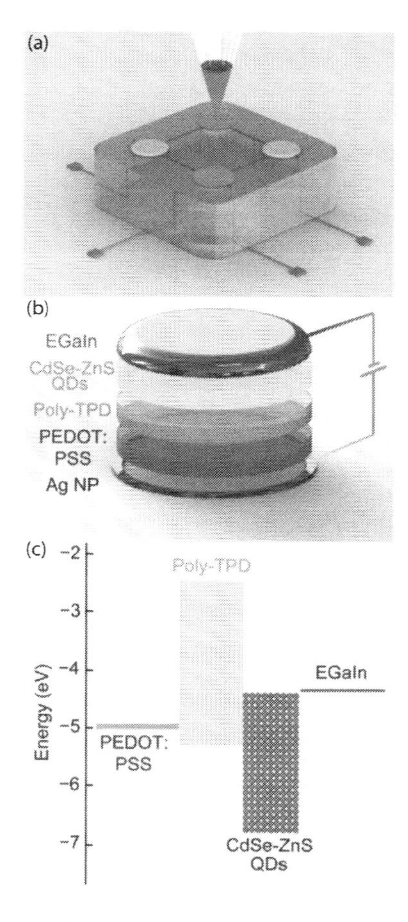

FIGURE 14.8 Schematic of fully printed QDLEDs. (a) Direct printing of QDLEDs on substrate, (b) layer by layer schematic of printed QDLEDs, and (c) QDLED energy level diagram. (Reprinted with permission from Kong, Y.L. et al., *Nano Lett.*, 14, 7017–7023, 2014. Copyright 2014 American Chemical Society.)

layer as can be seen in Figure 14.8. Next, a protective layer is printed on top of the PDOT:PSS layer in order to increase the non-polar solvent resistance of the entire structure and to reduce the residual organic emission electrical contact layer. For this reason, a thin layer of poly[N,N'-bis(4-butylphenyl)-N,N'-bis(phenyl) (poly-TPD) was used in order to maintain the color purity of the QDLED output. Next, the active layers of the LEDs were carefully printed by combining CdSe/ZnS QD emissive layers. In this case, 3D printing of the emissive layers possesses several advantages over traditional fabrication of QDLEDs by using spin coating due to the reduction of waste materials. For example, in traditional spin coating, nearly 94%–97% of the initial solution is wasted during the coating process which translate into a 20 times higher production cost of the LEDs. However, spin coating has the advantage of producing a much more uniform layer due to the suppression in the accumulation of suspended particles in a drying film. For this reason, the printing mechanism of the active layers had to be optimized by using a solutal Marangoni effect, which removes the accumulation of suspended particles and creates a uniform drying droplet without coffee ring pattern. Figure 14.2 shows the fabricated structure on top of a commercial contact lens. From the figure, it can be seen that the printed QDLED can be used in several applications such as transparent displays in a wide variety of substrates and with a cost reduction close to 20x due to the additive manufacturing nature of 3D printing. This demonstration represents a proof of concept that could open new doors towards the fabrication of fully 3D printed electronics. Furthermore, it represents a giant step forward in the fabrication of low cost electronic systems where all the components are printed and can be tuned to produce different functional structures without the need of sophisticated fabrication processes and tools. In terms of the QDLED performance, the tunability and pure color emission of the devices represent an excellent step forward for custom made displays. Finally, the authors anticipate that their process could allow the fabrication of topographically tailored devices onto curvilinear surfaces. However, there are still several key challenges that need to be resolved before this technology can be used in commercial electronics. Among them, increasing the resolution of the 3D printed structure, improving the performance and yield of the 3D printed devices, and incorporating additional functional blocks and devices in order to print a full system, are some of the most pressing ones. For this reason, in a near future, we expect to see co-printing of active flexible and stretchable electronics in biological constructs that could lead to new kinds of bionic devices such as prosthetic implants that could optically stimulate nerves.

14.4.2 Hybrid 3D Printed Flexible and Stretchable Electronics

Over the last few years, hybrid 3D printed electronics has become one of the most researched topics in the area of flexible and stretchable electronics. This is due to the extremely low cost advantage of 3D printing when compared with traditional electronic system fabrication. Also, the fast prototyping advantage of 3D printing makes it an excellent competitor for future fabrication of low cost systems. Here, we describe the use of different 3D printing techniques for the fabrication of hybrid 3D printed electronics, where active components can be placed manually or automatically using high precision pick and place tools on a wide variety of substrates, while the interconnects and isolation layers can be easily printed using traditional 3D printing technologies (Figure 14.9).

14.4.2.1 Hybrid 3D Printing of Soft Electronics

Recently, Professor Jennifer Lewis' group at Harvard University developed a state-of-the-art technique to "print" traditional electronics onto soft polymeric substrates [34]. Here, the main purpose is to combine the excellent performance of traditional silicon-based electronics with low cost additive manufacturing technologies in order to create complete systems that have the ability of being stretched and flexed without the loss of performance. The process started by developing a matrix of conductive and dielectric inks that could easily be 3D printed on different asymmetric surfaces. In order to print the different inks, different chemistries were used by mixing thermoplastic polyurethane (TPU) with silver nanoparticles. This helped to reduce delamination effects due to the robust interface between the silver TPU inks and the pure TPU substrates. Figure 14.10 shows the schematic process flow that was followed in order to produce

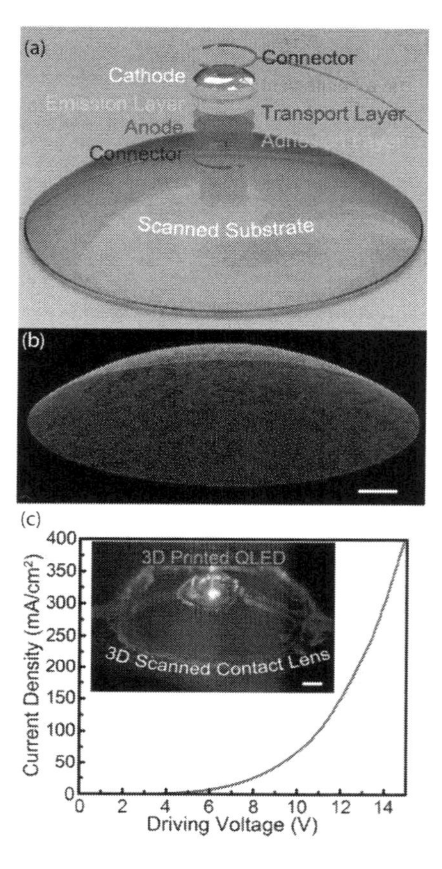

FIGURE 14.9 3D printed QDLED on a 3D scanned onto a curvilinear substrate. (a) CAD model showing the QDLED components and conformal integration onto a curvilinear substrate. (b) 3D model of a contact lens acquired by 3D structured-light scanning. (c) Current density versus voltage characteristics of the 3D printed QDLEDs on top of the 3D scanned contact lens. The inset shows the electro-luminescence output from the printed QDLED. Scale bars are 1 mm. (Reprinted with permission from Kong, Y.L. et al., *Nano Lett.*, 14, 7017–7023, 2014. Copyright 2014 American Chemical Society.)

high performance hybrid 3D printed electronic systems. In order to test the viability of the developed process, different kinds of conductive ink-based sensors were printed on top of a flexible 3D printed substrate. Figure 14.11 shows the performance of fully printed resistive strain and capacitive touch sensors. It can be seen that in both cases the sensors show excellent performance in terms of linearity, sensitivity, and repeatability. Next, the authors made use of the same process to create complete flexible systems that included traditional microcontrollers and a large area array of soft pressure sensors. Figure 14.12 shows the developed systems and the results obtained for the strain and pressure sensors. Here, the strain and pressure sensors were connected to the microcontroller with fully printed conductive interconnects. To evaluate the strain sensor, the complete system was placed posterior to the elbow joint and was tested at different joint angles. Once strain is detected by the sensor, the data are fed to the microcontroller which translates the analogue signal into its digital counterpart and outputs the level of detected strain to five LED indicators as shown in Figure 14.12. In the case of the pressure sensor, a 29 sensor array was printed on a custom made substrate for plantar sensing. As shown in Figure 14.12, the sensor array can be easily used to measure a unique plantar pressure profile. Due to the soft property and flexible form factor of the sensors, they can be readily integrated into different textiles and shoe insoles that can be used in several applications such as soft robotics, biomedical devices, and wearable electronics.

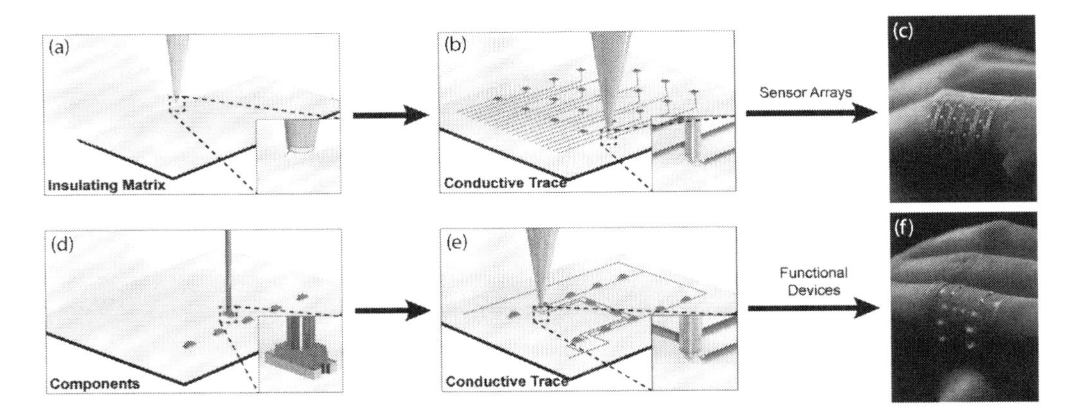

FIGURE 14.10 Hybrid 3D printing platform for soft electronics. Schematic images of (a) direct ink writing of a TPU matrix for the device body and (b) conductive AgTPU traces for the sensing elements. (c) Image of representative example of soft sensor array. Schematic images of (d) pick-and-place (P+P) of components using vacuum nozzle onto target positions and (e) direct writing of conductive AgTPU traces to interconnect the surface-mounted LEDs placed in the form of a soft, stretchable "H" LED array. (f) Image of a functional LED array wrapped around a human finger. (From Valentine A.D. et al.: Hybrid 3D Printing of Soft Electronics. *Advanced Materials*. 2017. 29. 1703817. Copyright Wiley-VCH Verlag GmbH & Co. KGaA. Reproduced with permission.)

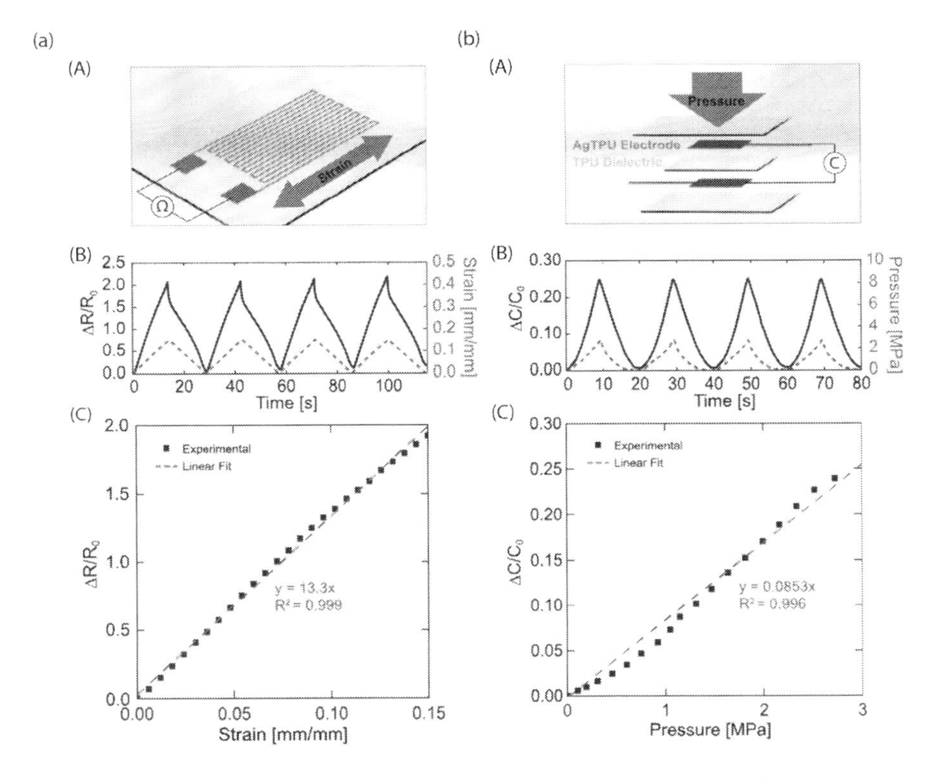

FIGURE 14.11 (a) Strain sensor device and performance: (A) device scheme, (B) plot of ΔR/R0 over time during a triangular strain cycle, and (C) plot of ΔR/R0 versus strain. (b) Capacitive pressure sensor and performance: (A) device scheme, (B) plot of ΔC/C0 over time during a triangular strain cycle, and (C) plot of ΔC/C0 versus pressure. (From Valentine A.D. et al.: Hybrid 3D Printing of Soft Electronics. *Advanced Materials*. 2017. 29. 1703817. Copyright Wiley-VCH Verlag GmbH & Co. KGaA. Reproduced with permission.)

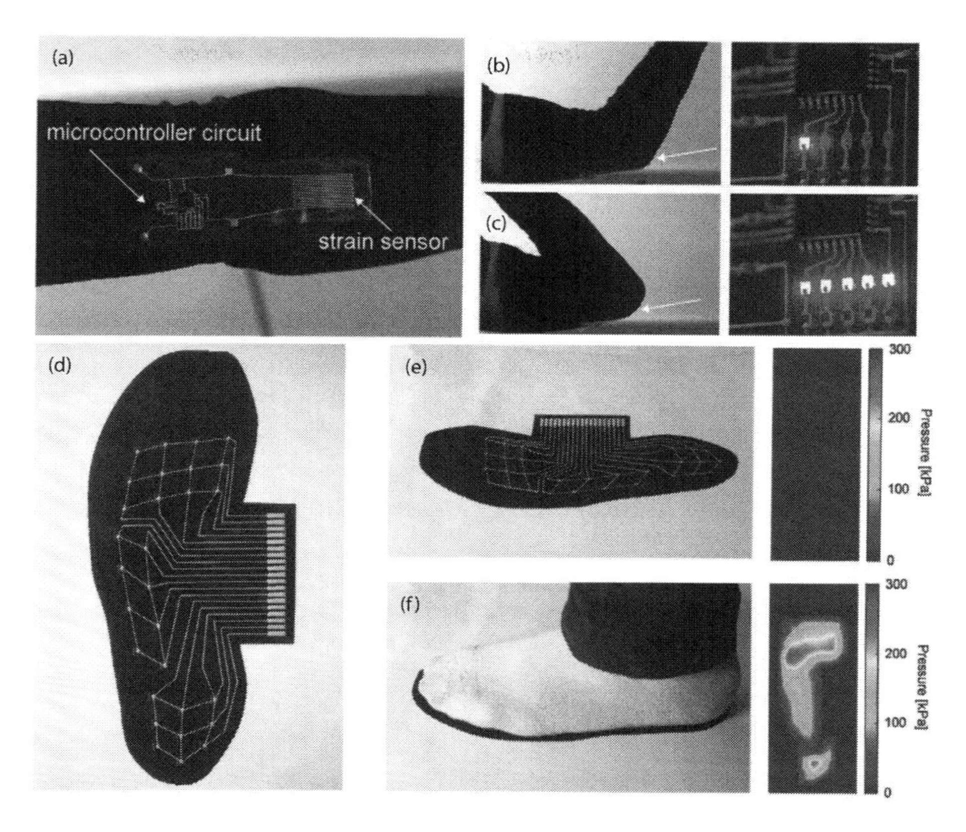

FIGURE 14.12 Wearable soft electronics fabricated by hybrid 3D printing: (a) image of textile-mounted printed strain sensor and microcontroller circuit. (b,c) Images of real-time function of the wearable strain device (identified with arrows) at modest and maximum joint bending, respectively, along with the corresponding LED readout. (d) Image of plantar sensor array. (e) Illustration of sensor array readout in the absence of applied pressure and (f) upon the application of pressure by a human foot. (From Valentine A.D. et al.: Hybrid 3D Printing of Soft Electronics. *Advanced Materials*. 2017. 29. 1703817. Copyright Wiley-VCH Verlag GmbH & Co. KGaA. Reproduced with permission.)

14.4.2.2 Hybrid 3D Printed Electronics for Radio-Frequency Applications

Radio-frequency (RF) electronics has been one of the most developed fields in the area of consumer electronics. This is due to the need of portable devices that can easily communicate with external peers without the need of cables or wires. For this reason, recently, Professor Jennifer Lewis' group at Harvard University developed an excellent process for direct writing of radio-frequency components using state-of-the-art 3D printing technologies [35]. Here, direct writing of visco-elastic silver nanoparticle-based inks were used to produce RF passive devices that are able to operate up to 45 GHz frequencies. Printed devices included inductors and capacitors, as well as wave-based devices such as transmission lines, resonant networks, and antennas. Also, these devices were combined with discrete traditional electronics transistors to produce self-sustained oscillators and synchronized oscillator arrays that are able to produce fully flexible 3D printed RF circuits. Figure 14.13 shows the different architectures of devices that can be fabricated using this technique. Here, miniaturization of the passive devices is an important aspect due to their regularly larger footprint when compared to state of the art complementary metal-oxide semiconductor (CMOS) transistors. In order to demonstrate the capabilities of the process, the authors created different RF circuits using printed passive devices and commercial RF active electronics mounted on top of the substrate using the same tool that was used for 3D printing. Here, an LC resonator was used in

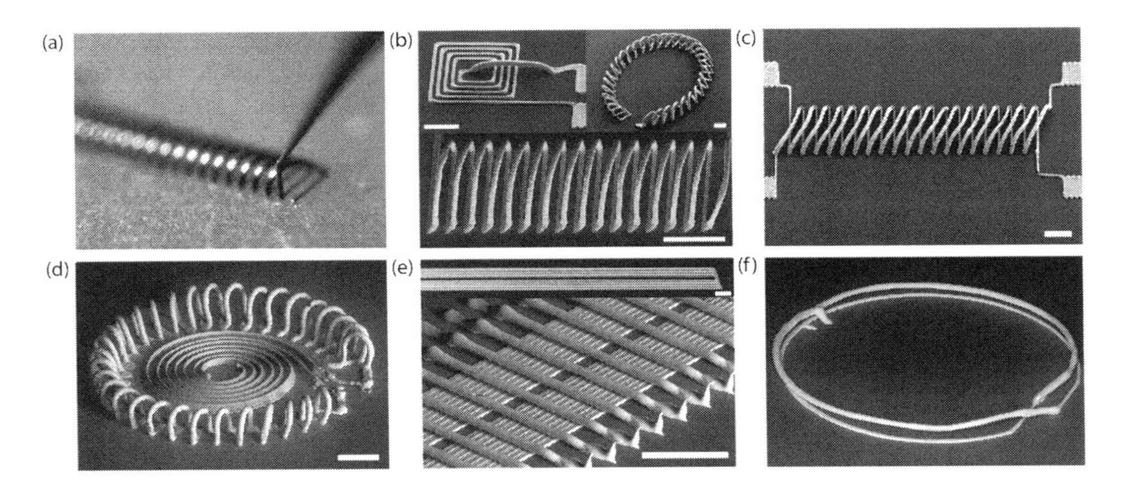

FIGURE 14.13 3D printed passive devices on glass substrates and their measured RF characteristics. (a) Direct writing of a viscoelastic silver ink through at 10 μm nozzle. (b) Top left: A 4-turn planar inductor; top right: a 33-turn toroidal inductor; bottom: a 16-turn solenoidal inductor. (c) 1:1 solenoidal transformer. (d) LC resonator consisting of a 32-turn toroidal inductor and a parallel-plate capacitor. (e) Floating metal strips (FMS)-free and FMS-loaded co-planar strip (CPS) standing wave resonators (SWR). (f) Möbius CPS ring (diameter: 5 mm). (From Zhou, N. et al.: Gigahertz Electromagnetic Structures via Direct Link Writing for Radio-Frequency Oscillator and Transmitter Applications. *Advanced Materials*. 2017. 29. 1605198. Copyright Wiley-VCH Verlag GmbH & Co. KGaA. Reproduced with permission.)

order to test the model. It is to be noted that no deterioration is seen due to the use of 3D printed passive devices as seen in Figure 14.14. Then, the oscillators were used to clock a 1.8 GH RF transmitter on different substrates such as glass and polyimide. The oscillator was used to drive a printed meander dipole antenna though a class-A power amplifier. Here, the power amplifier was also built using conventional RF electronics. In order to provide a strong direct current (DC) isolation between the main functional blocks of the RF transmitter, a 1:1 transformer was also 3D printed using the same conductive ink as in the case of the passive circuit components. Figure 14.15 shows the fabricated RF transmitter and the results obtained at two different antenna distances of 3 and 30 m. In summary, the authors have shown an excellent fabrication strategy to miniaturize and build passive RF electronic components on a wide variety of substrates using simple additive manufacturing technologies. All the passive structures such as resistors, inductors, and capacitors were easily printed using conductive inks and combined with conventional electronics to create high-performance flexible hybrid RF electronics. One of the main advantages of the process is the use of a single tool for both printing the passive components and mounting the active devices on top of the host substrate, showing the versatility of additive manufacturing tools for the creation of complete prototypes. Also, due to the ability of printing out-of-plane, RF components can be easily miniaturized in order to accommodate complete circuits with much lower footprints.

14.4.2.3 Embedded 3D Printed Electronic Sensors

Electronic sensors can be considered as an integral part of current wearable and future biomedical devices. This is due to the need of constant monitoring of different physiological functions of the human body in order to prevent or to early diagnose illnesses that could have extremely harmful effects on a person. However, due to the mechanical properties mismatch between conventional rigid electronics and soft objects such as human body tissue, it is difficult to provide seamless integration of soft and stretchable sensory systems. For this reason, recently, Professor Lewis' group at Harvard University developed a 3D printing method that allows printing and embedding of strain sensors onto flexible and stretchable polymeric substrates [36]. The method involves extruding viscoelastic

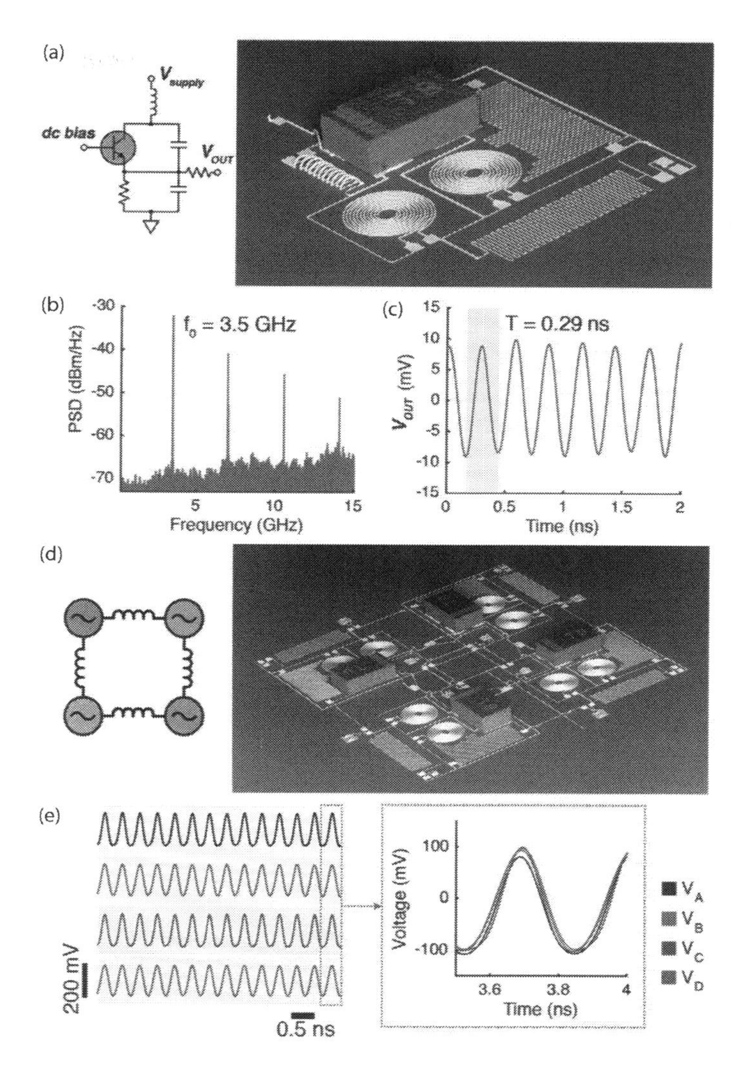

FIGURE 14.14 Lumped self-sustained oscillators and their synchronization network. (a–c) A Colpitts oscillator with a 3D printed LC resonator and its measured power spectral density (PSD) and time-domain signal. (d,e) An array of four injection-locked Colpitts oscillators and their measured time-domain signals. Each transistor in parts (a) and (d) has a footprint of 1 mm × 0.6 mm. Absolute time reference points in parts (c) and (e) are arbitrary. The power consumption of each Colpitts oscillator in parts (a) and (d) is ≈8 mW. (From Zhou, N. et al.: Gigahertz Electromagnetic Structures via Direct Link Writing for Radio-Frequency Oscillator and Transmitter Applications. *Advanced Materials*. 2017. 29. 1605198. Copyright Wiley-VCH Verlag GmbH & Co. KGaA. Reproduced with permission.)

ink through a deposition nozzle directly into a predefined elastomeric reservoir. Once the ink is positioned in the reservoir, it forms a resistive sensing element. This way, 3D printed sensors can be easily fabricated in a programmable and seamless manner. Figure 14.16 shows the developed printing mechanism. Here, material compatibility was one of the most important considerations for successful printing of soft sensor devices. For this reason, the materials were engineered to meet two different criteria: (I) the sensor ink should minimally dissociate or diffuse within the reservoir during and after printing and (II) the reservoir and the filler should be identical chemically in order to eliminate unnecessary interfaces in the elastomeric matrix after curing. Figure 14.17 shows the electrical

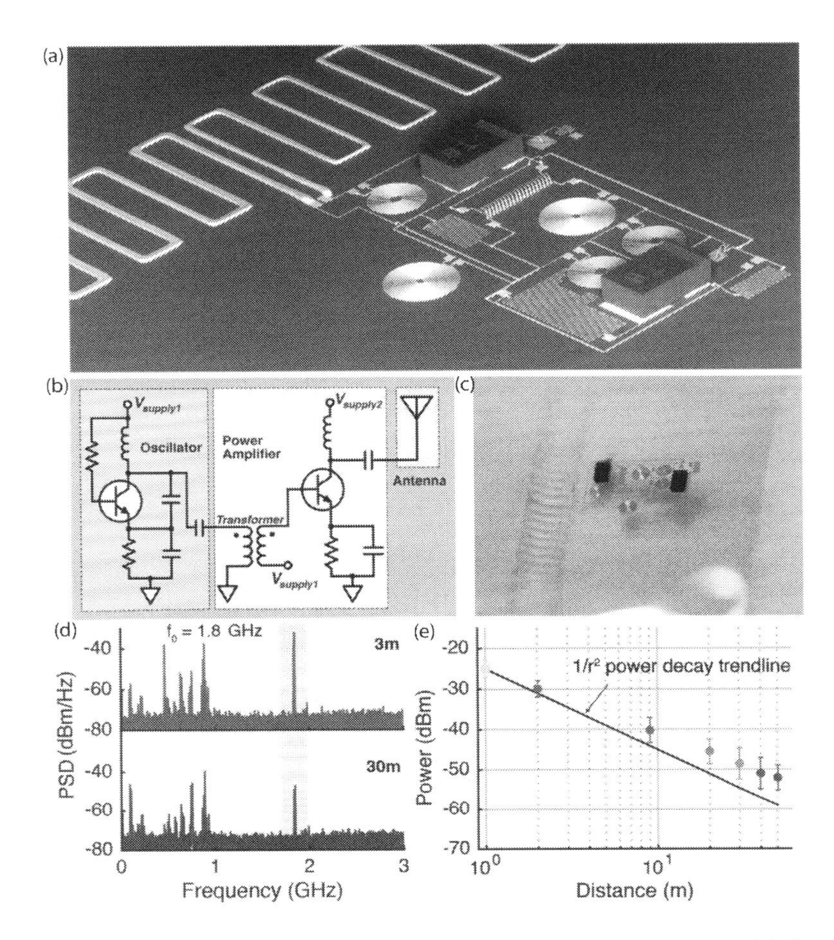

FIGURE 14.15 Wireless transmitter clocked by oscillator. (a–c) 1.8 GHz RF transmitter on (a) glass and (c) polyimide substrates. Each transistor shown has a footprint of 1 mm × 0.6 mm. The total power consumption of each transmitter in parts (a) and (c) is ≈72 mW. (d) PSD measured 3 and 30 m away from the transmitter on polyimide substrate. (e) 1.8 GHz power (colored dots) versus distance from the transmitter, up against the theoretical inverse square law. (From Zhou, N. et al.: Gigahertz Electromagnetic Structures via Direct Link Writing for Radio-Frequency Oscillator and Transmitter Applications. *Advanced Materials*. 2017. 29. 1605198. Copyright Wiley-VCH Verlag GmbH & Co. KGaA. Reproduced with permission.)

characteristics of the printed sensors under different test conditions. Here, to investigate the electrical performance of the different strain sensors as a function of cycling strain, the embedded devices were tested up to 100% strain and relaxed back to 0%. It can be seen that sensors with lower cross-sectional area produce a higher change in resistance indicating that the sensor sensitivity can be easily tuned by adjusting the printing speed of the devices. Although the performance of the sensors is comparable to commercial strain gauges, it can be seen that hysteresis arises due to the disparate time scales associated with breakdown and reformation of contacts between the carbon particles that comprise the resistive sensing material. To further demonstrate the capability of the fabricated devices, a large area array of strain sensors were embedded in a wearable objects which can be worn on different parts of the human body. Figure 14.18 shows the embedded devices in a pre-molded glove and the different responses of the embedded strain sensors. Here, by using the 3D printed devices, the demonstrations show how the sensors have the ability to sense different positions of the fingers. This could be useful in several emerging wearable devices where not only the sensors can be used to

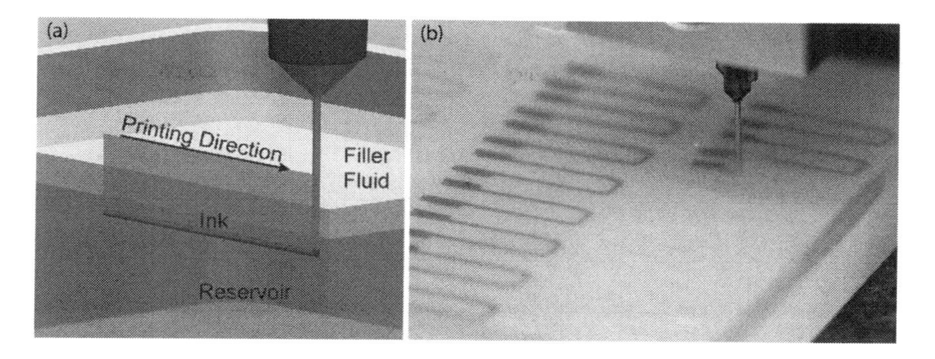

FIGURE 14.16 (a) Schematic illustration of the embedded 3D printing (e-3DP) process. A conductive ink is printed into an uncured elastomeric reservoir, which is capped by filler fluid. (b) Photograph of e-3DP for a planar array of soft strain sensors. (From Muth, J.T. et al.: Embedded 3D Printing of Strain Sensors within Highly Stretchable Elastomers. *Advanced Materials*. 2014. 26. 6307–6312. Copyright Wiley-VCH Verlag GmbH & Co. KGaA. Reproduced with permission.)

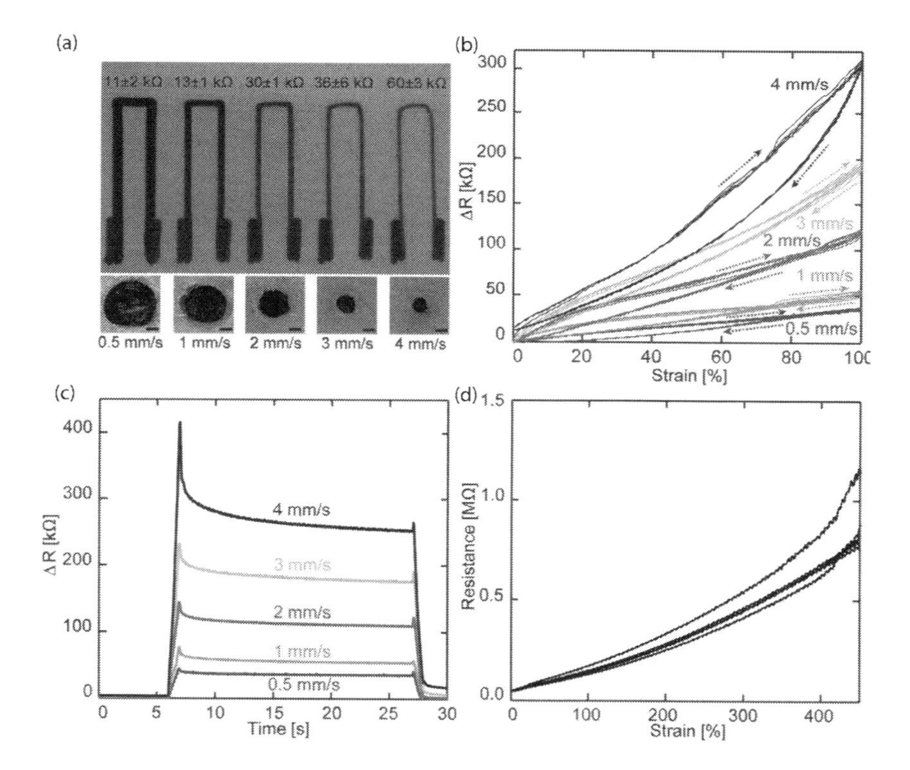

FIGURE 14.17 (a) Top and cross-sectional images of soft sensors printed by e-3DP at varying speeds. (b) Electrical resistance change as a function of elongation for sensors subjected to cyclic deformation, in which each sensor is cycled five times to 100% strain at a crosshead speed of 2.96 mm/s. (c) Electrical resistance change as a function of time for sensors subjected to step deformation to 100% strain at a crosshead speed of 23 mm/s. (d) Electrical resistance as a function of strain up to 450% strain for sensors strained to mechanical failure at a crosshead speed of 5 mm/s. (From Muth, J.T. et al.: Embedded 3D Printing of Strain Sensors within Highly Stretchable Elastomers. *Advanced Materials*. 2014. 26. 6307–6312. Copyright Wiley-VCH Verlag GmbH & Co. KGaA. Reproduced with permission.)

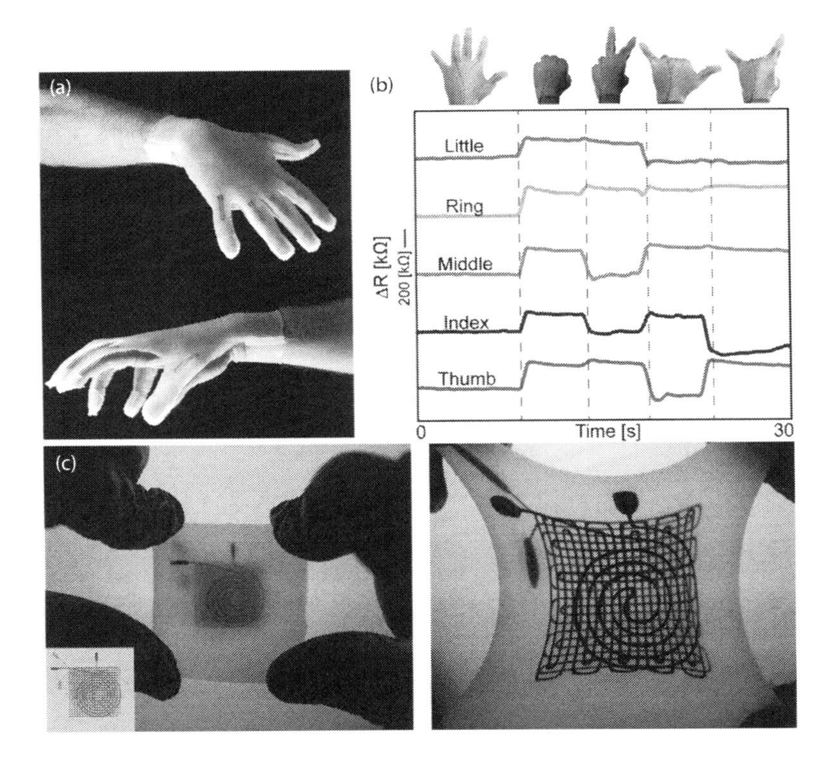

FIGURE 14.18 (a) Photograph of a glove with embedded strain sensors produced by e-3DP. (b) Electrical resistance change as a function of time for strain sensors within the glove at five different hand positions. (c) Photograph of a three-layer strain and pressure sensor in the unstrained state (left) and stretched state (right). The top layer consists of a spiral pressure sensor, below which lies a two-layer biaxial strain sensor that consists of two square meander patterns (20 mm × 20 mm) oriented perpendicular to each other [Inset: Shows the CAD model of the layered motif, in which each layer is depicted in a different color for visual clarity.]. (From Muth, J.T. et al.: Embedded 3D Printing of Strain Sensors within Highly Stretchable Elastomers. *Advanced Materials*. 2014. 26. 6307–6312. Copyright Wiley-VCH Verlag GmbH & Co. KGaA. Reproduced with permission.)

evaluate the physiological status of human body parts, but also in applications such as robotics where one can wear different sensors around the body to allow a robot to mimic the movements of a person. This shows that 3D printing of different sensors opens new doors in the creation of functional devices for human/machine interfaces, soft robotics, and beyond.

14.5 Conclusion

3D printing has become one of the most heavily researched topics around the world in the area of flexible and stretchable electronics due to its ability to produce low cost and fast prototypes without the need of specialized or expensive equipment. Although 3D printing has not reach an evolved stage where it could be adapted for mass manufacturing of commercial devices, we can see that many different approaches are being made in order to allow a wide range of electronics to be easily printed and tested. For this reason, here, we have described some of the most promising approaches for future 3D printing of high performance and low cost electronics. We can see that currently there are several 3D printing mechanisms that can be readily adapted to provide fast and reliable prototyping of different devices. Also, numerous materials with different properties are being developed in order to cover the wide range of electrical and mechanical properties needed to build a fully printed electronic system. Finally, several approaches have already been made in order to demonstrate the viability of 3D printing in many different branches of commercial electronics such as displays, robotics, and wearable devices that in the future can be used as the new generation of high performance devices. For this reason, we believe that going forward, 3D printing has the potential to become one of the best resources for people to design, evaluate, and fabricate devices that currently can only be made with specialized equipment in costly and inaccessible manufacturing plants.

References

1. A. P. Chandrakasan, N. Verma, and D. C. Daly, "Ultralow-power electronics for biomedical applications," *Annual Review of Biomedical Engineering*, vol. 10, pp. 247–274, 2008.
2. T. Eggers, C. Marschner, U. Marschner, B. Clasbrummel, R. Laur, and J. Binder, "Advanced hybrid integrated low-power telemetric pressure monitoring system for biomedical applications," In *Proceedings IEEE Thirteenth Annual International Conference on Micro Electro Mechanical Systems (Cat. No. 00CH36308)*, pp. 329–334, 2000.
3. E. Pickwell and V. P. Wallace, "Biomedical applications of terahertz technology," *Journal of Physics D: Applied Physics*, vol. 39, no. 17, pp. R301, 2006.
4. Y. Zhan, Y. Mei, and L. Zheng, "Materials capability and device performance in flexible electronics for the Internet of Things," *Journal of Materials Chemistry C*, vol. 2, no. 7, pp. 1220–1232, 2014.
5. S. Khan, L. Lorenzelli, and R. S. Dahiya, "Technologies for printing sensors and electronics over large flexible substrates: A review," *IEEE Sensors Journal*, vol. 15, no. 6, pp. 3164–3185, 2015.
6. Z. Bao and X. Chen, "Flexible and stretchable devices," *Advanced Materials*, vol. 28, no. 22, pp. 4177–4179, 2016.
7. H. Klauk, *Organic Electronics: Materials, Manufacturing, and Applications*. Weinheim, Germany, John Wiley & Sons, 2006.
8. Y. Sun and J. A. Rogers, "Inorganic semiconductors for flexible electronics," *Advanced Materials*, vol. 19, no. 15, pp. 1897–1916, 2007.
9. J. A. Rogers, T. Someya, and Y. Huang, "Materials and mechanics for stretchable electronics," *Science*, vol. 327, no. 5973, pp. 1603–1607, 2010.
10. J. Perelaer et al., "Printed electronics: The challenges involved in printing devices, interconnects, and contacts based on inorganic materials," *Journal of Materials Chemistry*, vol. 20, no. 39, pp. 8446–8453, 2010.
11. W. Clemens, W. Fix, J. Ficker, A. Knobloch, and A. Ullmann, "From polymer transistors toward printed electronics," *Journal of Materials Research*, vol. 19, no. 7, pp. 1963–1973, 2004.
12. S. Ready, F. Endicott, G. L. Whiting, T. N. Ng, E. M. Chow, and J. Lu, "3D printed electronics," In *NIP & digital Fabrication Conference*, vol. 2013, pp. 9–12, Society for Imaging Science and Technology, 2013.
13. E. Macdonald et al., "3D printing for the rapid prototyping of structural electronics," *IEEE Access*, vol. 2, pp. 234–242, 2014.
14. S. J. Leigh, R. J. Bradley, C. P. Purssell, D. R. Billson, and D. A. Hutchins, "A simple, low-cost conductive composite material for 3D printing of electronic sensors," *PloS One*, vol. 7, no. 11, pp. e49365, 2012.
15. B. Berman, "3-D printing: The new industrial revolution," *Business Horizons*, vol. 55, no. 2, pp. 155–162, 2012.
16. V. Mironov, T. Boland, T. Trusk, G. Forgacs, and R. R. Markwald, "Organ printing: Computer-aided jet-based 3D tissue engineering," *TRENDS in Biotechnology*, vol. 21, no. 4, pp. 157–161, 2003.
17. S. V. Murphy and A. Atala, "3D bioprinting of tissues and organs," *Nature Biotechnology*, vol. 32, no. 8, pp. 773–785, 2014.
18. S.-S. Yoo, "3D-printed biological organs: Medical potential and patenting opportunity," *Expert Opinion on Therapeutic Patents*, vol. 25, no. 5, pp. 507–511, 2015.

19. C. Schubert, M. C. Van Langeveld, and L. A. Donoso, "Innovations in 3D printing: A 3D overview from optics to organs," *British Journal of Ophthalmology*, vol. 98, no. 2, pp. 159–161, 2014.
20. P. F. Jacobs, *Rapid Prototyping & Manufacturing: Fundamentals of Stereolithography*. Dearborn, MI, Society of Manufacturing Engineers, 1992.
21. R. Liska et al., "Photopolymers for rapid prototyping," *Journal of Coatings Technology and Research*, vol. 4, no. 4, pp. 505–510, 2007.
22. I. Zein, D. W. Hutmacher, K. C. Tan, and S. H. Teoh, "Fused deposition modeling of novel scaffold architectures for tissue engineering applications," *Biomaterials*, vol. 23, no. 4, pp. 1169–1185, 2002.
23. J.-P. Kruth, P. Mercelis, J. Van Vaerenbergh, L. Froyen, and M. Rombouts, "Binding mechanisms in selective laser sintering and selective laser melting," *Rapid Prototyping Journal*, vol. 11, no. 1, pp. 26–36, 2005.
24. E. O. Olakanmi, "Selective laser sintering/melting (SLS/SLM) of pure Al, Al–Mg, and Al–Si powders: Effect of processing conditions and powder properties," *Journal of Materials Processing Technology*, vol. 213, no. 8, pp. 1387–1405, 2013.
25. B. Mueller and D. Kochan, "Laminated object manufacturing for rapid tooling and patternmaking in foundry industry," *Computers in Industry*, vol. 39, no. 1, pp. 47–53, 1999.
26. R. Bogue, "3D printing: The dawn of a new era in manufacturing?," *Assembly Automation*, vol. 33, no. 4, pp. 307–311, 2013.
27. J. W. Stansbury and M. J. Idacavage, "3D printing with polymers: Challenges among expanding options and opportunities," *Dental Materials*, vol. 32, no. 1, pp. 54–64, 2016.
28. S. Kumar and J. P. Kruth, "Composites by rapid prototyping technology," *Materials & Design*, vol. 31, no. 2, pp. 850–856, 2010.
29. T. A. Campbell and O. S. Ivanova, "3D printing of multifunctional nanocomposites," *Nano Today*, vol. 8, no. 2, pp. 119–120, 2013.
30. C. Shemelya et al., "3D printed capacitive sensors," In *SENSORS* 2013, pp. 1–4. IEEE.
31. Y. L. Kong et al., "3D printed quantum dot light-emitting diodes," *Nano Letters*, vol. 14, no. 12, pp. 7017–7023, 2014.
32. A. Curodeau, E. Sachs, and S. Caldarise, "Design and fabrication of cast orthopedic implants with freeform surface textures from 3-D printed ceramic shell," *Journal of Biomedical Materials Research Part A*, vol. 53, no. 5, pp. 525–535, 2000.
33. K. Kim et al., "3D optical printing of piezoelectric nanoparticle–polymer composite materials," *ACS Nano*, vol. 8, no. 10, pp. 9799–9806, 2014.
34. A. D. Valentine et al., "Hybrid 3D printing of soft electronics," *Advanced Materials*, vol. 29, no. 40, p. 1703817, 2017.
35. N. Zhou, C. Liu, J. A. Lewis, and D. Ham, "Gigahertz electromagnetic structures via direct ink writing for radio-frequency oscillator and transmitter applications," *Advanced Materials*, vol. 29, no. 15, p. 1605198, 2017.
36. J. T. Muth et al., "Embedded 3D printing of strain sensors within highly stretchable elastomers," *Advanced Materials*, vol. 26, no. 36, pp. 6307–6312, 2014.
37. J. Wallace et al., "Validating continuous digital light processing (cDLP) additive manufacturing accuracy and tissue engineering utility of a dye-initiator package," *Biofabrication*, vol. 6, no. 1, p. 015003, 2014.

15

Flexible and Stretchable Paper-Based Structures for Electronic Applications

Tongfen Liang,
Ramendra Kishor Pal, Xiyue Zou,
Anna Root, and
Aaron D. Mazzeo

15.1 Introduction

As the other contributors to this volume have undoubtedly highlighted, flexible and stretchable electronics are thin, lightweight, capable of managing large strains, and potentially inexpensive (Kim et al. 2015; Lu and Kim 2013). Flexible and stretchable devices can be wearable (Stoppa and Chiolerio 2014), function as electronic skins (Hammock et al. 2013), integrate seamlessly with humans (Kim et al. 2012), harvest energy (Hwang et al. 2015), and contribute to soft robotic technologies (Lu and Kim 2013). Such devices require materials with tunable flexibility, stretchability, and electrical properties (Rogers et al. 2010).

Paper has received refreshed attention as a material for advanced electronics because of its wide availability, sustainability, and small environmental footprint (Martins et al. 2011a; Tobjörk and Österbacka 2011). The field of paper-based electronics, papertronics, refers to electro-chemo-opto-mechanical devices with functionality patterned on or within paper-based substrates. Paper-based substrates are multi-scale, porous mats of fibers that manipulate the flow of fluids, heat, light, charges, magnetic fields, and forces/stresses.

TABLE 15.1 Properties of Paper and Corresponding Advantages or Opportunities for Electronics

Property	Advantages or Opportunities
Flexible, foldable, and stackable	Paper can adapt to 3D structures by molding; structural engineering techniques including origami and kirigami can endow stretchability.
Porous	The high surface-to-volume ratio of paper allows it to absorb and immobilize functional materials such as conductive additives or enzymes within cellulose-based fibrous networks; the porosity of paper can also facilitate a low-impedance interface for electroanalytical studies.
Light-weight	Ease of storage and transportation
Biodegradable	Potential to reduce produced electronic waste (E-waste)
Biocompatible	Cellulosic fibers may be suitable for implantable devices.
Low-cost and abundant	Paper is inexpensive and readily available; e-readers and digital media are shifting demand for paper-based products.
Renewable	Paper often comes from plants and trees, and although manufacturing often requires large quantities of water, these base materials are ecologically sustainable.
Manufacturable	Printing and writing techniques permit patterning of paper, but there are still issues in achieving resolution comparable to conventional electronics.

Source: Alam and Mandal, 2016; Lan et al., 2013; Lin, Y. et al. *ACS Appl. Mater. Interfaces*, 8, 20501–20515, 2016; Maxwell, E.J. et al., *MRS Bull.*, 38, 309–314, 2013; Tobjörk, D. and Österbacka, R., *Adv. Mater.*, 23, 1935–1961, 2011.

Paper already plays a significant role in writing, printing, packaging, currency, origami, personal hygiene, medicine, and construction. These applications are possible through a diverse set of manufacturing processes and chemical additives. Modification of the fibrous matrix of paper can tune its hydrophilicity or hydrophobicity, porosity, opacity or transparency, and surface roughness (Dodson and Sampson 1996; Alava and Niskanen 2006; Roberts 2007; Ras et al. 2017). Paper can be as soft as biological tissue or as strong as Kevlar (Habibi et al. 2010). Paper is also flexible, porous, lightweight, and biodegradable (see Table 15.1). Surface treatments and additives can modify the functional properties of paper to make it magnetic, photocatalytic, retardant to flames, antibacterial, deodorizing, electrically conductive, or thermally insulative (Shen et al. 2011).

Paper is a planar assembly of fibers that lie roughly horizontal or a mat formed by a non-woven network of randomly distributed fibers (Alava and Niskanen 2006; Dichiara et al. 2017). Printing paper usually contains fillers, such as clay, talc, rosin, or alum to improve brightness, printability, and resistance to water (Roberts 1996). Other non-cellulose-based 'paper-like films' include cellophane, sheets of carbon nanotubes, and sheets of graphene (Kim et al. 2006; Marinho et al. 2012; Yuan et al. 2014).

15.2 Cellulose in Plants

Cellulose mainly comes from wood, plants, tunicates, algae, and bacteria (Moon et al. 2011). Figure 15.1 shows the chemical structure of cellulose, which is a polysaccharide comprised of D-glucopyranose rings in a $^4C_1$-chair configuration. Two units of D-glucopyranose form cellobiose, a disaccharide with

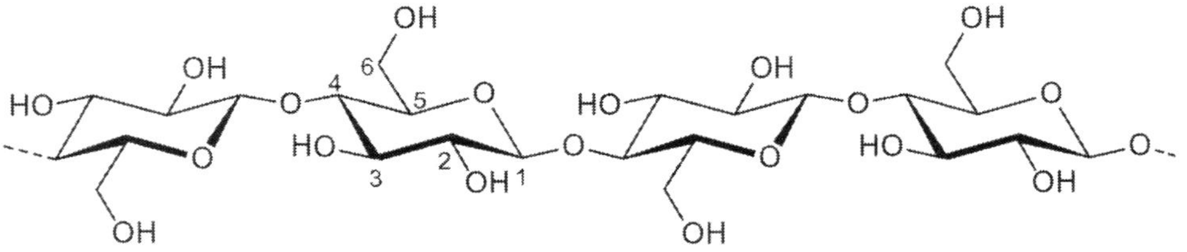

FIGURE 15.1 The chemical structure of cellulose. (Image taken from Heinze, T., Cellulose: Structure and Properties, In *Cellulose Chemistry and Properties: Fibers, Nanocelluloses and Advanced Materials*, O.J. Rojas, ed., Springer, Cham, Switzerland, pp. 1–52, 2015.)

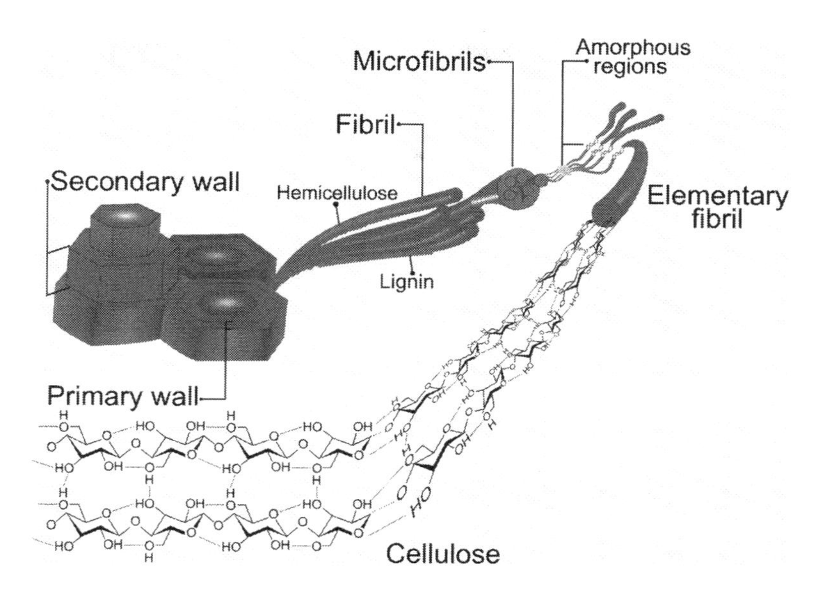

FIGURE 15.2 Hierarchical structure of a typical plant-based, cellulose fiber. (Image taken from Rojas et al. *Cellulose-Fundamental Aspects and Current Trends*, Chapter 8, 193–228, 2015.)

a length of 1.3 nm. These repeating units of cellobiose link together through β-1,4-glycosidic bonds, resulting in a chain of cellulose (French and Johnson 2007).

Figure 15.2 created by Rojas et al. shows that the chains of cellulose found in the secundary (secondary) walls of plant cells are the base of a hierarchical structure of elementary fibrils, microfibrils, and microfibrillar bands (fibrils). About 30 to 100 chains of cellulose bunch together to form elementary fibrils held together by intermolecular hydrogen bonding. These chains aggregate in both highly ordered, crystalline regions and disordered, amorphous regions (Rojas et al. 2015a). The elementary fibrils are ~100 nm in length with a characteristic lateral dimension of 1.5–3.5 nm, which aggregate to form fibrillar bundles (i.e., microfibrils with widths in the range of 10–30 nm). These microfibrils also have crystalline and amorphous regions (Klemm et al. 2005). The amorphous regions are present along the fibers and on the surfaces of the fibers (Kulachenko et al. 2012). Microfibrillar bands further bundle as fibrils of ~100 nm in width with length ranging from 100 nm to a few microns (Heinze 2015; Rojas et al. 2015a). Fibrils, along with hemicellulose and lignin, form an intertwined matrix within the secondary wall of plant cells (Heinze 2015). Hemicellulose strengthens the cell wall by forming hydrogen bonds with cellulose and lignin (Scheller and Ulvskov 2010). Lignin provides mechanical strength and rigidity to wood, facilitates the upward growth of the plants, and prevents the absorption of water due to its hydrophobicity. Lignin also protects plants from attacks by insects and fungi (Boerjan et al. 2003). Nonetheless, a sheet of paper containing lignin may become brittle as lignin degrades and discolors over time. Removing lignin during pulping improves the quality and mechanical properties of fabricated paper.

15.3 Technologies for the Fabrication of Papertronic Devices

Paper is an electrical insulator with a typical dielectric constant of 2.3 and electrical resistivity of ~10^9 Ω·cm (Josefowicz and Deslandes 1982). Nevertheless, add-on electronic functionality is attainable, as the fibrous networks and hydrophilic nature of cellulose-based paper enhance adhesion of functional fillers to its surface or in its porous interior. Electronic traces and components on the surface of a sheet of paper are papertronics *on* paper. Papertonics *in* paper tune the electromechanical properties within the volume of the paper substrate. In either class of papertronics, factors that influence electrical properties include conductive fillers, processing, surface roughness, bending stiffness, and absorbency of inks (Liang et al. 2016; Liu et al. 2017a; Martins et al. 2011a; Tobjörk and Österbacka 2011).

15.3.1 Electronics on Paper

In this category of papertronics, paper acts as a passive substrate to support functional electronic features. Fabrication involves coating, printing, and writing. For these techniques, formulating an appropriate ink is of prime importance. Inks may contain dispersed nanoparticles (NPs), dissolved organometallic compounds, or conductive polymers. Customized inks balance processability and desired physical properties including conductivity, optical transparency, mechanical strength, and adhesion (Kamyshny and Magdassi 2014). How binders and solvents in inks interact with paper is another consideration, as these interactions may lead to swelling, buckling, and shrinking of paper (Bollström et al. 2014).

Techniques for coating thin conductive layers on the outer surfaces of paper include rolling (Hu et al. 2009), spin coating (Zhong et al. 2013), sputtering (Martins et al. 2013), and evaporative deposition (Siegel et al. 2010). Multiple layers of conductive coatings may be necessary for surface homogeneity and uniform electrical conductivity. A typical conductive ink for coating consists of a binder and conductive fillers. Thus, the conductivity of coated paper follows the percolation theory in which there is a non-linear relationship between the concentration of a filler in a matrix and the bulk properties of the filler. The threshold of percolation depends on the applied binders and conductive fillers (Li et al. 2015a). The binders are typically electrical insulators with low conductivity. Conductive fillers include metals, metal oxides, silicon-based nanomaterials, carbon-based nanomaterials, and conductive polymers (Lin et al. 2016). Calendering or annealing can augment the electrical conductivity of the coating. Calendered samples are more compact, less porous, and less vulnerable to crease than uncalendered ones (Zhang et al. 2017b; Kumar et al. 2017; Nassar and Hussain 2017). Coating techniques *are* suitable where the application requires complete coverage of functional materials on a surface of paper with a controlled thickness (e.g., energy storage devices—supercapacitors, batteries, and solar cells) (Nogi et al. 2015; Yuan et al. 2012; Hu et al. 2010).

Coating techniques coupled with etching or patterned material removal can be effective techniques for patterning two-dimensional (2D) features or traces on the surface of a paper-based substrate (Maxwell et al. 2013). This technique is of particular use in capacitive sensing, when the patterned conductors benefit from having small gaps between them (Mazzeo et al. 2012; Xie et al. 2017; Zou et al. 2017, 2018). In lab settings with low volumes of produced devices, serial laser-based ablation is effective in patterning a device in a few minutes. Nonetheless, large-scale manufacturing may require the use of excimer lasers or chemical etching with high-resolution masks. In addition to electronic application, laser treatment can selectively modify hydrophobic paper (like parchment paper, wax paper, and palette paper) to create hydrophilic patterns with micro/nano-hybrid structures (Chitnis et al. 2011). This method provides a fast and cost-effective way to produce paper-based microfluidics.

To pattern electrical traces and interconnects on paper, techniques include inkjet, stencil, screen, and transfer printing (Figure 15.3). Inkjet printing has a micro-scale resolution with picoliter-sized droplets of ink ejected onto a substrate through one or more nozzles (Lessing et al. 2014; Derby 2010; Yin et al. 2010). Examples of printed functional devices include thin-film transistors, electronic circuits, biosensors, and energy storage devices (Costa et al. 2015; Mitra et al. 2017; Sundriyal and Bhattacharya 2017; Wang et al. 2016b; Zheng et al. 2013). Inkjet printing requires inks with low volatility and low viscosity to avoid clogging nozzles (Liu et al. 2017a). Therefore, it is necessary to keep the concentration of nanoparticles in ink at a sufficiently low concentration to ease the flow through a nozzle or make the ink thixotropic. A thixotropic or shear-thinning ink is one that has a low viscosity as it experiences high shear rates within a nozzle, and then has a high viscosity after exiting (Siqueira et al. 2017). To keep the ink from penetrating the surface of the substrate, pre-treating the paper is often necessary

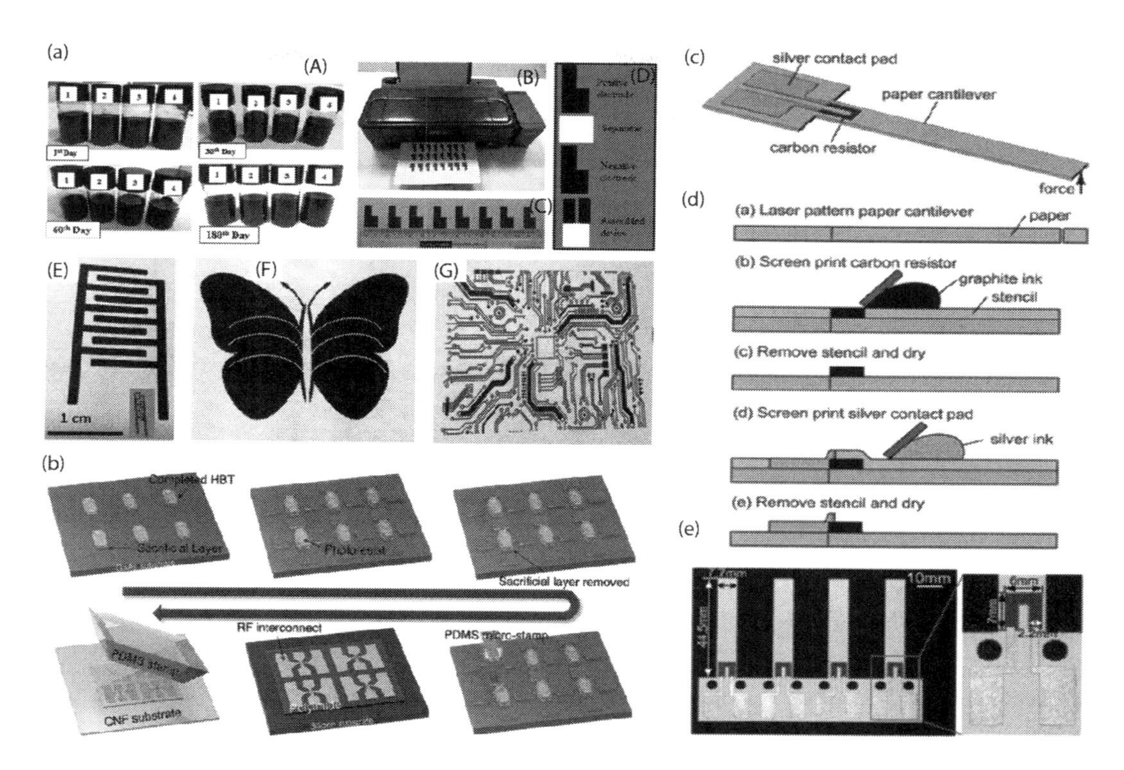

FIGURE 15.3 Techniques to fabricate electronics on paper. (a) Inkjet printing. (Images taken from Sundriyal, P., and Bhattacharya, S., *ACS Appl. Mater. Interfaces*, 9, 38507–38521, 2017.) (b) Transfer printing. (Images taken from Jung, Y.H. et al., *Nat. Commun.*, 6, 7170, 2015.) (c–e) Stencil-based printing. (Images taken from Liu, X. et al., *Lab. Chip*, 11, 2189–2196, 2011.)

(Wang et al. 2016b). Low-temperature thermal, plasma, and photonic sintering processes can reduce the electrical resistance by merging conductive nanoparticles and reducing the porosity of the printed traces (Sanchez-Romaguera et al. 2015).

Screen printing involves patterning conductive inks through a mask or mesh onto a substrate (Numakura 2008). This technique has shown applications in antennas for wireless communication on a nanopaper composite, thin-film transistors, and piezoresistive micro-electro-mechanical system (MEMS) sensors (Inui et al. 2015; Liu et al. 2011; Lu et al. 2011). Screen printing typically requires ink of a high viscosity (>1000 cP) to avoid excessive spreading and leakage through the mesh, and some techniques have even included cellulose-based binders to facilitate adhesion to paper (Hines et al. 2007).

Transfer printing is a technique for picking up traces or devices fabricated on one substrate and placing them on another (Hines et al. 2007). Using a soft transfer stamp, the technique leverages rate-dependent adhesion to pick up or deposit targeted components (Ahmed et al. 2015). Often, the transferred electronic components come from semiconductor-based rigid substrates, and the receptor is a soft, flexible substrate (Jung et al. 2015; Kim et al. 2009). There is also a non-contact laser-induced forward transfer (LIFT) technique to fabricate 3D microstructures (Piqué et al. 1999; Arnold et al. 2007). Wang et al. used this laser direct-write technique to transfer volumetric pixels of silver nanopastes from donor substrates to the receiving substrates in order to obtain ultra-fine pitch bonds (<10 μm) and interconnects for electronics (Wang et al. 2010).

Roll-to-roll (R2R) manufacturing is the ultimate strategy to mass-produce flexible electronics over large areas (Cheng and Wagner 2009). R2R manufacturing includes gravure printing (with engraved images) and flexography (with raised images). Serial rollers can work together to print images and

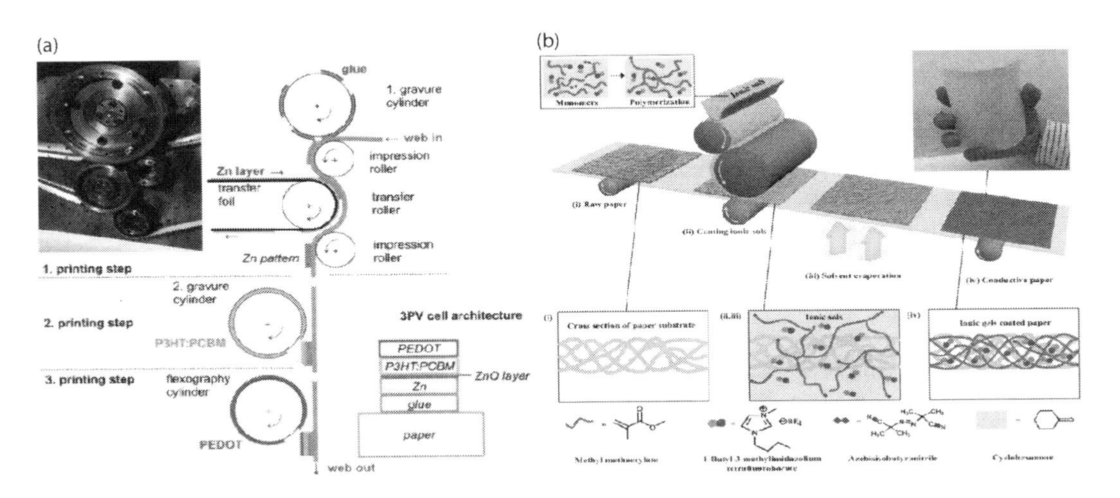

FIGURE 15.4 Large-scale transfer printing techniques: Gravure printing and flexography. (a) R2R process. (Image taken from Hübler, A. et al., *Adv. Energy Mater.*, 1, 1018–1022, 2011.) and (b) R2P process. (Image taken from He, M. et al., *ACS Appl. Mater. Interfaces*, 9, 16466–16473, 2017.)

structures layer-by-layer (Hübler et al. 2011). Alternatively, a roll-to-plate (R2P) technique can transfer ink continuously on individual flat sheets of paper (Figure 15.4) (He et al. 2017).

Handwriting is another strategy for fabricating *electronics on paper* (Figure 15.5). One technique employs pens compatible with highly conductive gallium-based liquid metal, copper ink, silver ink, carbon nanotube ink, enzymatic ink, and zinc oxide nanoparticle-based ink (Liu et al. 2017a; Grey et al. 2017). Another technique uses pencils to produce carbon-based *electronics on paper* in a solvent-free manner (Kurra and Kulkarni 2013). Pencils represent a simple and low-cost method suitable for rapid prototyping and educational purposes. However, there are many opportunities to explore the reproducibility and scalability of this wear-based (tribological) approach for the potential mass production of papertronics. Analogous to drawing with a pencil on paper, Mirica et al. mechanically abraded a pellet of compressed single-wall carbon nanotubes (SWCNTs) on the surfaces of paper to form gas sensors. They found that the sensors drawn on paper with smooth surfaces (e.g., weighing paper) had good sensitivity, high signal-to-noise ratio towards NH_3, and comparable reproducibility to SWCNT-based devices fabricated by solution-phase methods (Mirica et al. 2012).

While *electronics on paper* are straightforward and promising with their high spatial resolution, there are drawbacks to printing solely on the surface of a substrate. Poor adhesion between paper and conductive inks can affect the long-term stability of produced devices (Chen et al. 2016). For a folding substrate, the tensile or compressive stresses are highest at its surface away from its neutral bending axis. These bending stresses can lead to cracks in patterned circuit architectures or detachment of electronic components from flexible substrates (Thi Nge et al. 2013; Kim et al. 2018).

15.3.2 Electronics in Paper

In contrast to *electronics on paper*, *electronics in paper* incorporate conductive fillers within a cellulosic network to form intrinsically conductive composites that are less susceptible to cracking and abrasion than inks patterned on the surface of a substrate. Currently, there are three main methods for fabricating cellulose-based conductive paper (Figure 15.6a–c): (I) flow-directed filtration (Compton et al. 2010), (II) dip coating/soaking/deposition, and (III) additive manufacturing (or 3D printing).

Flow-directed filtration begins with the functionalization of fibers by coating treatments or adding fillers during pulping. After pulping, the fibers in solvent pass through a filtering screen or mesh

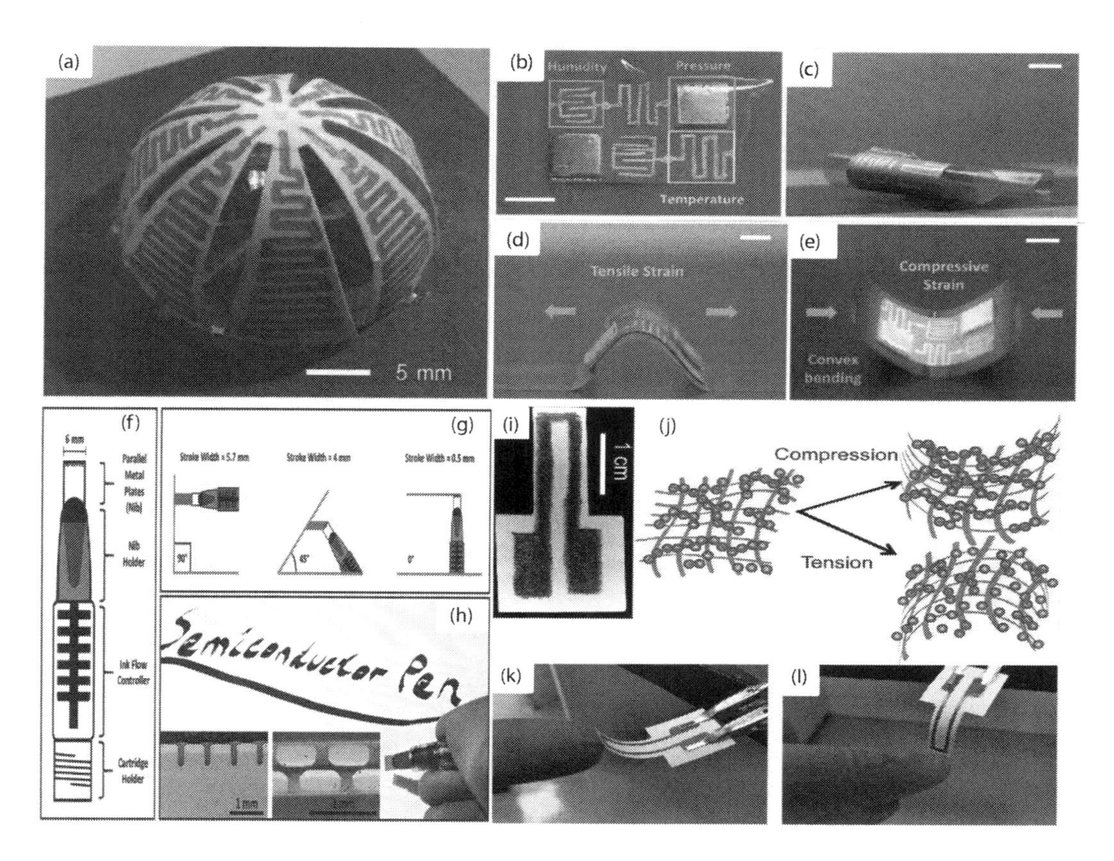

FIGURE 15.5 Writing techniques for papertronics. (a) A 3D antenna drawn by a silver-ink filled rollerball pen on a sticky paper. (Image taken from Russo, A. et al., *Adv. Mater.*, 23, 3426–3430, 2011.) (b–e) A flexible sensing circuit drawn by a silver-ink pen. (Images taken from Nassar, J.M. and Hussain, M.M., *IEEE Trans. Electron Devices*, 64, 2022–2029, 2017.) (f–h) A parallel metal-plate pen depositing ZnO nanoparticles on paper. (Images taken from Grey, P. et al., *Adv. Mater. Technol.*, 2, 1700009(1–7), 2017.) (i–l) A pencil-drawn strain gauge. (Images taken from Lin, C.-W. et al., *Sci. Rep.*, 4, 3812, 2014.)

to form a fibrous mat/sheet (e.g., handsheet forming or vacuum filtration). This wet mat then undergoes pressing and drying stages that tune its final morphology and resulting physical properties (Anderson et al. 2010; Sehaqui et al. 2011; Jabbour et al. 2012). The matted fibers function as a porous and hydrophilic framework to immobilize different types of fillers: (I) carbon-based fillers—carbon black, graphite, carbon fiber, carbon nanotube, reduced graphene oxide, and graphene (Dichiara et al. 2017; Anderson et al. 2010; Hamedi et al. 2014; Jabbour et al. 2012; Koehly et al. 2014; Koga et al. 2016; Luong et al. 2011; (II) conductive polymers [e.g., poly(3,4-ethylene dioxythiophene) polystyrene sulfonate (PEDOT: PSS) (Wang et al. 2016c)]; and (III) metal nanoparticles (e.g., silver nanowires) (Song et al. 2015). Flow-directed filtration ranges from low-volume fabrication with handsheet forming to medium-volume fabrication with dynamic sheet forming (Zhang et al. 2011) to large-scale manufacturing lines (Huang et al. 2016).

Simple dip coating, soaking, and deposition embed conductive inks/particles into the fibrous structure of paper (Liu et al. 2017b; Tao et al. 2017; Weng et al. 2011; Zhang et al. 2018a). Usually, it requires repetitive treatments by the selected process to make a uniformly conductive composite paper. These methods are suitable for substrates with varied porosities, such as tissue paper, filter paper, and copy paper (Gullapalli et al. 2010). With such paper, rapid prototyping is possible. However, dip coating, soaking, and deposition may create weak bonds between fillers and fibers and may also swell the paper. To overcome this, one

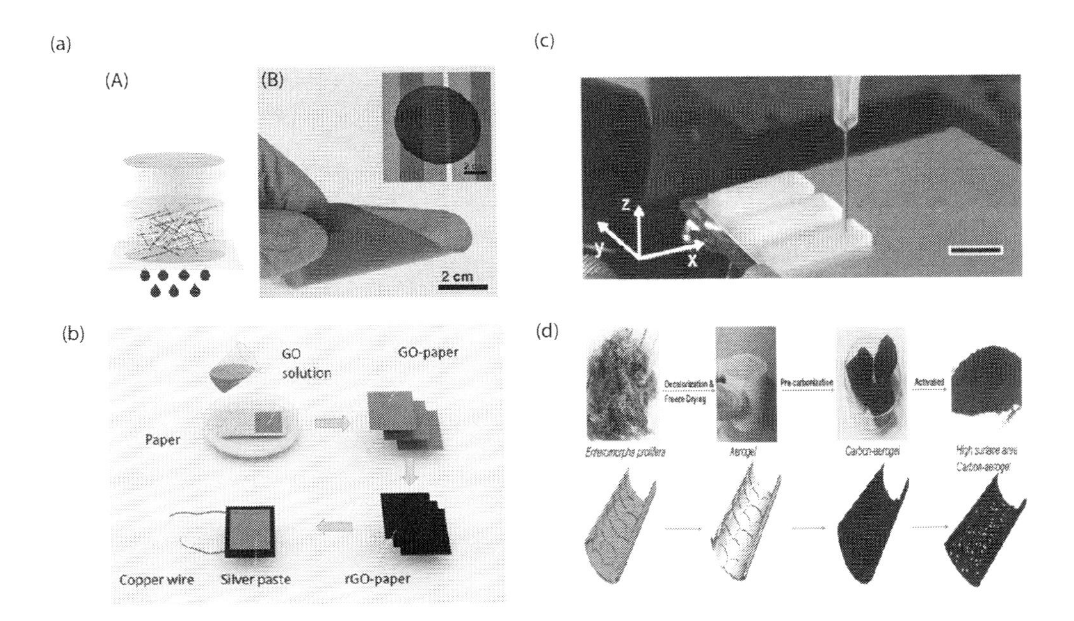

FIGURE 15.6 Three main techniques to fabricate cellulose-based conductive paper, including: (a) Flow-directed filtration. (Images taken from Hamedi, M.M. et al., *ACS Nano*, 8, 2467–2476, 2014.) (b) Soaking. (Images taken from Tao, L.-Q. et al., *ACS Nano*, 11, 8790–8795, 2017.) and (c) 3D printing. (Images taken from Siqueira, G. et al., *Adv. Funct. Mater.*, 27, 1604619, 2017.) (d) Carbonization that turns cellulose fibers into conductive carbon networks. (Images taken from Cui, J. et al., *Adv. Funct. Mater.*, 26, 8487–8495, 2016.)

unique form of deposition uses wax-based printing to form hydrophobic barriers, thus containing the lateral spread of aqueous conductive inks within paper (Hamedi et al. 2016).

Additive manufacturing (AM) is a technique for intricate patterning of three-dimensional structures. One AM-based process forms cellulosic structures through layer-by-layer deposition of a printable conductive composite paste (consisting of cellulose, conductive fillers, and a solvent). This technique achieved complex 3D geometric patterns for capacitors, batteries, and sensors (Li et al. 2017b). Previous studies have also demonstrated the printability of cellulose acetate (CA) (Pattinson and Hart 2017), cellulose nanocrystals (CNC) (Siqueira et al. 2017), composites of silver nanowires (Ag-NWs) and sodium carboxymethylcellulose (AgNW-CMC) (Park et al. 2017), and composites of carbon nanotube and nanofibrillated cellulose (CNT-NFC) (Li et al. 2017b).

Unlike some printed materials, hot-melt extrusion in typical AM-based processes is problematic for cellulosic pastes. Due to the intramolecular and intermolecular bonds, the hypothetical melting point of cellulose (467°C) is higher than its degradation temperature (around 315°C) (Ganster and Fink 2013; Krumm et al. 2016; Yang et al. 2007). Therefore, cellulose is not compatible with hot-melt extrusion as it degrades upon heating before melting/flowing sufficiently. To print cellulose effectively, current methods include: (I) using a low concentration of cellulose; (II) replacing the hydroxyl groups with other functional groups (e.g., acetate groups) to make the material soluble in volatile solvents, extruding the material, and then reverting the material to cellulose by restoring the hydroxyl groups (e.g., deacetylation) (Yamashita and Endo 2004; Pattinson and Hart 2017); and (III) extruding the spinning solution directly into a coagulating bath and then removing the extruded fibers (Li et al. 2017b).

The direction of extrusion in additive manufacturing can affect the mechanical properties of printed components. Components containing filaments of oriented particles will result in anisotropic composites. For instance, the reinforcing effects of CNC (an aspect ratio of about 18) in polymer nanocomposites are more significant in the longitudinal direction (LD) than the transverse direction (TD)

(Siqueira et al. 2017). Nonetheless, it is possible to minimize the effect of the printing direction with cellulose-based polymers. For printed CA, Young's moduli (2.2 ± 0.1 GPa in the LD and 2.2 ± 0.2 GPa in the TD) and yield strengths (45 ± 1.9 MPa in the LD and 44.7 ± 2.2 MPa in the TD) were similar in orthogonal directions (Pattinson and Hart 2017).

Apart from cellulose-based papertronics, another type of *electronics in paper* utilizes carbonized paper, or a hierarchical porous carbonaceous aerogel (HPCA), which is synthesized from commercial paper by freeze drying and high-temperature pyrolysis (Figure 15.6d). Carbonized paper shows promise in energy storage and strain sensors (Cui et al. 2016; Li et al. 2016; Liang et al. 2012; Ye et al. 2015), as it possesses high porosity (up to 99.56%) (Liang et al. 2012), high specific surface area (63.4 ~2200 $m^2 g^{-1}$) (Cui et al. 2016; Ye et al. 2015), and moderate electrical conductivity (13.6 S/m) (Li et al. 2016). When filled with elastomers, carbonized paper can have foldable and stretchable characteristics which allow it to function as a heater or a strain sensor (Li et al. 2016; Liang et al. 2012; Ye et al. 2015).

Overall, *electronics in paper* with inks/particles distributed among insulating cellulosic fibers enable tunable conductive, capacitive, piezoresistive, and piezoelectric properties. There are still opportunities to further understand the process-structure-property relationships of such volumetrically patterned materials. Furthermore, these composite sheets have customizable mechanical properties and keep patterned electrical regions away from exposed surfaces susceptible to scratching and high bending stresses.

15.3.3 Typical Paper-Based Substrates for Flexible Electronics

Paper-based electronics link distinct fiber morphologies, additives, and surface treatments to target applications. For *electronics on paper*, copy paper is a common choice for inkjet-based printed electronics due to its printability in conventional printers, availability, and low cost (Santhiago et al. 2017; Ahn et al. 2016). However, the absorption of ink in copy paper can lead to prints with low conductivity because the resulting tortuous conductive paths may lack complete inter-fiber connectivity along the surface of the substrate. One solution to this problem is to use paper with non-absorptive or hydrophobic surfaces such as glassine paper, parchment paper, or photo paper (Kim and Steckl 2010; Hyun et al. 2015; Zocco et al. 2014; Chitnis and Ziaie 2012). Another approach is to cut and/or etch metallized paper into conductive architectures and devices, as mentioned in Section 15.3.1 (Mazzeo et al. 2012; Xie et al. 2017; Zou et al. 2018).

In contrast, *electronics in paper* are functional conductive composites with inks/particles embedded in porous paper, such as filter paper and chromatography paper (Hamedi et al. 2016; Weng et al. 2011). However, the natural cellulose fibers in such paper have low strength due to their disrupted crystalline (amorphous) regions. In contrast, regenerated cellulose, bacterial cellulose, and nanocellulose fibers consist of only crystalline regions and will result in paper with superior mechanical attributes suitable for applications demanding high strength and endurance (Zhang et al. 2016a; Yun et al. 2009; Zhang et al. 2016b; Jung et al. 2015). The nanofibers with high crystallinity can form densely packed paper that scatters less light than regular paper with fiber bundles and/or pores, which makes such nanopaper transparent and useful for optoelectronic applications (Zhu et al. 2013; Xu et al. 2016; Nogi et al. 2015).

Different materials, grammages, and thicknesses of the paper substrate can achieve desirable mechanical properties for other specific electronic applications. Some paper-like materials made of cellulose derivatives (e.g., nitrocellulose fibers) and synthetic fibers of polyolefins (e.g., Teslin paper) are key components for microfluidics (e.g., lateral flow assays) and printed flexible electronics (Bhattacharyya and Klapperich 2006; Posthuma-Trumpie and Amerongen 2012; Bahadır and Sezgintürk 2016; Jenkins et al. 2015; Cook et al. 2013). These paper-like substrates have superior mechanical properties to cellulosic paper, fibrous structure, and porosity (Gao et al. 2016; Wang et al. 2016b; Fridley et al. 2013). Paper made of carbon-based materials (such as carbon fibers, carbonized cellulose, carbon nanotubes, and graphene) is suitable for energy storage applications due to its resulting conductive and porous morphology (Wang et al. 2008; Zhu et al. 2017; Zhang et al. 2016c; Li et al. 2017a). In addition to the type and material of paper, grammage (mass per unit area) and thickness also affect its overall tensile strength and elastic modulus. Typical tissue paper, copy paper, and filter paper have a grammage of 12–30 g m^{-2}, 80 g m^{-2}, and 84 g m^{-2}, respectively (Tobjörk and Österbacka 2011; Prambauer et al. 2015). For paper of similar grammage, filter paper is thicker, less compact, and mechanically weaker than copy paper (Prambauer et al. 2015).

15.4 Mechanical Flexibility, Strength, and Endurance of Paper-Based Substrates

Flexible and foldable paper-based substrates have gained significant attention in flexible electronics (Zhang et al. 2018b). This section focuses on the flexibility, mechanical strength, and bending stability of paper-based substrates. Flexibility is the inverse of stiffness, flexural rigidity, or bending stiffness. Carson and Worthington defined the stiffness of paper as the bending moment per unit width of the specimen producing per unit curvature, as shown in Equation (15.1):

$$S=\frac{M}{Kb}=\frac{MR}{b}=\frac{EI}{b}=\frac{Ed^3}{12}=\frac{ML^2}{3bF}=\frac{ML}{3b}f(\theta), \tag{15.1}$$

where M is the bending moment at the torque axis, b is the width of the specimen, K is the curvature, R is the radius of curvature at the torque axis ($K = 1/R$), E is the elastic modulus (bending modulus), I is the moment of inertia, d is the thickness of the specimen, L is the span or bending length (distance between the axes of the two clamps), F is the deflection of the free end (or loaded end), and θ is the bending angle (Carson and Worthington 1952).

The intrinsic chemistry, morphology, strength, and structure of individual fibers affect the mechanical properties of paper (Caulfield and Gunderson 1988; Page 1969). Properties of cellulosic fibers depend on their source (e.g., bacterial cellulose, wood-based plants, non-woody plants, regenerated cellulose, or refined cellulose). The Young's modulus of cotton fibers ranges from 5.5–13 GPa while that of ramie fibers varies from 44–128 GPa (Sathishkumar et al. 2013). Hardwood fibers are short, thin, and soft, while softwood fibers are long, thick, and strong (Rydholm 1965). Other contributing factors to the strength of fibers are fibrillar structure, microfibrillar angle, porosity, the percentage of associated polymers like hemicellulose and lignin, and mechanical deformation/defects created during the pulping process (Wathén 2006).

15.4.1 Methods for Improving the Strength of Cellulose Fibers and Paper

Mechanical refining and chemical processing alter the morphology of cellulosic fibers and thus the mechanical properties of paper. Mechanical refining beats, compresses, and shears the fibers (Gharehkhani et al. 2015). Compressive and shearing actions between the grooved surfaces of the refiners alter the structure of the trapped fibers through internal fibrillation, external fibrillation, shortening and straightening of fibers, and formation of fines. Fibrillation peels nanofibril bundles away from the fiber surfaces without detachment (Kang and Paulapuro 2006; Afra et al. 2013). The fibrillation can occur on both the internal and external surfaces of cellulose fibers. Internal fibrillation loosens the wall structure of fibers, making them flexible and collapsible, and increases inter-fiber bonding, bringing fibers together in close contact. Internal fibrillation also enhances the strength and bendability of fibers, but reduces tear resistance (Rusu et al. 2011). External fibrillation (i.e., fraying of nanofibrils on outer walls of fiber) improves mechanical interlocking between fibers, bonding, and tear resistance (Hirn and Schennach 2015). Fiber shortening is the breaking of fibers due to an applied strain. Fiber shortening reduces the tear resistance and leaves a smooth finish (Kerekes and Schell 1995). After straightening, fibers have better tensile strength and stress-distribution efficiency than curled fibers (of chemical pulp) in the fibrous network (El-Sharkawy et al. 2008). Fines in the pulp increase bonding, tear strength, folding endurance, wet strength, and dry strength of the constituted paper, but prolong the time to drain water in the process of papermaking, which limits production efficiency (Wistara and Young 1999; Taipale et al. 2010).

Chemical processing involves acid treatment or 2,2,6,6-tetramethylpiperidine-1-oxyl (TEMPO)-mediated oxidation of cellulosic fibers. Acid treatment hydrolyzes the structurally weak amorphous regions within the cellulosic microfibrils and thus cleaves them to form short nanocrystals that have better mechanical properties (Sofla et al. 2016). TEMPO-mediated oxidation of native cellulose converts hydroxyl groups present on fibril surfaces to carboxylate and aldehyde groups while maintaining its fibrous morphologies and crystallinities. The carboxylate and aldehyde groups disrupt tight bonds between inter-fiber hydroxyl groups. Usually, a mechanical treatment follows the oxidation to individualize cellulose fibrils (Saito et al. 2007).

These mechanical and chemical pulp-conditioning processes have led to the development of nanostructured cellulose—a new class of natural nanomaterials. Uniform and defect-free crystalline nanocelluloses possess unique properties: low density, biodegradability, high aspect ratio, high strength, high stiffness, optical transparency, low thermal expansion, gas impermeability, and adaptable surface chemistry (Trache et al. 2017; Abitbol et al. 2016). Based on morphology, function, and preparation protocols, nanocellulose has three categories: (I) cellulose nanocrystals (CNCs) (also known as nanocrystalline cellulose (NCC) or cellulose nanowhiskers (CNWs)), (II) cellulose nanofibrils (CNFs) (also known as nanofibrillated cellulose (NFC)), and (III) bacterial cellulose (BC). These three categories of nanocellulose use the preparation processes of acid hydrolysis/heat controlled techniques, mechanical/chemical/enzymatic treatment, and bacterial synthesis, respectively (Klemm et al. 2011; Abdul Khalil et al. 2014; Abitbol et al. 2016). The Young's moduli of the cellulose crystalline forms I and II (native and regenerated celluloses, respectively) are 167.5 GPa and 162.1 GPa theoretically and 138 GPa and 88 GPa experimentally (Tashiro and Kobayashi 1991; Nishino et al. 1995), greater than aluminum (70 GPa) (Gere and Goodno 2008).

Young's modulus or elastic modulus of the fibers, the three-dimensional inter-fiber hydrogen bonding, and the density of fibrous networks contributes to the elastic modulus of nanopaper (made of nanocellulose) (Kulachenko et al. 2012). Among these factors, the hydrogen bonds play a key role in the elastic modulus of paper. According to Alfred Nissan's hydrogen bond theory (Equation (15.2)), the elastic modulus of paper is proportional to the cube root of the number of effective hydrogen bonds (Nissan 1967),

$$E = k(n)^{1/3} \tag{15.2}$$

where E is the modulus of elasticity of paper, n is the number of effective hydrogen bonds per cubic volume, and k is the average spring constant for the hydrogen bond (Nissan 1967; Zauscher et al. 1996; Mortensen 2006). Because of hydrogen bonds, the mechanical properties of cellulosic nanopaper exhibit an anomalous scaling law between strength and toughness. Strength and toughness are typically mutually exclusive for a given material (e.g., metals and alloys) (Launey and Ritchie 2009). However, Zhu et al. demonstrated that strength and toughness of nanopaper increased together as the diameters of the cellulosic fibers decreased. Small fibers contributed to a significant increase in hydrogen bonding between cellulosic nanofibrils, which increased the shear stresses and energy required for inter-fiber sliding (Zhu et al. 2015). Similarly, well-aligned microfibers of CNT/NFC/CNT had significant hydrogen bonding and thus high tensile strength (247 ± 5 MPa) and toughness (Li et al. 2017b).

The wet strength—dictated primarily by fiber-fiber entanglement (Tejado and van de Ven 2010)—of paper is of particular interest for paper-based microfluidics (Cate et al. 2015). Soaking paper in water disrupts inter-fiber hydrogen bonding and results in a significant loss of dry strength (up to 90%) (Crisp and Riehle 2009). Introducing certain additives (e.g., polyaminopolyamide-epichlorohydrin (PAE) polymers) can increase the wet strength and integrity of paper (Crisp and Riehle 2009). In a different approach, TEMPO-based oxidation chemically introduces aldehyde groups to the fiber surface. The aldehyde and hydroxyl groups on the fiber surfaces form covalent bonds that also increase wet strength (Saito and Isogai 2005).

15.4.2 Bending and Folding Endurance of Paper

Paper is bendable, foldable, and capable of adapting to complicated geometries (Yang et al. 2015a; Lee et al. 2016a; Hamedi et al. 2016; Nogi et al. 2013). Figure 15.7 shows two examples of foldable paper-based electronic circuits; such flexible papertronics can operate under cyclic bending and folding conditions. The microscopic fragility and brittleness of paper mainly determine its endurance during bending and folding (Siegel et al. 2010; Lee et al. 2016a). Equation (15.3) gives the maximum curvature that a fiber can sustain under bending before fracture (Hull and Clyne 1996):

$$\kappa_{max} = \frac{2\sigma_f}{Ed}, \tag{15.3}$$

where E is Young's modulus, d is the diameter, and σ_f is the fracture strength of the fiber. For the same material, nanofibers can sustain a greater curvature before failure than microfibers. As a result, nanopaper composed of nanofibers and nanostructured networks has high foldability (Nogi et al. 2013; Sehaqui et al. 2011; Cox 1952). For instance, nanopaper-based circuits with two top layers of conductive nanoparticles (SWNT and silver nanowires) are stable under 500 folding cycles of ±180° (Kang et al. 2016).

15.4.3 Endurance of Electronics on Paper

Electronics on paper involve patterned electronic circuits and architectures on paper as described in Section 15.3. Such electronics experience failures of cracking and delamination under strain, due to mechanical mismatch of electronic materials and substrate (Park et al. 2008; Siegel et al. 2010; Harris et al. 2016). To prevent failures, printed electronics employ conductive nanomaterial-based inks (such as silver nanowire inks, graphene inks, carbon nanotube inks) to penetrate through the porous network of paper, forming a highly adherent fiber-ink network (Hu et al. 2009). Printed electronics using conductive inks

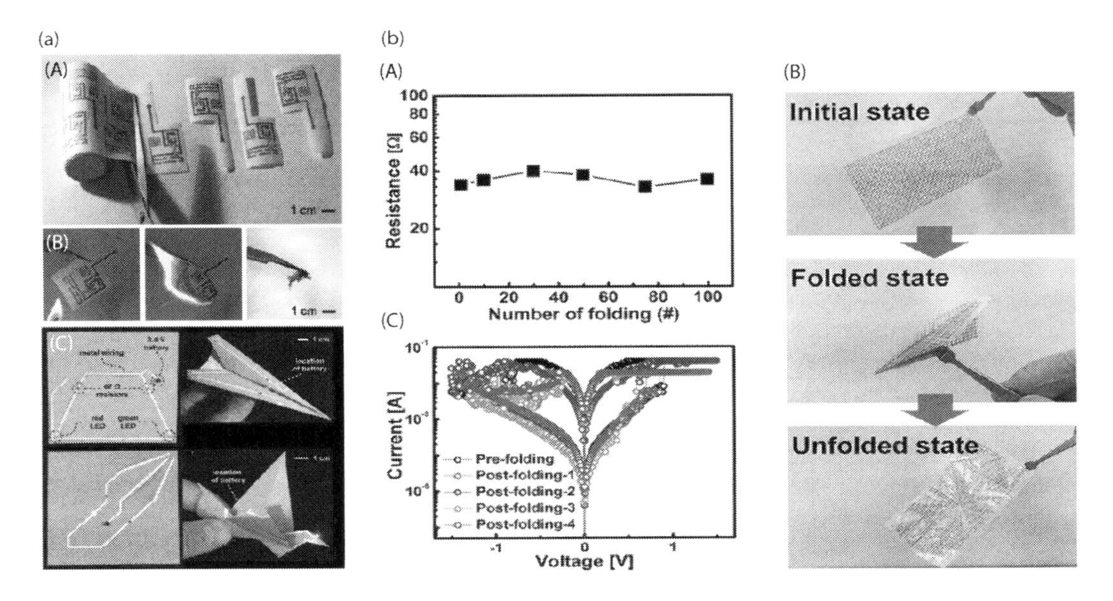

FIGURE 15.7 Bendable and foldable paper-based electronic circuits made by different techniques: (a) Paper-based electronic circuits made by evaporation, sputter deposition, and spray deposition. (Images taken from Siegel, A.C. et al., *Lab. Chip*, 9, 2775–2781, 2009.) and (b) paper-based memory made by inkjet printing and initiated chemical vapor deposition (iCVD). (Images taken from Lee, B.-H. et al., *Sci. Rep.*, 6, 38389, 2016a.)

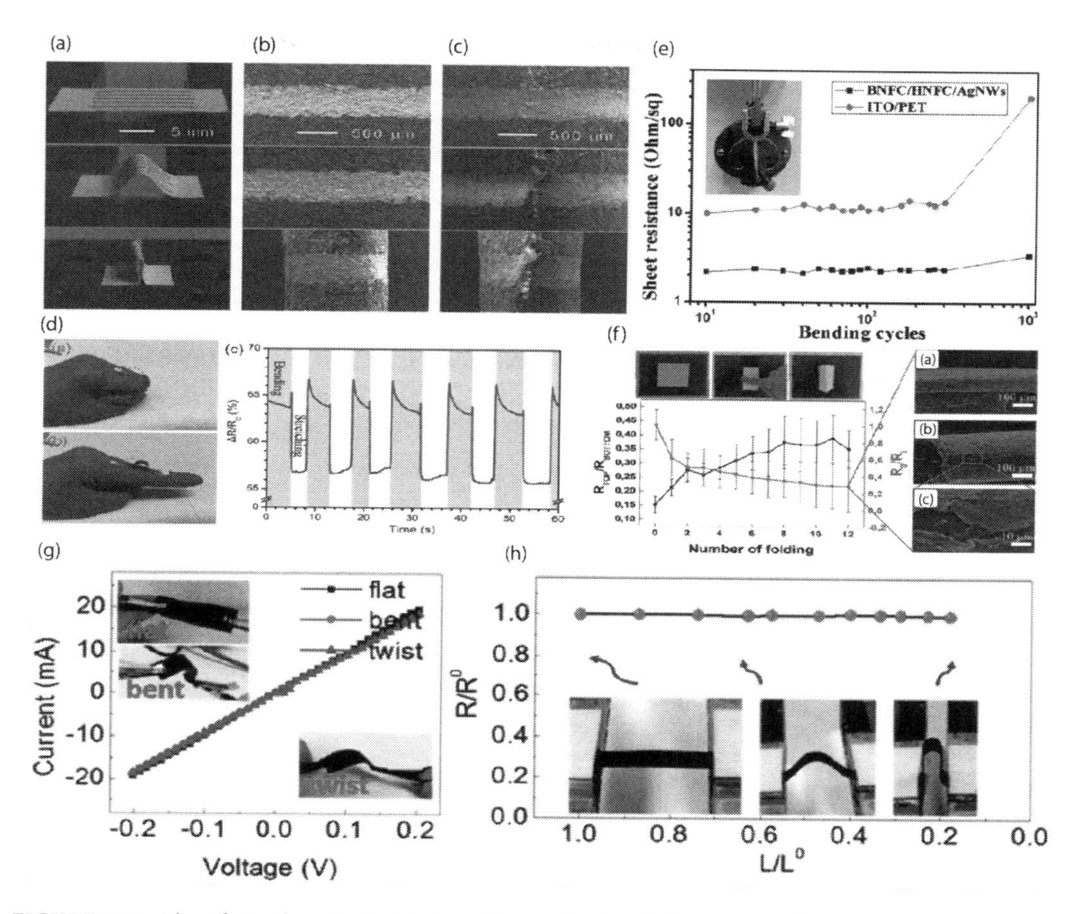

FIGURE 15.8 The effects of mechanical deformation on the electrical properties of papertronics. (i) Electronics on paper: Bending tests of the printed silver electrodes with different radii: (a, b) Images before bending cycles and (c) images after bending 10,000 bending cycles. (Images taken from Russo, A. et al., *Adv. Mater.*, 23, 3426–3430, 2011.) (d) Relative resistance changes of a pencil-drawn wearable sensor during finger bending tests. (Images taken from Liao, X. et al., *Adv. Funct. Mater.*, 25, 2395–2401, 2015.) (II) Electronics in paper: (e) Sheet resistance changes of bamboo NFC (BNFC)/hemp NFC (HNFC)/silver nanowires (AgNWs) hybrid nanopaper and the commercial ITO/PET film over 1000 bending cycles. (Images taken from Song, Y. et al., *Nanoscale*, 7, 13694–13701, 2015.) (f) Sheet resistance changes of graphene/bioplastic/cellulose over folding and unfolding. The SEM images shows crack after folding. (Images taken from Cataldi, P. et al., *Adv. Electron. Mater.*, 1, 1500224, 2015.) (g, h) Electrical properties of PEDOT nanopaper under bending and twisting. (Images taken from Wang, Z. et al., *J. Mater. Chem. A*, 4, 1714–1722, 2016c.)

were stable under 1000–18000 dynamic concave and convex bending cycles (Figure 15.8a–d) (He et al. 2017; Liu et al. 2017b; Russo et al. 2011; Guo et al. 2017b; Wei et al. 2015; Zhang et al. 2017a; Liao et al. 2015). Another approach to prevent delamination is to use coarse paper which provides a surface with greater adhesion than smooth paper. For instance, a strain sensor consisting of gold nanofilms on an abrasive paper (with many microcracks on the surface) was stable up to 18000 bending cycles (Liao et al. 2017).

15.4.4 Endurance of Electronics in Paper

As discussed in Section 15.3, conductive composite papers have embedded conductive fillers/inks Shen et al. 2011; Du et al. 2017; Li et al. 2015b; Jabbour et al. 2010; Heldt et al. 2013). Most conductive fillers/inks can disrupt the inter-fiber hydrogen bonding, weaken the paper, and limit its flexibility.

Available strategies to overcome this problem include: (I) introducing functional groups, (II) exfoliating conductive fillers with NFC, (III) reinforcing paper structure with conducting polymers, and (IV) using binder and cross-linking agents.

For the first strategy, adding functional groups to fillers or to cellulose establishes chemical bonds between fillers and cellulose to augment mechanical integrity (Dichiara et al. 2017; Luong et al. 2011). For the second strategy, NFC can exfoliate SWNTs into individual or small SWNT bundles or exfoliate graphite into single- or few-layer sheets to obtain a stable colloidal dispersion. The resulting composite nanopaper shows higher mechanical properties than pure nanopaper (Hamedi et al. 2014; Zhang et al. 2017b). For the third strategy, conductive polymers reinforce the paper structure by forming a polymer network in the cellulosic matrix. The composite paper made of PEDOT was stable under static bending and dynamic bending up to 250 cycles (Anothumakkool et al. 2015). Furthermore, mixing polypyrrole (PPy)-coated fibers and pristine fibers can preserve fiber-fiber bonding to produce composite paper with the same electrical conductivity but higher tensile strength than paper made exclusively of PPy-coated fibers (Huang et al. 2005). For the fourth strategy, binder and cross-linking agents improve the stability of conductive paper composites. Composite papers made from such strategies showed stability up to 1000 bending cycles and up to 12 folding cycles (Figure 15.8e–h) (Luong et al. 2011; Wan et al. 2017; Jabbour et al. 2012; Wang et al. 2016c; Yuan et al. 2013; Song et al. 2015; Cataldi et al. 2015).

15.5 Strategies for Making Paper-Based Electronics Stretchable

Health-monitoring flexible electronics must conform to curved surfaces such as human skin. Human skin—ignoring the folds at joints—is generally not extremely stretchable, with an elastic limit at 10% elongation and rupture point at 32%–35% elongation (Agache and Humbert 2004). However, the folds at knee joints can experience up to 55% strain upon stretch and contraction during walking (Yamada et al. 2011). To monitor human skin with papertronics, it is necessary to tailor the stretchability of papertronics according to the anticipated strain on different target regions of the human body/substrate.

Prevailing stretchable electronics come in two forms: stretchable structures and stretchable materials (Rogers et al. 2010). Conventional materials can accommodate applied strains of 100% or more by adopting stretchable structures (e.g., ‘wavy’ shapes, arc-shaped bridge structures, coplanar self-similar serpentine layouts) (Yan et al. 2014; Yang et al. 2015b; Dang et al. 2017). Additionally, substrates may gain stretchability by including mini-valleys, honeycomb lattice architectures, sponge-like structures, or nano-accordion structures in their surface topography (Lee et al. 2016b). Intrinsically stretchable materials equivalently play key roles in stretchable electronics. Stretchable materials can function either as a matrix to immobilize conductive dopants/fillers or as a substrate compatible with the direct printing of conductive traces. Unlike hard, brittle materials that tend to fail at small strains, intrinsically stretchable materials have a large critical strain of fracture, due to their elasticity with low Young's moduli (Suo 2012; Yang et al. 2016).

15.5.1 Stretchable Substrate for Papertronics

Paper and board products tend to suffer brittleness and exhibit small strain-to-failure (Zhang et al. 2016b). Paper is viscoelastic and shows delayed recovery after a strain-release cycle. The elastic modulus and tensile strength of paper increase with the strain rate. Paper becomes ductile and soft in humid conditions, as water disrupts the hydrogen-bonded network and decreases nanofibrillar friction (Alava and Niskanen 2006; Benítez et al. 2013). To gain stretchability in paper-based electronics, strategies include making paper intrinsically stretchable by changing its internal interfibril interactions and increasing its porosity.

First, weakening inter-fiber bonding can make paper ductile or increase its strain-to-failure (Dichiara et al. 2017; Sehaqui et al. 2012). As discussed in Section 15.4.1, the structure of paper changes upon contact with liquids, as liquids disrupt hydrogen bonds, relax fibers, and produce dimensional changes in pores and capillaries. Applying non-polar liquid during paper formation or supercritical drying weakens the interfibril interaction and results in softer and more ductile paper than that made from polar liquids (Sahin and Arslan 2008). For example, nanopaper made of TEMPO-oxidized-nano-fibrillated-cellulose (TO-NFC) prepared by supercritical CO_2 drying shows a moderate strain-to-failure of 17%, a porosity of 56%, Young's modulus of 1.4 GPa, and tensile strength of 84 MPa (Sehaqui et al. 2011).

Second, porous structures stretch more easily than non-porous structures (Fan et al. 2017). As for paper, the number of interfiber bonds per fiber decreases with increasing porosity, thus resulting in increased strain-to-failure (Sehaqui et al. 2012). For example, tissue paper is ductile and soft, due to its open and sparse structure (Alava and Niskanen 2006). Similarly, porous nanocomposites made of NFC coated with hydroxyethyl cellulose (HEC, a water-soluble cellulose derivative) after supercritical drying are soft with a modulus of 0.8–1.3 GPa. Such nanocomposites with 44%–63% porosity show a nominal strain-to-failure as high as 55% (Figure 15.9a), which is greater than those with low porosity (13%–27%) by nearly 100% (Figure 15.9b). This increased ductility results from high porosity and limited interfibril bonding. Low friction facilitating fibril-fibril slippage of HEC-coated NFC may also contribute to an increase in ductility. Such stretchable nanocomposites also show high tensile strengths of 80–93 MPa. However, there is a tradeoff between the maximum strain at break and the strength of the material. Pre-stretching the nanocomposite can increase modulus, strength, and the yield strength substantially while decreasing its maximum strain at break (Figure 15.9c and d) (Sehaqui et al. 2012).

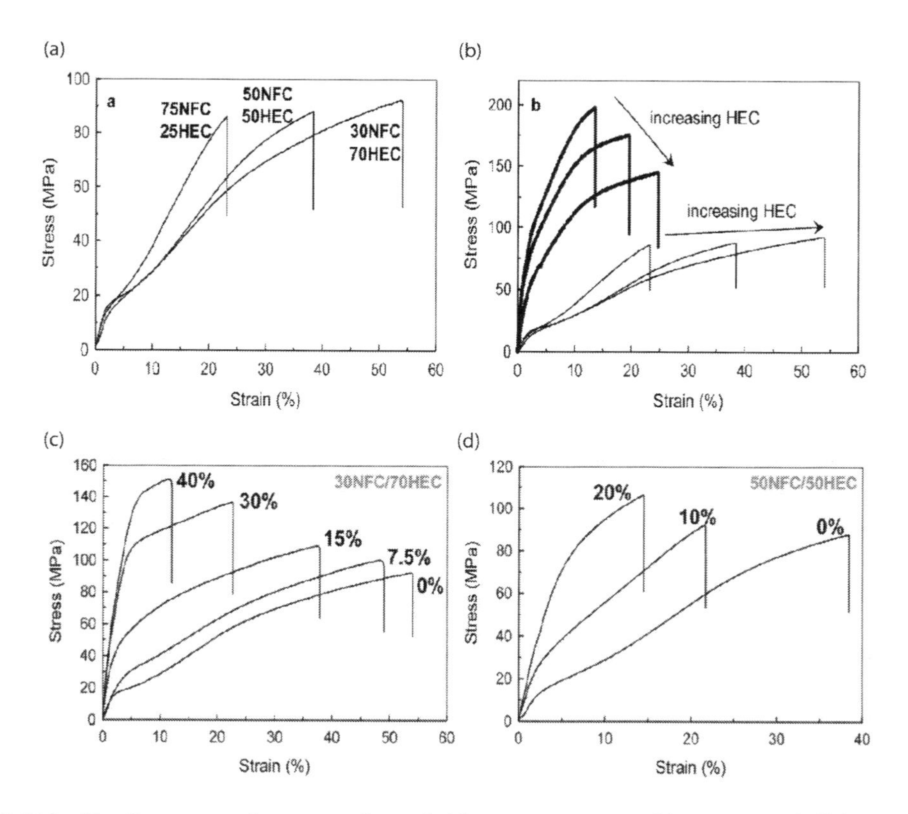

FIGURE 15.9 Tensile stress-strain curves of stretchable composite paper. (a) NFC-HEC (cellulose nanofiber/hydroxyethylcellulose) nanopaper of different ratios. The nanopaper has a porosity of 44%–63%, density of 525–770 kg m^{-3}. (b) Comparison high-porosity NFC-HEC in (a) and low-porosity NFC-HEC nanopaper (darker lines) (13%–27%, density of 980–1200 kg m^{-3}). High-porosity samples have double strain-to-failure, while increased HEC concentration improves the strain-to-failure of both paper. (c) Pre-stretched NFC/HEC non-woven composites (30:70 ratio). (d) Pre-stretched NFC/HEC non-woven composites (50:50 ratio). (Images taken from Sehaqui, H. et al., *Biomacromolecules*, 13, 3661–3667, 2012.)

Although the above approaches apply to stretchable electronic paper, there are currently few examples of stretchable conductive paper. Usually conductive composite paper cannot recover to its original resistance even within the mechanical elastic range. Zhang et al. fabricated a paper-based strain sensor by first prestraining the composite to remove some electron paths spanning over cellulosic fibers and facilitate good sensitivity. The resistance of their sensor increased linearly with stretch strains up to 6% (Zhang et al. 2018a). Hamedi et al. fabricated nanopaper out of SWNT and carboxymethylated NFC that was stretchable up to 8.9% with 3 wt% SWNT and to 7% with 10 wt% SWNT (Hamedi et al. 2014). Therefore, current examples of stretchable, conductive paper handless strain up to ~9%, and there may be future opportunities to augment the stretchability of conductive paper for skin-like sensing and other applications in stretchable electronics.

15.5.2 Structural Engineering for Papertronics

Despite the limited stretchability of paper itself, structural engineering allows large effective strains. Origami (by folding) and kirigami (by cutting) can achieve expandable structures. These techniques use cutting, bending, and folding to make complex three-dimensional hierarchical structures that handle large strains (Xu et al. 2017b).

15.5.2.1 Stretchable Papertronics Based on Origami

One straightforward, common origami to increase stretchability is to pre-strain paper to form pleats like an accordion. This wavy sheet can accommodate up to 1000% uniaxial strain upon stretching (Figure 15.10a) (Huang et al. 2017). Recently, the Innventia company has produced reels of stretchable paper by microscale origami. They created corrugated (or wavy) paper by micro-shrinking it consecutively along machine direction (MD) and cross direction (CD) before drying. The produced paper is stretchable up to 20% in two directions (Rushton 2017).

Miura-ori (or Miura folding), another origami structure, achieves omnidirectional stretchability and rigid foldability. Miura-ori generates parallelogram faces which preserve rigid materials or components on the substrate when the structure expands and collapses in four different directions. Miura-ori has proved to be an efficient packing method of solar sails and individual ZnO photodetectors (PDs) (Figure 15.10b) (Xu et al. 2017b; Callens and Zadpoor 2017; Lin et al. 2017).

15.5.2.2 Stretchable Papertronics Based on Kirigami

Kirigami is a powerful, customizable approach to fabricate mechanical metamaterials like stretchable electrodes and supercapacitors. Conventional kirigami cuts a pattern of slits in paper to obtain a desirable topology upon folding. Such patterns enable stiff sheets to acquire high extensibility

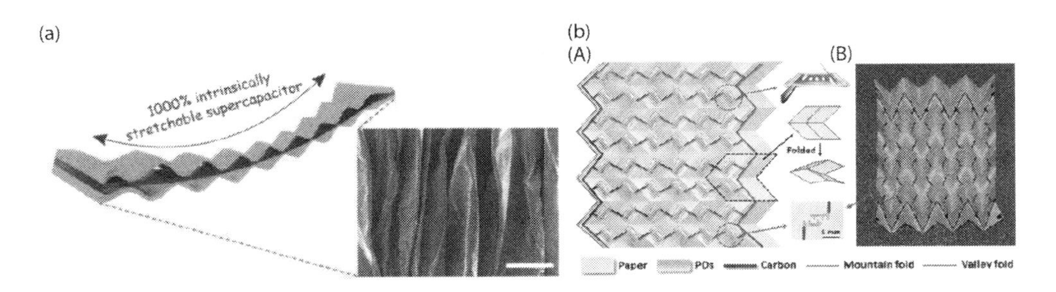

FIGURE 15.10 Two origami structures for stretchable papertronics. (a) Accordion structure. (Images taken from Huang, Y. et al., *Angew. Chem. Int. Ed.*, 56, 9141–9145, 2017.) (b) Miura-Ori structure. (Images taken from Lin, C.-H. et al., *ACS Nano*, 11, 10230–10235, 2017.)

(Blees et al. 2015; Shyu et al. 2015; Lv et al. 2018). In addition, kirigami enhances the adhesion of films to substrates that undergo large deformations such as human joints (Zhao et al. 2018).

The Föppl–von Kármán number γ is a crucial material parameter for kirigami. For a square sheet, Equation (15.4) gives):

$$\gamma = Y_{2D}L^2/K \approx \left(\frac{L}{t}\right)^2, \tag{15.4}$$

where Y_{2D} is the two-dimensional Young's modulus, K is the out-of-plane stiffness, L is the length, and t is the thickness of the square sheet. γ indicates the ratio between in-plane stiffness and out-of-plane bending stiffness. Membranes of high-value γ bend and crumple more easily than they stretch and shear. For example, γ of a sheet of graphene is in the order of 10^5–10^7, close to that of paper, thus stiff enough for kirigami (Blees et al. 2015).

Kirigami has a limitless variety of patterns: lines, squares, hexagons, circles, or an array of these units (Wu et al. 2016). The most simple, common case is a line-cut pattern consisting of straight lines in a centered rectangular arrangement. The remaining materials rotate like hinges to open the cuts upon stretching. The pre-set deformation points delocalize stress rather than concentrating it. Plastic rolling suppresses the fracture that is due to cutting and folding. Blunting the crack tip can further reduce the loads at the cuts.

The kirigami approach can provide predictive deformation mechanics for materials and a systematic means to engineer their elasticity. Shyu et al. showed that finite element modeling (FEM) can accurately predict the tensile behavior of sheets with linear kirigami patterns. From the analysis of beam deflection, the critical force f_c (or critical buckling load) relates geometric and material parameters as shown in Equation (15.5):

$$f_c \propto \frac{Eyt^3}{\left(L_c - x\right)^3}, \tag{15.5}$$

where E is Young's modulus, x is the spacing in the transverse direction, y is the spacing in the axial direction, L_c is the length of the cut, and t is the thickness of the sheet (Figure 15.11a). The extensibility of

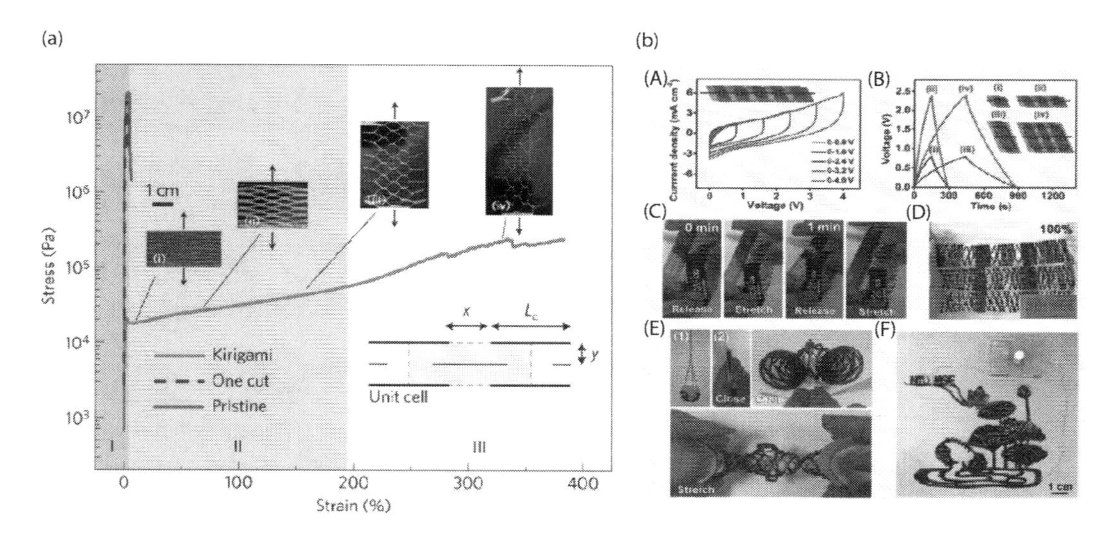

FIGURE 15.11 (a) Stress-strain curves for paper sheets with different patterns: Line-cut (green), a single notch in the middle (dashed blue), and without cut or pristine (grey curve). (Image taken from Shyu, T.C. et al., *Nat. Mater.*, 14, 785, 2015.) (b) Editable supercapacitors, which are connectable in series and parallel modes forming arrays and complicated stretchable structures. (Images taken from Lv, Z. et al., *Adv. Mater.*, 30, 1704531(1–9), 2018.)

the sheet increases with decreasing critical buckling load (achieved by decreasing the x- or y-axis spacing or increasing the cut length L_c) (Shyu et al. 2015). Such sheets with line-cut (or linear) patterns have a theoretical maximum stretchability (in % strain) calculated by Equation (15.6):

$$\% \text{ strain} = \frac{\Delta L}{L_i} \times 100\% = \frac{(L_c - x)}{2y} \times 100\% \tag{15.6}$$

where ΔL is the length increment after stretch, and L_i is the initial length before stretch (Lv et al. 2018).

The linear pattern transforms the paper sheet into a honeycomb-like structure upon stretching. Lv et al. applied such pattern to achieve a stretchable supercapacitor (composed of ultralong manganese dioxide nanowires (MNWs) and CNTs sandwiched by nanocellulose fibers (NCFs)), which showed a specific capacitance of 227.2 mF cm^{-2} and was stretchable up to 500% without degradation of the electrochemical performance. The supercapacitor exhibited excellent stability with 95% capacitance retention after 10,000 cycles of concurrent bending, folding, and twisting. Honeycomb-like supercapacitors can form parallel or serial arrays to enlarge output voltage (Figure 15.11b) (Lv et al. 2018).

Some sophisticated kirigami techniques like periodic serpentine shapes and 2D fractal iterations can facilitate conformability and stretchability of paper-based substrates (Fan et al. 2014). Typically, paper is likely to adopt unidirectional bending, but cannot support biaxial deformation. Pini et al. (2016) have confirmed that in-plane stresses occur when paper is subject to biaxial bending, which makes paper generate wrinkles instead of smooth curves. To conform papertronics to curvilinear biological surfaces including skins, paper-based substrates need to have skin-like mechanical properties (such as a low elastic modulus) and a low thickness (~5 μm) (Jeong et al. 2013). Sadri et al. fabricated paper-based sensors with open-mesh serpentine designs (effective Young's modulus = 0.2–9.1 MPa), which matched the mechanical impedance of most biological tissues, and had a maximum strain-to-failure of 68% (Sadri et al. 2018). Alternatively, fractal-cut patterns, or a set of simple cuts in a multi-level hierarchy with same or different motifs, can make rigid materials super-conformable to objects of non-zero Gaussian curvature (like a sphere or saddle), and expandable to more than 100% strain. The cutting divides a sheet into rotating units with interconnecting hinges, which therefore accommodates deformation without sacrificing the entire performance (Figure 15.12) (Yang et al. 2016).

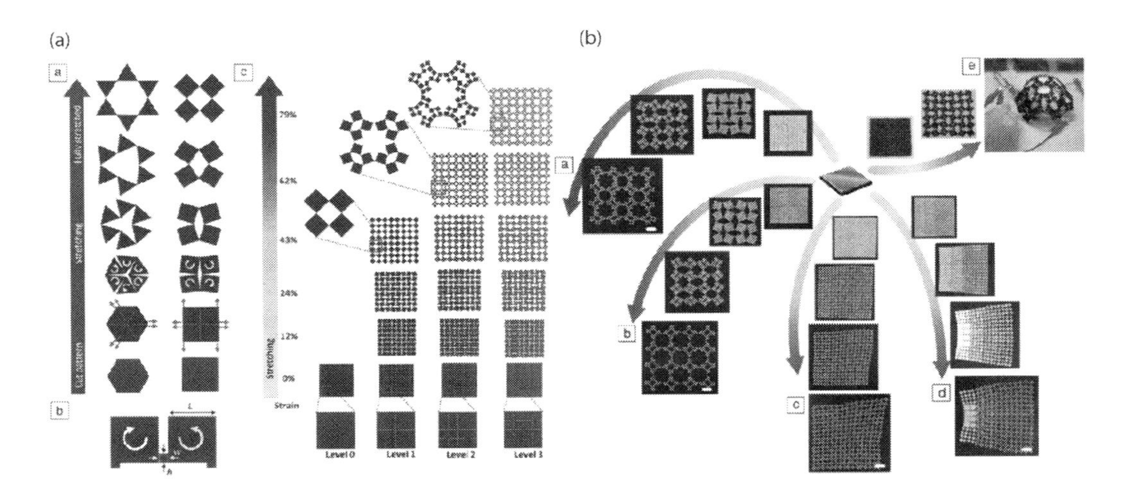

FIGURE 15.12 Fractal cuts that enable rigid materials to exhibit super conformality and biaxial expansion. (Image from Yang, S. et al., *MRS Bull.*, 41, 130–138, 2016.) (a) Cuts divide the material into rotating units, which enable the material to deform upon biaxial or uniaxial stretching. (b) silicone sheets stretch with fractural structures and become a stretchable electrode with conductive film of multiwall carbon nanotubes deposited on top.

Overall, origami and kirigami techniques can achieve expandable, shape-shifting, and reversible structures of papertronics. Such techniques also allow strain-invariant electrical properties of the devices. The scale-free geometric character of origami can extend to milli-, micro-, and nanometer-size systems (Silverberg et al. 2014). Kirigami is also applicable to multi-scale materials, but the notches/creases need to be larger than the typical variations in the materials. The characteristic dimensions of the patterns have to be ~100× smaller than that of the base materials, but ~100× larger than its granularity (Xu et al. 2017b).

15.6 Developments and Applications in Papertronics

The previous sections demonstrate the viability and conceivability of paper-based flexible and stretchable electronics. In this section, we discuss some representative applications of papertronics, including pressure sensors, transistors, energy storage devices, electroanalytical devices, memory devices, displays, antennas, sanitizers, and triboelectric nanogenerators.

15.6.1 Tactile/Pressure/Strain Sensors

Paper-based tactile/pressure/strain sensors usually function as input interfaces of flexible electronics. Paper-based tactile sensors are typically capacitive with a cellulosic layer as the passive substrate. Metallized paper (Figure 15.13a) and silver-nanowire printed paper are good candidates for capacitive touchpads (Mazzeo et al. 2012; Zou et al. 2018; Li et al. 2014).

Most paper-based force sensors rely on pressure-sensitive composite paper. Such composite paper utilizes the porous structure of paper by embedding conductive particles in cellulose fiber matrix. Applying pressure to this paper causes resistance changes (Gong et al. 2014). Tao et al. obtained a graphene-paper pressure sensor by soaking tissue paper in precursor ink. The sensitivity of the graphene-paper pressure sensor reached 17.2 kPa^{-1} in the pressure range of 0 to 20 kPa and was capable of pulse detection, respiratory detection, and voice recognition (Figure 15.13b). Such paper-based pressure sensors are stable under cyclic compressive loading/unloading with a slight reduction of sensitivity after 300 cycles (Tao et al. 2017).

Paper-based piezoelectric sensors can also detect pressure. Li and Liu grew zinc oxide NWs (ZnO-NWs) to cover the fibers in paper through a hydrothermal process. The ZnO-NWs length, growth

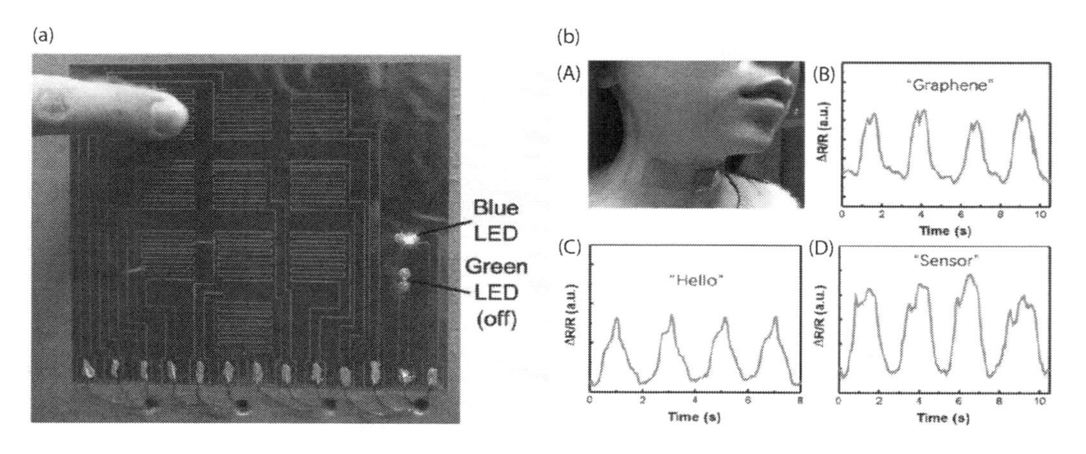

FIGURE 15.13 (a) A paper-based capacitive touchpad. (Image taken from Mazzeo, A.D. et al., *Adv. Mater.*, 24, 2850–2856, 2012.) (b) A graphene-paper pressure sensor used in voice recognition. (Images taken from Tao, L.-Q. et al., *ACS Nano*, 11, 8790–8795, 2017.)

rate, and the conductivity of the resulting paper are controllable by adjusting growth time. The ZnO-NWs paper is sensitive to pressure because the electrical centers displace under deformation, thus generating a piezoelectric potential (Li and Liu 2014).

Paper-based strain sensors function with changes in resistance under strain (Yan et al. 2014). Applying compressive and tensile strain to such sensors alter the network of conductive particles to be dense or sparse, respectively, thus decreasing or increasing the resistance (Figure 15.5j) (Lin et al. 2014). Paper-based strain sensors detected a change in bending radius down to 0.5 cm at a frequency up to 4 Hz for more than 3000 cycles (Wei et al. 2015). However, some electronic applications require constant electrical properties insensitive to strain under deformed states. In such cases, liquid metal and nanoconfined electronic inks are suitable due to their stable electrical properties under mechanical deformation (Zheng et al. 2013; Han et al. 2015; Xu et al. 2017a).

The performance of paper-based strain sensors depends on the type, structure, and concentration of conductive fillers, and the morphology of the conductive, fibrous network. Tuning these properties can result in a high-performance strain sensor that can sense small variations with significant sensitivity and reproducibility. Future paper-based tactile/pressure/strain sensors will work as an 'electronic skin' and adhere to the surface of robots for the environment-machine interfacing/interaction.

15.6.2 Transistors

Transistors are key components in logic gates of integrated circuits. The design and fabrication of paper-based transistors are emerging topics in papertronics. Some paper-based transistors use paper as the gate dielectric to separate the gate and drain/source terminals. Paper can offer a large capacitance at low frequencies due to its fiber-based, foam-like structure. Martins et al. fabricated a working, low-power, complementary metal-oxide semiconductor (CMOS) inverter on paper by reactive magnetron sputtering and electron beam evaporation (Figure 15.14a). The inverter achieved a gain of −4 and circuit leakage current of 150 μA. Other paper-based transistors have layers of functional materials stacked on the same side of paper. Zocco et al. have fabricated a paper-based organic thin-film transistor (OTFT) (by deposition of a copper gate electrode and a

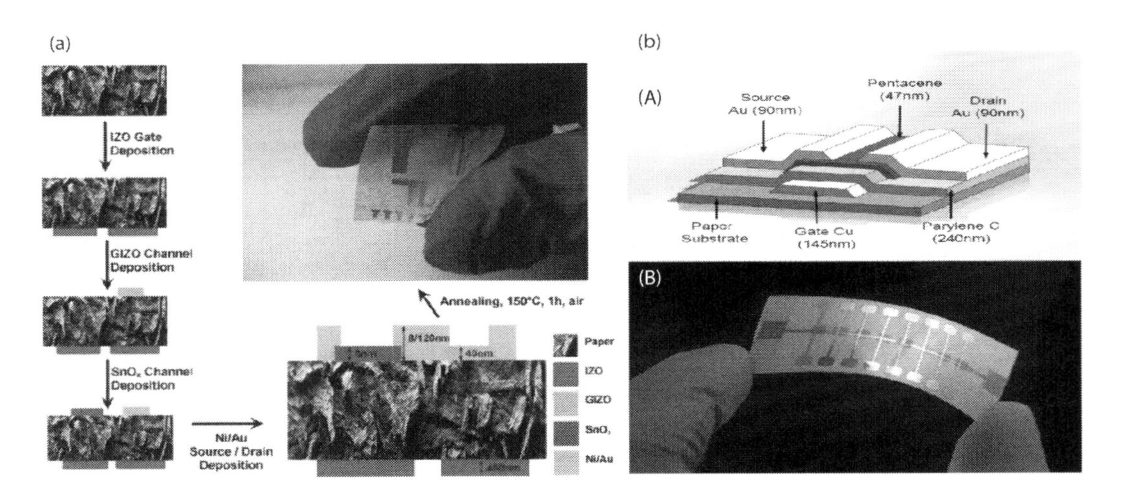

FIGURE 15.14 (a) A CMOS inverter fabricated on two sides of the paper substrate. (Images taken from Martins, R. et al., *Adv. Mater.*, 23, 4491–4496, 2011b.) (b) An organic thin-film transistor on the same side of paper substrate. (Images taken from Zocco, A.T. et al., *Nanotechnology*, 25, 094005, 2014.)

dielectric layer of Parylene C; and by thermal evaporation of a pentacene semiconductor layer and gold source/drain electrodes) on a piece of commercial photo paper. The transconductance, mobility, and on/off current ratio of the OTFTs reach 2.6×10^{-7} S, 0.09 $cm^2 V^{-1} s^{-1}$, and 1×10^5, respectively (Figure 15.14b) (Martins et al. 2011b; Zocco et al. 2014).

In addition to the sputtering, deposition, and evaporation, Rojas et al. (2015b) used a double-transfer printing method to fabricate printed thin-film transistors up-side-down on a silicon wafer, and then transferred them onto a carrier substrate, which could be paper, wood, stone, or other materials. The performance of paper-based transistors is comparable to that of conventional transistors fabricated on silicon or glass substrates (Liang et al. 2016). Future development needs to simplify the current complex fabrication process to propagate the usage of paper-based transistors.

15.6.3 Energy Storage Devices

Paper-based power sources are the essential components for functional papertronics independent of an external power supply. Disposable paper-based power devices can provide sufficient energy for small gadgets to function in resource-limited regions. Jabbour et al. and Nguyen et al. have comprehensively reviewed paper-based batteries, which include supercapacitors, electrochemical batteries, biofuel cells, microbial fuel cells, and Li-ion batteries (Jabbour et al. 2013; Nguyen et al. 2014). Here, we only list some examples. Pushparaj et al. fabricated nanocomposite units by embedding multi-walled nanotubes (MWNT) and an electrolyte in cellulosic paper, which serves as building blocks for supercapacitors, batteries, and their hybrids (Figure 15.15a). The adjustable porosity of the cellulosic layer promotes the intimate configuration and efficient packaging of MWNT and electrolyte. An extra cellulosic layer functioned as a spacer in the supercapacitor. Their supercapacitors showed specific capacitances of 36 F/g at the operating voltage of 2.3 V and a power density of 1.5 $kW \cdot kg^{-1}$ (energy density, $\approx$13 Wh/kg) with the electrolyte (1-butyl,3-methylimidazolium chloride ([bmIm][Cl])). Such supercapacitors can operate with bodily fluids (e.g., sweat and blood). Their batteries had a reversible capacity of 110 mAh/g over several tens of charge-discharge cycles between 3.6 V and 0.1 V at a constant current of 10 mA/g. When combining the supercapacitor and the battery in a hybrid system, the discharge of the battery charged the supercapacitor (Pushparaj et al. 2007).

As another example, paper-based microbial fuel cells (MFCs) generate power by transferring electrons to the anode via microorganism respiration. MFCs use microorganism-containing water as the media, which can come from wastewater, urine, or soiled water in a puddle. Fabrication of MFC benefits from easily patterned fluidic pathways and 3D origami structures of paper sheets. Fraiwan et al. fabricated a paper-based MFC stack of eight modular blades (Figure 15.15b). The device was retractable from sharp shuriken (closed) to round frisbee (opened) while it generated voltages from 0 V to 2.76 V. The shape of round frisbee connected the eight modules in series and exposed all air-cathodes for reactions. The device powered an light-emitting diode (LED) for 20 min. Alternatively, Gao and Choi fabricated a paper-based, bacteria-powered battery with different anode surface areas (Figure 15.15c). Anodes with larger areas enabled more bacteria to attach and provided a higher current density than those with smaller areas. Their battery provided a maximum current density of 0.87 $\mu A\ cm^{-2}$ and a power density of 55.2 $\mu W\ m^{-2}$ when connected with an external load of 1 kΩ. Overall, these biobatteries can be a simple, low-cost, disposable power supply for point-of-care (POC) diagnostic devices in resource-limited areas (Fraiwan et al. 2016; Gao and Choi 2017).

Paper-based batteries take advantage of the foldability of paper to simplify the fabrication process and improve their stackability and stability. Hamedi et al. made a foldable battery by wax-printing a counter electrode, a working electrode, and a separator on a single piece of paper. They folded this paper in such a way that the electrodes sandwiched the separator vertically to form a battery. When connecting six batteries in series, the device provided a voltage of 2.5 V at 12.5 μW (Figure 15.15d), which powered an LED for 15 s (Hamedi et al. 2016).

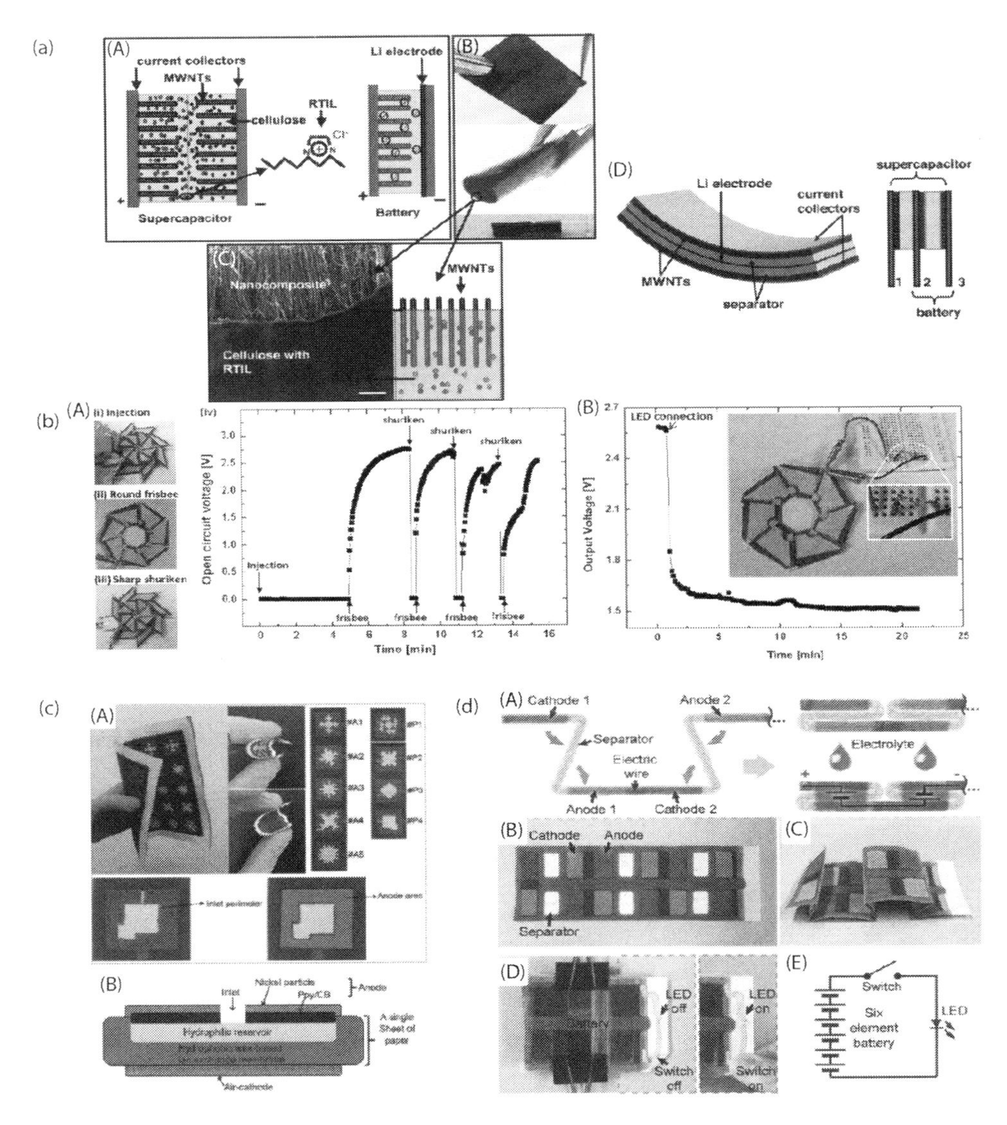

FIGURE 15.15 Paper-based batteries. (a) Nanocomposite paper serves as building blocks for supercapacitors, batteries, or their hybrids. (Images taken from Pushparaj, V.L. et al., *Proc. Natl. Acad. Sci.*, 104, 13574–13577, 2007.) (b) A retractable MFC stack to control voltage generation. The frisbee shape connect eight MFC modules in series, which provides a voltage up to 2.76 V and powers an LED for 20 min. (Images taken from Fraiwan, A. et al., *Biosens. Bioelectron.*, 85, 190–197, 2016.) (c) Biological fuel cells with different anode/inlet configuration. (Images taken from Gao, Y. and Choi, S., *Adv. Mater. Technol.*, 2, 1600194, 2017.) (d) Foldable paper-based batteries. (Images taken from Hamedi, M.M. et al., *Adv. Mater.*, 28, 5054–5063, 2016.)

15.6.4 Electroanalytical Devices

Porous cellulosic substrates promote the development of paper-based microfluidic devices. Among these devices, paper-based electrochemical analytical devices can hold biological fluids for reactions to diagnose potential diseases (Liang et al. 2016). Cellulose can provide a continuous capillary force for the transportation of fluids; for instance, a nanocellulose substrate in an electrolyte-sensing transistor absorbs and transports biofluid to sensing electrodes (Yuen et al. 2017). Additionally, cellulose with a specific porosity can filter out large particles to avoid interference in reactions. Some devices comprise paper-based microfluidic channels patterned by photolithography or wax printing and electrodes screen-printed from conductive inks. These devices can detect and quantify glucose by performing a chronoamperometric analysis based on glucose oxidase or can combine with a commercial glucometer to measure beta-hydroxybutyrate (BHB)—a biomarker for diabetic ketoacidosis (Nie et al. 2010; Wang et al. 2016a). Liu and Crooks fabricated a self-working paper-based electrochemical sensing platform with an integral battery and an electrochromic read-out. The integrated metal/air battery powers both the electrochemical sensor and the electrochromic read-out. The device detects glucose or hydrogen peroxide (H_2O_2) starting from 0.1 mM (Liu and Crooks 2012).

15.6.5 Memory Devices

Memory devices enable information storage by repetitive reading, writing, and erasing of data. Among disposable memory devices, paper-based substrates are useful for security systems against hacking. For non-volatile paper-based memory devices, resistive random-access memory (RRAM) can record data at the resistive switching layer. These memory devices function with changes in applied voltage. They are at the high-resistance state ('OFF' state) before setting; upon application of a positive voltage, the resistance of the memory decreases to the low-resistance state ('ON' state). Conversely, a negative voltage returns the memory to the high-resistance state ('OFF' state). Such paper-based memory devices are advantageous because they are bendable, foldable, and disposable by burning or shredding to permanently remove secure data. Lien et al. printed layers of silver/titanium dioxide/carbon ($Ag/TiO_2/C$) on a piece of copy paper to form a memory device. Their device showed high reliability under a bending radius of ~10 mm for more than 1000 cycles. They also tuned the ON/OFF memory window (up to 3 orders of magnitude) by changing the thickness of the TiO_2 layer. Lee et al. fabricated a paper-based memory device with stable conductivity after 100 folding cycles. Their devices operated under a low voltage near ±1 V with a memory window (ratio of high to low resistance states) of 10^2 (Lien et al. 2014; Lee et al. 2016a).

Another paper-based non-volatile memory device is a thin-film transistor-based organic flash memory, which includes a charge-storage layer (CSL) sandwiched between two dielectric bilayers (tunneling and blocking). The control gate (CG) bias controls charges injected or removed into the CSL through the tunneling dielectric layer (TDL), which programs or erases the memory. The blocking dielectric layer (BDL) constrains the charges stored within CSL to retain memory. Lee et al. fabricated such memories with polymeric layers grown by initiated chemical vapor deposition (iCVD) on dye-sublimation papers (DP), which have a significantly rough surface with bumps in both the micro-scale and nano-scale. Their devices have long memory retention (estimated to be of ~10 years), low-programming/erasing voltages (~±10 V), and good mechanical flexibility. The devices can maintain programming and erasing capabilities in the folded state as well as after 1200 folding cycles with a bending radius of 300 μm (Lee et al. 2017).

15.6.6 Displays

Paper-based displays are in the early developing stage. The most common paper-based displays rely on color changes of thermochromic ink upon Joule heating from electrically conductive wires (heaters) patterned on the back of the paper. Such displays normally work with a voltage less than 10 V and take 10–15 s to reach a steady state. Siegel et al. patterned thermochromic ink made of Leuco dye over pre-printed images on a photopaper (Figure 15.16a). Upon heating, the Leuco dye changed from opaque to transparent to reveal the images underneath. Their 100-μm thin display has a minimum resolution of approximately 200 μm × 200 μm. A key advantage of such displays is that they can present predefined complex messages including passages of text in multiple languages or intricate multi-color images (Siegel et al. 2009).

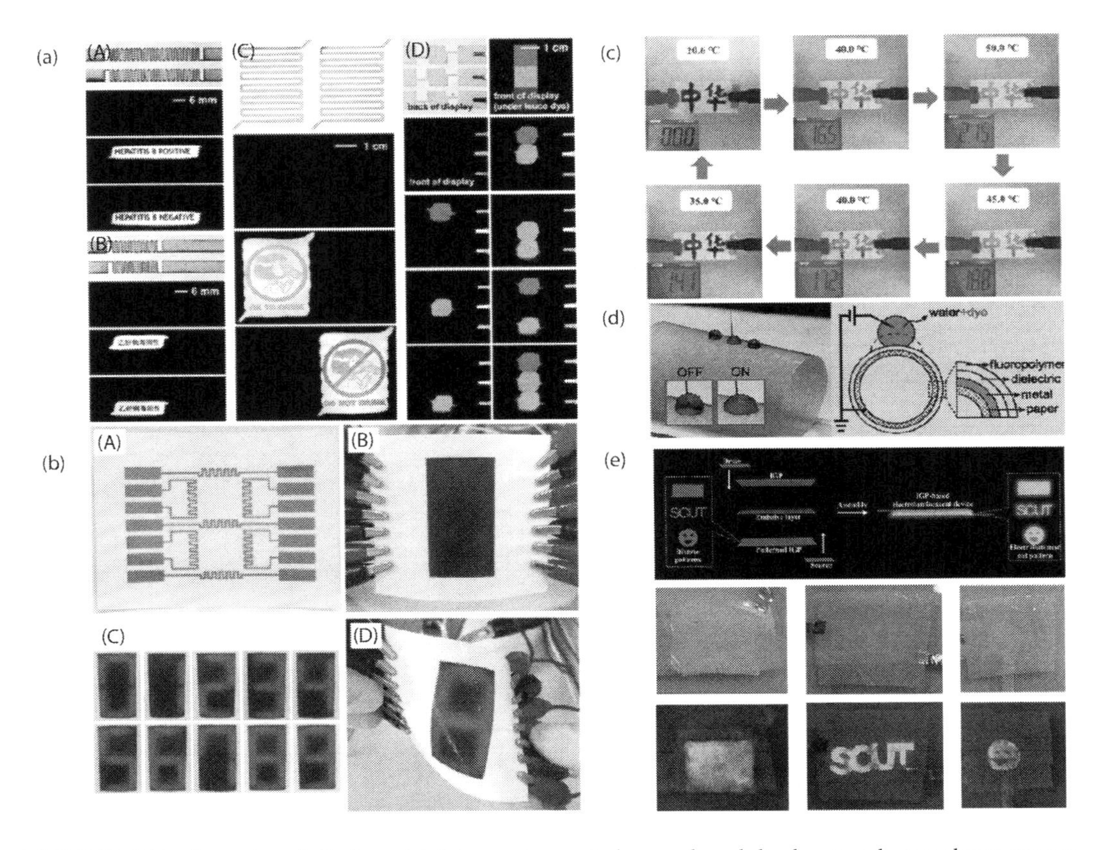

FIGURE 15.16 Paper-based displays. (a) A leuco-dye-coated paper-based display reveals complex messages or intricate multicolor images upon Joule heating. (Images taken from Siegel, A.C. et al., *Lab. Chip*, 9, 2775–2781, 2009.) (b) A seven-segment polydiacetylene (PDA) display. By heating the wires of different segments on the back of the photopaper, the PDA-coated side showed numeric digits from 0 to 9, due to the thermally induced blue-to red color change of PDAs. (Image taken from Shin, H. et al., *Nanotechnology*, 25, 094011, 2014.) (c) A simple display made by a composite paper with thermochromic ink painted on top. By applying voltage across the AgNW composite paper can heat the thermochromic paint to change from black to Gainsboro. (Images taken from Wang, G. et al., *Mater. Res. Express*, 4, 116405, 2017.) (d) Electrowetting (EW) effect of treated paper. Applying voltage to the water droplets on such paper, which leads to the change of water contact angle (~90° at most). (Images taken from Kim, D.Y. and Steckl, A.J., *ACS Appl. Mater. Interfaces*, 2, 3318–3323, 2010.) (e) A simple electroluminescent display by sandwiching an emissive polymer layer between two pieces of ionic-gel paper (IGP). When applying a high-frequency alternating voltage to the IGP, the emissive layer afforded most of the voltage and emitted a blue luminescent light. Cutting IGP into different shapes, their display can show patterns like a square, an alphabetic string, and a laughing face. (Images taken from He, M. et al., *ACS Appl. Mater. Interfaces*, 9, 16466–16473, 2017.)

Shin et al. further developed a display to present user-controllable patterns with a seven-segment polydiacetylene (PDA) display. Heating the wires of different segments on the back of the photopaper induced a blue-red color change of PDAs, showing numeric digits from 0 to 9 (Figure 15.16b). Instead of printing wires on the backside of the paper substrate, Wang et al. used a composite (with AgNWs embedded in a Chinese Xuan paper) as both a heater and a substrate for patterned thermochromic paint. The paint changed from black to Gainsboro with heat upon application of a voltage across the composite, and then returned to black when the composite cooled down to room temperature (Figure 15.16c). Future advanced thermochromic displays will progress by choosing different thermochromic paints, controlling the resistance of composite paper, and changing the input voltage (Shin et al. 2014; Wang et al. 2017).

Other approaches to make paper-based displays include harnessing the electrowetting (EW) effect, applying ionic conductors, and using electrochromic materials. Kim and Steckl deposited layers of metal, insulator, and hydrophobic fluoropolymer on top of paper. Then, they applied a voltage to the water droplets on the paper, which changed the water contact angle (~90° at most) (Figure 15.16d). These paper-based EW devices showed negligible hysteresis (~2°) and fast switching times of ~20 ms. Their work indicates the feasibility of applying paper as flexible substrates for e-paper and video displays. He et al. fabricated a simple electroluminescent display by sandwiching an emissive polymer layer between two pieces of ionic-gel paper (IGP). The emissive layer emitted a blue luminescent light under a high-frequency alternating voltage to the IGP. Their displays showed diverse patterns with different shapes of IGP, such as a square, an alphabetic string, and a laughing face (Figure 15.16e). Liana et al. fabricated an electrochromic paper-based read-out system compatible with pressure sensing. The read-out comprised several segments of gold-nanoparticle film separated by resistive graphite films on paper. The electrochromic coating of Prussian blue/polyaniline on gold-nanoparticle film changed color from green/blue to transparent when the voltage across the film increased. The read-out indicated the range of pressure by changing the color of a different number of segments (Kim and Steckl 2010; He et al. 2017; Liana et al. 2016).

15.6.7 Other Applications

In addition to the above applications, papertronics can also function as portable antennas, plasma generators, sanitizers, and triboelectric nanogenerators. We have shown a 3D antenna drawn by a silver-ink filled rollerball pen on a sticky paper in Figure 15.5a (Russo et al. 2011). Paper-based antennas have both a simple fabrication process and high performance. By inkjet printing, Cook and Shamim patterned antennas on a photopaper with small feature sizes (30–50 μm) and sintered them to high conductivities (1.2×10^7 [S/m]). Their antennas operated in the ultrawideband (UWB) (at frequencies from MHz to 12.5 GHz) and exhibited a high gain of up to 8 dBi (Cook and Shamim 2012). Paper-based antennas on smooth nanopaper are bendable and foldable. Folding such antennas leads to large shifts of resonance peak and enhances sensitivity at specific frequency bands (Nogi et al. 2013). To shorten the length of the antenna element while maintaining its radio-wave frequency, Inui et al. prepared nanopaper of high density (1.3 g cm^{-3}) and high dielectric constant ($k = 726.5$) by embedding silver nanowires at a non-percolative fraction (2.48 vol%). The resulting antenna was half-size of the printed on plain nanopaper, but with the same sensitivity and flexibility (Inui et al. 2015).

Paper-based plasma generators made of conductive paper sheets can discharge plasma under strain with kirigami patterns (Shyu et al. 2015). Xie et al. further extended paper-based plasma generators into paper-based sanitizers (Figure 15.17a). Their metallized-paper sanitizers were capable of generating a high level of ozone and deactivating greater than 99% bacteria with non-contact treatment for 30 s (Xie et al. 2017).

Paper-based triboelectric nanogenerators (TENG) harvest mechanical energy from the periodical shape-shifting of stretchable structures, such as origami or kirigami patterns. Yang et al. presented a slinky-shaped and doodlebug-shaped TENG with aluminum foil and polytetrafluoroethylene (PTFE) thin film glued sequentially on top of the paper. Periodical pressing and stretching seven units of such TENG in parallel connection generated a charge of 0.55 μC, which can light up commercial LEDs (Figure 15.17b). Wu et al. devised a paper-based kirigami TENG that consisted of a copper-coated paper and a film of fluorinated ethylene propylene (FEP) interlocked by kirigami patterns. When pressing, stretching, and twisting the TENG, the contact and separation of these two layers generated an electrical output; the outputs of pressing mode (an open-circuit voltage of 115.49 V and a maximum transferred charge of 39.87 nC) were the largest among three modes (Figure 15.17c) (Yang et al. 2015a; Wu et al. 2016; Guo et al. 2017a).

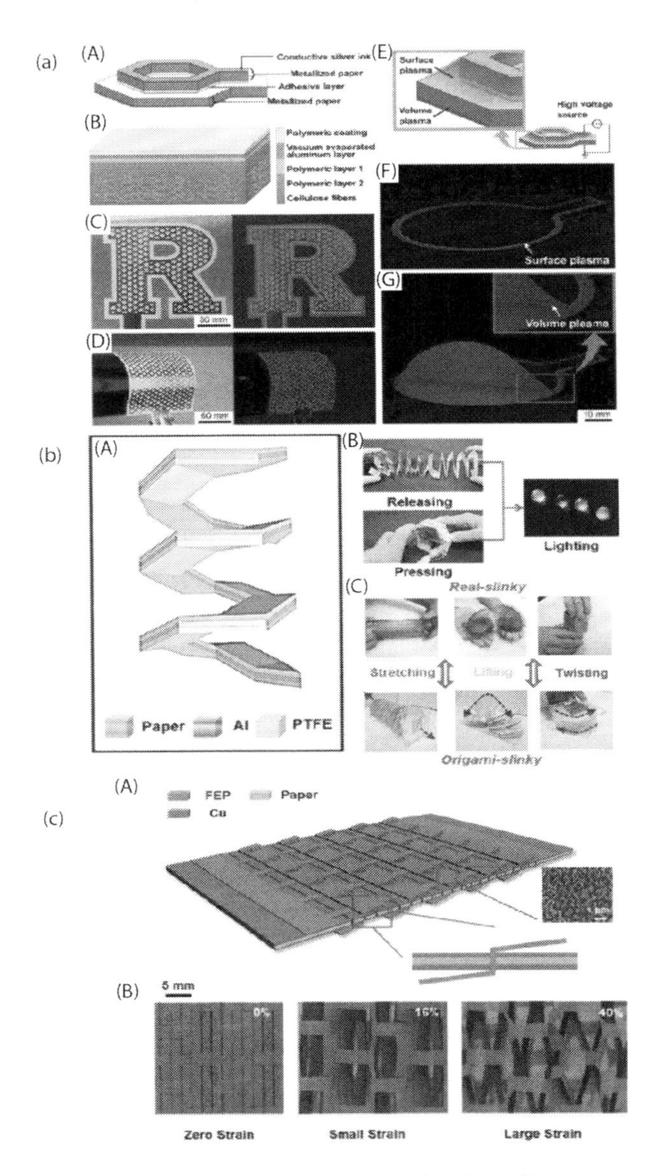

FIGURE 15.17 Other papertronics. (a) A plasma generator made of metallized paper with kirigami patterns. (Images taken from Xie, J. et al., *Proc. Natl. Acad. Sci.*, 114, 5119–5124, 2017.) (b) A triboelectric nanogenerators that can harvest mechanical energy to power LEDs. (Images taken from Yang, P.-K. et al., *ACS Nano*, 9, 901–907, 2015a.) (c) A paper-based triboelectric nanogenerator with stretchable interlocking kirigami patterns. (Images taken from Wu, C. et al., *ACS Nano*, 10, 4652–4659, 2016.)

15.7 Conclusion

This book chapter describes the building blocks of paper: its cellulosic fibers and multi-scale architecture. Approaches to tailoring the properties of paper include chemical and mechanical processing. Chemical methods break fibers into nanocrystals or modify their surface properties. Mechanical methods alter the physical morphology of the fibers. Both methods affect the structure-property relationships of the sheets of materials, which influence chemo-electro-opto-mechanical properties. *Electronics on paper* have already started to demonstrate their potential scalability and commercial viability with engineering and scientific advances in printing technologies and adhesion of electronic components to paper. *Electronics in paper* represent an alternative approach to papertronics with ongoing challenges and opportunities in tunable porosity, inter-fiber bonding, and fiber-filler interactions. Origami (folding) and kirigami (cutting) are additional strategies to creating stretchable and flexible devices.

Paper-based systems show promise in becoming stand-alone, sustainable products that will benefit society. In the short term, strategies that interface paper-based devices with conventional electronics will continue to develop, while future papertronic devices may also integrate multiple paper-based modules (e.g., battery, memory) into devices that work independently of external support. Paper-based electronics have the potential to become an eco-friendly class of flexible electronics for applications in sensing, computation, energy management, and human-machine interfaces. Fibrous, cellulose-based mats, while ancient, are a high-tech material with potential advantages in manipulating the flow of heat, mechanical stresses/forces, charges, light, and liquids. In addition to their environmental sustainability, low cost, and flexibility, such attributes will facilitate applications in dynamic delivery of information, smart packaging and labels, wearable and implantable healthcare sensors and actuators, robotics, and environmental monitoring systems.

Acknowledgment

The authors thank Prof. Jian Wen (School of Technology, Beijing Forest University, 100083, China) for helpful discussion and insightful suggestions. The authors also thank Jasmine Sawaged and Jingjin Xie for their assistance in editing this book chapter. The authors acknowledge support from the National Science Foundation Award Nos. 1610933 and 1653584. T.L. and X.Z. acknowledge fellowships from the China Scholarship Council.

References

Abdul Khalil, H.P.S., Davoudpour, Y., Islam, M.N., Mustapha, A., Sudesh, K., Dungani, R., and Jawaid, M. (2014). Production and Modification of Nanofibrillated Cellulose using Various Mechanical Processes: A Review. *Carbohydr. Polym. 99*, 649–665.

Abitbol, T., Rivkin, A., Cao, Y., Nevo, Y., Abraham, E., Ben-Shalom, T., Lapidot, S., and Shoseyov, O. (2016). Nanocellulose, a Tiny Fiber with Huge Applications. *Curr. Opin. Biotechnol. 39*, 76–88.

Afra, E., Yousefi, H., Hadilam, M.M., and Nishino, T. (2013). Comparative Effect of Mechanical Beating and Nanofibrillation of Cellulose on Paper Properties made from Bagasse and Softwood Pulps. *Carbohydr. Polym. 97*, 725–730.

Agache, P., and Humbert, P. (2004). *Measuring the Skin* (Springer-Verlag, Berlin, Heidelberg, Germany).

Ahmed, N., Rogers, J.A., and Ferreira, P.M. (2015). Microfabricated Instrumented Composite Stamps for Transfer Printing. *J. Micro Nano-Manuf. 3*, 021007–021012.

Ahn, J., Seo, J.-W., Lee, T.-I., Kwon, D., Park, I., Kim, T.-S., and Lee, J.-Y. (2016). Extremely Robust and Patternable Electrodes for Copy-Paper-Based Electronics. *ACS Appl. Mater. Interfaces 8*, 19031–19037.

Alava, M., and Niskanen, K. (2006). The Physics of Paper. *Rep. Prog. Phys. 69*, 669.

Anderson, R.E., Guan, J., Ricard, M., Dubey, G., Su, J., Lopinski, G., Dorris, G., Bourne, O., and Simard, B. (2010). Multifunctional Single-walled Carbon Nanotube–cellulose Composite Paper. *J. Mater. Chem. 20*, 2400–2407.

Anothumakkool, B., Soni, R., Bhange, S., and Kurungot, S. (2015). Novel Scalable Synthesis of Highly Conducting and Robust PEDOT Paper for a High Performance Flexible Solid Supercapacitor. *Energy Environ. Sci. 8*, 1339–1347.

Arnold, C.B., Serra, P., and Piqué, A. (2007). Laser Direct-Write Techniques for Printing of Complex Materials. *MRS Bull. 32*, 23–31.

Bahadır, E.B., and Sezgintürk, M.K. (2016). Lateral Flow Assays: Principles, Designs and Labels. *TrAC Trends Anal. Chem. 82*, 286–306.

Benítez, A.J., Torres-Rendon, J., Poutanen, M., and Walther, A. (2013). Humidity and Multiscale Structure Govern Mechanical Properties and Deformation Modes in Films of Native Cellulose Nanofibrils. *Biomacromolecules 14*, 4497–4506.

Bhattacharyya, A., and Klapperich, C.M. (2006). Design and Testing of a Disposable Microfluidic Chemiluminescent Immunoassay for Disease Biomarkers in Human Serum Samples. *Biomed. Microdevices 9*, 245.

Blees, M.K., Barnard, A.W., Rose, P.A., Roberts, S.P., McGill, K.L., Huang, P.Y., Ruyack, A.R. et al. (2015). Graphene Kirigami. *Nature 524*, 204.

Boerjan, W., Ralph, J., and Baucher, M. (2003). Lignin Biosynthesis. *Annu. Rev. Plant Biol. 54*, 519–546.

Bollström, R., Pettersson, F., Dolietis, P., Preston, J., Osterbacka, R., and Toivakka, M. (2014). Impact of Humidity on Functionality of On-paper Printed Electronics. *Nanotechnology 25*, 094003.

Callens, S.J.P., and Zadpoor, A.A. (2017). From Flat Sheets to Curved Geometries: Origami and Kirigami Approaches. *Mater. Today 21*, 241–264.

Carson, F.T., and Worthington, V. (1952). Stiffness of Paper. *J. Res. Natl. Bur. Stand. 49*, 385–391.

Cataldi, P., Bayer, I.S., Bonaccorso, F., Pellegrini, V., Athanassiou, A., and Cingolani, R. (2015). Foldable Conductive Cellulose Fiber Networks Modified by Graphene Nanoplatelet-Bio-Based Composites. *Adv. Electron. Mater. 1*, 1500224.

Cate, D.M., Adkins, J.A., Mettakoonpitak, J., and Henry, C.S. (2015). Recent Developments in Paper-Based Microfluidic Devices. *Anal. Chem. 87*, 19–41.

Caulfield, D.F., and Gunderson, D.E. (1988). Paper Testing and Strength Characteristics. In *TAPPI Proceedings of the 1988 Paper Preservation Symposium*.

Chen, Y., Chen, G., Cui, Y., and Yang, Y. (2016). Preparation and Performance Study of Paper-Based Resin Nano-silver Inkjet Conductive Ink. In *Advanced Graphic Communications, Packaging Technology and Materials* (Springer, Singapore), pp. 1001–1010.

Cheng, I.-C., and Wagner, S. (2009). Overview of Flexible Electronics Technology. In *Flexible Electronics* (Springer, Boston, MA), pp. 1–28.

Chitnis, G., and Ziaie, B. (2012). Waterproof Active Paper via Laser Surface Micropatterning of Magnetic Nanoparticles. *ACS Appl. Mater. Interfaces 4*, 4435–4439.

Chitnis, G., Ding, Z., Chang, C.-L., Savran, C.A., and Ziaie, B. (2011). Laser-treated Hydrophobic Paper: An Inexpensive Microfluidic Platform. *Lab. Chip 11*, 1161–1165.

Compton, O.C., Dikin, D.A., Putz, K.W., Brinson, L.C., and Nguyen, S.T. (2010). Electrically Conductive "Alkylated" Graphene Paper via Chemical Reduction of Amine-Functionalized Graphene Oxide Paper. *Adv. Mater. 22*, 892–896.

Cook, B.S., and Shamim, A. (2012). Inkjet Printing of Novel Wideband and High Gain Antennas on Low-Cost Paper Substrate. *IEEE Trans. Antennas Propag. 60*, 4148–4156.

Cook, B.S., Fang, Y., Kim, S., Le, T., Goodwin, W.B., Sandhage, K.H., and Tentzeris, M.M. (2013). Inkjet Catalyst Printing and Electroless Copper Deposition for Low-cost Patterned Microwave Passive Devices on Paper. *Electron. Mater. Lett. 9*, 669–676.

Costa, T.H. da, Song, E., Tortorich, R.P., and Choi, J.-W. (2015). A Paper-Based Electrochemical Sensor Using Inkjet-Printed Carbon Nanotube Electrodes. *ECS J. Solid State Sci. Technol. 4*, S3044–S3047.

Cox, H.L. (1952). The Elasticity and Strength of Paper and Other Fibrous Materials. *Br. J. Appl. Phys. 3*, 72.

Crisp, M.T., and Riehle, R.J. (2009). Wet-Strengthening of Paper in Neutral pH Papermaking Conditions. In *Applications of Wet-End Paper Chemistry* (Springer, Dordrecht, the Netherlands), pp. 147–169.

Cui, J., Xi, Y., Chen, S., Li, D., She, X., Sun, J., Han, W., Yang, D., and Guo, S. (2016). Prolifera-Green-Tide as Sustainable Source for Carbonaceous Aerogels with Hierarchical Pore to Achieve Multiple Energy Storage. *Adv. Funct. Mater. 26*, 8487–8495.

Dang, W., Vinciguerra, V., Lorenzelli, L., and Dahiya, R. (2017). Printable Stretchable Interconnects. *Flex. Print. Electron. 2*, 013003.

Derby, B. (2010). Inkjet Printing of Functional and Structural Materials: Fluid Property Requirements, Feature Stability, and Resolution. *Annu. Rev. Mater. Res. 40*, 395–414.

Dichiara, A.B., Song, A., Goodman, S.M., He, D., and Bai, J. (2017). Smart Papers Comprising Carbon Nanotubes and Cellulose Microfibers for Multifunctional Sensing Applications. *J. Mater. Chem. A 5*, 20161–20169.

Dodson, C.T.J., and Sampson, W.W. (1995). The Effect of Paper Formation and Grammage on Its Pore Size Distribution. *Journal of pulp and paper science* 22, no. 5 (1996): J165.

Du, X., Zhang, Z., Liu, W., and Deng, Y. (2017). Nanocellulose-based Conductive Materials and Their Emerging Applications in Energy Devices-A Review. *Nano Energy 35*, 299–320.

El-Sharkawy, K., Haavisto, S., Koskenhely, K., and Paulapuro, H. (2008). Effect of Fiber Flocculation and Filling Design on Refiner Loadability and Refining Characteristics. *BioResources 3*, 403–424.

Fan, J.A., Yeo, W.-H., Su, Y., Hattori, Y., Lee, W., Jung, S.-Y., Zhang, Y., Liu, Z., Cheng, H., Falgout, L., et al. (2014). Fractal Design Concepts for Stretchable Electronics. *Nat. Commun. 5*, 3266.

Fan, Y.J., Meng, X.S., Li, H.Y., Kuang, S.Y., Zhang, L., Wu, Y., Wang, Z.L., and Zhu, G. (2017). Stretchable Porous Carbon Nanotube-Elastomer Hybrid Nanocomposite for Harvesting Mechanical Energy. *Adv. Mater. 29*, 1603115(1–8).

Fraiwan, A., Kwan, L., and Choi, S. (2016). A Disposable Power Source in Resource-limited Environments: A Paper-based Biobattery Generating Electricity from Wastewater. *Biosens. Bioelectron. 85*, 190–197.

French, A.D., and Johnson, G.P. (2007). Cellulose Shapes. In *Cellulose: Molecular and Structural Biology* (Springer, Dordrecht, the Netherlands), pp. 257–284.

Fridley, G.E., Holstein, C.A., Oza, S.B., and Yager, P. (2013). The Evolution of Nitrocellulose as a Material for Bioassays. *MRS Bull. 38*, 326–330.

Ganster, J., and Fink, H.-P. (2013). Cellulose and Cellulose Acetate. In *Bio-Based Plastics* (Wiley-Blackwell, Hoboken, NJ), pp. 35–62.

Gao, Y., and Choi, S. (2017). Stepping Toward Self-Powered Papertronics: Integrating Biobatteries into a Single Sheet of Paper. *Adv. Mater. Technol. 2*, 1600194.

Gao, B., Liu, H., and Gu, Z. (2016). Patterned Photonic Nitrocellulose for Pseudo-Paper Microfluidics. *Anal. Chem. 88*, 5424–5429.

Gere, J.M., and Goodno, B.J. (2008). *Mechanics of Material*, 7th edition (Cengage Learning, Toronto, Canada).

Gharehkhani, S., Sadeghinezhad, E., Kazi, S.N., Yarmand, H., Badarudin, A., Safaei, M.R., and Zubir, M.N.M. (2015). Basic Effects of Pulp Refining on Fiber Properties—A Review. *Carbohydr. Polym. 115*, 785–803.

Gong, S., Schwalb, W., Wang, Y., Chen, Y., Tang, Y., Si, J., Shirinzadeh, B., and Cheng, W. (2014). A Wearable and Highly Sensitive Pressure Sensor with Ultrathin Gold Nanowires. *Nat. Commun. 5*, 3132.

Grey, P., Gaspar, D., Cunha, I., Barras, R., Carvalho, J.T., Ribas, J.R., Fortunato, E., Martins, R., and Pereira, L. (2017). Handwritten Oxide Electronics on Paper. *Adv. Mater. Technol. 2*, 1700009(1–7).

Gullapalli, H., Vemuru, V.S.M., Kumar, A., Botello-Mendez, A., Vajtai, R., Terrones, M., Nagarajaiah, S., and Ajayan, P.M. (2010). Flexible Piezoelectric ZnO–Paper Nanocomposite Strain Sensor. *Small 6*, 1641–1646.

Guo, H., Yeh, M.-H., Zi, Y., Wen, Z., Chen, J., Liu, G., Hu, C., and Wang, Z.L. (2017a). Ultralight Cut-Paper-Based Self-Charging Power Unit for Self-Powered Portable Electronic and Medical Systems. *ACS Nano 11*, 4475–4482.

Guo, R., Chen, J., Yang, B., Liu, L., Su, L., Shen, B., and Yan, X. (2017b). In-Plane Micro-Supercapacitors for an Integrated Device on One Piece of Paper. *Adv. Funct. Mater. 27*, 1702394(1–11).

Habibi, Y., Lucia, L.A., and Rojas, O.J. (2010). Cellulose Nanocrystals: Chemistry, Self-Assembly, and Applications. *Chem. Rev. 110*, 3479–3500.

Hamedi, M.M., Hajian, A., Fall, A.B., Håkansson, K., Salajkova, M., Lundell, F., Wågberg, L., and Berglund, L.A. (2014). Highly Conducting, Strong Nanocomposites Based on Nanocellulose-Assisted Aqueous Dispersions of Single-Wall Carbon Nanotubes. *ACS Nano 8*, 2467–2476.

Hamedi, M.M., Ainla, A., Güder, F., Christodouleas, D.C., Fernández-Abedul, M.T., and Whitesides, G.M. (2016). Integrating Electronics and Microfluidics on Paper. *Adv. Mater. 28*, 5054–5063.

Hammock, M.L., Chortos, A., Tee, B.C.-K., Tok, J.B.-H., and Bao, Z. (2013). 25th Anniversary Article: The Evolution of Electronic Skin (E-Skin): A Brief History, Design Considerations, and Recent Progress. *Adv. Mater. 25*, 5997–6038.

Han, Y.L., Liu, H., Ouyang, C., Lu, T.J., and Xu, F. (2015). Liquid on Paper: Rapid Prototyping of Soft Functional Components for Paper Electronics. *Sci. Rep. 5*, 11488.

Harris, K.D., Elias, A.L., and Chung, H.-J. (2016). Flexible Electronics Under Strain: A Review of Mechanical Characterization and Durability Enhancement Strategies. *J. Mater. Sci. 51*, 2771–2805.

He, M., Zhang, K., Chen, G., Tian, J., and Su, B. (2017). Ionic Gel Paper with Long-Term Bendable Electrical Robustness for Use in Flexible Electroluminescent Devices. *ACS Appl. Mater. Interfaces 9*, 16466–16473.

Heinze, T. (2015). Cellulose: Structure and Properties. In *Cellulose Chemistry and Properties: Fibers, Nanocelluloses and Advanced Materials*, O.J. Rojas, ed. (Springer, Cham, Switzerland), pp. 1–52.

Heldt, C.L., Sieloff, A.K., Merillat, J.P., Minerick, A.R., King, J.A., Perger, W.F., Fukushima, H., and Narendra, J. (2013). Stacked Graphene Nanoplatelet Paper Sensor for Protein Detection. *Sens. Actuators B Chem. 181*, 92–98.

Hines, D.R., Ballarotto, V.W., Williams, E.D., Shao, Y., and Solin, S.A. (2007). Transfer Printing Methods for the Fabrication of Flexible Organic Electronics. *J. Appl. Phys. 101*, 024503.

Hirn, U., and Schennach, R. (2015). Comprehensive Analysis of Individual Pulp Fiber Bonds Quantifies the Mechanisms of Fiber Bonding in Paper. *Sci. Rep. 5*, 10503.

Hu, L., Choi, J.W., Yang, Y., Jeong, S., Mantia, F.L., Cui, L.-F., and Cui, Y. (2009). Highly Conductive Paper for Energy-Storage Devices. *Proc. Natl. Acad. Sci. 106*, 21490–21494.

Hu, L., Wu, H., La Mantia, F., Yang, Y., and Cui, Y. (2010). Thin, Flexible Secondary Li-ion Paper Batteries. *ACS Nano 4*, 5843–5848.

Huang, B., Kang, G., and Ni, Y. (2005). Electrically Conductive Fibre Composites Prepared from Polypyrrole-Engineered Pulp Fibres. *Can. J. Chem. Eng. 83*, 896–903.

Huang, Q., Xu, M., Sun, R., and Wang, X. (2016). Large Scale Preparation of Graphene Oxide/Cellulose Paper with Improved Mechanical Performance and Gas Barrier Properties by Conventional Papermaking Method. *Ind. Crops Prod. 85*, 198–203.

Huang, Y., Zhong, M., Shi, F., Liu, X., Tang, Z., Wang, Y., Huang, Y., Hou, H., Xie, X., and Zhi, C. (2017). An Intrinsically Stretchable and Compressible Supercapacitor Containing a Polyacrylamide Hydrogel Electrolyte. *Angew. Chem. Int. Ed. 56*, 9141–9145.

Hübler, A., Trnovec, B., Zillger, T., Ali, M., Wetzold, N., Mingebach, M., Wagenpfahl, A., Deibel, C., and Dyakonov, V. (2011). Printed Paper Photovoltaic Cells. *Adv. Energy Mater. 1*, 1018–1022.

Hull, D., and Clyne, T.W. (1996). *An Introduction to Composite Materials* (Cambridge University Press, Cambridge, UK).

Hwang, G.-T., Byun, M., Jeong, C.K., and Lee, K.J. (2015). Flexible Piezoelectric Thin-Film Energy Harvesters and Nanosensors for Biomedical Applications. *Adv. Healthc. Mater. 4*, 646–658.

Hyun, W.J., Secor, E.B., Rojas, G.A., Hersam, M.C., Francis, L.F., and Frisbie, C.D. (2015). All-Printed, Foldable Organic Thin-Film Transistors on Glassine Paper. *Adv. Mater. 27*, 7058–7064.

Inui, T., Koga, H., Nogi, M., Komoda, N., and Suganuma, K. (2015). A Miniaturized Flexible Antenna Printed on a High Dielectric Constant Nanopaper Composite. *Adv. Mater. 27*, 1112–1116.

Jabbour, L., Gerbaldi, C., Chaussy, D., Zeno, E., Bodoardo, S., and Beneventi, D. (2010). Microfibrillated Cellulose–graphite Nanocomposites for Highly Flexible Paper-like Li-ion Battery Electrodes. *J. Mater. Chem. 20*, 7344–7347.

Jabbour, L., Chaussy, D., Eyraud, B., and Beneventi, D. (2012). Highly Conductive Graphite/Carbon Fiber/Cellulose Composite Papers. *Compos. Sci. Technol. 72*, 616–623.

Jabbour, L., Bongiovanni, R., Chaussy, D., Gerbaldi, C., and Beneventi, D. (2013). Cellulose-based Li-ion Batteries: A Review. *Cellulose 20*, 1523–1545.

Jenkins, G., Wang, Y., Xie, Y.L., Wu, Q., Huang, W., Wang, L., and Yang, X. (2015). Printed Electronics Integrated with Paper-based Microfluidics: New Methodologies for Next-generation Health Care. *Microfluid. Nanofluidics 19*, 251–261.

Jeong, J.-W., Yeo, W.-H., Akhtar, A., Norton, J.J.S., Kwack, Y.-J., Li, S., Jung, S.-Y. et al. (2013). Materials and Optimized Designs for Human-machine Interfaces via Epidermal Electronics. *Adv. Mater. Deerfield Beach Fla 25*, 6839–6846.

Josefowicz, J.Y., and Deslandes, Y. (1982). Electrical Conductivity of Paper: Measurement Methods and Charge Transport Mechanisms. In *Colloids and Surfaces in Reprographic Technology* (American Chemical Society, Washington, DC), pp. 493–530.

Jung, Y.H., Chang, T.-H., Zhang, H., Yao, C., Zheng, Q., Yang, V.W., Mi, H. et al. (2015). High-performance Green Flexible Electronics based on Biodegradable Cellulose Nanofibril Paper. *Nat. Commun. 6*, 7170.

Kamyshny, A., and Magdassi, S. (2014). Conductive Nanomaterials for Printed Electronics. *Small 10*, 3515–3535.

Kang, T., and Paulapuro, H. (2006). Effect of External Fibrillation on Paper Strength. *Pulp Pap. Can. 107*, 51–54.

Kang, W., Lin, M.-F., Chen, J., and Lee, P.S. (2016). Highly Transparent Conducting Nanopaper for Solid State Foldable Electrochromic Devices. *Small 12*, 6370–6377.

Kerekes, R., and Schell, C.J. (1995). Effects of Fiber Length and Coarseness on Pulp Flocculation. *TAPPI J. 78*, 133–139.

Kim, D.Y., and Steckl, A.J. (2010). Electrowetting on Paper for Electronic Paper Display. *ACS Appl. Mater. Interfaces 2*, 3318–3323.

Kim, D.-H., Kim, Y.-S., Wu, J., Liu, Z., Song, J., Kim, H.-S., Huang, Y.Y., Hwang, K.-C., and Rogers, J.A. (2009). Ultrathin Silicon Circuits With Strain-Isolation Layers and Mesh Layouts for High-Performance Electronics on Fabric, Vinyl, Leather, and Paper. *Adv. Mater. 21*, 3703–3707.

Kim, D.-H., Ghaffari, R., Lu, N., and Rogers, J.A. (2012). Flexible and Stretchable Electronics for Biointegrated Devices. *Annu. Rev. Biomed. Eng. 14*, 113–128.

Kim, J., Yun, S., and Ounaies, Z. (2006). Discovery of Cellulose as a Smart Material. *Macromolecules 39*, 4202–4206.

Kim, S., Yun, T.G., Kang, C., Son, M.-J., Kang, J.-G., Kim, I.-H., Lee, H.-J., An, C.-H., and Hwang, B. (2018). Facile Fabrication of Paper-Based Silver Nanostructure Electrodes for Flexible Printed Energy Storage System. *Mater. Des. 151*, 1–7.

Kim, S.J., Choi, K., Lee, B., Kim, Y., and Hong, B.H. (2015). Materials for Flexible, Stretchable Electronics: Graphene and 2D Materials. *Annu. Rev. Mater. Res. 45*, 63–84.

Klemm, D., Heublein, B., Fink, H.-P., and Bohn, A. (2005). Cellulose: Fascinating Biopolymer and Sustainable Raw Material. *Angew. Chem. Int. Ed Engl. 44*, 3358–3393.

Klemm, D., Kramer, F., Moritz, S., Lindström, T., Ankerfors, M., Gray, D., and Dorris, A. (2011). Nanocelluloses: A New Family of Nature-Based Materials. *Angew. Chem. Int. Ed. 50*, 5438–5466.

Koehly, R., Wanderley, M.M., Ven, T. van de, and Curtil, D. (2014). In-House Development of Paper Force Sensors for Musical Applications. *Comput. Music J. 38*, 22–35.

Koga, H., Tonomura, H., Nogi, M., Suganuma, K., and Nishina, Y. (2016). Fast, Scalable, and Eco-friendly Fabrication of an Energy Storage Paper Electrode. *Green Chem. 18*, 1117–1124.

Krumm, C., Pfaendtner, J., and Dauenhauer, P.J. (2016). Millisecond Pulsed Films Unify the Mechanisms of Cellulose Fragmentation. *Chem. Mater. 28*, 3108–3114.

Kulachenko, A., Denoyelle, T., Galland, S., and Lindström, S.B. (2012). Elastic Properties of Cellulose Nanopaper. *Cellulose 19*, 793–807.

Kumar, V., Forsberg, S., Engström, A.-C., Nurmi, M., Andres, B., Dahlström, C., and Toivakka, M. (2017). Conductive Nanographite–Nanocellulose Coatings on Paper. *Flex. Print. Electron. 2*, 035002.

Kurra, N., and Kulkarni, G.U. (2013). Pencil-on-paper: Electronic Devices. *Lab. Chip 13*, 2866–2873.

Launey, M.E., and Ritchie, R.O. (2009). On the Fracture Toughness of Advanced Materials. *Adv. Mater. 21*, 2103–2110.

Lee, B.-H., Lee, D.-I., Bae, H., Seong, H., Jeon, S.-B., Seol, M.-L., Han, J.-W., Meyyappan, M., Im, S.-G., and Choi, Y.-K. (2016a). Foldable and Disposable Memory on Paper. *Sci. Rep. 6*, 38389.

Lee, H.-B., Bae, C.-W., Duy, L.T., Sohn, I.-Y., Kim, D.-I., Song, Y.-J., Kim, Y.-J., and Lee, N.-E. (2016b). Mogul-Patterned Elastomeric Substrate for Stretchable Electronics. *Adv. Mater. 28*, 3069–3077.

Lee, S., Seong, H., Im, S.G., Moon, H., and Yoo, S. (2017). Organic Flash Memory on Various Flexible Substrates for Foldable and Disposable Electronics. *Nat. Commun. 8*, 725.

Lessing, J., Glavan, A.C., Walker, S.B., Keplinger, C., Lewis, J.A., and Whitesides, G.M. (2014). Inkjet Printing of Conductive Inks with High Lateral Resolution on Omniphobic "RF Paper" for Paper-Based Electronics and MEMS. *Adv. Mater. 26*, 4677–4682.

Li, X., and Liu, X. (2014). Hydrothermal Growth of ZnO Nanowires on Paper for Flexible Electronics. In *14th IEEE International Conference on Nanotechnology*, pp. 981–985.

Li, H., Qian, X., Li, T., and Ni, Y. (2015a). Percolation for Coated Conductive Paper: Electrical Conductivity as a Function of Volume Fraction of Graphite and Carbon Black. *BioResources 10*, 4877–4885.

Li, R.-Z., Hu, A., Zhang, T., and Oakes, K.D. (2014). Direct Writing on Paper of Foldable Capacitive Touch Pads with Silver Nanowire Inks. *ACS Appl. Mater. Interfaces 6*, 21721–21729.

Li, S., Ren, G., Hoque, M.N.F., Dong, Z., Warzywoda, J., and Fan, Z. (2017a). Carbonized Cellulose Paper as an Effective Interlayer in Lithium-sulfur Batteries. *Appl. Surf. Sci. 396*, 637–643.

Li, X., Zhao, C., and Liu, X. (2015b). A Paper-based Microfluidic Biosensor Integrating Zinc Oxide Nanowires for Electrochemical Glucose Detection. *Microsyst. Nanoeng. 1*, 15014.

Li, Y., Zhu, H., Wang, Y., Ray, U., Zhu, S., Dai, J., Chen, C., Fu, K., Jang, S.-H., Henderson, D., et al. (2017b). Cellulose-Nanofiber-Enabled 3D Printing of a Carbon-Nanotube Microfiber Network. *Small Methods 1*, 1700222(1–8).

Li, Y.-Q., Zhu, W.-B., Yu, X.-G., Huang, P., Fu, S.-Y., Hu, N., and Liao, K. (2016). Multifunctional Wearable Device Based on Flexible and Conductive Carbon Sponge/Polydimethylsiloxane Composite. *ACS Appl. Mater. Interfaces 8*, 33189–33196.

Liana, D.D., Raguse, B., Gooding, J.J., and Chow, E. (2016). An Integrated Paper-Based Readout System and Piezoresistive Pressure Sensor for Measuring Bandage Compression. *Adv. Mater. Technol. 1*, 1600143.

Liang, H.-W., Guan, Q.-F., Zhu, Z., Song, L.-T., Yao, H.-B., Lei, X., and Yu, S.-H. (2012). Highly Conductive and Stretchable Conductors Fabricated from Bacterial Cellulose. *NPG Asia Mater. 4*, e19.

Liang, T., Zou, X., and Mazzeo, A.D. (2016). A Flexible Future for Paper-based Electronics. *Micro-and Nanotechnology Sensors, Systems, and Applications VIII*. (International Society for Optics and Photonics), pp. 98361D(1–14).

Liao, X., Liao, Q., Yan, X., Liang, Q., Si, H., Li, M., Wu, H., Cao, S., and Zhang, Y. (2015). Flexible and Highly Sensitive Strain Sensors Fabricated by Pencil Drawn for Wearable Monitor. *Adv. Funct. Mater. 25*, 2395–2401.

Liao, X., Zhang, Z., Liang, Q., Liao, Q., and Zhang, Y. (2017). Flexible, Cuttable, and Self-Waterproof Bending Strain Sensors Using Microcracked Gold Nanofilms@ Paper Substrate. *ACS Appl. Mater. Interfaces 9*, 4151–4158.

Lien, D.-H., Kao, Z.-K., Huang, T.-H., Liao, Y.-C., Lee, S.-C., and He, J.-H. (2014). All-Printed Paper Memory. *ACS Nano 8*, 7613–7619.

Lin, C.-H., Tsai, D.-S., Wei, T.-C., Lien, D.-H., Ke, J.-J., Su, C.-H., Sun, J.-Y., Liao, Y.-C., and He, J.-H. (2017). Highly Deformable Origami Paper Photodetector Arrays. *ACS Nano 11*, 10230–10235.

Lin, C.-W., Zhao, Z., Kim, J., and Huang, J. (2014). Pencil Drawn Strain Gauges and Chemiresistors on Paper. *Sci. Rep. 4*, 3812.

Lin, Y., Gritsenko, D., Liu, Q., Lu, X., and Xu, J. (2016). Recent Advancements in Functionalized Paper-Based Electronics. *ACS Appl. Mater. Interfaces 8*, 20501–20515.

Liu, H., and Crooks, R.M. (2012). Paper-Based Electrochemical Sensing Platform with Integral Battery and Electrochromic Read-Out. *Anal. Chem. 84*, 2528–2532.

Liu, H., Qing, H., Li, Z., Han, Y.L., Lin, M., Yang, H., Li, A., Lu, T.J., Li, F., and Xu, F. (2017a). Paper: A Promising Material for Human-friendly Functional Wearable Electronics. *Mater. Sci. Eng. R Rep. 112*, 1–22.

Liu, H., Jiang, H., Du, F., Zhang, D., Li, Z., and Zhou, H. (2017b). Flexible and Degradable Paper-Based Strain Sensor with Low Cost. *ACS Sustain. Chem. Eng. 5*, 10538–10543.

Liu, X., Mwangi, M., Li, X., O'Brien, M., and Whitesides, G.M. (2011). Paper-based Piezoresistive MEMS Sensors. *Lab. Chip 11*, 2189–2196.

Lu, N., and Kim, D.-H. (2013). Flexible and Stretchable Electronics Paving the Way for Soft Robotics. *Soft Robot. 1*, 53–62.

Lu, A., Dai, M., Sun, J., Jiang, J., and Wan, Q. (2011). Flexible Low-Voltage Electric-Double-Layer TFTs Self-Assembled on Paper Substrates. *IEEE Electron Device Lett. 32*, 518–520.

Luong, N.D., Pahimanolis, N., Hippi, U., Korhonen, J.T., Ruokolainen, J., Johansson, L.-S., Nam, J.-D., and Seppälä, J. (2011). Graphene/cellulose Nanocomposite Paper with High Electrical and Mechanical Performances. *J. Mater. Chem. 21*, 13991–13998.

Lv, Z., Luo, Y., Tang, Y., Wei, J., Zhu, Z., Zhou, X., Li, W., Zeng, Y., Zhang, W., Zhang, Y., et al. (2018). Editable Supercapacitors with Customizable Stretchability Based on Mechanically Strengthened Ultralong MnO_2 Nanowire Composite. *Adv. Mater. 30*, 1704531(1–9).

Marinho, B., Ghislandi, M., Tkalya, E., Koning, C.E., and de With, G. (2012). Electrical Conductivity of Compacts of Graphene, Multi-wall Carbon Nanotubes, Carbon Black, and Graphite Powder. *Powder Technol. 221*, 351–358.

Martins, R., Ferreira, I., and Fortunato, E. (2011a). Electronics with and on Paper. *Phys. Status Solidi RRL – Rapid Res. Lett. 5*, 332–335.

Martins, R., Nathan, A., Barros, R., Pereira, L., Barquinha, P., Correia, N., Costa, R., Ahnood, A., Ferreira, I., and Fortunato, E. (2011b). Complementary Metal Oxide Semiconductor Technology With and On Paper. *Adv. Mater. 23*, 4491–4496.

Martins, R.F.P., Ahnood, A., Correia, N., Pereira, L.M.N.P., Barros, R., Barquinha, P.M.C.B., Costa, R., Ferreira, I.M.M., Nathan, A., and Fortunato, E.E.M.C. (2013). Recyclable, Flexible, Low-Power Oxide Electronics. *Adv. Funct. Mater. 23*, 2153–2161.

Maxwell, E.J., Mazzeo, A.D., and Whitesides, G.M. (2013). Paper-based Electroanalytical Devices for Accessible Diagnostic Testing. *MRS Bull. 38*, 309–314.

Mazzeo, A.D., Kalb, W.B., Chan, L., Killian, M.G., Bloch, J.-F., Mazzeo, B.A., and Whitesides, G.M. (2012). Paper-Based, Capacitive Touch Pads. *Adv. Mater. 24*, 2850–2856.

Mirica, K.A., Weis, J.G., Schnorr, J.M., Esser, B., and Swager, T.M. (2012). Mechanical Drawing of Gas Sensors on Paper. *Angew. Chem. Int. Ed. 51*, 10740–10745.

Mitra, K.Y., Polomoshnov, M., Martínez-Domingo, C., Mitra, D., Ramon, E., and Baumann, R.R. (2017). Fully Inkjet-Printed Thin-Film Transistor Array Manufactured on Paper Substrate for Cheap Electronic Applications. *Adv. Electron. Mater. 3*, 1700275(1–9).

Moon, R.J., Martini, A., Nairn, J., Simonsen, J., and Youngblood, J. (2011). Cellulose Nanomaterials Review: Structure, Properties and Nanocomposites. *Chem. Soc. Rev. 40*, 3941–3994.

Mortensen, A. (2006). *Concise Encyclopedia of Composite Materials* (Elsevier, Amsterdam, the Netherlands).

Nassar, J.M., and Hussain, M.M. (2017). Impact of Physical Deformation on Electrical Performance of Paper-Based Sensors. *IEEE Trans. Electron Devices 64*, 2022–2029.

Nguyen, T.H., Fraiwan, A., and Choi, S. (2014). Paper-based Batteries: A Review. *Biosens. Bioelectron. 54*, 640–649.

Nie, Z., Nijhuis, C.A., Gong, J., Chen, X., Kumachev, A., Martinez, A.W., Narovlyansky, M., and Whitesides, G.M. (2010). Electrochemical Sensing in Paper-based Microfluidic Devices. *Lab. Chip 10*, 477–483.

Nishino, T., Takano, K., and Nakamae, K. (1995). Elastic Modulus of the Crystalline Regions of Cellulose Polymorphs. *J. Polym. Sci. Part B Polym. Phys.* *33*, 1647–1651.

Nissan, A.H. (1967). The significance of hydrogen bonding at the surfaces of cellulose network structures. Surface and Coatings Related to Paper and Wood (Syrause, New York: Syracuse University Press), pp. 221–268.

Nogi, M., Komoda, N., Otsuka, K., and Suganuma, K. (2013). Foldable Nanopaper Antennas for Origami Electronics. *Nanoscale 5*, 4395–4399.

Nogi, M., Karakawa, M., Komoda, N., Yagyu, H., and Nge, T.T. (2015). Transparent Conductive Nanofiber Paper for Foldable Solar Cells. *Sci. Rep. 5*, 17254.

Numakura, D. (2008). Advanced Screen Printing "Practical Approaches for Printable & Flexible Electronics." In *2008 3rd International Microsystems, Packaging, Assembly Circuits Technology Conference*, pp. 205–208.

Page, D. (1969). A Theory for the Tensile Strength of Paper. *Tappi 52*, 674–681.

Park, J.S., Kim, T., and Kim, W.S. (2017). Conductive Cellulose Composites with Low Percolation Threshold for 3D Printed Electronics. *Sci. Rep. 7*, 3246.

Park, S.-I., Ahn, J.-H., Feng, X., Wang, S., Huang, Y., and Rogers, J.A. (2008). Theoretical and Experimental Studies of Bending of Inorganic Electronic Materials on Plastic Substrates. *Adv. Funct. Mater. 18*, 2673–2684.

Pattinson, S.W., and Hart, A.J. (2017). Additive Manufacturing of Cellulosic Materials with Robust Mechanics and Antimicrobial Functionality. *Adv. Mater. Technol. 2*, 1600084.

Pini, V., Ruz, J.J., Kosaka, P.M., Malvar, O., Calleja, M., and Tamayo, J. (2016). How Two-dimensional Bending Can Extraordinarily Stiffen Thin Sheets. *Sci. Rep. 6*, 29627.

Piqué, A., Chrisey, D.B., Auyeung, R.C.Y., Fitz-Gerald, J., Wu, H.D., McGill, R.A., Lakeou, S., Wu, P.K., Nguyen, V., and Duignan, M. (1999). A Novel Laser Transfer Process for Direct Writing of Electronic and Sensor Materials. *Appl. Phys. A 69*, S279–S284.

Posthuma-Trumpie, G.A., and Amerongen, A. van (2012). Lateral Flow Assays. In *Antibodies: Applications and New Developments* (Bentham Science Publishers, London, UK), pp. 175–183.

Prambauer, M., Paulik, C., and Burgstaller, C. (2015). The Influence of Paper Type on the Properties of Structural Paper–Polypropylene Composites. *Compos. Part Appl. Sci. Manuf. 74*, 107–113.

Pushparaj, V.L., Shaijumon, M.M., Kumar, A., Murugesan, S., Ci, L., Vajtai, R., Linhardt, R.J., Nalamasu, O., and Ajayan, P.M. (2007). Flexible Energy Storage Devices based on Nanocomposite Paper. *Proc. Natl. Acad. Sci. 104*, 13574–13577.

Ras, R.H.A., Tian, X., and Bayer, I.S. (2017). Superhydrophobic and Superoleophobic Nanostructured Cellulose and Cellulose Composites. In *Handbook of Nanocellulose and Cellulose Nanocomposites*, H. Kargarzadeh, I. Ahmad, S. Thomas, and A. Dufresne, eds. (Wiley-VCH Verlag GmbH & Co. KGaA, Weinheim, Germany), pp. 731–760.

Roberts, J.C. (1996). Chemical Additives in the Paper Formation Process. In *The Chemistry of Paper*, pp. 109–140.

Roberts, J.C. (2007). The Chemistry of Paper (Royal Society of Chemistry). (Royal Society of Chemistry, Cambridge, UK).

Rogers, J.A., Someya, T., and Huang, Y. (2010). Materials and Mechanics for Stretchable Electronics. *Science 327*, 1603–1607.

Rojas, J., Bedoya, M., and Ciro, Y. (2015a). Current Trends in the Production of Cellulose Nanoparticles and Nanocomposites for Biomedical Applications. *Cellulose-Fundamental Aspects and Current Trends* (Intech, Vigo, Spain) pp. 193–228.

Rojas, J.P., Torres Sevilla, G.A., Alfaraj, N., Ghoneim, M.T., Kutbee, A.T., Sridharan, A., and Hussain, M.M. (2015b). Nonplanar Nanoscale Fin Field Effect Transistors on Textile, Paper, Wood, Stone, and Vinyl via Soft Material-Enabled Double-Transfer Printing. *ACS Nano 9*, 5255–5263.

Rushton, M. (2017). Stretchable Paper—Yes, Really! Tappi *Paper 360°*, 20–21.

Russo, A., Ahn, B.Y., Adams, J.J., Duoss, E.B., Bernhard, J.T., and Lewis, J.A. (2011). Pen-on-Paper Flexible Electronics. *Adv. Mater. 23*, 3426–3430.

Rusu, M., Mörseburg, K., Gregersen, Ø., Yamakawa, A., and Liukkonen, S. (2011). Relation between Fibre Flexibility and Cross Sectional Properties. *BioResources 6*, 641–655.

Rydholm, S.A. (1965). *Pulping Processes*. (Interscience Publishers, New York).

Sadri, B., Goswami, D., Sala de Medeiros, M., Pal, A., Castro, B., Kuang, S., and Martinez, R.V. (2018). Wearable and Implantable Epidermal Paper-Based Electronics. *ACS Appl. Mater. Interfaces 10*, 31061–31068.

Sahin, H.T., and Arslan, M.B. (2008). A Study on Physical and Chemical Properties of Cellulose Paper Immersed in Various Solvent Mixtures. *Int. J. Mol. Sci. 9*, 78–88.

Saito, T., and Isogai, A. (2005). Novel Method to Improve Wet Strength of Paper. *Tappi J 4*, 3–8.

Saito, T., Kimura, S., Nishiyama, Y., and Isogai, A. (2007). Cellulose Nanofibers Prepared by TEMPO-Mediated Oxidation of Native Cellulose. *Biomacromolecules 8*, 2485–2491.

Sanchez-Romaguera, V., Wünscher, S., Turki, B.M., Abbel, R., Barbosa, S., Tate, D.J., Oyeka, D. et al. (2015). Inkjet Printed Paper Based Frequency Selective Surfaces and Skin Mounted RFID Tags: The Interrelation between Silver Nanoparticle Ink, Paper Substrate and Low Temperature Sintering Technique. *J. Mater. Chem. C 3*, 2132–2140.

Santhiago, M., Corrêa, C.C., Bernardes, J.S., Pereira, M.P., Oliveira, L.J.M., Strauss, M., and Bufon, C.C.B. (2017). Flexible and Foldable Fully-Printed Carbon Black Conductive Nanostructures on Paper for High-Performance Electronic, Electrochemical, and Wearable Devices. *ACS Appl. Mater. Interfaces 9*, 24365–24372.

Sathishkumar, T., Navaneethakrishnan, P., Shankar, S., Rajasekar, R., and Rajini, N. (2013). Characterization of Natural Fiber and Composites–A Review. *J. Reinf. Plast. Compos. 32*, 1457–1476.

Scheller, H.V., and Ulvskov, P. (2010). Hemicelluloses. *Annu. Rev. Plant Biol. 61*, 263–289.

Sehaqui, H., Zhou, Q., Ikkala, O., and Berglund, L.A. (2011). Strong and Tough Cellulose Nanopaper with High Specific Surface Area and Porosity. *Biomacromolecules 12*, 3638–3644.

Sehaqui, H., Morimune, S., Nishino, T., and Berglund, L.A. (2012). Stretchable and Strong Cellulose Nanopaper Structures Based on Polymer-Coated Nanofiber Networks: An Alternative to Nonwoven Porous Membranes from Electrospinning. *Biomacromolecules 13*, 3661–3667.

Shen, J., Song, Z., Qian, X., and Ni, Y. (2011). A Review on Use of Fillers in Cellulosic Paper for Functional Applications. *Ind. Eng. Chem. Res. 50*, 661–666.

Shin, H., Yoon, B., Park, I.S., and Kim, J.-M. (2014). An Electrothermochromic Paper Display Based on Colorimetrically Reversible Polydiacetylenes. *Nanotechnology 25*, 094011.

Shyu, T.C., Damasceno, P.F., Dodd, P.M., Lamoureux, A., Xu, L., Shlian, M., Shtein, M., Glotzer, S.C., and Kotov, N.A. (2015). A Kirigami Approach to Engineering Elasticity in Nanocomposites through Patterned Defects. *Nat. Mater. 14*, 785.

Siegel, A.C., Phillips, S.T., Wiley, B.J., and Whitesides, G.M. (2009). Thin, Lightweight, Foldable Thermochromic Displays on Oaper. *Lab. Chip 9*, 2775–2781.

Siegel, A.C., Phillips, S.T., Dickey, M.D., Lu, N., Suo, Z., and Whitesides, G.M. (2010). Foldable Printed Circuit Boards on Paper Substrates. *Adv. Funct. Mater. 20*, 28–35.

Silverberg, J.L., Evans, A.A., McLeod, L., Hayward, R.C., Hull, T., Santangelo, C.D., and Cohen, I. (2014). Using Origami Design Principles to Fold Reprogrammable Mechanical Metamaterials. *Science 345*, 647–650.

Siqueira, G., Kokkinis, D., Libanori, R., Hausmann, M.K., Gladman, A.S., Neels, A., Tingaut, P., Zimmermann, T., Lewis, J.A., and Studart, A.R. (2017). Cellulose Nanocrystal Inks for 3D Printing of Textured Cellular Architectures. *Adv. Funct. Mater. 27*, 1604619.

Sofla, M.R.K., Brown, R.J., Tsuzuki, T., and Rainey, T.J. (2016). A Comparison of Cellulose Nanocrystals and Cellulose Nanofibres Extracted from Bagasse using Acid and Ball Milling Methods. *Adv. Nat. Sci. Nanosci. Nanotechnol. 7*, 035004.

Song, Y., Jiang, Y., Shi, L., Cao, S., Feng, X., Miao, M., and Fang, J. (2015). Solution-processed Assembly of Ultrathin Transparent Conductive Cellulose Nanopaper Embedding AgNWs. *Nanoscale 7*, 13694–13701.

Stoppa, M., and Chiolerio, A. (2014). Wearable Electronics and Smart Textiles: A Critical Review. *Sensors 14*, 11957–11992.

Sundriyal, P., and Bhattacharya, S. (2017). Inkjet-Printed Electrodes on A4 Paper Substrates for Low-Cost, Disposable, and Flexible Asymmetric Supercapacitors. *ACS Appl. Mater. Interfaces 9*, 38507–38521.

Suo, Z. (2012). Mechanics of Stretchable Electronics and Soft Machines. *MRS Bull. 37*, 218–225.

Taipale, T., Österberg, M., Nykänen, A., Ruokolainen, J., and Laine, J. (2010). Effect of Microfibrillated Cellulose and Fines on the Drainage of Kraft Pulp Suspension and Paper Strength. *Cellulose 17*, 1005–1020.

Tao, L.-Q., Zhang, K.-N., Tian, H., Liu, Y., Wang, D.-Y., Chen, Y.-Q., Yang, Y., and Ren, T.-L. (2017). Graphene-Paper Pressure Sensor for Detecting Human Motions. *ACS Nano 11*, 8790–8795.

Tashiro, K., and Kobayashi, M. (1991). Theoretical Evaluation of Three-Dimensional Elastic Constants of Native and Regenerated Celluloses: Role of Hydrogen Bonds. *Polymer 32*, 1516–1526.

Tejado, A., and van de Ven, T.G.M. (2010). Why Does Paper Get Stronger as it Dries? *Mater. Today 13*, 42–49.

Thi Nge, T., Nogi, M., and Suganuma, K. (2013). Electrical Functionality of Inkjet-printed Silver Nanoparticle Conductive Tracks on Nanostructured Paper Compared with those on Plastic Substrates. *J. Mater. Chem. C 1*, 5235–5243.

Tobjörk, D., and Österbacka, R. (2011). Paper Electronics. *Adv. Mater. 23*, 1935–1961.

Trache, D., Hussin, M.H., Haafiz, M.K.M., and Thakur, V.K. (2017). Recent Progress in Cellulose Nanocrystals: Sources and Production. *Nanoscale 9*, 1763–1786.

Wan, C., Jiao, Y., and Li, J. (2017). Flexible, Highly Conductive, and Free-standing Reduced Graphene Oxide/polypyrrole/cellulose Hybrid Papers for Supercapacitor Electrodes. *J. Mater. Chem. A 5*, 3819–3831.

Wang, C.-C., Hennek, J.W., Ainla, A., Kumar, A.A., Lan, W.-J., Im, J., Smith, B.S., Zhao, M., and Whitesides, G.M. (2016a). A Paper-Based "Pop-up" Electrochemical Device for Analysis of Beta-Hydroxybutyrate. *Anal. Chem. 88*, 6326–6333.

Wang, D., Song, P., Liu, C., Wu, W., and Fan, S. (2008). Highly Oriented Carbon Nanotube Papers made of Aligned Carbon Nanotubes. *Nanotechnology 19*, 075609.

Wang, G., Xu, W., Xu, F., Shen, W., and Song, W. (2017). AgNW/Chinese Xuan Paper Film Heaters for Electro-thermochromic Paper Display. *Mater. Res. Express 4*, 116405.

Wang, J., Auyeung, R.C.Y., Kim, H., Charipar, N.A., and Piqué, A. (2010). Three-Dimensional Printing of Interconnects by Laser Direct-Write of Silver Nanopastes. *Adv. Mater. 22*, 4462–4466.

Wang, Y., Guo, H., Chen, J., Sowade, E., Wang, Y., Liang, K., Marcus, K., Baumann, R.R., and Feng, Z. (2016b). Paper-Based Inkjet-Printed Flexible Electronic Circuits. *ACS Appl. Mater. Interfaces 8*, 26112–26118.

Wang, Z., Tammela, P., Huo, J., Zhang, P., Strømme, M., and Nyholm, L. (2016c). Solution-processed poly(3,4-ethylenedioxythiophene) Nanocomposite Paper Electrodes for High-capacitance Flexible Supercapacitors. *J. Mater. Chem. A 4*, 1714–1722.

Wathén, R. (2006). Studies on Fiber Strength and its Effect on Paper Properties (Helsinki University of Technology, Espoo, Finland).

Wei, Y., Chen, S., Li, F., Lin, Y., Zhang, Y., and Liu, L. (2015). Highly Stable and Sensitive Paper-Based Bending Sensor Using Silver Nanowires/Layered Double Hydroxides Hybrids. *ACS Appl. Mater. Interfaces 7*, 14182–14191.

Weng, Z., Su, Y., Wang, D.-W., Li, F., Du, J., and Cheng, H.-M. (2011). Graphene–Cellulose Paper Flexible Supercapacitors. *Adv. Energy Mater. 1*, 917–922.

Wistara, N., and Young, R.A. (1999). Properties and Treatments of Pulps from Recycled Paper. Part I. Physical and Chemical Properties of Pulps. *Cellulose 6*, 291–324.

Wu, C., Wang, X., Lin, L., Guo, H., and Wang, Z.L. (2016). Paper-Based Triboelectric Nanogenerators Made of Stretchable Interlocking Kirigami Patterns. *ACS Nano 10*, 4652–4659.

Xie, J., Chen, Q., Suresh, P., Roy, S., White, J.F., and Mazzeo, A.D. (2017). Paper-based Plasma Sanitizers. *Proc. Natl. Acad. Sci. 114*, 5119–5124.

Xu, J., Wang, S., Wang, G.-J.N., Zhu, C., Luo, S., Jin, L., Gu, X. et al. (2017a). Highly Stretchable Polymer Semiconductor Films through the Nanoconfinement Effect. *Science 355*, 59–64.

Xu, L., Shyu, T.C., and Kotov, N.A. (2017b). Origami and Kirigami Nanocomposites. *ACS Nano 11*, 7587–7599.

Xu, X., Zhou, J., Jiang, L., Lubineau, G., Ng, T., Ooi, B.S., Liao, H.-Y., Shen, C., Chen, L., and Zhu, J.Y. (2016). Highly Transparent, Low-haze, Hybrid Cellulose Nanopaper as Electrodes for Flexible Electronics. *Nanoscale 8*, 12294–12306.

Yamada, T., Hayamizu, Y., Yamamoto, Y., Yomogida, Y., Izadi-Najafabadi, A., Futaba, D.N., and Hata, K. (2011). A Stretchable Carbon Nanotube Strain Sensor for Human-motion Detection. *Nat. Nanotechnol. 6*, 296–301.

Yamashita, Y., and Endo, T. (2004). Deterioration Behavior of Cellulose Acetate Films in Acidic or Basic Aqueous Solutions. *J. Appl. Polym. Sci. 91*, 3354–3361.

Yan, C., Wang, J., Kang, W., Cui, M., Wang, X., Foo, C.Y., Chee, K.J., and Lee, P.S. (2014). Highly Stretchable Piezoresistive Graphene–Nanocellulose Nanopaper for Strain Sensors. *Adv. Mater. 26*, 2022–2027.

Yang, H., Yan, R., Chen, H., Lee, D.H., and Zheng, C. (2007). Characteristics of Hemicellulose, Cellulose and Lignin Pyrolysis. *Fuel 86*, 1781–1788.

Yang, P.-K., Lin, Z.-H., Pradel, K.C., Lin, L., Li, X., Wen, X., He, J.-H., and Wang, Z.L. (2015a). Paper-Based Origami Triboelectric Nanogenerators and Self-Powered Pressure Sensors. *ACS Nano 9*, 901–907.

Yang, S., Chen, Y.-C., Nicolini, L., Pasupathy, P., Sacks, J., Su, B., Yang, R., Sanchez, D., Chang, Y.-F., Wang, P., et al. (2015b). "Cut-and-Paste" Manufacture of Multiparametric Epidermal Sensor Systems. *Adv. Mater. 27*, 6423–6430.

Yang, S., Choi, I.-S., and Kamien, R.D. (2016). Design of Super-conformable, Foldable Materials via Fractal Cuts and Lattice Kirigami. *MRS Bull. 41*, 130–138.

Ye, T.-N., Feng, W.-J., Zhang, B., Xu, M., Lv, L.-B., Su, J., Wei, X., Wang, K.-X., Li, X.-H., and Chen, J.-S. (2015). Converting Waste Paper to Multifunctional Graphene-decorated Carbon Paper: From Trash to Treasure. *J. Mater. Chem. A 3*, 13926–13932.

Yin, Z., Huang, Y., Bu, N., Wang, X., and Xiong, Y. (2010). Inkjet Printing for Flexible Electronics: Materials, Processes and Equipments. *Chin. Sci. Bull. 55*, 3383–3407.

Yuan, L., Xiao, X., Ding, T., Zhong, J., Zhang, X., Shen, Y., Hu, B., Huang, Y., Zhou, J., and Wang, Z.L. (2012). Paper-based Supercapacitors for Self-powered Nanosystems. *Angew. Chem. Int. Ed. 51*, 4934–4938.

Yuan, L., Yao, B., Hu, B., Huo, K., Chen, W., and Zhou, J. (2013). Polypyrrole-coated Paper for Flexible Solid-state Energy Storage. *Energy Environ. Sci. 6*, 470–476.

Yuan, Z., Peng, H.-J., Huang, J.-Q., Liu, X.-Y., Wang, D.-W., Cheng, X.-B., and Zhang, Q. (2014). Hierarchical Free-standing Carbon-Nanotube Paper Electrodes with Ultrahigh Sulfur-loading for Lithium-Sulfur Batteries. *Adv. Funct. Mater. 24*, 6105–6112.

Yuen, J.D., Walper, S.A., Melde, B.J., Daniele, M.A., and Stenger, D.A. (2017). Electrolyte-Sensing Transistor Decals Enabled by Ultrathin Microbial Nanocellulose. *Sci. Rep. 7*, 40867.

Yun, S., Jang, S.-D., Yun, G.-Y., Kim, J.-H., and Kim, J. (2009). Paper Transistor made with Covalently Bonded Multiwalled Carbon Nanotube and Cellulose. *Appl. Phys. Lett. 95*, 104102.

Zauscher, S., Caulfield, D.F., and Nissan, A.H. (1996). The Influence of Water on the Elastic Modulus of Paper. I. Extension of the H-bond Theory. *Tappi J. 79*, 178–182.

Zhang, G., Liao, Q., Zhang, Z., Liang, Q., Zhao, Y., Zheng, X., and Zhang, Y. (2016a). Novel Piezoelectric Paper-Based Flexible Nanogenerators Composed of $BaTiO_3$ Nanoparticles and Bacterial Cellulose. *Adv. Sci. 3*, 1500257.

Zhang, H., Yuan, Z., Gilbert, D., Ni, Y., and Zou, X. (2011). Use of a Dynamic Sheet Former (DSF) to Examine the Effect of Filler Addition and White Water Recirculation on Fine Papers Containing High-yield Pulp. *BioResources 6*, 5099–5109.

Zhang, J., Lee, G.-Y., Cerwyn, C., Yang, J., Fondjo, F., Kim, J.-H., Taya, M., Gao, D., and Chung, J.-H. (2018a). Fracture-Induced Mechanoelectrical Sensitivities of Paper-Based Nanocomposites. *Adv. Mater. Technol. 3*, 1700266.

Zhang, M., Hou, C., Halder, A., Wang, H., and Chi, Q. (2016c). Graphene Papers: Smart Architecture and Specific Functionalization for Biomimetics, Electrocatalytic Sensing and Energy Storage. *Mater. Chem. Front. 1*, 37–60.

Zhang, T., Cai, X., Liu, J., Hu, M., Guo, Q., and Yang, J. (2017a). Facile Fabrication of Hybrid Copper–Fiber Conductive Features with Enhanced Durability and Ultralow Sheet Resistance for Low-cost High-Performance Paper-based Electronics. *Adv. Sustain. Syst. 1*, 1700062.

Zhang, W., Jing, Z., Shan, Y., Ge, X., Mu, X., Jiang, Y., Li, H., and Wu, P. (2016b). Paper Reinforced with Regenerated Cellulose: A Sustainable and Fascinating Material with Good Mechanical Performance, Barrier Properties and Shape Retention in Water. *J. Mater. Chem. A 4*, 17483–17490.

Zhang, X., Lu, Z., Zhao, J., Li, Q., Zhang, W., and Lu, C. (2017b). Exfoliation/Dispersion of Low-Temperature Expandable Graphite in Nanocellulose Matrix by Wet Co-milling. *Carbohydr. Polym. 157*, 1434–1441.

Zhang, Y., Zhang, L., Cui, K., Ge, S., Cheng, X., Yan, M., Yu, J., and Liu, H. (2018b). Flexible Electronics Based on Micro/Nanostructured Paper. *Adv. Mater.* 1801588.

Zhao, R., Lin, S., Yuk, H., and Zhao, X. (2018). Kirigami Enhances Film Adhesion. *Soft Matter 14*, 2515–2525.

Zheng, Y., He, Z., Gao, Y., and Liu, J. (2013). Direct Desktop Printed-Circuits-on-Paper Flexible Electronics. *Sci. Rep. 3*.

Zhong, Q., Zhong, J., Hu, B., Hu, Q., Zhou, J., and Lin Wang, Z. (2013). A Paper-based Nanogenerator as a Power Source and Active Sensor. *Energy Environ. Sci. 6*, 1779–1784.

Zhu, H., Fang, Z., Preston, C., Li, Y., and Hu, L. (2013). Transparent Paper: Fabrications, Properties, and Device Applications. *Energy Environ. Sci. 7*, 269–287.

Zhu, H., Zhu, S., Jia, Z., Parvinian, S., Li, Y., Vaaland, O., Hu, L., and Li, T. (2015). Anomalous Scaling Law of Strength and Toughness of Cellulose Nanopaper. *Proc. Natl. Acad. Sci. 112*, 8971–8976.

Zhu, Y., Cheng, S., Zhou, W., Jia, J., Yang, L., Yao, M., Wang, M., Wu, P., Luo, H., and Liu, M. (2017). Porous Functionalized Self-Standing Carbon Fiber Paper Electrodes for High-Performance Capacitive Energy Storage. *ACS Appl. Mater. Interfaces 9*, 13173–13180.

Zocco, A.T., You, H., Hagen, J.A., and Steckl, A.J. (2014). Pentacene Organic Thin-film Transistors on Flexible Paper and Glass Substrates. *Nanotechnology 25*, 094005.

Zou, X., Liang, T., Lopez, N., Ahmed, M., Ajayan, A., and Mazzeo, A.D. (2017). Arrayed Force Sensors Made of Paper, Elastomer, and Hydrogel Particles. *Micromachines 8*, 356.

Zou, X., Chen, C., Liang, T., Xie, J., Gillette-Henao, E.-N., Oh, J., Tumalle, J., and Mazzeo, A.D. (2018). Paper-based Resistive Networks for Scalable Skin-like Sensing. *Adv. Electron. Mater. 4*, 1800131.

16

Reliability Assessment of Low-Temperature ZnO-Based Thin-Film Transistors

Chadwin D. Young,
Rodolfo A.
Rodriguez-Davila,
Pavel Bolshakov,
Richard A. Chapman,
and Manuel
Quevedo-Lopez

16.1 Introduction

Large-area/flexible electronics for Internet of Things (IoT) applications have garnered significant consideration recently. In addition, oxide-based semiconductors, namely, zinc oxide-based (i.e., ZnO and indium gallium zinc oxide, IGZO) materials, are attractive for use in flexible electronics due to their current or potential use in display technology, their transparency, and because they can be processed at low temperature, which is compatible with large-area/flex technology needs. Furthermore, these semiconductors have demonstrated noteworthy thin-film transistor (TFT) performance [1–3] when compared to organic-based semiconductors. Thus, ZnO-based semiconductors as potential candidates for use as an active layer in TFTs in flexible circuitry [3] are a viable option. Although this is the case, low-temperature deposition parameter impact of ultra-thin semiconducting ZnO-based layers on TFT performance is required. In addition, other critical materials are required to fabricate a TFT, such as the gate dielectric—a critical component of any metal-oxide-semiconductor field effect transistor (MOSFET). Currently in mainstream silicon-based transistor technologies, optimized high-k gate dielectrics, such as hafnium-based films are used. However, early hafnium dioxide (HfO_2) gate-first dielectric development in MOSFETs saw the impact of fast transient charge trapping and its effect on threshold voltage (V_t) instability. These high-k dielectrics were deposited and annealed at higher temperatures to achieve a low-leakage dielectric. This fast trapping phenomenon has been extensively investigated over the years, where channel electrons would be readily trapped in the HfO_2 rather than participate in the channel current [4–18]. This fast electron trapping can have trapping times on the order of microseconds and manifest itself as a shift in the threshold voltage (ΔV_t), and subsequently lower the drive current (Figure 16.1). The results demonstrate that electrically active defects that impact performance and reliability behave differently for low process temperature (100°C) HfO_2 gate dielectrics compared to HfO_2 processed at temperatures (~400°C + source/drain [S/D] anneal) for silicon complimentary

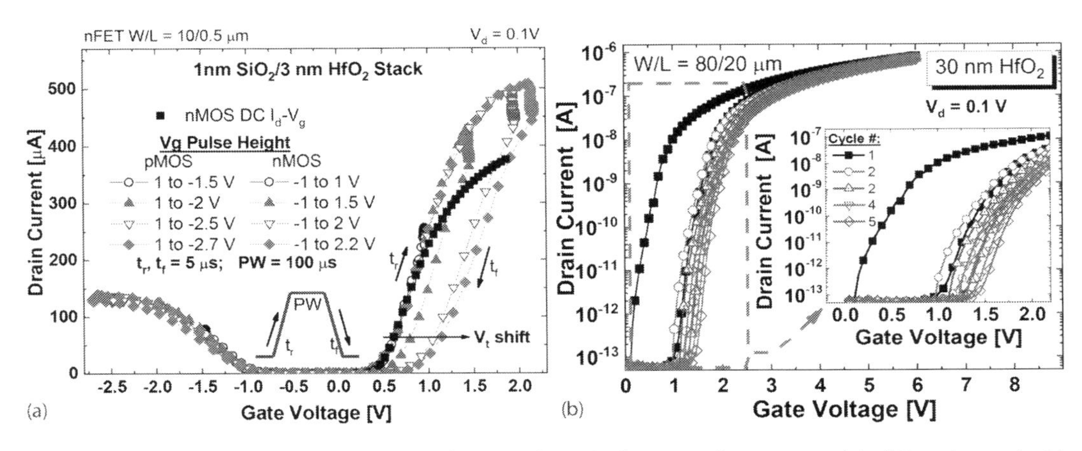

FIGURE 16.1 (a) 'Conventional' gate-first HfO_2 using [23,24] where complete recovery (t_r) of V_t is observed with repetitive measurements while (b) ZnO TFT HfO_2 using DC I-V does not recover after first measurement.

metal-oxide semiconductor (CMOS) gate-first technology [19–21]. The instability characteristics illustrate that, after an initially large shift for the TFT, the V_t remains shifted after each subsequent measurement (more details in this chapter), while a conventional Si transistor demonstrated large shifts as well, but demonstrated complete recovery with each subsequent measurement. These electrically active defects—also known as traps—affect device performance and reliability as a function time and the trap fundamental nature [22]; and therefore, need intense investigation to understand low temperature HfO_2 deposition's impact on non-Si, TFT V_t instability.

Considerable research and development have been done to enable Hf-based dielectric use in mainstream silicon technologies. Therefore, when high-k dielectrics are introduced in non-Si transistors or have major aspects of the deposition parameters changed—in this case, low temperatures for large-area/flex—the same critical investigation of electron trapping is important. Thus, several I-V methods have been employed to investigate relatively low temperature HfO_2 and Al_2O_3 gate dielectrics in TFTs.

16.2 Experimental

16.2.1 Device Fabrication

For TFTs evaluated in this work, bottom-gate, top-contact configured transistors were fabricated (Figure 16.2). The gate is defined after starting with an insulating glass substrate consisting of a conductive indium tin oxide (ITO) layer on top that is patterned. Then, multiple thickness values of HfO_2 or 15 nm of Al_2O_3 gate dielectrics are deposited by atomic layer deposition (ALD) at 100°C without patterning. Next, ZnO or IGZO at 100°C was deposited by pulsed laser deposition (PLD) under various conditions to evaluate the impact to device performance. Then, to serve as a hard-mask and passivation layer, Parylene-C (500 nm) is deposited on the ZnO or IGZO. Source and drain contacts are exposed through the hard-mask to deposit aluminum source-drain contacts. Further details are provided in [1,25]. Tables 16.1 through 16.3 show expanded fabrication details for the TFTs evaluated in this work. Each table contains a systematic study of either the gate dielectric or ZnO-based semiconductor.

16.2.2 Electrical Characterization and Reliability

For as-fabricated, initial instability (i.e., 'time zero' instability) current–voltage (I–V) measurements were executed using a Keithley 4200 analyzer. Three I_d–V_g measurement schemes were employed in the linear regime. Method #1 executed a dual sweep with a forward sweep to various maximum V_g bias

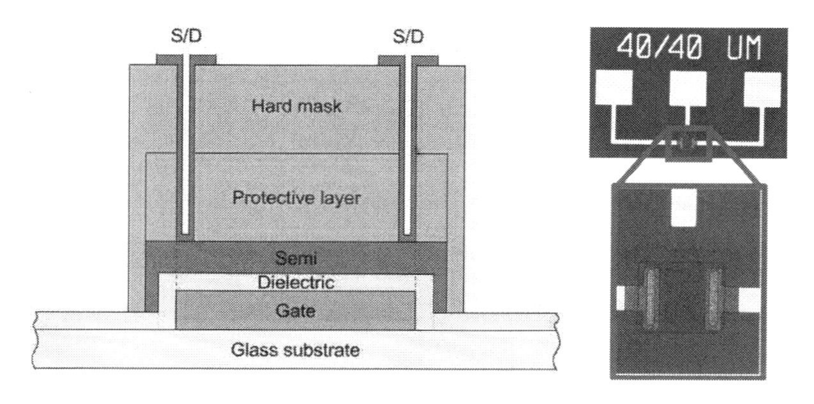

FIGURE 16.2 Cross-sectional schematic and plan-view image of the TFT structure.

TABLE 16.1 Expanded Fabrication Details for the Evaluation of Dielectric Type and Thickness

Gate Metal	100 nm of ITO		
Dielectric	90, 30, and 15 nm HfO_2	15 nm Al_2O_3	
Semiconductor	45 nm ZnO		
Zinc Oxide (ZnO) deposition conditions	**100°C** **50 nm thick**		
Protection layer	250 nm Parylene		
Hard-Mask	250 nm Parylene		
S-D Contacts	150 nm Al		
Forming Gas Anneal	No FGA	No FGA	150C FGA

TABLE 16.2 Expanded Fabrication Details for Evaluation of ZnO Deposition Pressure and Post Deposition Anneal

Gate Metal	100 nm of ITO		
Dielectric	15 nm Al_2O_3		
Semiconductor	45 nm ZnO		
Semiconductor deposition conditions	**100°C**		
	30 mTorr O_2	**30 mTorr O_2 + PDA**	**20 mTorr O_2**
Protection layer	250 nm Parylene		
Hard-Mask	250 nm Parylene		
S-D Contacts	150 nm Al		

TABLE 16.3 Expanded Fabrication Details for the Comparison of ZnO and IGZO

Gate Metal	100 nm of ITO	
Dielectric	15 nm Al_2O_3	
Semiconductor	45 nm ZnO	
Semiconductor deposition conditions	**100°C at 20 mTorr**	
	40 nm ZnO	**40 nm IGZO**
Protection layer	250 nm Parylene	
Hard-Mask	250 nm Parylene	
S-D Contacts	150 nm Al	

TABLE 16.4 Measurement Methods Used in this Work to Assess Electrically Active Defects/Traps

	Measurement Method	Schematic Representation	Notes
Time Zero	Method #1 (MD1)	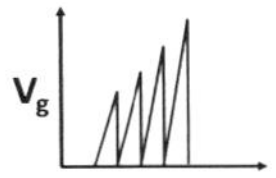 V_g	Repetitive, dual sweep with increasing $V_{g,\max}$
	Method #2 (MD2)	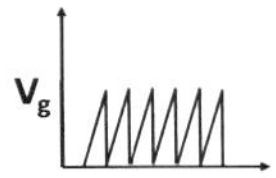 V_g	Repetitive, dual sweep with fixed $V_{g,\max}$
	Method #3 (MD3)	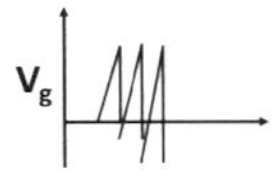 V_g	Repetitive, dual sweep with different discharge V_g (i.e., $-V_g$) and fixed $V_{g,\max}$ to assess recovery
	Method #4 (MD4)	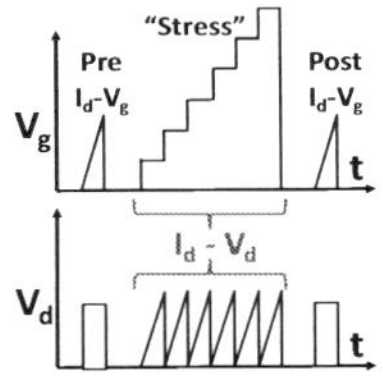 "Stress" Pre I_d-V_g Post I_d-V_g V_g t I_d - V_d V_d t	Saturation V_t Shift: • Execute an I_d–V_g (1) • Execute I_d–V_d sweeps (2) • Execute an I_d–V_g (3)
Time Dependent	CVS	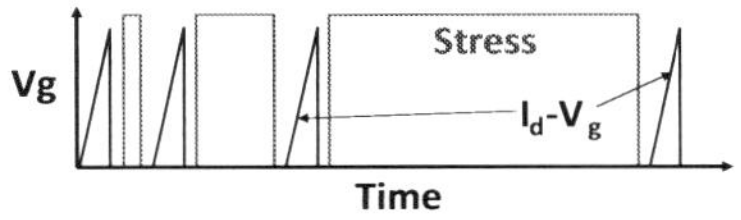 Stress I_d-V_g Vg Time	Essentially executed as positive bias temperature instability stress without the temperature
	HCI	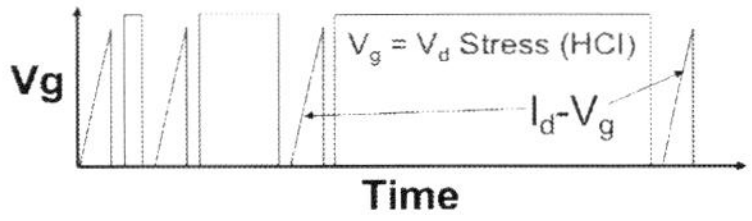 $V_g = V_d$ Stress (HCI) I_d-V_g Vg Time	Hot carrier injection where $V_{stress} = V_g = V_d$

values and back. Influence from the previous measurement during the sequential V_g increase is possible. Thus, method #2 executed 'cycling' several hysteresis forward and back sweeps at a fixed, maximum V_g bias. To study charge de-trapping, method #3 introduces a varying negative V_g start/stop dual sweeping approach [24]. For time-dependent degradation evaluation, typical constant voltage stress as well as typical hot carrier injection with interspersed I_d–V_g measurements throughout was executed to evaluate the charge trapping kinetics. Since there is no fast charge trapping experienced in these low-temperature devices, the stress sequences were conducted using the typical measure–stress–measure approach. These methods are summarized in Table 16.4.

16.3 Results and Discussion

Contrary to planar Si transistors with HfO_2 gate dielectrics, where complete recovery of the trapped charge [24,26] is observed (Figure 16.1a), these TFTs demonstrated non-recoverable trapped charge (Figures 16.3 through 16.9). For increasing V_g (Figure 16.3a), the I_d data continuously shifts due to electron trapping while maintaining a similarly detected V_t hysteresis (Figure 16.3b). Since a given V_g increase could be influenced from the previous measurement, a hysteresis 'cycling' measurement was executed on new sites at a particular bias. Figure 16.4a illustrates that after a large initial shift, the subsequent four cycles essentially lie on the back trace verifying that minimal charge de-trapping occurs,

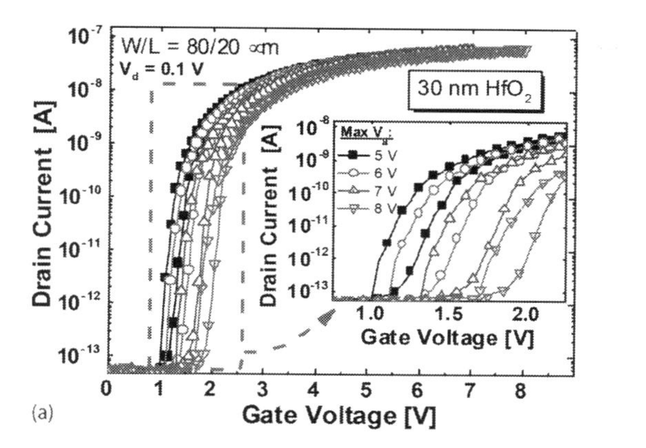

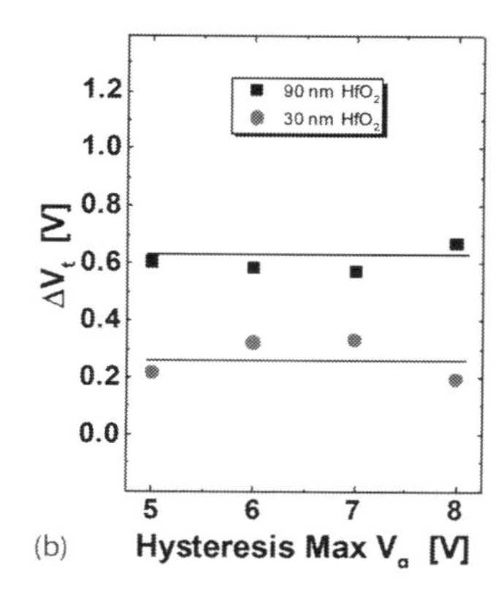

FIGURE 16.3 (a) Hysteresis sweeps starting from 0 V to 5 V, 6 V, 7 V, and 8 V. V_t continues to shift with only slight recovery. (b) While V_t shifts in (a), a relatively constant ΔV_t exists, although different maximum V_g values were measured. Results clearly demonstrate that ΔV_t reduces as HfO_2 thickness decreases. (From C.D. Young, et al., *2015 IEEE International Integrated Reliability Workshop (IIRW)*, 34–36, 2015. © 2015 IEEE. With permission.)

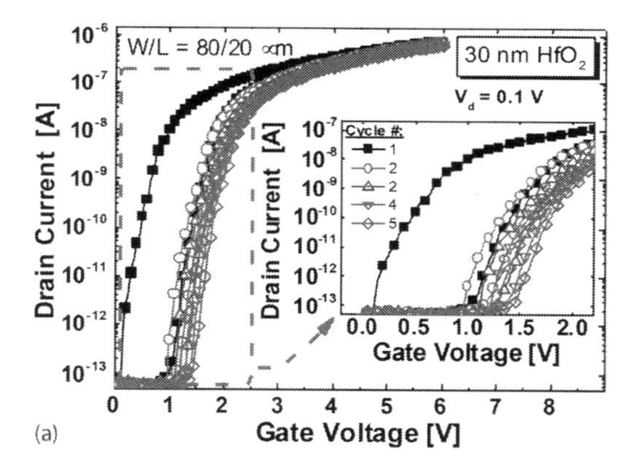

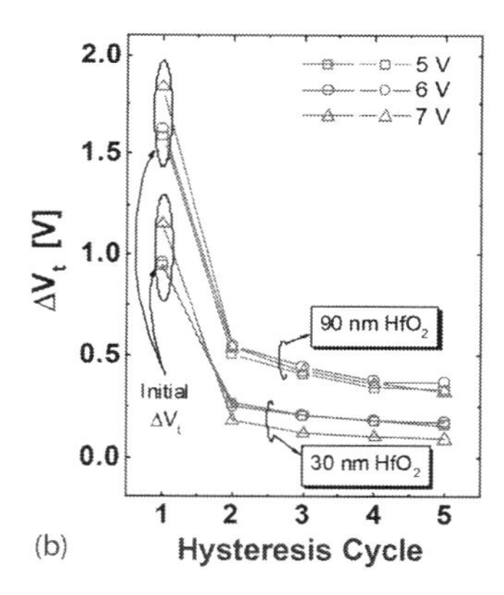

FIGURE 16.4 (a) Repetitive hysteresis 'cycling' from 0 V to a fixed V_g. After large initial shift, there is no recovery of the V_t. (b) Summary of hysteresis cycling at a given V_g showing significantly reduced ΔV_t after the initial cycle. Results clearly demonstrate that ΔV_t reduces as HfO_2 thickness decreases. (From C.D. Young, et al., *2015 IEEE International Integrated Reliability Workshop (IIRW)*, 34–36, 2015. © 2015 IEEE. With permission.)

and the ΔV_t remains fixed (Figure 16.4b). Next, applying a negative bias does not drastically de-trap the trapped charge which is in stark contrast to planar Si transistors (Figure 16.5) [24,26]. For the 15 nm HfO_2 case, a significant reduction in ΔV_t is observed. However, the hysteresis detected is counter clockwise (Figure 16.6). Since the charge trapping is significantly reduced, the ΔV_t does not mask a resistance capacitance (RC) delay measurement artifact that is observed [23].

The same measurement schemes were employed on TFTs fabricated with 15 nm of aluminum oxide (Al_2O_3) as the dielectric. Similar to HfO_2, for increasing V_g, the I_d data continuously shifts due to electron trapping while maintaining a similarly detected V_t hysteresis after the initial ΔV_t measurement (Figures 16.4 and 16.7a). When subjecting the 15 nm Al_2O_3 to the hysteresis cycling measurement, an overall significantly reduced ΔV_t is observed compared to 30 nm HfO_2, where, as before, there is a

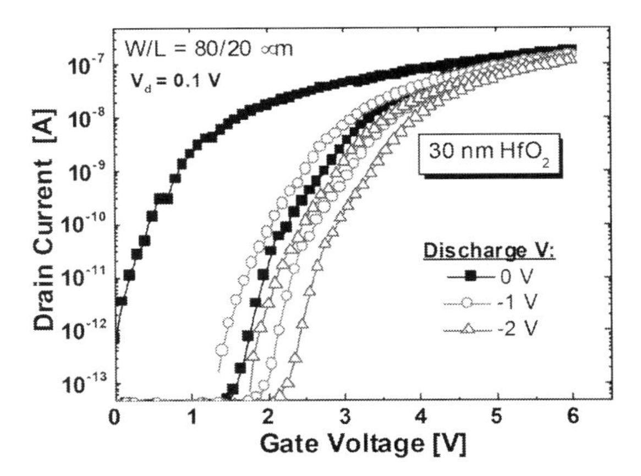

FIGURE 16.5 Applying discharge voltages has a minimal effect in recovering the threshold voltage back to the initial value.

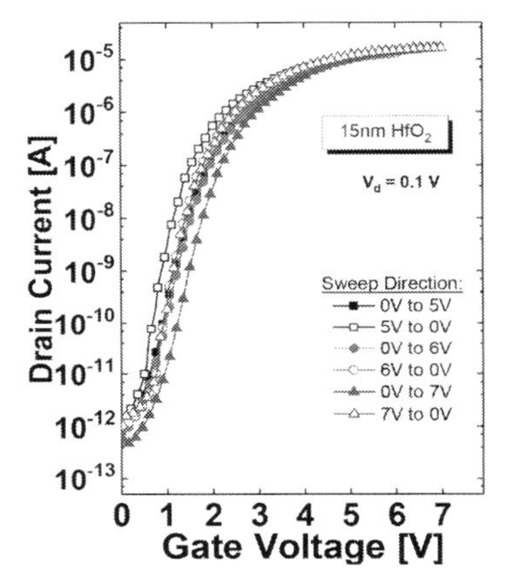

FIGURE 16.6 Hysteresis 'cycling' from 0 V to different $V_{g,max}$ values, where RC delay artifact occurs impacting ΔV_t. (From C.D. Young, et al., *2015 IEEE International Integrated Reliability Workshop (IIRW)*, 34–36, 2015. © 2015 IEEE. With permission.)

ΔV_t reduction after the initial cycle (Figure 16.7b). Finally, unlike the thin HfO_2 case, there was only a small hysteresis behavior observed (Figures 16.8 through 16.10) and applying a negative bias de-traps the trapped charge following the planar Si transistors behavior (Figure 16.10b). Further analysis of TFTs with HfO_2 as dielectric is required to determine if the ΔV_t, as a function of thickness, is due to the overall injected charge or the position of the charge centroid (Figure 16.11), and also to investigate more about the interface charge trapping in the Al_2O_3 case.

16.3.1 Process-Induced/'Time Zero' Device Instability

16.3.1.1 Threshold Voltage Instability

Since there is a V_t shift with a subsequent, non-recoverable component when conducting dual-sweep measurements, charge trapping is occurring during the first measurement after device fabrication. The previous results (Figures 16.3 through 16.10) demonstrated hysteresis and a V_t shift after $I_d - V_g$ measurements in the linear regime, where continuous shifting was usually observed. To study

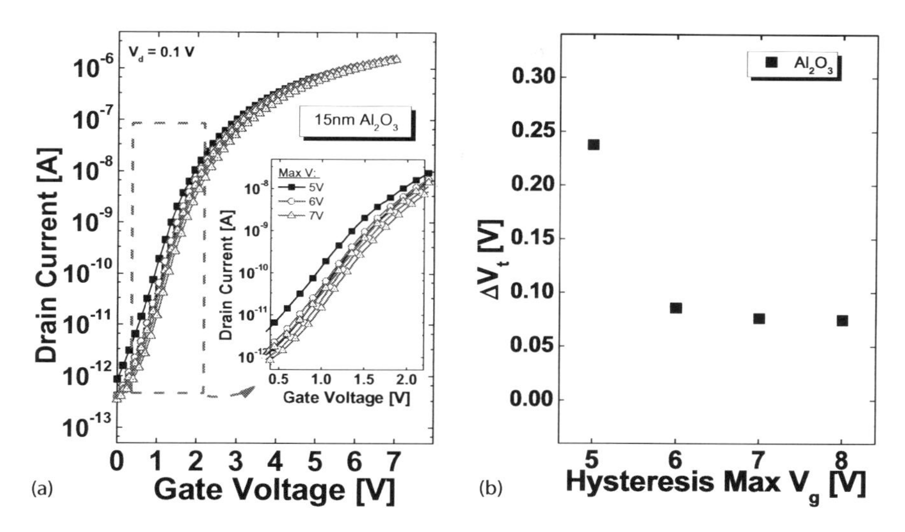

FIGURE 16.7 (a) Repetitive hysteresis sweeps starting from 0 V to 5 V, 6 V, 7 V, and 8 V, the ΔV_t reduced significantly with no major recovery of the V_t after each cycle. (b) Summary of hysteresis cycling at a given V_g showing significantly reduced ΔV_t after the initial cycle. (From C.D. Young, et al., *2015 IEEE International Integrated Reliability Workshop (IIRW)*, 34–36, 2015. © 2015 IEEE. With permission.)

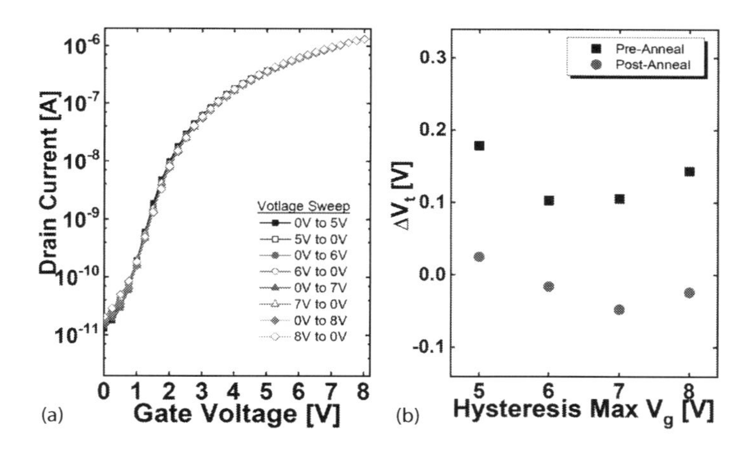

FIGURE 16.8 (a) Repetitive hysteresis sweeps starting from 0 V to 5 V, 6 V, 7 V, and 8 V 10 days after a FGA, the ΔV_t reduced further with the anneal with no major recovery of the V_t after each cycle. (b) ΔV_t during hysteresis testing for increasing maximum V_g for a TFT both before annealing and 10 days after annealing. (From C.D. Young, et al., *2015 IEEE International Integrated Reliability Workshop (IIRW)*, 34–36, 2015. © 2015 IEEE. With permission.)

charge-trapping in the saturation regime, method #4 (MD4) from Table 16.4 is performed. Figure 16.12 illustrates the threshold voltage shift before (trace 1) and after (trace 3) a collection of an $I_d - V_d$ family of curves. There is clear demonstration of electron trapping. Then, an additional $I_d - V_g$ measurement was made, to ascertain if any relaxation or further V_t shifting occurred (trace 3 again). However, no additional shift was detected. Thus, the measurement in the saturation regime (i.e., $I_d - V_d$) is the cause of V_t instability using MD4. The plausible origins of the shift could be from bulk Al_2O_3 oxide trapping [27,28] and/or trapping at the Al_2O_3 interface with the ZnO [29,30].

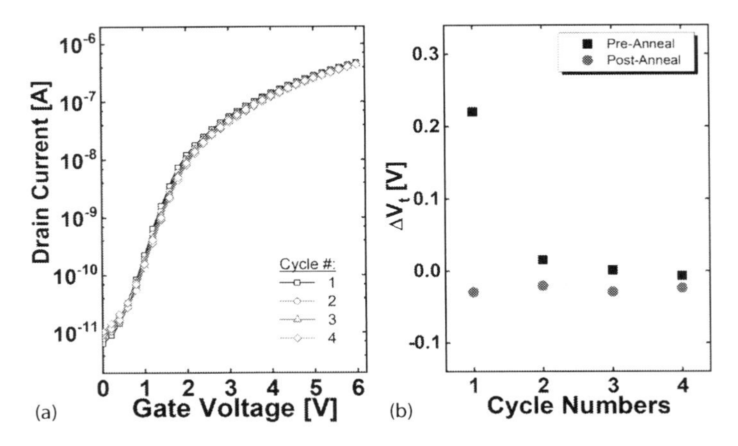

FIGURE 16.9 (a) Repetitive hysteresis 'cycling' at a fixed $V_g = 6$ V bias 10 days after an FGA. (b) Summary of hysteresis cycling for a TFT both before annealing and 10 days after annealing illustrating significantly reduced ΔV_t after anneal where an RC delay artifact occurs thereby affecting ΔV_t. (From C.D. Young, et al., *2015 IEEE International Integrated Reliability Workshop (IIRW)*, 34–36, 2015. © 2015 IEEE. With permission.)

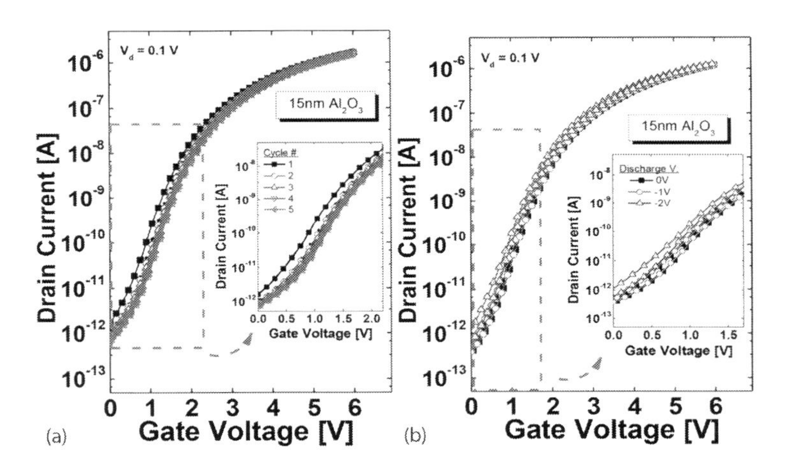

FIGURE 16.10 (a) Repetitive hysteresis 'cycling' from 0 V to $V_g = 6$ V. After an initial shift, there is no recovery of the V_t (inset). However, thin Al_2O_3 demonstrates significantly reduced ΔV_t compared to thicker HfO_2. (b) Applying discharge voltages does result in recovery of the threshold voltage for Al_2O_3.

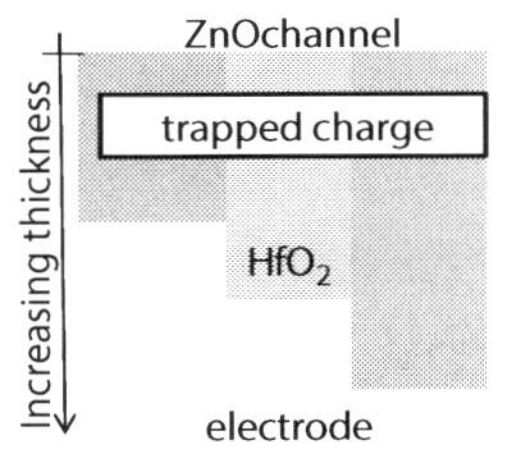

FIGURE 16.11 Ultimately, the question remains regarding correlation of injected charge to the extracted ΔV_t. The trapped charge location is also important, where a thicker dielectric could possible extract a larger ΔV_t compared to a thinner dielectric for a similar charge centroid (white box) due to image charge referenced from the gate electrode.

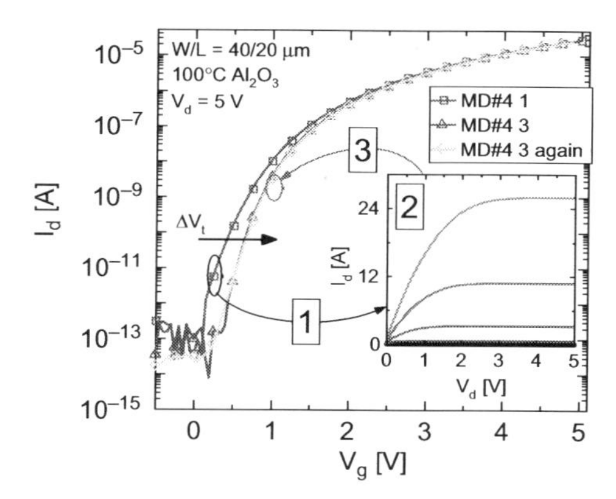

FIGURE 16.12 Typical pre- and post-I_d – V_d transfer curves using MD4 from Table 16.4 for the 100°C Al_2O_3 gate dielectric, where the numbers corresponding to the measurement sequence are outlined in MD4. Noticeable electron trapping is observed due to the positive ΔV_t.

16.3.2 Time Dependent Instability

Results from the previous sections demonstrate that scaled dielectrics and some annealing can significantly reduce the V_t instability and improve TFT performance. Therefore, investigating the time dependence of the V_t instability is the next step in evaluating relatively low temperature processing. In addition, studying the semiconductor deposition parameters and the interfacial impact on the underlying high-k dielectric is necessary. This can be achieved using these time-dependent stress methodologies (bottom of Table 16.4).

16.3.2.1 Constant Voltage Stress

In an effort to achieve large-area/flexible metal-oxide-semiconductor technology, low temperature high-k gate dielectric deposition is critical. Therefore, as part of this time-dependent V_t instability assessment, comparing 100°C Al_2O_3 to an expected higher quality Al_2O_3 at 250°C, is conducted. To ascertain robust comparison trends, multiple transistor channel width to length (W/L) = 400/20 μm TFTs at each respective stress bias were constant voltage stressed (CVS) using the methodology outlined in Table 16.4. The transformation of the 100°C Al_2O_3 I_d – V_g degradation is illustrated in Figure 16.13. Results demonstrate little to no V_t shift (ΔV_t) across the three sets of TFTs. However, Figure 16.14 provides a more descriptive picture of the (ΔV_t) with stress time. The ΔV_t monotonically increases during a 5.5 V stress; however, at 5.75 V and 6 V, there is a negative V_t shift during the latter portion of the stress time after an initially positive ΔV_t in the early stages of the stress. This observed 'turn-around' effect has been seen previously [31–35] in addition to a recent result in IGZO TFTs [36]. The higher stress voltages intensify the turn-around, which was also experienced before [36]. Thus, the results demonstrate that the trapping mechanism causing the negative V_t shift occurs at higher stress voltages, while a different trapping method, resulting in a positive shift, is evident at lower stress voltages. Since the mechanisms compete and begin to cancel each other out, an overall smaller ΔV_t occurs, thereby giving an impression of better stability in the 100°C Al_2O_3. While these competing mechanisms are taking place, a negligible, time-independent g_m or subthreshold slope (SS, extracted using an exponential fit between 10^{-11} and 10^{-9} A) change is observed which suggests minimal interface state generation (Figure 16.15).

The evolution of the V_t shift for the 250°C Al_2O_3 shows a typical positive V_t shift due to electron trapping during stress for the three stress voltages used (Figures 16.16 and 16.17). Meanwhile, similar to the 100°C Al_2O_3 case, there is essentially no g_m or SS change during the stress (Figure 16.18). To summarize,

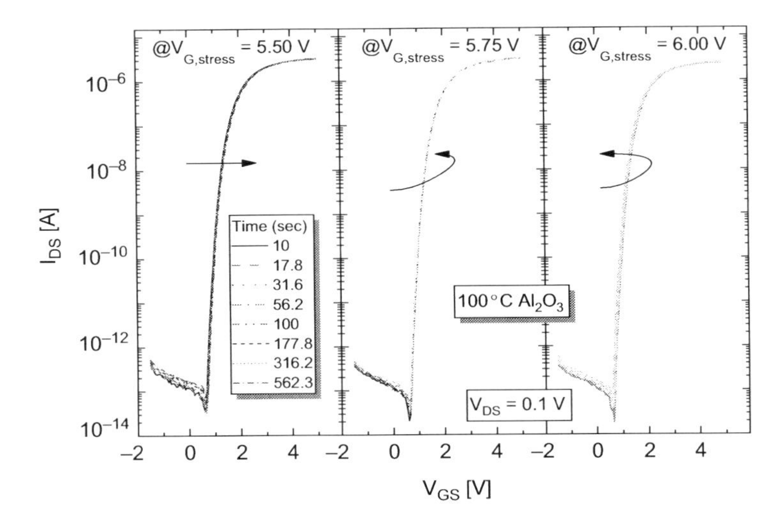

FIGURE 16.13 Example $I_{DS} - V_{GS}$ of TFTs with 100°C Al_2O_3 at three different stress biases. The threshold voltage (V_t) shift appears to be minor with no apparent degradation in the SS; however, a 'turn-around' effect is observed during stress.

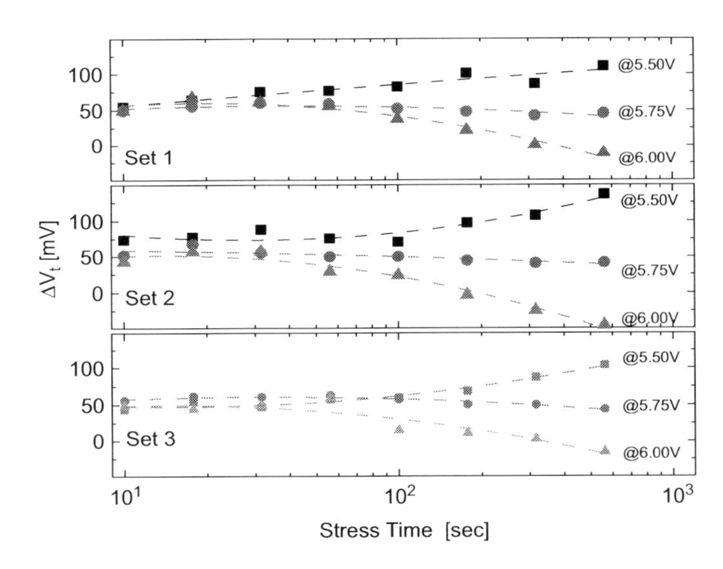

FIGURE 16.14 The ΔV_T versus stress time for different sets of TFTs with 100°C Al_2O_3 at three different stress biases. The fitted lines show the ΔV_T demonstrating the 'turn-around' effect shown in Figure 16.13.

there appears to be two competing charge trapping/generating mechanisms in 100°C Al_2O_3 that result in a smaller overall V_t shift, while the 250°C devices have only one of the mechanisms that induces the positive V_t shift, as summarized in Figure 16.19.

To ascertain a potential cause of the turn-around effect, further study provided two possible causes based on the device fabrication used for these TFTs. One cause could be more hydrogen in 100°C Al_2O_3 [37]. Previous temperature studies suggest that residual hydrogen content (>10 at.%) in low temperature Al_2O_3 may be the culprit [37], while the at.% of the present hydrogen decreases substantially with increasing ALD temperature. This would further indicate that the higher concentration of

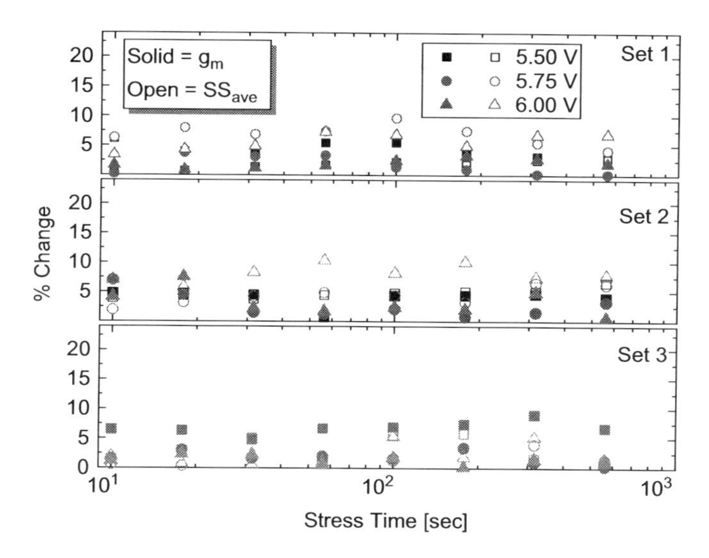

FIGURE 16.15 The % change in maximum transconductance (g_m) and average subthreshold swing (SS_{ave}) for TFTs with 100°C Al_2O_3 suggest that there is little to no interface state generation contributing to ΔV_T during stress.

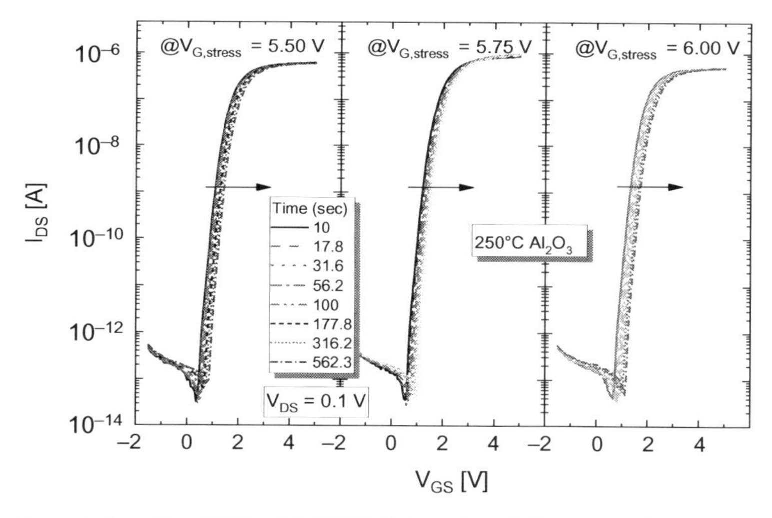

FIGURE 16.16 Example $I_{DS} - V_{GS}$ of TFTs with 250°C Al_2O_3 at three different stress biases. The threshold voltage (V_T) shift is significant in this dielectric with no apparent degradation in the SS. No 'turn-around' effect is present during stress.

hydrogen (>10 at.%) in 100°C Al_2O_3 is at least partially responsible for the observed 'turn-around' effect, whereas at 250°C, the low hydrogen content (< 2 at.%) does not appear to contribute significantly to the ΔV_t. Another cause, after evaluating elemental composition of the Al_2O_3 at both temperatures is shown in Figure 16.20. The X-ray photoelectron spectroscopy (XPS) spectra demonstrate hafnium content in the 100°C Al_2O_3, while hafnium is below the detection limit for 250°C. Meanwhile, carbon is detected in both ALD deposition temperatures, but to a lesser extent in 250°C. Therefore, Hf could be contributing to the turn-around effect. After revisiting the ALD processing sequence, the 100°C Al_2O_3 deposition occurred before the 250°C Al_2O_3. This resulted in Hf contamination from a prior hafnium oxide deposition done in a preceding run thereby providing the residual Hf in the 100°C Al_2O_3.

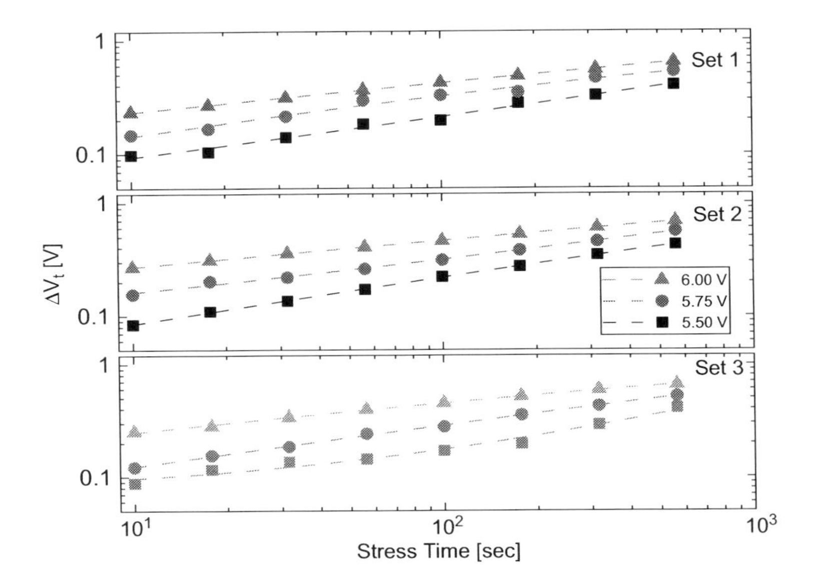

FIGURE 16.17 The ΔV_T versus stress time for different sets of TFTs with 250°C Al_2O_3 at three different stress biases. The fitted lines show trends consistent with traditional positive bias temperature instability (PBTI) in metal gate/high-k (MG/HK) n-channel metal-oxide-semiconductor (nMOS) degradation.

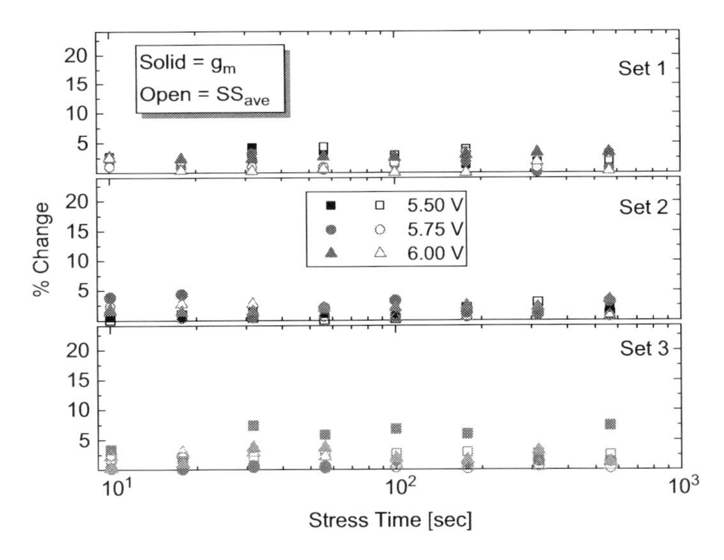

FIGURE 16.18 The % change in g_m and SS_{ave} for TFTs with 250°C Al_2O_3 suggests that there is little to no interface state generation contributing to ΔV_T, similar to the 100°C Al_2O_3 samples.

16.3.2.2 Hot Carrier Injection—ZnO

As previously stated, the TFT dielectric/semiconductor interface is exposed when the semiconductor layer is deposited by PLD. Thus, the possibility of unintentional effects at this interface need to be investigated. To study the PLD process on the underlying gate oxide, multiple TFTs from the three different thin-film semiconductor depositions in Table 16.2 were subjected to hot carrier injection (HCI) stress as outlined in Table 16.4. The data sets shown are representative of the trends observed when conducting the

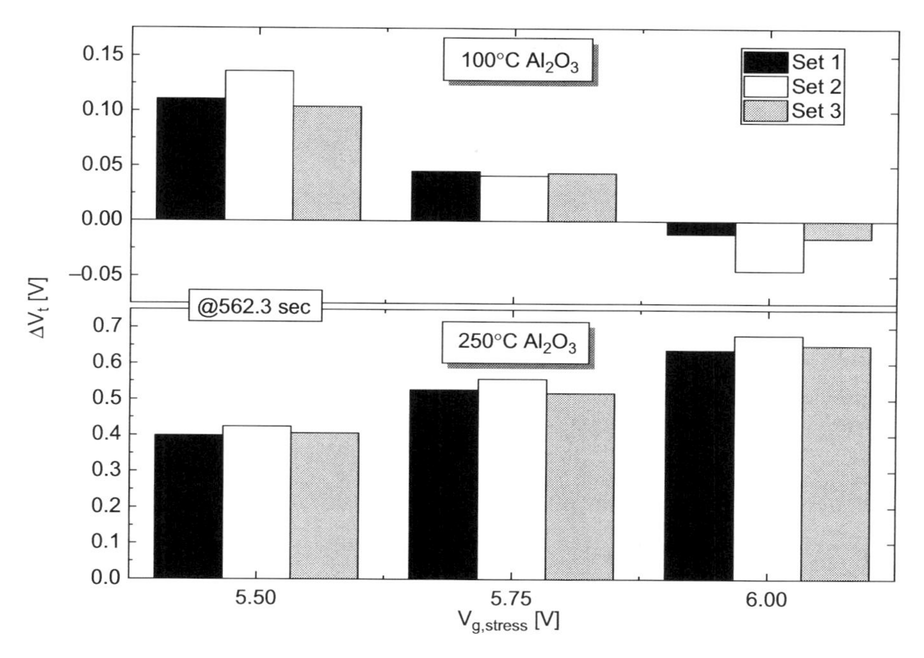

FIGURE 16.19 Comparison of ΔV_T at a fixed time between the TFTs with 100°C Al_2O_3 and 250°C Al_2O_3. Results demonstrate lower ΔV_T for the 100°C samples possibly caused by the proposed competing mechanisms of simultaneous electron trapping and de-trapping within stress duration. (From Sato, M. et al., *JPN J. Appl. Phys.*, 49, 04DC24, 2010; Jung, H.-S. et al., *Electrochem. Solid-State Lett.*, 13, G71–G74, 2010; Reimbold, G. et al., *Microelectron. Reliab.*, 47, 489–496, 2007; Sa, N. et al., *IEEE Electron Device Lett.*, 26, 610–612, 2005.)

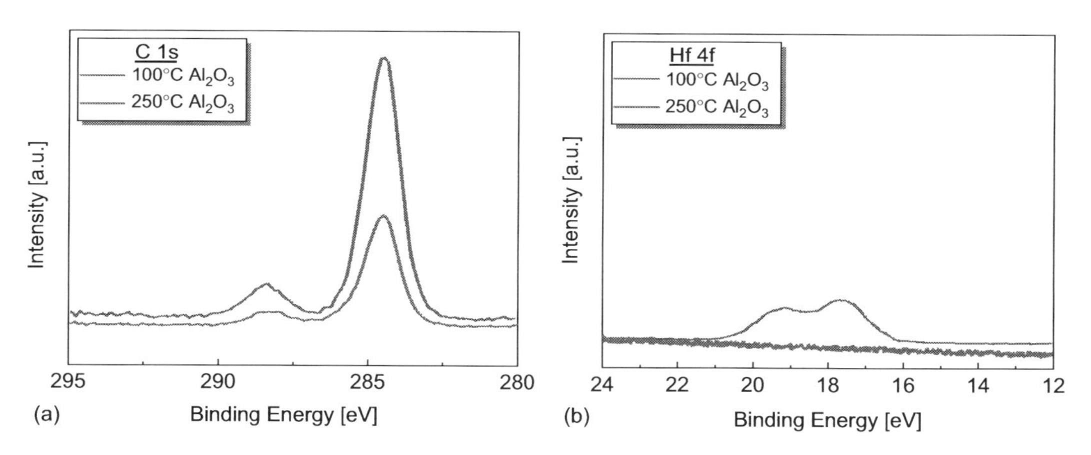

FIGURE 16.20 XPS spectra comparison of 100°C–250°C Al_2O_3 with C 1s (a) and Hf 4f (b). These spectra show that C content is present in both films while Hf content in oxide state, not metallic state, is in the 100°C Al_2O_3. For the 250°C Al_2O_3, the Hf is below the limit of detection. Therefore, Hf may contribute defect states that enable the 'turn-around' effect seen in the 100°C Al_2O_3.

HCI methodology on various TFTs. Figure 16.21 illustrates an example of the $I_d - V_g$ degradation (shifting $I_d - V_g$ curves with stress time) for the 20 mTorr sample. Here, this shifting towards the right signifies electron trapping for this accumulation mode TFT. From this data, the evolution of the degradation can be seen through a slight decrease in drain current, increasing threshold voltage shift (Figure 16.22a), and a degraded transconductance (Figure 16.22b). From the analyzed results, the 30 mTorr devices are

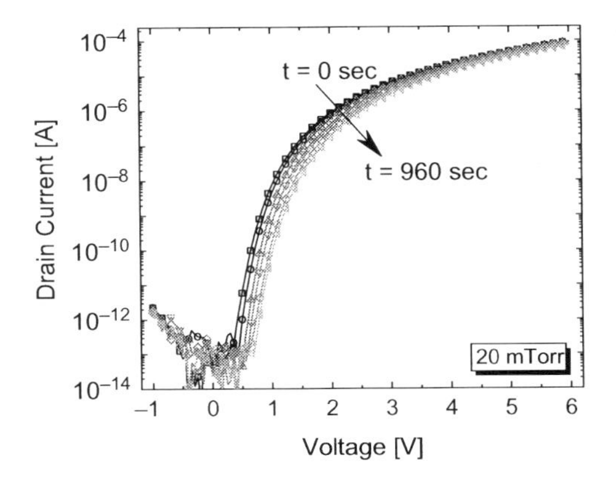

FIGURE 16.21 Example of the evolution of $I_d - V_g$ degradation for the 20 mTorr sample demonstrates a V_t shift when subjected to hot carrier stress. (From R.A. Rodriguez-Davila, et al., *2018 IEEE International Symposium on the Physical and Failure Analysis of Integrated Circuits (IPFA)*, 1–4, 2018. © 2018 IEEE. With permission.)

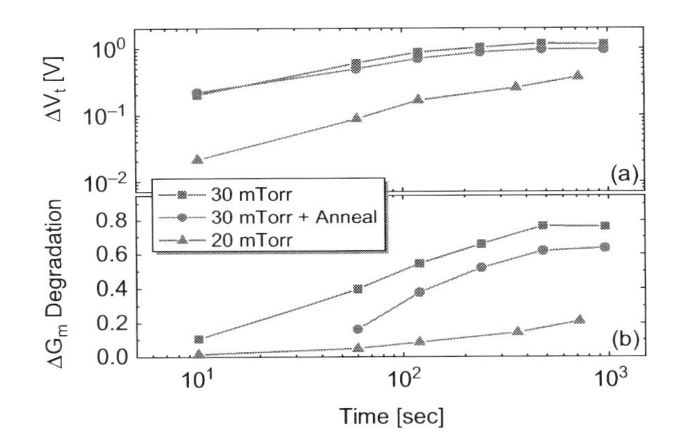

FIGURE 16.22 (a) The time evolution of the V_t shift illustrating that the 20 mTorr devices have reduced V_t instability compared to the 30 mTorr samples. (b) The time evolution of the change in g_m with respect to the pre-stress g_m, where the 20 mTorr sample demonstrated less degradation than 30 mTorr. (From R.A. Rodriguez-Davila, et al., *2018 IEEE International Symposium on the Physical and Failure Analysis of Integrated Circuits (IPFA)*, 1–4, 2018. © 2018 IEEE. With permission.)

worse than the 20 mTorr devices. Taking these findings a step further, continued analysis can potentially help understand where degradation is happening in the TFT gate stack. For instance, the transconductance, g_m, is often considered a measure of the quality that a semiconductor interface has with its dielectric. Thus, the degradation of g_m with stress time indicates charge trapping/generation in the interfacial region during stress. Meanwhile, the ΔV_t shift is a measure of all the trapped charge and trap generation in the bulk and at the interface of the gate dielectric. So, a correlation plot of g_m degradation with ΔV_t provides a semiquantitative assessment as to the contribution of interface traps to the overall V_t shift. This is demonstrated in Figure 16.23, where there is a clear correlation between g_m degradation and the threshold voltage shift with stress time suggesting that, indeed, interface trapping is a contributing factor. Furthermore, from the results in Figures 16.22 and 16.23, the 20 mTorr devices have the best overall reliability of the semiconductor conditions (Table 16.2) evaluated. A possible mechanism for this

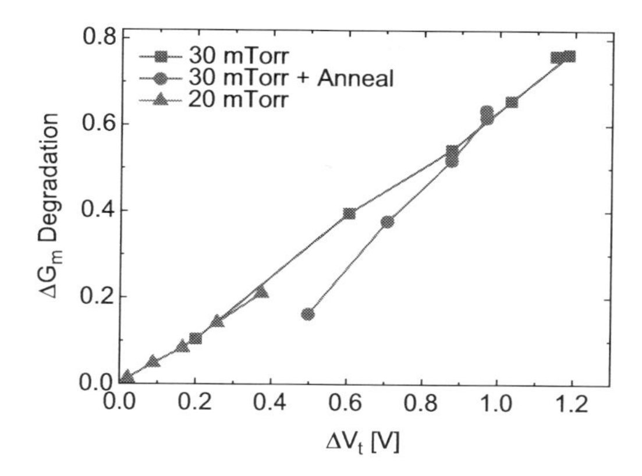

FIGURE 16.23 Transconductance to ΔV_t correlation that shows a significant amount of interface state generation—through g_m degradation—is contributing to the V_t shift. (From R.A. Rodriguez-Davila, et al., *2018 IEEE International Symposium on the Physical and Failure Analysis of Integrated Circuits (IPFA)*, 1–4, 2018. © 2018 IEEE. With permission.)

is the 30 mTorr deposition had a higher oxygen flow rate compared to the 20 mTorr deposition, which may have adversely altered the exposed surface of the Al_2O_3, which subsequently became the device channel interface.

As an attempt to improve the interface quality, forming gas anneals (FGAs) are typically done thereby making an FGA an important processing step in past and current device technologies. Therefore, an FGA, comprised of 95% N_2 and 5% H_2 at 400°C was done prior to 20 mTorr ZnO deposition, since our previous experimental work demonstrated that the ZnO cannot withstand temperatures above 200°C. This FGA temperature has been known to reduce interface state densities, thereby improving the dielectric/silicon interface. After the FGA, the HCI stress sequence was conducted, where representative data sets are shown in Figures 16.24 through 16.26. Figure 16.24 results illustrate a noticeable increase in off-state current compared to the non-annealed sample (Figure 16.21) coupled with an increased SS. Moreover, increased degradation is observed with greater V_t shifts (Figure 16.25a) and more g_m degradation

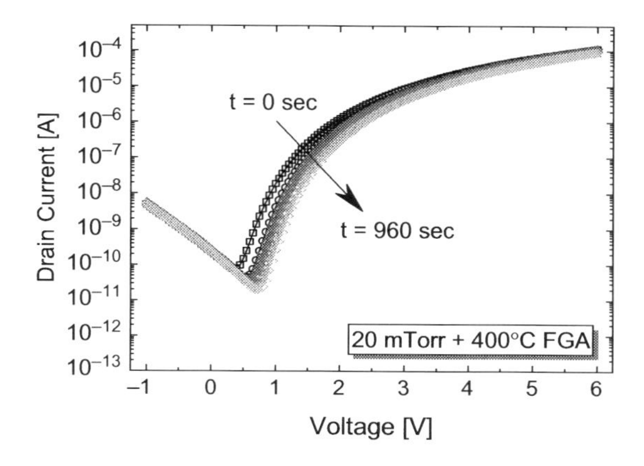

FIGURE 16.24 A 20 mTorr sample was forming gas annealed at 400°C, subjected to HCI stress, and compared with the unannealed results, where the evolution of $I_d - V_g$ degradation for the 20 mTorr + FGA sample demonstrates higher off-state current, along with V_t shifting with stress time. (From R.A. Rodriguez-Davila, et al., *2018 IEEE International Symposium on the Physical and Failure Analysis of Integrated Circuits (IPFA)*, 1–4, 2018. © 2018 IEEE. With permission.)

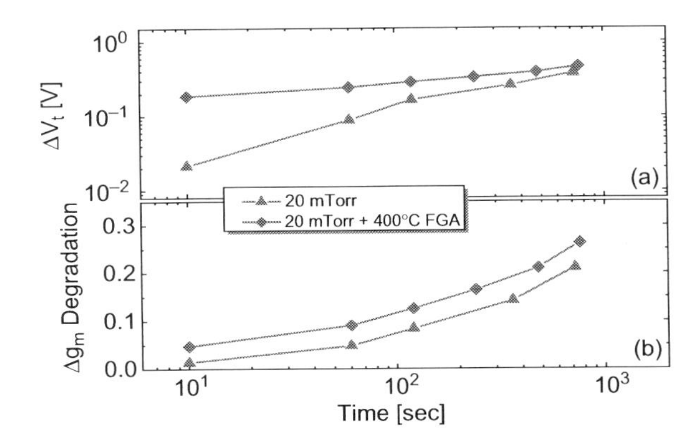

FIGURE 16.25 (a) The time evolution of the V_t shift illustrating that the 20 mTorr devices have reduced V_t instability compared to the 20 mTorr + 400°C FGA samples. (b) The time evolution of the change in g_m with respect to the pre-stress g_m, where the 20 mTorr sample demonstrated less degradation than 20 mTorr + 400°C FGA. (From R.A. Rodriguez-Davila, et al., *2018 IEEE International Symposium on the Physical and Failure Analysis of Integrated Circuits (IPFA)*, 1–4, 2018. © 2018 IEEE. With permission.)

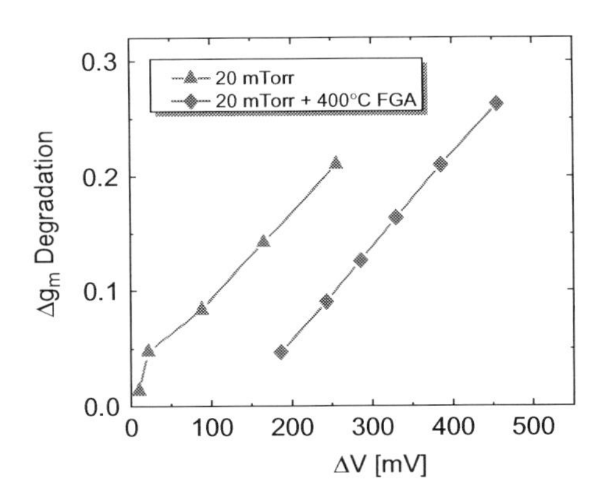

FIGURE 16.26 Transconductance to ΔV_t correlation that shows a significant amount of interface state generation—through g_m degradation—is contributing to the V_t shift, where the additional 400°C FGA exacerbates the interface degradation contribution to the overall V_t shift. (From R.A. Rodriguez-Davila, et al., *2018 IEEE International Symposium on the Physical and Failure Analysis of Integrated Circuits (IPFA)*, 1–4, 2018. © 2018 IEEE. With permission.)

(Figure 16.25b). Evaluating the interface degradation (i.e., g_m change) contribution to the total V_t shift is evident in Figure 16.26. Therefore, a 400°C FGA appears to be too aggressive, where more interface states are induced rather than improving the interface. A plausible mechanism is roughening of the surface [38] that leads to non-ideal dangling bonds at the Al_2O_3/ZnO interface [39]. This would certainly lead to a reduction in device performance and increased levels of degradation.

16.3.2.3 Hot Carrier Injection—ZnO vs IGZO

In addition to ZnO, IGZO is another commonly studied thin-film semiconductor. Therefore, to compare the 20 mTorr ZnO, IGZO was also deposited at 20 mTorr on the same 100°C Al_2O_3 gate dielectric and subjected to HCI stress at $V_g = V_d = 6$ V with $I_d - V_g$ (saturation regime) monitoring over many stressed devices. Example results for a typical device for each respective substrate is shown in Figure 16.27.

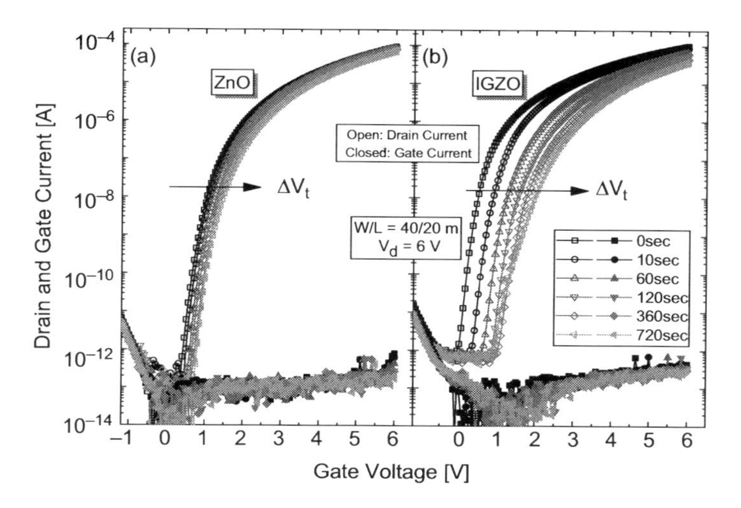

FIGURE 16.27 Representative $I_d - V_g$ data during HCI stress for (a) ZnO and (b) IGZO, where a greater detected V_t shift is observed for IGZO.

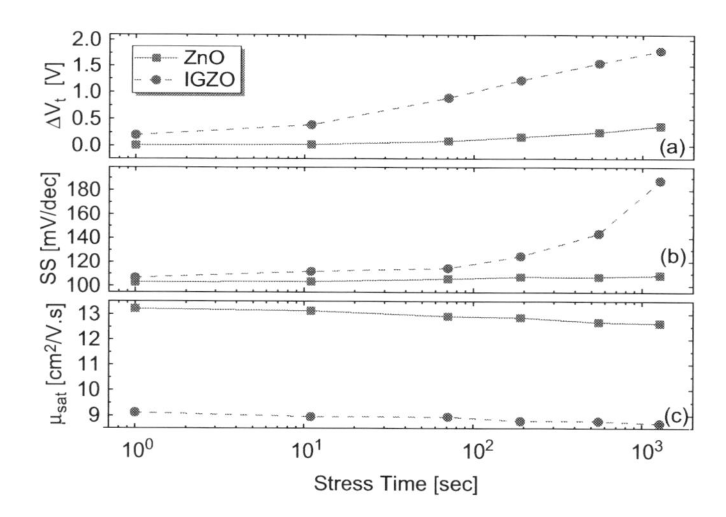

FIGURE 16.28 Monitored (a) threshold voltage shift (ΔV_t), (b) SS, and saturation mobility (μ_{sat}) for ZnO and IGZO during HCI stress.

Both semiconductor thin films demonstrate an expected positive V_t shift signifying electron trapping. Analysis of the monitored transfer characteristics also included SS and saturation mobility (μ_{sat}) as extracted in [41]. Figure 16.28 illustrates extracted $\Delta V_{t\text{-sat}}$, SS, and μ_{sat} with respect to stress time. Here, the $\Delta V_{t\text{-sat}}$ for both semiconductors steadily shifts positively with stress time. Meanwhile, the SS does not appear to deviate over time for ZnO; however, IGZO trends up after 740 seconds of stress. In addition, the saturation mobility remains constant with stress time. This lack of a change in SS suggests negligible interface state generation during the stress time (ZnO) or portion of the stress time (IGZO). This result is contrary to the previous section on PLD deposition parameters of ZnO because those $I_d - V_g$ sense measurements were done in the linear regime, while these TFT's sense data were collected in the saturation regime which masks the sensitivity to what could be occurring at the semiconductor/

dielectric interface until enough interface trap generation can become evident (i.e., IGZO result). As previously stated, charge trapping/generation at the interface [29,30] or near-interface deep states [27,28] of the Al_2O_3 could be plausible mechanisms for the $\Delta V_{t\text{-sat}}$ shifts reported herein. Furthermore, SS degradation for IGZO has been previously reported where oxygen vacancies could result in decreased resistivity of the film [42,43], thereby affecting the TFT subthreshold properties [44]. However, the true defect source is still elusive without direct confirmation. One report in the literature has recommended meta-stable oxygen vacancies at the interface after lengthy stress times [45] which puts this reference in line with our study.

16.4 Summary

ZnO-based thin-film transistors with an atomic layer deposited HfO_2 or Al_2O_3 gate dielectrics were evaluated using various *I–V* measurement methodologies and constant voltage stress to ascertain semi-quantitative understanding of V_t instability as a function of low-temperature fabrication conditions. The observed threshold voltage instability is caused by electron trapping from the accumulation layer in the transistor channel. The detected charge trapping was unable to de-trap for HfO_2. A reduction in dielectric thickness of HfO_2 or Al_2O_3 noticeably decreased the overall ΔV_t shift compared to thicker HfO_2. Furthermore, a 150°C forming gas anneal further reduced the ΔV_t to smaller values for Al_2O_3. Since Al_2O_3 demonstrated better stability compared to HfO_2, additional time-dependent stress was done to investigate the impact of low Al_2O_3 deposition temperature on longer-term V_t instability. Results show the Al_2O_3 deposited at 100°C demonstrated a 'turn-around' effect possibly caused by two competing defect mechanisms in Al_2O_3 resulting in $+V_t$ and $-V_t$ shifts that essentially cancel each either to some degree depending on the stress bias. Al_2O_3 deposited at 250°C experienced a more traditional, monotonic increase in V_t with stress time. Both deposition temperatures demonstrated negligible g_m and SS_{ave} degradation, which suggests little to no generated interface state contribution to the observed V_t instability. To investigate aspects of the pulse laser deposition of ZnO-based semiconductor films on underlying Al_2O_3, hot carrier injection stress was conducted while monitoring the threshold voltage and transconductance to semiquantitatively study the transistor instability. The PLD 30 mTorr ZnO films showed more degradation with a larger change in g_m, reduction in drain current (I_d), and an larger V_t shift. However, the 20 mTorr ZnO film demonstrated better performance and was more reliable with less V_t instability. A higher oxygen flow rate occurred during the 30 mTorr pressure deposition, and this may have had a detrimental impact on the exposed, underlying Al_2O_3, thereby creating a less robust dielectric interface compared to 20 mTorr devices. A correlation of the g_m degradation (i.e., interface degradation) to ΔV_t clearly demonstrated that interface traps played a significant role in the V_t shift under these HCI stress conditions. With 20 mTorr being the better deposition pressure, a comparison of ZnO and IGZO, representing polycrystalline and amorphous material, respectively, under HCI stress was executed. Results demonstrated that both semiconductors exhibit positive ΔV_t shifts caused by a negative trapped charge due to trap filling or generation. In addition, the amorphous IGZO experienced SS change after prolonged stress suggesting eventual interface state creation while the ZnO subthreshold slope remained constant during the same stress time duration. Overall, this work demonstrates the need to conduct further investigations on the instability of relatively low-temperature processed ZnO-based semiconductors in combination with gate insulating layers that form the gate stack channel region of thin-film transistors.

Acknowledgments

This work was supported in part by the NSF CAREER award ECCS-1653343, AFOSR project FA9550-18-1-0019, CONACYT, and RD Research Technology USA, LLC.

References

1. G. Gutierrez-Heredia, I. Mejia, M. E. Rivas-Aguilar, N. Hernandez-Como, V. H. Martinez-Landeros, F. S. Aguirre-Tostado, and M. A. Quevedo-Lopez, "Fully patterned and low temperature transparent ZnO-based inverters," *Thin Solid Films*, vol. 545, pp. 458–461, 2013.
2. Y. Kawamura, M. Horita, Y. Ishikawa, and Y. Uraoka, "Effects of gate insulator on thin-film transistors with ZnO channel layer deposited by plasma-assisted atomic layer deposition," *Journal of Display Technology*, vol. 9, no. 9, pp. 694–698, 2013.
3. M. S. Oh, W. Choi, K. Lee, D. K. Hwang, and S. Im, "Flexible high gain complementary inverter using n-ZnO and p-pentacene channels on polyethersulfone substrate," *Applied Physics Letters*, vol. 93, no. 3, p. 033510, 2008.
4. G. Bersuker, B. H. Lee, and H. R. Huff, "Novel dielectric materials for future transistor generations," *International Journal of High Speed Electronics and Systems*, vol. 16, no. 1, pp. 221–239, 2006.
5. G. Bersuker, B. H. Lee, H. R. Huff, J. Gavartin, and A. Shluger, "Mechanism of charge trapping reduction in scaled high-k gate stacks," in *Defects in High-k Gate Dielectric Stacks: Proceedings of the NATO Advanced Research Workshop on Defects in Advanced High-k Dielectric Nano-electronic Semiconductor Devices*, vol. 220, E. Gusev, Ed. (NATO Science Series, no. 220): Springer, Berlin, Germany, 2006, pp. 227–236.
6. G. Bersuker, J. Sim, C. S. Park, C. D. Young, S. Nadkarni, C. Rino, and L. Byoung Hun, "Mechanism of electron trapping and characteristics of traps in HfO_2 gate stacks," *IEEE Transactions on Device and Materials Reliability*, vol. 7, no. 1, pp. 138–145, 2007.
7. G. Bersuker, J. H. Sim, C. S. Park, C. D. Young, S. Nadkarni, R. Choi, and B. H. Lee, "Intrinsic threshold voltage instability of the HfO_2 NMOS transistors," in *International Reliability Physics Symposium*, Piscataway, NJ, 2006, pp. 179–183.
8. G. Bersuker, J. H. Sim, C. D. Young et al., "Effects of structural properties of Hf-based gate stack on transistor performance," in *2004 Spring Meeting of the Material Research Society*, 2004, vol. 811, pp. 31–35.
9. P. Broqvist and A. Pasquarello, "Oxygen vacancy in monoclinic HfO_2: A consistent interpretation of trap assisted conduction, direct electron injection, and optical absorption experiments," *Applied Physics Letters*, vol. 89, no. 26, p. 262904, 2006.
10. S. Chen, Y. Tian, L. Ming-Fu, W. Xinpeng, C. E. Foo, S. S. Ganesh, Y. Yee-Chia, and K. Dim-Lee, "Fast V_{th} instability in HfO_2 gate dielectric MOSFETs and its impact on digital circuits," *Electron Devices, IEEE Transactions on*, vol. 53, no. 12, pp. 3001–3011, 2006.
11. R. Choi, S. C. Song, C. D. Young, G. Bersuker, and B. H. Lee, "Charge trapping and detrapping characteristics in hafnium silicate gate dielectric using an inversion pulse measurement technique," *Applied Physics Letters*, vol. 87, no. 12, p. 122901, 2005.
12. B. H. Lee, C. Young, R. Choi, J. H. Sim, and G. Bersuker, "Transient charging and relaxation in high-k gate dielectrics and their implications," *Jap. J. of Applied Physics Part 1-Regular Papers Short Notes & Review Papers*, vol. 44, no. 4B, pp. 2415–2419, 2005.
13. B. H. Lee, C. D. Young, R. Choi et al., "Intrinsic characteristics of high-k devices and implications of fast transient charging effects (FTCE)," in *IEEE Intl. Electron Devices Meeting Tech. Digest*, 2004, pp. 859–862.
14. G. Ribes, S. Bruyere, D. Roy, C. Parthasarthy, M. Muller, M. Denais, V. Huard, T. Skotnicki, and G. Ghibaudo, "Origin of V_t instabilities in high-k dielectrics jahn-teller effect or oxygen vacancies," *IEEE Transactions on Device and Materials Reliability*, vol. 6, no. 2, pp. 132–135, 2006.
15. J. Robertson, "Band offsets of wide-band-gap oxides and implications for future electronic devices," *Journal of Vacuum Science & Technology B: Microelectronics and Nanometer Structures*, vol. 18, no. 3, pp. 1785–1791, 2000.
16. K. Xiong, J. Robertson, and S. J. Clark, "Defect energy states in high-K gate oxides," *Physica Status Solidi B-Basic Solid State Physics*, vol. 243, no. 9, pp. 2071–2080, 2006.

17. C. D. Young, G. Bersuker, G. A. Brown, P. Lysaght, P. Zeitzoff, R. W. Murto, and H. R. Huff, "Charge trapping and device performance degradation in MOCVD hafnium-based gate dielectric stack structures," in *42nd Annual IEEE International Reliability Physics Symposium Proceedings*, 2004, pp. 597–598.
18. G. Ribes, J. Mitard, M. Denais, S. Bruyere, F. Monsieur, C. Parthasarathy, E. Vincent, and G. Ghibaudo, "Review on high-k dielectrics reliability issues," *IEEE Transactions on Device and Materials Reliability*, vol. 5, no. 1, pp. 5–19, 2005.
19. D. Siddharth, G. Gutierrez-Heredia, I. Mejia, S. Benton, M. Quevedo-Lopez, and C. D. Young, "Investigation of V_t Instability in ZnO TFTs with an HfO_2 Dielectric," presented at the 18th International Workshop on Dielectrics in Microelectronics, Kinsale, Cork, 2014.
20. D. Siddharth, P. Zhao, I. Mejia, S. Benton, M. Quevedo-Lopez, and C. D. Young, "Threshold voltage instabilities in zinc oxide thin film transistors with high-k dielectrics," presented at the *International Integrated Reliability Workshop*, 2014.
21. C. D. Young, R. Campbell, S. Daasa, S. Benton, R. R. Davila, I. Mejia, and M. Quevedo-Lopez, "Effect of dielectric thickness and annealing on threshold voltage instability of low temperature deposited high-k oxides on ZnO TFTs," in *2015 IEEE International Integrated Reliability Workshop (IIRW)*, 2015, pp. 34–36.
22. T. Grasser, *Bias Temperature Instability for Devices and Circuits*. Springer Science & Business Media, New York, 2013.
23. C. D. Young, D. Heh, R. Choi, B.-H. Lee, and G. Bersuker, "The pulsed I_d–V_g methodology and its application to the electron trapping characterization of high-κ gate dielectrics," *Journal of Semiconductor Technology and Science (JSTS)*, vol. 10, no. 2, pp. 79–99, 2010.
24. C. D. Young, Y. Zhao, D. Heh, R. Choi, B. H. Lee, and G. Bersuker, "Pulsed I_d–V_g methodology and its application to electron-trapping characterization and defect density profiling," *IEEE Transactions on Electron Devices*, vol. 56, no. 6, pp. 1322–1329, 2009.
25. R. A. Chapman, R. A. Rodriguez-Davila, I. Mejia, and M. Quevedo-Lopez, "Nanocrystalline ZnO TFTs using 15-nm thick Al_2O_3 gate insulator: Experiment and simulation," *IEEE Transactions on Electron Devices*, vol. 63, no. 10, pp. 3936–3943, 2016.
26. A. Kerber, E. Cartier, L. Pantisano, R. Degraeve, T. Kauerauf, Y. Kim, A. Hou, G. Groeseneken, H. E. Maes, and U. Schwalke, "Origin of the threshold voltage instability in SiO_2/HfO_2 dual layer gate dielectrics," *IEEE Electron Device Letters*, vol. 24, no. 2, pp. 87–89, 2003.
27. C. v. Berkel and M. J. Powell, "Resolution of amorphous silicon thin-film transistor instability mechanisms using ambipolar transistors," *Applied Physics Letters*, vol. 51, no. 14, pp. 1094–1096, 1987.
28. M. J. Powell, C. v. Berkel, I. D. French, and D. H. Nicholls, "Bias dependence of instability mechanisms in amorphous silicon thin-film transistors," *Applied Physics Letters*, vol. 51, no. 16, pp. 1242–1244, 1987.
29. F. R. Libsch and J. Kanicki, "Bias-stress-induced stretched-exponential time dependence of charge injection and trapping in amorphous thin-film transistors," *Applied Physics Letters*, vol. 62, no. 11, pp. 1286–1288, 1993.
30. M. J. Powell, "Charge trapping instabilities in amorphous silicon-silicon nitride thin-film transistors," *Applied Physics Letters*, vol. 43, no. 6, pp. 597–599, 1983.
31. M. Houssa, *High k Gate Dielectrics*. Boca Raton, FL: CRC Press, 2003.
32. M. Sato, S. Kamiyama, T. Matsuki, D. Ishikawa, T. Ono, T. Morooka, J. Yugami, K. Ikeda, and Y. Ohji, "Study of a negative threshold voltage shift in positive bias temperature instability and a positive threshold voltage shift the negative bias temperature instability of yttrium-doped HfO_2 gate dielectrics," *Japanese Journal of Applied Physics*, vol. 49, no. 4, p. 04DC24, 2010.
33. H.-S. Jung, J.-M. Park, H. K. Kim et al., "The bias temperature instability characteristics of in situ nitrogen incorporated ZrO x N y gate dielectrics," *Electrochemical and Solid-State Letters*, vol. 13, no. 9, pp. G71–G74, 2010.

34. G. Reimbold, J. Mitard, X. Garros, C. Leroux, G. Ghibaudo, and F. Martin, "Initial and PBTI-induced traps and charges in Hf-based oxides/TiN stacks," *Microelectronics Reliability*, vol. 47, no. 4, pp. 489–496, 2007.
35. N. Sa, J. F. Kang, H. Yang, X. Y. Liu, Y. D. He, R. Q. Han, C. Ren, H. Y. Yu, D. S. H. Chan, and D. Kwong, "Mechanism of positive-bias temperature instability in sub-1-nm TaN/HfN/HfO/sub 2/ gate stack with low preexisting traps," *IEEE Electron Device Letters*, vol. 26, no. 9, pp. 610–612, 2005.
36. G. Baek, L. Bie, K. Abe, H. Kumomi, and J. Kanicki, "Electrical instability of double-gate a-IGZO TFTs with metal source/drain recessed electrodes," *IEEE Transactions on Electron Devices,* vol. 61, no. 4, pp. 1109–1115, 2014.
37. O. M. E. Ylivaara, X. Liu, L. Kilpi et al., "Aluminum oxide from trimethylaluminum and water by atomic layer deposition: The temperature dependence of residual stress, elastic modulus, hardness and adhesion," *Thin Solid Films*, vol. 552, pp. 124–135, 2014.
38. Z. Zurita, S. Subash, S. Sachin, K. Golap, and M. Tanemura, "Effect of annealing in hydrogen atmosphere on ZnO films for field emission display," *IOP Conference Series: Materials Science and Engineering*, vol. 99, no. 1, p. 012030, 2015.
39. M. Estrada, G. Gutierrez-Heredia, A. Cerdeira, J. Alvarado, I. Garduño, J. Tinoco, I. Mejia, and M. Quevedo-Lopez, "Temperature dependence of the electrical characteristics of low-temperature processed zinc oxide thin film transistors," *Thin Solid Films*, vol. 573, pp. 18–21, 2014.
40. R. A. Rodriguez-Davila, I. Mejia, M. Quevedo-Lopez, and C. D. Young, "Hot carrier stress investigation of zinc oxide thin film transistors with an Al_2O_3 gate dielectric," in *2018 IEEE International Symposium on the Physical and Failure Analysis of Integrated Circuits (IPFA)*, 2018, pp. 1–4.
41. R. A. Chapman, R. A. Rodriguez-Davila, W. G. Vandenberghe, C. L. Hinkle, I. Mejia, A. Chatterjee, and M. A. Quevedo-Lopez, "Quantum confinement and interface states in ZnO nanocrystalline thin-film transistors," *IEEE Transactions on Electron Devices*, vol. 65, no. 5, pp. 1787–1795, 2018.
42. K. Nomura, T. Kamiya, M. Hirano, and H. Hosono, "Origins of threshold voltage shifts in room-temperature deposited and annealed a-In–Ga–Zn–O thin-film transistors," *Applied Physics Letters*, vol. 95, no. 1, p. 013502, 2009.
43. H. Yabuta, M. Sano, K. Abe, T. Aiba, T. Den, H. Kumomi, K. Nomura, T. Kamiya, and H. Hosono, "High-mobility thin-film transistor with amorphous InGaZnO4 channel fabricated by room temperature rf-magnetron sputtering," *Applied Physics Letters*, vol. 89, no. 11, p. 112123, 2006.
44. K. Nomura, T. Kamiya, H. Ohta, M. Hirano, and H. Hosono, "Defect passivation and homogenization of amorphous oxide thin-film transistor by wet O_2 annealing," *Applied Physics Letters*, vol. 93, no. 19, p. 192107, 2008.
45. Y. Young-Soo, C. Sun-Hee, K. Joo-Sun, and N. Sang-Cheol, "Characteristics of parylene polymer and its applications," *Korean Journal of Materials Research*, vol. 14, no. 6, pp. 443–450, 2004.

III

Systems and Applications

17

Reconfigurable Electronics

Jhonathan Prieto Rojas

17.1 Introduction

Reconfigurable electronics has been born from the need of incorporating adaptability in electronic systems to any changes in their environment or new requirements mandated by the user's needs or comfort. A hardware that can adapt its digital circuit's interconnections dynamically to adapt to different conditions and improve performance was probably the first reconfigurable electronic system to be conceptualized (Estrin 2002). More recently, due to the rise of novel, mechanically challenging technologies, such as wearable electronics, bio-integrated systems, cybernetics, or robotics, mechanical reconfiguration and adaptability became physical requirements for this new generation of electronic devices. For instance, wearable and bio-integrated electronics have emerged as promising new technologies with growing markets and relevant commercial values, but they demand that devices demonstrate not only high electrical performance, but also high mechanical compliance and reconfigurability to continuously shape-shifting surfaces (Nassar et al. 2016). Other types of reconfiguration can be also identified and will be shortly discussed. Therefore, several types of reconfiguration can be recognized, either to exhibit flexibility in the electrical configuration at the device-level or circuit-level or to exhibit flexibility or adaptability in its mechanical configuration. The development of such systems with inherent flexibility either mechanically or electrically will be the center of discussion of this chapter.

Several approaches can be found as to how the reconfiguration might take place according to mechanical or electrical properties, as devices adapt to new conditions, configurations, positions, dimensions, etc. In the first part of this chapter, after a short historical background is described, a review will be done about some of the most note-worthy approaches to produce reconfigurable electronic systems or devices, ranging from the computing perspective to the concept of reconfiguration from big to small scales. Some important considerations will be discussed along the way, especially useful when designing and developing such systems. Next, a section about materials will discuss the need for a careful assessment of the mechanical and electrical requirements. Later, the role of different structures and geometries will be discussed and some of the most outstanding structures will be then described and studied in detail. Finally, the chapter ends with the main conclusions and references.

17.2 Beginnings of Reconfigurability

Gerald Estrin first suggested the concept of a system capable of reconfiguring its own hardware back in the 1960s. As opposed to standard, rigid computing systems, a more innovative and risky computer architecture was proposed. This architecture could be reconfigured by a main processor depending on the required task, such as pattern recognition, image processing, or others, so that flexibility is introduced, while the task is completed more efficiently by the recently adapted circuit. In this manner, the flexibility of the system would lead to higher speeds and performance, as if a dedicated hardware was used. After the task is completed, the main processor can reconfigure the hardware again to start a new task (Estrin 2002).

More recently, reconfiguration has evolved beyond the electrical circuit design into a much broader concept, which encompasses from material reconfiguration in transient electronics, to mechanical adaptability through structural reconfiguration. The following sections will cover in more detail all these novel concepts.

17.3 Reconfigurable Electronic Approaches

17.3.1 Reconfigurable Computing

Just as first introduced by Estrin, reconfigurable computing refers to a computer architecture capable of reconnecting the hardware's digital circuits to complete a specific task more efficiently. In a way, the system combines both the flexibility of a software with the efficiency and higher performance of a dedicated hardware. An ideal tool to implement such concept could be a field-programmable gate array (FPGA). In a FPGA, the user can reprogram the functionality of the system through a hardware description language (HDL), which contains the design and purpose of the required electronic system, so that the array of digital functional blocks can be interconnected accordingly and complete the task efficiently. The fact that the interconnections can be configured and reconfigured to meet the desired objective makes FPGAs an excellent candidate to implement a reconfigurable computing system (Hartenstein 2001).

Although there is no standard way to implement reconfigurable computing systems, there are several common parameters used to classify them, which will be shortly discussed below.

17.3.1.1 Granularity

In a reconfigurable computing system, we can define a logic block as the smallest functional unit. The size of such blocks will determine the granularity of the system. Thus, small logic blocks (fine-grains) mean a high granularity, which can lead to greater flexibility during the implementation process. Nonetheless, greater flexibility comes with the price of higher power consumption, larger required area, and increased delay. On the other hand, low granularity (coarse-grains) systems are used when larger computations are needed, in which data paths become more useful than single-bit-based operations. If, however, a low granularity system is required to complete operations of smaller size, the allocated resources would be under-utilized and thus the system becomes inefficient. A possible solution would be to design a system with both fine and coarse grain arrays for optimized flexibility and performance (Zain-ul-Abdin and Svensson 2009). A perfect example of a high granularity system would be the FPGA, whilst a low granularity system can be implemented with a reconfigurable data path array (rDPA), in which multiple-bits data paths are used instead of single-bit-based functional blocks, as used in FPGAs.

17.3.1.2 Reconfigurability

A reconfigurable computing system should be able to run its reconfiguration process before or even during operation. In the case of a high granularity system, a longer configuration time will be required due to more elements to be programmed, compared to a low granularity system. In that sense, one advantage of coarse-grained systems is less time and energy requirements during the initial configuration or future reconfiguration processes. This will directly influence power consumption. An alternative method to reduce the power consumed during reconfiguration would be to allow it to occur at the same time that another part of the system is running its normal operation. This is defined as partial-reconfiguration, whose main advantage is to reduce the instructions needed for the smaller portion of the system to be reprogrammed, thus also increasing the efficiency of the reconfiguration process. Alternatively, a compression scheme of the reconfiguration instructions could also be implemented with the same aim as partial-reconfiguration, although the need for decompression should also be taken into account when looking at the total power consumption (Zain-ul-Abdin and Svensson 2009).

17.3.1.3 Routing/Interconnects

Another very important parameter to determine the flexibility of a reconfigurable system is its interconnections or routing scheme. The most common routing scheme, used in FPGAs, is an array-based arrangement with horizontal and vertical interconnects. In any case, a balanced routing is always the best way to ensure the best reconfiguration performance. For instance, insufficient or deficient routing can lead to less flexibility and poor utilization of resources, or too many interconnects can lead to higher resources and power usage (Zain-ul-Abdin and Svensson 2009).

17.3.1.4 Applications

The signal processing in many radar sensor systems can be very demanding. The large number of sensor channels, the need for a fast analogue-to-digital conversion-sampling rate, and the rigorous requirements on the filter design lead to very high processing requirements reaching a computational demand of up to hundreds of billion operations per second (GOPS). Even though custom very large scale integration (VLSI) designs can perform this kind of computations, the use of a reconfigurable computing system could bring many advantages. First, the design and manufacturing cost of FPGAs-based solutions is less than developing custom VLSI chips. Next, substituting the FPGA device with newer and more capable models can make the upgrade of the system very simple, for example, to achieve higher analog-to-digital converter (ADC) sampling rates. Finally, it is possible to implement different filtering functions thanks to the inherent flexibility of FPGA devices. In order to demonstrate the feasibility of a reconfigurable computing system in this practical scenario, the hardware implementation of a high-performance radar signal-processor was completed on a Xilinx Virtex chip (XCV1000), showing the same system performance as a custom VLSI design (Martinez et al. 2001).

17.3.2 Reconfigurability from Macro to Micro and Nanoelectronics

According to the size of the electronic system, there are different methodologies and opportunities to achieve certain levels of reconfiguration either at the electrical or mechanical level. Subsequently, a great diversity of applications can emerge from these reconfigurable systems. In the following subsections, a discussion about reconfigurability at different sizes will show several different approaches, with their own current challenges and applications.

17.3.2.1 Macro-size Reconfigurable Electronics: Flexible and Stretchable Printed Circuit Boards

Nowadays, we use printed circuit boards (PCBs) to implement almost all electronic systems. Being able to add new mechanical properties to the PCB will enable the development of novel wearable and ubiquitous technologies. For example, a much less stiff PCB that can exhibit flexibility and even stretchability would be appropriate for applications in on-body continuous measurements such as health monitoring devices or other scenarios in which mechanical decoupling is needed or preferred. Thus, mechanical reconfiguration can be introduced at the macro level to electronic systems of all kinds and open the door to novel applications such as e-textiles (Figure 17.1) and smart clothing (Hamedi et al. 2007, Cherenack and Van Pieterson 2012, Tao et al. 2017).

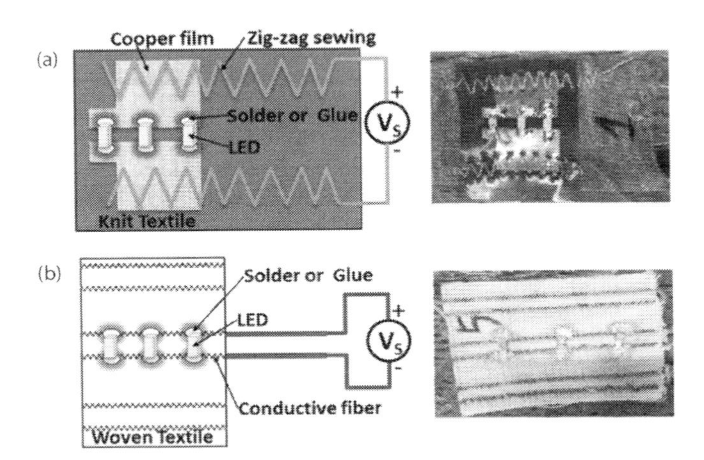

FIGURE 17.1 e-textiles implemented on a flexible PCB with three LEDs integration by (a) soldering and (b) by silver adhesive on conductive thread. (Reprinted with permission from Tao, X. et al., *Sensors*, 17 (4), 673, 2017. Copyright 2017 MDPI AG, Basel, Switzerland.)

Flexibility in a PCB can be easily obtained from commercially available products with a great diversity of thicknesses for copper and polymeric supporting material, usually polyimide. How the copper is attached to the polymeric material can also have different approaches ranging from direct metal growth through electrodeposition (ED) to rolled and annealed (RA). In both cases good adhesion can be achieved, although direct growth usually leads to a stronger attachment, and it is suitable and compatible with standard PCB manufacturing processes. Nevertheless, more defects in the ED copper might lead to lower conductivity, although this can be partially improved by heat treatment (McLeod 2002).

In the case of stretchable PCB, the relation between the metallic material for electric interconnects and the elastomeric material used as structural support needs to be special due to the distinct difference between their mechanical characteristics. Therefore, since the copper itself is not a naturally elastomeric material, it should be structured in an appropriate way; the most common structures being serpentines or horseshoes, which can elongate as the elastomeric material is stretched (Chtioui et al. 2016, Plovie et al. 2017). Section 3.4 elaborates further on possible structures useful for mechanical reconfiguration of the material itself.

17.3.2.2 Silicon-Based Reconfigurable Microelectronics

The world of electronics is constantly evolving and nowadays we are looking at a new paradigm in which electronic device's performance is no longer measured only by computing capability, but also by functionality, adaptability, and its capacity to be conformably integrated with mechanically demanding systems. Such new systems can be curvilinear, irregular, contain complex shapes and surfaces, and be in constant movement or change (Rojas et al. 2015, Nassar et al. 2016). From this new exchange between electronics and mechanics, the concept of mechanically reconfigurable electronics is leveraging the rise of novel technologies such as wearable electronics, bio-integrated devices, and active modules for the Internet of Things (IoT) (Swan 2012, Kim et al. 2015, Xu, Gutbrod et al. 2015, Koh et al. 2016, Haghi et al. 2017). For these new technologies to be part of our daily life, there exists still a big gap to be bridged. Current conventional electronics are mechanically rigid and brittle, and therefore their full integration with constantly shape-changing systems, such the human body, is still not possible. Although novel and ingenious ways to bridge this gap have been proposed and demonstrated at the research level, electric and mechanical performance and reliability needs to be improved a lot.

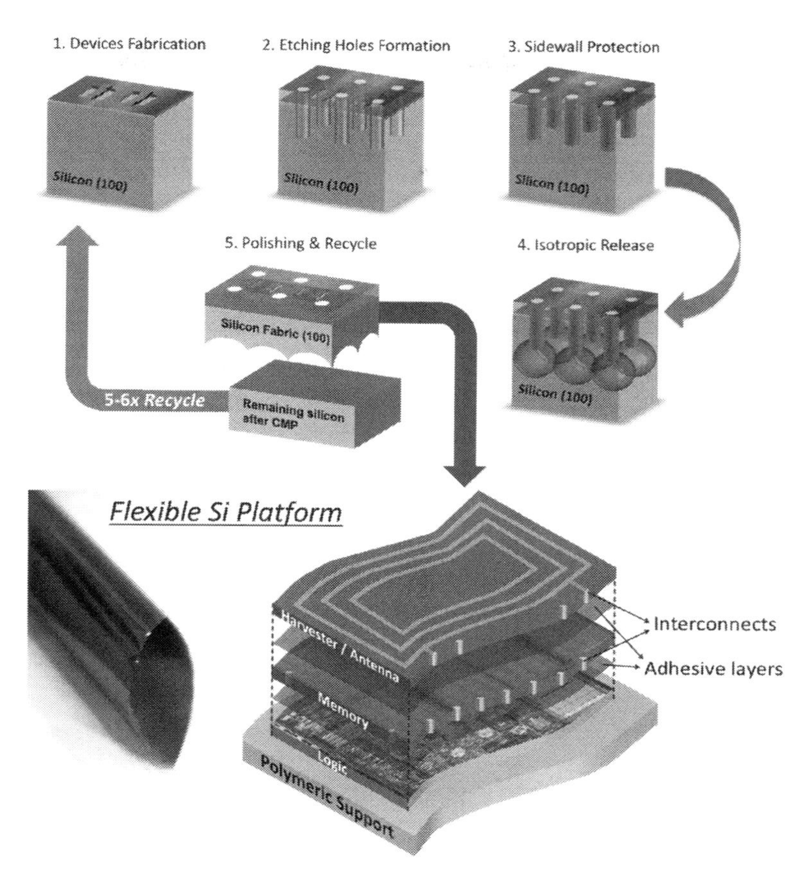

FIGURE 17.2 Fabrication process flow of flexible silicon platform and schematic of 3D flexible system implementation with digital photograph of curled-thin silicon sheet.

An interesting and highly manufacturable alternative technique consists of the formation of mechanically flexible, optically semitransparent, post-processed silicon sheets with pre-fabricated electronic devices. The technique is based on an ingenious sequence of dry anisotropic and isotropic etching steps. In summary, the peel-off process consists of deep trench formation followed by isotropic etching-based release, and it has been demonstrated with several devices and thicknesses (5–50 μm) by controlling various design parameters, such as the trenches' depth. Additionally, this technique is cost attractive since it allows the reutilization of the remaining substrate to release a total of 5–6 thin-Si-sheets out a single wafer (Rojas et al. 2012, Sevilla, Rojas et al. 2013). Figure 17.2 illustrates the process and the resulting flexible and semi-transparent silicon sheet, on top of which different applications can be implemented, from logic to energy harvesting modules (Rojas et al. 2013, Sevilla, Inayat et al. 2013, Alfaraj et al. 2015, Ghoneim et al. 2015) that could be interconnected using through silicon vias (TSV). Furthermore, this technique can be extended to other materials as has been previously demonstrated (Rojas, Torres, Sevilla et al. 2014).

It is important to clarify that flexibility is achieved due to the strong dependency of the flexural rigidity of a material to its thickness, as can be seen on the following expression (Symon 1971),

$$D = \frac{Et^3}{12\left(1-\nu^2\right)}, \tag{17.1}$$

where D is the flexural rigidity of a plate, E is the Young's modulus, t is the thickness, and ν is the Poisson's ratio. As can be observed, the rigidity, or resistance offered by a structure while undergoing bending, depends on the thickness to the power of three. Consequently, the thinner the structure, the much less resistance to bending it presents.

An extended implication of this concept allows us to form stretching structures out of rigid materials by clever use of enfolded ultra-thin arms in diverse shapes as will be elaborated in Section 3.4. For instance, an all-silicon-based ultra-stretchable platform has been demonstrated based on a double-arm spiral design. In such development, structural modifications were applied on a rigid material, such as silicon, in order to achieve not only flexibility, but also ultra-high, controllable stretchability (Rojas, Arevalo et al. 2014, Rehman and Rojas 2017).

17.3.2.2.1 **Silicon Crystal Orientation Consideration**

It is well known that single crystalline silicon has a highly anisotropic nature, thus, depending on the crystal orientation, its electrical and mechanical properties might vary. Consequently, it is important to bear this in mind while designing a silicon-based device, whose Young's modulus (or elastic modulus), for example, can vary as much as 45% depending on the crystalline plane. For the most common case, a silicon (100) wafer, the parallel and perpendicular directions to its wafer's flat (⟨110⟩ direction), exhibit a Young's modulus of 169 GPa, which can be used for calculations, simulations, and design purposes. In the case that a different wafer orientation must be used, it should be remembered that the ⟨111⟩ direction exhibits the highest Young's modulus value as 188 GPa, and that the ⟨100⟩ direction has the smallest Young's modulus as 130 GPa (Hopcroft et al. 2010).

17.3.2.2.2 **Drastic Reconfiguration: Transient or Destructible Electronics**

Drastic reconfiguration can be considered in cases where there is a sudden and complete destruction of the electronic device or when its component materials slowly dissolve until operation stops. The latter case is denominated transient electronics as its operation is constrained to a certain lifetime, which can be designed depending on the application. This scheme is especially useful for temporary functional systems such as medical implants, degradable environmental monitors, and hardware-secure systems. Such systems can be designed to fully or partially degrade its constituting materials in a controllable way through chemical or physical processes. When selecting materials for such technologies, it is important to consider that they must be not only electrically active when needed, but also biocompatible and biodegradable in general. Clear examples would be selecting single-crystalline silicon nanosheets or thin films of zinc oxide (ZnO) as semi-conductors, Zn, Mg, W, Mo, or Fe as metallic interconnects, SiO_2, SiN_x, or MgO as encapsulation layers and dielectrics, and poly(lactic acid) (PLA), polycaprolactone (PCL), or silk fibroin as mechanical support and packaging or as substrates (Hwang et al. 2015). Several examples of devices using these kinds of materials can be found ranging from simple electric devices and logic gates (such as transistors and inverters) (Hwang et al. 2014), to diverse sensors and solar cells (Hwang et al. 2012), and energy harvesters (Dagdeviren et al. 2013). In all these examples, the lifetime of devices can be designed by controlling the dissolution time of the encapsulating layers and substrate itself. Further control can be introduced by adding variances in thicknesses and types of materials to establish specific time sequences for dissolution of the system's components, thus enabling multi-staged reconfiguration. Such is the case of the reconfigurable transient electronic system, in which harmless, water-soluble materials were used to build diverse biodegradable electronic devices (Hwang et al. 2015).

For the case of destructible electronics, again chemical or physical processes can be used to fully or partially destroy the devices and stop their operation in a more abrupt way. Ultra-secure systems would benefit from this capability as long as the destruction can be triggered in a very controllable manner. For instance, Gao et al. reported an ultra-thin device fabricated on top of a polymeric substrate, which can be drastically deformed by heating application, and thus the ultra-thin device is destructed alongside (Figure 17.3). Furthermore, the thermal trigger can be controlled easily through an integrated resistive heater (Gao et al. 2017).

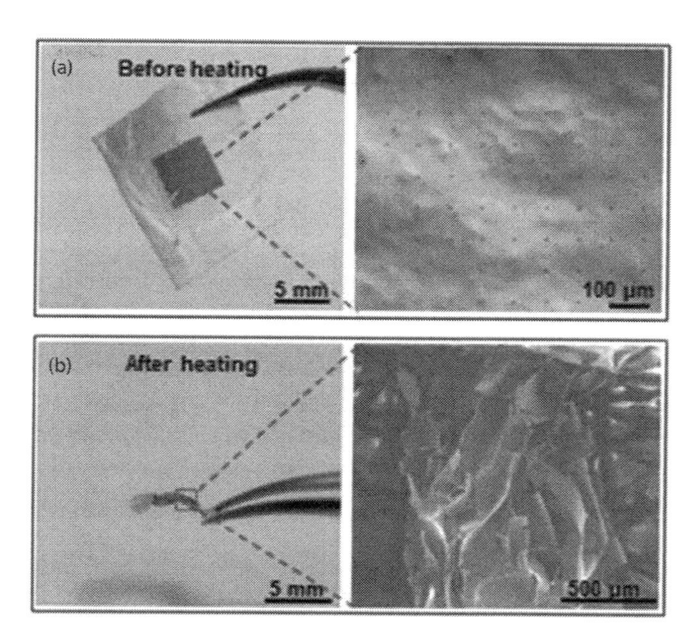

FIGURE 17.3 Digital photographs of thermally triggered destruction of Si membrane on polymer film (a) before and (b) after heating. (With kind permission from Springer Science+Business Media: *Scientific Reports*, Thermally triggered mechanically destructive electronics based on electrospun poly(ϵ-caprolactone) nanofibrous polymer films, 7, 2017, 947, Gao, Y. et al.)

17.3.2.2.3 Nanotechnology-Enabled Reconfigurability

Even at the nanometric-scale, reconfigurability appears in novel electronic devices such as nanowire-based transistors that are capable of reconfiguring their electronic characteristics to exhibit either n- or p-type semi-conductive properties. These devices integrate a fourth terminal that uses an additional electric signal to dynamically establish their charge carrier polarity. This new feature of single electronic devices opens up the door to dynamical reconfiguration with ultra-fine-tuning capabilities, allowing higher circuit design flexibility from the device level. For instance, Heinzig et al. demonstrated a nanowire-based transistor with heterostructure (metal/intrinsic-silicon/metal) and double gate on each Schottky junction. It showed excellent electrical characteristics with an outstanding on/off ratio, up to 9 decades for a Schottky transistor. In this case, the charge carrier polarity is selectively controlled by the charge carrier injections at each Schottky junction (Heinzig et al. 2012). Figure 17.4 shows a schematic of the proposed structure described above.

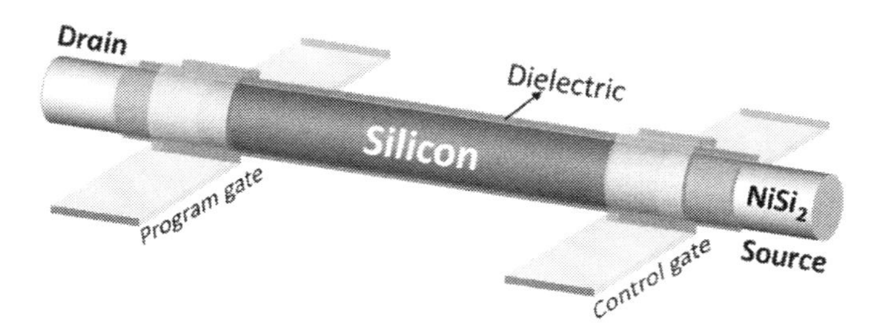

FIGURE 17.4 Schematic diagram of a reconfigurable silicon nanowire field effect transistor (FET) with two independent gates on top of the junctions (source and drain).

17.3.3 Materials in Mechanical Reconfigurability

Flexible and stretchable electronics have evolved into novel technologies with growing markets and huge potential such as wearable and bio-integrated systems. Promising new devices are coming to market, such as smart watches, glasses, jewelry and fashion, as well as fitness and health trackers, with increasing revenue every year (Delabrida Silva et al. 2017). The challenge remains with the mechanical characteristics needed to integrate the rigid, silicon-based electronics of nowadays with the soft, mobile, and stretchy nature of the human body. Less bulky and more compliant electronics would increase comfort and be appealing for consumers, hence boosting their market even more. Clever approaches to close this mechanical gap have been studied by several research groups. A first important approach consists of the use low cost and flexible organic materials, which have been already introduced to market with commercial flexible light-emitting diode (LED) screens (Thejo Kalyani and Dhoble 2012). Nevertheless, their intrinsic electrical properties are still well behind the current silicon-based devices, which limits their use to a smaller application niche (Nassar et al. 2016). Alternative materials involve novel 2-dimensional (2D) and 1-dimensional (1D) carbon-based structures such as carbon nanotubes (CNTs) and graphene (Kang et al. 2007, Lee, Kim et al. 2011). Even medium-scale integrated circuits (ICs) have been demonstrated using CNT random networks, but with a lower performance (Cao et al. 2008).

Although these demonstrations and novel materials can be encouraging, compared to the more complex and dense silicon-based electronics, such as the silicon-based VLSI and ultra large scale integration (ULSI) technologies, the techniques and integration levels of these demonstrations are still behind the current capability of packing a billion or trillion devices in a small area. More recently, several works have been completed to integrate the mechanical bendability and support of polymeric materials with the outstanding electric performance of silicon and other semi-conductors (Ahn et al. 2006, Ko et al. 2006, Sun et al. 2010). In this way, small-scale transistors can be fabricated from silicon micro/nano-membranes or ribbons (Lee et al. 2010, Kim et al. 2013), which then can be transferred on top of polymeric materials such as polydimethylsiloxane (PDMS), polyethylene terephthalate (PET) or polymide (PI) whose function is to be a mechanical flexible support to the electronics. Circuits with medium complexity have been already demonstrated through this approach (Ahn et al. 2007), although there is still room for improvement regarding the kind of substrates usually used to produce such ultra-thin silicon membranes, namely, silicon-on-insulator (SOI) or silicon (111), which are more expensive than the common silicon (100). Furthermore, a silicon (111) substrate also exhibits higher defect density than silicon (100) (Kato et al. 2004). Finally, and once again, the challenge remains in trying to achieve the ultra-high integration device density and nano-alignment of nowadays' electronics, which is not yet possible through these techniques.

17.3.4 Structural Reconfiguration

A crucial property of nowadays electronics is the ability to integrate three very different materials to manufacture a useful device. We are talking about conductors, semiconductors, and insulators. Besides their very different electric properties, these materials also differ from each other in their mechanical properties. In reconfigurable electronics, the challenge remains on how these materials can be integrated together given the dissimilar induced stress and its distribution for each material type. Although diverse integration methods have been proposed and demonstrated (Sanda et al. 2005, Kim, Won et al. 2009, Zhai et al. 2012), a common approach is to attach thin sheets of rigid inorganic materials (electrically active, but brittle and rigid materials such as silicon or other crystalline semiconductors) on top of soft substrates (usually an elastomer that would enable flexibility of the system). Just as described in the previous section, by transferring a less rigid version of the rigid material onto a bendable supporting layer, induced mechanical stress is sharply reduced in the overall structure, while preserving the exceptional electrical properties of the inorganic material on top (Kim et al. 2013). Although the concept of this technique is ingenious and practical, different mechanical failure modes might emerge depending on some characteristics of the materials, and more importantly, the adhesion means between them. For instance, if there is poor adhesion between inorganic and soft materials, delamination can occur upon bending. Even with a stronger adhesion, slipping and even cracking could also occur, depending on thicknesses and other material's characteristics, resulting in complete failure of devices (Park et al. 2008). On the other hand, if further stress reduction is needed, an additional layer of soft material can be placed on top, thus fully encapsulating the electrically active material. An added advantage to this scheme is the protection of the electrically active material from environmental conditions. Furthermore, if the electrically active material is placed at the neutral mechanical plane of the structure stack, which is neither under tensile nor compressive stress, then the mechanical performance can improve by effectively reducing the stress on the mechanically more rigid, electrically active material sheet (Park et al. 2008).

Concerning stretchable-capable reconfigurable electronics, it is possible to identify two main schemes, either: (i) engineer novel materials that incorporate both, advantageous mechanical and electrical properties or (ii) engineer structural modifications for established inorganic and rigid materials to reconfigure them into stretchable platforms (Rogers et al. 2010). The last approach makes use of the fact that flexural rigidity is drastically reduced with thickness. Hence, commonly rigid materials can be reconfigured to a certain degree of mechanical flexibility (Nassar et al. 2016). Such is the case of silicon, which is inherently rigid and brittle, but can become flexible after its thickness is reduced down to micro- or nano-size as discussed in Section 3.2.2. Likewise, going beyond flexibility, additional structural modifications to a thin sheet of a rigid material can further reconfigure its mechanical characteristic to make it exhibit even stretchable capabilities. For instance, ultra-thin silicon sheets transferred onto a pre-strained PDMS can exhibit an advantageous wavy configuration. In such a case, after the transfer process and the applied strain is released, the PDMS returns to its original size and the silicon sheets form the wavy structures as shown in Figure 17.5. The formed rippling pattern offers unidirectional stretchability as the structure is deformed and elastic response at strain release (Choi et al. 2007). An enhanced approach can be prepared by restructuring the straight-thin silicon sheets with serpentine shapes. In such case, the level of stretchability can be further improved thanks to the higher elongation that can be achieved by the twisting/buckling nature of the serpentine structures and their resulting unfolding (Kim et al. 2008). Another example of such serpentine structures can be seen in Figure 17.6 (Kim et al. 2009).

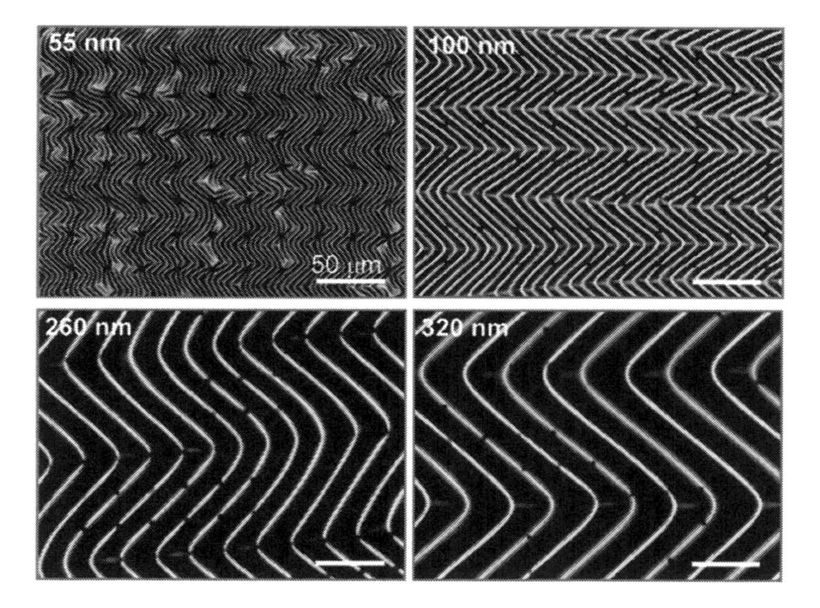

FIGURE 17.5 Optical micrographs of silicon wavy nano-membranes with various thickness (55, 100, 260, 320 nm) on pre-strained PDMS. (Reprinted with permission from Choi, W.M. et al., *Nano Letters*, 7, 1655–1663, 2007. Copyright 2007 American Chemical Society.)

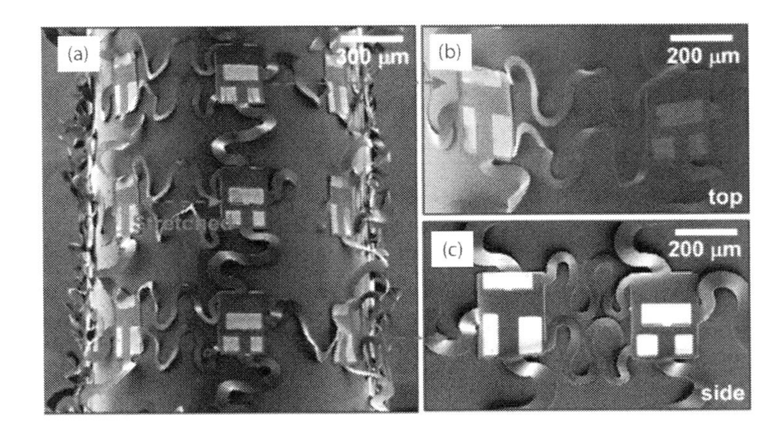

FIGURE 17.6 (a) Scanning electron microscopy image of array of complementary metal-oxide semiconductor (CMOS) inverters joined with serpentine interconnects. Zoom in views of serpentine (b) being elongated and (c) being compressed. (From Kim, H.S.: Self-assembled nanodielectrics and silicon nanomembranes for low voltage, flexible transistors, and logic gates on plastic substrates. *Applied Physics Letters*. 2009. 95. 183504. Copyright Wiley-VCH Verlag GmbH & Co. KGaA. Reproduced with permission.)

17.3.4.1 Interconnecting Rigid Islands on Soft Substrates

A direct extension of the approaches discussed earlier consist of forming an array of rigid islands of an electrically active material, where the electronic devices will be fabricated on, on top of a soft, elastomeric material. The key principle for the success of this scheme is to be able to form stretchable interconnects between the regularly distributed rigid islands over the elastic substrate, so that they can maintain an electrical connection between them, as shown in Figure 17.7 (Lee, Wu et al. 2011). Moreover, the final goal

FIGURE 17.7 Optical microscope images of stretchable GaAs photovoltaic modules in (a) un-stretched state and (b) 20% biaxial strain. (From Lee, J. et al., 2011. Stretchable GaAs photovoltaics with designs that enable high areal coverage. *Advanced Materials*. 2011. 23. 986–991. Copyright Wiley-VCH Verlag GmbH & Co. KGaA. Reproduced with permission.)

behind this scheme is to be able to isolate the brittle and rigid islands, containing the electronics, from experiencing any stress. On the other hand, in order to lessen the induced stress on the island's interconnects during deformation of the bottom soft substrate (stretching, bending, or even twisting), the design of the interconnects must be based on mechanically optimal structures. Unlike the fixed, rigid islands, these interconnects must be able to expand, upon substrate deformation, thanks to their unusual structure. Although commonly serpentine structures are used, some other structures have been proposed such as horseshoes and other similar derivatives of serpentines (Widlund et al. 2014). Even spirals have been demonstrated as well, which are able to withstand much larger strain deformations, and will be discussed further later on. One final characteristic of this approach is that it allows the possibility of arranging and organizing different components of a system (power management, sensor modules, communication, etc.) in different islands, which then can be interconnected for optimal distribution of a fully stretchable and reconfigurable system.

Unlike rigid islands, interconnects can be restructured to be able to stretch to even large deformations. Kim et al. proposed an array of islands on top of an elastomeric substrate, which were interconnected through buckled-arch-shaped structures as the serpentines shown in Figure 17.6 (Kim, Kim et al. 2009). Hence, the interconnections can move out-of-plane to move along with the elongated elastomer upon deformation. If the straight-arch-shaped interconnects are replaced by effectively longer serpentine structures, then higher deformation can be applied, with less impact on the stress experienced by the island and the interconnects themselves.

17.3.4.2 Fractals, Spirals, and Other Structures

A powerful alternative to serpentine designs comes from naturally occurring self-repeating structures known as fractals (Figure 17.8), with the potential to reduce the induced stress or to show even higher stretchability. Fan et al. demonstrated their use with a stretchable health monitor and communication device. In this work, it is shown that higher order fractal structures (higher self-repetition) display better stretchability properties (Fan et al. 2014). More recently, complex 3D out-of-plane structures were demonstrated by using multi-layer 2D precursors on a pre-strained polymeric substrate (Yan et al. 2016).

FIGURE 17.8 (a) 2D fractals of Peano structures with different iterations. (b) Optical image of Peano fractal-based wires on skin and scanning electron microscopy image of zoom-in view on skin-replica. (With kind permission from Springer Science+Business Media: *Nat. Commun.*, Fractal design concepts for stretchable electronics, 5, 2014, 3266, Fan, J. et al.)

A great variety of geometries were fabricated including entangled wavy arcs, blooming flowers, and circular cages. The applications are not merely for aesthetic appeal, but also original out-of-plane and stretchable applications were demonstrated, such as a helix-based tunable inductor for wireless communication. Xu et al. fabricated similar structures by using compressive buckling of 2D thin films into 3D out-of-plane structures, such as spirals, toroids, and helices (Xu, Yan et al. 2015).

Spirals can have actually advantageous mechanical characteristics among the aforementioned structures. In fact, it has been shown that spirals can provide larger stretchability as compared to serpentines with the same used area (plastic deformation was reached at 200% applied strain with spiral-based structures, compared to ~100% for serpentines) (Lv et al. 2014). Earlier, Huang et al. made use of double-arm spirals to build an ultra-stretchable, silicon-based array. It consisted of interconnected silicon spirals, whose centre can host electronic devices, and the spring-arms of the spirals act as physical and electrical interconnections. Very large area expansion ratios can be achieved with these structures (51 times the original size), a property that could have enormous potential in macro-electronics applications (Huang et al. 2007). Later on, an array of silicon-based hexagonal islands with double-arm spirals as physical interconnects were simulated and fabricated, as shown in Figure 17.9. An unprecedented stretchability higher than 1000% was demonstrated with these structures. Furthermore, the ultra-high stretch ratio can be controlled and designed, without requiring much more additional area, by controlling the number of the spiral turns along the inner circle (Rojas, Arevalo et al. 2014, Qaiser et al. 2017). Recently, also based on spiral structures, a stretchable thermoelectric generator (TEG) was demonstrated with the interesting characteristic of higher electric power generation while stretching. This happens due to the higher temperature difference between the two extremes of the stretched TEG (Rojas et al. 2016).

Further mechanical optimization of structures can be achieved through the efficient combination of structures, in a similar fashion as the self-repetition of fractals. Compound structures, can be then designed with the aim of reducing stress localization and improving the overall mechanical robustness. For instance, it has been shown that combining spiral structures with serpentine arms (Figure 17.10) can improve strain and stress distribution along the structure, while drastically reducing the maximum peaks of both stress and strain. In fact, after the integration of serpentine-arms in the spirals, the stress was reduced more than half the initial value with straight arms (Rehman and Rojas 2017).

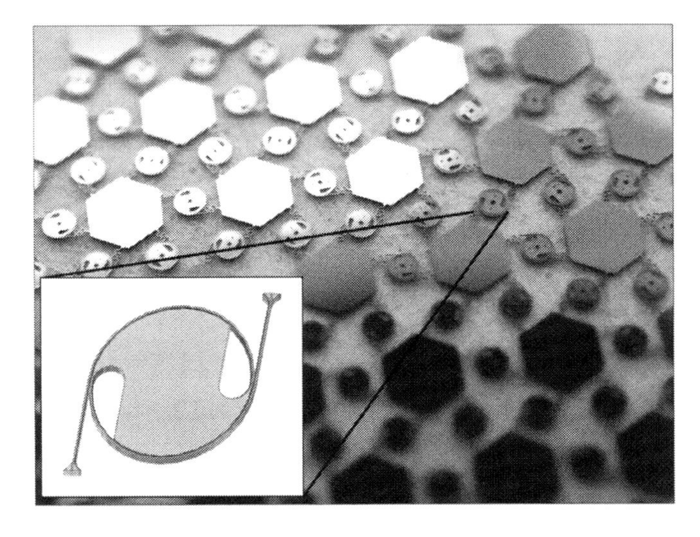

FIGURE 17.9 Digital photograph of array of hexagons interconnected by double-arm spirals. Inset shows schematic of spiral structure.

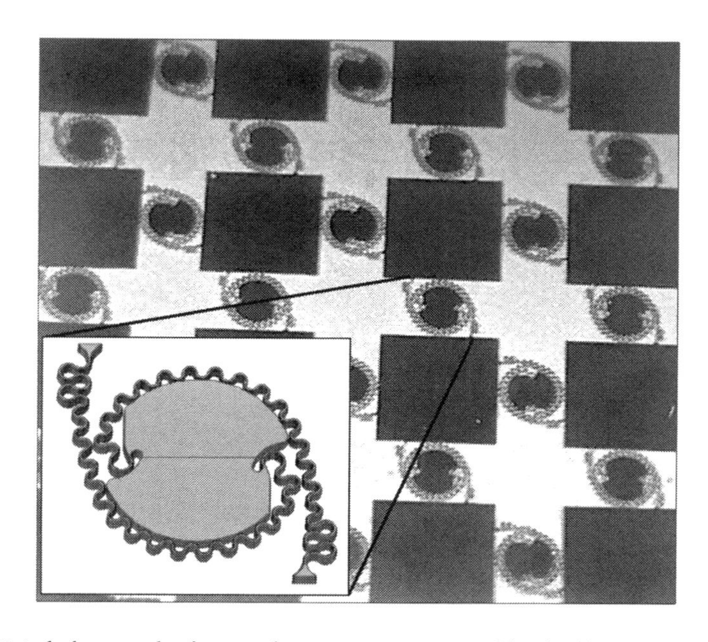

FIGURE 17.10 Digital photograph of array of squares interconnected by double-arm compound spirals. Inset shows schematic of serpentine-spiral structure.

17.4 Conclusions

Reconfigurable electronics have been around for a long time, but it wasn't until recently that novel concepts started to appear to extend the reconfigurability to new horizons. Initially, reconfigurable electronic systems were born from the need of incorporating adaptability to changes in the environment or new requirements mandated by the user's needs and comfort. More recently, the concept of reconfiguration has evolved beyond electrical circuit design towards new possibilities, such as material reconfiguration in transient electronics, and even mechanical adaptability through structural reconfiguration. The continuous rise of ingenious techniques looking for alternative ways to integrate materials together will keep redefining the concept of reconfiguration, as well as opening the door for more innovative solutions and still more novel applications.

References

Ahn, J.H., Kim, H.S., Lee, K.J., Jeon, S., Kang, S.J., Sun, Y., Nuzzo, R.G., and Rogers, J.A., 2006. Heterogeneous three-dimensional electronics by use of printed semiconductor nanomaterials. *Science*, 314 (5806), 1754–1757.

Ahn, J.-H., Kim, H.-S., Menard, E., Lee, K.J., Zhu, Z., Kim, D.-H., Nuzzo, R.G. et al., 2007. Bendable integrated circuits on plastic substrates by use of printed ribbons of single-crystalline silicon. *Applied Physics Letters*, 90 (21), 213501.

Alfaraj, N., Hussain, A.M., Torres Sevilla, G.A., Ghoneim, M.T., Rojas, J.P., Aljedaani, A.B., and Hussain, M.M., 2015. Functional integrity of flexible n-channel metal-oxide-semiconductor field-effect transistors on a reversibly bistable platform. *Applied Physics Letters*, 107 (17), 174101.

Cao, Q., Kim, H., Pimparkar, N., Kulkarni, J.P., Wang, C., Shim, M., Roy, K., Alam, M.A., and Rogers, J.A., 2008. Medium-scale carbon nanotube thin-film integrated circuits on flexible plastic substrates. *Nature*, 454 (7203), 495–500.

Cherenack, K. and Van Pieterson, L., 2012. Smart textiles: Challenges and opportunities. *Journal of Applied Physics*, 112 (9), 091301.

Choi, W.M., Song, J., Khang, D.Y., Jiang, H., Huang, Y.Y., and Rogers, J.A., 2007. Biaxially stretchable 'Wavy' silicon nanomembranes. *Nano Letters*, 7 (6), 1655–1663.

Chtioui, I., Bossuyt, F., de Kok, M., Vanfleteren, J., and Bedoui, M.H., 2016. Arbitrarily shaped rigid and smart objects using stretchable interconnections. *IEEE Transactions on Components, Packaging and Manufacturing Technology*, 6 (4), 533–544.

Dagdeviren, C., Hwang, S.-W., Su, Y., Kim, S., Cheng, H., Gur, O., Haney, R., Omenetto, F.G., Huang, Y., and Rogers, J.A., 2013. Transient, biocompatible electronics and energy harvesters based on ZnO. *Small*, 9 (20), 3398–3404.

Delabrida Silva, S.E., Rabelo Oliveira, R.A., and Loureiro, A.A.F., 2017. *Examining Developments and Applications of Wearable Devices in Modern Society*. Hershey, PA: IGI Global.

Estrin, G., 2002. Reconfigurable computer origins: The UCLA fixed-plus-variable (F+V) structure computer. *IEEE Annals of the History of Computing*, 24 (4), 3–9.

Fan, J., Yeo, W.-H., Su, Y., Hattori, Y., Lee, W., Jung, S.-Y., Zhang, Y. et al., 2014. Fractal design concepts for stretchable electronics. *Nature Communications*, 5, 3266.

Gao, Y., Sim, K., Yan, X., Jiang, J., Xie, J., and Yu, C., 2017. Thermally triggered mechanically destructive electronics based on electrospun poly(ϵ-caprolactone) nanofibrous polymer films. *Scientific Reports*, 7 (1), 947.

Ghoneim, M.T., Rojas, J.P., Young, C.D., Bersuker, G., and Hussain, M.M., 2015. Electrical analysis of high dielectric constant insulator and metal gate metal oxide semiconductor capacitors on flexible bulk mono-crystalline silicon. *IEEE Transactions on Reliability*, 64 (2), 579–585.

Haghi, M., Thurow, K., and Stoll, R., 2017. Wearable devices in medical internet of things: Scientific research and commercially available devices. *Healthcare Informatics Research*, 23 (1), 4–15.

Hamedi, M., Forchheimer, R., and Inganäs, O., 2007. Towards woven logic from organic electronic fibres. *Nature Materials*, 6, 357.

Hartenstein, R., 2001. A decade of reconfigurable computing: A visionary retrospective. In: *Proceedings of the Design, Automation and Test in Europe*, Piscataway, NJ: IEEE Press, 642–649.

Heinzig, A., Slesazeck, S., Kreupl, F., Mikolajick, T., and Weber, W.M., 2012. Reconfigurable silicon nanowire transistors. *Nano Letters*, 12 (1), 119–124.

Hopcroft, M.A., Nix, W.D., and Kenny, T.W., 2010. What is the Young's modulus of silicon? *Journal of Microelectromechanical Systems*, 19 (2), 229–238.

Huang, K., Dinyari, R., Lanzara, G., Jong, Y.K., Feng, J., Vancura, C., Chang, F.K., and Peumans, P., 2007. An approach to cost-effective, robust, large-area electronics using monolithic silicon. *Technical Digest: International Electron Devices Meeting, IEDM. IEEE*, 217–220.

Hwang, S.-W., Kang, S.-K., Huang, X., Brenckle, M.A., Omenetto, F.G., and Rogers, J.A., 2015. Materials for programmed, functional transformation in transient electronic systems. *Advanced Materials*, 27 (1), 47–52.

Hwang, S.-W., Song, J.-K., Huang, X., Cheng, H., Kang, S.-K., Kim, B.H., Kim, J.-H., Yu, S., Huang, Y., and Rogers, J.A., 2014. High-performance biodegradable/transient electronics on biodegradable polymers. *Advanced Materials*, 26 (23), 3905–3911.

Hwang, S.-W., Tao, H., Kim, D.-H., Cheng, H., Song, J.-K., Rill, E., Brenckle, M.A. et al., 2012. A physically transient form of silicon electronics. *Science*, 337 (6102), 1640.

Kang, S.J., Kocabas, C., Ozel, T., Shim, M., Pimparkar, N., Alam, M.A., Rotkin, S. V, and Rogers, J.A., 2007. High-performance electronics using dense, perfectly aligned arrays of single-walled carbon nanotubes. *Nature Nanotechnology*, 2 (4), 230–236.

Kato, Y., Takao, H., Sawada, K., and Ishida, M., 2004. The characteristic improvement of Si (111) metal-oxide-semiconductor field-effect transistor by long-time hydrogen annealing. *Japanese Journal of Applied Physics, Part 1: Regular Papers and Short Notes and Review Papers*, 43 (10), 6848–6853.

Kim, D.H., Kim, Y.S., Wu, J., Liu, Z., Song, J., Kim, H.S., Huang, Y.Y., Hwang, K.C., and Rogers, J.A., 2009. Ultrathin silicon circuits with strain-isolation layers and mesh layouts for high-performance electronics on fabric, vinyl, leather, and paper. *Advanced Materials*, 21 (36), 3703–3707.

Kim, D.-H., Song, J., Choi, W.M., Kim, H.-S., Kim, R.-H., Liu, Z., Huang, Y.Y., Hwang, K.-C., Zhang, Y.-W., and Rogers, J.A., 2008. Materials and noncoplanar mesh designs for integrated circuits with linear elastic responses to extreme mechanical deformations. *Proceedings of the National Academy of Sciences*, 105 (48), 18675–18680.

Kim, H.S., Won, S.M., Ha, Y.G., Ahn, J.H., Facchetti, A., Marks, T.J., and Rogers, J.A., 2009. Self-assembled nanodielectrics and silicon nanomembranes for low voltage, flexible transistors, and logic gates on plastic substrates. *Applied Physics Letters*, 95 (18), 183504.

Kim, J., Banks, A., Cheng, H., Xie, Z., Xu, S., Jang, K.-I., Lee, J.W. et al., 2015. Epidermal electronics with advanced capabilities in near-field communication. *Small*, 11 (8), 906–912.

Kim, T., Hwan Jung, Y., Chung, H.-J., Jun Yu, K., Ahmed, N., Corcoran, C.J., Suk Park, J., Hun Jin, S., and Rogers, J.A., 2013. Deterministic assembly of releasable single crystal silicon-metal oxide field-effect devices formed from bulk wafers. *Applied Physics Letters*, 102 (18), 182104.

Ko, H.C., Baca, A.J., and Rogers, J.A., 2006. Bulk quantities of single-crystal silicon micro-/nanoribbons generated from bulk wafers. *Nano Letters*, 6 (10), 2318–2324.

Koh, A., Kang, D., Xue, Y., Lee, S., Pielak, R.M., Kim, J., Hwang, T. et al., 2016. A soft, wearable microfluidic device for the capture, storage, and colorimetric sensing of sweat. *Science Translational Medicine*, 8 (366), 366ra165.

Lee, J., Wu, J., Shi, M., Yoon, J., Park, S.-Il, Li, M., Liu, Z., Huang, Y., and Rogers, J.A., 2011. Stretchable GaAs photovoltaics with designs that enable high areal coverage. *Advanced Materials*, 23 (8), 986–991.

Lee, K.J., Ahn, H., Motala, M.J., Nuzzo, R.G., Menard, E., and Rogers, J.A., 2010. Fabrication of microstructured silicon (ms-Si) from a bulk Si wafer and its use in the printing of high-performance thin-film transistors on plastic substrates. *Journal of Micromechanics and Microengineering*, 20 (7), 75018.

Lee, S.-K., Kim, B.J., Jang, H., Yoon, S.C., Lee, C., Hong, B.H., Rogers, J.A., Cho, J.H., and Ahn, J.-H., 2011. Stretchable graphene transistors with printed dielectrics and gate electrodes. *Nano Letters*, 11 (11), 4642–4646.

Lv, C., Yu, H., and Jiang, H., 2014. Archimedean spiral design for extremely stretchable interconnects. *Extreme Mechanics Letters*, 1, 29–34.

Martinez, D.R., Moeller, T.J., and Teitelbaum, K., 2001. Application of reconfigurable computing to a high performance front-end radar signal processor. *Journal of VLSI Signal Processing Systems for Signal, Image and Video Technology*, 28 (1), 63–83.

McLeod, P., 2002. *A Review of Flexible Circuit Technology and Its Applications*. Loughborough, UK: PRIME Faraday Partnership.

Nassar, J.M., Rojas, J.P., Hussain, A.M., and Hussain, M.M., 2016. From stretchable to reconfigurable inorganic electronics. *Extreme Mechanics Letters*, 9, 245–268.

Park, S.-Il, Ahn, J.H., Feng, X., Wang, S., Huang, Y., and Rogers, J.A., 2008. Theoretical and experimental studies of bending of inorganic electronic materials on plastic substrates. *Advanced Functional Materials*, 18 (18), 2673–2684.

Plovie, B., Yang, Y., Guillaume, J., Dunphy, S., Dhaenens, K., Van Put, S., Vandecasteele, B., Vervust, T., Bossuyt, F., and Vanfleteren, J., 2017. Arbitrarily shaped 2.5D circuits using stretchable interconnects embedded in thermoplastic polymers. *Advanced Engineering Materials*, 19 (8), 1700032.

Qaiser, N., Khan, S.M., Nour, M., Rehman, M.U., Rojas, J.P., and Hussain, M.M., 2017. Mechanical response of spiral interconnect arrays for highly stretchable electronics. *Applied Physics Letters*, 111 (21), 214102.

Rehman, M.U. and Rojas, J.P., 2017. Optimization of compound serpentine–spiral structure for ultra-stretchable electronics. *Extreme Mechanics Letters*, 15, 44–50.

Rogers, J.A., Someya, T., and Huang, Y., 2010. Materials and mechanics for stretchable electronics. *Science*, 327 (5973), 1603–1607.

Rojas, J.P., Arevalo, A., Foulds, I.G., and Hussain, M.M., 2014. Design and characterization of ultra-stretchable monolithic silicon fabric. *Applied Physics Letters*, 105 (15), 154101.

Rojas, J.P., Hussain, A.M., Arevalo, A., Foulds, I.G., Torres Sevilla, G.A., Nassar, J.M., and Hussain, M.M., 2015. Transformational electronics are now reconfiguring. In: *Proceedings of SPIE*. Baltimore, MD, 946709.

Rojas, J.P., Sevilla, G.A.T., and Hussain, M.M., 2013. Can we build a truly high performance computer which is flexible and transparent? *Scientific Reports*, 3, 2609.

Rojas, J.P., Singh, D., Conchouso, D., Arevalo, A., Foulds, I.G., and Hussain, M.M., 2016. Stretchable helical architecture inorganic-organic hetero thermoelectric generator. *Nano Energy*, 30, 691–699.

Rojas, J.P., Syed, A., and Hussain, M.M., 2012. Mechanically flexible optically transparent porous mono-crystalline silicon substrate. In: *Proceedings of the IEEE International Conference on Micro Electro Mechanical Systems (MEMS)*. Paris, France, 281–284.

Rojas, J.P., Torres Sevilla, G.A., Ghoneim, M.T., Inayat, S. Bin, Ahmed, S.M., Hussain, A.M., and Hussain, M.M., 2014. Transformational silicon electronics. *ACS Nano*, 8 (2), 1468–1474.

Sanda, H., McVittie, J., Koto, M., Yamagata, K., Yonehara, T., and Nishi, Y., 2005. Fabrication and characterization of CMOSFETs on porous silicon for novel device layer transfer. In: *Electron Devices Meeting, 2005. IEDM Technical Digest. IEEE International*. Washington, DC, 679–682.

Sevilla, G.A.T., Inayat, S. Bin, Rojas, J.P., Hussain, A.M., and Hussain, M.M., 2013. Flexible and semi-transparent thermoelectric energy harvesters from low cost bulk silicon (100). *Small*, 9 (23), 3916–3921.

Sevilla, G.T., Rojas, J.P., Ahmed, S., Hussain, A., Inayat, S.B., and Hussain, M.M., 2013. Silicon fabric for multi-functional applications. In: *The 17th International Conference on Solid-State Sensors, Actuators and Microsystems, Transducers And Eurosensors 2013*. Barcelona, Spain, 2636–2639.

Sun, L., Qin, G., Seo, J.-H., Celler, G.K., Zhou, W., and Ma, Z., 2010. 12-GHz thin-film transistors on transferrable silicon nanomembranes for high-performance flexible electronics. *Small*, 6 (22), 2553–2557.

Swan, M., 2012. Sensor mania! The internet of things, wearable computing, objective metrics, and the quantified self 2.0. *Journal of Sensor and Actuator Networks*, 1 (3), 217–253.

Symon, K.R., 1971. *Mechanics*. 3rd ed. Reading, UK: Addison Wesley.

Tao, X., Koncar, V., Huang, T.H., Shen, C.L., Ko, Y.C., and Jou, G.T., 2017. How to make reliable, washable, and wearable textronic devices. *Sensors (Switzerland)*, 17 (4), 673.

Thejo Kalyani, N. and Dhoble, S.J., 2012. Organic light emitting diodes: Energy saving lighting technology - A review. *Renewable and Sustainable Energy Reviews*, 16 (5), 2696–2723.

Widlund, T., Yang, S., Hsu, Y.Y., and Lu, N., 2014. Stretchability and compliance of freestanding serpentine-shaped ribbons. *International Journal of Solids and Structures*, 51 (23–24), 4026–4037.

Xu, L., Gutbrod, S.R., Ma, Y., Petrossians, A., Liu, Y., Webb, R.C., Fan, J.A. et al., 2015. Materials and fractal designs for 3D multifunctional integumentary membranes with capabilities in cardiac electrotherapy. *Advanced Materials*, 27 (10), 1731–1737.

Xu, S., Yan, Z., Jang, K.-I., Huang, W., Fu, H., Kim, J., Wei, Z. et al., 2015. Assembly of micro/nanomaterials into complex, three-dimensional architectures by compressive buckling. *Science*, 347 (6218), 154.

Yan, Z., Zhang, F., Liu, F., Han, M., Ou, D., Liu, Y., Lin, Q. et al., 2016. Mechanical assembly of complex, 3D mesostructures from releasable multilayers of advanced materials. *Science Advances*, 2 (9), e1601014.

Zain-ul-Abdin and Svensson, B., 2009. Evolution in architectures and programming methodologies of coarse-grained reconfigurable computing. *Microprocessors and Microsystems*, 33 (3), 161–178.

Zhai, Y., Mathew, L., Rao, R., Xu, D., and Banerjee, S.K., 2012. High-performance flexible thin-film transistors exfoliated from bulk wafer. *Nano Letters*, 12 (11), 5609–5615.

18

Flexible and Stretchable Devices for Human-Machine Interfaces

Irmandy Wicaksono
and
Canan Dagdeviren

18.1 Introduction

The discovery of the transistor in the mid-twentieth century has revolutionized modern electronics and transformed the way we live, experience, and interact with the world today. Vast developments in semiconductor materials and fabrication technologies enabled exponential growth of transistor counts and miniaturization of electronic systems, down from logic gates to integrated circuits (IC), as predicted by Moore's law. With further advances in large-scale integration, compact packaging, high durability, and low-power design of microelectronics, electronic devices with many functionalities have become ubiquitous and embedded in all aspects of our life. They have now been applied to a broad range of applications, from entertainment, health-care, business development, to information technology, all converging to augment the way we interact with digital information and support our daily needs and goals.

The miniaturization, reduced cost, and improved performance of these electronic devices and systems have also sparked various technological breakthroughs in computing interfaces and pushed forward their

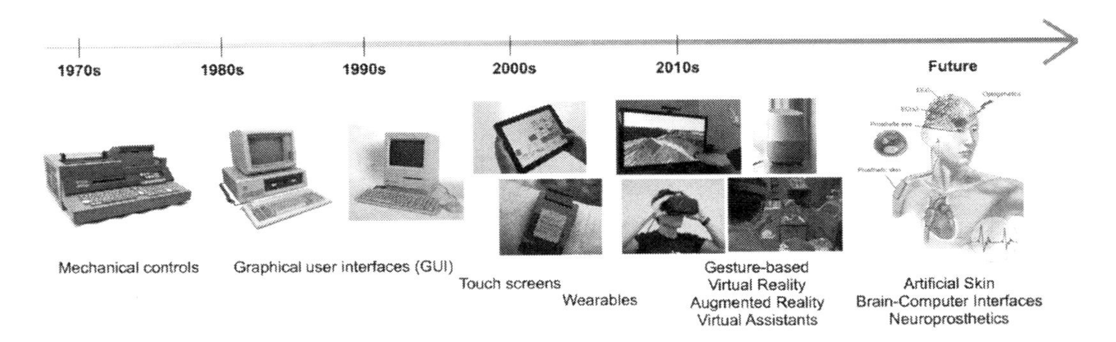

FIGURE 18.1 Timeline of human-computer interface technologies. Mechanical controls in Olivetti P6060. (Reprinted from Museoscienza.org. 1976, Olivetti P6066, 1976.) GUIs in IBM5150. (Reprinted from Ruben de Rijcke, IBM PC 5150, 2010.) and Apple Macintosh Classic. (Reprinted from Schaelss, Alexander, Apple Macintosh Classic, 2004.) capacitive touch sensing in consumer tablet, wearable smartwatch. (Reprinted from Peeble Technology, Peeble Watch Email 1, 2012.) gesture vision sensing in Kinect. (Reprinted from Open Exhibits, Open. 2011, 360 Degree VR Image Viewer Controlled by Kinect, 2011.) virtual reality headset, augmented reality experience in 1st person view. (Reprinted from Sweden, Microsoft, win10-HoloLens-Minecraft, 2015.) Google Home virtual assistant. (Reprinted from Exhibits, Open. 2011, 360 Degree VR Image Viewer Controlled by Kinect, 2011.) and the vision of bio-integrated electronics for human-machine interfaces. (Adapted from Choi, Suji et al., *Adv. Mater.*, 28, 4203–4218, 2016; With kind permission from Springer Science+Business Media: *Nature*, A Hemispherical Electronic Eye Camera Based on Compressible Silicon Optoelectronics, 454, 2008, 748–753, Ko, Heung Cho et al.)

commercialization as consumer electronics. Human-computer or human-machine interface (HMI) requires an input mechanism that sends information or requests from a human to the machine. The machine then processes this command and relays feedback in the form of output. This information can be ergonomically communicated through various means, either through visual, auditory, or other cognitive or physical interaction channels (Jaimes and Sebe 2007). As illustrated in Figure 18.1, the first computing devices required simple mechanical switches for user controls. After the appearance of the first series of personal computers in the 1980s, there has been a rapid development in new interface technologies that enable users to interact with the digital world intuitively. Several user interface technologies have become matured and widespread, such as the mouse and keypad set to control a graphical user interface (GUI) in a personal computer, touch-sensitive devices that have become default controls of today's' tablets and smartphones, as well as wearables that actively respond to the activity or physiological state of the users.

Continuous innovation in sensing, display, hardware processing, and software developments around the early 21st century have also triggered advances in technologies for immersive sensory experiences, such as virtual and augmented reality. With haptic feedback, gesture-sensing, voice-recognition devices, and head-mounted displays, these technologies attempt to radically change the way users perceive and interact with digital information. It is done by bringing them into a computer-simulated reality or by overlaying digital information to the physical world respectively. They also aim to reduce users' memory load and provide intuitive, seamless interaction between human and computers. However, they are still under early developmental stage before becoming universal. Their relatively large size, limited mobility, and in some cases, additional setup for the controllers limit the applications and restrain them from becoming truly wearable and being used continuously. Extensive research in brain-computer interfaces and neuro-prosthetics has also been recently conducted to ultimately accomplish a direct bi-directional control between the nervous system and external devices. Current research trends show various invasive technologies for a high-quality sensing, actuation, and direct access to the neural tissues. Nonetheless, invasive neural interfacing approaches provide several technological challenges. Standard electronic devices are not compatible with the human body, as they are commonly planar, rigid, and several orders of magnitude stiffer than the body tissue.

In this light, recent advances in new materials, device designs, and fabrication strategies have established a new form of soft electronics that are biocompatible and can be flexed and stretched to bridge the biological, geometrical, and mechanical mismatch between electronics and the human body (Rogers 2015). They enable a myriad of novel wearable and implantable applications, from physiological and activity monitoring, physical interaction media, prosthetics, to robotics. As discussed in the previous chapters of this book and illustrated in Figure 18.2, the performance and applications of these flexible and stretchable devices are defined by their material properties, choice of substrates, and fabrication techniques. Most of the state-of-the-art flexible electronics are developed by fabricating (Ahn et al. 2006) or transfer-printing (Meitl et al. 2006) devices on flexible substrates or by thinning down silicon wafers with methods such as dry etching, wet chemical etching, grinding, chemical-mechanical polishing, and exfoliating (Feil et al. 2003; Gumus et al. 2017; Zhai et al. 2012). On the other hand, stretchable electronics are realized by developing intrinsically stretchable materials (Wang et al. 2017), designing serpentine interconnect architecture that allows stretching of rigid structures on an elastomeric substrate (Zhang et al. 2013a), or leveraging buckling mechanisms by pre-stretching an elastomeric substrate (Sun et al. 2006).

To bolster technological translation, using materials and substrates that are imperceptible, transparent, and have self-healing properties is also essential (Benight et al. 2013; Salvatore et al. 2014). Self-healing ability allows certain electronic devices to regain back their mechanical and electrical characteristics upon minor damage and is particularly useful for devices that experience accidental cuts or scratches. In addition, ultra-thin and transparent designs allow intimate integration, enhance comfortability, and support a widespread user acceptance.

This chapter will give an overview of existing flexible and stretchable sensors, actuators, and transducers systems with a focus on their applications for the next generation of HMI. The mechanically adaptive features of these artificial skin, neural implants, and sensory prostheses facilitate conformal and intimate integration to different regions of the human body. Depending on their modalities and locations, as listed in Table 18.1, these collective devices can give insights to various physiological and biomechanical changes of the human body. They can also further realize the ultimate goal of seamless bi-directional communication between machines and the human nervous system. This effort promises to restore and augment our physical and mental capabilities and will radically transform the way we interact with computers in the future.

18.2 On-body Interfaces

The rise of ubiquitous computing and the Internet of Thing has brought us into an era where our physical and digital worlds start to blend, seamlessly coupling without boundaries. However, current wearable devices, mobile phones, and remote displays are still limited in their active surface and accessibility in particular scenarios. Given the fact that human skin and its appendages are easily accessible, intimate to us, and they provide a vital sensory barrier between ourselves and the environments, visions start to emerge of using on-body electronics, or electronics that merge with the body as extensions of self (Leigh et al. 2017). Leveraging our skin as mobile input and output surfaces that are with us and available all the time can result in compelling human-computer interaction (HCI) applications (Steimle 2016). This vision is further supported by the current developments of flexible and stretchable devices, which are light-weight, seamless, aesthetically pleasing, and compatible with the skin, making them comfortable and socially acceptable to everyone.

18.2.1 On-body Inputs

Several techniques have been adapted for the development of touch panels, including capacitive, resistive, surface acoustic wave, and infrared-based sensing. Capacitive and resistive methods, in particular, are widely used for on-body applications. Compared to the others, these methods are more approachable

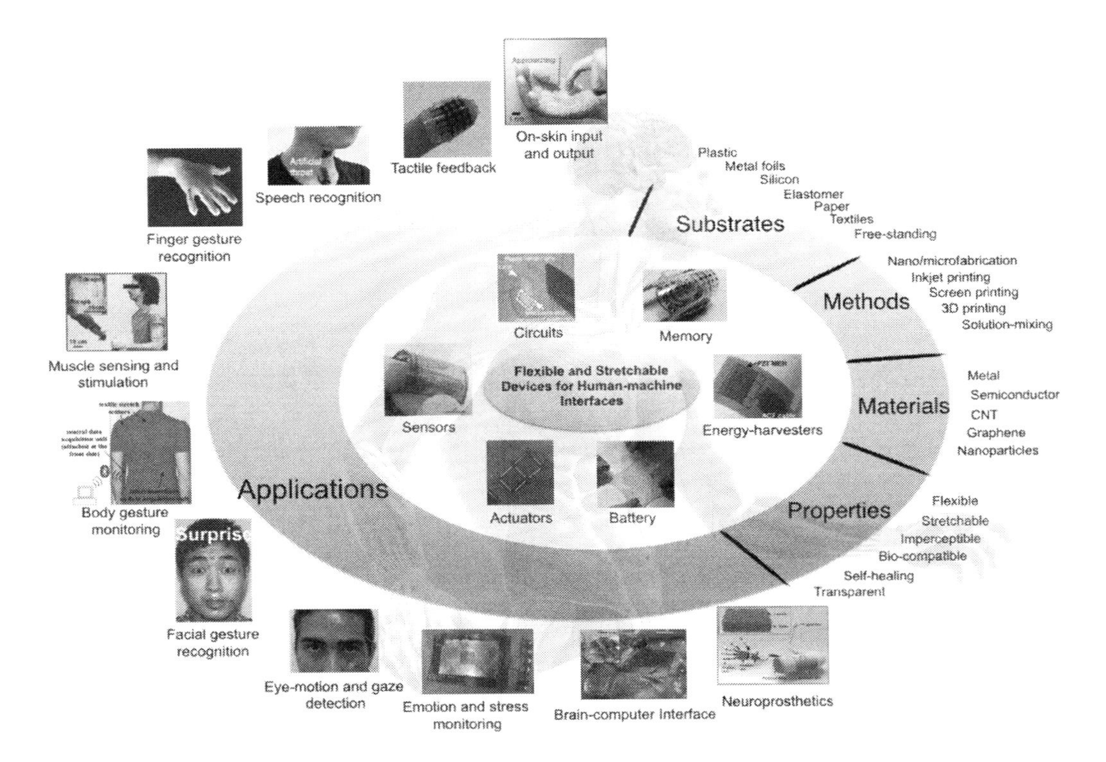

FIGURE 18.2 Flexible and stretchable devices for human-machine interfaces. Various types of electronic devices (clockwise): circuits, memory, energy-harvesters, battery, actuators, and sensors on unconventional substrates for flexible and stretchable systems. (With kind permission from Springer Science+Business Media: *Nat. Commun.*, Wafer-Scale Design of Lightweight and Transparent Electronics That Wraps around Hairs, 5, 2014, 2982, Salvatore, Giovanni A. et al.; With kind permission from Springer Science+Business Media: *Nat. Commun.*, Stretchable Batteries with Self-Similar Serpentine Interconnects and Integrated Wireless Recharging Systems, 4, 2013, Xu, Sheng et al.); (Reprinted with permission from Dagdeviren, Canan et al. 2014, Conformal Piezoelectric Energy Harvesting and Storage from Motions of the Heart, Lung, and Diaphragm, *Proc. Natl. Acad. Sci.*,111, 1927–1932, Copyright (2014) National Academy of Sciences, USA.); Reprinted with permission from Kim, Soo Jin, and Jang Sik Lee. *Nano Lett.*, 10, 2884–2890, 2010. Copyright 2010 American Chemical Society; Yokota, Tomoyuki et al., Ultraflexible Organic Photonic Skin, *Sci. Adv.*, 2, e1501856–e1501856, 2016. Reprinted with permission of AAAS). Applications of flexible and stretchable sensors and actuators for (anti-clockwise) on-skin input and output, tactile feedback, voice recognition, data glove, muscle sensing and stimulation, body gesture recognition, facial gesture recognition, eye-motion and gaze detection, stress and emotion monitoring, brain-computer interface, and neuroprosthetics. (Reprinted with permission from Kang, Minpyo. et al., *ACS Nano*, 11, 7950–7957. Copyright 2017 American Chemical Society; Ko, Heung Cho et al., *Nature*, 454, 748–753, 2008. Copyright 2014. IEEE with permission; With kind permission from Springer Science+Business Media: *Nat. Commun.*, An Intelligent Artificial Throat with Sound-Sensing Ability Based on Laser Induced Graphene, 8, 2017, 14579, Tao, Lu-Qi et al.; With kind permission from Springer Science+Business Media: *Scientific Reports*, A Flexible and Wearable Human Stress Monitoring Patch, 6, 2016, 23468, Yoon, Sunghyun et al.; With kind permission from Springer Science+Business Media: *Nature Neuroscience*, Flexible, Foldable, Actively Multiplexed, High-Density Electrode Array for Mapping Brain Activity in Vivo, 14, 2011, 1599–1605, Viventi, Jonathan et al.; With kind permission from Springer Science+Business Media: *Nat. Commun.*, Stretchable Silicon Nanoribbon Electronics for Skin Prosthesis, 5, 2014, 5747, Kim et al.; From Muth, Joseph T., et al., Embedded 3D Printing of Strain Sensors within Highly Stretchable Elastomers. *Adv. Mater.* 2014. 26. 6307–6312. Copyright Wiley-VCH Verlag GmbH & Co. KGaA. Reproduced with permission.; Su, Meng et al., *Adv. Mater.*, 28, 1369–1374, 2016; Mattmann et al. (2008), Copyright MDPI; Reprinted from *Biosens. Bioelectron.*, 91, Mishra, Saswat et al., Soft, Conformal Bioelectronics for a Wireless Human-Wheelchair Interface, 796–803. Copyright 2017 with permission for Elsevier.)

TABLE 18.1 Various HMI Applications Enabled by Flexible and Stretchable Devices

Application	Sensing	Actuation	Location	References
On-body Interfaces	Touch Pressure Bend	Display Tactile	Skin surface Clothing	Weigel et al. (2015) Wang et al. (2013) Yokota et al. (2016) Poupyrev et al. (2016)
Body Gesture Recognition and Activity Monitoring	Acceleration Electromyography Strain		Body segments Hand, Arm, Forearm Body joints	Varga et al. (2017) Jeong et al. (2013) Muth et al. (2014)
Voice and Speech Recognition	Vibration Electromyography Strain Pressure		Neck	Liu et al. (2016) Wang et al. (2015) Yang et al. (2015)
Facial Gesture Recognition	Electromyography Electrooculography Strain		Facial skin	Lee et al. (2017) Paul et al. (2014) Su et al. (2016)
Eye Motion and Gaze Detection	Electrooculography Strain Magnetic Field		Upper, lower eyelid, and canthus Cornea	Mishra et al. (2017) Lee et al. (2014) Whitmire et al. (2016)
Emotion and Stress Recognition	Electrocardiography Electrodermal Activity Electrooculography Electromyography Blood Pressure Blood Oxygenation		Chest Arm Forehead	Yoon et al. (2016) Jang et al. (2014)
Brain-Computer Interface and Neuroprosthetics	Electroencephalography Electrocorticography Local Field Potential	Electrical Chemical Optical	Transcranial Intracranial	Norton et al. (2015) Viventi et al. (2011) Toda et al. (2011) Minev et al. (2015) Canales et al. (2015) Tee et al. (2015)

and scalable. One of the first explorations of on-body touch-sensing using stretchable sensors for user interface applications was conducted by Weigel et al. (2015). iSkin is a flexible and stretchable array of touch sensors fabricated by laser-patterning of carbon black and polydimethylsiloxane composite (PDMS). A spacing layer is made by making holes across a PDMS layer. This layer is then sandwiched between two conductive carbon PDMS layers to a make touch-sensitive sensor. Possible applications of this device range from a headset control by laminating the stretchable touch sensing skin at the back of the ear, a one-handed slider or touch control by wrapping the sensor skin on the fingers, to a rollout and on-skin keyboard control by attaching the sensor skin on a smartwatch and surface of the forearm.

Another approach, as shown in Figure 18.3a, utilizes a multi-layer of graphene grown by chemical vapor deposition (CVD) and transferred onto a polyethylene (PET) substrate (Kang et al. 2017). The study, using similar approach as Weigel et al. (2015), leveraged the mutual capacitive coupling in between two electrodes to detect a presence of finger touch. As our fingers get close to the sensing layer, they disrupt the electromagnetic field due to the role of our body as a ground conductor and reduce the total capacitance of the sensor. Depending on the sensor design and active area, this method could also detect near-proximity gestures of up to 7 cm (Figure 18.3b). The row-column architecture enables multi-touch gesture capability (Figure 18.3c), and the addition of the third bottom ground layer eliminates the noise caused by the change in surface charge of the skin underneath due to sweat and other physical conditions. DuoSkin presents a similar concept of on-body touch sensing by using a low-cost material, which is gold leaf, to rapidly fabricate temporary tattoo (Kao et al. 2016). Instead of using a mutual capacitive method, a floating or self-capacitive touch sensing mechanism is applied in this work where a finger strike in contrast increases the total capacitance of the sensor. By experimenting with the electrode design, one can design not only

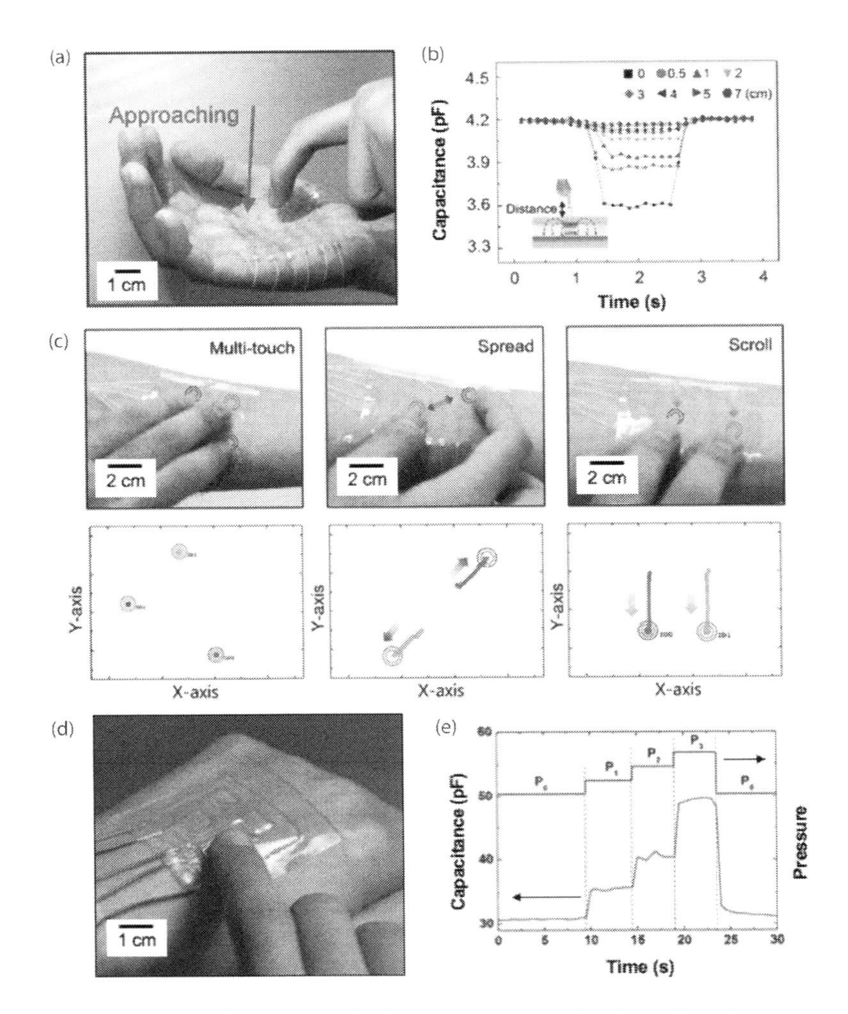

FIGURE 18.3 On-skin touchpad (a) image of a finger approaching the 3D conformable sensor on the palm. (b) Capacitance change *versus* time for finger touch and proximity of various distances between 0 and 7 cm. (c) Demonstration of graphene-based wearable capacitive touch sensor for multi-touch, spread, and scroll user interaction. (Reprinted with permission from Kang, Minpyo. et al., *ACS Nano*, 11, 7950–7957. Copyright 2017 American Chemical Society.) (d) Ionic pressure sensor laminated on the back of the hand with a pressing interaction. (e) Capacitance change versus time in one location as finger pressure gets stronger and back to initial condition. (Adapted with permission from Sun, Jeong Yun et al., *Adv. Mater.*, 26, 7608–7614, 2014.)

a touch button, but also a a continuous slider and an XY pad. The ease of DuoSkin fabrication and the use of low-cost material allow people to design and personalize the shape and functionality of their own on-skin touch sensor.

Other researchers have also explored the use of biocompatible, hyper-elastic, and transparent materials, such as hydrogels, to develop a mechanically invisible and stretchable ionic sensory skin (Sun et al. 2014). Figure 18.3d illustrates a pressure sensor array that consists of a stretchable dielectric layer sandwiched in between two stretchable ionic conductors. The hydrogel ionic conductors are fabricated by mixing polyacrylamide powder, sodium chloride salts, and cross-linker agents. The pressure exerted on the sensor will reduce the distance between the two conductors and result in the increase of capacitance (Figure 18.3e). In order to create a high resolution touchpad, Kim et al. (2016) developed a stretchable transparent ionic touch panel for a continuous, two-dimensional (2D) gesture by using similar materials. The processing method in this ionic touch panel is based on 4-wire sensing, where a signal source is applied to these 4-points. As a finger couples the active area to the ground, it closes the electrical network through the resistance of the active hydrogel. The ratio in between the flowing currents can then be analyzed to find the location of a single touch. The device works reliably on the skin, even though calibration algorithm would improve its accuracy and precision when the device is pre-stretched. They further demonstrated various gestural control applications using the touchpad, including tapping, holding, and dragging to perform certain tasks, such as writing words, performing music, and playing a game.

To extend our perception in interacting with physical and virtual objects in mixed reality, Bermúdez et al. (2018) explored the possibility of touchless manipulation through magnetic field sensing. They fabricated a conformal magnetosensitive electronic skin (e-skin), which consists of 2D giant magnetoresistive (GMR) spin valve sensors in a full-bridge configuration (Figure 18.4a). The full-bridge Wheatstone configuration eliminates the temperature dependence and intrinsic output offset of each meander spin valve stack sensor. The stack is comprised of alternating ferromagnetic and non-ferromagnetic thin-films. The GMR e-skin calculates the angle of a magnetic field input by comparing cosine and sine voltage outputs of the inner and outer bridge, respectively (Figure 18.4b, c). As shown in Figure 18.4d, laminating the GMR e-skin on a palm and interacting with a magnet enables a touchless 2D control of virtual objects. The GMR e-skin, therefore, could continuously monitor the body movement wirelessly with respect to an external magnetic field. It allows various novel applications in navigation and motion tracking for robotics, augmented, and virtual reality.

Textiles are soft and conformable materials, as opposed to conventional electronics built on rigid structures or flexible substrates that can only be deformed along one axis. Supported by recent advances in intelligent textiles, its pervasion in our daily life, particularly in clothing, presents many exciting applications for on-body interactions. Intelligent textiles are fabricated by fusing electronic materials with common textile fibers, threads, yarns, and fabrics. They allow fabrics to not only passively

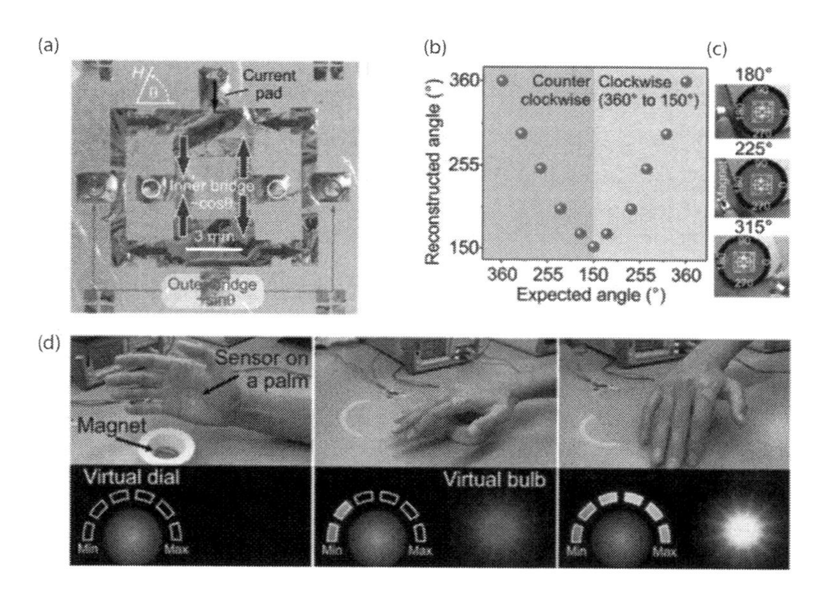

FIGURE 18.4 Magnetosensitive e-skin for augmented reality (a) 2D magnetic field sensor based on spin valve sensors in Wheatstone bridge configuration. (b) Magnetic field angle reconstruction based on flexible 2D magnetic field sensor on a flat surface in respect to the orientation of a permanent magnet. The experiment setup is shown in (c). (d) Demonstration of augmented reality-based, virtual object touchless manipulation using the on-skin magnetic field sensor. (Adapted from Bermúdez et al. (2018), Copyright American Association for the Advancement of Science.)

sense, but also to react and adapt its behavior to the user through computation (Post and Orth 1997). One example is FabricKeyboard, a multi-sensory textile-based interface that attempts to enrich physical interaction modalities by embedding a multi-layer of smart and common fabrics (Wicaksono and Paradiso 2017). By using a combination of Ag-plated conductive fabrics and polypyrrole (PPy)-coated piezoresistive fabrics, the textile sensors stack can simultaneously detect proximity, touch, pressure, stretch, and electromagnetic fields. This effort enables new tactile experiences and novel interactions with both physical and non-contact gestures, particularly for deformable interfaces that can conform to any 3D structure.

A functional garment, Levi's Commuter X Jacket by Google Jacquard has a woven touchpad with gesture sensing capability on its sleeve (Poupyrev et al. 2016). The woven touchpad consists of an XY matrix of conductive threads. Similar to how a touchpad works, projected capacitive sensing is applied to the woven matrix to detect the capacitance in between the threads and calculate XY position of finger gestures. Conductive threads commonly consist of purely metal filaments or a combination of metal filaments and common yarns or fibers (Stoppa and Chiolerio 2014). Metal filaments are fabricated by wire-drawing and twisted with the core fibers using a spinning machine. Another approach to making conductive threads is by metal-coating common fibers. Google's Project Jacquard, in particular, developed special conductive yarns that are customizable and can be easily soldered by twisting insulated copper core with ordinary threads. The project demonstrated a significant effort to large-scale manufacture and commercialize intelligent clothing: from customizing conductive yarns, weaving them into fabrics, to integrating all electronic components. Current research endeavors are directed to an emerging area of fiber electronics (fibertronics), which explores methods to fabricate electronic devices such as sensors, actuators, and transistors directly onto fibers for a high-density and seamless integration between electronics and textiles (Cherenack et al. 2010; Hamedi et al. 2007).

18.2.2 On-body Sensing and Display

Going further beyond sensing, the mechanical properties and the appeal of our skin presents several exciting tactile and visual interaction possibilities. SkinMarks explores the applications of conformal electronics for novel on-skin input and output devices (Weigel et al. 2017). Thin layers of sensing and display electrodes are fabricated by screen-printing poly(3,4-ethylenedioxythiophene)-poly(styrenesulfonate) (PEDOT:PSS). The electroluminescent (EL) display is implemented by printing thin layers of phosphor and dielectric resin paste composite in between the electrodes. The stretchability of PEDOT:PSS prevents structural damages to the devices as our skin deforms. Unique patterns of conductive traces can form input sliders, by leveraging interpolation of neighboring electrodes or squeeze sensors, by the serpentine design of strain gauges. Moreover, the placement of SkinMarks around body landmarks such as fingers could also enable novel input gestures, such as finger bending. Visual feedback via on-skin display further enhances the functionality and interactivity of the SkinMarks.

An exploration of coupling on-skin sensor and display is also demonstrated by Yokota et al. (2016). The ultra-flexible light-emitting diodes (LEDs) are fabricated using polymeric light-emitting materials, sandwiched in-between two thin electrodes. The integration of an ultra-flexible organic pulse oximeter and display allows simultaneous sensing and display of physiological signals in digital format directly from the skin, as illustrated in Figure 18.5a. Further efforts have also been conducted to develop 2D stretchable displays by incorporating serpentine interconnects to bridge inorganic LED-islands on an elastomeric substrate (Hu et al. 2011) or developing novel polymer-based organic LEDs that are intrinsically stretchable (Liang et al. 2013; Sekitani et al. 2009). The thin and deformable nature of these stretchable displays permits their integration onto the surface of the skin for on-body interface applications.

Another muscle-actuated modality besides touch, squeeze, bend, or strain sensing for on-skin interfaces is pressure sensing. The previous examples of on-skin interfaces, however, have limited input-output modalities and resolution. To enable a truly interactive, multi-gesture, and programmable on-skin

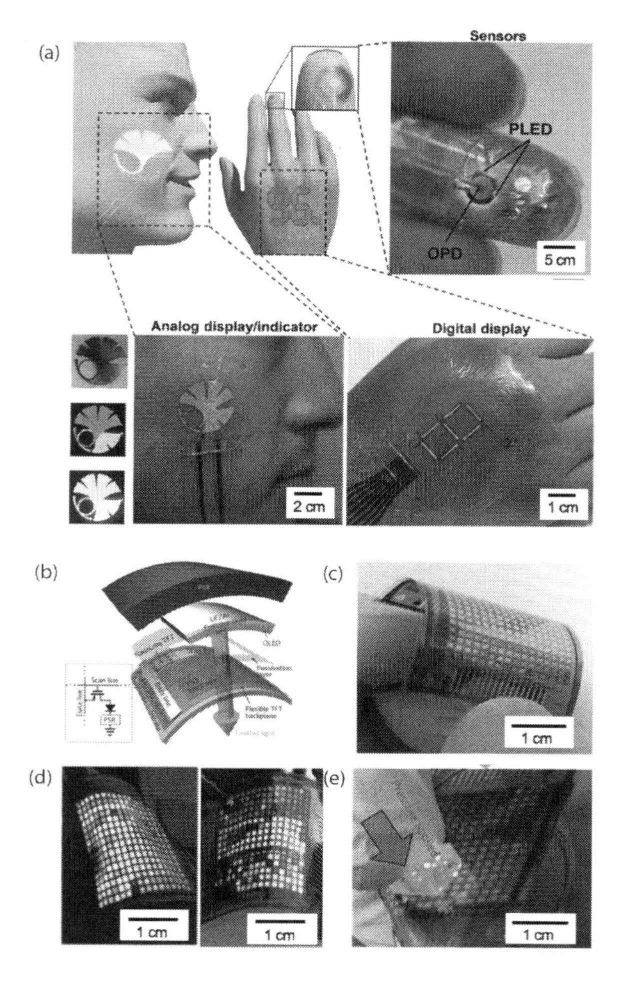

FIGURE 18.5 On-skin interactive sensing and display (a) illustration of an optoelectronic system on the skin with individual organic photodiode as a sensor, polymer LED as an actuator, analogue, and digital display prototype. (Yokota, Tomoyuki et al., Ultraflexible Organic Photonic Skin, *Sci. Adv.*, 2, e1501856–e1501856, 2016. Reprinted with permission of AAAS.) (b) Exploded view of a single pixel of pressure sensitive display, comprising a thin film transistor (TFT), organic LED, and a pressure sensor all integrated on a polyimide substrate. (c) Digital image of the pressure sensitive electronic skin with 16 × 16 pixels. (d) Digital image of bendable single-color (green) and tricolor (red, green, and blue) matrix of active matrix organic light emitting diodes (AMOLEDs) being turned on. (e) Digital image of a finger pressure applied to the interactive e-skin display. The exerted pressure correlates to the brightness of the display. (With kind permission from Springer Science+Business Media: *Nature Mater.*, User-Interactive Electronic Skin for Instantaneous Pressure Visualization, 12, 2013, 899–904, Wang, Chuan et al.)

interface, multi-array and multi-layer architecture of sensing and display electronics must be adopted (Wang et al. 2013). Figure 18.5c illustrates an interactive electronic skin that optically responds simultaneously to pressure. In the 2D array (Figure 18.5b), a single device stack comprises a pressure-sensitive rubber, an organic LED, and a carbon nanotube thin-film transistor (CNT-TFT) for matrix-addressing, are all fabricated, on a polyamide substrate. Each pressure-sensitive element is hard-wired to each of the LEDs, resulting in a higher flow of current and brightness as a stronger force is applied to the electronic skin (Figure 18.5d,e). The progress mentioned above shows how innovation in new materials and fabrication strategies can provide opportunities of functionalizing our skin as future sensing and display surfaces for HMI applications. Further efforts in miniaturization of each sensing, display, and circuit component, as well as a large-scale integration of dense array, are still required for them to compete with current wearable technologies.

18.2.3 Tactile Stimulation

With the rapid growth of technologies for immersive sensory experience, there has also been an increasing interest in the development of wearable tactile feedback to provide users with a physical representation of information beyond visual and auditory. Tactile feedback can trigger several somatosensory receptors in the human skin. To they could relay various sensory information including light stroke, touch, vibration, stretch, texture, and pain (Delmas et al. 2011). Having a programmable tactile stimulation platform facilitates a more seamless, realistic, and meaningful interaction in an HMI system. Possible areas that can benefit from tactile feedback include sensory augmentation, motion training, rehabilitation therapy, telepresence, robotics, assistive technology, to virtual and augmented reality interaction.

The rigidness and lack of resolution of current tactile devices limit them from being highly wearable and practical. The ability of these devices to adapt to the curvilinear structure of our skin surface is paramount, as intimate contact improves the effectiveness of sensation. Any tactile sensation can be theoretically recreated with an adequate spatiotemporal resolution, which is around 1.5–3.0 mm and 0–1 kHz, respectively (Yem and Hiroyuki Kajimoto 2017). These challenges call for microfabricated, soft, and conformal tactile stimulators.

Figure 18.6 shows three types of fingertip soft tactile stimulators. The first approach (Figure 18.6a), which is based on electro-tactile stimulation, employs a flowing current to excite mechanoreceptors transcutaneously (Ying et al. 2012). The device consists of an array of gold (Au) circular electrodes on polyimide and is multiplexed by silicon (Si) nanomembrane (NM) diodes for programmable addressing. The entire structure, except active areas, is encapsulated with polyimide (PI) and transfer-printed to an elastomeric substrate (Ecoflex). Experimental results show that the voltage required to achieve sensation declines with an increasing actuation frequency, which agrees with the skin impedance model (Figure 18.6b).

Another approach, as shown in Figure 18.6c, leverages electroactive polymers (EAP) to develop a wearable soft-actuator tactile display (Koo et al. 2008). Commonly used as artificial muscles, capacitive-type EAPs deform with respect to a voltage potential, thus creating mechanical actuation force (Figure 18.6d). The fabrication process starts with spin-coating and curing a dielectric elastomer layer. A mask is then used to pattern electrodes deposited through spray printing of carbon powder solution on both sides of the elastomer layer. A multi-layer structure of electrodes can be applied to reduce the required actuation voltages. The simple fabrication process and low-cost material enable a fast production of large matrix arrays of EAP soft-actuators. The intrinsic mechanical properties of the dielectric and conductive composite layers also facilitate a realization of entirely soft and deformable actuators that can conform to any geometrical shapes. One possible application from this soft-actuator array is a wearable, finger-tip Braille device that can be used for visually impaired individuals.

A soft actuator-sensor skin for vibrotactile feedback has also been demonstrated by Sonar and Paik (2016) (Figure 18.6e,f). The tactile feedback relies on pneumatic actuation, where compressed air is converted into mechanical force. An array of hollow chambers for shape inflation is fabricated by embedding a masking layer in between two silicone layers. At the bottom of each pneumatic actuator, there is a piezoelectric ceramic encapsulated in silicone layers. These piezoelectric sensors allow precise control of actuation amplitude and vibration through a closed-loop system as well as detection of external interaction forces. The actuator is capable of providing 0.3 N force with a frequency of 5–100 Hz. This device is useful not only for wearable tactile feedback in virtual reality systems, but also for rehabilitation devices requiring a haptic feedback.

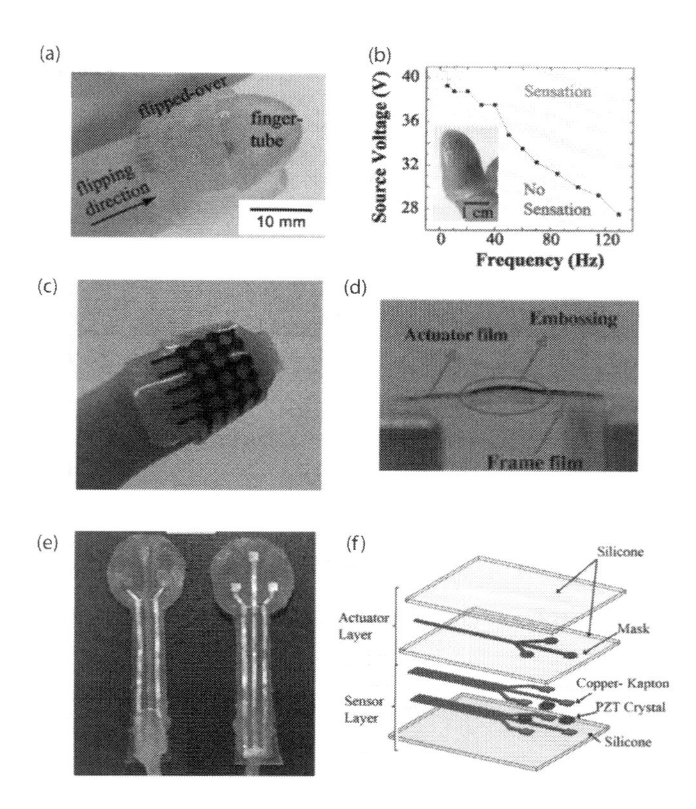

FIGURE 18.6 Wearable tactile stimulators. (a) Tactile electrode array for muscle stimulation on a finger-tube substrate and worn, (b) voltage applied as a function of stimulation frequency, showing the required range for perceived sensation on a fingertip. (Ying, Ming et al., Silicon Nanomembranes for Fingertip Electronics, *Nanotechnology*, 23, 2012. Copyright 2012, IOP Publishing.) (c) A wearable soft tactile display array based on dielectric elastomer and carbon-based conductor worn on a fingertip. (d) Side-view of a tactile pixel assembly. (From Ko, Heung Cho et al., *Nature*, 454, 748–753, 2008. Copyright 2014. IEEE with permission.) (e) Soft pneumatic actuator skin integrated with piezoelectric sensors. (f) Design schematic of the SPA skin with three independent sensor-actuator elements. (Adapted from Sonar, Harshal Arun, and Jamie Paik. 2016. "Soft Pneumatic Actuator Skin with Piezoelectric Sensors for Vibrotactile Feedback." *Frontiers in Robotics and AI* 2. https://doi.org/10.3389/frobt.2015.00038.)

18.3 Body Gesture Recognition and Activity Monitoring

The advance of multimedia technologies, such as in mobile devices, virtual, and augmented reality, as well as in robotics necessitate a richer set of alternative control modalities that enable a human to intuitively and seamlessly interact with the digital and physical world. Our gesture is a universal body language that enables us to communicate and express our intentions and ideas naturally. The recognition, classification, and interpretation of body gestures, particularly the hand motions have, therefore, been extensively studied and used in the HCI and augmentative and alternative communication (AAC) field (Higginbotham et al. 2007; Sharma and Verma 2015). Continuously monitoring the body gesture and posture is also useful for gait analysis and activity recognition, particularly in prosthetic control, rehabilitation, elderly care, sports, surgical, and industrial applications (Tao et al. 2012).

Most of the current gesture recognition platforms used in consumer electronics or research prototypes rely on vision-sensing (Cheng et al. 2015). Even though vision-based sensors are less cumbersome, non-contact, and allow full-body motion tracking, they commonly require a bulky set-up, color-contrast markers, and a high computational power. Moreover, they could not accurately sense micromotions and have a confined working space. These challenges restrain vision-based sensors from being used for long-term and mobile gesture recognition. Flexible and stretchable electronic devices that can be seamlessly laminated on our body or integrated into textiles for high-quality sensing and maximum comfort could solve these issues. In this section, we will separate these novel devices for human-motion detection into three main categories: inertial, electrical muscle activity, and strain sensing.

18.3.1 Inertial Sensing

Extensive studies have been conducted to perform activity recognition and gait analysis with micro-electro-mechanical systems (MEMS) inertial sensors attached to multiple body segments (Tao et al. 2012). To sufficiently classify body motion in normal daily activities, inertial measurement units (IMUs) must be able to sense accelerations with amplitude from −12 to +12 g and frequency of up to 20 Hz (Bouten et al. 1997). An IMU typically comprises single or multiple accelerometers, a gyroscope, and a magnetometer. It collectively measures 3-dimensional linear acceleration, angular rate, and magnetic field, respectively. Body-worn inertial sensors enable the calculation of angles around the body joints based on their orientations relative to one another. These combined joint angles and other markers from the IMUs can be collected and processed in real-time to characterize gestures. Ultimately, users activity can be recognized or even predicted from this subsequent set of motions (Seel et al. 2014, Yang and Hsu 2010).

One of the initial attempts in developing a flexible format of inertial sensors began by printing silver nano-ink on a paper substrate (Zhang et al. 2013b). A flexible paper accelerometer consists of two conductive layers, in which one of them is a floating membrane supported by two hinges. Due to gravity, there is a force exerted by the membrane mass on the floating top electrode towards or against the bottom electrode as the sensor is accelerated. The distance in between these two electrodes correlates to the capacitance of the accelerometer. As the acceleration towards the bottom electrode gets stronger, the gap between these two electrodes becomes smaller, resulting in a higher capacitance from the equilibrium. The plate size and structure of the hinges can be modified to control the sensitivity and resonant frequency of the paper accelerometer.

An alternative form of a deformable accelerometer is demonstrated by Yamamoto et al. (2016) in their multi-functional on-skin electronics for physiological and motion sensing. The device has reusable and disposable parts; in the disposable part, a printed 3-axis accelerometer is fabricated on a PET substrate with kirigami structure that allows sufficient strain limit against skin deformation (Figure 18.7a–c). The inertial sensor consists of three silver-based strain sensors as hinges and an acrylic plate as a centre membrane (Figure 18.7b). The whole structure is then supported by a silicone rubber. Due to the gravitational force, the centre membrane exerts a force and pulls the hinges. Z-direction acceleration exerts a strain to all of the sensors, while x-direction motion influences sensor #2, and y-direction influences sensor #1 and #3, respectively. The resistance change is affected by the conductive network in the silver-based strain sensors. As illustrated in Figure 18.7d, the acceleration in sensor #1 due to x, y, and z movements begins to increase in a relatively linear fashion from around 5 to 12 m/s^2, depending on the direction of movement and except for the x axis. Thus, the flexible accelerometer can detect various high-intensity activities when intimately attached to the chest (Figure 18.7e).

Instead of using a floating membrane, free-moving smart materials such as liquid metal can also be utilized to develop deformable inertial sensors (Varga et al. 2017). As presented in Figure 18.7f,g, the device consists of a glycerol chamber encapsulated in silicone, in which an embedded droplet of eutectic gallium indium (eGaIn) could tilt to modulate the capacitance in between two electrodes. A planar eGaIn coil as an extension of the two electrodes is spray-painted on the silicone substrate; the inductor-capacitor (LC) resonant frequency formed by the tilt sensor and the coil correlates to the movements of the eGaIn droplet and can be analyzed wirelessly. Figure 18.7h shows both experimental and theoretical capacitance response of the tilt sensor due to various tilting angles. The tilt sensors, for example, can be worn as a bracelet

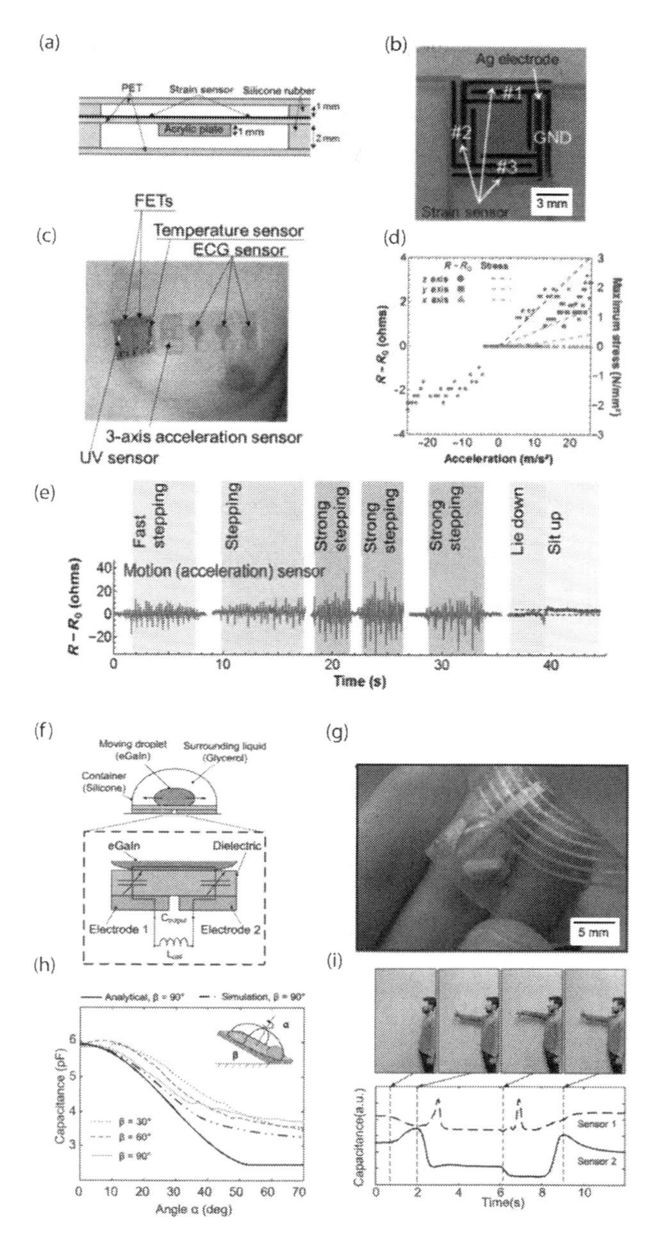

FIGURE 18.7 On-skin accelerometer based on strain sensors bridge. (a) Cross-sectional device structure and (b) digital image of the strain-based on-skin accelerometer. (c) Photograph of the acceleration sensor along with the others attached to a chest. (d) Resistance change and FEM simulation of stress of sensor #1 under a range of acceleration in x, y, and z-axis. (e) Motion sensing demonstration of the on-skin accelerometer during various physical activities. (With kind permission from Springer Science+Business Media: *Science Advances*, Printed Multifunctional Flexible Device with an Integrated Motion Sensor for Health Care Monitoring, 2, 2016, e1601473–e1601473, Yamamoto, Y. et al.) Wearable and flexible tilt sensor based on eGaIn droplet. (f) Cross-sectional view of device structure with its electrical model. (g) Digital image of the moving droplet inside a chamber with two electrodes. (h) Measured and simulated capacitance response of the tilt sensor over a range of tilt angles α and inclinations β. (i) Demonstration of output capacitance through different arm gestures as two tilt sensors are mounted on the wrist. (Varga, Matija et al., *Lab Chip*, 17, 3272–3278, 2017. Copyright 2017 Royal Society of Chemistry.)

to recognize many gestures by processing their distinct output capacitances (Figure 18.10i). Another approach by Persano et al. (2013) attempts to develop a poly(vinylidenefluoride-co-tri-fluoroethylene) or P(VDF-TrFe) nanofiber-based device and explore its piezoelectric behavior in order to measure acceleration and orientation. An electrospinning process is used to fabricate the free-standing, highly aligned, and pressure-sensitive (0.1 Pa) P(VDF-TrFe) fibers. By placing the fiber array as a diaphragm in a chamber inside a box, vibrations and accelerations on the box induce bending on the P(VDF-TrFE) device and generate a voltage. Adding a test mass in the structure allows the device to function as an orientation sensor, as tilting the device would exert lower. Therefore, it generates a small voltage output. Even though the flexible accelerometers discussed here possess lower sensitivity and are relatively larger in size than state-of-the-art MEMS inertial sensors, they demonstrate the future possibility of large-scale manufacturing of soft wearable IMUs for human motion detection.

18.3.2 Electromyography

The second principle of detecting body gestures is to read bio-potential signals generated from the human body. Electromyography (EMG), electrooculography (EOG), and electroencephalography (EEG) are the most commonly used modalities for human-machine interface applications. In this part, we will focus on EMG signals from the body as they are correlated to our physical gestures. EMG detects the change in potentials produced by the muscle cells when they relax and contract. The signal amplitude typically ranges from 100 μv to 90 mv with frequencies from DC to around 10 kHz (Ferreira et al. 2008).

The traditional approach of EMG is to attach wet Ag/AgCl surface electrodes on the skin and record the electric signals generated from muscle activities. However, this approach is cumbersome and uncomfortable, as it requires the application of electrolyte gels, attachment of straps or adhesives, and connection through bulky wires to the electronic systems. The developments of electronic packaging, system design, and signal processing techniques have mitigated the low electrical coupling of skin-electrode interfaces and the influence of motion artifacts, which result in an improved signal-to-noise performance of dry and non-contact electrodes for bio-potential sensing. A commercially-available wearable EMG device such as the Myo armband relies on an array of dry surface electrodes to detect bio-potential signals on the arm and are integrated with on-board IMU for wireless gesture recognition (Torres 2015).

Advances in flexible and stretchable electronics have also further transformed the bio-potential sensing to intimately conform to our skin, especially in curvy and dynamic regions of the human body. An initial example is the epidermal electronic system (EES) proposed by Kim et al. (2011). The EES consists of an assortment of electronic devices, including EMG electrodes integrated with circuitry for signal amplification and processing. The devices, as well as the interconnects, are fabricated in a serpentine structure and encapsulated in PI to withstand skin deformations. The EES is transferred to the skin with a temporary substrate such as polyester or water-soluble polyvinyl alcohol (PVA) and strongly adheres via van der Waals force. By mounting the EES on different parts of the body such as legs, the EES can detect EMG signals of a person walking and standing.

A comprehensive study of the materials, mechanics, and geometric designs of the EES for surface EMG (sEMG) is required for a high signal-to-noise ratio (SNR) output by minimizing contact impedance, motion artifacts, and crosstalk. For this purpose, electrode size and design, inter-electrode distance, and conformity to the skin are the main optimization parameters in the design of on-skin electrophysiological sensors (Figure 18.8a). A study by Jeong et al. (2013) concluded an optimum electrode spacing of 20 mm for minimum crosstalk contamination and larger electrode area for a maximum integrative signal and thus, optimum SNR. The filamentary serpentine (FS) mesh design provides intimate and adaptive contact to the skin, while avoiding interface stress and delamination. The contact resistance of FS electrodes can be reduced by compromising trace width with the curvature radius of the serpentine structure. The mechanics model of skin-surface interaction to estimate the required van der Waals forces for a conformal contact reveals a threshold thickness of 25 μm (Figure 18.8b), with Figure 18.8c showing an excellent contact of

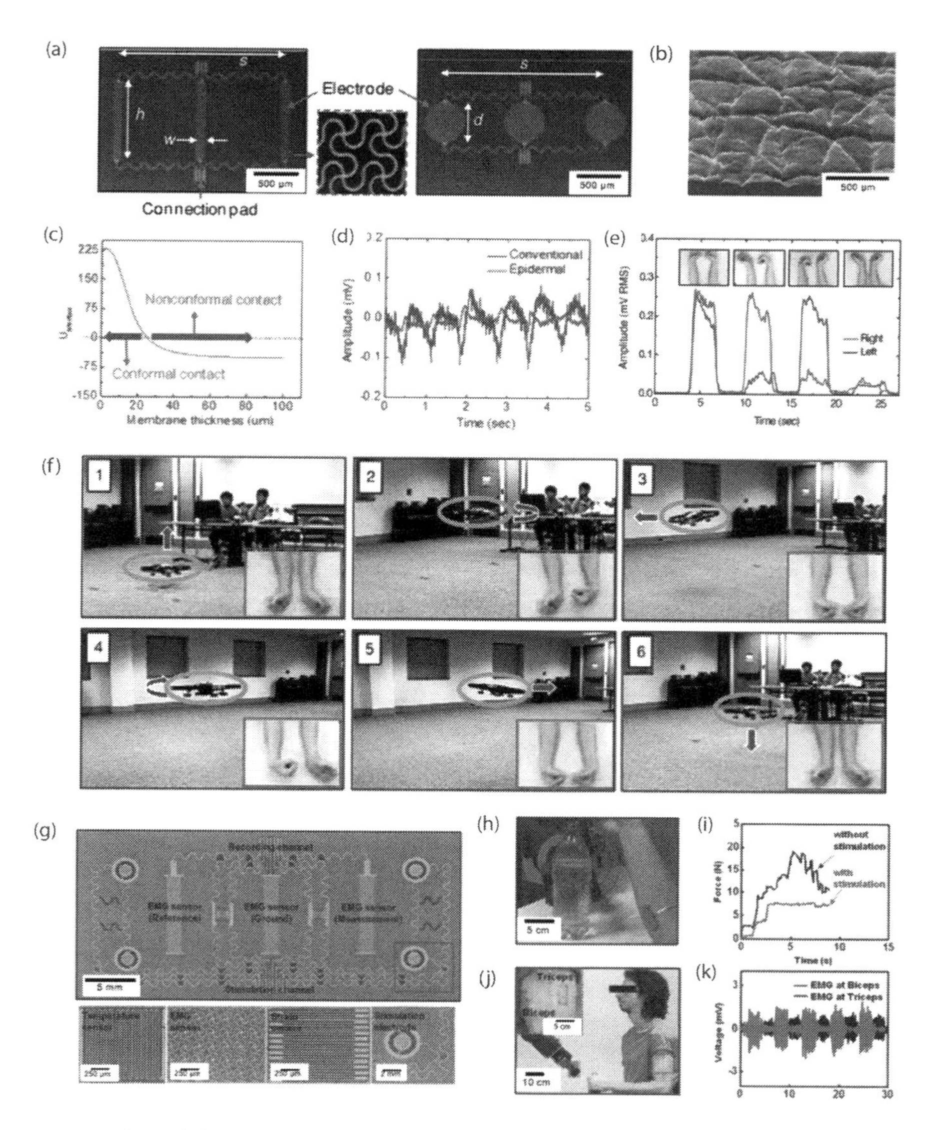

FIGURE 18.8 Epidermal electronics for sensing electrical muscle activity through sEMG. (a) Two configurations (bar and disk-type) of sEMG device with measurement, ground, and reference electrodes. (b) SEM image showing a conformal contact between thin silicone layer and the surface of skin. (c) Analytical plot of interfacial contact energy ($U_{interface}$) through various substrate thicknesses. Conformal contact can be deduced for the thickness with positive $U_{interface}$. (d) Comparison of sEMG voltage output between epidermal sensor and conventional electrode with gels. (e–f) Quadrotor control by various arm gestures using epidermal sEMG device. (From Jeong, Jae Woong, et al. Materials and Optimized Designs for Human-Machine Interfaces via Epidermal Electronics. *Adv. Mater.* 2014. 25. 6839–6846. Copyright Wiley-VCH Verlag GmbH & Co. KGaA. Reproduced with permission.) Integrated with sensory feedback from stimulation electrodes for prosthetic control. (g) Multifunctional EES device for simultaneous muscle sensing and stimulation among others. (h–i) Controlling a force on a robotic gripper using sEMG with and without muscle stimulation from the EES. (j–k) sEMG signals on both triceps and biceps while controlling a robotic gripper through the flexion and extension of an arm. (Adapted from Xu et al. (2016), Copyright John Wiley & Sons.)

5 μm silicone membrane with the skin. A membrane thickness over this threshold value will lead to the formation of air gaps, thus low conformability and SNR.

Figure 18.8d compares epidermal and conventional sEMG signal response as the skin is mechanically deformed. The result proves that EES provides a more robust signal for sEMG from noise caused by skin deformations. As illustrated in Figure 18.8e,f, the FS electrodes are then attached to various locations of the body such as the forearm, cheek, forehead, neck, and finger to perform sEMG reading, and both forearms to control movements of a quadrotor via pattern recognition. Further work also explores the possibility of non-contact, capacitive sensing of bio-signals through the electrical coupling between tissues and electrodes (Jeong et al. 2014). In this work, the epidermal electrodes are separated from the skin, with an insulation barrier made out of silicone. Given that there is no direct contact between the metal electrodes and the skin, this technique enables a more durable and long-term reading of bio-signals.

A conformal bioelectronic device that combines muscle electrical activity sensing and stimulation in one platform is also practical in assistive and rehabilitative technology. Xu et al. (2016) presented an EES with closed-loop sensing and stimulation capability for lower back exertion and sensorimotor prosthetics. The skin-like, multi-functional device consists of EMG, strain, and temperature sensors as well as electro-muscle stimulation (EMS) electrodes (Figure 18.8g). The close proximity of EMS and EMG electrodes enables further studies of muscle response to varying stimulation voltages. Sensorimotor control of a robot arm through the EES in the case of a paralyzed limb is also demonstrated. Stable control of gripping force can be achieved by real-time monitoring of biceps and triceps EMG signals and applying a local EMS feedback in concert (Figure 18.8h,i). Another application is to provide a proprioceptive feedback, by laminating two EES around the biceps and triceps brachii muscles (Figure 18.8j). As shown in Figure 18.8k, EMG signals from the EES are classified by linear discriminant analysis (LDA)to give insights to the arm movements. Two stimulation electrodes create a tactile funneling illusion that makes the user to perceive a phantom sensation in a point between the electrodes.

18.3.3 Strain Sensing

The third principle of wearable gesture sensing is by measuring strain change across the body using active skin sensors. Stretchable strain sensors can be used to measure the flexion of the fingers, wrist, arms, legs, and other body joints. They can be separated into three main classes: resistive, capacitive, and piezoelectric-based. Resistive or capacitive strain sensors are mainly developed by employing conductive materials such as CNTs, graphene, and metal in nanowires or nanoparticles, as well as liquid-metal alloys. These materials are combined with a stretchable polymeric base or substrate such as Ecoflex or PDMS through various techniques including filtration, deposition, transferring, coating, printing, and solution mixing (Amjadi et al. 2016; J. Park et al. 2015). The selection of materials, structural design, and fabrication methods will influence the strain sensor characteristics, such as stretchability, sensitivity, linearity, response and recovery time, and durability.

Conventional resistive strain sensors, commonly known as strain gauges, rely on a change in geometrical structure and inherent piezoresistive effects. The resistance of a conductor is given by $R = \rho L/A$, where ρ is the electrical resistivity, L is the length, and A is the cross-sectional area. Taking into account the intrinsic change of resistivity in the material due to piezoresistive effects, the relative change of the resistance can be written as $\Delta R/R = (1 + 2\upsilon)\varepsilon + \Delta\rho/\rho$, where υ is the material Poisson's ratio (Mohammed et al. 2008). Semiconducting materials such as silicon and germanium with high piezo-resistivity can be used to fabricate highly sensitive strain gauges. Nevertheless, these devices can typically withstand a strain of up to 1%, unless buckling mechanism or a horseshoe pattern is incorporated into the design (Kim, Xiao et al. 2010; J. Park et al. 2015). As a result, alternative materials and fabrication techniques based on conductive percolation network, micro-crack propagation, or electron tunneling effect have been investigated to develop highly stretchable and sensitive piezoresistive strain sensors, particularly for human motion detection (Amjadi et al. 2016).

The dexterity of our hand motivates a large number of studies to demonstrate the use of strain sensors in monitoring wrist and finger gestures. Yamada et al. (2011), for example, developed CNT-PDMS strain sensors that can be integrated into a glove for the detection of multiple hand gestures (Figure 18.12b, c). The sensors are fabricated by growing thin-films of aligned single walled (SW) CNTs and transferring them to a PDMS substrate (Figure 18.9a). These sensors can detect and withstand a strain of up to 280%, are highly durable when tested in 150% strain for 10,000 cycles. Another example of material that can be transferred or deposited onto an elastomeric substrate for micro-crack propagation strain sensing is gold nanosheet (Lim et al. 2016). Gerratt et al. (2015) used direct deposition of gold thin-films onto elastomeric substrates in order to develop a strain sensor integrated data glove (Figure 18.9d, e). To create the smart glove, strain sensors are connected to the read-out circuit through printed elastic liquid metal (EGaIn) interconnects. The sensor placements on the

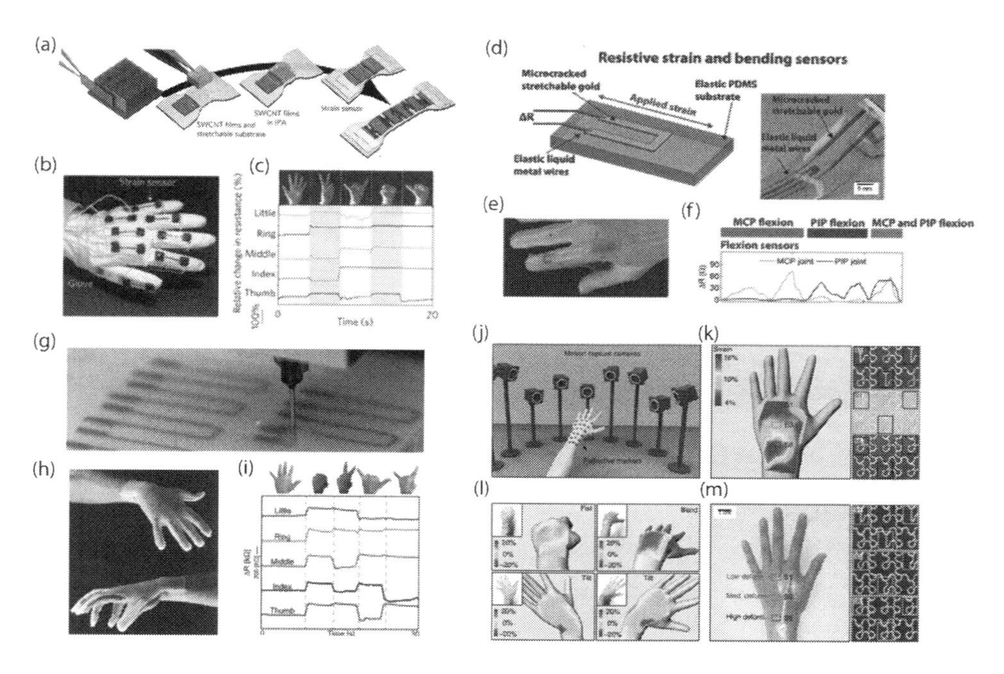

FIGURE 18.9 Data glove or strain-sensing skin fabricated by resistive strain sensors. (a) Fabrication steps of a SWCNT strain sensor. (b) Photograph of SWCNT strain sensors attached on finger joints of a glove. (c) Relative change in resistance of each strain sensor for multiple hand gestures. (With kind permission from Springer Science+Business Media: *Nat. Nanotechnol.*, A Stretchable Carbon Nanotube Strain Sensor for Human-Motion Detection, 6, 296–301, Yamada, Takeo et al.) (d) Device configuration and operation of micro-cracked stretchable gold strain or flexion sensor. (e) Photograph of a glove integrated with stretchable gold strain sensors. (f) Resistance change of the strain sensors with separate and simultaneous metacarpophalangeal (MCP) and interphalangeal (PIP) finger joint flexions. (From Gerratt, Aaron P., et al., 2015. Elastomeric Electronic Skin for Prosthetic Tactile Sensation. *Adv. Funct. Mater.* 2015. 25. 2287–2295. Copyright Wiley-VCH Verlag GmbH & Co. KGaA. Reproduced with permission.) (g) Embedded 3D-printing (e-3DP) process to pattern and embed stretchable strain sensors on an uncured elastomer. (h) A prototype of data glove embedded with strain sensors from e-3DP process. (From Muth, Joseph T., et al., Embedded 3D Printing of Strain Sensors within Highly Stretchable Elastomers. *Adv. Mater.* 2014. 26. 6307–6312. Copyright Wiley-VCH Verlag GmbH & Co. KGaA. Reproduced with permission.) (i) Resistance change of the embedded strain sensors through multiple hand gestures. (j) A motion-capture system to reflect strain distribution on the skin surface. (k–m) Results of motion capture system inform the array design and placement of three silicon nanoribbon strain gauges with different strain limit. S1, S3, and S6 strain sensor design for low, medium, and high deformation location, respectively. (With kind permission from Springer Science+Business Media: *Nat. Commun.*, Stretchable Silicon Nanoribbon Electronics for Skin Prosthesis, 5, 2014, 5747, Kim et al.)

metacarpophalangeal and proximal interphalangeal joints enable accurate measurements of finger flexions (Figure 18.9f). Combined with additional pressure sensors, the smart glove is able to sense the compressibility of an object and guide users to maintain their grasping strength for prostheses tactile feedback.

As shown in Figure 18.9g, instead of integrating strain sensors onto a textile-based material, Muth et al. (2014) explored the applications of 3D-printing of conductive ink (Figure 18.9g) to make a soft glove that detects various finger gestures (Figure 18.9h). The conductive ink consists of silicone oil filled with carbon black. A 3D-printing fluid nozzle, programmed to pattern a strain gauge, prints the conductive ink inside an uncured silicone elastomer and filler fluid layers. Due to the intrinsic conductive network, each strain gauge increases in resistance as it is being stretched (Figure 18.9i). The strain sensor has been tested to reasonably detect a strain of up to 400% and undergo percolation network breakdown at 700%. A high-resolution spatiotemporal strain sensing by a miniaturized array of strain sensors has also been demonstrated by Kim Jaemin et al. (2014). Because of its high piezo-resistivity and fracture strength, p-type doped single crystalline silicon nanoribbon (SiNR) is employed as the active layer of the strain sensing component. The SiNR array is fabricated on a PI and then encapsulated in a PDMS substrate. A motion-capture system is used to study the strain distribution of multiple hand motions (Figure 18.9j). The results are used to respectively, design the mechanical structure of a strain sensor array based on the maximum local strain (Figure 18.9k, l). As shown in Figure 18.9m, the SiNR serpentine design allows higher deformations with the compromise of lower sensitivity. SiNR6 can withstand a strain of up to 30%, which is suitable for strain sensing on the area close to the wrist, whereas SiNR1 can only handle up to 10% strain. This design strategy permits a robust implementation of high-resolution strain sensing with site-specific sensitivity.

Mattmann et al. (2008) designed a sensor-integrated shirt that can measure the strain distribution across multiple locations of the torso to distinguish different body gestures. The strain sensor is fabricated by mixing thermoplastic elastomer (TPE) and carbon black particles before processing them into a fiber. The sensor's linearity and sensitivity can be engineered by experimenting the composition of carbon black particles in the mix. As depicted in Figure 18.10a, 21 strain sensor fibers are attached to a shirt and connected to the data acquisition unit by sewing conductive threads. The strain sensor could work with a maximum strain of around 80%. Strain rate testing is conducted by subsequently increasing the strain velocity from 50 to 600 mm/min, and a slight mean error of 5.5% is observed (Figure 18.10b).

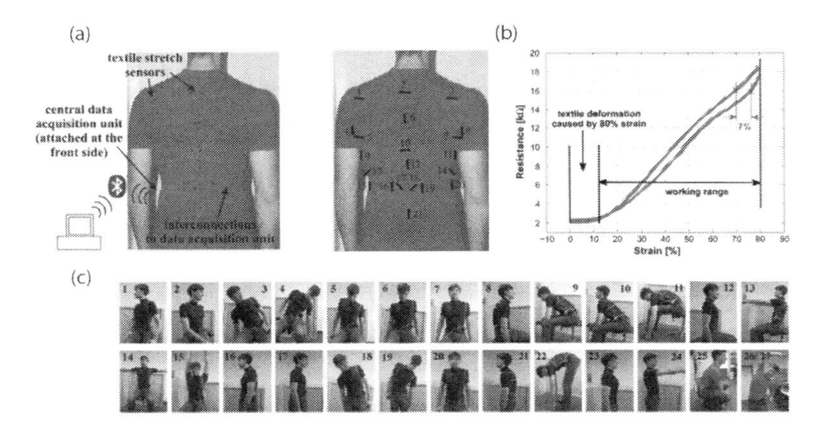

FIGURE 18.10 Sensor-integrated shirt for upper body gesture and posture monitoring. (a) Position of 21 textile resistive strain sensors connected to a data acquisition unit through conductive thread interconnects. (b) Typical resistance change due to cyclic strain to 80% with an observed hysteresis. (c) Demonstration of pattern recognition training study where a participant performs 27 upper body gestures while wearing the sensor-rich shirt. (Reprinted from Mattmann, Corinne et al., *Sensors*, 8, 3719–3732. Copyright 2008 MDPI.)

The strain sensor also retains its functionality under extreme long-term, ageing, and washability tests. With a pattern recognition algorithm and prior user-specific training, the shirt successfully classifies 27 upper-body gestures with 97% accuracy (Figure 18.10c).

For monitoring lower-body gestures, Menguc et al. (2013) demonstrated a soft wearable motion sensing suit for lower-limb biomechanics measurements. The sensors are fabricated by injecting EGaIn liquid metal into a silicone mould. They are then attached to the hip, knee, and ankle parts of the lower body. Experiment results reveal a non-linear resistance change consistent up to 200% and fracture at 364% strain. The wearable motion sensing suit is used to approximate joint angles on the lower-limb, allowing the analysis of a dynamic range of motions and gait of the user for rehabilitation training and sports applications. Using other elastomeric materials with self-healing properties can also significantly improve the elasticity of a strain sensor. A recent effort led by Cai et al. (2017) leveraged a self-healing property of hydrogels to develop a SWCNT/hydrogel strain sensor that can work with up to 1000% strain. The strain sensor is fabricated by dispersing SWCNT and adding PVA solution. The SWCNT/PVA solution is then cross-linked by the addition of an aqueous borax solution and stirred until a hydrogel is formed. This ultra-stretchable strain sensor retains stable electrical and mechanical performance under multiple stretching cycles. Applying the sensors to different parts of the body such as finger, knee, neck, and arm allows them to monitor a diverse set of human motions continuously.

A capacitive strain sensor employs conductive and stretchable materials as active layer and substrate. Its structure consists of a stretchable dielectric layer in between two stretchable conductive electrodes, resembling the parallel plate architecture of a capacitor. The capacitance can then be written as $C = \varepsilon_0 \varepsilon_r A/d$, where ε_0 is the electric constant, ε_r is the dielectric constant of the material between the plates, A is the overlapping area of the two plates, and d is the separation distance between the plates. Based on this underlying equation, tension on the strain sensor changes the effective area of the two electrodes and reduces the dielectric thickness in between, with respect to Poisson's ratio. This will increase the sensor's capacitance and *vice versa*. In comparison to resistive strain sensors, capacitive strain sensors exhibit better linearity, stretchability, recovery, and hysteresis performance, but with a trade-off of low sensitivity (J. Park et al. 2015; Amjadi et al. 2016). Thus, developing a strain sensor with an excellent performance on all of these parameters is still a challenge.

Some studies have been conducted to develop stretchable capacitive strain sensors for human motion detection. One approach is by depositing conductive films on an elastomeric substrate (L. Cai et al. 2013). Thin carbon nanotube films are initially grown using a CVD process. They are then transferred to the front and back surface of a stretchable PDMS base through PET frames. Because the CNT film is thin, the strain sensor exhibits a high optical transparency of 80%. It can also detect strain as high as 300% with an excellent durability. As an alternative approach, Atalay et al. (2017) investigated metal deposition and laser rastering to develop a capacitive strain sensor. Before depositing metal layers (Al and Ag) on a silicone substrate, surface modification techniques are executed. The surface modification through rapid laser rastering forms microgrooves. Pre-straining of the silicone substrate avoids cracks on the metal electrodes and ensures high repeatability of the sensor. Figure 18.11a shows the working mechanism of the laser-treated capacitive strain sensor and its cross-sectional view. The strain sensor remains linear and repeatable up to 85% strain and can withstand a maximum strain of up to 250%. The sensor successfully detects arm movements when attached to the elbow joint (Figure 18.11b, c).

The aforementioned capacitive strain sensors are customized to only allow single axis detection. To finely detect strain distributions on a large surface of the skin, an array of strain sensors must be developed. Zhao et al. (2015) reported a stretchable multi-functional electronic skin capable of mapping static and dynamic strain for HMI application (Figure 18.11d). This stretchable sensor consists of a silicone (Ecoflex) dielectric in between two layers of 7 by 7 islands or arrays of silver metal electrodes on a PET substrate and encapsulated in a PDMS layer. As a specific area of the sensor is touched, bent, stretched, or pressed, the strain distribution across this electronic skin can be analyzed through the change of

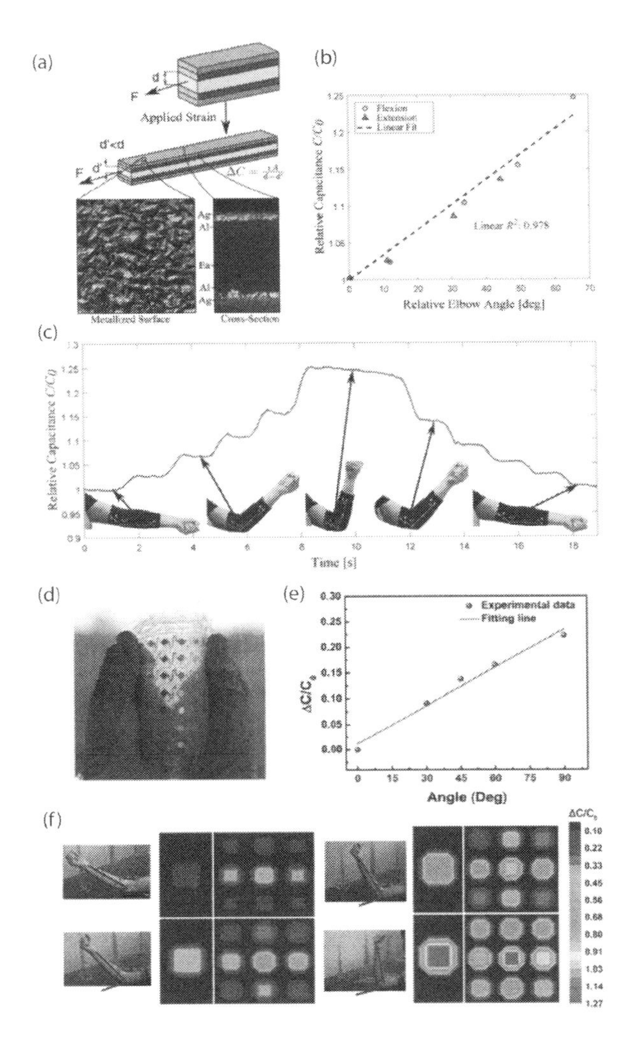

FIGURE 18.11 Capacitive strain sensors for recognizing elbow movements. (a) Design principle of capacitive soft strain sensor with surface and cross-sectional SEM image of the metal-elastomer sandwich stack. (b) Linear relative capacitance change of the strain sensor in response to elbow angle. (c) Relative capacitance change of the strain-sensor integrated on a sleeve to detect elbow joint flexion and extension. (From Atalay, Ozgur, et al., A Highly Stretchable Capacitive-Based Strain Sensor Based on Metal Deposition and Laser Rastering. *Adv. Mater. Technol.* 2017. 1700081:1700081. Copyright Wiley-VCH Verlag GmbH & Co. KGaA. Reproduced with permission.) (d) Digital image of a wearable, flexible, and stretchable pressure and strain sensor array. (e) Linear relative capacitance change of a pixel with various elbow angles. (f) Relative capacitance distribution (3 × 3 array) of the device laminated on the arm with 30°, 45°, 60°, and 90° flexion. (From Cai, Le et al., Super-Stretchable, Transparent Carbon Nanotube-Based Capacitive Strain Sensors for Human Motion Detection. *Scientific Reports*. 2013. 3. 1–9. Copyright Wiley-VCH Verlag GmbH & Co. KGaA. Reproduced with permission.)

capacitance in individual islands. Bending tests on the sensor reveal a linear response of relative capacitance in respect to various bending angles (Figure 18.11e). Figure 18.11f demonstrates spatiotemporal strain mapping around the elbow joint with respect to multiple flexion angles.

Piezoelectric nanogenerators (PENGs) convert mechanical energy into electrical charges and *vice versa*. Since they can produce a relatively large output with a minimal strain, piezoelectric materials exhibit a high sensitivity in response to strain. They also generate electrical energy, making them consume little or no power and requiring a relatively simple read-out circuitry to digitize the sensor signals. The fact that PENGs and other nanogenerators intrinsically generate power also opens up the possibility of a self-powered system, where the output from the sensor is used not only to detect motion, but also to output power to the read-out and communication interface circuit simultaneously. However, PENGs also have their limitations; it can only be used to measure changing or dynamic strain instead of static change, limiting its application for continuous strain sensing (Kon et al. 2007). As a consequence, piezoelectric-based sensors can only be used in certain applications that do not require static load sensing.

An example of a piezoelectric-based gesture recognition device is a ZnO conformal homojunction nanowire film that can be laminated directly to the skin, as shown in Figure 18.12a (Pradel et al. 2014). The nanogenerator device fabrication starts with sputtered Al-doped ZnO (AZO) on a PET substrate as the bottom electrode. The hydrothermally grown, undoped n-type ZnO and Sb-doped p-type ZnO are then spin-coated with polymethylmethacrylate (PMMA) for current leakage intervention and finally sputtered with another AZO layer as the top electrode. Experimental results prove that the bilayer, homojunction structure of pn-ZnO gives the best piezoelectric performance compared to single-layer or other homojunction pairs. The piezoelectric sensor is then encapsulated in silicone and mounted on a wrist to detect multi-finger gestures. By monitoring the PENGs voltage output due to the movement of flexor tendon, this device can distinguish multiple complex finger gestures as demonstrated in Figure 18.12b. Another work by Lim et al. (2015) utilized transparent and ultrathin nanomaterials to develop an invisible and interactive human-machine interface based on a piezoelectric motion sensor and electrotactile stimulator or EMS. The piezoelectric sensor consists of a polylactic acid (PLA)/SWNT composite film as the active element and is sandwiched in between transparent graphene/PMMA films. The EMS electrodes are composed of a graphene (GP)/silver nanowire (AgNW)/GP layers on top of a PDMS substrate (Figure 18.12d). Material selection, thin structure, and serpentine design allow the device to be transparent and stretchable. The addition of an electro-tactile stimulator enables the demonstration of interactive HMI systems such as in a prosthetic arm, where the sensor and stimulator are linked to control an external robot arm in a closed-loop sensory feedback mechanism (Figure 18.12e). As shown in Figure 18.12f, bending and pressure tests confirm the electrical property of the sensor in detecting dynamic motions of the wrist and providing real-time tactile stimulation.

Another type of nanogenerator that can be used for human motion sensing is triboelectric nanogenerator (TENG). TENG produces electricity from mechanical energy through a combination of electrostatic induction and contact electrification due to frictions between different materials. TENGs are particularly attractive, as they can be fabricated with a wide option of materials, are relatively low-cost, light-weight, and easily scalable (Wang 2014). Lai et al. (2017), as shown in Figure 18.12g, developed a TENG single-thread that can be sewn into textiles. The device consists of a single stainless-steel conductive thread coated with silicone rubber. Another thread configuration, such as the helix-belt structure of an inner electrode with an outer electrode can be explored to maximize the output energy of thread-based TENGs (Wang et al. 2016).

The silicone rubber is used to attract electrons since it has a high electron affinity. Sewing the thread in a serpentine pattern on an elastic textile allows stretchability of the cloth-based TENG up to 100%. The entire cloth can be folded, twisted, and crumpled without any degradation in structural mechanics and performance of the TENG. The active, self-powered sensing device operates by the triboelectrification and electrostatic induction during its regular contact with the skin. As the TENG thread is compressed and separated to and from the skin, electrons travel from the skin to the silicone rubber surface and *vice versa* (Figure 18.12h). The recurred induction of charges results in a successive flow of positive and negative currents. By sewing the TENG-thread on a textile glove, one can demonstrate an application of self-powered gesture sensing (Figure 18.12i) to detect specific finger actions based on its unique signal patterns, as well as self-powered wearable touch sensing (Figure 18.12j).

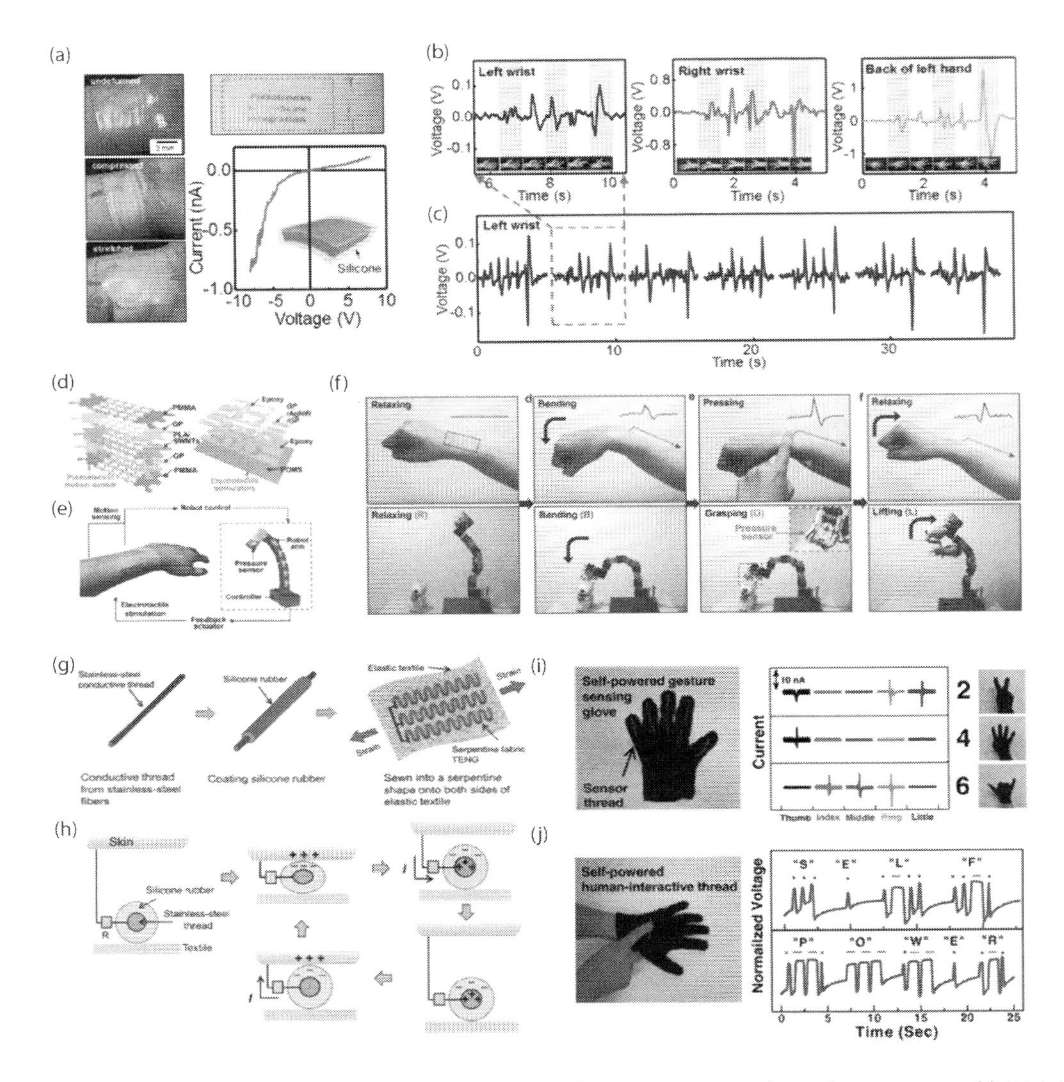

FIGURE 18.12 Conformal and transparent piezoelectric device on a wrist to detect finger gestures. (a) Digital image of the conformal ZnO nanowire device laminated and deformed on a wrist. The device is also highly transparent. A pn junction characteristic can be observed through current-voltage (IV) measurements. (b–c) Different wrist movements and finger gestures result in unique voltage outputs from the device. The unique patterns are consistent, and therefore, can be used as an input for hand gesture-based human-machine interfaces. (From Pradel, Ken C. et al., *Nano Lett.*, 14, 6897–6905, 2014, Copyright 2014 American Chemical Society.) Thin and transparent piezoelectric sensor integrated with electrotactile stimulator for interactive HMI. (d) Exploded view of patterned GP heterostructures with PLA/SWNT as a sensing (left) and AgNW as a stimulation (right) layer. (e) The devices are then attached to the wrist for robotic arm control with closed-loop sensory feedback. (f) Relaxing, bending, and pressing the piezoelectric sensing element actuates the robotic arm to bend, grasp with stimulation feedback, and lift correspondingly. (Adapted from Lim et al. (2015), Copyright John Wiley & Sons.) Thread-based triboelectric nanogenerator for self-powered interactive HMI. (g) Fabrication process of single-thread triboelectric nanogenerator and its integration into textile. (h) Electricity generation principle of the triboelectric thread. Demonstration of triboelectric-integrated glove for (i) self-powered finger gesture sensing and (j) interactive touchpad showing signals from tapping Morse code patterns. (Adapted from Lai et al. (2017), Copyright John Wiley & Sons.)

18.4 Speech and Voice Recognition

Human vocal folds are integral media for human interaction through communication. They oscillate to generate audible waves, enabling us to express ourselves and interact with one another effectively (Traunmüller and Eriksson 1994). With the vast advances in hardware systems, big data, and machine learning algorithms, it is now possible for computers to efficiently process audio signals for applications in voice recognition, particularly for speaker identification, authentication, and speech recognition. Typically, we interface with these audio technologies through a microphone. However, the distance, indirect contact, and environmental noise sometimes misinterpret our commands and force us to speak louder than usual. Additionally, these technologies are also ineffective for people with speech disorders. In this respect, we will discuss the role of flexible and stretchable devices that can be intimately attached to sense vibrations or muscle movements and recognize the sound generated by the vocal cords for the next generation of personal virtual assistants. These devices can be as a media for people with speech disorders to communicate or as novel voice-based HCI devices such as for silent speech interfaces. Several techniques can be used to detect human voice and speech with on-skin electronics: sensing the electrical activity around the thyroarytenoid muscle or monitoring the physical change of skin surface around the neck with vibration, strain, and pressure sensors.

18.4.1 Electromyography and Vibration Sensing

A wireless, on-skin stethoscope by Liu et al. (2016) offers both EMG and vibration sensing for listening to the internal sounds of the human body. As shown in Figure 18.13a, the mechano-acoustic system consists of a collection of electronic devices with their read-out chips embedded in an elastomeric shell. The elastomeric shell, for the structural protection of the electronic chips, is made out of outer and inner silicone that can withstand a biaxial strain of up to 25%. The stretchable system is laminated on the neck (Figure 18.13b) to allow simultaneous sensing of EMG signals through serpentine conductive electrodes and acoustic vibrations through the accelerometers for sensor fusion (Figure 18.13c). As shown in Figure 18.13d, compared to a standard microphone system, the on-skin stethoscope performs better in capturing sound signals in noisy conditions due to the intimate contact between the sensor and the skin. Hence, this device is particularly useful for first responders, security agents, or ground controllers to communicate in noisy environments. They demonstrated an HMI application of the system for speech recognition, which is to control a game by detecting in real-time particular words (left, right, up, down) based on the sensor outputs. With preprocessing of signals to reduce noise and classification techniques based on linear discriminant analysis, a recognition accuracy of 90% can be achieved (Figure 18.13e). This speech recognition system can also be used in many other applications, such as in unmanned aerial vehicles and robotics remote control.

EMG signals generated from swallowing by submental muscles have also been utilized as an HMI input (Lee et al. 2017). Thin gold electrodes, resting on a PI and elastomer substrate, are designed in an island-bridge configuration with serpentine bridge structures. The device can withstand a biaxial strain of up to 150%, which is sufficient to accommodate maximum deformation of the human skin of up to 70% (Pawlaczyk et al. 2013). The swallowing detection is done by initially employing a 30–150 Hz bandpass filter to the raw EMG signals from swallowing, then smoothing the filtered signals by taking root-mean-square values of 250 data points. Finally, the processed signals are compared to a pre-calibrated baseline and threshold value for classification. A biofeedback game, customized for dysphagia rehabilitation exercises is also demonstrated. The biofeedback game takes binary data from the EMG sensor to trigger a ball that jumps through obstacles. Evaluation of the results concludes that the skin-like electrodes have a comparable and better performance than rigid electrodes, with a false positive rate of 3%, which is 2% lower than using rigid electrodes.

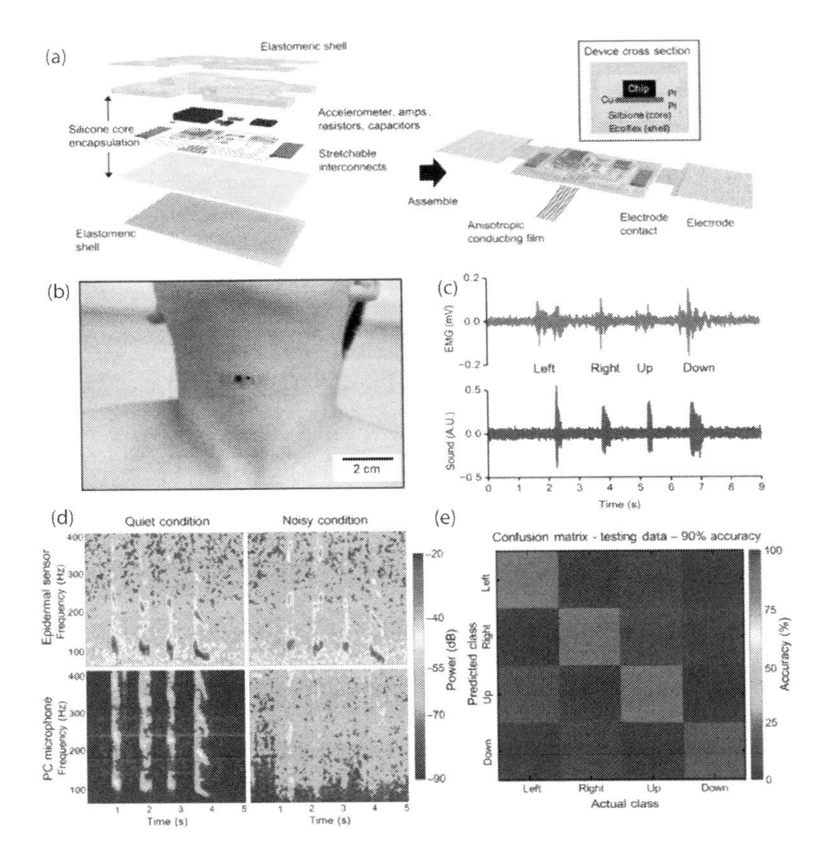

FIGURE 18.13 Epidermal mechano-acoustic sensing system for human-machine interfaces. (a) Exploded layout view of the encapsulated system showing accelerometer with various elements connected to accessible EMG electrodes and anisotropic conductive ribbon for power and communication interface. (b) Digital image of the epidermal mechano-acoustic sensing system attached on a neck around the vocal cords. (c) Simultaneous sensing of speech vibration with EMG and accelerometer data. (d) Comparison of speech output from the epidermal device with commercial microphone in quiet and noisy conditions. (e) Confusion matrix that shows the accuracy of a speech classification algorithm in distinguishing different words. (Adapted from Liu, Y. et al. *Sci. Adv.*, 2, e1601185–e1601185, Copyright 2016 American Association for the Advancement of Science.)

18.4.2 Pressure Sensing

Inspired by the architecture of human eardrum, a flexible bionic membrane sensor (BMS) that measures dynamic pressure of cardiovascular and vocal cords activities was developed (Yang et al. 2015). The BMS, as shown in Figure 18.14a, leverages TENG effects from polytetrafluoroethylene (PTFE) nanowires to sense a high range of vibration frequency from 0.1 to 3.2 kHz with a pressure detection limit of 2.5 Pa and sensitivity of 51 $mVPa^{-1}$. By attaching the BMS on the throat, the system can simultaneously and sensitively detect the low-frequency vibration components from the arterial pulse and the high-frequency components from muscle movements stimulated by the vocal cords; applying a band-pass filter (BPF) in the range of 45–1.5 kHz reveals the vocal signals without the low-frequency arterial pulse component (Figure 18.14b,c). Given this capability, a novel multi-modal biometric authentication can be realized by correlating both arterial pulse and throat sound to find a unique signal pattern among users.

Dagdeviren, Su, et al. (2014) presented a PENG-based conformal sensor with transistor amplification for cutaneous pressure monitoring. An audio speaker system is used to observe the piezoelectric sensor response to audible tones under high frequencies and low pressures. The flexible device can detect

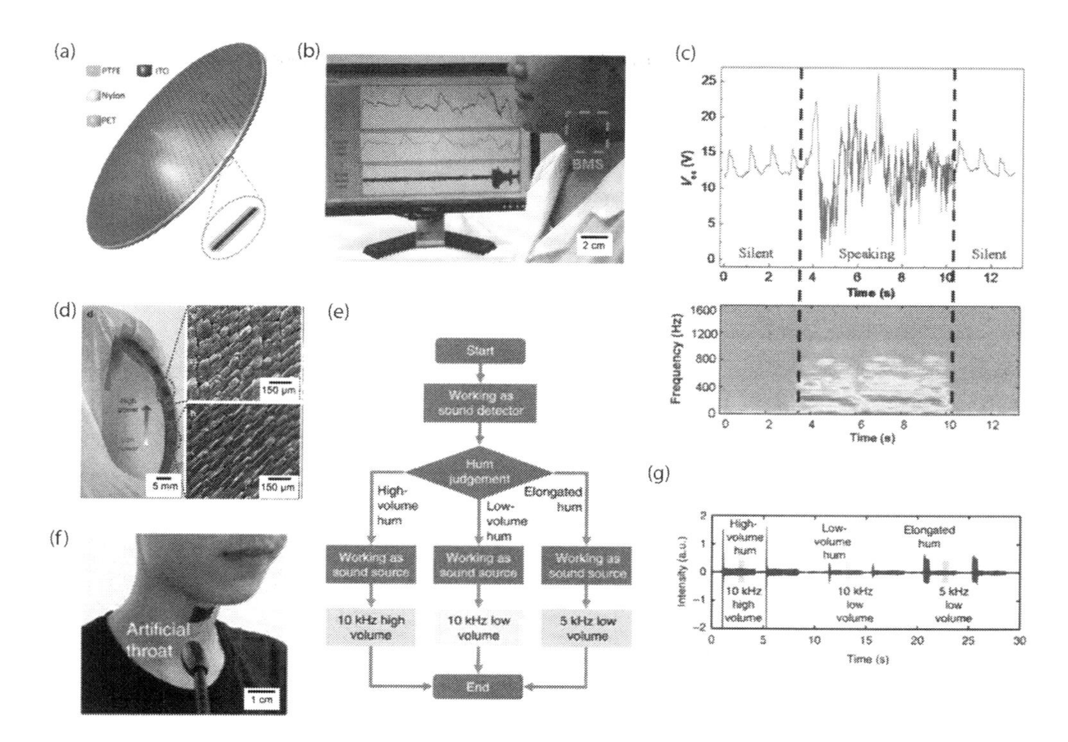

FIGURE 18.14 Bionic membrane sensor for speech recognition. (a) Design stack of the self-powered piezoelectric membrane sensor. (b) A participant with BMS attached on her throat with a screen showing the signal output and its components (arterial pulse and sound wave). (c) Voltage response of the BMS during silent and speaking with its corresponding Fourier transform. The Fourier transform indicates bandwidth of the voltage output from user's speech which ranges from 45 Hz–1.5 kHz. (From Yang, Jin et al., Eardrum-Inspired Active Sensors for Self-Powered Cardiovascular System Characterization and Throat-Attached Anti-Interference Voice Recognition. *Adv. Mater.* 27. 2015. 1316–1326. Copyright Wiley-VCH Verlag GmbH & Co. KGaA. Reproduced with permission.) Artificial throat for bi-directional voice sensing and actuation system. (d) SEM image of laser-induced graphene morphology with different power (290 and 125 mW). (e) State-machine of the LIG artificial throat. (f) A participant wearing the LIG artificial throat on the neck. (g) A graph showing several detected hums and the artificial throat response by outputting high-volume 10 kHz, low-volume 10 kHz, and low-volume 5 kHz sound. (With kind permission from Springer Science+Business Media: *Nat. Commun.*, An Intelligent Artificial Throat with Sound-Sensing Ability Based on Laser Induced Graphene, 8, 2017, 14579, Tao, Lu-Qi et al.)

pressure as low as ~0.005 Pa with a sensitivity of ~1 μAPa^{-1} thanks to the 2D array configuration of sensing elements and SiNM n-channel metal-oxide semiconductor field effect transistor (n-MOSFET). The IV response of the SiNM n-MOSFET enables amplification of the piezoelectric voltage output from the array to the gate of the n-MOSFET, converted in the form of drain-source current. The device can sensitively detect pressure variations of blood flow in the near-surface arteries when attached around the wrist, arm, and throat.

Figure 18.14f presents a bi-directional, flexible artificial throat based on laser-induced graphene (LIG) (Tao et al. 2017). Instead of using the principle of piezoelectricity or triboelectricity, the LIG harnesses the thermo-acoustic effect to both detect and generate sound in the audible range. A direct laser writing for rapid and low-cost prototyping is used to transform a PI substrate into a graphene layer (Figure 18.14d). A periodic joule heating, induced by the applied alternating current (AC) signals, results in an expansion of air and thus, generation of sound. On the other hand, vibrations on the LIG surface change its electrical behavior. The LIG could sensitively detect human-generated sound, including a cough, hum, scream, swallow, and nod when placed on the throat. Simultaneous sound sensing and generation are also demonstrated by training specific sound inputs via a machine learning model and correlating it with a particular, higher-volume sound output (Figure 18.14e–g). The LIG artificial throat, therefore, enables people particularly with speech disorders to express themselves through this bidirectional technology.

In a related application, Wang et al. (2014) explored the possibility of monitoring muscle movements during speech by developing an ultra-sensitive pressure sensor. Silk-based textiles are used as a mould to create a PDMS layer with a microstructure pattern on its surface. Free-standing, thin-films of SWCNT are then transferred to this PDMS layer and annealed. Two layers of the micro-structured PDMS are then sandwiched to serve as a resistive pressure sensor (Figure 18.15a). The density of

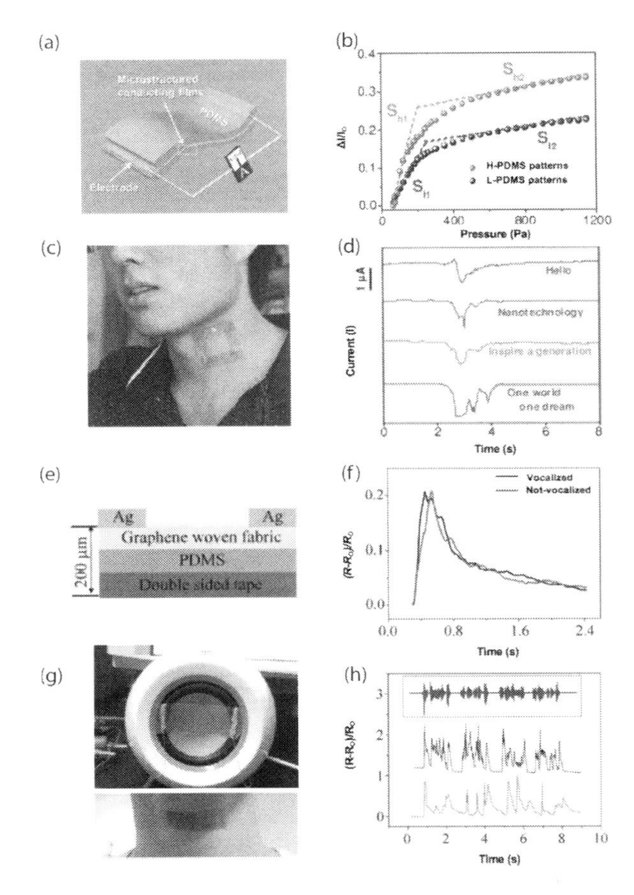

FIGURE 18.15 Flexible sound recognition devices based on pressure sensor. (a) Device structure of a conductive micro-structured PDMS pressure sensor. (b) Comparison of sensitivities between high-density polydimethylsiloxane (H-PDMS) and low-density polydimethylsiloxane (L-PDMS) sensor patterns. (c) A user wearing the pressure sensor on his neck to recognize muscle movements during speech activity. (d) IV curves of the H-PDMS pressure sensor as the user speaks different words. (From Wang, Xuewen et al. Silk-Molded Flexible, Ultrasensitive, and Highly Stable Electronic Skin for Monitoring Human Physiological Signals. *Adv. Mater.* 2014. 26. 1336–1342. Copyright Wiley-VCH Verlag GmbH & Co. KGaA. Reproduced with permission.) Based on strain sensor. (e) Device structure of graphene-woven fabric on a PDMS substrate. (f) Resistance response of graphene strain sensor during vocalization and non-vocalization as the user performs a similar speech action. (g) The graphene strain sensor attached to a loudspeaker's membrane and the neck. (h) Resistance change of the strain sensor attached to the throat (red) and the loudspeaker (black) to the same speech action with the recording data (blue). (With kind permission from Springer Science+Business Media: *Nano Res.*, Ultra-Sensitive Graphene Strain Sensor for Sound Signal Acquisition and Recognition, 8, 2015, 1627–1636, Wang, Yan et al.)

micro-structure patterns ultimately gives a boost of sensitivity to the device (Figure 18.15b). Under 300 Pa, the sensitivity of the high-performance high-density PDMS sensor is as high as 1.8 kPa^{-1}. As demonstrated in Figure 18.15c,d, attaching this pressure-sensitive skin on the neck reveals unique resistance patterns when the user speaks a different set of words and allows the application of speech recognition.

18.4.3 Strain Sensing

The last approach of an on-skin device for wearable speech recognition is based on strain-sensing (Wang et al. 2015). The device incorporates a graphene woven fabric (GWF), embedded in a PDMS, and acting as a microstructured conducting film. The GWF is developed by a CVD process on a seed copper mesh. The resistance of the GWF-based sensor changes when the user speaks, which corresponds to the muscle motions around the neck and strain across the sensor. The signal peaks of the GWF-based sensor when laminated on a throat and a vibrating membrane of a loudspeaker confirm with the original audio waveform, as shown in Figure 18.15h. In addition, testing data show a similar change in resistance as a user performs the same speech action with and without vocalization, revealing the usability of the sensor to speech-impaired individuals or for a silent-speech interface (Figure 18.15f).

18.5 Facial Gesture Recognition

Muscular movements of the face are triggered verbally by our speech and non-verbally by our facial expression and gestures. A facial gesture conveys valuable information of our emotional states and is a key parameter in the study of social interactions. Monitoring facial gestures is, therefore, essential in human-machine interfaces. It enables machines to understand human behaviors and to proactively respond by providing contextual and emotionally aware decisions. It has also extended its applications to AAC and medical rehabilitation.

The most commonly used technique to detect a facial gesture is vision-based sensing. This method requires high complexity processing, is not robust to environmental noise, and is relatively bulky. Researchers have also explored other modalities such as EMG, EEG, EOG, capacitive, and electromagnetic sensing for facial gesture recognition (Matthies et al. 2017). However, current research mostly used off-the-shelf components or commercial gel electrodes, restraining the system from giving high-quality signals while also being wearable and comfortable. As a solution, flexible and stretchable devices that can be easily worn, seamless, and conform to the skin should be developed.

18.5.1 Electromyography and Electrooculography

Paul et al. (2014) proposed a reusable fabric patch, which consists of EMG and EOG electrode arrays for facial electrical muscle activity monitoring. Twenty silver polymer electrodes are fabricated by using a screen-printing technique. A textile-based substrate, treated with an interfacial layer is used in this work. The interfacial layer ensures a smooth surface on the textile and prevents discontinuities when printing the conductive materials. The fabric patch, in the form of a headband, is then tested on a human subject by corresponding facial movements, such as raising eyebrows (frontalis muscle), clenching jaw (temporalis muscle), and horizontal EOG (lateral rectus muscle) to a cursor or controller. Testing procedures showed a significant change of voltage amplitude in four of the electrodes around the top of user's forehead.

Amplification and filtering system is constructed to process the data from these four electrodes to three output channels. 25–125 Hz filter is used to process EMG signal from the jaw and eyebrow while 10 Hz filter is used to process EOG signal from the eyes. With feature extractions and baseline calibrations, the user interface system can classify eyebrows and jaw movements to control 'up' and 'down' commands and eyeball movements to control 'right' and 'left' commands. This work is particularly useful for AAC, since the facial gesture recognition system can support speech and writing for those with dysarthria or other severe neuromotor disorders, such as cerebral palsy, amyotrophic lateral sclerosis, traumatic brain injury, and Parkinson's disease.

18.5.2 Strain Sensing

On-skin electronics that are attachable to the human body especially around the face or the neck should also be designed to look seamless and natural in order to gain widespread user acceptance. As a result, optical transparency or invisibility is a critical feature in the development of wearable electronics on the facial skin. As shown in Figure 18.16a,b, a transparent and stretchable piezoresistive strain sensor has been developed to monitor facial gestures for emotion monitoring during daily activities

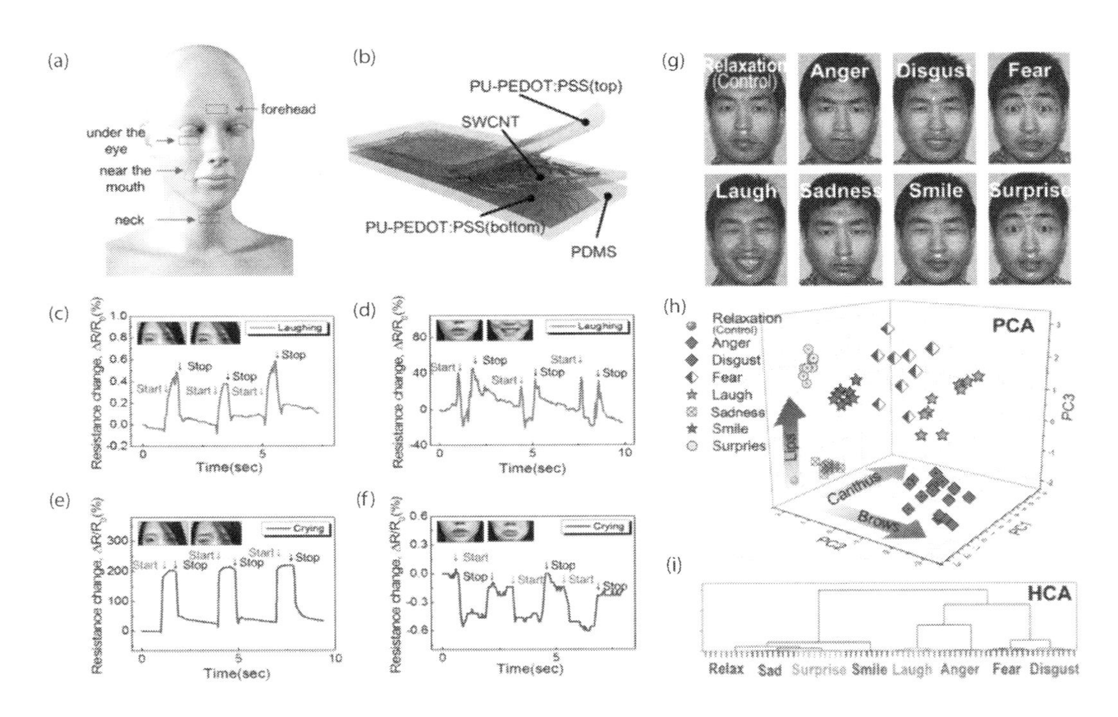

FIGURE 18.16 Resistive strain sensors for facial gesture monitoring. (a) Placement of transparent, stretchable, and ultrasensitive around the forehead, near the mouth, under the eye, and on the neck to measure skin micromotion induced by movements of the facial muscle movements during emotional response and daily activities. (b) A cross-sectional structure of the strain sensor comprising a three-layer stack of PU-PEDOT:PSS/SWCNT/PU-PEDOT:PSS on a PDMS substrate. The relative resistance change of the strain sensor when attached to the (a, e) forehead and (b, f) near the mouth as the subject was laughing and crying, respectively. (Roh, Eun et al., *ACS Nano*, 9, 6252–6261, 2015. Copyright 2015 American Chemical Society.) Facial expression recognition. (g) Using the AgNP curved array strain sensors previously shown in Figure 18.16g and placing six of them around the facial skin basic muscle groups allow feature extraction for classification. (h) PCA and (i) HCA results show a significant separation between each cluster when the strain sensors are placed at the corner of the lips, canthus, and in between the eyebrows. (From Su, Meng et al., Nanoparticle Based Curve Arrays for Multirecognition Flexible Electronics. *Adv. Mater.*, 2016. 28. 1369–1374. Copyright Wiley-VCH Verlag GmbH & Co. KGaA. Reproduced with permission.)

(Roh et al. 2015). The strain sensor has a three-layer stack structure consisting of SWCNTs sandwiched in between two polyurethane (PU)-PEDOT:PSS layers on a PDMS substrate. The PU-PEDOT:PSS and SWCNT solutions are deposited and developed onto the PDMS substrate by spin-coating, followed by thermal annealing and a chemical functionalization process. The sensor gauge factor and transmittance are dependent on the SWCNT concentrations. Mechanical testing confirmed the sensor can operate with a strain of up to 100%. Placing the sensor around various regions of the face reveal a maximum of 0.6% and 40% strain on the forehead and skin near the mouth, respectively. Figure 18.16c–f shows the sensor response as the subject was laughing and crying.

Another research of on-skin devices for facial skin micro-motion detection was conducted by Su et al. (2016). The sensing device is designed by the self-assembly of Ag nanoparticles (NPs)-containing liquid in a curve-patterned micropillar (Figure 18.18g). The AgNP array is then printed on a PDMS film and sintered. Cr/Au interdigitated electrodes are then deposited on the PDMS film and provide the base connections to the AgNP curves array. The strain sensor gives consistent results when stretched to 5% its length in 1000 cycles. Six of these sensors are then mounted on basis muscle groups around the face to recognize eight facial expressions, which are relaxed or default, angry, disgusted, scared, laugh, sad, smile, and surprised (Figure 18.16g). Jack-knifed classification procedures were performed on each of the six sensors and gave insights and validations on the top three sensor locations, which are on the corner of the lip, canthus, and in between the brows. As illustrated in Figure 18.16h i, principal component analysis (PCA) and hierarchical cluster analysis (HCA) results show a clear separation between each cluster of the facial expressions. It can be concluded that most of the negative emotions correlate with the movements of the brows, while the contraction of muscles around the canthus and lips exemplify delightful and strong expressions, such as smiling, laughter, fear, and surprise.

18.6 Eye-motion and Gaze Detection

Eye motions and gaze contain a natural, high-bandwidth source of voluntary and involuntary inputs for human-machine interfaces. They are one of the main markers that correspond to our attention. Eye tracking devices that measure eye movements and positions have been used in a broad range of applications. These include visual behavior monitoring in psychology (Gidlöf et al. 2013), oculomotor rehabilitation (Sharma 2011), activity recognition (Bulling et al. 2011), gaze-based human-computer interaction (Bulling and Gellersen 2010), sleep monitoring (Keenan and Hirshkowitz 2010), and augmentative and alternative communication (Al-Rahayfeh and Faezipour 2013).

There are three well-known technologies for eye tracking: optical, electric potential, and eye-attached tracking. Most efforts on eye tracking are predominantly vision-based, with a typical spatial resolution of 0.5–1 (Whitmire et al. 2016). Some of these vision-based systems used optical instruments that can be relatively worn as a light head-mounted gear. High-resolution optical instruments are currently expensive and computationally demanding. They also require a complex processing setup. Moreover, movements of the user and varying environmental conditions could heavily add noise and reduce the accuracy of the tracking. The recent advances in mobile systems, particularly for the emerging augmented reality (AR) and virtual reality (VR) technologies, hence, demand a novel solution for highly accurate, low-power, and wearable eye tracking.

18.6.1 Electrooculography

EOG measures electrical potentials caused by the movements of cornea in between multiple electrodes around the eye. Compared to the optical tracking, this method works under various lighting conditions, requires a low computational power, and can be worn non-obtrusively (Bulling et al. 2011). EOG systems can even be implemented to record eye motions with the eyes closed, which is useful during sleep study for the detection of rapid eye movements. Although EOG can be effectively used to monitor saccadic eye movement and blinks, due to its low SNR and potential drifts, it is challenging to use it for detecting gaze direction and monitoring slow eye motions with high accuracy.

Mishra et al. (2017) demonstrated soft, bio-signal electrodes to perform wireless EOG that can be attached intimately to the skin (Figure 18.17a). The electrode fabrication is similar to the EES

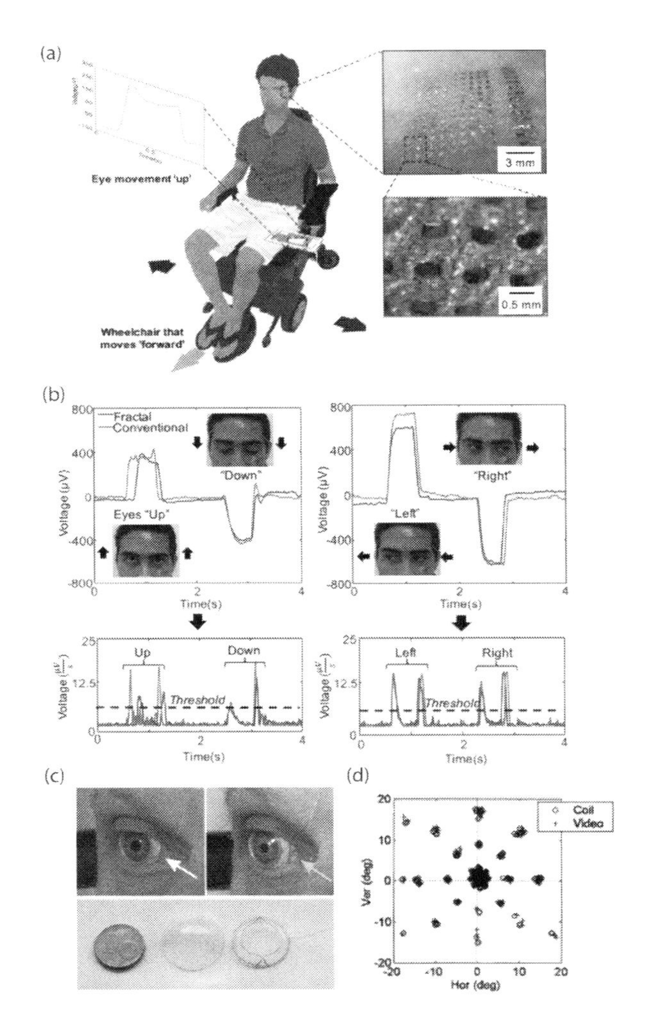

FIGURE 18.17 Monitoring eye movements through conformal EOG sensor. (a) Illustration of a user with the EES attached around the facial skin and close to the eyes to measure EOG and control a wheelchair. The exploded digital images of the fractal electrodes are also presented. (b) EOG voltage signals in four eye movements and its bandpass filtered (0.1–20 Hz) forms showing the possibility of classification. (Adapted from Mishra et al. (2017), Copyright Elsevier.) Scleral search coil for eye gaze tracking. (c) Image of a user wearing a scleral search coil contact lens (left) and with an additional bandage (right). The arrow shows the wire exits with and without the bandage. (Reprinted from *J. Neurosci. Methods*, 170, Sprenger, Andreas et al.,"Long-Term Eye Movement Recordings with a Scleral Search Coil-Eyelid Protection Device Allows New Applications, 305–309, Copyright 2008, with permission for Elsevier.) (d) The performance of scleral search coil method in comparison with video recognition showing both high accuracies in detecting eyeball fixation. (Reprinted from *J. Neurosci. Methods.*, 114, van Der Geest, Josef. N. Van Der, and Maarten A. Frens, Recording Eye Movements with Video-Oculography and Scleral Search Coils: A Direct Comparison of Two Methods, 185–195, Copyright 2002, with permission for Elsevier.)

previously mentioned, with a fractal structure of electrodes transferred to an elastomeric substrate (Kim et al. 2011). They compared the performance between fractal and conventional electrodes in measuring EOG potentials, as shown in Figure 18.17b. Based on the filtered and processed derivative peaks, five features are defined: amplitude, velocity, mean, wavelet energy, and definite integral. A machine learning algorithm is then used to classify the EOG signals in real-time. In the end, they showed an application of human-wheelchair interface for individuals with a disability. The soft bioelectronic sensor with the integrated system can detect four eyeball directions (up, down, right, and left) with 94% of accuracy in order to control the movements of a wheelchair.

18.6.2 Scleral Search Coils

A more high-resolution method to detect eye motions and gaze direction that is comparable to video recognition is magnetic field tracking with scleral search coils (van der Geest and Frens 2002) (Figure 18.17c,d). This semi-invasive technique provides a high temporal (>1 kHz) tracking with a spatial resolution of under 0.1° and is still functional with the eyes closed (Sprenger et al. 2008; Whitmire et al. 2016). The setup consists of torsion coil(s) embedded in a silicone rubber as a contact lens. The lens adheres to the sclera of the eye around the iris. Multiple Helmholtz coils that uniformly produce a magnetic field are placed around the head. This magnetic field induces a voltage in the torsion coil(s) according to their 2-/3D orientations. The voltage change given by these scleral coils in respect to the transmitters, therefore, corresponds to the eye motions (Robinson 1963). Since the torsion coil is semi-invasive and uncomfortable to the eye, this approach currently cannot be worn for long-term applications and is mostly used in medical settings.

18.6.3 Pressure and Strain Sensing

Several investigations of on-skin eye tracking involved novel materials and device designs. Lee et al. (2014), for example, presented a flexible, light-weight, ultra-thin PENG for detecting small skin deformations. The PENG is developed by depositing a ZnO seed layer onto Al foil with Al_2O_3 and then growing ZnO NW through a hydrothermal process. The high sensitivity of this thin PENG to bending enables the application of eyeball motion tracking when the device is laminated around the eyelid. The output voltage and current results generated by the piezoelectric device during the strain influenced by the eye movements showed the functionality of the PENG to detect slow to rapid (0.4–1.6 Hz) eye movements. The proposed system can be utilized to monitor emotion, alertness, and sleeping pattern. A TENG patch can also be attached near the canthus to sensitively detect mechanical micromotion of the skin due to the eye motions and blinks (Pu et al. 2017). This flexible and transparent TENG patch consists of fluorinated ethylene propylene (FEP) coated with indium tin oxide and laminated to a PET substrate. A cavity wall formed by PET then separates a natural latex layer with the FEP. As the users' eye is closed, the muscle contraction pushes the latex towards the FEP and electrons flow from the external circuit to the indium tin oxide (ITO) and *vice versa*. The voltage generation by the patch due to blinking (~750 mV) surpasses the typical voltage generated by EOG (1 mV). This performance reduces the possibility of false-positive detections due to low signal level and high noise. The interface system could also be self-powered and does not require complex circuitry and signal processing.

Another approach explores the use of piezoresistive strain sensors as previously discussed in detail in the previous section of this chapter (Roh et al. 2015; Su et al. 2016). Attaching these micro-motion strain sensors around the upper eyelid, lower eyelid, and canthus of the eye allows tracking of four eyeball directions through the resistance change of the sensors (Figure 18.18h). Further work of incorporating more nodes of micro-fabricated strain sensor with high-level signal processing could improve the eye-tracking resolution and enable a higher accuracy, point-of-gaze detection.

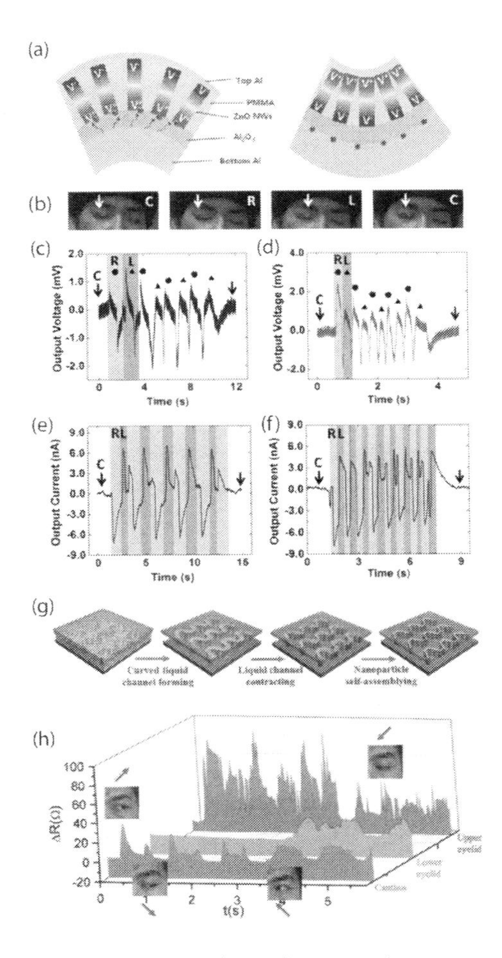

FIGURE 18.18 Detecting eyeball movements with on-skin piezoelectric sensor. (a) ZnO NWs layer under tensile and compressive strain, showing the screening effect that makes the device produces three times higher output signals under compressive than tensile strain. (b) The super-flexible ZnO NG attached around the eyelid for detecting eyeball motion. Output voltage of the ZnO NG device under (c) slow and (d) rapid eyeball motion going from the centre (C), to the left (L), and right (R). Output current is also measured during (e) slow and (f) rapid eye movements. (From Lee, Sangmin, Ronan Hinchet et al., Ultrathin Nanogenerators as Self-Powered/active Skin Sensors for Tracking Eye Ball Motion. *Adv. Funct. Mater.* 2014. 24. 1163–1168. Copyright Wiley-VCH Verlag GmbH & Co. KGaA. Reproduced with permission.) Eyeball motion recognition using on-skin strain sensor. (g) Silver nanoparticle self-assembly mechanism on a PDMS substrate through a pillar template to create a curved array strain sensor. (h) Multiple strain sensors attached to the upper eyelid, lower eyelid, and canthus of facial skin to real-time monitor eyeball motion in four directions. (From Su, Meng et al., Nanoparticle Based Curve Arrays for Multirecognition Flexible Electronics. *Adv. Mater.*, 2016. 28. 1369–1374. Copyright Wiley-VCH Verlag GmbH & Co. KGaA. Reproduced with permission.)

18.7 Emotion and Stress Monitoring

Emotions play a vital role in our daily life as they enable us to express and understand each other's feelings. They are represented by external physical expressions and internal mental processes that may be imperceptible to us. The ability to recognize human emotions and simulate empathy has become an important aspect in human-machine interaction systems, prompting the field of affective computing or artificial emotional intelligence (emotion AI) (Picard 1997). Recognizing emotions enables machines to adapt and react depending on the user's behaviors, allowing a more natural and efficacious mutual relationship between human and computers. Multiple methods have been explored in the past years to monitor and classify human emotions. The most widely used approach involves the detection of facial expressions, speech, body gestures, and physiological signals (Castellano et al. 2008). Except for physiological monitoring, which uses wearable sensors, current approaches of emotion recognition mainly use an external camera in order to recognize facial gestures or microphone to process voice signals. As we have covered recent developments of flexible and stretchable devices for body gesture, speech, and facial expression recognition (Sections 3 through 5), in this section, we will mainly discuss the development of these devices for physiological sensing. The fact that individuals cannot easily control their physiological signals makes sensing them extremely useful, as manipulating these signals to hide our emotions is challenging.

18.7.1 Electrodermal Activity

Several physiological parameters can give insights into human emotions. These include skin conductance, temperature, respiration rate, cardiac function, and electrical activity of the muscle, heart, and brain (Wioleta 2013). Electro-dermal activity (EDA) relates to the conductance of the skin, commonly known as galvanic skin response (GSR) that can be influenced by psychological arousal (Boucsein 2012). This skin conductance change is primarily influenced by sweat from stimulation of sweat glands through the autonomic nervous system. The water and electrolytes in the sweat provide conductive pathways through the skin, thus increasing the overall conductance. The EDA can be measured by measuring the change in impedance or potential in between two electrodes or using a combination of them. Several researchers designed EDA monitoring systems that are wearable and conformable. This prevents the use of wet or gel electrodes, which are uncomfortable and not reusable, or dry electrodes, which have low performance and are not adaptive to the curvature of the skin.

Figure 18.19a shows a flexible GSR sensor with conductive polymer foam, designed by Kim, Jeehon et al. (2014). The use of the soft conductive foam for the electrode enhances the sensor's comfort, reusability, and contact to the skin upon pressure. To test the system, they strapped the wearable GSR sensing system on the back of a body and compared it to a conventional GSR system. The results showed a high correlation between the two GSR signals when three auditory stimuli, which are screaming, glass breaking, and gun blasting were triggered without the subject's knowledge (Figure 18.19b). Further tests to 19 participants revealed a 94.7% success rate of GSR reading.

18.7.2 Multi-modal Physiological Sensing

The use of more than one modality of physiological parameters has been proven to improve classification number and accuracy of emotion recognition (Lisetti and Nasoz 2004). Yoon et al. (2016) proposed a multi-modal flexible human stress monitoring patch that incorporates not only GSR, but also skin temperature and arterial pulse-wave sensors in a multi-layer structure (Figure 18.19c). Two silver electrodes (A and D) serve as the GSR sensor while a meander-line structure of interconnect (B and C) acts as the skin temperature sensor. A Parylene C layer separates the skin contact layer with the pulse wave sensing layer. The arterial pulse is sensed by a thin flexible piezoelectric membrane in between two silver electrodes (E and F). Figure 18.19d–f shows the experimental and theoretical response of each sensor to its corresponding stimulus. They concluded that with a machine learning algorithm such as support vector machine (SVM), the physiological parameters given by this patch could be potentially used to classify four types of human emotion, such as surprised, angry, stressed, and sad.

A study to observe the electrophysiological response given by an individual in an unforeseen situation was conducted by Jang et al. (2014). For this purpose, a stretchable electronic system that includes electrophysiological (EP) sensing electrodes and an optical blood oximeter is fabricated (Figure 18.20a). Filamentary serpentine mesh geometry serves as the three EP electrodes for EOG, EMG, or ECG. A micro-scale inorganic light-emitting diode (μILED) composed of AlInGaP is used as the blood

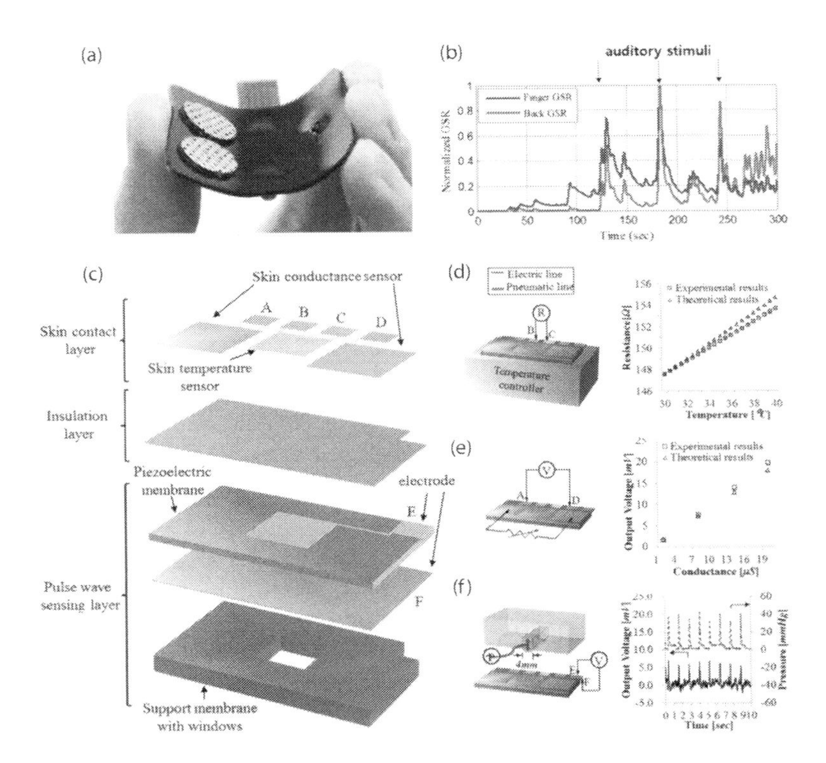

FIGURE 18.19 Flexible and wearable GSR sensor. (a) Digital image of the soft GSR sensor using conductive foam as electrodes for maximum comfort and reliable contact to the skin. (b) Finger and back GSR response of a subject which can be observed to react upon several unexpected auditory stimuli. (From Kim, Jaemin et al., *Nat. Commun.* 5, 5747, 2014. Copyright 2014. IEEE with permission.) Multi-sensory flexible patch for wearable stress and emotion monitoring. (c) Schematic of the device stack consisting of a piezoelectric membrane for pulse-wave sensing, two electrodes for skin conductance/GSR sensing, and a meander line for temperature sensing with support membrane and an insulation layer. Experimental and theoretical characterization of (d) temperature sensor, (e) GSR sensor, and (f) pulse wave sensor confirming the functionality of the multi-sensory flexible patch. (With kind permission from Springer Science+Business Media: Yoon, Sunghyun et al., *Scientific Reports*, A Flexible and Wearable Human Stress Monitoring Patch, 6, 2016, 23468.)

oximetry actuation element. The whole system including the interconnects is supported by a PI layer and transferred to a breathable, silicone-coated textile substrate. Based on the intensity of the scattered lights, they showed a comparison between a human subject performing meditation and a stressful task and concluded an inversely proportional relationship between the intensity of task and the light absorption due to the blood flow rate (Figure 18.20b,c). Further tests in a virtual driving simulator also revealed an abrupt change in EP signals (Figure 18.20d) around the eye, chest, and arm during sudden braking and merging events, as shown in the spectrograms of Figure 18.20e. There is a sudden increase of eye activity level through the EOG signals after the event. EMG signals collected from the EP sensor on the arm showed a brake reflex of the subject, as the subject reacted to avoid the collision by turning the wheel. Regional activations in two different frequencies can be observed on the EP sensor attached to the chest. These activations correlate to the surprised and nervous state of the subject, represented by an increase of ECG and EMG response evoked by the unexpected event.

The physiological parameters discussed in this chapter can also be used not only to monitor stress and cognitive states, but also detect pattern abnormalities that may lead to mental disorders, such as schizophrenia and depression. Given the fact that flexible and stretchable microfabricated sensors allow long-term, comfortable, high-sensitivity monitoring of physiological signals, as these devices are

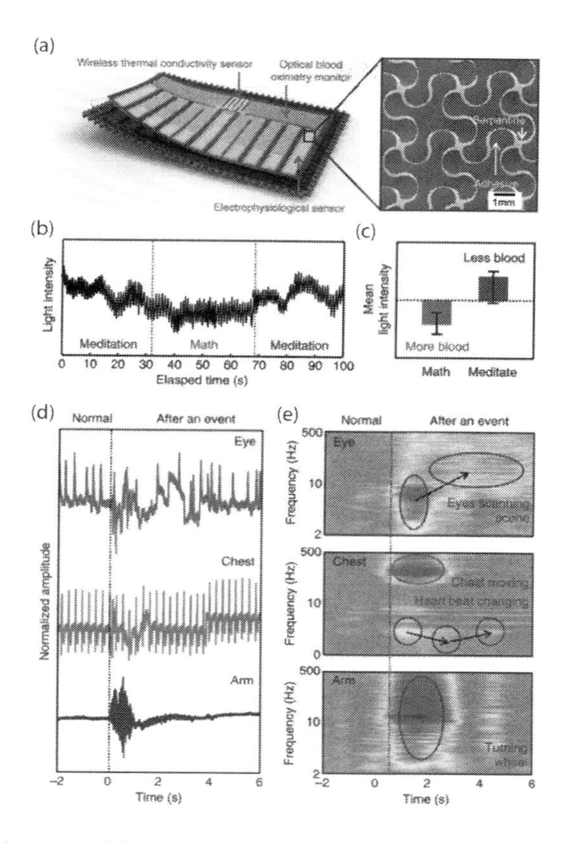

FIGURE 18.20 Breathable, stretchable electronics patch for transcutaneous physiological and cognitive load monitoring. (a) Illustration of the electronic patch comprising an electrophysiological sensor, temperature sensor, and blood oximeter on an elastomeric (~100 μm thick) and stretchable fabric substrate (~1 mm thick; 90% nylon, 10% spandex). SEM image shows the serpentine design of EP electrodes on a silicone-coated stretchable fabric. (b) Light intensity averaged response during mental math test and relaxation period. (c) It can be deduced that the mean light intensity correlates to the mental activity of the subject, with increased blood flow during high cognitive load stress. (d) EP signals including from EOG, ECG, and EMG from various locations (eye, chest, and arm) of the subject body during a realistic virtual driving test in a car simulator with a frequency spectrum shown in (e). An unexpected accident (braking and merging events) can be observed to trigger electrical muscle activity around the eye, chest, and arm as the subject is stunned, reacting by braking and turning the car, and becoming more attentive and focused. (With kind permission from Springer Science+Business Media: *Nat. Commun.*, Rugged and Breathable Forms of Stretchable Electronics with Adherent Composite Substrates for Transcutaneous Monitoring, 5, 2014, 4779, Jang, Kyung-In et al.)

becoming more accessible to researchers in artificial emotional intelligence, we are transitioning to an era where machines will start to gain a sense of empathy as they process affective information from their users in ambient.

18.8 Brain-Computer Interfaces and Neuroprosthetics

The brain is a primary coordinator of the nervous system. It receives and sends sensory and motoric signals from and to the rest of the body through the spinal cord. It is a powerful tool that allows us to perceive, process, and store information. Studying the brain dynamics will help us to understand how the brain responds to particular stimuli or how changes in brain structure can influence a persons' personality, cognitive behavior, or well-being. It could also potentially enable future computers to precisely identify our thoughts, predict our intentions, and recognize our emotions. Neuroimaging tools such as functional magnetic resonance imaging (fMRI), functional near-infrared spectroscopy (fNIRS), positron emission tomography, functional transcranial Doppler sonography (fTCD), magnetoencephalography (MEG), and EEG have consequently been used to map and monitor brain activity by measuring localized changes in hemodynamic response or recording electrical currents and magnetic fields due to neuronal activations (Duschek and Schandry 2003; Shibasaki 2008). Brain stimulation techniques, such as transcranial magnetic stimulation (TMS), deep brain stimulation (DBS), electroconvulsive therapy (ECT), and transcranial direct current stimulation (tCDS) have also been used in research and medical settings, mainly for the treatment of mental disorders (George et al. 2002; Fregni et al. 2006).

Most of these tools are currently operator-dependent and require a large setup. These challenges limit their usage for wearable, real-time, and long-term applications. Researchers have recently developed non-invasive systems such as screen-printed flexible receiver coils for magnetic resonance imaging (Corea et al. 2016) and wearable fNIRS with tCDS (McKendrick et al., 2015). Commercial and personal EEG headsets such as NeuroSky and Emotiv also exist and have been widely used in brain-computer interface (BCI) research and clinical trials (Swan 2012). However, to achieve a high SNR and direct contact with the brain, spinal cord, peripheral nerves, or even specific neurons, device miniaturization, modification efforts, and its translation into implants are required. Multi-modal, neuromodulation capability also needs to be incorporated to facilitate bi-directional communication. This section will cover recent development of various flexible and stretchable devices for neural reading, stimulation, and neuroprosthetics. The mechanically compliant structure and intrinsic property of these devices bridge the biological, mechanical, and geometrical mismatch between electronics and our body tissue, facilitating seamless and intimate coupling between the two.

18.8.1 Electroencephalography

In EEG, the electrical signals induced from the neural activity are spatially read by a set of electrodes, attached around the scalp. The electrical signals, which can be classified into several frequency bands, could give insights to human cognition, emotion, and attention (Ray and Cole 1985). Even though current EEG systems are wearable, they are still bulky and uncomfortable, restraining them from daily use for long-term monitoring of brain signals. Norton et al. (2015) explored the applications of an EES attached to the auricle for BCI applications. The Au electrodes are designed with stretchable filamentary serpentine traces, deposited on a PI, and then transferred to an elastomer. Placing these electrodes around the curvilinear region of the ear (Figure 18.21b) presents a novel and comfortable alternative than attaching them around the scalp, which is enabled by the conformal design of the electrodes. Biocompatibility and long-term wearing tests of the device both show no adverse effects on the skin. A second structure, with tripolar concentric ring electrodes in a capacitive configuration, is also designed for an improved spatial resolution and durability.

The EEG system, as shown in Figure 18.21a, consists of three electrodes: recording (REC), reference, (REF), and ground (GND) with stretchable interconnects to the read-out and processing system. The signal acquisition system will then amplify the EEG signals, and the computer will process these signals to extract suitable features for classification (Figure 18.21c). To demonstrate its application for BCI, steady-state visually evoked potential (SSVEP) approach is tested on several subjects. The SSVEP test triggers optical stimulation that flicker around 6–10 Hz to the retina. The electrical signals generated by the brain during the visual stimulus can be used to monitor eye gaze. The SSVEP BCI ultimately allows the subjects to interact with a text speller software and populate a word from a group of letters. Other applications of EES are also demonstrated by Kim et al. (2011). By applying an EES on the forehead, alpha brain waves of around 10 Hz can be observed to reflect opening, closing, and blinking of the subject's eyes. A Stroop test to study human cognition is also investigated, by performing demanding tasks and analyzing reaction time of the subjects based on their EEG signals.

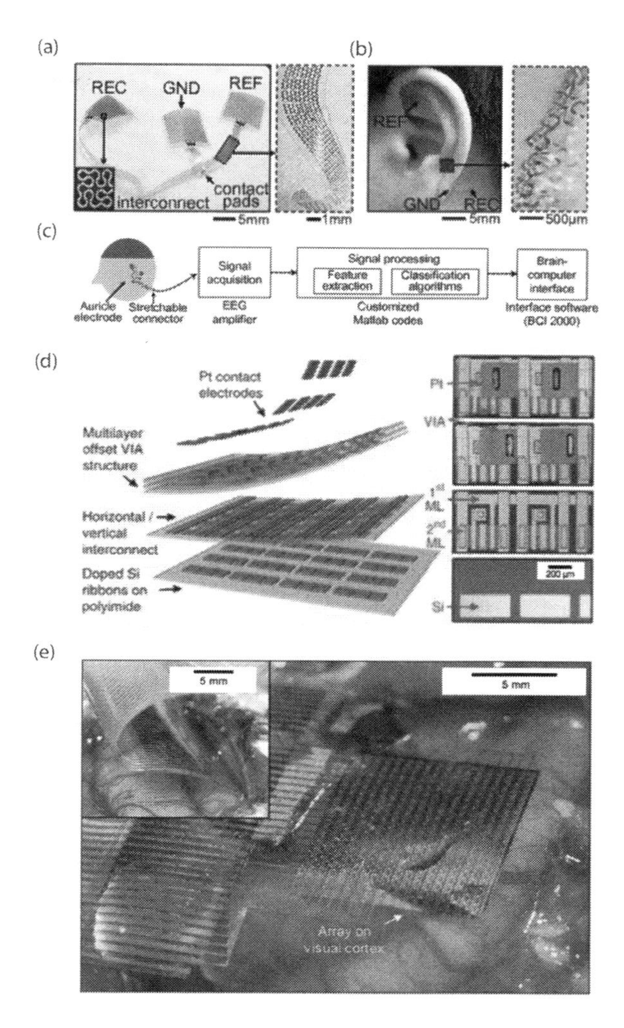

FIGURE 18.21 Novel conformal device for EEG sensing on the surface of the auricle. (a) Images of a fractal design of three electrodes (REC, GND, and REF) and interconnects with a magnified image (right). (b) The conformal EEG system laminated on the auricle and mastoid with its magnified image (right). (c) Flow-chart of signal acquisition and processing of EEG-enabled brain-computer interface. (Reprinted with permission from Norton, James J. S., et al. 2015, Soft, Curved Electrode Systems Capable of Integration on the Auricle as a Persistent Brain–computer Interface, *Proc. Natl. Acad. Sci.*, 112, 3920–3925. Copyright 2015 National Academy of Sciences.) Flexible 360-channel high-density electrode array for ECoG sensing. (d) Exploded view of the array stack consisting of outer Pt electrodes, multi-layer via, first and second metal layers for interconnects, and at the bottom, doped silicon ribbons on polyimide for multiplexing circuits. (e) Photograph of high-density ECoG array on the visual cortex for high spatiotemporal mapping of neural activity. (With kind permission from Springer Science+Business Media: *Nat. Neurosci.*, Flexible, Foldable, Actively Multiplexed, High-Density Electrode Array for Mapping Brain Activity In Vivo, 14, 2011, 1599–1605, Viventi, Jonathan et al.)

18.8.2 Electrocorticography

Due to noise signals and attenuation caused by the skin and bone barriers, research has also been conducted to fabricate an array of electrodes that can be laminated directly on the surface of the brain to perform electrocorticography (ECoG). Conventional invasive neural recordings and stimulations have mainly relied on rigid materials, such as silicon or metals which have a mechanical, geometrical, and biological mismatch to the brain tissues (Jog et al. 2002; Kim et al. 2009). These incompatibility issues could result in hemorrhage and inflammatory response. They could damage target tissues and influence the clarity of neural reading (Polikov et al. 2005; Saxena et al. 2013). Flexible and stretchable, minimally invasive implantable electronics could solve these challenges with a greater precision, sensitivity, as well as spatial and temporal resolution. Several researchers used a polymer layer as a substrate for micro-fabricated sensors and circuits. One example is a flexible multi-channel electrode array to simultaneously monitor ECoG and local field potential (LFP) in the visual cortex of a rat (Toda et al. 2011). The device consists of gold electrodes encapsulated in a Parylene C substrate. Visual stimulus from a monitor evoked a neural response from the rat, and recording results showed reliable data throughout the 2 weeks implantation.

Another work by Viventi et al. (2011) produced a high-density, flexible electrode array integrated with multiplexing circuitry for high resolution (micro-ECoG or μECoG) neural reading. Transistors are fabricated from silicon nanomembranes and transferred into a PI substrate. Interconnects and vias are then deposited layer by layer with PI encapsulation that connects to the outermost platinum (Pt) electrodes. The multiplexer enables addressing mechanisms that minimize connection numbers to access a total of 360 active electrodes. As illustrated in Figure 18.21e, this compliant, foldable, and flexible device allows high-resolution spatiotemporal neural activity reading even in a commonly inaccessible region of the brain. Other researchers have also incorporated bioresorbable and transient materials such as silicon and silk in their design strategies. The usage of these materials enable ECoG systems that can dissolve with the cerebrospinal fluid (Kim, Viventi et al. 2010; Kang et al. 2016).

18.8.3 Multi-modal Implantable Interfaces

The previous examples of flexible 2D devices provide non- to minimally invasive approaches to neural reading. Yet, there are some cases when it is necessary to penetrate further to target specific regions for simultaneous neural reading and modulation. New materials and fabrication strategies have motivated researchers to develop a 3D out-of-plane structure of probes or injectrodes that are bio-compatible and compliant with the mechanical properties of the nerves. Figure 18.22 shows many existing platforms and techniques of multi-modal neural interfacing that can penetrate to delicate regions or conform to the curvilinear surface of the brain, spinal cord, and peripheral nerves while alleviating tissue damage and foreign-body reaction (Lacour et al. 2016). Besides conventional electrical stimulation that uses an electric current to excite surrounding neurons, and delivery of biochemical agents to affect neurotransmission, a new modality for neuronal activations through the flow of photons has emerged. Optical stimulation through optogenetics provides a means to activate specific neurons that exhibit light-responsive ion channels by genetically modifying them with reagents such as channelrhodopsin (Grill et al. 2009; Fenno et al. 2011).

One example of multi-modal neural interfaces is a flexible injectrode for stimulation, sensing, and actuation of the brain soft tissues developed by Kim et al. (2013). The injectrode consists of several layers of microdevices such as an electrode, inorganic photo-diode, and LEDs, as well as a bi-functional temperature sensor/heater on a plastic strip. They are then wrapped in a removable, bio-degradable substrate. The needle-type devices discussed here are commonly used for deep brain recording and stimulation; therefore, they should be sharp and pointed to be able to move freely around the cerebrospinal fluid. The micro-scale size and structure of these devices proved to improve spatial targeting and reduce gliosis after implantation. An injectable micro-LED with near-field radio-frequency (RF) coil also serves as a stand-alone, wireless optoelectronic subdermal implant (Shin et al. 2017). The system consists of several devices with Au-interconnects on PI that are encapsulated in Parylene or PDMS. A soft and stretchable version of this optoelectronic implant is demonstrated in Figure 18.23a,b (S.I. Park et al. 2015). The deformable feature of the device allows it to be intimately attached to the epidural space of the spinal cord or peripheral nerves (Figure 18.23c). An external coil induces an electromagnetic field that couples with the receiver coil to power and communicate with the implantable system. Through wireless capacitive or inductive coupling, these devices bypass the use of batteries or wired connections and correspondingly facilitate free-moving and long-term *in vivo* trials.

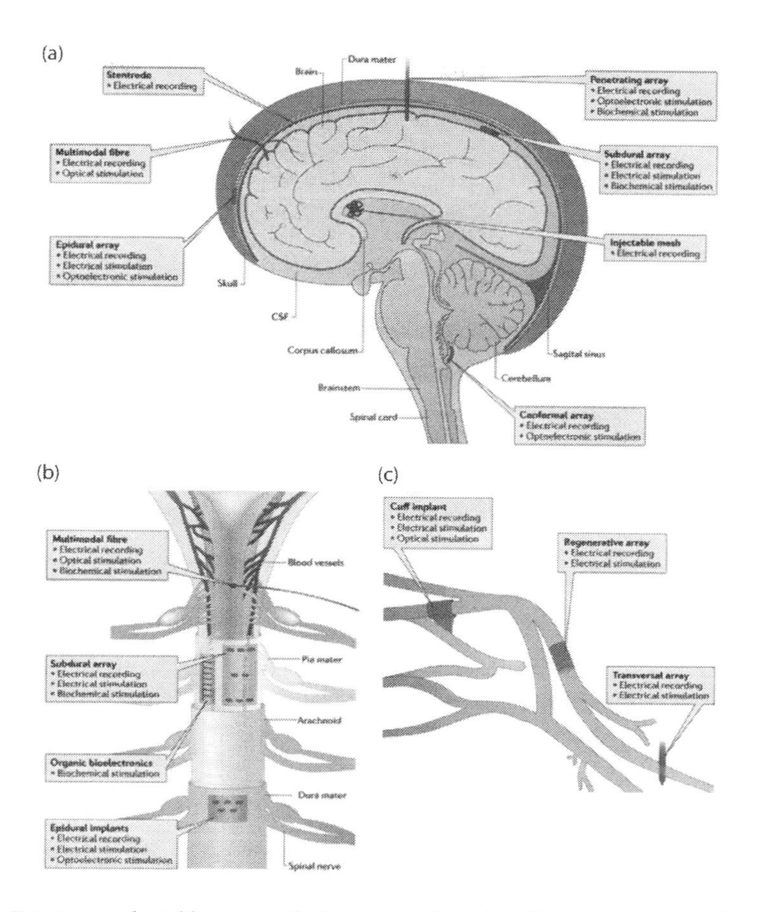

FIGURE 18.22 Existing implantable systems for human-machine interfaces. Recording and stimulation efforts using novel microfabricated devices to bi or uni-directionally interface with (a) brain, (b) spinal cord, and (c) peripheral nerves through various means (electrical, optical, and/or chemical). (With kind permission from Springer Science+Business Media: *Nat. Rev. Mater.*, Materials and Technologies for Soft Implantable Neuroprostheses, 1, 2016, 16063, Lacour, Stéphanie P. et al.)

Instead of using an LED with patterned metal interconnects on a long strip of polymer, research performed by Lu et al. (2017) and Canales et al. (2015) exploited thermal drawing techniques to develop flexible and stretchable fiber-based neural probes. The thermal drawing of a polycarbonate and cyclic olefin copolymer (COC) with the introduction of a conductive polymer composite resulted in a flexible multi-modal all-polymer fiber probe (Figure 18.23d,e). The implantable, fiber-based neural probe facilitates *in vivo* optical stimulation, drug delivery, and electrophysiological recording through the delicate regions of the brain with pinpoint accuracy. Figure 18.23f shows a long-term trial of the fiber probes in freely moving *Thy1-ChR2-YFP* mice. They retain both recording and stimulation capabilities for 2 months. These multi-modal injectrodes have the potentials to advance tools for manipulation and analysis of brain circuits, particularly for the treatment of neurodegenerative disorders. For instance, they could be integrated with wirelessly controlled pumps that could deliver micro-liters dose of drugs on-demand (Dagdeviren et al. 2018).

As an alternative approach to high spatiotemporal monitoring of neural activity in the deep brain, Liu et al. (2015) introduced a novel, syringe-injectable electronic device. This device consists of an array of Pt-electrodes with metal interconnects, encapsulated in a polymer material, and fabricated in a mesh structure. The size and flexibility of the device facilitate compact storing of a dense, large-volume mesh

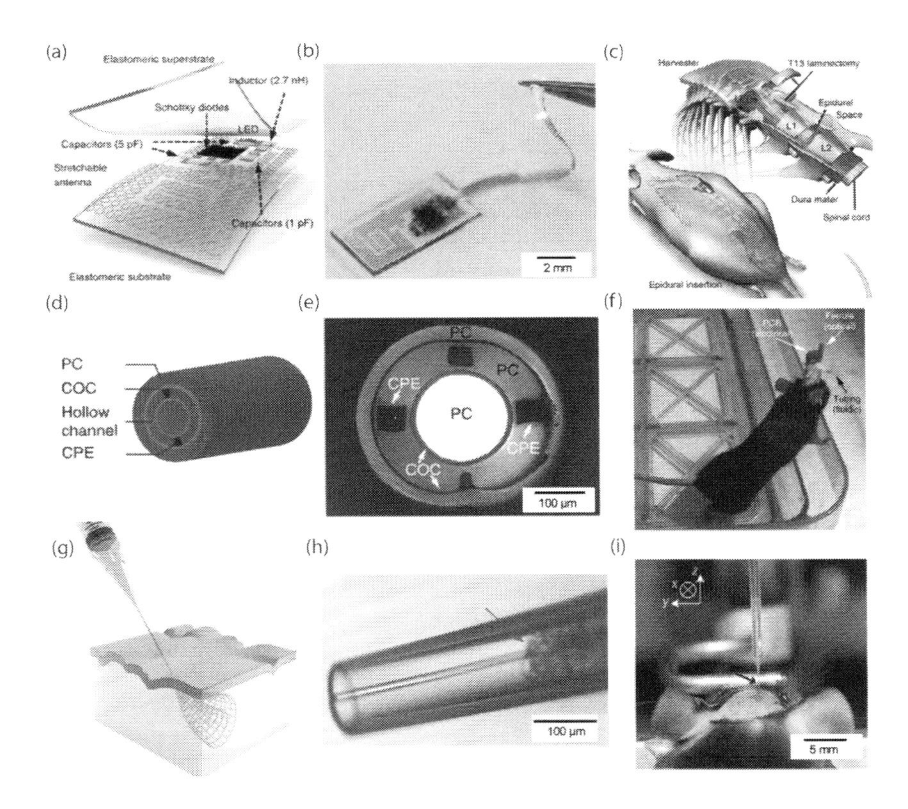

FIGURE 18.23 RF-powered stretchable and implantable optoelectronic system for wireless optogenetics. (a) Illustration of the device consisting of an RF system integrated with LED. (b) Photograph of the soft epidural device with stretchable interconnects leading to an LED at the tip. (c) Location of the implanted soft optoelectronic epidural device relative to the spinal cord of a rat. (With kind permission from Springer Science+Business Media: Park, Jaeyoon et al., 2015. *ChemPhysChem*, Material Approaches to Stretchable Strain Sensors, 2015; With kind permission from Springer Science+Business Media: *Nat. Biotechnol.*, Soft, Stretchable, Fully Implantable Miniaturized Optoelectronic Systems for Wireless Optogenetics, 33, 2015, 1280–1286, Park, Sung Il et al.) Flexible multi-modal brain probe for simultaneous electrical, optical, and chemical signaling. (d) Cross-sectional structure and (e) image of the fiber with polycarbonate (PC) and cyclic olefin copolymer (COC) as light transmission channels, conductive polyethylene (CPE) as recording electrodes, and hollow channel for chemical transport. (f) Transgenic *Thy1-ChR2-YFP* mice implanted with the multi-modal probe on the medial prefrontal cortices. (With kind permission from Springer Science+Business Media: *Nat. Biotechnol*, Multifunctional Fibers for Simultaneous Optical, Electrical and Chemical Interrogation of Neural Circuits in Vivo, 33, 2015, 277–284, Canales et al.) Syringe-injectable electronics for dense sensing and stimulation array. (g) A principle of injectable electronics. Red-orange lines indicate passivating polymer mesh substrate. Yellow lines represent metal interconnects from the recording nodes (blue circles) to I/O pads (green circles). (h) Photograph of the electronic mesh (red arrow) inside a glass needle (ID = 95 μm) before injection. (i) Stereotaxic injection of the electronic mesh into an anaesthetized mouse brain. (With kind permission from Springer Science+Business Media: *Nat. Nanotechnol.*, Syringe-Injectable Electronics, 10, 2015, 629–635. Liu et al.)

network in a syringe (Figure 18.23h), and after injection, enables the mesh structure to unfold and three-dimensionally conform to the surrounding tissue (Figure 18.23g). Injection of the mesh into the hippocampus of anaesthetized mice confirmed the ability of 16 channels to record brain electrical activity (Figure 18.23i). The mesh structure of this device provides seamless, unobtrusive interface with brain tissue and results in a better chronic immune response compared to the other flexible thin-film devices (Zhou et al. 2017).

18.8.4 Soft Neuroprosthetics

Soft implantable neuroprostheses aim to substitute or restore sensory, motor, or cognitive function that could have been damaged due to injury, ageing, or neurodegenerative diseases. Due to their adaptive structures, they function while alleviating foreign body reaction and maintaining long-term stability. One of the most recent applications of soft neural implants is to restore motor or sensory pathways due to spinal cord injury. Electronic dura mater (E-dura) is a multi-functional subdural device that can simultaneously perform electrical sensing, stimulation, and biochemical drug delivery (Minev et al. 2015). This soft implant has been clinically tested in a spinal cord injury case of a mouse. By precisely delivering electrical stimulation around the lumbosacral segment of the spine and injecting serotonergic agents, the E-dura can restore paralyzed rats and control their locomotion behavior.

The progression of soft electronics for the human body has also promoted a new generation of artificial organs integrated with a neural feedback. One example is an organic skin prosthesis that mimics how a biological mechanoreceptor functions (Tee et al. 2015). The skin-inspired digital mechanoreceptor consists of a micro-structured resistive pressure sensor with sensitivity approaching the human skin capability, a printed organic voltage-controlled oscillator circuit, and an optogenetic neural interface system. The CNT-based, piezoresistive pressure sensor converts physical inputs to electrical signals that oscillate and change in frequency through a three-stage ring oscillator. To model similar action potentials response during human tactile stimulations, the sensor and circuitry are engineered to output a maximum frequency of 200 Hz. The pressure-dependent frequency signals given by the digital mechanoreceptor successfully stimulate neurons in the primary somatosensory cortex region of a mouse either electrically or optically.

Another research, as shown in Figure 18.24a, presents a multi-sensory prosthetic skin integrated with platinum(Pt)-NWs on a stretchable multi-electrode array (MEA) to interface with the peripheral nerves (Kim Jaemin et al. 2014). The stretchable MEA allows the relay of sensation from each sensor to its corresponding peripheral nerve through electrical stimulation. A microcontroller unit processes pressure sensor signals and as a feedback, controls stimulation voltage of an electrode on a sciatic nerve (Figure 18.24b,c). Amplified and filtered EEG signals from the ventral posterolateral nucleus (VPL) in the thalamus are simultaneously read to confirm a successful afferent signal transmission from the sciatic nerve to the brain.

For intimate, effective, and robust long-term interfacing with the peripheral nerves, bio-compatible compliant electrodes with low impedance must be developed. Metals such as tungsten, gold, platinum, and iridium have been adopted as the electrodes, typically deposited or embedded in an insulating substrate such as PET or PI (Geddes and Roeder 2003; HajjHassan et al. 2008). Figure 18.24d shows conformal lamination of the PtNWs/Au MEA on peripheral nerve and muscle tissues of a rat model. To reduce the impedance of the electrodes, PtNWs are grown on the MEA using an electrochemical method. Ceria nanoparticles are also introduced on the PtNWs to suppress and prevent inflammation caused by reactive oxygen species. The PtNWs/Au MEA is then transfer-printed onto a stretchable and thin PDMS substrate. The prosthetic skin has a potential for burnt victims or individuals with a prosthesis to restore their skin sense of touch, pressure, and temperature (Figure 18.24e).

Inspired by the structure of the human eye, Choi et al. (2017) proposed a soft retinal implant. As illustrated in Figure 18.24a, the working mechanism of a human eye starts with a lens that captures incoming lights. The retina then converts these lights into action potentials and transmits them to the brain through optic nerves. To overcome the mechanical mismatch between soft retina and rigid electronic devices, an array of optoelectronic devices is configured in such a way to mimic the hemispherical structures of the human eye. The system induces minimal stress and deformation to an artificial eye model (0.61 MPa), which is orders-of-magnitude lower than the other approach (Figure 18.24g). The image sensing component consists of hemispherical curved arrays of MoS_2–graphene phototransistors. The property and high-density design of the phototransistors eliminate infrared (IR) noises and enable high-quality imaging.

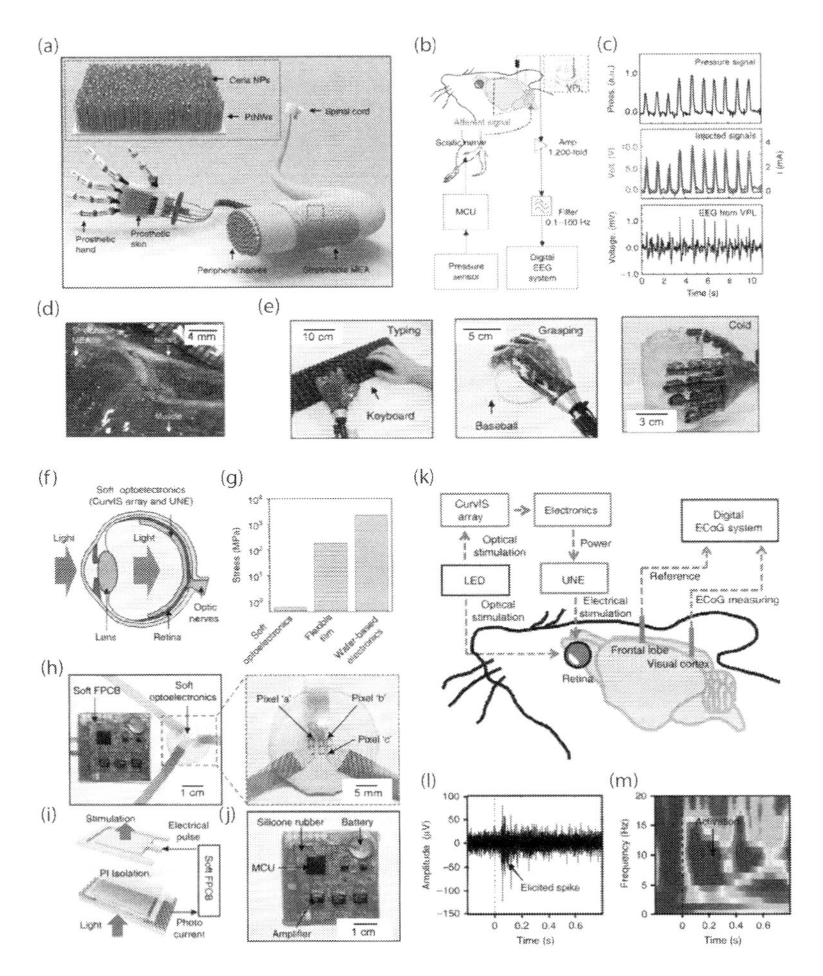

FIGURE 18.24 Silicon nanoribbon electronics for skin prosthesis. (a) System illustration of the multi-sensory prosthetic skin interfacing to the peripheral nerves through MEA. Inset shows the stretchable MEA, which consists of PtNWs with adsorbed ceria NPs grown on Au electrodes. (b) Flowchart of peripheral nerve simulations based on pressure sensor signals processed by microprocessor with cross-validation by EEG recordings of the ventral posterolateral nucleus (VPL) of the thalamus in the right hemisphere (inset). (c) Signals from pressure sensor (top) trigger and affect the stimulation amplitude to the peripheral nerve (centre, red, and blue). Correspondingly, neural activations can be observed from the VPL through EEG signals (bottom). (d) Optical image of the conformal stretchable MEA on peripheral nerves (blue arrows) in a rat model. (e) Several applications demonstrated with the multi-sensory prosthetic limb including typing on a keyboard, grasping a ball, and grabbing a cold mug. (With kind permission from Springer Science+Business Media: *Nat. Commun.*, Stretchable Silicon Nanoribbon Electronics for Skin Prosthesis, 5, 2014, 5747, Kim et al.) Eye-inspired soft optoelectronic system as retinal implant. (f) Illustration of the ocular structure of human integrated with the soft optoelectronic device. (g) Stress induced by the soft optoelectronic device compared to other technologies. (h) Optical image of CurvIS array and UNE connected by soft flexible printed circuit board (FPCB). Magnified image shows three pixels array. (i) Layer configuration of phototransistor (bottom) and stimulation electrode (top) separated by a PI isolation and connected to the soft FPCB. (j) The soft FPCB consists of a system that receives optical signals from CurvIS array, amplifies them, and electrically stimulate through the UNE to the optic nerves (k) of a rat correspondingly. Digital ECoG system is used as a cross-validation of the neural stimulation. Transient voltage graph of the evoked spike (l) and LFP changes (m) in the visual cortex by electrostimulation. (With kind permission from Springer Science+Business Media: *Nat. Commun.*, Human Eye-Inspired Soft Optoelectronic Device Using High-Density MoS2-Graphene Curved Image Sensor Array, 8, 2017, 1664, Choi, C. et al.)

A system to interface with the curved image sensor (CurvIS) array as a soft implantable optoelectronics is shown in Figure 18.24h. A soft and flexible printed circuit board (PCB) processes optical information from the LED to the image sensor and correspondingly applies electrical pulses to a set of Au neural-interfacing electrodes (UNE). The UNE is located on the back side of the image sensor array and facing towards the retina (Figure 18.24i). The soft PCB consists of an amplifier and microprocessor unit that processes the signals from the CurvIS array, makes decisions, and actuates the UNE (Figure 18.24j). ECoG reading is conducted to validate the functionality of this soft optoelectronic implant by simultaneously reading the LFP changes at the visual cortex of the rat as optical stimulation is pulsed to the CurvIS array (Figure 18.24k). From these discoveries, it can be observed that soft implantable devices could pave the way towards the future of electronic implants, digital nervous systems, and thought-controlled prostheses for human augmentation and assistive technology.

18.9 Conclusion

In the last few decades, we have witnessed the emergence of new technologies that transform personal electronics and reconfigure the way we understand ourselves and interact with the world. With the rapid advances in flexible and stretchable bio-integrated devices, we have come to a new age where electronics are becoming much closer to the body, breaking boundaries between the synthetic and biological systems. This chapter highlights the role of various novel sensors, actuators, and transducers that can be conformably laminated to the curvilinear surface of the skin, embedded into textiles, or implanted into the body for seamless body augmentation, as well as intimate physical activity recognition and physiological monitoring. New materials, device designs, and microfabrication techniques allow these electronic devices to be flexible, stretchable, imperceptible, bio-compatible, self-healable, and transparent which improves their sensitivity, comfortability, robustness, long-term usage, and social acceptability. Integrating these devices into different regions of the human body internally and externally enables various applications of human-machine interfaces from on-skin sensing and display, tactile feedback, body gesture monitoring, speech recognition, emotion and stress recognition, brain-computer interfaces, to neuroprosthetics.

Future challenges include system-level efforts of heterogeneous integration of sensors, actuators, and transducers with other electronic modules, such as wireless communication, circuits, memory, energy storage, and energy harvesters. Large-scale fabrication will enable high-coverage of active areas that could entirely cover skin or brain surface, for example. Multi-modal, dense array of sensing and actuation elements will provide a high-resolution spatiotemporal mapping and actuations with high recognition rate and pinpoint accuracy. In addition to these, extensive clinical studies must be conducted, particularly for implantable devices to gain regulatory approval. It is, therefore, an exciting time for materials scientists, engineers, medical practitioners, computer scientist, and interaction designers to work together and push forward wearable and implantable electronics through flexible and stretchable devices and revolutionize the future of human-machine interfaces.

References

Ahn, Jong Hyun, Hoon Sik Kim, Keon Jae Lee, Zhengtao Zhu, Etienne Menard, Ralph G. Nuzzo, and John A. Rogers. 2006. "High-Speed Mechanically Flexible Single-Crystal Silicon Thin-Film Transistors on Plastic Substrates." *IEEE Electron Device Letters* 27 (6):460–462. https://doi.org/10.1109/LED.2006.874764.

Al-Rahayfeh, Amer, and Miad Faezipour. 2013. "Eye Tracking and Head Movement Detection: A State-of-Art Survey." *IEEE Journal of Translational Engineering in Health and Medicine* 1:2100212–2100212. https://doi.org/10.1109/JTEHM.2013.2289879.

Amjadi, Morteza, Ki Uk Kyung, Inkyu Park, and Metin Sitti. 2016. "Stretchable, Skin-Mountable, and Wearable Strain Sensors and Their Potential Applications: A Review." *Advanced Functional Materials* 26 (11):1678–1698. https://doi.org/10.1002/adfm.201504755.

Atalay, Ozgur, Asli Atalay, Joshua Gafford, Hongqiang Wang, Robert Wood, and Conor Walsh. 2017. "A Highly Stretchable Capacitive-Based Strain Sensor Based on Metal Deposition and Laser Rastering." *Advanced Materials Technologies* 1700081:1700081. https://doi.org/10.1002/admt.201700081.

Benight, Stephanie J., Chao Wang, Jeffrey B.H. Tok, and Zhenan Bao. 2013. "Stretchable and Self-Healing Polymers and Devices for Electronic Skin." *Progress in Polymer Science*. https://doi.org/10.1016/j.progpolymsci.2013.08.001.

Bermúdez, Cañón, Gilbert Santiago, Dmitriy D. Karnaushenko, Daniil Karnaushenko, Ana Lebanov, Lothar Bischoff, Martin Kaltenbrunner, Jürgen Fassbender, Oliver G. Schmidt, and Denys Makarov. 2018. "Magnetosensitive E-Skins with Directional Perception for Augmented Reality." *Science Advances* 4 (1):eaao2623. https://doi.org/10.1126/sciadv.aao2623.

Boucsein, Wolfram. 2012. *Electrodermal Activity*. New York: Springer Science & Business Media, 1–8. https://doi.org/10.1007/978-1-4614-1126-0.

Bouten, Carlijn, Karel Koekkoek, Maarten Verduin, Rens Kodde, and Jan Janssen. 1997. "A Triaxial Accelerometer and Portable Data Processing Unit for the Assessment of Daily Physical Activity." *IEEE Transactions on Bio-Medical Engineering* 44 (3):136–147. https://doi.org/10.1109/10.554760.

Bulling, Andreas, and Hans Gellersen. 2010. "Toward Mobile Eye-Based Human-Computer Interaction." *IEEE Pervasive Computing* 9 (4):8–12. https://doi.org/10.1109/MPRV.2010.86.

Bulling, Andreas, Jamie A. Ward, Hans Gellersen, and Gerhard Tröster. 2011. "Eye Movement Analysis for Activity Recognition Using Electrooculography." *IEEE Transactions on Pattern Analysis and Machine Intelligence* 33 (4):741–753. https://doi.org/10.1109/TPAMI.2010.86.

Cai, Guofa, Jiangxin Wang, Kai Qian, Jingwei Chen, Shaohui Li, and Pooi See Lee. 2017. "Extremely Stretchable Strain Sensors Based on Conductive Self-Healing Dynamic Cross-Links Hydrogels for Human-Motion Detection." *Advanced Science* 4 (2). https://doi.org/10.1002/advs.201600190.

Cai, Le, Li Song, Pingshan Luan, Qiang Zhang, Nan Zhang, Qingqing Gao, Duan Zhao et al. 2013. "Super-Stretchable, Transparent Carbon Nanotube-Based Capacitive Strain Sensors for Human Motion Detection." *Scientific Reports* 3:1–9. https://doi.org/10.1038/srep03048.

Canales, Andres, Xiaoting Jia, Ulrich P. Froriep, Ryan A. Koppes, Christina M. Tringides, Jennifer Selvidge, Chi Lu et al. 2015. "Multifunctional Fibers for Simultaneous Optical, Electrical and Chemical Interrogation of Neural Circuits in Vivo." *Nature Biotechnology* 33 (3):277–284. https://doi.org/10.1038/nbt.3093.

Castellano, Ginevra, Loic Kessous, and George Caridakis. 2008. "Emotion Recognition through Multiple Modalities: Face, Body Gesture, Speech." In *Lecture Notes in Computer Science (Including Subseries Lecture Notes in Artificial Intelligence and Lecture Notes in Bioinformatics)*, 4868 LNCS:92–103. https://doi.org/10.1007/978-3-540-85099-1_8.

Cheng, Hong, Lu Yang, and Zicheng Liu. 2015. "A Survey on 3D Hand Gesture Recognition." *IEEE Transactions on Circuits and Systems for Video Technology* (99):1. https://doi.org/10.1109/TCSVT.2015.2469551.

Cherenack, Kunigunde, Christoph Zysset, Thomas Kinkeldei, Niko Münzenrieder, and Gerhard Tröster. 2010. "Woven Electronic Fibers with Sensing and Display Functions for Smart Textiles." *Advanced Materials* 22 (45):5178–5182. https://doi.org/10.1002/adma.201002159.

Choi, Changsoon, Moon Kee Choi, Siyi Liu, Min Sung Kim, Ok Kyu Park, Changkyun Im, Jaemin Kim et al. 2017. "Human Eye-Inspired Soft Optoelectronic Device Using High-Density MoS2-Graphene Curved Image Sensor Array." *Nature Communications* 8 (1):1664. https://doi.org/10.1038/s41467-017-01824-6.

Choi, Suji, Hyunjae Lee, Roozbeh Ghaffari, Taeghwan Hyeon, and Dae Hyeong Kim. 2016. "Recent Advances in Flexible and Stretchable Bio-Electronic Devices Integrated with Nanomaterials." *Advanced Materials* 28 (22):4203–4218. https://doi.org/10.1002/adma.201504150.

Corea, Joseph R., Anita M. Flynn, Balthazar Lechêne, Greig Scott, Galen D. Reed, Peter J. Shin, Michael Lustig et al. 2016. "Screen-Printed Flexible MRI Receive Coils." *Nature Communications* 7. https://doi.org/10.1038/ncomms10839.

Dagdeviren, Canan, Byung Duk Yang, Yewang Su, Phat L. Tran, Pauline Joe, Eric Anderson, Jing Xia et al. 2014. "Conformal Piezoelectric Energy Harvesting and Storage from Motions of the Heart, Lung, and Diaphragm." *Proceedings of the National Academy of Sciences* 111 (5):1927–1932. https://doi.org/10.1073/pnas.1317233111.

Dagdeviren, Canan, Khalil B. Ramadi, Pauline Joe, Kevin Spencer, Helen N. Schwerdt, Hideki Shimazu, Sebastien Delcasso et al. 2018. "Miniaturized Neural System for Chronic, Local Intracerebral Drug Delivery." *Science Translational Medicine* 10 (425). https://doi.org/10.1126/scitranslmed.aan2742.

Dagdeviren, Canan, Yewang Su, Pauline Joe, Raissa Yona, Yuhao Liu, Yun Soung Kim, Yongan Huang et al. 2014. "Conformable Amplified Lead Zirconate Titanate Sensors with Enhanced Piezoelectric Response for Cutaneous Pressure Monitoring." *Nature Communications* 5. https://doi.org/10.1038/ncomms5496.

Delmas, Patrick, Jizhe Hao, and Lise Rodat-Despoix. 2011. "Molecular mechanisms of mechanotransduction in mammalian sensory neurons." *Nature Reviews Neuroscience* 12(3): 139.

Duschek, Stefan, and Rainer Schandry. 2003. "Functional Transcranial Doppler Sonography as a Tool in Psychophysiological Research." *Psychophysiology* 40 (3):436–454. https://doi.org/10.1111/1469-8986.00046.

Exhibits, Open. 2011. "360 Degree VR Image Viewer Controlled by Kinect." 2011. https://www.flickr.com/photos/openexhibits/5448542351. (Acessed April, 2018).

Feil, Michael, Cliff Adler, Gerhard Klink, Martin Konig, Christof Landesberger, Sabine Scherbaum, Gregor Schwinn, et al. 2003. "Ultra Thin ICs and MEMS Elements: Techniques for Wafer Thinning, Stress-Free Separation, Assembly and Interconnection." *Microsystem Technologies* 9 (3):176–182. https://doi.org/10.1007/s00542-002-0223-5.

Fenno, Lief, Ofer Yizhar, and Karl Deisseroth. 2011. "The Development and Application of Optogenetics." *Annual Review of Neuroscience* 34 (1):389–412. https://doi.org/10.1146/annurev-neuro-061010-113817.

Ferreira, Andre, Wanderley C. Celeste, Fernando A. Cheein, Teodiano F. Bastos-Filho, Mario Sarcinelli-Filho, and Ricardo Carelli. 2008. "Human-Machine Interfaces Based on EMG and EEG Applied to Robotic Systems." *Journal of NeuroEngineering and Rehabilitation* 5 (1):10. https://doi.org/10.1186/1743-0003-5-10.

Fregni, Felipe, Paulo S. Boggio, Michael A. Nitsche, Marco A. Marcolin, Sergio P. Rigonatti, and Alvaro Pascual-Leone. 2006. "Treatment of Major Depression with Transcranial Direct Current Stimulation." *Bipolar Disorders* 8 (2):203–204. https://doi.org/10.1111/j.1399-5618.2006.00291.x.

Geddes, Leslie A., and Rebecca A. Roeder. 2003. "Criteria for the Selection of Materials for Implanted Electrodes." *Annals of Biomedical Engineering* 31 (7):879–890. https://doi.org/10.1114/1.1581292.

George, Mark S., Ziad Nahas, Xiangbao Li, F. Andrew Kozel, Berry Anderson, Kaori Yamanaka, Jeong-Ho Chae, and Milton J. Foust. 2002. "Novel Treatments of Mood Disorders Based on Brain Circuitry (ECT, MST, TMS, VNS, DBS)." *Seminars in Clinical Neuropsychiatry* 7 (4):293–304. https://doi.org/S108436120250040X [pii].

Gerratt, Aaron P., Hadrien O. Michaud, and Stéphanie P. Lacour. 2015. "Elastomeric Electronic Skin for Prosthetic Tactile Sensation." *Advanced Functional Materials* 25 (15):2287–2295. https://doi.org/10.1002/adfm.201404365.

Gidlöf, Kerstin, Annika Wallin, Richard Dewhurst, and Kenneth Holmqvist. 2013. "Using Eye Tracking to Trace a Cognitive Process: Gaze Behaviour during Decision Making in a Natural Environment.." *Journal of Eye Movement Research* 6(1):3 (1):1–14. https://doi.org/10.16910/jemr.6.1.3.

Grill, Warren M., Sharon E. Norman, and Ravi V. Bellamkonda. 2009. "Implanted Neural Interfaces: Biochallenges and Engineered Solutions." *Annual Review of Biomedical Engineering* 11 (1):1–24. https://doi.org/10.1146/annurev-bioeng-061008-124927.

Gumus, Abdurrahman, Arsalan Alam, Aftab M. Hussain, Kush Mishra, Irmandy Wicaksono, Galo A. Torres Sevilla, Sohail F. Shaikh et al. 2017. "Expandable Polymer Enabled Wirelessly Destructible High-Performance Solid State Electronics." *Advanced Materials Technologies* 2 (5):1600264. https://doi.org/10.1002/admt.201600264.

HajjHassan, Mohamad, Vamsy Chodavarapu, and Sam Musallam. 2008. "NeuroMEMS: Neural Probe Microtechnologies." *Sensors*. https://doi.org/10.3390/s8106704.

Hamedi, Mahiar, Robert Forchheimer, and Olle Inganäs. 2007. "Towards Woven Logic from Organic Electronic Fibres." *Nature Materials* 6 (5):357–362. https://doi.org/10.1038/nmat1884.

Higginbotham, D. Jeffery, Howard Shane, Susanne Russell, and Kevin Caves. 2007. "Access to AAC: Present, Past, and Future." *AAC: Augmentative and Alternative Communication* 23 (3):243–257. https://doi.org/10.1080/07434610701571058.

Hu, Xiaolong, Peter Krull, Bassel De Graff, Kevin Dowling, John A. Rogers, and William J. Arora. 2011. "Stretchable Inorganic-Semiconductor Electronic Systems." *Advanced Materials* 23 (26):2933–2936. https://doi.org/10.1002/adma.201100144.

Jaimes, Alejandro, and Nicu Sebe. 2007. "Multimodal Human-Computer Interaction: A Survey." *Computer Vision and Image Understanding* 108 (1–2):116–134. https://doi.org/10.1016/j.cviu.2006.10.019.

Jang, Kyung-In, Sang Youn Han, Sheng Xu, Kyle E. Mathewson, Yihui Zhang, Jae-Woong Jeong, Gwang-Tae Kim et al. 2014. "Rugged and Breathable Forms of Stretchable Electronics with Adherent Composite Substrates for Transcutaneous Monitoring." *Nature Communications* 5. Nature Publishing Group:4779. https://doi.org/10.1038/ncomms5779.

Jeong, Jae Woong, Min Ku Kim, Huanyu Cheng, Woon Hong Yeo, Xian Huang, Yuhao Liu, Yihui Zhang et al. 2014. "Capacitive Epidermal Electronics for Electrically Safe, Long-Term Electrophysiological Measurements." *Advanced Healthcare Materials* 3 (5):642–648. https://doi.org/10.1002/adhm.201300334.

Jeong, Jae Woong, Woon Hong Yeo, Aadeel Akhtar, James J.S. Norton, Young Jin Kwack, Shuo Li, Sung Young Jung et al. 2013. "Materials and Optimized Designs for Human-Machine Interfaces via Epidermal Electronics." *Advanced Materials* 25 (47):6839–6846. https://doi.org/10.1002/adma.201301921.

Jog, Mandar, Christopher Ian Connolly, Yasuo Kubota, D. R. Iyengar, Leoncio Garrido, Ray Harlan, and Ann M. Graybiel. 2002. "Tetrode Technology: Advances in Implantable Hardware, Neuroimaging, and Data Analysis Techniques." *Journal of Neuroscience Methods* 117 (2):141–152. https://doi.org/10.1016/S0165-0270(02)00092-4.

Kang, Minpyo, Jejung Kim, Bongkyun Jang, Youngcheol Chae, Jae Hyun Kim, and Jong Hyun Ahn. 2017. "Graphene-Based Three-Dimensional Capacitive Touch Sensor for Wearable Electronics." *ACS Nano* 11 (8):7950–7957. https://doi.org/10.1021/acsnano.7b02474.

Kang, Seung Kyun, Rory K.J. Murphy, Suk Won Hwang, Seung Min Lee, Daniel V. Harburg, Neil A. Krueger, Jiho Shin et al. 2016. "Bioresorbable Silicon Electronic Sensors for the Brain." *Nature* 530 (7588):71–76. https://doi.org/10.1038/nature16492.

Kao, Hsin-Liu (Cindy), Christian Holz, Asta Roseway, Andres Calvo, and Chris Schmandt. 2016. "DuoSkin: Rapidly Prototyping On-Skin User Interfaces Using Skin-Friendly Materials." *Proceedings of the 2016 ACM International Symposium on Wearable Computers—ISWC'16*, 16–23. https://doi.org/10.1145/2971763.2971777.

Keenan, Sharon, and Max Hirshkowitz. 2010. "Monitoring and Staging Human Sleep." In *Principles and Practice of Sleep Medicine: Fifth Edition*, 1602–1609. https://doi.org/10.1016/B978-1-4160-6645-3.00141-9.

Kim, Chong-Chan, Hyun-Hee Lee, Kyu Hwan Oh, and Jeong-Yun Sun. 2016. "Highly Stretchable, Transparent Ionic Touch Panel." *Science* 353 (6300):682–687. https://doi.org/10.1126/science.aaf8810.

Kim, Dae-Hyeong, Nanshu Lu, Rui Ma, Yun-Soung Kim, Rak-Hwan Kim, Shuodao Wang, Jian . Wu et al. 2011. "Epidermal Electronics." *Science* 333 (6044):838–843. https://doi.org/10.1126/science.1206157.

Kim, Dae Hyeong, Jianliang Xiao, Jizhou Song, Yonggang Huang, and John A. Rogers. 2010. "Stretchable, Curvilinear Electronics Based on Inorganic Materials." *Advanced Materials*. https://doi.org/10.1002/adma.200902927.

Kim, Dae Hyeong, Jonathan Viventi, Jason J. Amsden, Jianliang Xiao, Leif Vigeland, Yun Soung Kim, Justin A. Blanco et al. 2010. "Dissolvable Films of Silk Fibroin for Ultrathin Conformal Bio-Integrated Electronics." *Nature Materials* 9 (6):1–7. https://doi.org/10.1038/nmat2745.

Kim, Jaemin, Mincheol Lee, Hyung Joon Shim, Roozbeh Ghaffari, Hye Rim Cho, Donghee Son, Yei Hwan Jung et al. 2014. "Stretchable Silicon Nanoribbon Electronics for Skin Prosthesis." *Nature Communications* 5:5747. https://doi.org/10.1038/ncomms6747.

Kim, Jeehoon, Sungjun Kwon, Sangwon Seo, and Kwangsuk Park. 2014. "Highly Wearable Galvanic Skin Response Sensor Using Flexible and Conductive Polymer Foam." *Conference Proceedings:... Annual International Conference of the IEEE Engineering in Medicine and Biology Society. IEEE Engineering in Medicine and Biology Society. Annual Conference* 2014 (Figure 18.1):6631–6634. https://doi.org/10.1109/EMBC.2014.6945148.

Kim, S., R. Bhandari, M. Klein, S. Negi, L. Rieth, P. Tathireddy, M. Toepper, H. Oppermann, and F. Solzbacher. 2009. "Integrated Wireless Neural Interface Based on the Utah Electrode Array." *Biomedical Microdevices* 11 (2):453–466. https://doi.org/10.1007/s10544-008-9251-y.

Kim, Soo Jin, and Jang Sik Lee. 2010. "Flexible Organic Transistor Memory Devices." *Nano Letters* 10 (8):2884–2890. https://doi.org/10.1021/nl1009662.

Kim, Tae-il, Jordan G. McCall, Yei. Hwan Jung, Xian Huang, Edward. R. Siuda, Yuhang Li, Jizhou Song et al. 2013. "Injectable, Cellular-Scale Optoelectronics with Applications for Wireless Optogenetics." *Science* 340 (6129):211–216. https://doi.org/10.1126/science.1232437.

Ko, Heung Cho, Mark P. Stoykovich, Jizhou Song, Viktor Malyarchuk, Won Mook Choi, Chang Jae Yu, Joseph B. Geddes et al. 2008. "A Hemispherical Electronic Eye Camera Based on Compressible Silicon Optoelectronics." *Nature* 454 (7205):748–753. https://doi.org/10.1038/nature07113.

Kon, Stanley, Kenn Oldham, and Roberto Horowitz. 2007. "Piezoresistive and Piezoelectric MEMS Strain Sensors for Vibration Detection" 6529:65292V. https://doi.org/10.1117/12.715814.

Koo, Ig Mo, Kwangmok Jung, Ja Choon Koo, Jae Do Nam, Young Kwan Lee, and Hyouk Ryeol Choi. 2008. "Development of Soft-Actuator-Based Wearable Tactile Display." *IEEE Transactions on Robotics* 24 (3):549–558. https://doi.org/10.1109/TRO.2008.921561.

Lacour, Stéphanie P., Grégoire Courtine, and Jochen Guck. 2016. "Materials and Technologies for Soft Implantable Neuroprostheses." *Nature Reviews Materials* 1 (10):16063. https://doi.org/10.1038/natrevmats.2016.63.

Lai, Ying Chih, Jianan Deng, Steven L. Zhang, Simiao Niu, Hengyu Guo, and Zhong Lin Wang. 2017. "Single-Thread-Based Wearable and Highly Stretchable Triboelectric Nanogenerators and Their Applications in Cloth-Based Self-Powered Human-Interactive and Biomedical Sensing." *Advanced Functional Materials* 27 (1). https://doi.org/10.1002/adfm.201604462.

Lee, Sangmin, Ronan Hinchet, Yean Lee, Ya Yang, Zong Hong Lin, Gustavo Ardila, Laurent Montès, Mireille Mouis, and Zhong Lin Wang. 2014. "Ultrathin Nanogenerators as Self-Powered/active Skin Sensors for Tracking Eye Ball Motion." *Advanced Functional Materials* 24 (8):1163–1168. https://doi.org/10.1002/adfm.201301971.

Lee, Yongkuk, Benjamin Nicholls, Dong Sup Lee, Yanfei Chen, Youngjae Chun, Chee Siang Ang, and Woon-Hong Yeo. 2017. "Soft Electronics Enabled Ergonomic Human-Computer Interaction for Swallowing Training." *Scientific Reports* 7 (April). 46697. https://doi.org/10.1038/srep46697.

Leigh, Sang-won, Harpreet Sareen, Hsin-Liu Kao, Xin Liu, and Pattie Maes. 2017. "Body-Borne Computers as Extensions of Self." *Computers* 6 (1):12. https://doi.org/10.3390/computers6010012.

Liang, Jiajie, Lu Li, Xiaofan Niu, Zhibin Yu, and Qibing Pei. 2013. "Elastomeric Polymer Light-Emitting Devices and Displays." *Nature Photonics* 7 (10):817–824. https://doi.org/10.1038/nphoton.2013.242.

Lim, Guh-Hwan, Nae-Eung Lee, and Byungkwon Lim. 2016. "Highly Sensitive, Tunable, and Durable Gold Nanosheet Strain Sensors for Human Motion Detection." *Journal of Materials Chemistry C* 4 (24):5642–5647. https://doi.org/10.1039/C6TC00251J.

Lim, Sumin, Donghee Son, Jaemin Kim, Young Bum Lee, Jun-Kyul Song, Suji Choi, Dong Jun Lee et al. 2015. "Transparent and stretchable interactive human machine interface based on patterned graphene heterostructures." *Advanced Functional Materials* 25 (3): 375–383.

Lisetti, Christine Lætitia, and Fatma Nasoz. 2004. "Using Noninvasive Wearable Computers to Recognize Human Emotions from Physiological Signals." *Eurasip Journal on Applied Signal Processing*. https://doi.org/10.1155/S1110865704406192.

Liu, Jia, Tian Ming Fu, Zengguang Cheng, Guosong Hong, Tao Zhou, Lihua Jin, Madhavi Duvvuri et al. 2015. "Syringe-Injectable Electronics." *Nature Nanotechnology* 10 (7):629–635. https://doi.org/10.1038/nnano.2015.115.

Liu, Y., J. J. S. Norton, R. Qazi, Z. Zou, K. R. Ammann, H. Liu, L. Yan et al. 2016. "Epidermal Mechano-Acoustic Sensing Electronics for Cardiovascular Diagnostics and Human-Machine Interfaces." *Science Advances* 2 (11):e1601185–e1601185. https://doi.org/10.1126/sciadv.1601185.

Lu, Chi, Seongjun Park, Thomas J. Richner, Alexander Derry, Imogen Brown, Chong Hou, Siyuan Rao et al. 2017. "Flexible and Stretchable Nanowire-Coated Fibers for Optoelectronic Probing of Spinal Cord Circuits." *Science Advances* 3 (3):e1600955. https://doi.org/10.1126/sciadv.1600955.

Matthies, Denys J.C., Bernhard A. Strecker, and Bodo Urban. 2017. "EarFieldSensing: A Novel In-Ear Electric Field Sensing to Enrich Wearable Gesture Input through Facial Expressions." *Chi 2017*, 1911–1922. https://doi.org/10.1145/3025453.3025692.

Mattmann, Corinne, Frank Clemens, and Gerhard Tröster. 2008. "Sensor for Measuring Strain in Textile." *Sensors* 8 (6):3719–3732. https://doi.org/10.3390/s8063719.

McKendrick, Ryan, Raja Parasuraman, and Hasan Ayaz. 2015. "Wearable Functional near Infrared Spectroscopy (fNIRS) and Transcranial Direct Current Stimulation (tDCS): Expanding Vistas for Neurocognitive Augmentation." *Frontiers in Systems Neuroscience* 9. https://doi.org/10.3389/fnsys.2015.00027.

Meitl, Matthew A., Zheng Tao Zhu, Vipan Kumar, Keon Jae Lee, Xue Feng, Yonggang Y. Huang, Ilesanmi Adesida, Ralph G. Nuzzo, and John A. Rogers. 2006. "Transfer Printing by Kinetic Control of Adhesion to an Elastomeric Stamp." *Nature Materials* 5 (1):33–38. https://doi.org/10.1038/nmat1532.

Menguc, Yigit, Yong Lae Park, Ernesto Martinez-Villalpando, Patrick Aubin, Miriam Zisook, Leia Stirling, Robert J. Wood, and Conor J. Walsh. 2013. "Soft Wearable Motion Sensing Suit for Lower Limb Biomechanics Measurements." *Proceedings—IEEE International Conference on Robotics and Automation*, 5309–5316. https://doi.org/10.1109/ICRA.2013.6631337.

Minev, I. R., P. Musienko, A. Hirsch, Q. Barraud, N. Wenger, E. M. Moraud, J. Gandar et al. 2015. "Electronic Dura Mater for Long-Term Multimodal Neural Interfaces." *Science* 347 (6218):159–163. https://doi.org/10.1126/science.1260318.

Mishra, Saswat, James J.S. Norton, Yongkuk Lee, Dong Sup Lee, Nicolas Agee, Yanfei Chen, Youngjae Chun, and Woon Hong Yeo. 2017. "Soft, Conformal Bioelectronics for a Wireless Human-Wheelchair Interface." *Biosensors and Bioelectronics* 91 (December 2016). Elsevier:796–803. https://doi.org/10.1016/j.bios.2017.01.044.

Mohammed, Ahmed, Walied Moussa, and Edmond Lou. 2008. "High Sensitivity MEMS Strain Sensor: Design and Simulation." *Sensors* 8 (4):2642–2661. https://doi.org/10.3390/s8042642.

Museoscienza.org. 1976. "Olivetti P6066." 1976 . https://commons.wikimedia.org/wiki/Category:Olivetti_computers#/media/File:Computer_minipersonal_-_Museo_scienza_tecnologia_Milano_09298.jpg. (Acessed April, 2018).

Muth, Joseph T., Daniel M. Vogt, Ryan L. Truby, Yiḡit Mengüç, David B. Kolesky, Robert J. Wood, and Jennifer A. Lewis. 2014. "Embedded 3D Printing of Strain Sensors within Highly Stretchable Elastomers." *Advanced Materials* 26 (36):6307–6312. https://doi.org/10.1002/adma.201400334.

Norton, James J. S., Dong Sup Lee, Jung Woo Lee, Woosik Lee, Ohjin Kwon, Phillip Won, Sung-Young Jung et al. 2015. "Soft, Curved Electrode Systems Capable of Integration on the Auricle as a Persistent Brain–computer Interface." *Proceedings of the National Academy of Sciences* 112 (13):3920–3925. https://doi.org/10.1073/pnas.1424875112.

Park, Jaeyoon, Insang You, Sangbaie Shin, and Unyong Jeong. 2015. "Material Approaches to Stretchable Strain Sensors." *ChemPhysChem*. https://doi.org/10.1002/cphc.201402810.

Park, Sung Il, Daniel S. Brenner, Gunchul Shin, Clinton D. Morgan, Bryan A. Copits, Ha UK Chung, Melanie Y. Pullen et al. 2015. "Soft, Stretchable, Fully Implantable Miniaturized Optoelectronic Systems for Wireless Optogenetics." *Nature Biotechnology* 33 (12). Nature Publishing Group:1280–1286. https://doi.org/10.1038/nbt.3415.

Paul, Gordon Mark, Fan Cao, Russel Torah, Kai Yang, Steve Beeby, and John Tudor. 2014. "A Smart Textile Based Facial Emg and Eog Computer Interface." *IEEE Sensors Journal* 14 (2):393–400. https://doi.org/10.1109/JSEN.2013.2283424.

Pawlaczyk, Mariola, Monika Lelonkiewicz, and Michal Wieczorowski. 2013. "Age-Dependent Biomechanical Properties of the Skin." *Postepy Dermatologii I Alergologii*. https://doi.org/10.5114/pdia.2013.38359.

Peeble Technology. 2012. "Peeble Watch Email 1." 2012. https://commons.wikimedia.org/wiki/File:Pebble_watch_email_1.png . (Acessed April, 2018)

Persano, Luana, Canan Dagdeviren, Yewang Su, Yihui Zhang, Salvatore Girardo, Dario Pisignano, Yonggang Huang, and John A. Rogers. 2013. "High Performance Piezoelectric Devices Based on Aligned Arrays of Nanofibers of Poly(vinylidenefluoride-Co-Trifluoroethylene)." *Nature Communications* 4. https://doi.org/10.1038/ncomms2639.

Photos, NDB. 2017. "Google's Smart Speaker, the Google Home." https://commons.wikimedia.org/wiki/File:Google_Home_sitting_on_table.jpg. (Acessed April, 2018)

Picard, Rosalind W. 1997. *Affective Computing. Pattern Recognition*. Vol. 73. https://doi.org/10.1007/BF01238028.

Polikov, Vadim S., Patrick A. Tresco, and William M. Reichert. 2005. "Response of Brain Tissue to Chronically Implanted Neural Electrodes." *Journal of Neuroscience Methods*. https://doi.org/10.1016/j.jneumeth.2005.08.015.

Post, E. Rehmi., and Maggie Orth. 1997. "Smart Fabric, or Wearable Clothing." In *Digest of Papers. First International Symposium on Wearable Computers*, 167–168. IEEE.

Poupyrev, Ivan, Nan-Wei Gong, Shiho Fukuhara, Mustafa Emre Karagozler, Carsten Schwesig, and Karen E. Robinson. 2016. "Project Jacquard: Interactive Digital Textiles at Scale." In *Proceedings of the 2016 CHI Conference on Human Factors in Computing Systems*, 4216–4227. https://doi.org/10.1145/2858036.2858176.

Pradel, Ken C., Wenzhuo Wu, Yong Ding, and Zhong Lin Wang. 2014. "Solution-Derived ZnO Homojunction Nanowire Films on Wearable Substrates for Energy Conversion and Self-Powered Gesture Recognition." *Nano Letters* 14 (12):6897–6905. https://doi.org/10.1021/nl5029182.

Pu, Xianjie, Hengyu Guo, Jie Chen, Xue Wang, Yi Xi, Chenguo Hu, and Zhong Lin Wang. 2017. "Eye Motion Triggered Self-Powered Mechnosensational Communication System Using Triboelectric Nanogenerator." *Science Advances* 3 (7):e1700694. https://doi.org/10.1126/sciadv.1700694.

Ray, W. J., and H. W. Cole. 1985. "EEG Alpha Activity Reflects Attentional Demands, and Beta Activity Reflects Emotional and Cognitive Processes." *Science (New York, N.Y.)* 228 (4700):750–752. https://doi.org/10.1126/science.3992243.

Robinson, David A. 1963. "A Method of Measuring Eye Movement Using a Scieral Search Coil in a Magnetic Field." *IEEE Transactions on Bio-Medical Electronics* 10 (4):137–145. https://doi.org/10.1109/TBMEL.1963.4322822.

Rogers, John A. 2015. "Electronics for the Human Body." *JAMA* 313 (6):561. https://doi.org/10.1001/jama.2014.17915.

Roh, Eun, Byeong Ung Hwang, Doil Kim, Bo Yeong Kim, and Nae Eung Lee. 2015. "Stretchable, Transparent, Ultrasensitive, and Patchable Strain Sensor for Human-Machine Interfaces Comprising a Nanohybrid of Carbon Nanotubes and Conductive Elastomers." *ACS Nano* 9 (6):6252–6261. https://doi.org/10.1021/acsnano.5b01613.

Ruben de Rijcke. 2010. "IBM PC 5150." https://commons.wikimedia.org/wiki/File:Ibm_pc_5150.jpg. (Acessed April, 2018)

Salvatore, Giovanni A., Niko Münzenrieder, Thomas Kinkeldei, Luisa Petti, Christoph Zysset, Ivo Strebel, Lars Büthe, and Gerhard Tröster. 2014. "Wafer-Scale Design of Lightweight and Transparent Electronics That Wraps around Hairs." *Nature Communications* 5:2982. https://doi.org/10.1038/ncomms3982.

Saxena, Tarun, Lohitash Karumbaiah, Eric A. Gaupp, Radhika Patkar, Ketki Patil, Martha Betancur, Garrett B. Stanley, and Ravi V. Bellamkonda. 2013. "The Impact of Chronic Blood-Brain Barrier Breach on Intracortical Electrode Function." *Biomaterials* 34 (20):4703–4713. https://doi.org/10.1016/j.biomaterials.2013.03.007.

Schaelss, Alexander. 2004. "Apple Macintosh Classic." 2004. https://commons.wikimedia.org/wiki/File:Macintosh_classic.jpg.

Seel, Thomas, Jörg Raisch, and Thomas Schauer. 2014. "IMU-Based Joint Angle Measurement for Gait Analysis." *Sensors* 14 (4):6891–6909. https://doi.org/10.3390/s140406891.

Sekitani, Tsuyoshi, Hiroyoshi Nakajima, Hiroki Maeda, Takanori Fukushima, Takuzo Aida, Kenji Hata, and Takao Someya. 2009. "Stretchable Active-Matrix Organic Light-Emitting Diode Display Using Printable Elastic Conductors." *Nature Materials* 8 (6):494–499. https://doi.org/10.1038/nmat2459.

Sharma, Rakesh. 2011. "Oculomotor Dysfunction in Amyotrophic Lateral Sclerosis." *Archives of Neurology* 68 (7):857. https://doi.org/10.1001/archneurol.2011.130.

Sharma, Ram Pratap, and Gyanendra K. Verma. 2015. "Human Computer Interaction Using Hand Gesture." In *Procedia Computer Science*, 54:721–727. https://doi.org/10.1016/j.procs.2015.06.085.

Shibasaki, Hiroshi. 2008. "Human Brain Mapping: Hemodynamic Response and Electrophysiology." *Clinical Neurophysiology*. https://doi.org/10.1016/j.clinph.2007.10.026.

Shin, Gunchul, Adrian M. Gomez, Ream Al-Hasani, Yu Ra Jeong, Jeonghyun Kim, Zhaoqian Xie, Anthony Banks et al. 2017. "Flexible Near-Field Wireless Optoelectronics as Subdermal Implants for Broad Applications in Optogenetics." *Neuron* 93 (3):509–521.e3. https://doi.org/10.1016/j.neuron.2016.12.031.

Sonar, Harshal Arun, and Jamie Paik. 2016. "Soft Pneumatic Actuator Skin with Piezoelectric Sensors for Vibrotactile Feedback." *Frontiers in Robotics and AI* 2. https://doi.org/10.3389/frobt.2015.00038.

Sprenger, Andreas, Birte Neppert, Sabine Köster, Steffen Gais, Detlef Kömpf, Christoph Helmchen, and Hubert Kimmig. 2008. "Long-Term Eye Movement Recordings with a Scleral Search Coil-Eyelid Protection Device Allows New Applications." *Journal of Neuroscience Methods* 170 (2):305–309. https://doi.org/10.1016/j.jneumeth.2008.01.021.

Steimle, Jurgen. 2016. "Skin-The Next User Interface." *Computer* 49 (4):83–87. https://doi.org/10.1109/MC.2016.93.

Stoppa, Matteo, and Alessandro Chiolerio. 2014. "Wearable Electronics and Smart Textiles: A Critical Review." *Sensors (Switzerland)*. https://doi.org/10.3390/s140711957.

Su, Meng, Fengyu Li, Shuoran Chen, Zhandong Huang, Meng Qin, Wenbo Li, Xingye Zhang, and Yanlin Song. 2016. "Nanoparticle Based Curve Arrays for Multirecognition Flexible Electronics." *Advanced Materials* 28 (7):1369–1374. https://doi.org/10.1002/adma.201504759.

Sun, Jeong Yun, Christoph Keplinger, George M. Whitesides, and Zhigang Suo. 2014. "Ionic Skin." *Advanced Materials* 26 (45):7608–7614. https://doi.org/10.1002/adma.201403441.

Sun, Yugang, Won Mook Choi, Hanqing Jiang, Yonggang Y. Huang, and John A. Rogers. 2006. "Controlled Buckling of Semiconductor Nanoribbons for Stretchable Electronics." *Nature Nanotechnology* 1 (3):201–207. https://doi.org/10.1038/nnano.2006.131.

Swan, Melanie. 2012. "Sensor Mania! The Internet of Things, Wearable Computing, Objective Metrics, and the Quantified Self 2.0." *Journal of Sensor and Actuator Networks* 1 (3):217–253. https://doi.org/10.3390/jsan1030217.

Sweden, Microsoft. 2015. "win10-HoloLens-Minecraft." 2015. https://www.flickr.com/photos/microsoftsweden/15716942894.

Tao, Lu-Qi, He Tian, Ying Liu, Zhen-Yi Ju, Yu Pang, Yuan-Quan Chen, Dan-Yang Wang et al. 2017. "An Intelligent Artificial Throat with Sound-Sensing Ability Based on Laser Induced Graphene." *Nature Communications* 8. Nature Publishing Group:14579. https://doi.org/10.1038/ncomms14579.

Tao, Weijun, Tao Liu, Rencheng Zheng, and Hutian Feng. 2012. "Gait Analysis Using Wearable Sensors." *Sensors* 12 (2):2255–2283. https://doi.org/10.3390/s120202255.

Tee, B. C.- K., A. Chortos, A. Berndt, A. K. Nguyen, A. Tom, A. McGuire, Z. C. Lin et al. 2015. "A Skin-Inspired Organic Digital Mechanoreceptor." *Science* 350 (6258):313–316. https://doi.org/10.1126/science.aaa9306.

Toda, Haruo, Takafumi Suzuki, Hirohito Sawahata, Kei Majima, Yukiyasu Kamitani, and Isao Hasegawa. 2011. "Simultaneous Recording of ECoG and Intracortical Neuronal Activity Using a Flexible Multichannel Electrode-Mesh in Visual Cortex." *NeuroImage* 54 (1):203–212. https://doi.org/10.1016/j.neuroimage.2010.08.003.

Torres, Timothy. 2015. "Myo Gesture Control Armband." *PC Magazine*. https://www.pcmag.com/review/335036/myo-gesture-control-armband. (Acessed April, 2018).

Traunmüller, Hartmut, and Anders Eriksson. 1994. "The Frequency Range of the Voice Fundamental in the Speech of Male and Female Adults." *Department of Linguistics, University of Stockholm* 97:1905191–1905195.

van Der Geest, Josef. N., and Maarten A. Frens. 2002. "Recording Eye Movements with Video-Oculography and Scleral Search Coils: A Direct Comparison of Two Methods." *Journal of Neuroscience Methods* 114 (2):185–195. https://doi.org/10.1016/S0165-0270(01)00527-1.

Varga, Matija, Collin Ladd, Siyuan Ma, Jim Holbery, and Gerhard Tröster. 2017. "On-Skin Liquid Metal Inertial Sensor." *Lab Chip* 17 (19):3272–3278. https://doi.org/10.1039/C7LC00735C.

Viventi, Jonathan, Dae-Hyeong Kim, Leif Vigeland, Eric S. Frechette, Justin A. Blanco, Yun-Soung Kim, Andrew E. Avrin et al. 2011. "Flexible, Foldable, Actively Multiplexed, High-Density Electrode Array for Mapping Brain Activity in Vivo." *Nature Neuroscience* 14 (12):1599–1605. https://doi.org/10.1038/nn.2973.

Wang, Chuan, David Hwang, Zhibin Yu, Kuniharu Takei, Junwoo Park, Teresa Chen, Biwu Ma, and Ali Javey. 2013. "User-Interactive Electronic Skin for Instantaneous Pressure Visualization." *Nature Materials* 12 (10):899–904. https://doi.org/10.1038/nmat3711.

Wang, Jie, Shengming Li, Fang Yi, Yunlong Zi, Jun Lin, Xiaofeng Wang, Youlong Xu, and Zhong Lin Wang. 2016. "Sustainably Powering Wearable Electronics Solely by Biomechanical Energy." *Nature Communications* 7. https://doi.org/10.1038/ncomms12744.

Wang, Xuewen, Yang Gu, Zuoping Xiong, Zheng Cui, and Ting Zhang. 2014. "Silk-Molded Flexible, Ultrasensitive, and Highly Stable Electronic Skin for Monitoring Human Physiological Signals." *Advanced Materials* 26 (9):1336–1342. https://doi.org/10.1002/adma.201304248.

Wang, Yan, Tingting Yang, Junchao Lao, Rujing Zhang, Yangyang Zhang, Miao Zhu, Xiao Li et al. 2015. "Ultra-Sensitive Graphene Strain Sensor for Sound Signal Acquisition and Recognition." *Nano Research* 8 (5):1627–1636. https://doi.org/10.1007/s12274-014-0652-3.

Wang, Yue, Chenxin Zhu, Raphael Pfattner, Hongping Yan, Lihua Jin, Shucheng Chen, Francisco Molina-Lopez et al. 2017. "A Highly Stretchable, Transparent, and Conductive Polymer." *Science Advances* 3 (3):e1602076. https://doi.org/10.1126/sciadv.1602076.

Wang, Zhong Lin. 2014. "Triboelectric Nanogenerators as New Energy Technology and Self-Powered Sensors: Principles, Problems and Perspectives." *Faraday Discuss.* 176:447–458. https://doi.org/10.1039/C4FD00159A.

Weigel, Martin, Aditya Shekhar Nittala, Alex Olwal, and Jürgen Steimle. 2017. "SkinMarks: Enabling Interactions on Body Landmarks Using Conformal Skin Electronics." *Proceedings of the 2017 CHI Conference on Human Factors in Computing Systems—CHI'17*, 3095–3105. https://doi.org/10.1145/3025453.3025704.

Weigel, Martin, Tong Lu, Gilles Bailly, Antti Oulasvirta, Carmel Majidi, and Jürgen Steimle. 2015. "iSkin." In *Proceedings of the 33rd Annual ACM Conference on Human Factors in Computing Systems—CHI'15*, 2991–3000. https://doi.org/10.1145/2702123.2702391.

Whitmire, Eric, Laura Trutoiu, Robert Cavin, David Perek, Brian Scally, James O Phillips, and Shwetak Patel. 2016. "EyeContact: Scleral Coil Eye Tracking for Virtual Reality." *Proceedings of the 2016 ACM International Symposium on Wearable Computers—ISWC'16*, 184–191. https://doi.org/10.1145/2971763.2971771.

Wicaksono, Irmandy, and Joseph A. Paradiso. 2017. "FabricKeyboard: Multimodal Textile Sensate Media as an Expressive and Deformable Musical Interface." In *NIME*, 348–353. http://homes.create.aau.dk/dano/nime17/papers/0066/paper0066.pdf.

Wioleta, Szwoch. 2013. "Using Physiological Signals for Emotion Recognition." In *2013 6th International Conference on Human System Interactions (HSI)*, 556–561. https://doi.org/10.1109/HSI.2013.6577880.

Xu, Baoxing, Aadeel Akhtar, Yuhao Liu, Hang Chen, Woon Hong Yeo, Sung II Park, Brandon Boyce et al. 2016. "Flexible Electronics: An Epidermal Stimulation and Sensing Platform for Sensorimotor Prosthetic Control, Management of Lower Back Exertion, and Electrical Muscle Activation (Adv. Mater. 22/2016)." *Advanced Materials* 28 (22):4563. https://doi.org/10.1002/adma.201670154.

Xu, Sheng, Yihui Zhang, Jiung Cho, Juhwan Lee, Xian Huang, Lin Jia, Jonathan A. Fan et al. 2013. "Stretchable Batteries with Self-Similar Serpentine Interconnects and Integrated Wireless Recharging Systems." *Nature Communications* 4. https://doi.org/10.1038/ncomms2553.

Yamada, Takeo, Yuhei Hayamizu, Yuki Yamamoto, Yoshiki Yomogida, Ali Izadi-Najafabadi, Don N. Futaba, and Kenji Hata. 2011. "A Stretchable Carbon Nanotube Strain Sensor for Human-Motion Detection." *Nature Nanotechnology* 6 (5):296–301. https://doi.org/10.1038/nnano.2011.36.

Yamamoto, Y., S. Harada, D. Yamamoto, W. Honda, T. Arie, S. Akita, and K. Takei. 2016. "Printed Multifunctional Flexible Device with an Integrated Motion Sensor for Health Care Monitoring." *Science Advances* 2 (11):e1601473–e1601473. https://doi.org/10.1126/sciadv.1601473.

Yang, Che Chang, and Yeh Liang Hsu. 2010. "A Review of Accelerometry-Based Wearable Motion Detectors for Physical Activity Monitoring." *Sensors*. https://doi.org/10.3390/s100807772.

Yang, Jin, Jun Chen, Yuanjie Su, Qingshen Jing, Zhaoling Li, Fang Yi, Xiaonan Wen, Zhaona Wang, and Zhong Lin Wang. 2015. "Eardrum-Inspired Active Sensors for Self-Powered Cardiovascular System Characterization and Throat-Attached Anti-Interference Voice Recognition." *Advanced Materials* 27 (8):1316–1326. https://doi.org/10.1002/adma.201404794.

Yem, Vibol, and Hiroyuki Kajimoto. 2017. "Wearable Tactile Device Using Mechanical and Electrical Stimulation for Fingertip Interaction with Virtual World." *IEEE Virtual Reality (VR)*.

Ying, Ming, Andrew P. Bonifas, Nanshu Lu, Yewang Su, Rui Li, Huanyu Cheng, Abid Ameen, Yonggang Huang, and John A. Rogers. 2012. "Silicon Nanomembranes for Fingertip Electronics." *Nanotechnology* 23 (34). https://doi.org/10.1088/0957-4484/23/34/344004.

Yokota, Tomoyuki, Peter Zalar, Martin Kaltenbrunner, Hiroaki Jinno, Naoji Matsuhisa, Hiroaki Kitanosako, Yutaro Tachibana, Wakako Yukita, Matsuhisa Koizumi, and Takao Someya. 2016. "Ultraflexible Organic Photonic Skin." *Science Advances* 2 (4):e1501856–e1501856. https://doi.org/10.1126/sciadv.1501856.

Yoon, Sunghyun, Jai Kyoung Sim, and Young-Ho Cho. 2016. "A Flexible and Wearable Human Stress Monitoring Patch." *Scientific Reports* 6 (1):23468. https://doi.org/10.1038/srep23468.

Zhai, Yujia, Leo Mathew, Rajesh Rao, Dewei Xu, and Sanjay K. Banerjee. 2012. "High-Performance Flexible Thin-Film Transistors Exfoliated from Bulk Wafer." *Nano Letters* 12 (11):5609–5615. https://doi.org/10.1021/nl302735f.

Zhang, Yihui, Haoran Fu, Yewang Su, Sheng Xu, Huanyu Cheng, Jonathan A. Fan, Keh Chih Hwang, John A. Rogers, and Yonggang Huang. 2013a. "Mechanics of Ultra-Stretchable Self-Similar Serpentine Interconnects." *Acta Materialia* 61 (20):7816–7827. https://doi.org/10.1016/j.actamat.2013.09.020.

Zhang, Yuanfeng, Chupeng Lei, and Woo Soo Kim. 2013b. "Design Optimized Membrane-Based Flexible Paper Accelerometer with Silver Nano Ink." *Applied Physics Letters* 103 (7):1–4. https://doi.org/10.1063/1.4818734.

Zhao, Xiaoli, Qilin Hua, Ruomeng Yu, Yan Zhang, and Caofeng Pan. 2015. "Flexible, Stretchable and Wearable Multifunctional Sensor Array as Artificial Electronic Skin for Static and Dynamic Strain Mapping." *Advanced Electronic Materials* 1 (7):1–7. https://doi.org/10.1002/aelm.201500142.

Zhou, Tao, Guosong Hong, Tian-Ming Fu, Xiao Yang, Thomas G. Schuhmann, Robert D. Viveros, and Charles M. Lieber. 2017. "Syringe-Injectable Mesh Electronics Integrate Seamlessly with Minimal Chronic Immune Response in the Brain." *Proceedings of the National Academy of Sciences* 114 (23):5894–5899. https://doi.org/10.1073/pnas.1705509114.

19

Wearable Electronics

Sherjeel M. Khan
and Muhammad
Mustafa Hussain

19.1 Introduction

We are witnessing an era of exponential growth in the field of microelectronics and wireless devices leading to the rise of small devices capable of performing complex tasks with exceptional performance. These advancements can largely be contributed to performance improvement in microprocessors due to increased transistor speeds, reduced energy consumption, and increased transistor density per unit area from Moore's Law. However, due to challenges involved with further transistor scaling, the improvements in the processing are becoming slow due to increased power consumption. Moving ahead improvements in software and architecture can help keep up this fast-paced growth [1]. This ability to create high performing microprocessors in smaller sizes gave rise to the market of wearables which is expected to grow to a staggering 150 billion by 2027. Wearables are electronic devices that can be worn on the body as an accessory or embedded in the clothes worn by humans. Jiang H. et al. declared five characteristics that a wearable device needs to have: usable while the wearer is in motion, usable when one or both hands are being used, become a part of the user, allow the user to keep control, and be constantly available [2]. The ability of wearables to connect to the Internet has enabled big data analysis and subsequently brought wearables to the forefront of the Internet of Things. Currently, the wearables market is dominated by smartwatches and fitness trackers followed by smart eyewear, smart clothing, and medical devices [3]. Thus, the application of wearables is not only limited to individuals who want to keep track of their fitness, but also by doctors and patients alike for disease diagnostics, and even treatment of diseases in the near future. Moving ahead, we envision a paradigm shift from bulky high power-consuming electronics to small flexible and conformal wearables with human-friendly electronics that can be worn or attached to any part of the body to carry out both disease diagnostics and treatments. Adam Greenfield (Fitbit CEO) once said is 2006, 'We want to get to the point where not only are we addressing lifestyle conditions but more chronic conditions as well, whether it's heart disease, obesity, etc.'

Before the advent of wearables, people relied on hospitals and clinics to keep track of their heath. For measuring mere blood pressure or blood glucose levels, patients were required to go to a clinic, as a result of which, people were not aware of their physical health unless they reached a point when the situation became adverse. On the other hand, if doctors wanted to keep track of a patient's health post-treatment, they needed to rely on self-reporting from patients [4]. The proliferation of adaptation of wireless technologies enabled the transfer of information from one device to another without the limit of wires and distances. Wireless connectivity has become a necessity for wearables which come laden with sensors like accelerometers, heart rate, and blood oxygen level allowing the monitoring of patient health remotely. This allowed continuous gathering of physiological data and vital signs of individuals over long intervals of time among a diverse range of patients rendering early patient identification, treatment customization, and intervention implementation plausible [4].

Depending on the target application, a wearable can be equipped with movement, biometric, or environmental sensors [5]. A wearable may have a combination of any of these, for example, for early detection of asthma, the wearable can be used to sense the sounds coming from lungs while at the same time monitoring the surroundings to identify potential triggers of an asthma attack [6–9]. Wearables have a vast potential in health monitoring and prognosis [10]. However, currently, the wearable market is dominated by smartwatches and fitness trackers which have resulted in a stagnant growth of the wearable market owing to their limited functionality, as the consumers see them as merely a tool for entertainment and secondary notification system to their smartphone [11]. In the US, the growth of wearables has been much slower in the past 3 years [12]. Together with the lack of functionality, these wearables have remained bulky and rigid which limits the application of such wearables. For the same reason, we see most wearables are usually wrist-worn, and having a wearable on the wrist does not cover the vast range of physiological parameters that can be measured. Currently, wrist-worn wearables can only sense activities like walking and exercising, and physiological parameters like heart rate. Even with such limited functionality, we see a large discrepancy in the data obtained from different wearables [13,14]. In contrast, the interest in flexible systems has been on the rise to enable wearables with a wide variety of flexible sensors in compliant flexible platforms that can measure a vast range of body vitals beyond heart rate while being able to be worn unobtrusively on the human body or integrated seamlessly into the clothing. Significant research has been carried out on the fabrication techniques to make flexible and stretchable electronics [15–19]. In order to realize flexible wearables, all of the components that are a part of the wearable need to be flexible which include; integrated circuits (IC), circuit boards, display, energy storage or harvesting elements, antennas, and sensors.

19.2 Key Components of a Flexible Wearable

Recent advances in telecommunications, microelectronics, sensor manufacturing, and data analysis techniques have allowed the materialization of wearable devices. Any wearable device needs to have three functional systems working together in harmony: (1) sensors, signal conditioning circuits, and microprocessor for data collection and preliminary interpretation, (2) a transceiver for wireless transfer of data from the wearable device to a remote centre, and (3) advanced signal processing and data analysis to make sense of collected data [20]. In the past, the size of sensors and front-end electronics made it too difficult to use them in wearable technology to gather physiological and movement data. With smaller sized circuits, microcontroller functions, front-end amplification, and wireless data transmission, wearable sensors can be packaged together in a miniaturized wearable that can be worn unobtrusively on a human body or embedded in textile materials seamlessly. On one hand, researchers have been pushing the limits to achieve miniaturization of sensors and electronics, while on the other hand there has been a significant improvement in data processing algorithms for fast performance at lower energy costs. In the following subsections, we will discuss the key components of a fully compliant wearable and discuss the flexibility of each component.

19.2.1 Sensors and Actuators

Every wearable comes equipped with certain kinds of sensors, the choice of which depends on the intended application. Fitness trackers come equipped with biometric and activity sensors. Virtual augmented and mixed reality devices usually contain various sensors in different combinations of cameras, inertial measurement units, depth sensing, and force/pressure sensors for user's interaction with the content and the environment. Wearables intended for medical and healthcare-based applications are built in such a way that they can monitor and interact directly with bodily processes. Sensors can be loaded into different accessories such as garments, hats, wrist bands, socks, shoes, eyeglasses, and other devices such as wristwatches, headphones, and smartphones. The types of sensors being used in wearable applications include temperature, force, pressure, stretch, optical, microphones, global positioning system (GPS), chemical and gas, IMUs, electrodes (electrocardiography (ECG)), and many others.

These sensors usually fall under two domains: wearable sensors and ambient sensors. Wearable sensors are worn on the human body to measure certain physiological parameters as indicators of health which can potentially be used for disease diagnostics. These parameters include heart rate, body temperature, touch, blood oxygen level, blood pressure, respiratory rate, and muscle activity [20]. These sensors can be worn on any part of the body depending on the application and type of sensor. Blood pressure sensors are generally attached near the wrist or the elbow, while heart rate sensors are attached on the wrist. Sometimes researchers need to come up with innovative designs for the sensors to be worn on different parts of the body. Asada et al. combined a heart rate and blood oxygen level sensor in a ring-shaped design to be worn on the finger [21]. Here, we see that a body-worn wearable system typically requires flexible and stretchable sensors so that they can be easily wrapped around parts of the body for various physiological measures. Paterson et al. designed such a flexible photoplethysmogram (PPG) sensor that can be worn on the ear for measuring heart rate unobtrusively for long intervals of time [22]. The sensors may not be limited to contact with the outer surface as some applications demand interaction of sensor with body parts underneath the skin. As an example, the glucose level in the blood, unlike blood oxygen level, cannot be measured from the outer skin and thus requires the sensor to come in contact with the blood to collect the sample. Dudde et al. used a microperfusion technique to find out blood glucose levels and further deliver insulin into the body as required for diabetes patients [23]. Here, we see a combination of sensors with actuators. The actuator is the component responsible for the movement of a mechanism in a system. Here, the actuator is in the form of microneedles injecting the insulin into the human body.

On the other hand, ambient sensors are aimed at monitoring the environment around humans. Such sensors include light, heat, chemical, gas, and motion sensors. Ambient sensors find use in wearables aimed at monitoring environmental conditions around a subject. This may include checking the quality of air around an asthma patient to warn patients of a possible asthma attack. In any case, we need flexible and stretchable sensors that can conform to the human skin or clothes so that they can be worn for long periods of time without causing any discomfort. A plethora of other flexible and stretchable sensors and actuators have been discussed in previous chapters.

19.2.2 Microcontroller

The microcontroller is the IC which contains the processor, memory, and input/output peripherals. It acts as the brain of any electronic system as it performs necessary computations to collect, process, and store data from the sensors. For a long time, the size of electronics has been a hurdle towards making a flexible wearable that can be worn unobtrusively on the human body, but researchers have been working hard towards making the electronics smaller and smaller by reducing the size of individual transistors and thus increasing the number of transistors per unit area. This led to the achievement of the same performance, or even better, in a much smaller area. By having a smaller microcontroller IC, we are left with more space to integrate other components of a wearable system as will be discussed in later subsections. To visualize the drastic size transformation of a single transistor, imagine the size of a transistor 4 decades ago was the size of a red blood cell (10 μm). Now the industry, led by Intel and TSMC, has pushed the transistor size down to 10 nm which is less than the size of an average virus [24,25]. This thousand-fold size also allowed a decrease in cost where we can see a linear trend in price reduction for size reduction [26]. This allowed manufacturers to make small size ICs by packing as many as 100 million transistors in a millimeter squared area. With the growing interest and demand for flexible systems, researchers are working towards making these chips flexible so that they can be seamlessly integrated into a fully compliant and flexible wearable platform. One approach is to use inherently flexible printed or organic thin-film transistors (TFTs) to make flexible microprocessors, but these transistors have 2–3 orders of magnitude less mobility compared to complementary metal-oxide semiconductor (CMOS)-based circuits. Transistor performance is generally evaluated by the switching speed which is directly proportionally related to mobility [27]. It is for the same reason that, by far, printed and organic TFTs have only found applications in areas of flexibility: displays, radio-frequency identification (RFID) tags, and large areas sensors [28]. In addition to being faster, CMOS-based circuits dissipate much less power which is an important characteristic for a wearable device as wearables have smaller spaces for energy storage. Furthermore, the sizes of printed and organic TFTs have only been brought down to a mere 1 μm, thus unable to reach the smaller size of CMOS-based chips [29]. It has been established that CMOS-based electronics have a significantly higher performance compared to other non-CMOS-based technologies, thus we see flexible CMOS-based devices being used as the microcontroller of a flexible wearable system. The flexibility of CMOS-based IC can be achieved by thinning down the underlying silicon wafer that carries the electronic circuitry. The average thickness of a silicon bare die IC is 200–700 μm. By bringing down the thickness to less than 25 μm, the IC becomes flexible while the performance remains similar. Different techniques for thinning down silicon wafers to make them flexible have been discussed in Chapter 17.

19.2.3 Analog Front End

Generally, a wearable system may require more than just one microcontroller IC. Most sensors need to be attached to an Analog Front End (AFE) which in turn is connected to the microcontroller. The analogue output signals from a sensor can be very small making them impossible to be detectable by any microcontroller. The first function of an AFE is to amplify the input analogue signals using operational amplifiers. The input signal can be a voltage, current, or even an alternating signal like a sine wave. Once the signal is amplified, an AFE further converts the analogue data from the sensors into digital data, so that the data can be read by a microcontroller using analogue to digital convertor. These components can be part of an AFE as individual ICs or they can all be embedded in a single AFE IC. Apart from the amplification of the signal, many sensors require further filtering in order to remove unwanted noise signals. As an example, the audio system is prone to electric power line 50–60 Hz noise which can be heard as a hum on the audio signals. In such a case, an AFE must contain a high pass filter to remove unwanted low-frequency noise signals.

The individual components of an AFE differ by application. For example, in the case of chemical sensing, a specific chemical or gas can alter the impedance of an electrolyte once it comes in contact with the sensor. Typically, these sensors have three connections: the counter electrode (CE), the reference electrode (RE), and the working electrode (WE). Voltage is monitored at the RE when current is fed into the CE. The voltage at the RE is kept constant by using closed-loop feedback such that return current changes at the WE. This current is converted to a voltage using a trans-impedance amplifier (TIA). Companies have integrated all these components and circuitry of an AFE into a signal chip, an example of which is LMP91000. For an ultrasound sensor, the AFE is composed of various components: voltage controlled amplifier (VCA), analogue-to-digital converter (ADC), and continuous wave mixer. The VCA includes a low noise amplifier (LNA), voltage controlled attenuator (VCAT), programmable gain amplifier (PGA), and low-pass filter (LPF). Companies like TI and Maxim have several single IC ultrasound AFE chips available. In the case of a CMOS image sensor, the AFE is merely composed of an analogue-to-digital converter to convert the raw voltage coming from the photodetectors of a CMOS image sensor to a digital value. However, CMOS image sensors contain thousands of photodetectors where each pixel (photodetector) contains information about the image. Such a large amount of data cannot be read by a general-purpose microcontroller, and thus require a digital application-specific integrated circuit (ASIC) which is a specialized image processing unit (IPU). This ASIC may also contain timing generation circuitry to control the timing of data acquisition from an AFE since the AFE has to acquire data from thousands of individual pixels, and it cannot collect data from all of these pixels simultaneously. Timing circuits help an AFE take data from rows of pixels in serial fashion and convert them to digital data for them to be read by the ASIC to make a processed digital image. Temperature, light, and heat sensors generally require a simpler AFE comprised only of an ADC, with digital data directly fed to a microcontroller. Since an AFE and ASIC come in the form of an IC, we can use the same techniques of thinning down silicon wafers to make them flexible as discussed in the previous section.

19.2.4 Wireless Connectivity

Wearables are stand-alone devices that can perform sensing and data processing by itself. Wearable like smartwatches and fitness trackers come equipped with display modules to show alerts and relevant data. This comes at the cost of additional power consumption as display consumes much more power than a microcontroller and an AFE. Wearables targeted towards medical applications demand a long term operation and thus would favor a system in which the wearable is connected to a peripheral like a smartphone to display and store large sets of data. Additionally, a wearable attached to one part of the body may need to communicate with a wearable on another part of the body. Moreover, a wearable needs to have wireless functionality in order to be able to be integrated into an Internet of Things (IoT)-based system [30,31]. Healthcare-based wearables are largely intended for real-time intervention systems which make wireless connectivity a necessity so that the wearable can inform relevant authorities or physicians in case of emergencies [32,33].

A wearable is limited in terms of space and a processing power computer compared to remote server stations and desktop computers. The onboard processor and memory suffice for data collection, low-level processing (amplification and filtering), and short term data storage, but they are not enough to undergo artificial intelligence (AI)-based analysis, big data analysis, or even run complex signal processing algorithms. In order for any device to perform AI-based tasks, it needs to be equipped with specialized processors like a neural processing unit (NPU) which can perform matrices-based calculation. The conventional microcontroller does not have the processing power to perform matrix-based calculations and are mainly limited to performing basic multiplications and division tasks. An NPU has the ability to perform computation in the floating-point unit (FPU). An FPU is a real number like fractions, irrational numbers, etc. A conventional microcontroller can only perform computations in decimal format limited only to a 32-bit number. This ability to operate with a floating-point unit allows an NPU to perform machine learning. Machine learning is the process where the device teaches itself to differentiate between different things like conversations, pictures, or words. Just recently, smartphone industries have started to equip a smartphone with a secondary, higher-level processor than the normal microprocessor. Google introduced an image processing unit inside its Pixel 2 phone which is dedicated to applying the AI-based algorithm on large sets of image data [34]. Huawei pioneered in introducing the first NPU HiSilicon Kirin 970 in their latest smartphone Mate 10 [35]. However, so far this processor use has only been limited to improving the picture quality of the camera, but as we move on, developers will start using these processors to undergo machine learning and AI. Thus, we imply from this discussion that wearables still need a central hub to which they can periodically send data to be processed for complex analysis.

Having a central hub doesn't solve the problem, as the microcontroller still needs to acquire and transmit data which in certain applications could become a strenuous task. As an example, any sensing application employing ECG electrodes suffers from the issue of extremely large sets of data. Generally, ECG sampling takes place at 200 samples per second or more which results in a huge amount of data that are hard to store, analyze, and transmit [36]. Significant research is being carried out to compress these large sets of data so that they can be stored and transmitted by a low-level processor like the ones found in wearables [37–41]. Low-level processing can only detect the heart rate, but deeper analysis on ECG data can help us predict heart attacks, get an indication of heart muscle thickness, detect weakened blood flow, or find patterns of anomalous cardiac rhythm, but a wearable powered by a microcontroller cannot handle even such tasks [42].

Almost all wearables use a standard wireless protocol like Bluetooth, ZigBee, NFC, IrDA, or IEEE 802.11 g for data transmission [43]. Each of the protocols has certain characteristics that are favorable in different circumstances. 802.11 g has the largest range of communication, typically around 200 m with high data rates of 54 Mbps, but it comes at the cost of large power consumption of 1 W and higher cost of purchase around $9. On the other hand, Bluetooth is the most energy-efficient method with power consumption as low as 2.5 mW, but it comes with a limited range of 50 m and data rates of up to 3 Mbps. NFC and IrDA consume extremely low power of up to 25 μW, but with a range of only 1–2 m. These communication methods are targeted for low power consumption and longer-lasting wearable devices as it requires the central node to be in a nearby range. Technologies like WLAN, GSM, GPRS, and WiMAX offer extremely long ranges and ubiquitous network access, but the power consumption and cost of running such systems are much higher. Any wireless communication system will require a transceiver IC, an antenna, and an antenna matching circuit. We have seen significant research conducted in making all of these components flexible (and stretchable) as discussed in Chapter 8. Looking ahead, we can see a widespread adaptation of wearables which each person can be wearing one or more wearables. For high-density deployment scenarios, like connecting a central hub to all users on a train or a bus, the short-range low energy communications methods fail as they tend to have a smaller number of user per cluster. Figure 19.1 shows the ranges and an allowable number of users for the popular communication protocols in wearables. On the other hand, 60 Hz protocols can handle high user density communication requirements, but they tend to lack mechanisms in handling situations where neighboring networks overlap each other [44].

19.2.5 Battery

The battery is another essential component of a wearable device. There has been a proposition of powering wearable devices using energy harvesting devices, but such methods generally fail to provide a continuous flow of energy which is essential for a wearable, especially a wearable targeted towards healthcare-based applications. Solar energy is a large source of free energy, but it cannot provide a continuous flow of energy at night times or during cloudy days. Similarly, piezoelectric energy harvesting relies on the movement of the subject, but a person cannot keep moving at all times. Thus, having an onboard battery becomes a necessity where the energy harvesting sources can recharge batteries when they can to increase the lifetime of the wearable. The battery can be either lithium-ion polymer (LiPo) battery, alkaline, or nickel metal hybrid. Alkaline batteries are the ubiquitous AA and AAA batteries that have been around since the mid-1900s. For wearables, LiPo batteries are the most popular choice due to their long-lasting life and ease of recharge. Most wearables use Li-ion batteries in the form of a coin cell that ranges from 10 to 20 mm so that it can easily fit inside a wearable. Figure 19.2 shows different battery technologies available for use in wearables. So far, batteries have proved to be the bottleneck in the proliferation of flexible wearable devices.

Even when all other components of an electronic system have been made flexible, the batteries remain rigid, presenting a redundancy in the claims of a fully compliant wearable system [30,46–53]. These batteries are electrolyte batteries which pose a hazard risk which makes the manufacturers hesitant in

	Rate (Mb/s)	Range (m)	Users per cluster
Bluetooth	0.5–20	10–15	8
IEEE 802.11g/n	50–120	10–50	5–10
IEEE 802.11ac	up to 400	10–30	10–15
ZigBee	0.5	10–100	5–10

FIGURE 19.1 Comparison of current short-range radio technologies. (From Pyattaev, A. et al., *IEEE Wirel. Commun.*, 22, 12–18, 2015.)

Type	Characteristics	Devices	Limitations
Lithium Coin	Small in size, lightweight, inexpensive, and fairly safe.	Small and low-power wearables.	Non-rechargeable and have low current draw capability.
Lithium-ion (Li-ion)	Rechargeable, lightweight, have higher power density and cheaper in cost than Li-Poly, do not require priming when first used and have a low self-discharge.	Many wearables including wrist-worn and head-worn.	Suffer from aging even when not in use.
Lithium Polymer (Li-Poly)	Rechargeable, lightweight and flexible (more than Li-Ion) and have improved safety.	Many wearables including wrist-worn and head-worn.	Cost more to manufacture and have a worse energy density than Li-ion batteries.
Graphene	Promised to be more efficient, more flexible, have higher energy density and storage capacity than all of the current battery types.	Still under development.	Expected to be more expensive than other batteries.

FIGURE 19.2 Wearable battery types. (From Seneviratne, S. et al., *IEEE Commun. Surv. Tut.*, 19, 2573–2620, 2017.)

using such batteries when the wearables come in contact with humans especially for wearables targeted towards healthcare applications. Solid-state batteries are a potential solution, and they were first used in pacemakers [54]. A large number of manufacturers like Toyota are investing in the commercialization of solid-state batteries. Since solid-state batteries can usually present on a silicon wafer, this provides us with a potential of making them flexible using different silicon thinning techniques as discussed in Chapter 7. Flexible batteries are coming out as an emerging field where researchers are working on either making the Li-ion batteries flexible or coming out with newer inherently flexible materials to make batteries [55–60].

19.2.6 Display

A display can be added to a wearable device for a visual representation of data and alerts. The most common type of displays is LCD, LED, and organic LED display. For example, a display can show the heart rate of a patient and show an alert if the patient is detected to be in a chemically hazardous environment. The displays are the most power-consuming components in a wearable electronic system where a small 1.5 inch TFT-based LCD display can consume the power of about 272 mW. The amount of power consumed by a display is directly proportional to the size of the same technology display [61]. Displays are an optional component of a wearable device as we discussed that wireless connectivity has become a necessity for a wearable, and a wearable can use the display on the central node to display relevant data. It is for the same reason that we see smartphones and wearables with displays need to recharge many times a week.

LCD and LED displays area are rigid and thick, and since the display covers a larger surface area compared to other electronic components, a rigid display would be unusable in a flexible wearable. Research on flexible displays has been ongoing for ages, and now we see the commercialization of flexible displays in many devices such as smartphones, TVs, and even wearables. An organic LED display uses OTFT LED to form a transparent and flexible display [62]. In addition, they consume less power, have higher contrast, better response time, and excellent color reproduction [61]. The flexible displays have been discussed in detail in Chapter 18.

19.2.7 Circuit Board

A circuit board, commonly referred to as printed circuit board (PCB), is the substrate that holds the electronic components on insulating material and connects the components to each other using electrically conductive metal tracks. A flexible circuit board is very similar to a rigid circuit board where the only difference is that the substrate is now flexible instead of being rigid. The metal interconnects are inherently flexible as their thickness is less than 500 μm. This allows the circuit board to possess any shape provided it stays within the bending limits of the substrate material. In previous subsections, we have discussed the flexibility of all the components in a wearable, and their integration on a flexible circuit board can help us envision a fully compliant wearable system. Commercial companies like Flexible Circuit© Technologies are dedicated to making flexible circuit boards for companies in need of making a flexible wearable device. The circuit boards need to be tested for performance under long term bending conditions.

On a rigid board, packaged ICs are soldered on the metal pads, but for a flexible wearable system, a rigid packaged IC cannot be used. A flexed bare die is the potential solution as discussed before, but bare dies are not compatible with soldering. Bare dies are generally connected to metal pads using wire bonding where the chip is placed on a substrate with metal pads facing upwards and the wire bonding tool forms a metal interconnect from the pad on the die to a pad on the substrate. However, wire bonding is known to not perform well under bending conditions [63,64]. In such cases, unconventional bonding techniques need to be used. Anisotropic conductive films (ACFs) are widely used to bond flexible display modules to a circuit board [65]. An ACF is an adhesive with small embedded metal particles inside the adhesives. When the ACF is sandwiched between two conductors, the embedded metal particles allow the contact formation between the two conductors. This allows the conduction of electricity in one direction which is why they are called anisotropic conductive tapes. An ACF can also be used to bond an IC chip to a flexible substrate providing a viable alternative to wire bonding and soldering for attaching ICs on flexible circuit boards [66]. Furthermore, Sohail et al. showed a Lego-based assembly of IC on a physically compliant system which is compatible with large scale manufacturing [67]. Each bare die with a certain geometry can fit on a destination site with the same dimensions to allow targeted site binding. Flip-chip bonding is another popular technique to bond bare dies on a substrate. Solder bumps are deposited on a metal pad on the top of the bare die IC and then flipped on top of a pattern of pad aligned such that each metal pad on bare dies lands on its respective metal pad on the circuit board. The solder is then reflowed to form an electrical connection. Behnam et al. printed silver pillar arrays on top of a metal pad on a bare die using a silver nanoparticle piezoelectric inkjet printer [68]. This increased the contact reliability of flip-chip bonding.

19.3 Design Issues and Considerations

For a wearable to be considered as practical, it has to be non-invasive, reliable, user-friendly and, able to give feedback to the user [5]. In the following sections, we will discuss the considerations for wearable design and the issues that arise consequently.

19.3.1 Sensor Placement

Different parts of the human body can provide a different set of physiological parameters. The preferred place of wearing for commercially available wearables is the wrist as such wearables are easy to install and remove. Furthermore, having worn watches and wrist-worn jewelry for ages, people find it as the least obtrusive position for wearing an accessory (wearable) [69]. Besides, wearables on the wrist laden with appropriate sensors can be used to measure physiological parameters like movements, heart rate, and sleep activity, as well as environmental variables like temperature, light, sounds, location (GPS-based), and altitude [5,14,70]. Another popular way of attaching wearables to the body is to wear the sensors in a chest strap or a belt. Chest straps and belts are easy to install, and they can remain sturdy in a single place for reliable data collections [5]. A wearable worn on the chest can measure the same physiological parameters as the wrist with the addition of ECG. Wearable ECG devices are popular tools to understand heart conditions in order to detect heart diseases [71]. For human activity monitoring, each place on the body or a combination of them can give us information about a vast range of movements. The most commonly used activity sensor is an accelerometer and researchers have used them in a position such as hip, wrist, ankle, arm, thigh, waist, back, head, and ears [72]. The head is another place on the human body that can be used to gather information about head injuries and impacts. Many researchers have placed sensors on the head to monitor falls and posture monitoring systems [73–77]. There are some niche applications for which the sensors have to be placed in certain places outside (e-skin) or inside of the human body (implantable). A vast array of such sensors has been discussed in Chapters 10 and 20. Figure 19.3

Signal	Sensor	Measurement site	Fs
Altitude	Air Pressure (Suunto X6HR)	Wrist	0.5
Audio	Microphone (AKG C417)	Chest, on rucksack strap	22000, mono, 16 bit
Body Position	Metal ball moves between resistors (ProTech Position)	Chest	200
Chest Accelerations	3D acceleration (2 x Analog Devices ADXL202)	Chest, on rucksack strap	200
Chest Compass	3D compass (Honeywell HMC-1023)	Chest, on rucksack strap	200
EKG	Voltage between EKG electrodes (Blue Sensor VL, Embla A10)	Below left armpit, on breastbone	200
Environmental Humidity	Humidity (Honeywell, HIH-3605-B)	Chest, on rucksack strap	200
Environmental Light Intensity	Light sensor with two output dynamics (Siemens SFH 203P)	Chest, on rucksack strap	200
Environmental Temperature	Temperature sensor (Analog Devices TMP36)	Chest, on rucksack strap	200
Event Button	Switch (Embla XN Oximeter)	Chest, on rucksack strap	-
Heart Rate	IR light absorption (Embla XN oximeter)	Finger	1
Heart Rate	IR light reflectance (Nonin XPOD)	Forehead	3
Heart Rate	Voltage between chest belt electrodes (Suunto X6HR)	Chest	0.5
Location	GPS satellite receiver (Garmin eTrex Venture)	Shoulder, on rucksack strap	Based on location
Pulse Plethysmogram	IR light reflectance (Nonin XPOD)	Forehead	75
Respiratory Effort	Piezo sensor (Pro-Tech Respiratory Effort)	Chest	200
SaO2	IR light absorption (Embla XN Oximeter)	Finger	1
SaO2	IR light reflectance (Nonin XPOD)	Forehead	3
Skin Resistance	Resistance between two metal leads (Custom-made)	Chest	200
Skin Temperature	Resistive temperature sensor (YSI 409B)	Upper back, below neck	200
Wrist Accelerations	3D acceleration (Analog Devices, ADXL 202E)	Wrist, dominant hand	40
Wrist Compass	2D compass (Honeywell HMC-1022)	Wrist, dominant hand	40

FIGURE 19.3 Signals and sensors of the data collection system. (From Parkka, J. et al., *IEEE Trans. Inf. Technol. Biomed.*, 10, 119–128, 2006.)

shows a variety of signals that can be acquired by different sensors placed in different positions of the body [78]. Antenna placement also becomes critical as the radiation pattern and efficiency of the antenna is affected by the proximity to human skin. Wearables are generally worn on or in close proximity to the skin. Researchers observed the efficiency of an antenna increased 18% when moved 3 mm away from the skin [79]. Thus, a sufficient gap between the place of the antenna and the skin is essential.

19.3.2 Obtrusiveness

The number of sensors and the related circuitry should be minimal in order to form a wearable in order for it to be acceptable by everyday users [10]. The numbers of sensors can be minimized by using a software algorithm. For instance, Bao et al. put accelerometers on more than five parts of the human body and found out that just two accelerometers placed on the wrist and thigh are enough to monitor and identify movement activities [80]. Reducing the number of sensors also decreased the complexity of the system and hence the size of the wearable can be smaller. A smaller sized wearable can better be able to stay on the body unobtrusively. Additionally, a lesser number of sensors and the analogue front will warrant a reduced power consumption resulting in longer-lasting battery life.

19.3.3 Processing and Performance

A wearable device tends to have low power-consuming microcontrollers which in turn have low processing power, as it is more important for a wearable to have a longer-lasting battery, especially in healthcare oriented applications. This may render the wearable handicapped when handling a large set of data and complex signal processing tasks. In such a case, a wearable saves the raw data in the wearable memory and sends the data to a remote server wirelessly. However, for applications requiring continuous monitoring, like glucose monitoring, the wearable will then be forced to continuously send raw data to the central server which results in increased power consumption. In applications like ECG-based heart monitoring, the amount of data is so large that it becomes difficult for a wearable to process or store the data. In those cases, software-based data compression techniques are employed as discussed in Section 19.2. Thus, one solution for all may not exist, and the designer needs to make a compromise between energy consumption and processing power depending upon the targeted applications. Thus, we see there are many variables involved in designing wearables and the choice of strategy will depend upon the nature of the application. For any algorithm designed for wearables, there will always be a trade-off between algorithm performance (correct detections), algorithm cost (false detections), and power consumption [42]. Moving forward, researchers are empowering wearables to perform data mining tasks such that they can do much more than mere data reporting. Data mining-based healthcare services are now able to perform prediction-based analysis so that a wearable can predict a future event on the basis of data collected over time [81]. Meanwhile, healthcare providers are using wearables to detect anomalies and perform diagnosis of diseases. Figure 19.4 shows the main data mining tasks undertaken by wearables in comparison to different health monitoring systems. It is generally observed that applications targeted towards home monitoring perform anomaly detection and prediction systems while clinical applications are focused on diagnosis [81]. A standard approach used towards mining data from wearable sensors can be seen in Figure 19.5. Raw data fed to the wearable undergoes some pre-processing for feature extraction, and then the data are fed to models for learning and making decisions.

19.3.4 Flexibility

Many new applications are arising which demand wearables to be fully flexible and conformal so that they can be placed in areas where rigid and bulky wearables cannot be worn especially for implantable and e-skins discussed in Chapters 10 and 20. An implantable intended for eyes or brains needs to be conformal such that it can be placed seamlessly onto the desired organ such that it does not interfere with the

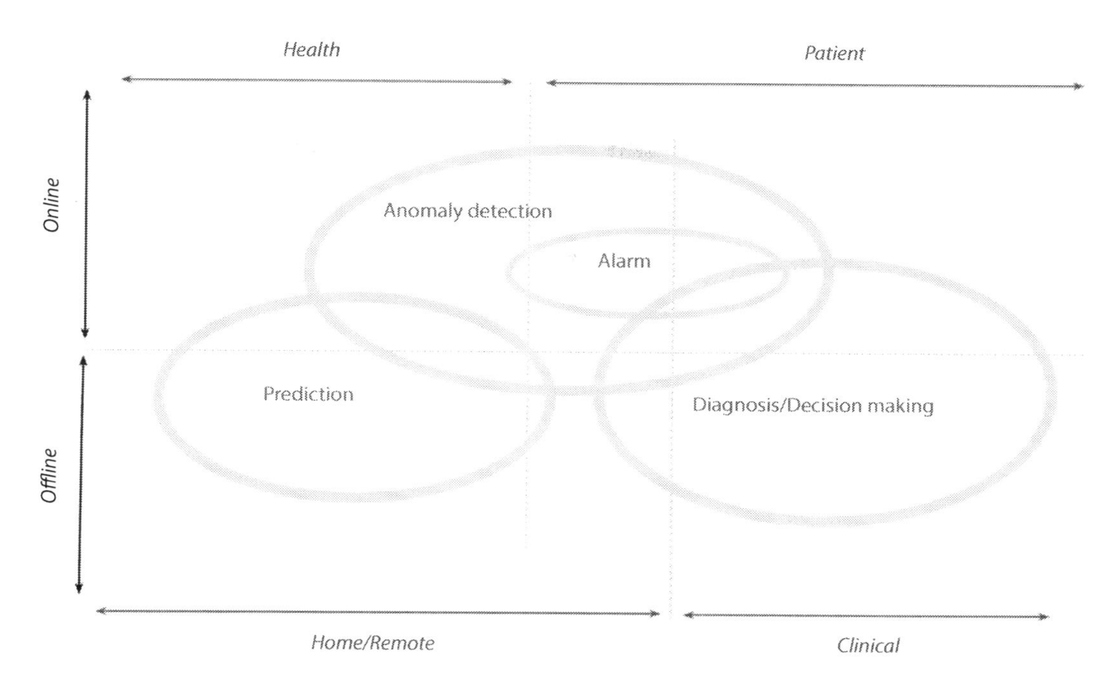

FIGURE 19.4 A schematic overview of the position of the main data mining tasks (anomaly detection, prediction, and diagnosis/decision making) in relation to the different aspects of wearable sensing in the health monitoring systems. (From Banaee, H. et al., *Sensors*, 13, 17472–17500, 2013.)

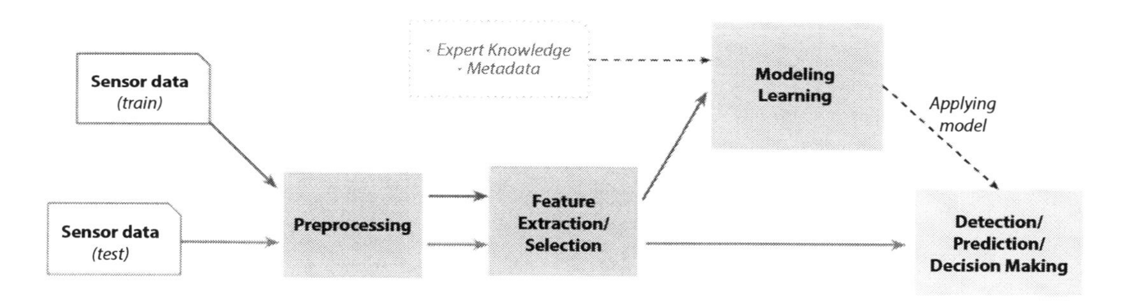

FIGURE 19.5 A generic architecture of the main data mining approach for wearable sensor data. (From Banaee, H. et al., *Sensors*, 13, 17472–17500, 2013.)

functionality of the respective organ [82,83]. The eye is a very delicate organ where the external layer of the eyeball in protecting delicate internal structures [84]. Glaucoma is an eye disease where the patient suffers from impaired peripheral visual function and caused by increased intraocular pressure [85]. Researchers have developed a specialized system on a chip that can be placed inside the eye to measure the intraocular pressure while being powered wirelessly from an external source like a smartphone [86]. A flexible IC for the glaucoma monitor can allow unobtrusive and conformal attachment on the spherical structure of an eye such that it doesn't affect the delicate internal structures of the eye. Cardiac physiological mapping allows the understanding of the functions of the heart, but it will require a characterization device to have a uniform contact across the whole surface of the heart. Using elastic and flexible membranes, some researchers have been able to adhere sensors to the complete surface of the heart by making a carrier substrate using 3D printed elastic membranes in the form of a jacket such that it wraps around the whole heart [87]. We see that by making flexible devices, we can reach places inside the human body to study and identify diseases or replace the skin to provide similar functionality after the skin may have completely burned away [88,89].

19.3.5 Energy Consumption

Battery life becomes critical when the application is targeted towards healthcare-based applications. Display and wireless transmission are generally the most power-consuming elements in a wearable [90]. As discussed in Section 19.2, wireless connectivity is a necessity since the wearable needs to store and process a complex set of data which cannot be done locally owing to the small size of the wearable platform. With the availability of wireless functionality, wearables can take advantage of the already available displays on smartphones or computers to store and process large sets of data, and then display the relevant information to the user. It was further discussed in the same section that there are different communication protocols available with each one having its own advantages and disadvantages. Researchers and manufacturers tend to use Bluetooth Low Energy technology as it offers high data rates, acceptable range, and comparatively low power consumption. Computational complexity is the main factor of higher power consumption as higher complexity demands higher CPU usage and ultimately increases the power consumption. As we move to intelligent processors that can make decisions based on the previous sets of data available, large sets of data are now available in order to process, which increases the computational complexity significantly, which pushed researchers to develop algorithms that reduce the number of features in the learning tasks to only include the necessary tasks for achieving a minimum acceptable accuracy [45].

On one hand, the miniaturized form factor of most wearables leaves a small space available for the battery, while on the other hand, wearables are expected to run at all times [91]. This presents designers with a serious challenge of reducing power consumption to a bare minimum. When the usage of high power-consuming elements becomes a necessity, smart power-saving techniques are utilized in order to reduce power consumption. The microcontroller and AFE are mostly kept in sleep mode, wherein the electronics are taken to a low processing state most of the time. The electronics are taken back to active mode at regular time intervals to gather the new set of data, perform necessary computations, and then go back to sleep mode again. For the sensors, they can be completely turned off when the controller is in the sleep state as they do not have volatile memory, whereas a microcontroller has a volatile memory (RAM) which gets erased whenever it is turned off. Microcontrollers need RAM in order to function properly. For instance, GPS consumes a significant amount of power, and thus they are kept in the off state when not in use [92]. Another method is to reduce the sampling time when the user is in a state wherein the relevant data may not be available. As an example, a sensor monitoring exercise and movement can be taken to a low sampling interval state when the user is sleeping. System designers employ several other software-based power-saving techniques in order to increase the battery lifetime of wearable devices. Figure 19.6 shows how the current consumption of various sensors increases with the increasing sampling rate [91].

19.3.6 Adaptability and Reliability

Wearables are generally designed as a one fit for all, but some researchers believe that having just one recognition model for one activity is not enough as people's activities vary among individuals due to having different age, gender, weight, and so on [10]. Another example could be an average woman's walking posture is different from an average man. This led to the formation of two types of analyses for activity recognitions as subject-dependent and subject-independent analyses [93]. In some cases, it becomes hard to train the systems for every single user, including having too many activities, lack of data for certain activities that are not preferred to be re-enacted (for example, people falling down the stairs), or collecting data from people who do not/cannot cooperate (for example, patients sufferings from mental illnesses) [10].

The push for the integration of wearables in the healthcare system has been a little difficult despite thousands of publications available in academic journals on new and advanced wearable systems. Non-reliable systems have been the obstruction towards the commercialization and adaptation of these

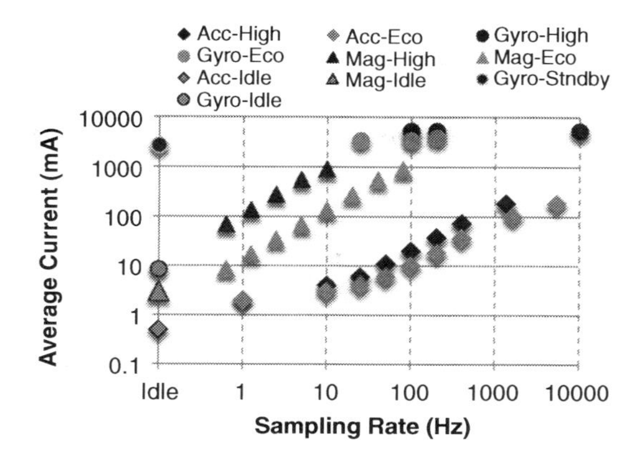

FIGURE 19.6 Average current of MEMS IMUs at different frequencies and operation modes. (From Williamson, J., et al., *Design Automation Conference (ASP-DAC), 2015 20th Asia and South Pacific*, 2015. © 2015 IEEE. With permission.)

wearables. Researchers do not focus on the practical issues, most importantly long term operation, while designing their devices. Results are published based on the short term performance of the wearables [94]. When the industry tries to adopt the technology, they often deem the technology unfeasible for manufacturing due to many reasons like higher cost of production, incompatibility with large scale production, being non-user friendly, unreliable performance in real life, being bulky and uncomfortable to wear, not complying with regulations set by regulating bodies, high power consumption, etc. Researchers need to understand these issues and prove the practicality of the wearables while publishing the research in order to improve the chances of seeing their research being adapted by wearable manufacturing organizations. Just recently, Samsung partnered with the University of California, San Francisco (UCSF) to use an optical-based sensor for measuring blood pressure in their flagship Samsung Galaxy S9 instead of the conventional strap-based blood pressure monitor. Although the results look promising, the error in reading greatly increases when the user is doing exercise or any rigorous activity [95].

19.3.7 Privacy and Security

Wearables have access to vital personal information related to the subject's health status which should not be available to anyone other than the physician and the patient itself [43]. Wearables generally transmit data wirelessly which need to be encrypted so that they cannot be read by an unauthorized user. Standard protocols like bluetooth low energy (BLE) and ZigBee come equipped with several high to low-level encryption and security capabilities. In some cases, wearables gain access to conversations and location of a user which raises concerns over the confidentiality of such data [96]. This calls for jotting down some legal terms to overcome the arising concerns with privacy and security, and researchers studied a few wearable projects and put forward some proposals to overcome the legal complications [97]. In the US, an application that needs access to a phone's wireless modules can be subject to regulations laid down by the Federal Communications Commission (FCC). Similarly, Food and Drug Administration (FDA) has strict regulation over any medical device, and recently they have laid down further regulations even for mobile applications that are used with medical devices [98]. These commissions help ensure laying down a standardized model for legal and safety issues arising from the manufacturing and usage of wearables is done. Since the wearable and even the connected phone lack the processing power to handle computationally intensive tasks, cloud-based servers present hubs of enormous computational resources. The question arises on how to ensure data privacy on these servers which are mainly controlled by third-party service providers unrelated to the patient, physician, or hospital. Researchers have been working towards identifying the challenges and solutions associated with security and privacy in cloud-assisted wireless wearable communications [99].

19.4 Applications

A new wave of wearable applications is rising where we see wearables being used beyond the confines of entertainment and smartwatches. As we move closer to the realization of flexible wearables, the scope of the applications will become more diverse. Wearables that have not seen widespread adoption due to their large size and rigidity will be easily accepted as they become conformal and unobtrusive. Most notable new contributions of wearables have been in sports and the healthcare industry. The application is briefly discussed below along with the need for wearable devices to be conformal and flexible in order to prove more effective for the intended purpose.

19.4.1 Sports

The wearable has been gaining interest among athletes and coaches in order to improve their performance in the field. Wearables have been designed to improve the posture and technique of an athlete by tracking hand and leg movements. Beginners use such technologies to compare their movements and technique to that of a professional athlete in order to correct their shortcomings. Wearables that tracks body movements have been used to monitoring lower limb activity during rowing [100], movement assessment during sports sessions [101], quantify the tackling demands for an (American) football player [102], monitor rehabilitation procedures of running athletes [103], measurement of ski jumping phases [104], and many others [105,106]. Training load monitoring is gaining popularity in sports science. In training load programs, athletes are subject to endure the progressively increasing amount of loads for certain muscles in order to keep improving the muscle strength. Wearables have been used to both monitor the internal physiological and psychological stimulation/stress and external load measures like speed, distance, acceleration, and metabolic power during load monitoring programs to gain insight on how the program is improving the muscle strength [107].

Athletes will benefit greatly from the use of flexible electronics. In most arm-based sports like badminton and baseball, the posture and weight of the hand has a major effect on the performance. Large bulky sensors attached to the arms of an athlete will surely alter their actual posture and movement, thus, making it counter-intuitive to track and monitor that movement. For some sports, even the seemingly small effects on aerodynamics of the arm due to these sensors can greatly change the performance of the athlete. We have seen in the previous chapters that flexible wearables in the form of e-skins practically feel like part of the skin. They are lightweight and conformal. Athletes equipped with flexible wearables can potentially perform unaltered which opens up the possibility of monitoring athletes in real-time during a game rather than experimenting in the confines of a laboratory. In the game of cricket, strict regulations are enforced as to how much the elbow of a bowler can bend during the delivery. According to the rules, 'a ball is fairly delivered in respect of the arm if, once the bowler's arm has reached the level of the shoulder in the delivery swing, the elbow joint is not straightened partially or completely from that point until the ball has left the hand' [108]. The angle between the bowler's biceps and forearms shouldn't exceed 15 degrees. In the field, it is extremely difficult for the field referees to judge the angle of bowling to such a low degree of elbow angle. Moreover, the bent elbow can injure the bowler [109]. Thus, a flexible wearable that can provide real-time field data can accurately identify the elbow angle of the bowlers to assist umpires and bowling coaches ensure that the bowlers stay within the assigned limits.

19.4.2 Health, Safety, Rehabilitation, and Post-treatment Monitoring

Advancements in technology are pushing for a new era of health monitoring where the patients can be monitored in the confines of their homes. The long term remote monitoring of physiological data and performing machine learning algorithms on them can provide for improvements in diagnosis and treatment of various chronic conditions [20]. Many researchers have shown different potentials of remote health monitoring, diagnosis, and treatment of diseases: to make more efficient therapeutic treatments for Autism spectrum disorders, provide first aid in battlefields and disaster-struck areas, continuously monitor the health of elderly remotely, assist behavioral therapy by monitoring the physiology of patients suffering from behavioral and anxiety disorders, and monitor glucose levels of diabetes patients and inject insulin as needed [32]. Safety monitoring has become a topic of rising interest in the field of remote health monitoring. Wearables are now commercially available that can be worn by patients which monitor their movements and alert authorities in case of the detection of an irregular activity like a fall. These wearables are potential lifesavers for the elderly and people suffering from epilepsy. Wearables have further shown to assist in remote rehabilitation by keeping a check on the patient's regularity with the prescribed exercises and monitoring the movements to help the patients to follow the correct posture [20]. However, to ensure the widespread adoption of these technologies, it can be said with confidence that only flexible wearables can be considered feasible for long term monitoring of patients. A device that monitors human movements needs to stay on the body round the clock. Thus, it becomes imperative that the wearable is flexible and conformal so that the patients can keep up with their daily activities without being affected by the presence of a monitoring device attached to the body.

Beyond monitoring the health of humans, wearables have also proven to be a valuable asset in overseeing the rehabilitation process of patients and observing vital signs of patients once they leave the confines of clinics or hospitals. Wearable devices keep track of movements and vital signs of patients that are sent wirelessly to a medical facility, thus, allowing doctors to monitor their patients remotely. Patients enjoy the luxury of going home early while opening up space for incoming patients. By reducing the time spent at the hospitals, patients can lower the cost of the treatments and doctors can entertain a larger number of patients. Some companies have adopted a clever way to support the rehabilitation process by using virtual reality and games. The games and the virtual reality environment direct patients to perform the movements and exercises as needed to undergo the process of physiotherapy. Honda Stride Management Assist technology was revealed in 2008. Supported by their ASIMO robot technology, Honda has configured a girdle designed for robots such that it's metal braces wrap around the human legs while being supported by a hip piece. The brace lowers the amount of load on the legs of the wearer, allowing patients suffering from leg injuries to walk on their own and undertake therapeutical activities while avoiding any physical damage. General Electric Healthcare developed a wireless medial monitoring system that tracks body movements and sends them to the caregiver allowing the patients to undergo physiotherapy from their homes [110]. As with monitoring devices, a rehabilitation tracking device will stay attached to the human body while performing the therapeutical activities. It becomes ever so important that the wearable remains lightweight and unobtrusive to allow the patient to follow natural movements. A device that restricts the natural flow of movement can alter the efficacy of the rehabilitation process.

Wearables also provide a great way to track the efficacy of a treatment by tracking relevant bodily signs between outpatient visits. Doctors can alter and improve upon the treatments depending upon the patients' response to the treatments allowing patients to get personalized medical care. Towards the later stages of Parkinson's disease, it becomes hard to observe the motor symptoms of the patients as their occurrence becomes erratic, making it difficult for the physician to observe the symptoms during the patient's visit. Many researchers use accelerometer sensors to keep track of the hand movement of Parkinson's disease patients throughout the day. The data can later be analyzed by the physicians to prescribe the optimum treatment and medicine dosage [111–113]. The rehabilitation is usually a long term process and it takes months of physiotherapy to completely recover. The use of such sensors and technology makes the process effective and cost-effective. The timings between doctor appointments can vary from days to weeks which means that activity monitoring devices are to be worn for extremely long intervals of time. Thus, it is vital that the activity monitoring devices are lightweight and conformal so that the patients can carry on with their normal lives without noticing the device.

References

1. Borkar, S. and A.A. Chien, The future of microprocessors. *Communications of the ACM*, 2011. **54**(5): 67–77.
2. Jiang, H. et al. Software for wearable devices: Challenges and opportunities. in *Computer Software and Applications Conference (COMPSAC), 2015 IEEE 39th Annual*. 2015. IEEE.

3. Hayward, J., G. Chansin and H. Zervos, Wearable technology 2016–2026: Markets, players and 10-year forecasts. *IDTechEx*, 2016.
4. Appelboom, G. et al. The promise of wearable activity sensors to define patient recovery. *Journal of Clinical Neuroscience*, 2014. **21**(7): 1089–1093.
5. King, R.C. et al. Application of data fusion techniques and technologies for wearable health monitoring. *Medical Engineering & Physics*, 2017. **42**: 1–2.
6. Firrincieli, V. et al. Decreased physical activity among headstart children with a history of wheezing: Use of an accelerometer to measure activity. *Pediatric Pulmonology*, 2005. **40**(1): 57–63.
7. Oletic, D., B. Arsenali and V. Bilas. Towards continuous wheeze detection body sensor node as a core of asthma monitoring system. in *International Conference on Wireless Mobile Communication and Healthcare*. 2011. Springer.
8. Seto, E.Y., et al. *A wireless body sensor network for the prevention and management of asthma*. in *Industrial Embedded Systems, 2009. SIES 09. IEEE International Symposium on*. 2009. IEEE.
9. Vilar, M.R. et al. Development of nitric oxide sensor for asthma attack prevention. *Materials Science and Engineering: C*, 2006. **26**(2–3): 253–259.
10. Lara, O.D. and M.A. Labrador. A survey on human activity recognition using wearable sensors. *IEEE Communications Surveys and Tutorials*, 2013. **15**(3): 1192–1209.
11. Kumar, D. and V. Venkateshwarlu. Consumer perception and purchase intention towards smartwatches. *IOSR Journal of Business and Management*, 2017. **19**(01): 26–28.
12. Deloitte. Smart watch adoption rate among consumers in the United States from 2013 to 2018 [Graph]. [cited August 7, 2019]; Available from: https://www.statista.com/statistics/949230/united-states-smart-watch-adoption-rate/.
13. Evenson, K.R., M.M. Goto and R.D. Furberg. Systematic review of the validity and reliability of consumer-wearable activity trackers. *International Journal of Behavioral Nutrition and Physical Activity*, 2015. **12**(1): 159.
14. de Arriba-Pérez, F., M. Caeiro-Rodríguez and J.M. Santos-Gago. Towards the use of commercial wrist wearables in education. in *Experiment@ International Conference (exp. at'17), 2017 4th*. 2017. IEEE.
15. Ghaffari, R., et. al. Reinventing biointegrated devices: Roozbeh Ghaffari discusses silicon-based nanomaterials configured in flexible and stretchable formats, and their potential to rapidly transform the medical landscape. *Materials Today*, 2013. **16**(5): 156–157.
16. Qaiser, N. et al. Mechanical response of spiral interconnect arrays for highly stretchable electronics. *Applied Physics Letters*, 2017. **111**(21): 214102.
17. Qaiser, N., S.M. Khan and M.M. Hussain. Understanding the stretching mechanism of spiral-island configurations for highly stretchable electronics. in *2018 International Flexible Electronics Technology Conference (IFETC)*. 2018. IEEE.
18. Qaiser, N. et al. 3D printed robotic assembly enabled reconfigurable display with higher resolution. *Advanced Materials Technologies*, 2018. **3**(12): 1800344.
19. Qaiser, N., S. Khan and M.M. Hussain. In-plane and out-of-plane structural response of spiral interconnects for highly stretchable electronics. *Journal of Applied Physics*, 2018. **124**(3): 034905.
20. Patel, S. et al. A review of wearable sensors and systems with application in rehabilitation. *Journal of Neuroengineering and Rehabilitation*, 2012. **9**(1): 21.
21. Asada, H.H. et al. Mobile monitoring with wearable photoplethysmographic biosensors. *IEEE Engineering in Medicine and Biology Magazine*, 2003. **22**(3): 28–40.
22. Patterson, J.A., D.C. McIlwraith and G.-Z. Yang. A flexible, low noise reflective PPG sensor platform for ear-worn heart rate monitoring. in *Wearable and Implantable Body Sensor Networks, 2009. BSN 2009. Sixth International Workshop on*. 2009. IEEE.
23. Dudde, R. et al. Computer-aided continuous drug infusion: Setup and test of a mobile closed-loop system for the continuous automated infusion of insulin. *IEEE Transactions on Information Technology in Biomedicine*, 2006. **10**(2): 395–402.

24. Graphics, Mentor. Mentor Graphics and TSMC collaborate to deliver IC design and signoff infrastructure for 10 nm. Available from: http://www.mentor.com/company/news/mentor-tsmc-design-10nm.
25. Merritt, R., TSMC Preps 10 nm, Tunes 16 nm. 2015. http://www.eetimes.com/document.asp?doc_id=1327725, Accessed on March 2019.
26. Bohr, M. 14 nm process technology: Opening new horizons. *Intel Development Forum*. 2014. https://www.intel.com.tw/content/dam/www/public/us/en/documents/technology-briefs/bohr-14nm-idf-2014-brief.pdf, Accessed on March 2019.
27. Wu, Q., J. Zhang and Q. Qiu. Design considerations for digital circuits using organic thin film transistors on a flexible substrate. in *2006 IEEE International Symposium on Circuits and Systems*. 2006. IEEE.
28. Takeda, Y. et al. Fabrication of ultra-thin printed organic TFT CMOS logic circuits optimized for low-voltage wearable sensor applications. *Scientific Reports*, 2016. **6**: 25714.
29. Noh, J. et al. Key issues with printed flexible thin film transistors and their application in disposable RF sensors. *Proceedings of the IEEE*, 2015. **103**(4): 554–566.
30. Khan, S. and M.M. Hussain. IoT enabled plant sensing systems for small and large scale automated horticultural monitoring. in *2019 IEEE 5th World Forum on Internet of Things (WF-IoT)*. 2019. IEEE.
31. El-Atab, N. et al. Bi-facial substrates enabled heterogeneous multi-dimensional integrated circuits (MD-IC) for Internet of Things (IoT) applications. *Advanced Engineering Materials*, 2019: 1900043.
32. Fletcher, R.R., M.-Z. Poh and H. Eydgahi. Wearable sensors: Opportunities and challenges for low-cost health care. in *Engineering in Medicine and Biology Society (EMBC), 2010 Annual International Conference of the IEEE*. 2010. IEEE.
33. Khan, S.M. et al., CMOS enabled microfluidic systems for healthcare based applications. *Advanced Materials*, 2018. **30**(16): 1705759.
34. Shacham, O. and M. Reynders. Pixel Visual Core: Image processing and machine learning on Pixel 2. [cited March 8, 2018]; Available from: https://www.blog.google/products/pixel/pixel-visual-core-image-processing-and-machine-learning-pixel-2/.
35. Low, C. A dedicated AI chip is squandered on Huawei's Mate 10 Pro. https://www.engadget.com/2017/11/10/huawei-s-mate-10-p-review-ai-chip/, Accessed on March 2019.
36. Abenstein, J.P. and W.J. Tompkins. A new data-reduction algorithm for real-time ECG analysis. *IEEE Transactions on Biomedical Engineering*, 1982(1): 43–48.
37. Cho, G.-Y., S.-J. Lee and T.-R. Lee. An optimized compression algorithm for real-time ECG data transmission in wireless network of medical information systems. *Journal of Medical Systems*, 2015. **39**(1): 161.
38. Deepu, C.J. and Y. Lian. A joint QRS detection and data compression scheme for wearable sensors. *IEEE Transactions on Biomedical Engineering*, 2015. **62**(1): 165–175.
39. Manikandan, M.S. and S. Dandapat. Wavelet-based electrocardiogram signal compression methods and their performances: A prospective review. *Biomedical Signal Processing and Control*, 2014. **14**: 73–107.
40. Suo, Y. et al. Energy-efficient multi-mode compressed sensing system for implantable neural recordings. *IEEE Transactions on Biomedical Circuits and Systems*, 2014. **8**(5): 0–0.
41. Zhang, J. et al. Energy-efficient ECG compression on wireless biosensors via minimal coherence sensing and weighted ℓ_1 minimization reconstruction. *IEEE Journal of Biomedical and Health Informatics*, 2015. **19**(2): 520–528.
42. Casson, A.J. Opportunities and challenges for ultra low power signal processing in wearable healthcare. in *Signal Processing Conference (EUSIPCO), 2015 23rd European*. 2015. IEEE.
43. Pantelopoulos, A. and N.G. Bourbakis. A survey on wearable sensor-based systems for health monitoring and prognosis. *IEEE Transactions on Systems, Man, and Cybernetics, Part C (Applications and Reviews)*, 2010. **40**(1): 1–12.
44. Pyattaev, A. et al. Communication challenges in high-density deployments of wearable wireless devices. *IEEE Wireless Communications*, 2015. **22**(1): 12–18.

45. Seneviratne, S. et al. A survey of wearable devices and challenges. *IEEE Communications Surveys & Tutorials*, 2017. **19**(4): 2573–2620.
46. Nassar, J.M. et al. Compliant lightweight non-invasive standalone "Marine Skin" tagging system. *npj Flexible Electronics*, 2018. **2**(1): 13.
47. Shaikh, S.F. et al. Noninvasive featherlight wearable compliant "Marine Skin": Standalone multi-sensory system for deep-sea environmental monitoring. *Small*, 2019. **15**(10): 1804385.
48. Nassar, J.M. et al. Compliant plant wearables for localized microclimate and plant growth monitoring. *npj Flexible Electronics*, 2018. **2**(1): 24.
49. Khan, S.M. et al. Flexible lightweight CMOS-enabled multisensory platform for plant microclimate monitoring. *IEEE Transactions on Electron Devices*, 2018. **65**(11): 5038–5044.
50. Khan, S.M., N. Qaiser and M.M. Hussain. Do-It-Yourself (DIY) based flexible paper sensor based electronic system for pill health monitoring. in *2018 International Flexible Electronics Technology Conference (IFETC)*. 2018. IEEE.
51. Khan, S. and M.M. Hussain. Low-cost foil based wearable sensory system for respiratory sound analysis to monitor wheezing. in *2019 IEEE 16th International Conference on Wearable and Implantable Body Sensor Networks (BSN)*. 2019. IEEE.
52. Khan, S. and M.M. Hussain. Integration strategy for standalone compliant interactive systems for add-on IoT based electronics. in *2019 IEEE 5th World Forum on Internet of Things (WF-IoT)*. 2019. IEEE.
53. Khan, S., N. Qaiser and M.M. Hussain. An inclinometer using movable electrode in a parallel plate capacitive structure. *AIP Advances*, 2019. **9**(4): 045118.
54. Gradne, L., *Solid-State and Polymer Batteries 2017–2027: Technology, Markets, Forecasts.* Available online: http://www.idtechex.com/research/reports/solid-state-and-polymer-batteries-2017-2027-technology-markets-forecasts-000498.asp, Accessed on January 2019.
55. Nishide, H. and K. Oyaizu. Toward flexible batteries. *Science*, 2008. **319**(5864): 737–738.
56. Zhou, G., F. Li and H.-M. Cheng. Progress in flexible lithium batteries and future prospects. *Energy & Environmental Science*, 2014. **7**(4): 1307–1338.
57. Hu, L. et al. Thin, flexible secondary Li-ion paper batteries. *ACS Nano*, 2010. **4**(10): 5843–5848.
58. Li, N. et al. Flexible graphene-based lithium ion batteries with ultrafast charge and discharge rates. *Proceedings of the National Academy of Sciences*, 2012. **109**(43): 17360–17365.
59. Koo, M. et al. Bendable inorganic thin-film battery for fully flexible electronic systems. *Nano Letters*, 2012. **12**(9): 4810–4816.
60. Park, M.H. et al. Flexible dimensional control of high-capacity Li-ion-battery anodes: From 0D hollow to 3D porous germanium nanoparticle assemblies. *Advanced Materials*, 2010. **22**(3): 415–418.
61. Fernández, M.R., E.Z. Casanova and I.G. Alonso. Review of display technologies focusing on power consumption. *Sustainability*, 2015. **7**(8): 10854–10875.
62. Hack, M.G. et al. Flexible low-power-consumption OLED displays for a universal communication device. in *Cockpit Displays X*. 2003. International Society for Optics and Photonics.
63. Chan, Y.H. et al. Comparative performance of gold wire bonding on rigid and flexible substrates. *Journal of Materials Science: Materials in Electronics*, 2006. **17**(8): 597–606.
64. Schafft, H.A. *Testing and Fabrication of Wire-Bond Electrical Connections—A Comprehensive Survey*. 1972, National Bureau of Standards. Washington, DC, Electronic Technology Div.
65. Lai, Y.H. et al. 4.2: Study of ACF bonding technology in flexible display module packages. in *SID Symposium Digest of Technical Papers*. 2015. Wiley Online Library.
66. Chang, S.-M. et al. Characteristic study of anisotropic-conductive film for chip-on-film packaging. *Microelectronics Reliability*, 2001. **41**(12): 2001–2009.
67. Shaikh, S.F. et al. Modular lego-electronics. *Advanced Materials Technologies*, 2017. **3**: 1700147.
68. Khorramdel, B., T.M. Kraft and M. Mäntysalo. Inkjet printed metallic micropillars for bare die flip-chip bonding. *Flexible and Printed Electronics*, 2017. **2**(4): 045005.
69. Stein, P. Why We Wear Wellness Wearables on Our Wrist. [cited December 3, 2018]; Available from: https://blog.philipstein.com/wear-wellness-wearables-wrist/.

70. de Arriba-Pérez, F., M. Caeiro-Rodríguez and J.M. Santos-Gago. Collection and processing of data from wrist wearable devices in heterogeneous and multiple-user scenarios. *Sensors*, 2016. **16**(9): 1538.
71. Gargiulo, G. et al. An ultra-high input impedance ECG amplifier for long-term monitoring of athletes. *Medical Devices (Auckland, NZ)*, 2010. **3**: 1.
72. Atallah, L. et al. Sensor positioning for activity recognition using wearable accelerometers. *IEEE Transactions on Biomedical Circuits and Systems*, 2011. **5**(4): 320–329.
73. Chen, J. et al. Wearable sensors for reliable fall detection. in *Engineering in Medicine and Biology Society, 2005. IEEE-EMBS 2005. 27th Annual International Conference of the*. 2006. IEEE.
74. Li, Q. et al. Accurate, fast fall detection using gyroscopes and accelerometer-derived posture information. in *Wearable and Implantable Body Sensor Networks, 2009. BSN 2009. Sixth International Workshop on*. 2009. IEEE.
75. Noury, N., et al. Fall detection-principles and methods. in *Engineering in Medicine and Biology Society, 2007. EMBS 2007. 29th Annual International Conference of the IEEE*. 2007. IEEE.
76. Rougier, C. and J. Meunier. Fall detection using 3d head trajectory extracted from a single camera video sequence. *Journal of Telemedicine and Telecare*, 2005. **11**(4): 37–42.
77. Yu, X. Approaches and principles of fall detection for elderly and patient. in *e-health Networking, Applications and Services, 2008. HealthCom 2008. 10th International Conference on*. 2008. IEEE.
78. Parkka, J. et al. Activity classification using realistic data from wearable sensors. *IEEE Transactions on Information Technology in Biomedicine*, 2006. **10**(1): 119–128.
79. Sojuyigbe, S. and K. Daniel. Wearables/IOT devices: Challenges and solutions to integration of miniature antennas in close proximity to the human body. in *Electromagnetic Compatibility and Signal Integrity, 2015 IEEE Symposium on*. 2015. IEEE.
80. Bao, L. and S.S. Intille. Activity recognition from user-annotated acceleration data. in *International Conference on Pervasive Computing*. 2004. Springer.
81. Banaee, H., M.U. Ahmed and A. Loutfi. Data mining for wearable sensors in health monitoring systems: A review of recent trends and challenges. *Sensors*, 2013. **13**(12): 17472–17500.
82. Mercanzini, A. et al. Demonstration of cortical recording using novel flexible polymer neural probes. *Sensors and Actuators A: Physical*, 2008. **143**(1): 90–96.
83. Viventi, J. et al. Flexible, foldable, actively multiplexed, high-density electrode array for mapping brain activity in vivo. *Nature Neuroscience*, 2011. **14**(12): 1599.
84. Mescher, A.L. *Junqueira's Basic Histology: Text and Atlas*. 2013. Mcgraw-Hill Education, New York.
85. Quigley, H.A. Number of people with glaucoma worldwide. *British Journal of Ophthalmology*, 1996. **80**(5): 389–393.
86. Marnat, L. et al. On-chip implantable antennas for wireless power and data transfer in a glaucoma-monitoring SoC. *IEEE Antennas and Wireless Propagation Letters*, 2012. **11**: 1671–1674.
87. Xu, L. et al. 3D multifunctional integumentary membranes for spatiotemporal cardiac measurements and stimulation across the entire epicardium. *Nature Communications*, 2014. **5**: ncomms4329.
88. Nassar, J.M. et al. Paper skin multisensory platform for simultaneous environmental monitoring. *Advanced Materials Technologies*, 2016. **1**(1): 1600004.
89. Nassar, J.M. et al. A CMOS-compatible large-scale monolithic integration of heterogeneous multi-sensors on flexible silicon for IoT applications. in *Electron Devices Meeting (IEDM), 2016 IEEE International*. 2016. IEEE.
90. Mukhopadhyay, S.C. Wearable sensors for human activity monitoring: A review. *IEEE Sensors Journal*, 2015. **15**(3): 1321–1330.
91. Williamson, J., et al. Data sensing and analysis: Challenges for wearables. in *Design Automation Conference (ASP-DAC), 2015 20th Asia and South Pacific*. 2015. IEEE.
92. Reddy, S. et al. Using mobile phones to determine transportation modes. *ACM Transactions on Sensor Networks (TOSN)*, 2010. **6**(2): 13.

93. Tapia, E.M. et al. Real-time recognition of physical activities and their intensities using wireless accelerometers and a heart rate monitor. in *Wearable Computers, 2007 11th IEEE International Symposium on*. 2007. IEEE.
94. McAdams, E. et al. Wearable sensor systems: The challenges. in *Engineering in Medicine and Biology Society, EMBC, 2011 Annual International Conference of the IEEE*. 2011. IEEE.
95. Petrov, D. Can the Galaxy S9 measure your blood pressure? We put its new optical sensor to the test. https://www.phonearena.com/news/We-put-the-new-Galaxy-S9-optical-pulse-reader-to-the-test-as-blood-pressure-monitor_id103241, Accessed on April 2019.
96. Chan, M. et al. Smart wearable systems: Current status and future challenges. *Artificial Intelligence in Medicine*, 2012. **56**(3): 137–156.
97. Hill, J.W. and P. Powell. The national healthcare crisis: Is eHealth a key solution? *Business Horizons*, 2009. **52**(3): 265–277.
98. Munos, B. et al. Mobile health: The power of wearables, sensors, and apps to transform clinical trials. *Annals of the New York Academy of Sciences*, 2016. **1375**(1): 3–18.
99. Zhou, J. et al. Security and privacy in cloud-assisted wireless wearable communications: Challenges, solutions, and future directions. *IEEE Wireless Communications*, 2015. **22**(2): 136–144.
100. Tesconi, M. et al. Wearable sensorized system for analyzing the lower limb movement during rowing activity. in *Industrial Electronics, 2007. ISIE 2007. IEEE International Symposium on*. 2007. IEEE.
101. Ahmadi, A. et al. Automatic activity classification and movement assessment during a sports training session using wearable inertial sensors. in *Wearable and Implantable Body Sensor Networks (BSN), 2014 11th International Conference on*. 2014. IEEE.
102. Gastin, P.B. et al. Quantification of tackling demands in professional Australian football using integrated wearable athlete tracking technology. *Journal of Science and Medicine in Sport*, 2013. **16**(6): 589–593.
103. Glaros, C. et al. A wearable intelligent system for monitoring health condition and rehabilitation of running athletes. in *Information Technology Applications in Biomedicine, 2003. 4th International IEEE EMBS Special Topic Conference on*. 2003. IEEE.
104. Chardonnens, J. et al. Automatic measurement of key ski jumping phases and temporal events with a wearable system. *Journal of Sports Sciences*, 2012. **30**(1): 53–61.
105. Düking, P. et al. Comparison of non-invasive individual monitoring of the training and health of athletes with commercially available wearable technologies. *Frontiers in Physiology*, 2016. **7**: 71.
106. Chambers, R. et al. The use of wearable microsensors to quantify sport-specific movements. *Sports Medicine*, 2015. **45**(7): 1065–1081.
107. Cardinale, M. and M.C. Varley Wearable training-monitoring technology: applications, challenges, and opportunities. *International Journal of Sports Physiology and Performance*, 2017. **12**(Suppl 2): S2-55–S2-62.
108. Marshall, R. and R. Ferdinands. Cricket: The effect of a flexed elbow on bowling speed in cricket. *Sports Biomechanics*, 2003. **2**(1): 65–71.
109. Marshall, R.N. and R. Ferdinands. The biomechanics of the elbow in cricket bowling. *International SportMed Journal*, 2005. **6**(1): 1–6.
110. Saini, P., et al. Philips stroke rehabilitation exerciser: a usability test. *Proceedings of the IASTED International Conference on Telehealth/Assistive Technologies, ser. Telehealth/AT*, 2008. **8**: 116–122.
111. Thielgen, T. et al. Tremor in Parkinson's disease: 24-hr monitoring with calibrated accelerometry. *Electromyography and Clinical Neurophysiology*, 2004. **44**(3): 137–146.
112. Paquet, J. et al. Analysis of gait disorders in Parkinson's disease assessed with an accelerometer. *Revue Neurologique*, 2003. **159**(8–9): 786–789.
113. Weiss, A. et al. Can an accelerometer enhance the utility of the Timed Up & Go Test when evaluating patients with Parkinson's disease? *Medical Engineering & Physics*, 2010. **32**(2): 119–125.

20

Flexible Electronic Technologies for Implantable Applications

Sohail Faizan Shaikh

20.1 Introduction

The majority of the current medical treatments are focused towards "fixing" the patients during the evolution of their diseases to a significant level leading to detrimental consequences on patients both physically and mentally. It would not be wrong to accept that we fail to take care of our body as much as we do care for our owned vehicles. Periodic check-up of the vehicles result in cheaper maintenance in addition to significantly prolonging the life of the vehicle by virtue of improving its quality. A question then arises, why don't we keep track of our body and pay attention to maintenance far beyond our vehicles? One of the prevalent reasons is lack of tools or gauges to continuously or intermittently monitor our health condition. We do not have sophisticated tools which direct us personally to take more sleep, rehydrate, eat healthier, reduce the stress levels, or when to visit a doctor for preventive measures. People often rely on their feelings and other underlying or obvious physical signals, pain, etc., that trigger an alarm for visiting a doctor, who normally performs diagnostics based on pathological or other electronic diagnostic equipment available at the centre. These devices mainly gather a few data points from which a medical condition is inferred and then treatment or control measures begin. Often the visits to the doctors are at the stages when the disease is already progressed and create many complications which eventually increase the costs, treatment time, and difficulties. Thus, if we have the tools that monitor the body vitals and health conditions on a regular basis, we can take preventive measures to lead an improved and healthier life. The data points required can be provided by tiny or micro implantable devices that can be used to monitor wellness and early diagnosis of any upcoming ailment.

The human body and biomedical healthcare has always been the frontier area since the advent of miniaturized state-of-the-art semiconducting technologies—complementary metal-oxide semiconductor (CMOS).

CMOS technology has matured over decades to provide the information of the universe all in our palm, blinks away from accessing. For decades now, electronics technology has been used for biomedical applications in different ways, from diagnostic instruments, to treatments and surgical tools. In addition, implantable devices have also been in practice for more than 6 decades, although earlier implantable devices do not necessarily involve electronic components. A medical device that is intended to remain inside the body after being inserted partly or completely when a procedure is performed by surgical or medical means can be termed as an implantable device. A significant percentage of the world population has been using implantable medical devices for expanding longevity, achieving better life quality, and regaining body functions after impairments. Since the first implantable device—a cardiac pacemaker implanted in 1958, tremendous efforts have been put towards development of advanced medical devices such as implantable cardiac defibrillator, pacemakers, cochlear implants, wireless pressure sensors, glucose monitoring, deep-brain simulators (DBSs), etc. Over the last 60 years, implantable electronic systems and devices have undergone a significant transformation, becoming a valuable biomedical tool for monitoring, measuring, and soliciting physiological responses *in vivo* (Bazaka and Jacob 2012; Pang et al. 2013; Gutruf and Rogers 2018).

The invention and subsequent advancement of implantable devices fundamentally rely heavily on the growing knowledge regarding various aspects of the human physiology and neuro-motor system and the tremendous achievements of electronics technologies that have intercepted every aspect of our life, capable of interfacing with living tissues and organs at micro and nanoscale. Increased *in vivo* stability, miniaturization, and lower energy requirements of modern electronics led to a multitude of miniature wireless electronic devices, such as sensors, intelligent gastric and cardiac pacemakers, cochlear implant, implantable cardioverter defibrillators, and deep-brain, nerve, and bone stimulators being implanted in patients worldwide. Furthermore, ultra-miniaturization, well matured CMOS technologies, and advancing design and packaging techniques in addition to improvements in low power electronics have driven the onset of the wearable and implantable real-time monitoring devices that can sense a physiological response and trigger decisive events by actuating a certain stimuli to a specific organ. For example, diseases with neural origin currently do not have any chemical or drug therapies, neuro-stimulation is the only existing stimulation today. Incidentally, electrical stimulation of certain regions of the brain have remarkable responsivity in treating debilitating effects of chronic disorders like essential tremor, Parkinson's disease, major depression, Tourette syndrome, and chronic pain without causing any permanent damage to physiological or anatomical structure of the relevant organ (Poon 2014; Miller et al. 2015; Fekete and Pongrácz 2017; Keskinbora and Keskinbora 2018).

Thus, implantable real-time monitoring can potentially shift the traditional medical system from fixing/repairing the diseased portion after the diagnosis, to controlling the onset of the disease before it triggers and take decisive preventive measures. People who are genetically predisposed to or those who have already developed certain medical conditions can also benefit from diagnostic and therapeutic implantable devices. A very widely successful example for such device in practice is a Cardiac pacemaker—widely used implantable therapeutic device that has a tremendous impact on prolonging lives of people with chronic heart diseases.

In this chapter, we discuss different types of implantable devices in their current commercial rigid forms, current cutting-edge research, and what necessitates emerging compliant devices towards advanced flexible high-performing devices. In addition, we also discuss the recent advances in different proposed implantable devices for a multitude of applications from brain-mapping, retinal implants, to neural interfaces/prosthetics and their challenges and outlook.

20.2 Present State of Implants

Implantable medical devices have had a presence of over 6 decades in the commercial and medical field in different forms. Here, we present the devices which have been available in commercial or are recently emerging as technologies for future applications depending on their deployment inside the body. The devices presented are cardiac implants (pacemakers and defibrillators), retinal implants, brain interfacing implants, and other miscellaneous applications like glucose monitoring, bladder, and cochlear implants, etc.

20.2.1 Cardiac Implants

The heart is a small four-chambered muscular lump that receives and pumps blood to the systemic and pulmonary circulatory system of the human body. The pumping mechanism of the heart is regulated by cardiac muscle cells that are of two kinds—pacemaker cells and non-pacemaker cells. The pacemaker cells have no resting cell membrane potential, but depolarize instantaneously, whereas non-pacemaker cells have resting potential that depolarizes quickly on application of electrical stimuli. Pacemaker cells in the heart are responsible for causing the heart muscles to rhythmically contract and relax at regular intervals, providing the necessary pumping force to pump blood around the circulating system of body. Electrocardiogram (ECG) measurements allow physicians to have a closer look at the patients' heart, and it can be used to detect arrhythmias and heart attacks (myocardial infarctions). For instance, a leadless cardiac pacemaker has been presented by Reddy et al. which has notable extremely low power consumption of 64 nW, which raises the bar in terms of power budgets (Miller et al. 2015; Neuzil and Reddy 2015; Sideris et al. 2017). Nevertheless, it does not allow for continuous monitoring, as it only stores abnormal events into the memory for posterior wireless relaying. Status-quo of implantable devices used for continuous monitoring are discussed in later sections.

20.2.1.1 Pacemakers and Cardioverter Defibrillators

Irregular heart rhythms also known as cardiac arrhythmias need significant attentions in a timely manner as the chronic condition leads to weakening of the cardiac muscles over a prolonged time. A regular heartbeat rhythm in adults is considered between 60 and 80 bpm, and divergence from these numbers on a consistent basis are taken care of by usage of pacemakers and implantable cardioverter defibrillators (ICDs) that restore the rhythm (Bazaka and Jacob 2012; Lee 2014). Implanted electrodes pave a path for providing an electrical impulse to heart muscles which effectively brings back the pace of the heart, and these are achieved by pacemakers that sense and provide a stimuli for conditions like bradycardia arrhythmias (slow rhythms) and impulse conduction blocks in one or more chambers of the heart.

Dual chamber pacemakers are typically used in significant atrioventricular (AV) blocks, in which one lead is inserted in the right atrium and right ventricle each. It monitors intrinsic activity and delivers the impulses to individual or both chambers as needed. Biventricular pacing, also known as cardio resynchronization therapy (CRT) uses an electrical stimulus to provide normal and more balanced contractions of the ventricles in the patients with heart failure cases. All modern pacemakers have the ability to continuously monitor electrical activity of the heart and deliver artificial stimuli on demand to pace the heart. These pacemakers are surgically inserted in a subcutaneous pouch located in the upper chest region below collarbone, while the electrode is normally landed in the heart chamber through cephalic or subclavian veins. These rigid devices typically use a small battery to keep powering over the years which needs to be replaced by another surgery in few years.

ICDs are another type of cardiac implant used to detect and correct for tachycardia arrhythmias (fast rhythms) by applying a low-energy, high-rate electrical stimulus or a high-energy shock to the heart muscle. Modern state-of-the-art ICDs are tiny and incorporate pacemaker functionality together with sophisticated techniques, providing monitoring of electrical activities of the heart, providing necessary stimuli shocks while storing and transmitting ECG data using wireless communication. ICDs differ from pacemakers in that they sense for the ventricular arrhythmias; ventricular tachycardia (VT) and ventricular fibrillation (VF), both of which are life threatening and associated with sudden death (Alhammad et al. 2016; Cha et al. 2017) (Figure 20.1).

20.2.2 Retinal Implants

Glaucoma—a progressive disease that eventually can lead to sight loss, is a result of an increasing internal pressure of eyeball that can damage the optic nerve at the back of the eye and have irreversible effects. Onset of such disease can be managed with different treatments if detected at an early stage. Early diagnosis of the disease and avoiding the danger of sight loss is possible by measuring the intraocular pressure (IOP) of the

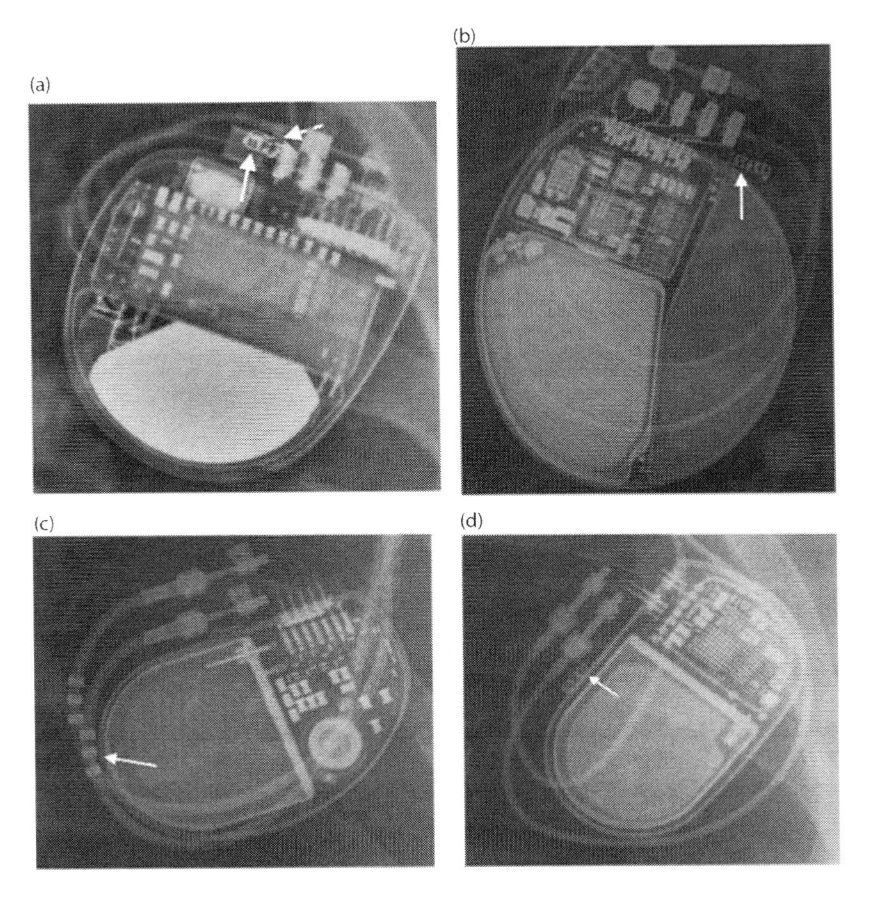

FIGURE 20.1 Different commercial cardiac implants over the years from leading organizations. (a) Radiograph of the pacemaker Adapta™ from Medtronic, inside the heart and (b) Viva® XT CRT-D (Copyright © Medtronic. Reprinted with permission.). (c) Evia® DR-T pacemaker. (Copyright © Biotronik. Reprinted with permission.), and (d) Unify Quadrat™ CRT-D system. (Copyright © St Jude Medical. Reprinted with permission.)

eye (Fitzpatrick 2015c). Different approaches for measuring IOP are possible and range from non-invasive devices, such as contact lenses, smart lenses that measure the deformation of the cornea curvature due to the extra pressure, to invasive, implantable sensors that directly measure the IOP inside the eye. The device presented can be viewed as the state-of-the-art of IOPs which occupies 1.5 mm^2 sensor area with wireless communication capabilities requiring only 7 μW power (Lewis et al. 2016). Other examples of such devices are the Triggerfish IOP sensor from Sensimed—a contact lens that is designed to monitor IOP changes in the eye over a 24 hour period by detecting changes in the curvature of the cornea. A microfabricated strain gauge sensor, antenna, and dedicated telemetry circuits (application specific integrated circuit [ASIC]) are embedded into the soft contact lens, arranged in such a way that they do not interfere with normal vision Figure 20.2 (Fitzpatrick 2015d). One must reflect that retinal implants will not benefit people who have blindness from birth, because their optical circuit and visual processing centers are not synchronized and conditioned to perceive the vision. Inner nerve cells of the retina are stimulated electrically in an ordered pattern by using an array of electrodes implanted in the retina. Micro photodiodes can be used to convert the incident light energy of images on the retina into the electrical stimuli as the lens of the eye is functional. Symmetrical alignment of micro photodiodes and microelectrode arrays (MEAs) effectively bypass the outer damaged photoreceptor cells and stimulate inner cells directly. These retinal prostheses can also be toned to evoke a pattern of light

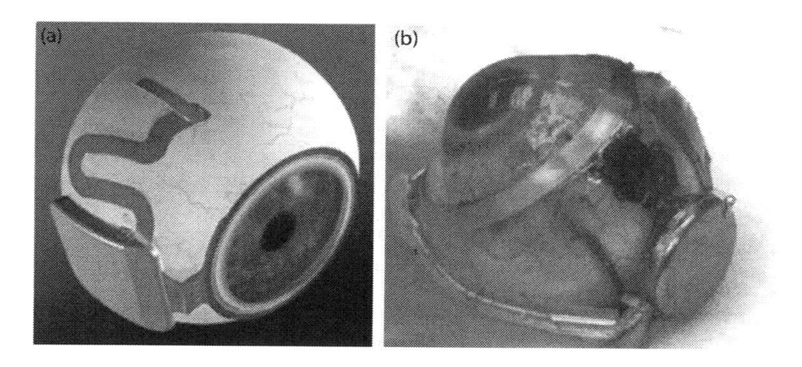

FIGURE 20.2 Boston retinal implant with enclosed receiver coil and electrode array in (a) model concept, (b) after deployment on a human eye, and (c) detailed components of the implant (Adapted with permission from Kelly et al. 2013)

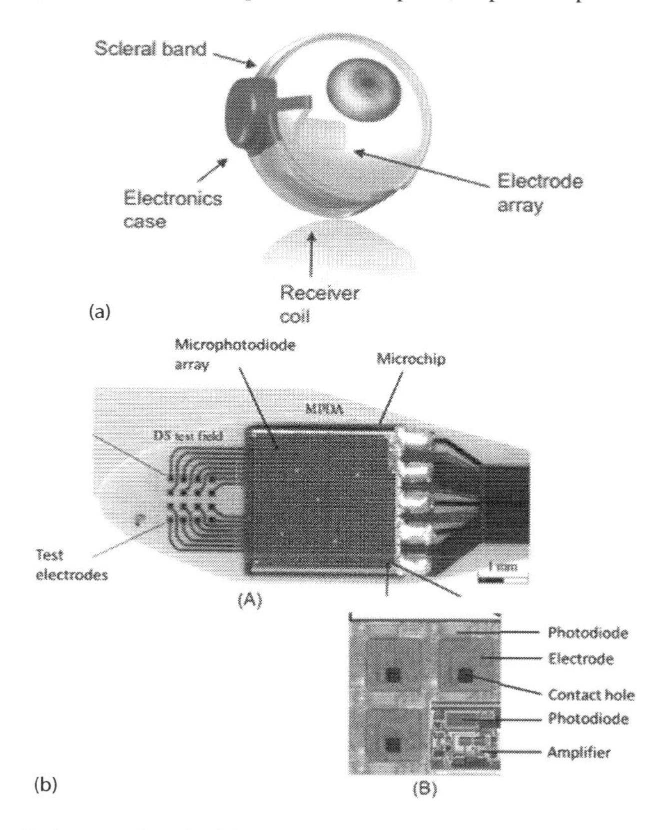

FIGURE 20.3 Digital photographs of Alpha-IMS prosthesis (a) microchip and simulation electrodes, and (b) imaging elements of micro-photodiodes. (Copyright © Retina Implant AG. Reprinted with permission)

dots representing Braille characters by conditioning electrical impulses. Figure 20.3a and b shows different retinal implants developed and under research by different organizations (Lewis et al. 2016).

20.2.3 Brain Implants

The brain is the single most complex organ of the human body, it controls all the activities through electrical signals often known as neural or synaptic signals. These signals have a specific path and run throughout the body within interweaved path of neurons. The brain is often termed as the powerhouse of the body mainly because of the fact that it consumes a lot of blood oxygen which is the main source of power through the mitochondria. However, when it comes to the processing of the actions, making decisions, analyzing the objects, or any other activity, it has the best performance in terms of power consumption for any kind of data processing, a normalized power consumption per task/activity. Neurons generate conductive electrical signals forming the main component of the nervous system which helps us in making decisions, move our bodies, and think. The term action potential (AP) refers to these electrical signals. On average, over 100 billion neurons are working in a synchronous manner to form more than 100 trillion synapses. A synaptic signal is transferred from the point of generation (neurons) and spreads through the cell membrane via axons and is relayed to another neuron at the synaptic junction using neurotransmitters. Scientists have been spending tremendous effort to understand and unravel these networks and neural circuits to learn about how our brain communicates and makes decisions. For any such activity, interrogation of a neural circuit is performed by measuring AP from multiple sites related to a specific command or function of the brain that requires an array of MEAs. These MEAs can also address individual neurons for acquiring single neuron AP, which is a direct measurement technique in contrast to measurement of extra-cellular field potential measurements. Extracellular recording techniques give an advantage of reduced damage to a single cell and allow prolonged signal acquisition making it a potential technique of investigating variation in signaling or firing pattern during development of a command from the brain to the organ.

20.2.3.1 Electrodes

Lately, the concept of electroceuticals which essentially means stimulation of the peripheral nervous system to modulate the organ physiology and treatment of neurodegenerative diseases like rheumatoid arthritis, Parkinson's disease, and other sensory motor related issues has evolved (Lacour et al. 2016). Broadly speaking, the recording electrodes for brain activity mapping can be classified in three categories depending on the location of deployment as: non-invasive electroencephalography (EEG), extra-cortical electrocorticography (ECoG), and invasive intracortical electrodes or MEAs. MEAs are used for micro-stimulation by injecting charges through electrodes at the same time they measure extracellular field potentials generated as a result of membrane currents and AP. Figure 20.4 represents bulky DBSs from different medical device makers which are commercially used for the advanced treatment of Parkinson's disease and dystonia for electrical stimulation therapy of the subthalamic nucleus or globous pallidus region of the brain. For all brain implants, conventional electrodes are implanted in the brain and an extension of the leads travels subcutaneously to the neuro-stimulator which is surgically

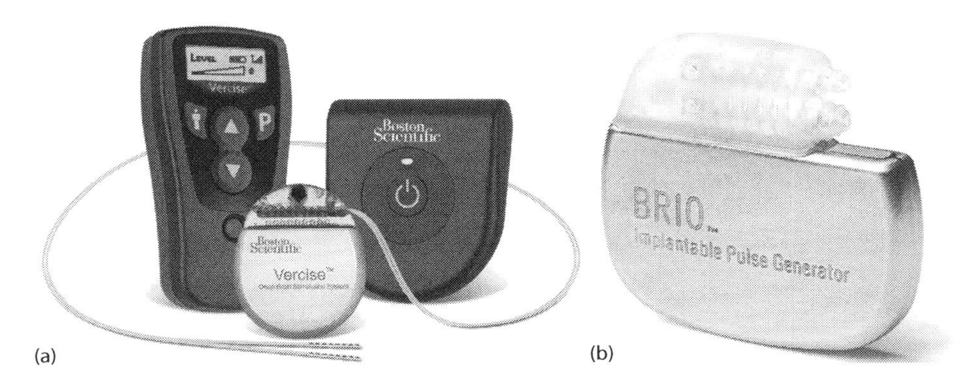

FIGURE 20.4 Rigid and bulky implantable neural devices. (a) Vercise™ DBS (Copyright © Boston Scientific. Reprinted with permission.) and (b) Brio™ Neurostimulator from St. Jude Medicals (Copyright © St Jude Medical. Reprinted with permission.)

inserted in the chest close to clavicle subcutaneously. These kinds of implants allow control of external stimulation, keeping track of the battery, and monitoring the condition manually.

A good variety of electrodes have been developed in the past for use with brain implantable devices and have continued to be explored as technology advances over time. The prominent types of neural recording electrodes include: (i) cuff electrodes that surround the nerve completely with an insulator and coaxial electrode in direct contact with the nerve, they provide minimum damage to the nerve as there is no puncture of nerves. (ii) Flat interface nerve electrodes (FINE) and longitudinal implanted intrafascicular electrodes (LIFE) are variants of cuff electrodes that are Teflon-insulated, platinum-iridium wires recording from sites at the end of the wire inserted parallel to the fascicle inside nerve (Fitzpatrick 2015b). Electrode designs that lack flexibility cause increased inflammation, hence, (iii) transverse intrafascicular multichannel electrodes record from multiple sites simultaneously which can be fabricated on a flexible polyimide shank using semiconductor processes hosting multiple conducting electrodes. (iv) ECoGs consist of a flexible grid of a planar electrode array placed directly on a surface of the brain, residing within subdural space not penetrating the brain. Localization of epileptogenic foci during surgeries for severe diseases is extremely critical, and, in such cases, ECoGs are temporarily implanted in a patient to aid surgery. Using a 32-electrode ECoG and associated training, successful preliminary demonstrations of computer mouse control in three-dimensional space by differentiating the upper limb of a paralyzed patient was shown (Fitzpatrick 2015b).

Unlike non-invasive EEG and ECoG, signals acquired from close proximity to the neurons provide more accurate information with less noise and localized single-unit recording than is feasible by using intracortical electrodes that directly penetrate the brain. Michigan electrodes, microwires, and Utah electrodes are most prominent and dominate invasive technologies in the intracortical section.

Michigan electrodes are fabricated using semiconductor processes, as an array of silicon pillars with high aspect ratio, typical length of shank is 3–5 mm, and the diameter (thickness) of electrode lies under 15 μm. *Microwire electrodes* are the oldest form of recording neural activity and are made from conductive stainless steel or tungsten core insulated with biocompatible polyamide or Teflon, and generally practiced in non-human primates. The only electrodes that secured FDA approval are *Utah electrodes* and have been implemented in the human body. These are boron-doped silicon electrodes fabricated using CMOS technologies, with the first demonstration performed on the cortex of a cat to record a neural signal for 13 months. Subsequently, it led to an international collaboration with Applied Physics Laboratory, and the first mind-controlled prosthetic for the patient of tetraplegia. In these clinical trials, neural signals were recorded for more than 1000 days, and notably the paralyzed patient was able to feed herself from a robotic arm connected to the interface using Utah electrodes (Falcone et al. 2014).

20.2.4 Miscellaneous Applications

In addition to the cardiac and brain implants, several other internal organs and body vitals need maintenance and monitoring on a timely basis. One of the most important organs for metabolism and proper functioning of the overall body is the bladder. For the diagnosis of bladder dysfunctions, bladder pressure monitoring can be used as an essential tool, more significantly due to lack of obvious symptoms that are similar to normal daily activities and go unnoticed during activities, and acute symptoms may disappear momentarily when visiting the physician. In such cases, chronic readings are necessary by means of implantable devices preferably without discomfort to patient (Ivanova et al. 2014; Fitzpatrick 2015a).

Diabetes is one of the most prevalent diseases globally, and its effects on the human body are not hidden. Glucose monitoring devices have been in the consumer market for a long time and are nowadays available as point-of-care diagnostic sticks that give instant readings by pricking the fingertips and drawing a small blood sample. For chronic monitoring and automated control of insulin pumping in the blood, implantable alternatives have been researched over the years and have been successfully demonstrated that trigger the alarm for manual injection of insulin or automatically infuse the required quantity in the blood on the onset of changed blood sugar levels (Montornes et al. 2008; Juanola-Feliu et al. 2014).

Loss of any sensory organ of the body leads to several challenges in one's life. Hearing loss is one such example where it can be either conductive, sensorineural, or a combination of both. The mechanism of hearing is through conduction of bones that transfer sound waves to the inner diaphragm and nerves translate these signals to the brain. Hearing loss can be due to age (presbycusis) which is a conductive loss as sound signals are not able to pass to the inner ear from the outer ear. The sound diminishing factors can be stiffening of the eardrum, thereby losing its elasticity, loss of mobility of conducting bones becoming rigid in their action. Whereas the damage of inner ear sensory hair cells or auditory nerve damage causes sensorineural deafness. For all such cases, cochlear implants are used externally for external applications. Preliminary work on a cochlear implant had begun in the 1960s, the first neural interface designed to revive the hearing for clinically deaf representing the first successful integration of stimulation electrodes within the peripheral nervous system (PNS) (Macherey and Carlyon 2014; Jalili et al. 2017). Cochlear implants are essentially used to bypass the sound signals from hair cells by inserting electrodes into the inner ear and directly stimulating the auditory nerve in response to the incoming audio signal from a microphone placed in the external ear area.

20.3 Flexible Implants

CMOS technology has been at the heart of the digital revolution with micro-and-nanofabrication processes inarguably the most reliable and advanced existing technologies which continue to be perfected. Living species have irregular and asymmetric contours, soft tissues, and irregular skin surfaces on which status-quo rigid and bulky electronics and ICs do not comply. Hence, a holistic approach to deploy these electronic systems on the human body or internal organs necessitates flexibility in these systems not only at the individual module (logic, processor, memory, communication, and sensing) level, but also at the entire system level (Hussain et al. 2016, 2018a; Gumus et al. 2017; Shaikh et al. 2017, 2018).

Flexible and stretchable electronics have tremendous potential to augment and enhance the quality of life by connecting people, processes, data, and devices. Significant progress has been achieved in this emerging field using naturally flexible and stretchable materials in combination with organic materials, low dimensional materials like nanowires, nanoribbons, nanotubes, graphene, and other 2D/1D materials (Lu et al. 2018; Park et al. 2018; Won et al. 2018). For continuous real-time healthcare monitoring or chronic diagnosis, implantable and wearable devices are a must which will be deployed either inside or on the soft tissues, skin, joints, neck, and other curvilinear locations that not only require flexible, conformal devices, but also stretchable to adapt to the motions of these organs (twisting, bending, contracting, or stretching). Furthermore, implantable applications present a whole new level of challenges among which biocompatibility and biodegradability are just a couple to name (Hussain and Hussain 2016; Park et al. 2017, 2018; Shaikh et al. 2017; Hussain et al. 2018b; Almuslem et al. 2019).

Soft electrodes mimicking mechanical properties like soft tissues play a pivotal role in implantable monitoring applications. The following paragraphs provide the insight on the recent advances in the field of flexible implants focusing on brain-machine interfaces with specific focus on neural interfaces and spinal cord related implantable technologies.

20.3.1 Neural Interfaces

The action potential (AP) is the main source of electrical signal from neurons that is conveyed to different parts of the neural network through synaptic junctions and neural circuits. Understanding the fundamental principle of how the human brain works and its PNS poses humongous challenges with greater intellectual gains in academic research in addition to practical importance in human healthcare. In the early 1990s, cranial electrodes have been successfully implanted to provide electrical stimulation to neurons that were coined as cranial pacemakers or DBSs. This led to subsequent advances in therapeutic modalities for epileptic and neurodegenerative disorders over the years and they have continued to improve. Concomitantly, rehabilitation became possible in otherwise untreatable conditions by means of stimuli generated by micro devices, implanted electrodes, and tiny batteries connected around the organ of interest (Rogers et al. 2016). This lead to a phenomenal rush in activities towards obtaining prosthetic devices. Optical stimulation by means of optical fibers as a simple waveguide has become an excellent mean of light transport which evolved as a multimodal probing technique to deliver light, fluid, and electrical stimulus with simultaneous *in situ* reading and computing capabilities. Furthermore, syringe injectable electronics or capsule/edible electronics has unfolded many unprecedented application areas where targeted delivery and monitoring systems become evident in the near future. These tiny devices can record electrical signals by virtue of ultrathin organic transistor arrays for *in situ* myoelectric signal recording of optogenetically evoked spikes in muscle and multifunctional electrodes for recordings in non-human primates and wireless ultrasonic powered electrical devices for electromyogram and electroneurogram measurements (Rogers et al. 2016; Gutruf and Rogers 2018).

Among neural interfacing devices, MEAs technology, materials, and processing techniques are at the forefront of runners that act as bridging between recording sites to the computational units. Experimental platforms have been recently demonstrated for electrophysiological studies on stem cells, dissociated cultures, to slices of the brain, thanks to the planar passive MEAs. These neural interfacing electrodes provide spatiotemporal confinement and recording of the signals from tens of microelectrodes providing simultaneous stimuli as well. The measurement can be non-invasive, allowing long-term recording and stimulation even for months. A huge reliance of MEAs is on electrode material such primarily platinum, gold, nanostructures, and aluminum. In addition, materials for flexible neural interfacing MEAs include conductive polymers such as PEDOT (poly 3,4-ethylenedioxythio-phene), transparent conductive oxides such as indium tin oxide (ITO), nanomaterials such carbon nanotube (CNT) or nanowire, and 2D materials such as graphene are dominant. Silicon dioxide, photoresists, dielectric polymers like polyamide, polydimethylsiloxane (PDMS), and Parylene are used as substrate and encapsulating materials for packaging of these devices in order to increase the lifetime and protection from harsh biological environment (Minev and Lacour 2016; Won et al. 2018).

Neural interfacing electrodes are typically measuring the extracellular field potential which can be modeled as a time-varying voltage source. A fully fabricated and packaged MEA interface is connected to a multichannel amplifier unit that includes signal conditioning, filtering (e.g., 300 Hz to 5 kHz for spikes, 10–200 Hz for local field potentials), and low noise amplifiers shown in Figure 20.5a. These huge data acquisitions and processing systems provide the much needed processing, however, in these systems, only a small portion of the interface is flexible and other systems are still bulky which restricts the usage for day-to-day stand-alone neural interfaces for chronic monitoring. Optical neuron stimulation can be achieved using genetically encoded light sensitive ion channels in the cell membrane, this technique is coined as optogenetic. Because communication at synapses is chemical, activity can be elicited by the targeted delivery of chemicals such as neurotransmitters released from an implanted device. Other forms of energy delivered are heat and ultrasound agitation, these have been explored as alternative routes for less invasive solutions for DBS and neural interfaces. Figure 20.5b represents flexible neural implants for deep-brain interfacing. The use of polyimide as a substrate allows the probe to bend up to a 1 mm radius without altering device performance. The electrode interconnects are made from thin metal in the neutral plane, and active sites of electrodes are fabricated using glossy carbon

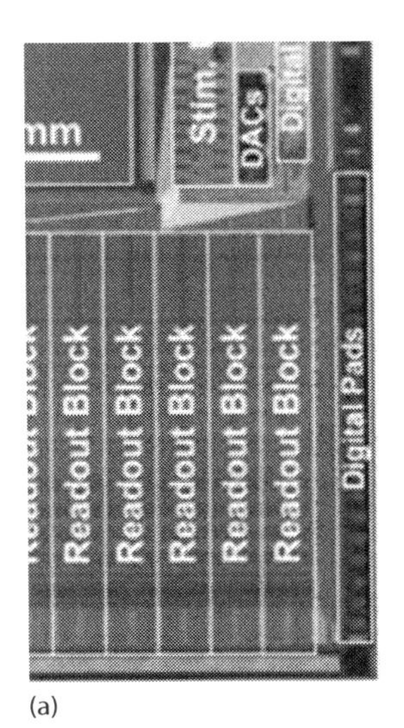

(a)

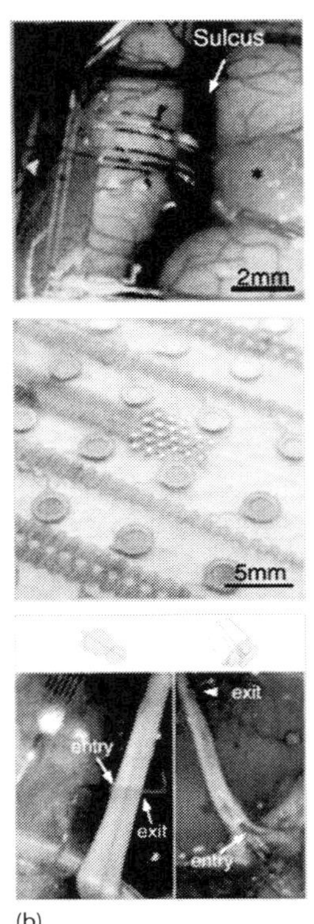

(b)

FIGURE 20.5 Neural interfaces in two different forms rigid and flexible. (a) Rigid high-density MEA neural interface showing the multichannel amplifiers, low-noise amplifiers and other building blocks of the read-out circuitry and the overall fabricated platform for culture and testing (Reproduced with permission from Ballini et al. 2014), (b) shows a schematic of different electrode types for neural probe deep brain stimulation (Reproduced with permission from Badia et al. 2011), and (c) microvessels under observation during surgery for implanting a branch of the electrocorticographic (ECoG) probe inserted into the sulcus. (Reproduced with permission from Matsuo et al. *2014*).

which can isolate electrical functionality from chemical sensing of dopamine that may be required in some diseases. A stable and reliable neural interface to the spinal cord requires maintaining a fixed relative position between tissue and implant. Lacour et al. recently demonstrated that implants with elastic properties similar to the outermost meningeal layer (dura mater) integrated seamlessly on the surface of the spinal cord in rats for extended periods of time. This enabled a functional interface closer to the target neural structures by placing the implant in the intrathecal space (over the spinal cord, but bellow dura mater), where we have demonstrated topical delivery of drugs and selective electrical stimulation (Lacour et al. 2005, 2016; Minev and Lacour 2016).

Commercially available multichannel systems have produced significant increased electrode density with 1024 channel microelectrodes systems that are used to actively multiplex 26,400 electrodes that were developed through standard photolithography. The conductor line is ITO, and the electrode material is TiN. Electrode density of 1264 electrodes per mm^2 (= 0.126 electrodes per 100 μm^2) present a case for huge potential recording of individual neural spikes which have later been shown to further increase density (Ballini et al. 2014). The line width and electrode size are the limiting factors in coplanar increasing density of MEAs, while multilayer 3D interconnects can be approached for further developments as proposed by Shaikh et al. (Hussain et al. 2016; Shaikh et al. 2017, 2018; Hussain et al. 2018a, 2018b).

Intraneural interfaces have been reported to be inserted transversally or longitudinally with respect to the long axis of the nerve and rely on the formation of a fibrous capsule that binds nerve and implant together. Inability to transect nerves poses a definite requirement on conformability of implants. Similarly, peripheral nerve implants rely on nerve regeneration, flexible 2D materials allow rolling of structure into 3D implant for guidance, and interfacing for regenerating exons. Integration of these neural active or passive electrodes for applications beyond incorporating stimulating signal and recording electrodes holds the key for future developments.

20.4 Challenges

Establishing the interface of communication with received signals has great challenges. Properties of the received signal like non-linearity, noise, instabilities, small data set, and dimensionality are just a few issues related to recorded signals. For brain implants, EEG signals can be better characterized by non-linear dynamic methods while non-stationary attributes of brain signals present issues on brain computer interfaces (Mavoori et al. 2005; Falcone et al. 2014; Abdulkader et al. 2015; Hughes 2016; Won et al. 2018). Several general design considerations to be addressed are size, weight, reliability, biocompatibility, minimal toxicity, high data rate, and low power consumption. Minimum invasiveness is the significant challenge when high signal quality and high spatiotemporal resolution are needed, however, for implants it is much less of a concern. For the long life-time of the device and safety of the patient, low power consumption is an inevitable condition. Heat dissipation due to high density electrodes in close proximity to living tissues demands strict restrictions on the amount of dissipation in powering an implanted electronic system which can inflict damage onto these soft tissues can have. In battery-less devices that are powered by a radio-frequency (RF) link, low power restrictions also apply to enforce electromagnetic energy radiation or backscattering during communication in line with IEEE exposure standards.

A flexible electronic system that is composed of an electrode interface, signal conditioning, communication, power management either through battery or wireless energy harvesting, and stimulation, faces major challenges in the field of communication, miniaturization, conformability, and power/energy management.

20.4.1 Power Management

To comply with the demand of the growing field of flexible electronics, huge efforts are made towards realizing the potential of flexible energy sources such as flexible lithium-ion batteries (LIBs), supercapacitors, solar cells, fuel cells, etc. (Kutbee et al. 2017; Bahabry et al. 2018a, 2018b; El-Atab et al. 2019). Different research groups have demonstrated the flexible energy sources, however, the capacities of these sources are extremely low when compared to the demands of the implantable system like a neural prosthesis or even a normal cardiac pacemaker. For real-time monitoring, where there is a constant data flow from the measuring site to the acquisition/analysis device, the power consumed is 100 s of mW in each hour, which reduces the battery life and has constraints on the energy sources. As an alternatives, RF, microwave, and other wireless energy harvesters are being investigated. An example of such a device is demonstrated by Ballini et al. who developed a 64-channel minimally invasive micro-ECoG implant that can provide wireless energy to operate the IC (Ballini et al. 2014). The wireless energy harvester though provides 800 μW of power which is reported to be enough to power the ASIC designed in this application, the point to note here is the communication of the data to the outer analysis system is done using the standard batteries outside the body or using a data acquisition system (DAQ) which is connected to a computer. Thus, the true wireless communication needs a highly efficient energy harvester coupled with high capacity flexible storage systems to have a genuine flexible implantable system (Worms 2002; Fitzpatrick 2015b; Jalili et al. 2017; Neely et al. 2018).

20.4.2 High-Density Electrodes

MEAs have been the standard for brain related chronic disease monitoring like ECoG and to understand fundamentals of different neurodegenerative and spinal cord injury related diseases. The highest number of electrodes reported in the recent past is from Muller et al. who have reported 59,760 electrodes with 2048 channels to measure different parameters. The group has demonstrated the acquisition of data from these 2048 channels using an external DAQ *in vitro*, which again restricts the fact that when it needs to be implemented for *in vivo* applications, the power consumption and acquisition of signals through a wired system has to be implemented. The key challenge in having high density electrodes is the multiplexing abilities, which increases the complexity of the overall integrated circuitry for the stand-alone IC. Electrode array design, multistage AP read out circuit, analogue to digital converters (ADCs), and multiplexers will eventually increase the footprint. Even if state-of-the-art nm scale gate transistors are used, then the fan-out makes the foot-print larger which is usually connected through a ribbon cable to the DAQ system. This concludes that a truly wireless platform is needed for the better implementation of the high-density electrode flexible implantable system (Dragas et al. 2017; Ballini et al. 2014).

Nanoscale, nanoelectronics devices such as nanowire field-effect transistors (FETs) (NWFETs) have emerged in order to reduce the active device area for neural recording devices. The active recording area can be reduced by several orders of magnitude by means of nanoscale devices compared to conventional MEAs. It subsequently increases spatial resolution allowing intracellular activity recording. These can be fabricated on non-planar flexible substrates as well (Park et al. 2017, 2018; El-Atab, Shaikh and Hussain 2019). Furthermore, flexible organic FETs (OFETs) have also been suggested for neural signal recording and stimulation, unlike the previous FETs that use gate capacitance coupling for the extracellular signal to be converted to drain currents, in these OFETs, neurons are placed far away from the channel of the transistors on the opposite side of the gate electrode. Nevertheless, relative poor mobility could limit signal-to-noise ratio (SNR) and bandwidth of the recording system made from organic devices (Lacour et al. 2004,

2005, 2016). Typical circuits designed for high density MEAs are comprised of buffer, amplifiers, active (or passive) filters, multiplexers, processors, ADCs and memory elements. Different architectures of MEAs are well classified and summarized, and the highest number of channel electrodes reported to date is 26,400 on 3.85×2.1 mm^2 sensing area from Ballini et al. (2014).

20.4.3 Materials and Integration

The passive or active materials choice in flexible electronics in general is fundamental to advancing the status-quo. Active or functional materials must integrate with a substrate in a seamless manner to avoid delamination, cracking, leakages (e.g., of current or light), or release of harmful substances into tissue. Roles of the passive substrate materials are to hold the functional elements together, facilitate insertion, and ensure biomechanical properties compatible with the host tissues.

Bulk metals have a very high elastic moduli on the order of GPa (6 orders higher than neural tissues) and have a narrow (1% strain) elastic region (~20% for tissues); it restricts their usage in soft neural interfaces in the form of foil, thin films, and microwires. Delamination of metals in elastic electronic devices due to their poor adhesion to a substrate necessitates the encapsulation layers of polymers. In addition, the interconnection of an electrode with rigid ICs using solder bumps is still a challenge as cracks and breaks occur at these high stress points of mechanical mismatch. The electronic conductivity of conductive polymers still remains low (1000 S.cm^{-1}) which currently hinders their integration as interconnects in microelectrodes. Also, the properties of these active materials are way below par compared to the CMOS status-quo which has been explained in the review article by Shaikh et al. justifying the necessity of flexing CMOS devices rather than continuing with organic materials for computing applications (Park et al. 2017; Shaikh et al. 2017; Alcheikh et al. 2018, 2019; Torres Sevilla et al. 2018).

The substrate materials are often dielectric organic polymers and are chosen for their bio-inertness, process compatibility, and mechanical stability. It contributes to the bulk of implantable devices that serves to hold functional elements with integrity, isolate them electrically, and facilitate insertion in the tissue. Parylene and polyimide are the most common used thermoplasts with elastic moduli of 3.2 and 2.5 GPa, respectively. A comprehensive review of the two materials and their use in neural implants can be found in reviews. Silicones and polyurethanes are preferred in applications demanding strains in several percent. Silicone is compatible with a majority of lithographical and spin coating processes in semiconductor processes excluding high temperature processes and micromachining. Though the materials used in general are biocompatible, however, in neural interfaces or brain related implantable devices, revascularization and complete healing of the blood-brain barrier, invasion of cells, and free diffusion of the metabolites are still not possible. Other materials which permit these are not CMOS compatible (Gumus et al. 2017; Nassar et al. 2018; Almuslem et al. 2019; El-Atab et al. 2019; Hussain 2019).

The true hybrid integration approach for successful implementation of implantable and wearable devices has been recently proposed by Shaikh et al. which combines CMOS ICs, modular-Lego approach for interconnecting, and 3D integration for increased area efficiency. This approach will also facilitate the isolation of electrodes and sensing area from the processing unit, while communicating wirelessly is being powered by the micro-batteries in flexible forms on the rear side of coin architecture. Thus, hybrid approaches and 3D integration can lead to true stand-alone flexible electronic system developments. An example of a wearable marine tagging device using a process flow is shown in Figure 20.6 (Park et al. 2017; Hussain et al. 2018a; Shaikh et al. 2018, 2019b; Khan et al. 2019).

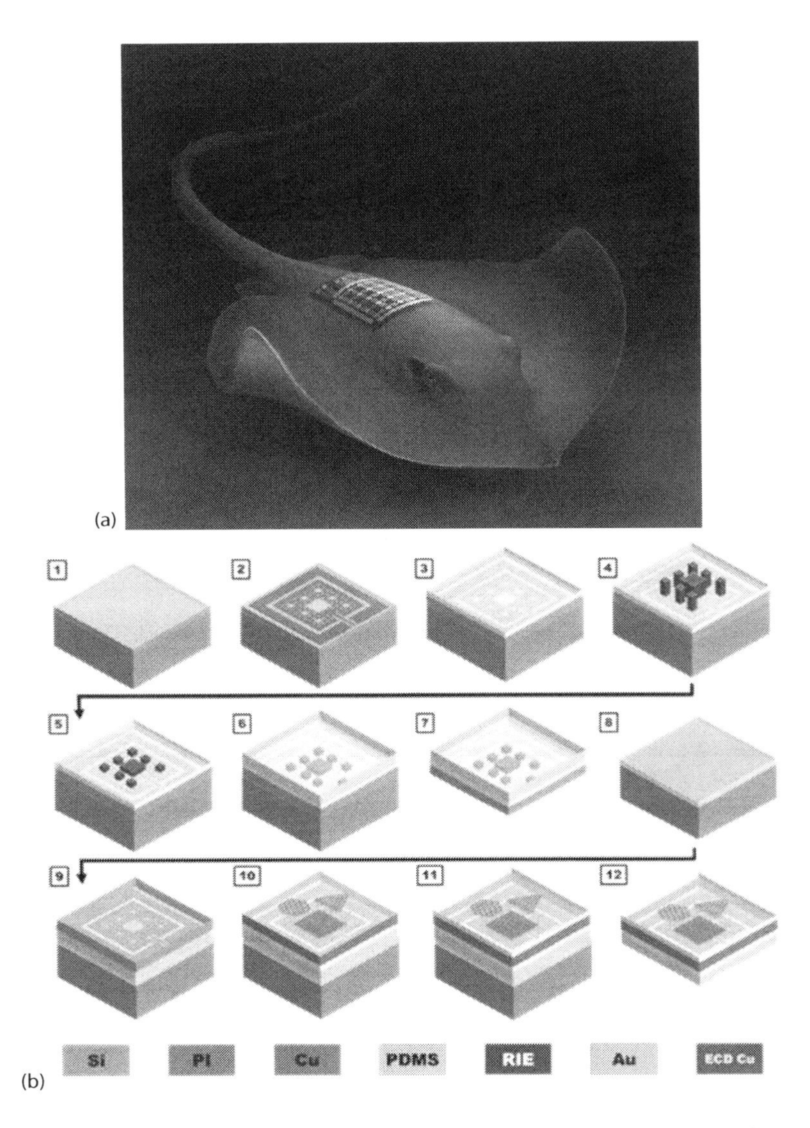

FIGURE 20.6 Fully compliant electronics system approaches. (a) Featherlight non-invasive "Marine-Skin" tagging device for deep sea environment attached to a sting ray (Reproduced with permission from Shaikh et al. 2019) and (b) 3D schematic illustration of heterogeneous integration flow for obtaining hybrid flexible high-performance electronics (Reproduced with permission from Shaikh et al. 2018).

20.5 Summary

In this chapter, we have presented an overview of the implantable electronics which includes both flexible and non-flexible devices. Cardiac pacemakers and cochlear implants have been used ever since their first inception in the 1960s, and since the 1990s, implantable electrodes for diagnosis and treatment of neurodegenerative diseases have been widely accepted. The earlier generations of the implantable devices especially the cardiac pacemakers and cochlear implants were still rigid, however, there has been a dramatic decrease in the form-factor as well as huge improvements in the packaging, biocompatibility, and shelf-life. The current state-of-the-art commercially available implantable devices rely on batteries to power them, resulting in bulky form factors. The size of the devices makes them difficult to implant, requiring expensive invasive surgery. Moreover, the batteries only last for 3–5 years, requiring the patient to undergo a surgery to replace the batteries with the recovery period lasting up to several days. To make these devices accessible for battery replacement, the pacemaker is placed under the patient's skin somewhere on the chest with long leads running subcutaneously to the region where the actual stimulation has to be performed. In the case of a DBS device, the lead runs to the top of a patient's head and is inserted deep inside the brain through an opening in the skull. The leads, their placement, and implantation surgeries can cause complications and significantly increase the risk of infection. Therefore, replacing batteries with alternative energy sources can help to dramatically reduce the device sizes and thus alleviate these serious problems, and will be the future direction for implantable devices. Implantable systems might soon be an integral part of minimally invasive diagnostic, therapeutic, and surgical treatments to attain a more accurate diagnosis and enhance the success rate of complex procedures. Many of the essential components have already been developed and demonstrated such as locomotion in fluid medium, energy harvesting for miniaturized implants, efficient communication, actuation and drug delivery, and low-power diagnostics (Hussain et al. 2015; Rojas et al. 2015; Nassar et al. 2016, 2017; Torres Sevilla et al. 2018; Shaikh et al. 2019a).

Trivially, when it comes to the conformability, the devices that show flexibility and stretchability are still under research and they present cases for brain related implants. These devices presented are mostly the electrodes only, which is not a complete system on its own and they still need bulky DAQ, ribbons for connection, and large power supply. The important factors to be considered when designing any free-form compliant system have been reviewed and extensively presented in this chapter. Moreover, we have also shown why we must pursue the approach of a hybrid integration strategy to integrate CMOS-based high-performance devices that outperform on every scale with flexible sensors and energy management devices using different technologies to truly ante the game. Going forward, we think that the absolute hybrid heterogeneous integration of the materials, interconnect technology, and CMOS-based computing will be the key enabler in developing a true flexible compliant implantable standalone system.

References

Abdulkader, S. N., Atia, A. and Mostafa, M. S. M. (2015) 'Brain computer interfacing: Applications and challenges', *Egyptian Informatics Journal*. Ministry of Higher Education and Scientific Research, 16(2), pp. 213–230. doi:10.1016/j.eij.2015.06.002.

Alcheikh, N., Shaikh, S. F. and Hussain, M. M. (2018) 'Ultra-stretchable archimedean interconnects for stretchable electronics', *Extreme Mechanics Letters*. 24, pp. 6–13. doi:10.1016/j.eml.2018.08.005.

Alcheikh, N., Shaikh, S. F. and Hussain, M. M. (2019) 'In-plane deformation mechanics of highly stretchable Archimedean interconnects', *AIP Advances*. 9(1). doi:10.1063/1.5053967.

Alhammad, N. J. et al. (2016) 'Cardiac implantable electronic devices and end-of-life care: An Australian perspective', *Heart Lung and Circulation*. Australian and New Zealand Society of Cardiac and Thoracic Surgeons (ANZSCTS) and the Cardiac Society of Australia and New Zealand (CSANZ), 25(8), pp. 814–819. doi:10.1016/j.hlc.2016.05.103.

Almuslem, A. S., Shaikh, S. F. and Hussain, M. M. (2019) 'Flexible and stretchable electronics for harsh-environmental applications', *Advanced Materials Technologies*. p. 1900145. doi:10.1002/admt.201900145.

Badia, J. et al. (2011) 'Comparative Analysis of Transverse Intrafascicular Multichannel , Longitudinal Intrafascicular and Multipolar Cuff Electrodes for the Selective Stimulation of Nerve Fascicles', *Journal of Neural Engineering*, 8, p. 036023. doi.org/10.1088/1741-2560/8/3/036023.

Bahabry, R. R. et al. (2018a) 'Corrugation architecture enabled ultraflexible wafer-scale high-efficiency monocrystalline silicon solar cell', *Advanced Energy Materials*. 8(12), p. 1702221. doi:10.1002/aenm.201702221.

Bahabry, R. R. et al. (2018b) 'Corrugation architecture enabled ultra-flexible mono-crystalline silicon solar cells via plasma etching and laser ablation', in *2018 IEEE 7th World Conference on Photovoltaic Energy Conversion (WCPEC) (A Joint Conference of 45th IEEE PVSC, 28th PVSEC & 34th EU PVSEC)*. IEEE, pp. 0289–0292. doi:10.1109/PVSC.2018.8547699.

Ballini, M. et al. (2014) 'A 1024-channel CMOS microelectrode array with 26,400 electrodes for recording and stimulation of electrogenic cells in vitro', *IEEE Journal of Solid-State Circuits*. Europe PMC Funders, 49(11), pp. 2705–2719. doi:10.1109/JSSC.2014.2359219.

Bazaka, K. and Jacob, M. (2012) *Implantable Devices: Issues and Challenges, Electronics*. doi:10.3390/electronics2010001.

Cha, Y.-M., Lee, B. K. and Chung, M. K. (2017) 'Advances in cardiac implantable electronic devices 2016—American college of cardiology', pp. 1–5. Available at: https://www.acc.org/latest-in-cardiology/articles/2017/03/27/15/15/advances-in-cardiac-implantable-electronic-devices-2016.

Dragas, J., et al. (2017) 'In Vitro Multi-Functional Microelectrode Array Featuring 59 760 Electrodes, 2048 Electrophysiology Channels, Stimulation, Impedance Measurement, and Neurotransmitter Detection Channels', *IEEE Journal of Solid-State Circuits* 52(6), pp. 1576–1590. doi.org/10.1109/JSSC.2017.2686580.

El-Atab, N. et al. (2019) 'Bi-facial substrates enabled heterogeneous multi-dimensional integrated circuits (MD-IC) for Internet of Things (IoT) applications', *Advanced Engineering Materials*, 1900043, pp. 1–6. doi:10.1002/adem.201900043.

El-Atab, N., Shaikh, S. F. and Hussain, M. M. (2019) 'Nano-scale transistors for interfacing with brain: Design criteria, progress and prospect', *Nanotechnology* (December 2016), pp. 11–14. doi:10.1088/1361-6528/ab3534.

Falcone, J. D. et al. (2014) 'Neural interfaces: From human nerves to electronics', *Implantable Bioelectronics*, 9783527335251, pp. 87–113. doi:10.1002/9783527673148.ch6.

Fekete, Z. and Pongrácz, A. (2017) 'Multifunctional soft implants to monitor and control neural activity in the central and peripheral nervous system: A review', *Sensors and Actuators, B: Chemical*. 243, pp. 1214–1223. doi:10.1016/j.snb.2016.12.096.

Fitzpatrick, D. (2015a) 'Bladder implants', *Implantable Electronic Medical Devices*, pp. 99–106. doi:10.1016/b978-0-12-416556-4.00007-3.

Fitzpatrick, D. (2015b) 'Electrical stimulation therapy for Parkinson's disease and dystonia', *Implantable Electronic Medical Devices*, pp. 111–116. doi:10.1016/b978-0-12-416556-4.00009-7.

Fitzpatrick, D. (2015c) 'Retinal implants', *Implantable Electronic Medical Devices*, pp. 1–18. doi:10.1016/b978-0-12-416556-4.00001-2.

Fitzpatrick, D. (2015d) 'Smart contact lens', *Implantable Electronic Medical Devices*, pp. 19–25. doi:10.1016/b978-0-12-416556-4.00002-4.

Gumus, A. et al. (2017) 'Expandable polymer enabled wirelessly destructible high-performance solid state electronics', *Advanced Materials Technologies*, 2(5), pp. 1–6. doi:10.1002/admt.201600264.

Gutruf, P. and Rogers, J. A. (2018) 'Implantable, wireless device platforms for neuroscience research', *Current Opinion in Neurobiology*. 50, pp. 42–49. doi:10.1016/j.conb.2017.12.007.

Hughes, M. A. (2016) 'Insinuating electronics in the brain', *Surgeon*, 14(4), pp. 213–218. doi:10.1016/j.surge.2016.03.003.

Hussain, A. M. et al. (2015) 'Metal/polymer based stretchable antenna for constant frequency far-field communication in wearable electronics', *Advanced Functional Materials*, 25(42), pp. 6565–6575. doi:10.1002/adfm.201503277.

Hussain, A. M. and Hussain, M. M. (2016) 'CMOS-technology-enabled flexible and stretchable electronics for internet of everything applications', *Advanced Materials*, 28(22), pp. 4219–4249. doi:10.1002/adma.201504236.

Hussain, A. M., Shaikh, S. F. and Hussain, M. M. (2016) 'Design criteria for XeF 2 enabled deterministic transformation of bulk silicon (100) into flexible silicon layer', *AIP Advances*, 6(7), p. 075010. doi:10.1063/1.4959193.

Hussain, M. M. et al. (2018a) 'Manufacturable heterogeneous integration for flexible CMOS electronics', in *2018 76th Device Research Conference (DRC)*. IEEE, pp. 1–2. doi:10.1109/DRC.2018.8442163.

Hussain, M. M. (2019) 'Marine IoT: Non-invasive wearable multisensory platform for oceanic environment monitoring', *2019 IEEE 5th World Forum on Internet of Things (WF-IoT)*. IEEE, pp. 309–312.

Hussain, M. M., Ma, Z. (Jack) and Shaikh, S. F. (2018b) 'Flexible and stretchable electronics—Progress, challenges, and prospects', *The Electrochemical Society Interface*, 27(4), pp. 65–69. doi:10.1149/2.f08184if.

Ivanova, E. P., Bazaka, K. and Crawford, R. J. (2014) *Introduction to Biomaterials and Implantable Device Design, New Functional Biomaterials for Medicine and Healthcare*. doi:10.1533/9781782422662.1.

Jalili, R. et al. (2017) 'Implantable electrodes', *Current Opinion in Electrochemistry*, 3(1), pp. 68–74. doi:10.1016/j.coelec.2017.07.003.

Juanola-Feliu, E. et al. (2014) 'Nano-enabled implantable device for glucose monitoring', *Implantable Bioelectronics*, 9783527335251, pp. 247–263. doi:10.1002/9783527673148.ch12.

Keskinbora, K. H. and Keskinbora, K. (2018) 'Ethical considerations on novel neuronal interfaces', *Neurological Sciences*, 39(4), pp. 607–613. doi:10.1007/s10072-017-3209-x.

Khan, S. M. et al. (2019) 'Do-It-Yourself integration of a paper sensor in a smart lid for medication adherence', *Flexible and Printed Electronics*, 4(2), p. 025001. doi:10.1088/2058-8585/ab10f5.

Kutbee, A. T. et al. (2017) 'Flexible and biocompatible high-performance solid-state micro-battery for implantable orthodontic system', *npj Flexible Electronics*, 1(1), p. 7. doi:10.1038/s41528-017-0008-7.

Lacour, S. P. et al. (2005) 'Stretchable interconnects for Elastic.pdf', *Proceedings of the IEEE*, 93(8), pp. 1459–1467.

Lacour, S. P., Courtine, G. and Guck, J. (2016) 'Materials and technologies for soft implantable neuroprostheses', *Nature Reviews Materials*, 1(10). doi:10.1038/natrevmats.2016.63.

Lacour, S. P., Tsay, C. and Wagner, S. (2004) 'An elastically stretchable TFT Circuit', *IEEE Electron Device Letters*, 25(12), pp. 792–794. doi:10.1109/LED.2004.839227.

Lee, C. (2014) 'Pacemakers and implantable cardioverter defibrillators', *Update in Anaesthesia*, 29(December), pp. 5–9. doi:10.1016/B978-0-12-416556-4.00006-1.

Lewis, P. M. et al. (2016) 'Advances in implantable bionic devices for blindness: A review', *ANZ Journal of Surgery*, 86(9), pp. 654–659. doi:10.1111/ans.13616.

Lu, Y., Liu, X. and Kuzum, D. (2018) 'Graphene-based neurotechnologies for advanced neural interfaces', *Current Opinion in Biomedical Engineering*, 6, pp. 138–147. doi:10.1016/j.cobme.2018.06.001.

Macherey, O. and Carlyon, R. P. (2014) 'Cochlear implants', *Current Biology*, 24(18), pp. R878–R884. doi:10.1016/j.cub.2014.06.053.

Matsuo, T. et al. (2011) 'Intrasulcal Electrocorticography in Macaque Monkeys with Minimally Invasive Neurosurgical Protocols', *Frontiers in Neuroscience* 5, pp. 1–9. doi.org/10.3389/fnsys.2011.00034.

Mavoori, J. et al. (2005) 'An autonomous implantable computer for neural recording and stimulation in unrestrained primates', *Journal of Neuroscience Methods*, 148(1), pp. 71–77. doi:10.1016/j.jneumeth.2005.04.017.

Miller, M. A. et al. (2015) 'Leadless cardiac pacemakers back to the future', *Journal of the American College of Cardiology*. 66(10), pp. 1179–1189. doi:10.1016/j.jacc.2015.06.1081.

Minev, I. R. and Lacour, P. (2016) 'Stretchable bioelectronics for medical devices and systems', pp. 257–273. doi:10.1007/978-3-319-28694-5.

Montornes, J. M., Vreeke, M. S. and Katakis, I. (2008) 'Glucose biosensors', *Bioelectrochemistry: Fundamentals, Experimental Techniques and Applications*, pp. 199–217. doi:10.1002/9780470753842.ch5.

Nassar, J. M. et al. (2016) 'From stretchable to reconfigurable inorganic electronics', *Extreme Mechanics Letters*. doi:10.1016/j.eml.2016.04.011.

Nassar, J. M. et al. (2017) 'Recyclable nonfunctionalized paper-based ultralow-cost wearable health monitoring system', *Advanced Materials Technologies*, 2(4), pp. 1–11. doi:10.1002/admt.201600228.

Nassar, J. M. et al. (2018) 'Compliant lightweight non-invasive standalone "Marine Skin" tagging system', *npj Flexible Electronics*. 2(1), p. 13. doi:10.1038/s41528-018-0025-1.

Neely, R. M. et al. (2018) 'Recent advances in neural dust: Towards a neural interface platform', *Current Opinion in Neurobiology*. 50, pp. 64–71. doi:10.1016/j.conb.2017.12.010.

Neuzil, P. and Reddy, V. Y. (2015) 'Leadless cardiac pacemakers: Pacing paradigm change', *Current Cardiology Reports*, 17(8), pp. 1–8. doi:10.1007/s11886-015-0619-3.

Pang, C., Lee, C. and Suh, K. Y. (2013) 'Recent advances in flexible sensors for wearable and implantable devices', *Journal of Applied Polymer Science*, 130(3), pp. 1429–1441. doi:10.1002/app.39461.

Park, W. et al. (2017) 'Stable MoS_2 Field-effect transistors using TiO_2 interfacial layer at metal/MoS_2 contact', *Physica Status Solidi (A) Applications and Materials Science*, 214(12). doi:10.1002/pssa.201700534.

Park, W. et al. (2018) 'Contact resistance reduction of ZnO thin film transistors (TFTs) with saw-shaped electrode', *Nanotechnology*, 29(32). doi:10.1088/1361-6528/aac4b9.

Poon, A. S. Y. (2014) 'Miniaturized biomedical implantable devices', *Implantable Bioelectronics*, 9783527335251, pp. 45–64. doi:10.1002/9783527673148.ch4.

Rogers, J. A., Ghaffari, R. and Kim, D.-H. (2016) *Microsystems and Nanosystems Stretchable Bioelectronics for Medical Devices and Systems*. Springer, Switzerland.

Rojas, J. P. et al. (2015) 'Nonplanar nanoscale fin field effect transistors on textile, paper, wood, stone, and vinyl via soft material-enabled double-transfer printing', *ACS Nano*, 9(5), pp. 5255–5263. doi:10.1021/acsnano.5b00686.

Shaikh, S. F. et al. (2017) 'Freeform compliant CMOS electronic systems for Internet of everything applications', *IEEE Transactions on Electron Devices*, 64(5). doi:10.1109/TED.2016.2642340.

Shaikh, S. F. et al. (2018) 'Modular Lego-electronics', *Advanced Materials Technologies*, 3(2), p. 1700147. doi:10.1002/admt.201700147.

Shaikh, S. F. et al. (2019a) 'Environmental monitoring: Noninvasive featherlight wearable compliant "Marine Skin": Standalone multisensory system for deep-sea environmental monitoring (Small 10/2019)', *Small*, 15(10), p. 1970051. doi:10.1002/smll.201970051.

Shaikh, S. F. et al. (2019b) 'Noninvasive featherlight wearable compliant "Marine Skin": Standalone multisensory system for deep-sea environmental monitoring', *Small*, 15(10), p. 1804385. doi:10.1002/smll.201804385.

Sideris, S. et al. (2017) 'Leadless cardiac pacemakers: Current status of a modern approach in pacing', *Hellenic Journal of Cardiology*, 58(6), pp. 403–410. doi:10.1016/j.hjc.2017.05.004.

Torres Sevilla, G. A. et al. (2018) 'Fully spherical stretchable silicon photodiodes array for simultaneous 360 imaging', *Applied Physics Letters*, 113(13). doi:10.1063/1.5049233.

Won, S. M. et al. (2018) 'Recent advances in materials, devices, and systems for neural interfaces', *Advanced Materials*, 30(30), pp. 1–19. doi:10.1002/adma.201800534.

Worms, J. G. (2002) 'Direction finding with array sensors in broadband applications', *Frequenz*, 56(9–10), pp. 220–228. doi:10.1515/FREQ.2002.56.9-10.220.

21

Bioresorbable Electronics

Joong Hoon Lee,
Gwan-Jin Ko,
Huanyu Cheng, and
Suk-Wong Hwang

21.1 Introduction

Recent transitions in electronic systems have been represented by various technologies based on a soft platform, and such an unusual format allows electronics to integrate with biology for a wide scope of clinical, medical applications.[1–13] Another approach of bio-applicable technologies, often referred to as "transient electronics," has a unique advantage that all constituent materials/components completely dissolve or disappear via hydrolysis in the body, which eliminates the procedure for collection and recovery associated with medical devices implanted into the human body.[14–18] Candidate materials involve monocrystalline silicon nanomembranes (Si NMs),[14,19–23] zinc oxide (ZnO),[24] and Si-germanium for semiconductors, magnesium (Mg), iron (Fe), Zn, tungsten (W), and molybdenum (Mo) for contacts and interconnects,[25,26] silicon oxides/nitrides (SiO_2/Si_3N_4) for gate and interlayer dielectrics,[27] and poly lactic acid (PLA), poly glycolic acid (PGA), poly lactic-co-glycolic acid (PLGA), polycaprolactone (PCL), and silk fibroin for substrates and encapsulators.[19,28–31] One of the key features in this type of technology includes an ultrathin sheet of silicon, allowing exploitation of established aspects of manufacturing strategies and design layouts with a high fidelity of operational characteristics that could compete with those of non-transient counterparts built on bulky wafer substrates. Sophisticated, miniaturized electronic components include different classes of sensors (e.g., strain, temperature, pH, hydration, solar cells, and photodetectors),[14,19,32] actuators (e.g., mechanical energy harvesters), collections of power supply and wireless technologies,[20,24] and a complementary metal-oxide semiconductor (CMOS).[19] In the following, we describe recent advances in the field of emerging transient technology that can broadly cover fundamental aspects of bioresorbable electronics, as well as other features towards potential applications.

21.2 Materials for Bioresorbable, Transient Electronics

Figure 21.1a shows a representative example of bioresorbable electronic systems that includes transistors, diodes, capacitors, inductors, and resistors with interconnects.[14] All electronic components consist of diverse biocompatible and bioresorbable electronic materials, including mono-crystalline Si NMs for semiconductors, Mg for conductors, MgO as dielectrics, and silk as substrates and encapsulants. Beyond these materials described here, other degradable materials are also possible. Silicon germanium (Si-Ge), germanium (Ge), and ZnO serve as semiconducting components,[24,33] iron (Fe), Zn, tungsten (W), molybdenum, and polycrystal silicon (p-Si) act as metallic contacts[26] and interconnects, SiO_2 and Si_3N_4 provide gate dielectrics and passivation elements,[27] and biodegradable and synthetic polymers

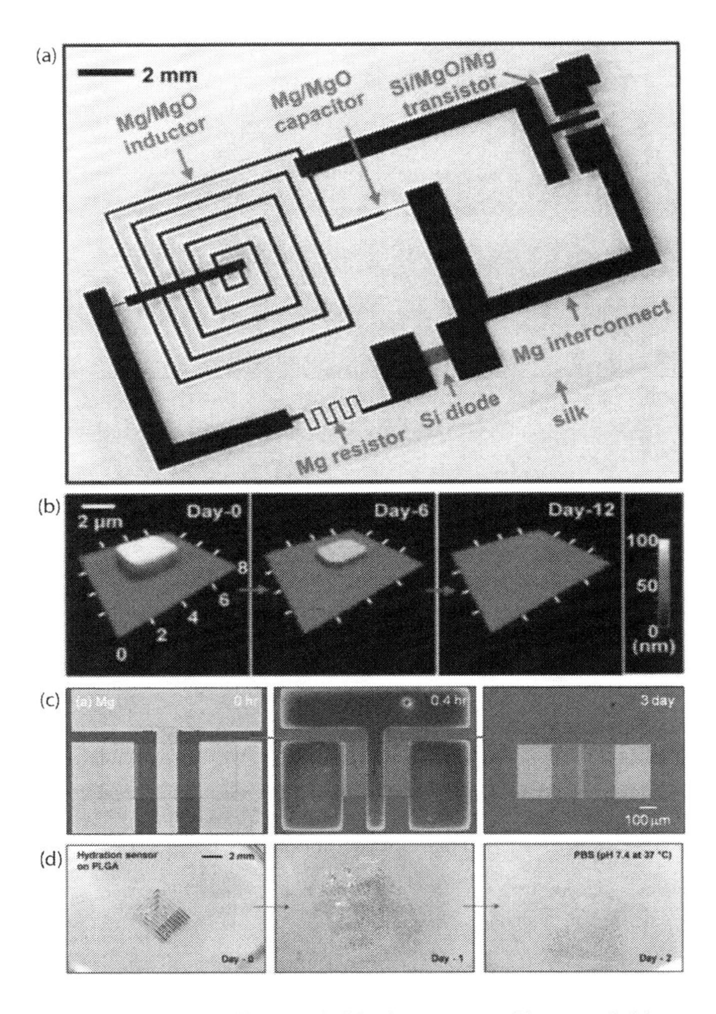

FIGURE 21.1 Representative examples of bioresorbable electronics and key resorbable materials. (a) Image of a representative bioresorbable electronic system that includes transistors, diodes, inductors, capacitors, and resistors, with interconnects and interlayer dielectrics, all on a thin silk substrate. (b) Atomic force microscope topographical images of single crystalline Si NMs at various stages of hydrolysis in PBS at 37°C. (c) Dissolution images of Mg metal electrodes in transient n-channel MOSFETs as time goes on. (d) Images of a bioresorbable hydration sensor on a thin PLGA film as a substrate showing various stages of dissolution in PBS (1M, pH 7.4) at physiological temperature (37°C) after 1 day and 2 days, respectively.

such as PLA, PCL, poly-glycolide (PGA), PLGA, and poly(1,8-octanediol-co-citrate) (POC) can be used for supporting and protection layers.[19,32] A series of atomic force microscope (AFM) topographical images in Figure 21.1b present the dissolution process of a most critical constituent of transient systems, semiconducting grade of Si NMs, with a dimension of 3 μm × 3 μm × 70 nm, at various temporal stages in phosphate-buffered saline (PBS; pH of 7.4) solution at 37°C.[14] The small Si NM dots had completely disappeared within 12 days via hydrolysis. Similar dissolution behaviors occurred in magnesium contacts in an n-channel metal-oxide-semiconductor field-effect transistor (MOSFET) induced by immersion in deionized (DI) water at room temperature (Figure 21.1c).[26] The degradation rate of Mg electrodes is relatively faster compared to that of silicon mainly due to microstructure, surface topology, film density, and others.[34–37] Hydrolysis of other materials with similar experimental approaches can be found in previous reports.[22,27] Serial images of a water-soluble hydration sensor exhibit the time sequence of physical/chemical transience during immersion in PBS (pH 7.4) at body temperature (Figure 21.1d).[19] A thin substrate of PLGA dominantly reacted with the bio-fluid, leading to swelling and degradation, and each of individual elements dissolved at their own rates.[15,16]

21.3 Dissolution Behaviors of Essential Materials, and Encapsulation Strategies

Fundamental kinetics of hydrolysis of bioresorbable electronic materials are dependent on pH values, chemical composition of solutions, doping type and concentration of materials, and temperature and external conditions. Figure 21.2a presents a summary of dissolution rates of Si NMs by use of pH dependence (pH 6 to 14) of thickness measurements performed by an AFM at room (red) and body temperatures (blue, 37°C).[23] The results indicate that hydrolysis with temperature and pH only considers surface reaction without diffusion of solution into the Si NMs, leading to a simple, linear behavior of dissolution. Differences of the dissolution rate even in similar pH levels reveal that ionic content and concentration in solution have an impact on the chemical reaction, for example, tap water (black, pH ~ 7.8), deionized water (blue, pH ~ 8.1), and spring water (red, pH ~ 7.4) in Figure 21.2b.[22] Other examples can be found somewhere else in previous articles.[33,38] External interventions with intensive light exposure to semiconducting elements have been studied to accelerate the speed of etching of semiconducting elements as a photoelectrochemical etching approach[39–41] that might be capable of varying the dissolution rate of Si NMs (Figure 21.2c).[22] Examination of dissolution results in PBS (0.1 M, pH ~7.4) at room temperature while exposing natural daylight (red) and ultraviolet light (blue, UV, λ = 365 nm, I = 590 μW/cm^2 at a distance of 7 cm) reveals no observation of significant dissolution rate changes. Such results arise from low levels of illumination compared to conventional high levels of illumination source (~1 to ~500 mW/cm^2). The concentrations of dopants in the Si NMs can have an effect on the dissolution rate. Three different doping concentrations (10^{17} cm^{-3}, black; 10^{19} cm^{-3}, red; 10^{20} cm^{-3}, blue) of boron-doped Si NMs were prepared and dissolved in the buffer solution (0.1 M, pH 7.4) at body temperature (Figure 21.2d).[22] The results indicate a strong reduction of dissolution rate at dopant concentrations that exceed a certain level, such as 10^{20} cm^{-3}, and those behaviors are consistent with previous studies on extensive silicon electrochemical etching in various conditions.[27] Similar trends on phosphorus-doped Si NMs are described in previous reports.[22]

The ultimate, desired behavior for dissolvable electronics is to operate in a stable mode as that of conventional electronics within a certain period in aqueous solutions and/or the body, then completely dissolve and disappear. For the purpose of this type of ideal behavior, encapsulation and passivation techniques with various inactive materials are used for programming their certain life span that can be controlled through protecting electronics against aqueous solutions by covering thick protection layers with insulating elements and/or forming a multilayer with diverse dielectrics including SiO_2, Si_3N_4, and MgO. Figure 21.2e presents measurements of changes in resistance of Mg traces (300 nm thick) encapsulated with different combinations of silicon oxide- and silicon nitride-based materials and their different thicknesses while immersed in DI water at room temperature.[27] At the single layer, the atomic

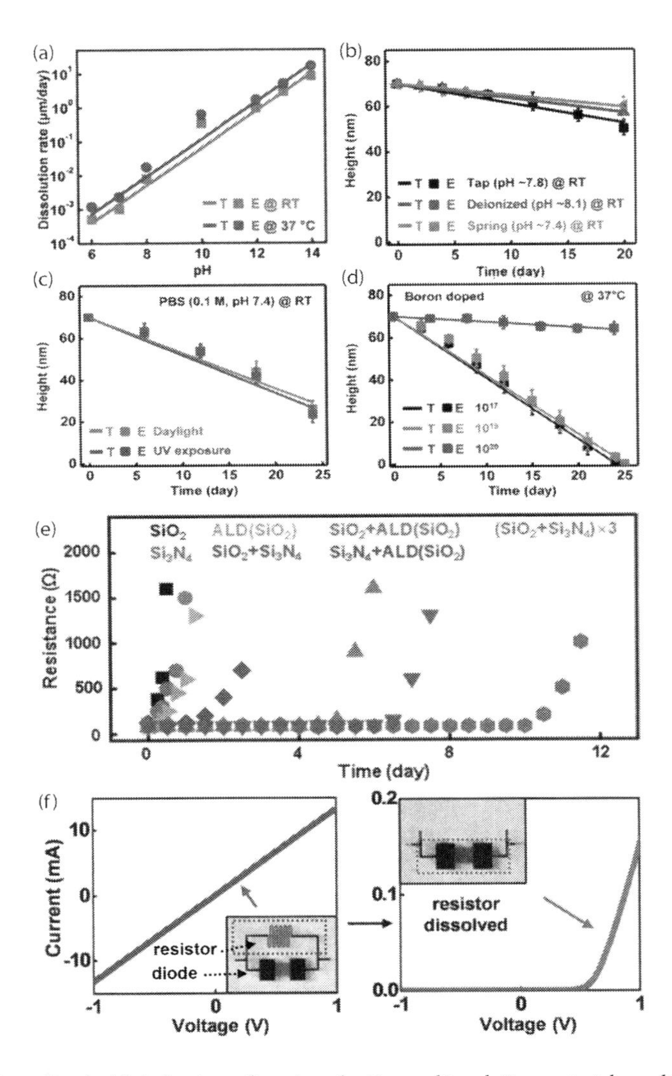

FIGURE 21.2 Various dissolvable behaviors of semiconducting and insulating materials, and functional devices. (a) Theoretical (T, lines) and experimental (E, symbols) dissolution rates of Si NMs in different pH levels (pH 6 to 14) at room (red) and body temperatures (blue, 37°C). (b) Calculated (T, lines) and measured (E, symbols) dissolution of Si NMs during immersion in different contents of waters at similar pH levels (black, tap water, pH ~7.8; blue, deionized water, pH ~8.1; red, spring water, pH ~7.4), at room temperature. (c) Theoretical (T, lines) and experimental (E, symbols) values of Si dissolution rate with the dependence of external light exposures (red, daylight; blue, UV light) in PBS (0.1 M, pH 7.4 at room temperature [RT]). (d) Calculated model (T, lines) and measured data (E, symbols) of silicon dissolution rate at different doping concentrations with boron (black, 10^{17} cm^{-3}; red, 10^{19} cm^{-3}; blue, 10^{20} cm^{-3}) while immersed in PBS (0.1 M, pH 7.4) at physiological temperature (37°C). (e) Measurements of changes in resistance of Mg traces (~300 nm thick) encapsulated with different types and thicknesses of insulating materials while immersed in DI water at room temperature. A single layer of plasma enhanced chemical vapor deposition (PECVD) SiO_2 (black, 1 μm), PECVD-low frequency (LF) Si_3N_4 (red, 1 μm), and ALD SiO_2 (orange, 20 nm), a double layer of PECVD SiO_2/PECVD-LF Si_3N_4 (blue, 500/500 nm), PECVD SiO_2/ALD SiO_2 (magenta, 500/20 nm), PECVD-LF Si_3N_4/ALD SiO_2 (purple, 500/20 nm), and a triple layer of PECVD SiO_2/PECVD-LF Si_3N_4 (Cyan, 200/200/200/200/100/100 nm) were used for the encapsulation. (f) Functional transformation of an integrated system into differentiated sub-systems via predefined spatial dissolution. Current-voltage (I-V) characteristics and inset images showing a combined system of a Mg resistor and a Si NM diode can be functionally transformed from the metal resistor with a low resistance (left, blue) to the silicon diode with a high resistance (right, red) through dissolution of the Mg trace.

layer deposition (ALD) method presents a better performance of encapsulation than that of the plasma enhanced chemical vapor deposition (PECVD) approach. The multiple stacks of protection materials present an effective improvement of encapsulation performance, compared to the single layer strategy. Similar to single component cases, a combination of ALD SiO_2 and PECVD Si_3N_4 layers (purple) show enhanced performance than that of PECVD SiO_2 and Si_3N_4 layers (blue). These results revealed that the films grown by the ALD have much lower defects than those grown by the PECVD.

Advanced encasements and well-controlled dissolution strategies define the operational lifetimes of transient electronic circuits, enabling functional transformations in various modes, such as conversion of integrated circuits into individual components and/or equivalent circuits, but with different functions. Figure 21.2f shows a simple example of functional transformation of transient circuits that combines a serpentine Mg resistor with a silicon p-n diode encapsulated with a thin layer of MgO.[42] At the initial mode, the whole system operates as a resistor due to a much lower electrical resistance than that of the silicon diode at low voltages (left). Pre-defined, local dissolution of the Mg resistor allows the system to functionally transform from the resistor to the semiconductor diode (right). Similar strategies can be used in complicated logic gates and wireless antenna systems as well.[43]

21.4 Analytical Models of Dissolvable Materials

The hydrolysis of transient materials with relatively closed packing structures can be explained by considering the reaction only at the surface of transient materials, which the surface of transient material is continuously saturated with reactive substances in aqueous solution, and a surface reaction with a constant dissolution rate can account for the dissolution kinetics of such materials.[14,23] The remaining thickness is predicted to be linearly proportional to the time, and the slope of the linear plots is the dissolution rate. The dependence of dissolution rate on temperature can be captured by the Arrhenius' equation.

In addition to the surface reaction, porous transient materials allow molecules in solutions to reactively diffuse into the materials inside, and the reaction between diffused molecules and their surrounding transient material need to be considered. Based on a previous study that focused on absorption by simultaneous diffusion and chemical reaction,[44] a model of reactive diffusion has been established to analytically study the dissolution process of the porous transient materials.[45] For the case of near neutral solutions, the model considered the diffusion of water molecules and hydroxyl ions into the porous material, where water molecules react with the transient material with hydroxyl ions as the catalyst, leading to a high reaction rate with increased reactive surface regions. In this reactive diffusion model, the reaction is assumed to be irreversible first order or pseudo first order.

In the experiment, the initial thickness of the transient porous material is much smaller than lateral dimensions, i.e., width and length. Therefore, a one-dimensional model has been developed to adequately capture the dissolution behavior. Setting the y-axis in the thickness direction with $y = 0$ at the bottom of the porous transient material in Figure 21.3a, the water concentration $w(y, t)$ at location y and time t satisfies the reactive diffusion equation,

$$D\frac{\partial^2 w}{\partial y^2} - kw = \frac{\partial w}{\partial t}, \quad 0 \le y \le h_0, \tag{21.1}$$

where D is the diffusivity of water in the porous transient material and k is the rate of the reaction between water and the transient material. For a negligible rate k, Equation (21.1) reduces to the diffusion equation, where the ideal diffusion law holds for the diffusion of unreacted solution through the porous material. Assuming that the surface of the transient material is continuously saturated with the reactive substance (e.g., water molecules and hydroxyl ions) in the solution, the boundary condition at the solution/porous material interface is given by a constant water concentration, i.e., $w\big|_{y=h_0} = w_0$. Because of a nearly perfect bonding of the porous material to the bottom substrate, the boundary condition at

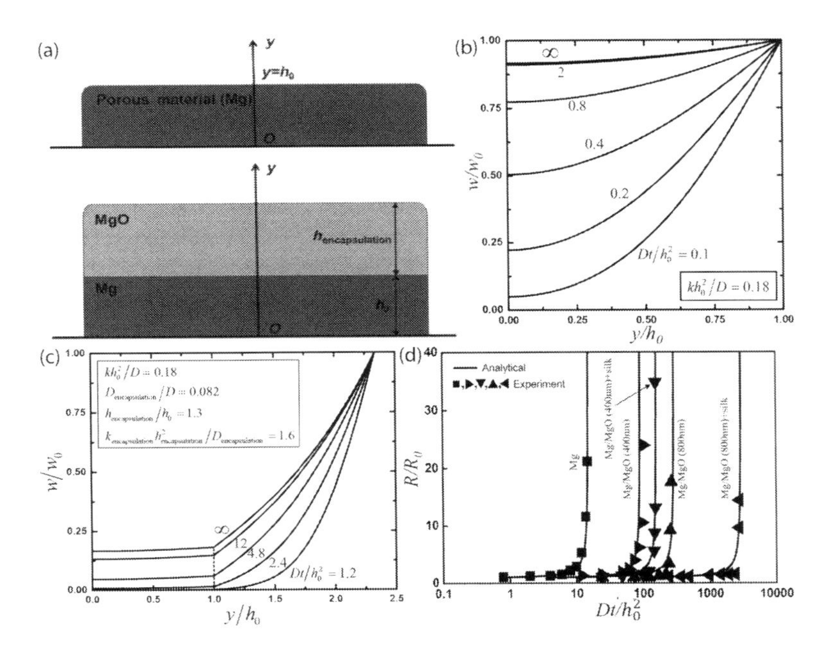

FIGURE 21.3 Schematic illustrations for the model of reactive diffusion and predictions of the normalized water concentration as well as the normalized electrical resistance from the model. (a) Single-layered structure (top) and bi-layered structure (bottom) in models of reactive diffusion for porous transient materials. (b) Distribution of the normalized water concentration as a function of the normalized position predicted from the model of reactive diffusion for the normalized time, $Dt/h_0^2 = 0.1$, 0.2, 0.4, 0.8, 2 and ∞, in the case of Mg layer ($kh_0^2/D = 0.18$) without encapsulation layer. (c) Same approach, but an Mg conductor layer encapsulated by an MgO layer ($kh_0^2/D = 0.18$, $D_{\text{encapsulation}}/D = 0.082$, $h_{\text{encapsulation}}/h_0 = 1.3$, and $k_{\text{encapsulation}}h^2_{\text{encapsulation}}/D_{\text{encapsulation}} = 1.6$). (d) Experimental and modeling results of the normalized electric resistance of the Mg layer and Mg with different encapsulation strategies (e.g., MgO encapsulation layers and/or silk overcoats).

the porous material/substrate interface is a zero water flux, i.e., $\partial w/\partial y|_{y=0} = 0$. Taken together with a zero water concentration of $w|_{t=0} = 0$ $(0 \le y < h_0)$ for the initial condition, the solution of the reactive diffusion equation in Equation (21.1) can be obtained, by introducing a new variable $\theta = w - w_0$ and then solving the corresponding inhomogeneous equation, as:

$$w(y,t) = w_0 \left\{ \frac{\cosh\left(\sqrt{\frac{kh_0^2}{D}}\frac{y}{h_0}\right)}{\cosh\sqrt{\frac{kh_0^2}{D}}} + 2\sum_{n=1}^{\infty} \frac{(-1)^n\left(n-\frac{1}{2}\right)\pi}{\frac{kh_0^2}{D} + \left(n-\frac{1}{2}\right)^2\pi^2} e^{-\left[\frac{kh_0^2}{D} + \left(n-\frac{1}{2}\right)^2\pi^2\right]\frac{Dt}{h_0^2}} \cos\left[\left(n-\frac{1}{2}\right)\pi\frac{y}{h_0}\right] \right\}. \tag{21.2}$$

The water concentration w normalized by its initial concentration w_0 is governed by a scaling law $w/w_0 = \overline{w}\left(y/h_0, Dt/h_0^2, kh_0^2/D\right)$, in which the normalized water concentration w/w_0 depends on the normalized position y/h_0, the normalized time Dt/h_0^2, and a non-dimensional ratio of reaction rate to diffusivity kh_0^2/D. For $D = 6.0 \times 10^{-12}$ cm²/s and $k = 1.2 \times 10^{-3}$ s^{-1} extracted from the experimental measurement of Mg with an initial thickness of 300 nm,[14] Figure 21.3b shows the normalized water concentration as a function of the normalized position for the normalized time of $Dt/h_0^2 = 0.1$, 0.2, 0.4, 0.8, 2 and ∞. The normalized time $Dt/h_0^2 = \infty$ is the steady-state limit of the normalized water concentration in the Mg layer, $w(y, t \to \infty) = w_0 \cosh\left(\sqrt{kh_0^2/D}\, y/h_0\right)/\cosh\sqrt{kh_0^2/D}$.

The mass of water reacted in an element of volume (unit cross-sectional area) is $kwdtdy$. Stoichiometric ratio q of water molecules to the transient material gives the mass of the dissolved porous material of $kwM/(qM_{H_2O})dtdy$, where M and $M_{H_2O} = 18\ \text{g}\cdot\text{mol}^{-1}$ are the molar masses of transient material and water, respectively. Integration with the thickness and time yields the total dissolved mass of the porous material, and it in turn gives the remaining thickness h normalized by its initial thickness h_0 as:

$$\frac{h}{h_0} = 1 - \frac{w_0 M}{q\rho M_{H_2O}} \frac{kh_0^2}{D} \left\{ \frac{Dt}{h_0^2} \cdot \frac{\tanh\sqrt{\frac{kh_0^2}{D}}}{\sqrt{\frac{kh_0^2}{D}}} - 2\sum_{n=1}^{\infty} \frac{1 - e^{-\left[\frac{kh_0^2}{D} + \left(n-\frac{1}{2}\right)^2 \pi^2\right]\frac{Dt}{h_0^2}}}{\left[\frac{kh_0^2}{D} + \left(n - \frac{1}{2}\right)^2 \pi^2\right]^2} \right\}, \tag{21.3}$$

where ρ is the mass density of the porous material.

The dissolution rates of transient materials can be affected by their physical and chemical properties from different deposition/growth methods and conditions. Density is one of the key parameters, and its variation can drastically change the dissolution rate. The reduced density is associated with increased porosity, and it also reduces the amount of materials that need to be dissolved. The effective density ρ_{eff} of porous material is related to the density ρ_s of the fully dense materials as $\rho_{eff} = \rho_s V_s/(V_s + V_{air})$, where V_s and V_{air} are the volumes of the porous material and air cavity, respectively. To account for the density variation, the reactive diffusion equation in Equation (21.1) can be modified by replacing the diffusivity D with an effective diffusivity D_e.[46] The effective diffusivity of water in a porous medium is linearly proportional to its pore fraction, i.e., $D_e \propto V_{air}/(V_{air} + V_s) = (\rho_s - \rho_{eff})/\rho_s$. When the air pores are filled with water at time $t = 0$, the initial condition can also be updated to $w|_{t=0} = w_0(\rho_s - \rho_{eff})/\rho_s\ (0 \le y < h_0)$. With the same boundary conditions as those for Equation (21.1), the normalized thickness is obtained as[46]:

$$\frac{h}{h_0} = 1 - \frac{w_0 M}{q\rho_{eff} M_{H_2O}} \frac{kh_0^2}{D_e} \left\{ \frac{D_e t}{h_0^2} \frac{\tanh\sqrt{\frac{kh_0^2}{D_e}}}{\sqrt{\frac{kh_0^2}{D_e}}} - 2\sum_{n=1}^{\infty} C_n \frac{1 - e^{-\left[\frac{kh_0^2}{D_e} + \left(n-\frac{1}{2}\right)^2 \pi^2\right]\frac{D_e t}{h_0^2}}}{\left[\frac{kh_0^2}{D_e} + \left(n - \frac{1}{2}\right)^2 \pi^2\right]^2} \right\}, \tag{21.4}$$

where C_n is $C_n = \left[kh_0^2/D_e + (n-1/2)^2 \pi^2\right]^{-1} + (\rho_{eff}/\rho_s - 1)/(n-1/2)^2 \pi^2$. The dissolution rate is then approximated as $-dh/dt = kh_0 w_0 M/(q\rho_{eff} M_{H_2O}) \tanh\sqrt{kh_0^2/D_e}/\sqrt{kh_0^2/D_e}$.

Application of transient electronics requires a stable operation of devices in a certain timeframe before they completely disappear. Encapsulation or packaging layers are typically used on top of the device towards such goal. For instance, MgO can serve as an encapsulation layer for Mg. To provide physical insight for such a layered system, an analytical model has also been established.[45] In the model, zero initial condition at $t = 0$ applies to both Mg and MgO layers. The reactive diffusion equation in Equation (21.1) and the zero water flux boundary condition at the bottom surface of Mg still hold for the Mg layer (Figure 21.3a). As for the MgO encapsulation with an initial thickness of $h_{\text{encapsulation}}$, the reactive diffusion equation in Equation (21.1) becomes: $D_{\text{encapsulation}} \frac{\partial^2 w}{\partial y^2} - k_{\text{encapsulation}} w = \frac{\partial w}{\partial t}\ (h_0 \le y \le h_0 + h_{\text{encapsulation}})$, where $D_{\text{encapsulation}}$ and $k_{\text{encapsulation}}$ are the diffusivity of water in MgO and the rate of the reaction between MgO and water, respectively.[45] The constant water concentration boundary condition at the MgO/water interface is $w|_{y=h_0+h_{\text{encapsulation}}} = w_0$. At the MgO/Mg interface, the continuity conditions of water concentration and flux are $w|_{y=h_0-0} = w|_{y=h_0+0}$ and $D\partial w/\partial y|_{y=h_0-0} = D_{\text{encapsulation}}\, \partial w/\partial y|_{y=h_0+0}$. The analytical solution of this bi-layered model also reveals a scaling law. In addition to the three non-dimensional parameters (y/h_0, Dt/h_0^2, and kh_0^2/D) as in the single-layer solution, the normalized water concentration w/w_0 also depends on the normalized reaction constant $k_{\text{encapsulation}} h_{\text{encapsulation}}^2 / D_{\text{encapsulation}}$ of the encapsulation layer, the diffusivity ratio $D_{\text{encapsulation}}/D$, and initial thickness ratio $h_{\text{encapsulation}}/h_0$. When the diffusivity ratio is small ($D_{\text{encapsulation}}/D \ll 1$ as in the Mg/MgO bi-layered structure), the water concentration in Mg is relatively uniform, whereas that in MgO is not (Figure 21.3c). The water concentration in the Mg layer increases much slower as opposed to that in Figure 21.3b, which effectively extends the lifetime of Mg.

In contrast to the intermittent thickness measurement, the measurement of electrical properties can be continuous. Such electrical measurements can also be used to quantify the dissolution behavior of a conductive transient material below a non-conductive layer (e.g., Mg below MgO). Considering the fact that a very significant change occurs in the thickness direction during dissolution as opposed to those in the width and length directions, a simple expression of the electric resistance of the conductive transient material is given as $R = R_0 h_0/h$, where R_0 is the initial resistance. As shown in Figure 21.3d, the normalized electrical resistance R/R_0 as a function of the normalized time Dt/h_0^2 predicted from the analytical model agrees reasonably well with the experimental measurements. The applicability of the analytical model goes beyond Mg to the other transient metals that include Mg alloy, zinc, tungsten (W), and molybdenum.[26] The prediction from the model can reproduce the dissolution behaviors of these transient metals in both DI water and simulated body fluids (e.g., Hanks' solution with pH from 5 to 8).

21.5 A Variety of Water-Soluble Electronic Devices and Systems

Various types of electronic devices, sensors, and systems including those from simple electronic passive and active elements to sensors and power supply elements can be fabricated by using dissolvable materials. Figure 21.4 presents images and electrical characteristics of diverse biodegradable electronic devices, involving physical/chemical sensors, optoelectronic systems, and essential components for wireless communication such as an inductor-capacitor (L-C) circuit, rectifiers, and oscillators.[14,20] A narrow-strip type of Si NMs slightly doped with boron senses physical movements in bi-directional modes with dissolvable Mg contact pads and interconnects on a thin sheet of silk for an applicable strain senor as shown in the left frame of Figure 21.4a.[14] Tensile and compressive loading to the sensors induced by mechanically repetitive deformations can be detected by the fractional changes of electrical resistance, and gauge factor is about ~40 (right). Figure 21.4b shows an optical image and properties of an array of a transient electrochemical sensor to evaluate hydration levels as a physiological parameter.[19] Phosphorous-doped Si NMs served as an active electrode and Mg, SiO_2, and PLGA are used as interconnects, interlayer dielectrics, and substrate, respectively (left). Comparison of hydration levels with three different types of electrodes (blue, circular electrode; red, interdigitated electrode; black, a commercial moisture meter [CMM]) indicates that detection capability of transient hydration sensors are comparable to that of a standard tool (right). Other types of Si NMs-based electrochemical sensors are demonstrated in previous reports.[32] An 8 × 8 array of Si p-n diodes can produce a system level of a photodetector as appears in Figure 21.4c (left).[14] Individual photodiodes combined with a blocking diode operates as a single pixel of a digital imaging system, enabling obtainment of various kinds of images through passive matrix addressing with external interfaces. The inset shows the original pattern (right). Figure 21.4d shows an image of an inductor-capacitor oscillator built with planar spiral coils in several turns for an inductor and metal-insulator-metal (MIM) structure for a capacitor.[14] Here, metallic electrodes and interlayer dielectrics used Mg and MgO, respectively (left). These radio-frequency (RF) passive elements operate as an antenna through near-field mutual inductance coupling to separately powered, external primary coils, and the operating frequencies vary up to ~3 GHz (right). Inductors and capacitors with different types and sizes in high-speed designs are possible.[20] A silicon-based transient RF diode in Figure 21.4e exhibits a high forward current of 1.5 mA at 1 V with a turn-on voltage of 0.7 V, including a representative image of the device in the inset (left).[20] Design parameters of the diodes exploit short channel lengths and wide channel widths, and cover most silicon contact areas with Mg electrodes to maximize the speed of the RF device, enabling RF diodes to operate up to a few GHz as a rectifier (right). Figure 21.4f shows CMOS-based three-stage ring oscillators built using n-type and p-type MOSFETs based on transient Si NMs (left).[20] The oscillation frequency in time-dependent output voltage can be controlled up to 4.1 MHz with different applied voltages (V_{dd}), 10 V (black), 15 V (red), and 20 V (blue) (right).

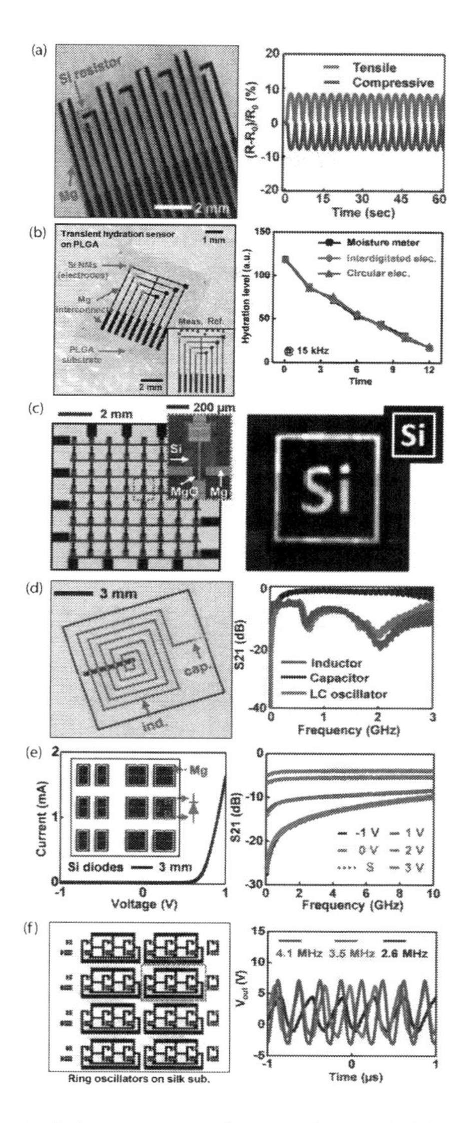

FIGURE 21.4 Images and electrical characteristics of various bioresorbable devices, sensors, and systems. (a) An optical photography of a resistive type of a Si strain sensor with Mg interconnects (left), and their functional changes in electrical resistance as a function of time under repetitive tensile (red) and compressive (blue) bending (right). (b) Image of a bioresorbable hydration sensor, comprised of doped Si NMs as measurement electrodes, Mg contacts for interconnects and reference electrodes, and PLGA for the substrate (left). Electrical performance of fabricated sensors with different electrode types, e.g., interdigitated (red) and circular (blue) electrodes, was compared to a commercial, standard moisture meter (black) at a frequency of 15 kHz (right). (c) An 8 × 8 array of individually addressable Si diodes-based photodetectors with blocking diodes, including detailed device layout in the inset (left). An image obtained using 8 × 8 passive matrix of transient photodetectors, with the original pattern in the inset (right). (d) A transient inductor-capacitor oscillator fabricated with Mg electrodes and MgO dielectric layers (left). Measurements of the S21 scattering parameter of an inductor (blue), capacitor (black), and inductor-capacitor oscillator (red) at frequencies up to 3 GHz (right). (e) Current-voltage characteristics of a silicon RF diode, including a device image in the inset (left). Experimental values (lines) and simulations (dots) of the S21 scattering parameter at frequencies up to 10 GHz with different DC biases (right). (f) Image of Si CMOS-based three-stage ring oscillators on a water-soluble silk substrate (left). Measured output responses of a ring oscillator at different frequencies (black, 2.6 MHz; red, 3.5 MHz; blue, 4.1 MHz) in time domain (right).

21.6 Transient Energy Harvesters and Batteries

Recent breakthroughs of transient energy systems in several ways could power transient electronic devices. Arrays of Si diodes in Figure 21.5a represent photovoltaic cells using a single crystal silicon micro ribbon with the thickness of a few microns (left).[14] Measured current density (red) and power (blue) indicate that the power conversion efficiency is ~3% (right). ZnO, an alternative water-soluble semiconductor, is interesting due to its property of the generation of electricity.[24] Figure 21.5b

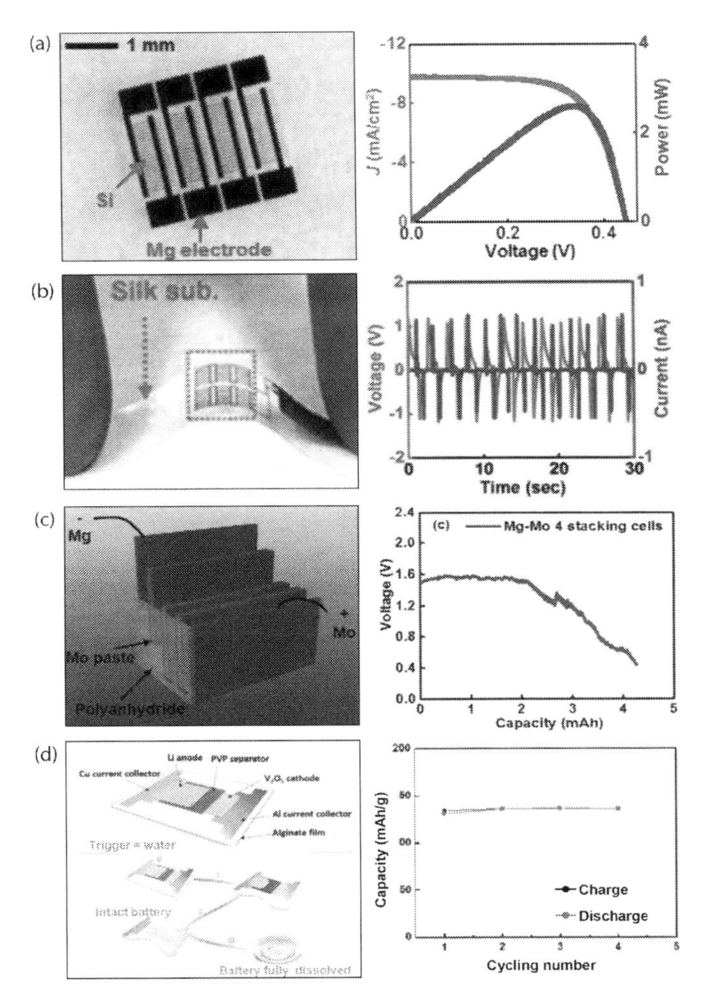

FIGURE 21.5 Examples of bioresorbable energy components and their electrical characteristics. (a) An array of water-soluble photovoltaic cells based on long, narrow pieces of thin Si membranes (~3 μm thick) with Mg contacts/interconnects (left). Measurements of current density (red) and power (blue) as a function of voltage while illuminated by a solar simulator (right). (b) Image of a bioresorbable ZnO-based energy harvester on a silk substrate while a deformed configuration (left). Output electrical performance of voltage and current values during cycles of bending in time domain (right). (c) Schematic illustration of a completely biodegradable primary battery that consists of four Mg-Mo cells in series (left). Discharging behavior of a primary magnesium battery as a function of capacity (right). (d) Description of structural arrangement of a rechargeable bioresorbable battery, their triggered dissolution procedure through a cascade chemical reaction (left). Electrical behaviors of charge (black) and discharge (red) cycling performance of the transient battery (right).

shows an array of transient mechanical energy harvesters (MEH) on a flexible silk substrate that utilize a capacitor type of geometry using Mg conductors as top and bottom electrodes to the ZnO layer.[24] Mechanical motions produce electrical energy as expected, generating positive and negative output voltages (red) and currents (blue) from ~1.14 V and ~0.55 nA under tensile and compressive strains. A wireless, long-range power scavenging system with a complex, advanced layout is possible with a combination of RF rectifiers, inductors, resistors, capacitors, and an antenna.[20] Figure 21.5c illustrates a fully biodegradable primary battery based on four pairs of Mg-Mo cells in a series for multiplying the output power (left).[25] The four stacking battery cells use Mg foils as the anode material, Mo foils for the cathode electrodes, and a sheet of polyanhydride for spacers between individual cells as well as barriers of the electrolytes between a couple of Mg-Mo cells. Evaluations on combinations of Mg with other dissolvable metals provide more details of characteristics and underlying chemistry of the primary battery.[25] Discharging behavior of the battery at a constant current density of 0.1 mA cm^{-2} presents a stable output voltage of 1.6 V for up to 3 hours. As opposed to a disposable or primary battery, a transient form of rechargeable battery is designed to undergo several chemical reactions for complete dissolution once triggered by water.[47] Figure 21.5d illustrates materials, a device structure, and a configuration of a transient rechargeable battery, and its multi-step reactions after activation by water (left).[47] The rechargeable battery consists of vanadium oxide (V_2O_5) as a cathode, lithium (Li) as an anode, polyvinylpyrrolidone (PVP) for a separator, and sodium alginate for a battery encasement. Such battery supplies an energy of ~0.29 mWh at the operating voltage of ~2.8 V and represents a high Coulombic efficiency of 99% and stable electrochemical performance during repeated cycles of a charging (black) and discharging (red) test (right). The rechargeable battery can be completely dissolved in water triggered by cascaded reactions within minutes after useful timeframes.

21.7 Biocompatibility of Key Materials for Bioresorbable Electronics

Various potential applications for bioresorbable electronic implants with silicon-based elements require thorough *in vitro/vivo* investigations of constituent materials for their toxicity and degradability.[22] For *in vitro* evaluations of dissolution behaviors and cell toxicity of Si NMs, a patterned array of Si NMs were fabricated and installed to culture metastatic breast cancer cells (MDA-MB-231) that are useful for rapid growth and culture. The series of differential contrast images illustrate the proliferation and propagation behaviors of cells during the cell culture, and the array of dot patterns of Si NMs were no longer observed after 4 days due to complete dissolution in the cell medium (Figure 21.6a). Figure 21.6b presents the time sequence of fluorescent images during growth and proliferation for viability through analysis of numbers of live and dead cells at various time stages of 1, 5, and 10 days, respectively. Here, live/dead assays exhibited cell viability, i.e., green indicates living cells and red indicates dead cells. Evaluations on *in vivo* cytotoxicity of Si NMs (semiconductor) as well as other transient electronic materials such as Mg (conductor), MgO (insulator), and silk (substrate) appear in Figure 21.6c. Histological examinations from implanted skin sections stained with hematoxylin and eosin (H&E) revealed no significant harmful responses. Numbers of individual cells, polymorphonuclear cells (PMNs), lymphocytes, plasma cells, and fibrosis, near implanted regions of transient electronic materials are comparable to those of a control material, high-density polyethylene (HDPE), although the fibrosis somewhat increased at the HDPE-implanted area due to permeation of the protein constituent (collagen) forming fibroblasts.[22,23]

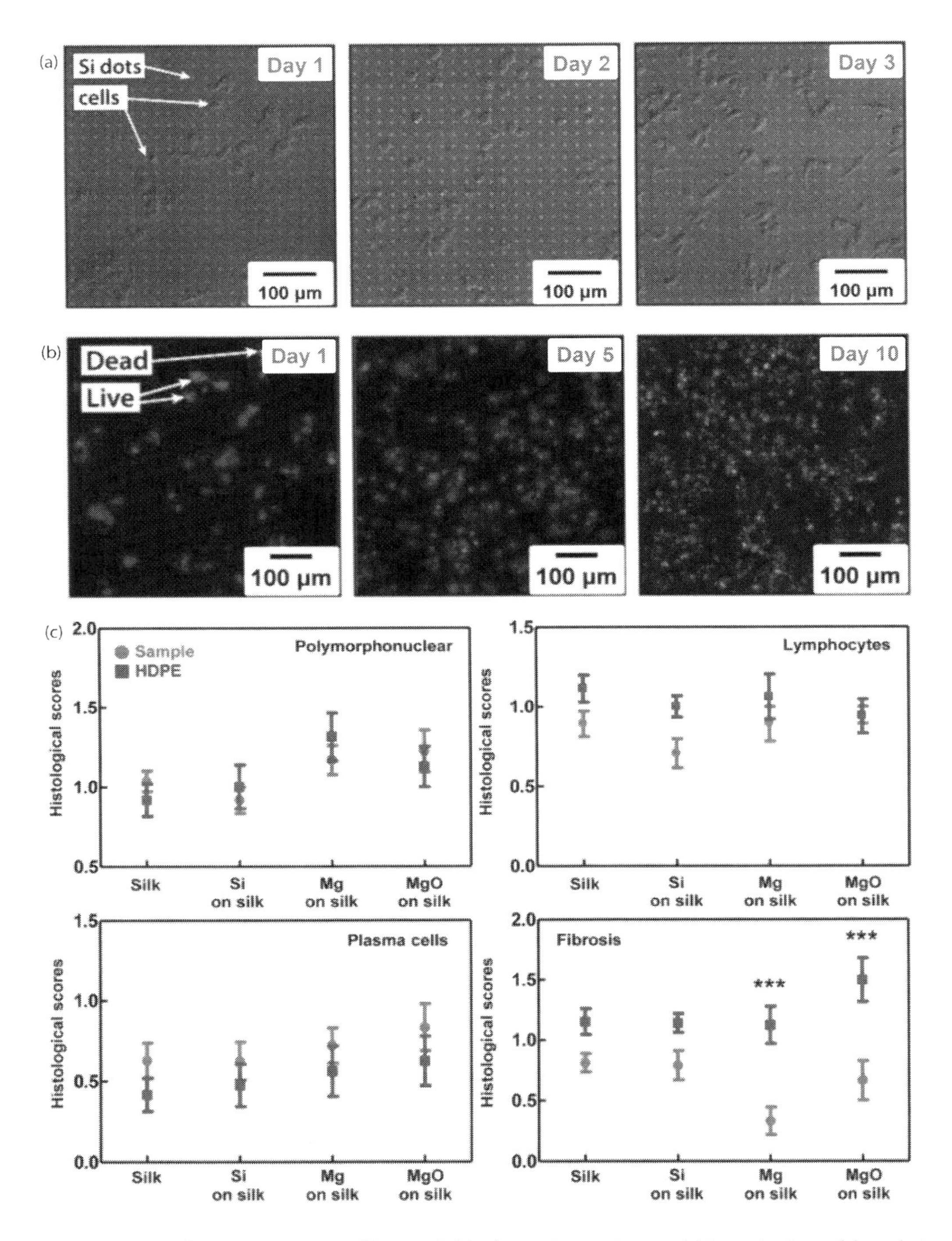

FIGURE 21.6 Evaluations on toxicity of bioresorbable electronic constituents. (a) Investigations of degradation and cytotoxicity associated with patterned Si NMs using *in vitro* cell culture. Differential interference contrast images describing dissolution behaviors of an array of Si dots with adhered cells over 1, 2, and 3 days. (b) A series of fluorescent images indicating cell viability using live/dead assay on Si NMs at 1, 5, and 10 days, respectively. (c) Histological scores of various tissues at 5 weeks of implantation, based on haematoxylin and eosin (H&E) staining of implanted skin sections from different groups of animals.

21.8 Biomedical Applications for Transient, Bioresorbable Electronic System

Collection of stable, reliable electrophysiological signal measurements for accurate diagnoses/treatments of diseases would be important. Figure 21.7a represents a platform of passive transient sensors associated with electrophysiological signals for skin-mount systems, capable of providing

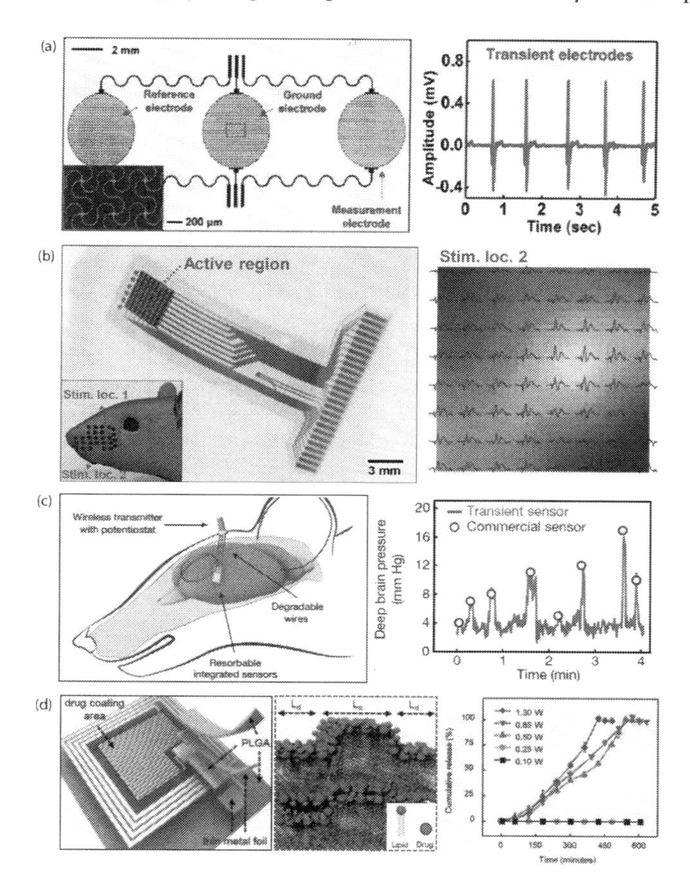

FIGURE 21.7 Various medical applications for transient, bioresorbable electronics. (a) An optical image of a soft, biocompatible electrophysiological (EP) sensor for an epidermal electronic system, including magnified view of Mg contacts and interconnects in stretchable configurations in the inset (left). Temporal measurement of electrocardiogram (ECG) using a transient device mounted on the chest (right). (b) Photograph of an actively addressable, multiplexed bioresorbable brain electrode for monitoring high spatiotemporal electrical activity, electrocorticography (ECoG). The inset describes stimulation locations on the whisker in an animal model (left). Spatial color map indicates measured locations of relative, evoked potential through the active matrix array after stimulating the barrel cortex, exhibiting good agreement in between stimulated coordinates of the barrel cortex and locations of evoked potentials via a bioresorbable ECoG array. (c) Schematic illustration of a wireless bioresorbable electronic system for recording ICP/ICT, with degradable wires as electrical interconnects that provide an interface to an external wireless data-transmission unit for a long-range operation (left). Comparison of measured intracranial pressures through a commercial sensor with a transient monitor. The recorded pressure from both sensors indicates that the performance of implanted wireless, transient brain sensors is comparable to that of a standard, commercial sensor (blue, commercial ICP sensor; red, transient ICP sensor). (d) Exploded view schematic illustration of bioresorbable, wirelessly programmable drug delivery vehicle that exploits thermally triggered lipid membranes loaded with drug molecules (left). *In vitro* assessments of a cumulative amount of programmed drug release (doxorubicin), activated by various wireless power values (right).

real-time, wireless data recording from the body, and reducing skin irritations because of utility of non-toxic materials.[32] Materials used for such a standard type sensor include a combination of thin layers of a conductor (Mg, 300 nm) and a dielectric (SiO_2, 100 nm), patterning into the design of filamentary serpentine (FS) structure to obtain conformal contact to any types of skins and/or tissues. Minimized physical/mechanical mismatches at interfaces are capable of improving electrical behaviors of skin-like electronic patches, for example, measured performance of electrocardiogram (ECG) with a capacitive transient sensor could excel over that of a commercial, wet electrode. Device implants into critical regions such as the brain would maximize the effectiveness of bioresorbable electronics. Figure 21.7b shows an array of actively multiplexed neural electrodes, consisting of 128 metal-oxide-MOSFETs based on Si NMs (~300 nm) as the active semiconductor materials and the neural sensing interfaces, and other conductors and gate/interlayer dielectrics rely on thin layers of Mo (~300 nm) and SiO_2 (~100 nm) and/or combinations of SiO_2 and Si_3N_4.[16] Demonstration of *in vivo* microscale electrocorticography (μECoG) recordings with a transient, multiplexed recording array presents spatial distribution of the amplitude of the evoked potential measured at the cortical surface by the array, which coincides with relative positions of activated whiskers on the cortex as appeared in the color map. In addition to spatial distributions, epileptiform activity can be captured using the same device. A device, module, or subsystem that intends to monitor and detect particular events and/or changes in the environment of the brain can expand neuroscientific researches beyond the traditional field of electrophysiology measurements. Assembly of electronic components and circuitry with consideration of physical structures creates wireless, bioresorbable intracranial pressure/temperature (ICP/ICT) monitors that enable recording real-time changes in the brain (Figure 21.7c).[15] Implanted sensors were connected to a wireless radio module mounted on the top of the skull via a conductive, dissolvable Mo wire. Here, the implanted intracranial pressure sensor was designed to maximize sensitivity by creating a cavity configuration, which collected measurements recorded with a transient ICP sensor, well-matched the data obtained using a standard, conventional ICP sensor in similar places. The ideal approach of pharmacological treatment is to deliver drugs through on-demand, localized release in accurate and controlled patient-specific time sequences.[48] Figure 21.7d presents a biodegradable, wirelessly programmable drug delivery system (DDS) with a lipid membrane loaded with drug molecules (left). A DDS device was equipped with a micro-heater to thermally activate a periodic drug release, a resonant receiver coil for wireless power transmission, and multilayers of uniform biological lipid membranes (1,2-dilauroyl-sn-glycero-3-phosphoethanolamine, dipalmitoylphosphatidylcholine, dipalmitoylphosphatidylglycerol, or 1,2-dioleoyl-3-trimethylammonium-propane and cholesterol) loaded with hydrophilic drug particles. The detailed formation of lipid membranes (middle) involves two coexisting phases: an ordered phase (L_o) with rigid all-trans tails and an unordered phase (L_d) with alkyl tails for swift changes in arrangement by a critical temperature. Profiles of a cumulative released amount of doxorubicin demonstrated wireless, on-demand, and tunable operating characteristics based on various external powers, allowing this type of bioresorbable drug system to envision the treatment of chronic disease with time-controlled pulsatile delivery of multi-dosage for a therapeutic regimen.

21.9 Potential Applications of Transient Electronics

Besides biomedicine, transient electronics have a wide scope of potential applications for technology involving hardware secure systems, vanishing environmental monitors/sensors, and compostable "eco-friendly" electronics. Examples of approaching to ensure electronic information in Figure 21.8a[49] and b[50] utilized principles of destroying critical components and/or entire systems through dissolution in a liquid form and ignition in a flammable format. The former constitutes a transient activation circuitry integrated with a microfluidic system that ceased operation of a radio-frequency identification (RFID) device whose disappearance of a near field communication (NFC) chip appeared in the inset of Figure 21.8a. The transient triggering system was initiated by turning on individual heaters to inflate

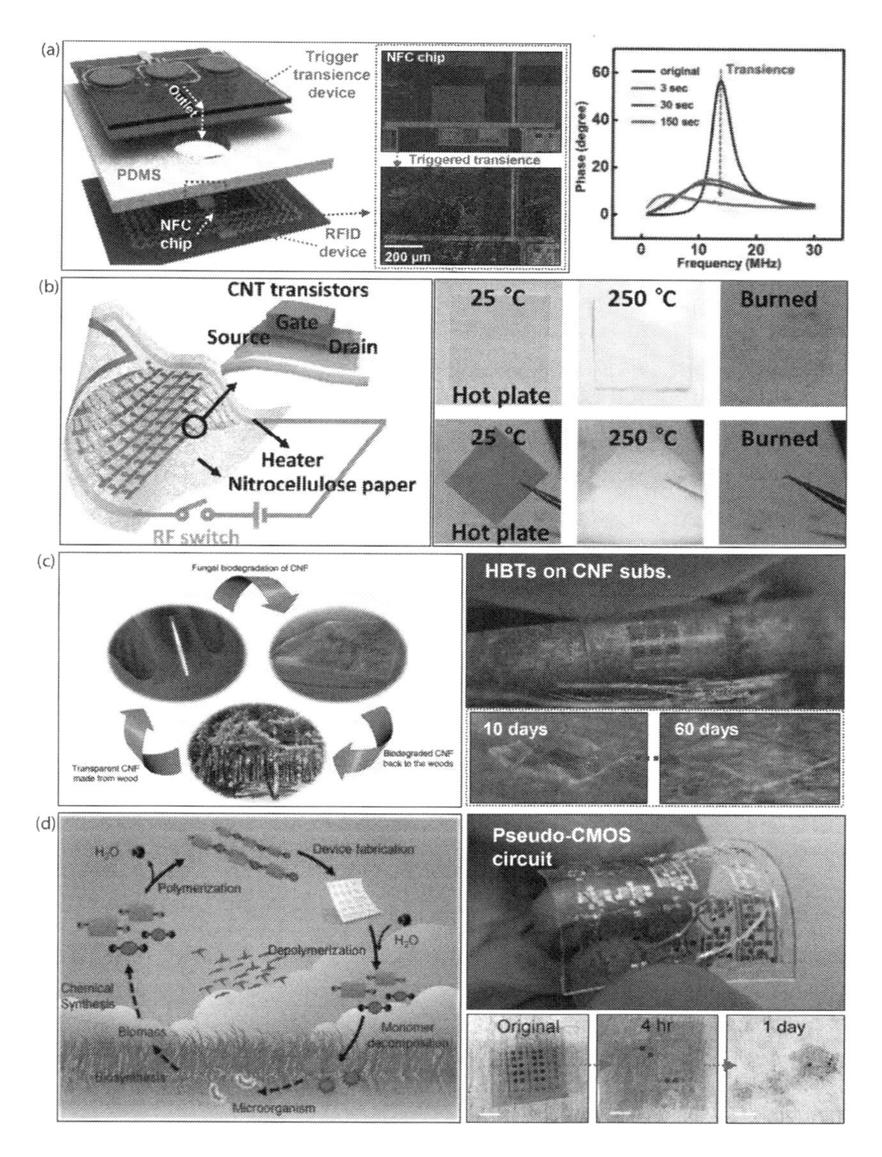

FIGURE 21.8 Potential applications for transient electronic system. (a) Exploded view drawing of a wireless, transient RFID device, capable of triggering release of etchant through a microfluidic system to physically/chemically vanish all components. The inset exhibits images of a commercial NFC chip before (top) and after (bottom) chemical dissolution of NFC chips by the triggered release of solution (left). Temporal transience of phase (degree) under the continuous release of a chemical solvent (KHF), on a range of frequencies up to 30 MHz (right). (b) Schematic illustration of flammable transient electronics based on a top-gate carbon nanotube (CNT) transistors on a nitrocellulose paper, with a resistive heater built in behind paper substrate. The voltage induced by the heater triggers the RF switch (left). Thermal trigger test where the entire area (top) and the edge of the nitrocellulose paper (bottom) rest on the hot plate. The flammable paper burned completely and quickly at 250°C with no residue. (c) Illustration of a cyclic life of a bio-based and compostable CNF paper (left). A photograph of an array of heterojunction bipolar transistors (HBTs) on a CNF substrate wrapped around a tree stick (right), and the inset image shows decomposed CNF-based electronics by a microorganism during 60 days. (d) Description of various procedures of organic materials-based electronic system in nature for green electronics (left). A fabricated disintegrable pseudo-CMOS circuit based on a degradable organic semiconductor, polydiketopyrrolopyrrole-*p*-phenylene diamine (PDPP-PD), onto an elastomer, and their optical images of disintegration at various time stages (right).

thermal expansion polymer-based reservoirs, enabling water to flow into fluidic channels containing powders of KOH, KHF, and NaOH for destruction of the system elements. Electrical characteristics of the vanishing electronics exhibited that phases of RFID were drastically changed in a few seconds after the triggered flow of etchants in the right frame of Figure 21.8a. The use of a thermo-responsive polymer contributes to construct a heat-triggered, combustible transient device (Figure 21.8b).[50] Single-walled carbon nanotubes (SCNTs)-based transistors incorporated with a RF heater were built on a flammable nitrocellulose paper that was very stable at room temperature (~25°C), but abruptly burned while exposed to excessive heat by the RF heater, without any residue remaining. Biodegradable elements could break down or decompose into minimum units by biological means or be consumed by microorganisms. Such nature-inspired collections of chemical compounds and materials provide critical components to create a new class of electronics that can offer undiscovered properties and/or characteristics. Cellulose nanofibril (CNF) paper originated from woods, an environmentally friendly product that is able to decompose back into natural elements, has transparent, flexible, and good insulating performance (Figure 21.8c).[51] A set of microwave GaAs-based active and passive components, ranging from heterojunction bipolar transistors (HBTs), digital logic circuits to inductors and capacitors, can be integrated onto a CNF substrate for their superior high-frequency performance and power management although their tiny design layouts and high temperature procedures limit building those systems on alien substrates. A completed array of microwave electronic components were gradually degraded when exposed to fungi, eventually fungus completely covered the device after 60 days. Despite the drawback of inferior electrical performance, organic materials derived from living organisms or lab-synthesized matters have advantages over inorganic elements in terms of natural biodegradation. Figure 21.8d presents an ultrathin, flexible transient electronic using a biocompatible and disintegrable semiconducting polymer of polydiketopyrrolopyrrole-*p*-phenylene diamine (PDPP-PD) that synthesized diketopyrrolopyrrole (DPP) and *p*-toluenesulfonic acid (PTSA). The synthesized semiconductor served the channel material for a pseudo-CMOS circuit on a cellulose substrate, and such components can be disintegrated in a pH 4.6 buffer solution, most constituents were fully degraded within 30 days.[52]

21.10 Conclusion

Bioresorbable electronics described in this chapter introduce the idea of the system operation, various types of dissolvable materials and electronic components, and widespread potential applications. In the materials context, dissolution experiments and theoretical studies established the underlying chemistry and kinetics of key elements of bioresorbable semiconductors, conductors, insulators, and substrates in extensive sorts of aqueous solutions and fluids. A variety of manufactured resorbable devices provided a baseline of an elaborated, sophisticated system level of electronics including typical passive sensors, silicon-based active components, and power management systems. *In vitro* and *in vivo* inspections support the non-toxicity of essential electronic constituents of active/passive bioresorbable electronics, offering an opportunity for applying such technology to unrevealed fields of areas. Numbers of potential applications were exploited, ranging from medical implants and epidermal devices, to security systems and compostable/green/eco-friendly electronic components.

Future research directions would include an effort to find innovative materials or synthetic substances for expanding the options of available resources. A cost-effective, optimized fabrication process will be required to accomplish a large-scale, assembly-line, mass production of technologies using standardized designs. A research field that is interesting, but challenging or needs to be focused on involve development of energy transmission or management technologies to provide sufficient power supply for an ultimate wireless digital communication.

References

1. D.-H. Kim, N. Lu, R. Ma, Y.-S. Kim, R.-H. Kim, S. Wang, J. Wu et al., Epidermal electronics. *Science* **2011**, *333*, 838.
2. J. Viventi, D.-H. Kim, L. Vigeland, E. S. Frechette, J. A. Blanco, Y.-S. Kim, A. E. Avrin et al., Flexible, foldable, actively multiplexed, high-density electrode array for mapping brain activity *in vivo*. *Nat. Neurosci.* **2011**, *14*, 1599.
3. J. Kim, M. Lee, H. J. Shim, R. Ghaffari, H. R. Cho, D. Son, Y. H. Jung et al., Stretchable silicon nanoribbon electronics for skin prosthesis. *Nat. Commun.* **2014**, *5*, 5747.
4. A. Koh, D. Kang, Y. Xue, S. Lee, R. M. Pielak, J. Kim, T. Hwang et al., A soft, wearable microfluidic device for the capture, storage, and colorimetric sensing of sweat *Sci. Transl. Med.* **2016**, *8*, 366ra165.
5. J. Park, S. Choi, A. H. Janardhan, S.-Y. Lee, S. Raut, J. Soares, K. Shin et al., Electromechanical cardioplasty using a wrapped elasto-conductive epicardial mesh. *Sci. Transl. Med.* **2016**, *8*, 344ra86.
6. C. Dagdeviren, B. D. Yang, Y. Su, P. L. Tran, P. Joe, E. Anderson, J. Xia et al., Conformal piezoelectric energy harvesting and storage from motions of the heart, lung, and diaphragm. *Proc. Natl. Acad. Sci.* **2014**, *111*, 1927.
7. J. Viventi, D.-H. Kim, J. D. Moss, Y.-S. Kim, J. A. Blanco, N. Annetta, A. Hicks et al., A Conformal, Bio-Interfaced Class of Silicon Electronics for Mapping Cardiac Electrophysiology. *Sci. Transl. Med.* **2010**, *2*, 24ra22.
8. R. C. Webb, Y. Ma, S. Krishnan, Y. Li, S. Yoon, X. Guo, X. Feng et al., Epidermal devices for noninvasive, precise, and continuous mapping of macrovascular and microvascular blood flow. *Sci. Adv.* **2015**, *1*, e1500701.
9. C. Dagdeviren, Y. Shi, P. Joe, R. Ghaffari, G. Balooch, K. Usgaonkar, O. Gur et al., Conformal piezoelectric systems for clinical and experimental characterization of soft tissue biomechanics. *Nat. Mater.* **2015**, *14*, 728.
10. L. Xu, S. R. Gutbrod, A. P. Bonifas, Y. Su, M. S. Sulkin, N. Lu, H.-J. Chung et al., 3D multifunctional integumentary membranes for spatiotemporal cardiac measurements and stimulation across the entire epicardium. *Nat. Commun.* **2014**, *5*, 3329.
11. T. Yokota, Y. Inoue, Y. Terakawa, J. Reeder, M. Kaltenbrunner, T. Ware, K. Yang et al., Ultraflexible, large-area, physiological temperature sensors for multipoint measurements. *Proc. Natl. Acad. Sci.* **2015**, *112*, 14533.
12. W. Gao, S. Emaminejad, H. Y. Y. Nyein, S. Challa, K. Chen, A. Peck, H. M. Fahad et al., Fully integrated wearable sensor arrays for multiplexed in situ perspiration analysis. *Nature* **2016**, *529*, 509.
13. D.-H. Kim, N. Lu, R. Ghaffari, Y.-S. Kim, S. P. Lee, L. Xu, J. Wu et al., Materials for multifunctional balloon catheters with capabilities in cardiac electrophysiological mapping and ablation therapy. *Nat. Mater.* **2011**, *10*, 316.
14. S.-W. Hwang, H. Tao, D.-H. Kim, H. Cheng, J.-K. Song, E. Rill, M. A. Brenckle et al., Materials for multifunctional balloon catheters with capabilities in cardiac electrophysiological mapping and ablation therapy. *Science (80-.).* **2012**, *337*, 1640.
15. S.-K. Kang, R. K. J. Murphy, S.-W. Hwang, S. M. Lee, D. V. Harburg, N. A. Krueger, J. Shin et al., Bioresorbable silicon electronic sensors for the brain. *Nature* **2016**, *530*, 71.
16. K. J. Yu, D. Kuzum, S.-W. Hwang, B. H. Kim, H. Juul, N. H. Kim, S. M. Won et al., Bioresorbable silicon electronics for transient spatiotemporal mapping of electrical activity from the cerebral cortex. *Nat. Mater.* **2016**, *15*, 782.
17. H. Tao, S.-W. Hwang, B. Marelli, B. An, J. E. Moreau, M. Yang, M. A. Brenckle et al., Silk-based resorbable electronic devices for remotely controlled therapy and in vivo infection abatement. *Proc. Natl. Acad. Sci.* **2014**, *111*, 17385.

18. M. Irimia-Vladu, P. A. Troshin, M. Reisinger, L. Shmygleva, Y. Kanbur, G. Schwabegger, M. Bodea et al., Edible Electronics: Biocompatible and Biodegradable Materials for Organic Field-Effect Transistors. *Adv. Funct. Mater.* **2010**, *20*, 4069.
19. S.-W. Hwang, J.-K. Song, X. Huang, H. Cheng, S.-K. Kang, B. H. Kim, J.-H. Kim, S. Yu, Y. Huang, J. A. Rogers, High-Performance Biodegradable/Transient Electronics on Biodegradable Polymers. *Adv. Mater.* **2014**, *26*, 3905.
20. S.-W. Hwang, X. Huang, J.-H. Seo, J.-K. Song, S. Kim, S. Hage-Ali, H.-J. Chung et al., Materials for Bioresorbable Radio Frequency Electronics. *Adv. Mater.* **2013**, *25*, 3526.
21. S.-W. Hwang, D.-H. Kim, H. Tao, T.-I. Kim, S. Kim, K. J. Yu, B. Panilaitis et al., Materials and Fabrication Processes for Transient and Bioresorbable High-Performance Electronics. *Adv. Funct. Mater.* **2013**, *23*, 4087.
22. S.-W. Hwang, G. Park, C. Edwards, E. A. Corbin, S.-K. Kang, H. Cheng, J.-K. Song et al., Dissolution Chemistry and Biocompatibility of Single-Crystalline Silicon Nanomembranes and Associated Materials for Transient Electronics. *ACS Nano* **2014**, *8*, 5843.
23. S.-W. Hwang, G. Park, H. Cheng, J.-K. Song, S.-K. Kang, L. Yin, J.-H. Kim et al., Materials for High-Performance Biodegradable Semiconductor Devices. *Adv. Mater.* **2014**, *26*, 1992.
24. C. Dagdeviren, S.-W. Hwang, Y. Su, S. Kim, H. Cheng, O. Gur, R. Haney, F. G. Omenetto, Y. Huang, J. A. Rogers, Transient, Biocompatible Electronics and Energy Harvesters Based on ZnO. *Small* **2013**, *9*, 3398.
25. L. Yin, X. Huang, H. Xu, Y. Zhang, J. Lam, J. Cheng, J. A. Rogers, Materials, Designs, and Operational Characteristics for Fully Biodegradable Primary Batteries. *Adv. Mater.* **2014**, *26*, 3879.
26. L. Yin, H. Cheng, S. Mao, R. Haasch, Y. Liu, X. Xie, S.-W. Hwang et al., Transient Electronics: Dissolvable Metals for Transient Electronics. *Adv. Funct. Mater.* **2014**, *24*, 645.
27. S.-K. Kang, S.-W. Hwang, H. Cheng, S. Yu, B. H. Kim, J.-H. Kim, Y. Huang, J. A. Rogers, Dissolution Behaviors and Applications of Silicon Oxides and Nitrides in Transient Electronics. *Adv. Funct. Mater.* **2014**, *24*, 4427.
28. Y. Gao, K. Sim, X. Yan, J. Jiang, J. Xie, C. Yu, Thermally Triggered Mechanically Destructive Electronics Based On Electrospun Poly(ε-caprolactone) Nanofibrous Polymer Films. *Sci. Rep.* **2017**, *7*, 947.
29. C. J. Bettinger, Z. Bao, Organic thin-film transistors fabricated on resorbable biomaterial substrates. *Adv. Mater.* **2010**, *22*, 651.
30. D.-H. Kim, J. Viventi, J. J. Amsden, J. Xiao, L. Vigeland, Y.-S. Kim, J. A. Blanco et al., Dissolvable films of silk fibroin for ultrathin conformal bio-integrated electronics. *Nat. Mater.* **2010**, *9*, 511.
31. M. A. Brenckle, H. Cheng, S.-W. Hwang, H. Tao, M. Paquette, D. L. Kaplan, J. A. Rogers, Y. Huang, F. G. Omenetto, Modulated Degradation of Transient Electronic Devices through Multilayer Silk Fibroin Pockets. *ACS Appl. Mater. Interfaces* **2015**, *7*, 19870.
32. S.-W. Hwang, C. H. Lee, H. Cheng, J.-W. Jeong, S.-K. Kang, J.-H. Kim, J. Shin et al., Biodegradable Elastomers and Silicon Nanomembranes/Nanoribbons for Stretchable, Transient Electronics, and Biosensors. *Nano Lett.* **2015**, *15*, 2801.
33. S.-K. Kang, G. Park, K. Kim, S.-W. Hwang, H. Cheng, J. Shin, S. Chung et al., Dissolution Chemistry and Biocompatibility of Silicon- and Germanium-Based Semiconductors for Transient Electronics. *ACS Appl. Mater. Interfaces.* **2015**, *7*, 9297.
34. N. T. Kirkland, N. Birbilis, M. P. Staiger, Assessing the corrosion of biodegradable magnesium implants: A critical review of current methodologies and their limitations. *Acta Biomater.* **2012**, *8*, 925.
35. H. Wang, Z. Shi, *In vitro* biodegradation behavior of magnesium and magnesium alloy. *J. Biomed. Mater. Res. Part B Appl. Biomater.* **2011**, *98*, 203.
36. J. Walker, S. Shadanbaz, N. T. Kirkland, E. Stace, T. Woodfield, M. P. Staiger, G. J. Dias, Magnesium alloys: Predicting in vivo corrosion with in vitro immersion testing. *J. Biomed. Mater. Res. Part B Appl. Biomater.* **2012**, *100*, 1134.

37. W. F. Ng, K. Y. Chiu, F. T. Cheng, Effect of pH on the in vitro corrosion rate of magnesium degradable implant material. *Mater. Sci. Eng. C.* **2010**, *30*, 898.
38. S.-K. Kang, S.-W. Hwang, S. Yu, J.-H. Seo, E. A. Corbin, J. Shin, D. S. Wie, R. Bashir, Z. Ma, J. A. Rogers, Biodegradable Thin Metal Foils and Spin-On Glass Materials for Transient Electronics. *Adv. Funct. Mater.* **2015**, *25*, 1789.
39. H. Maher, D. W. DiSanto, G. Soerensen, C. R. Bolognesi, H. Tang, J. B. Webb, Smooth wet etching by ultraviolet-assisted photoetching and its application to the fabrication of AlGaN/GaN heterostructure field-effect transistors. *Appl. Phys. Lett.* **2000**, *77*, 3833.
40. M. S. Minsky, M. White, E. L. Hu, Room-temperature photoenhanced wet etching of GaN. *Appl. Phys. Lett.* **1996**, *68*, 1531.
41. H. Cho, K. H. Auh, J. Han, R. J. Shul, S. M. Donovan, C. R. Abernathy, E. S. Lambers, F. Ren, S. J. Pearton, UV-photoassisted etching of GaN in KOH. *J. Electron. Mater.* **1999**, *28*, 290.
42. S.-W. Hwang, S.-K. Kang, X. Huang, M. A. Brenckle, F. G. Omenetto, J. A. Rogers, Materials for Programmed, Functional Transformation in Transient Electronic Systems. *Adv. Mater.* **2015**, *27*, 47.
43. K. Sim, X. Wang, Y. Li, C. Linghu, Y. Gao, J. Song, C. Yu, Destructive electronics from electrochemical-mechanically triggered chemical dissolution. *J. Micromech. Microeng.* **2017**, *27*, 6.
44. J. Shibasaki, T. Koizumi, W. Higuchi, Drug Absorption, Metabolism, and Excretion. IV. Pharmacokinetic Studies on Renal Transport. (1). Simultaneous Chemical Reaction and Diffusion (SCRD) Model for Uphill Transport. *Chem. Pharm. Bull.* **1968**, *16*, 2273.
45. R. Li, H. Cheng, Y. Su, S.-W. Hwang, L. Yin, H. Tao, M. A. Brenckle et al., An Analytical Model of Reactive Diffusion for Transient Electronics. *Adv. Funct. Mater.* **2013**, *23*, 3106.
46. X. Huang, Y. Liu, S.-W. Hwang, S.-K. Kang, D. Patnaik, J. F. Cortes, J. A. Rogers, Biodegradable Materials for Multilayer Transient Printed Circuit Boards. *Adv. Mater.* **2014**, *26*, 7371.
47. K. Fu, Z. Liu, Y. Yao, Z. Wang, B. Zhao, W. Luo, J. Dai et al., Transient Rechargeable Batteries Triggered by Cascade Reactions. *Nano Lett.* **2015**, *15*, 4664.
48. C. H. Lee, H. Kim, D. V. Harburg, G. Park, Y. Ma, T. Pan, J. S. Kim et al., Biological lipid membranes for on-demand, wireless drug delivery from thin, bioresorbable electronic implants. *NPG Asia Mater.* **2015**, *7*, e227.
49. C. H. Lee, J.-W. Jeong, Y. Liu, Y. Zhang, Y. Shi, S.-K. Kang, J. Kim et al., Materials and Wireless Microfluidic Systems for Electronics Capable of Chemical Dissolution on Demand. *Adv. Funct. Mater.* **2015**, *25*, 1338.
50. J. Yoon, J. Lee, B. Choi, D. Lee, D. H. Kim, D. M. Kim, D. Il Moon, M. Lim, S. Kim, S. J. Choi, Flammable carbon nanotube transistors on a nitrocellulose paper substrate for transient electronics. *Nano Res.* **2017**, *10*, 87.
51. Y. H. Jung, T.-H. Chang, H. Zhang, C. Yao, Q. Zheng, V. W. Yang, H. Mi et al., High-performance green flexible electronics based on biodegradable cellulose nanofibril paper. *Nat. Commun.* **2015**, 6, 7170.
52. T. Lei, M. Guan, J. Liu, H.-C. Lin, R. Pfattner, L. Shaw, A. F. McGuire et al., Biocompatible and totally disintegrable semiconducting polymer for ultrathin and ultralightweight transient electronics. *Proc. Natl. Acad. Sci.* **2017**, *114*, 5107.

Index

Note: Page numbers in italic and bold refer to figures and tables, respectively.

M

N

O

S

T